OXYGEN SENSING

LUNG BIOLOGY IN HEALTH AND DISEASE

Executive Editor

Claude Lenfant
Director, National Heart, Lung, and Blood Institute
National Institutes of Health
Bethesda, Maryland

1. Immunologic and Infectious Reactions in the Lung, *edited by C. H. Kirkpatrick and H. Y. Reynolds*
2. The Biochemical Basis of Pulmonary Function, *edited by R. G. Crystal*
3. Bioengineering Aspects of the Lung, *edited by J. B. West*
4. Metabolic Functions of the Lung, *edited by Y. S. Bakhle and J. R. Vane*
5. Respiratory Defense Mechanisms (in two parts), *edited by J. D. Brain, D. F. Proctor, and L. M. Reid*
6. Development of the Lung, *edited by W. A. Hodson*
7. Lung Water and Solute Exchange, *edited by N. C. Staub*
8. Extrapulmonary Manifestations of Respiratory Disease, *edited by E. D. Robin*
9. Chronic Obstructive Pulmonary Disease, *edited by T. L. Petty*
10. Pathogenesis and Therapy of Lung Cancer, *edited by C. C. Harris*
11. Genetic Determinants of Pulmonary Disease, *edited by S. D. Litwin*
12. The Lung in the Transition Between Health and Disease, *edited by P. T. Macklem and S. Permutt*
13. Evolution of Respiratory Processes: A Comparative Approach, *edited by S. C. Wood and C. Lenfant*
14. Pulmonary Vascular Diseases, *edited by K. M. Moser*
15. Physiology and Pharmacology of the Airways, *edited by J. A. Nadel*
16. Diagnostic Techniques in Pulmonary Disease (in two parts), *edited by M. A. Sackner*
17. Regulation of Breathing (in two parts), *edited by T. F. Hornbein*
18. Occupational Lung Diseases: Research Approaches and Methods, *edited by H. Weill and M. Turner-Warwick*
19. Immunopharmacology of the Lung, *edited by H. H. Newball*
20. Sarcoidosis and Other Granulomatous Diseases of the Lung, *edited by B. L. Fanburg*
21. Sleep and Breathing, *edited by N. A. Saunders and C. E. Sullivan*
22. *Pneumocystis carinii* Pneumonia: Pathogenesis, Diagnosis, and Treatment, *edited by L. S. Young*
23. Pulmonary Nuclear Medicine: Techniques in Diagnosis of Lung Disease, *edited by H. L. Atkins*
24. Acute Respiratory Failure, *edited by W. M. Zapol and K. J. Falke*
25. Gas Mixing and Distribution in the Lung, *edited by L. A. Engel and M. Paiva*

26. High-Frequency Ventilation in Intensive Care and During Surgery, *edited by G. Carlon and W. S. Howland*
27. Pulmonary Development: Transition from Intrauterine to Extrauterine Life, *edited by G. H. Nelson*
28. Chronic Obstructive Pulmonary Disease: Second Edition, *edited by T. L. Petty*
29. The Thorax (in two parts), *edited by C. Roussos and P. T. Macklem*
30. The Pleura in Health and Disease, *edited by J. Chrétien, J. Bignon, and A. Hirsch*
31. Drug Therapy for Asthma: Research and Clinical Practice, *edited by J. W. Jenne and S. Murphy*
32. Pulmonary Endothelium in Health and Disease, *edited by U. S. Ryan*
33. The Airways: Neural Control in Health and Disease, *edited by M. A. Kaliner and P. J. Barnes*
34. Pathophysiology and Treatment of Inhalation Injuries, *edited by J. Loke*
35. Respiratory Function of the Upper Airway, *edited by O. P. Mathew and G. Sant'Ambrogio*
36. Chronic Obstructive Pulmonary Disease: A Behavioral Perspective, *edited by A. J. McSweeny and I. Grant*
37. Biology of Lung Cancer: Diagnosis and Treatment, *edited by S. T. Rosen, J. L. Mulshine, F. Cuttitta, and P. G. Abrams*
38. Pulmonary Vascular Physiology and Pathophysiology, *edited by E. K. Weir and J. T. Reeves*
39. Comparative Pulmonary Physiology: Current Concepts, *edited by S. C. Wood*
40. Respiratory Physiology: An Analytical Approach, *edited by H. K. Chang and M. Paiva*
41. Lung Cell Biology, *edited by D. Massaro*
42. Heart–Lung Interactions in Health and Disease, *edited by S. M. Scharf and S. S. Cassidy*
43. Clinical Epidemiology of Chronic Obstructive Pulmonary Disease, *edited by M. J. Hensley and N. A. Saunders*
44. Surgical Pathology of Lung Neoplasms, *edited by A. M. Marchevsky*
45. The Lung in Rheumatic Diseases, *edited by G. W. Cannon and G. A. Zimmerman*
46. Diagnostic Imaging of the Lung, *edited by C. E. Putman*
47. Models of Lung Disease: Microscopy and Structural Methods, *edited by J. Gil*
48. Electron Microscopy of the Lung, *edited by D. E. Schraufnagel*
49. Asthma: Its Pathology and Treatment, *edited by M. A. Kaliner, P. J. Barnes, and C. G. A. Persson*
50. Acute Respiratory Failure: Second Edition, *edited by W. M. Zapol and F. Lemaire*
51. Lung Disease in the Tropics, *edited by O. P. Sharma*
52. Exercise: Pulmonary Physiology and Pathophysiology, *edited by B. J. Whipp and K. Wasserman*
53. Developmental Neurobiology of Breathing, *edited by G. G. Haddad and J. P. Farber*
54. Mediators of Pulmonary Inflammation, *edited by M. A. Bray and W. H. Anderson*
55. The Airway Epithelium, *edited by S. G. Farmer and D. Hay*

56. Physiological Adaptations in Vertebrates: Respiration, Circulation, and Metabolism, *edited by S. C. Wood, R. E. Weber, A. R. Hargens, and R. W. Millard*
57. The Bronchial Circulation, *edited by J. Butler*
58. Lung Cancer Differentiation: Implications for Diagnosis and Treatment, *edited by S. D. Bernal and P. J. Hesketh*
59. Pulmonary Complications of Systemic Disease, *edited by J. F. Murray*
60. Lung Vascular Injury: Molecular and Cellular Response, *edited by A. Johnson and T. J. Ferro*
61. Cytokines of the Lung, *edited by J. Kelley*
62. The Mast Cell in Health and Disease, *edited by M. A. Kaliner and D. D. Metcalfe*
63. Pulmonary Disease in the Elderly Patient, *edited by D. A. Mahler*
64. Cystic Fibrosis, *edited by P. B. Davis*
65. Signal Transduction in Lung Cells, *edited by J. S. Brody, D. M. Center, and V. A. Tkachuk*
66. Tuberculosis: A Comprehensive International Approach, *edited by L. B. Reichman and E. S. Hershfield*
67. Pharmacology of the Respiratory Tract: Experimental and Clinical Research, *edited by K. F. Chung and P. J. Barnes*
68. Prevention of Respiratory Diseases, *edited by A. Hirsch, M. Goldberg, J.-P. Martin, and R. Masse*
69. *Pneumocystis carinii* Pneumonia: Second Edition, *edited by P. D. Walzer*
70. Fluid and Solute Transport in the Airspaces of the Lungs, *edited by R. M. Effros and H. K. Chang*
71. Sleep and Breathing: Second Edition, *edited by N. A. Saunders and C. E. Sullivan*
72. Airway Secretion: Physiological Bases for the Control of Mucous Hypersecretion, *edited by T. Takishima and S. Shimura*
73. Sarcoidosis and Other Granulomatous Disorders, *edited by D. G. James*
74. Epidemiology of Lung Cancer, *edited by J. M. Samet*
75. Pulmonary Embolism, *edited by M. Morpurgo*
76. Sports and Exercise Medicine, *edited by S. C. Wood and R. C. Roach*
77. Endotoxin and the Lungs, *edited by K. L. Brigham*
78. The Mesothelial Cell and Mesothelioma, *edited by M.-C. Jaurand and J. Bignon*
79. Regulation of Breathing: Second Edition, *edited by J. A. Dempsey and A. I. Pack*
80. Pulmonary Fibrosis, *edited by S. Hin. Phan and R. S. Thrall*
81. Long-Term Oxygen Therapy: Scientific Basis and Clinical Application, *edited by W. J. O'Donohue, Jr.*
82. Ventral Brainstem Mechanisms and Control of Respiration and Blood Pressure, *edited by C. O. Trouth, R. M. Millis, H. F. Kiwull-Schöne, and M. E. Schläfke*
83. A History of Breathing Physiology, *edited by D. F. Proctor*
84. Surfactant Therapy for Lung Disease, *edited by B. Robertson and H. W. Taeusch*
85. The Thorax: Second Edition, Revised and Expanded (in three parts), *edited by C. Roussos*

86. Severe Asthma: Pathogenesis and Clinical Management, *edited by S. J. Szefler and D. Y. M. Leung*
87. *Mycobacterium avium*–Complex Infection: Progress in Research and Treatment, *edited by J. A. Korvick and C. A. Benson*
88. Alpha 1–Antitrypsin Deficiency: Biology • Pathogenesis • Clinical Manifestations • Therapy, *edited by R. G. Crystal*
89. Adhesion Molecules and the Lung, *edited by P. A. Ward and J. C. Fantone*
90. Respiratory Sensation, *edited by L. Adams and A. Guz*
91. Pulmonary Rehabilitation, *edited by A. P. Fishman*
92. Acute Respiratory Failure in Chronic Obstructive Pulmonary Disease, *edited by J.-P. Derenne, W. A. Whitelaw, and T. Similowski*
93. Environmental Impact on the Airways: From Injury to Repair, *edited by J. Chrétien and D. Dusser*
94. Inhalation Aerosols: Physical and Biological Basis for Therapy, *edited by A. J. Hickey*
95. Tissue Oxygen Deprivation: From Molecular to Integrated Function, *edited by G. G. Haddad and G. Lister*
96. The Genetics of Asthma, *edited by S. B. Liggett and D. A. Meyers*
97. Inhaled Glucocorticoids in Asthma: Mechanisms and Clinical Actions, *edited by R. P. Schleimer, W. W. Busse, and P. M. O'Byrne*
98. Nitric Oxide and the Lung, *edited by W. M. Zapol and K. D. Bloch*
99. Primary Pulmonary Hypertension, *edited by L. J. Rubin and S. Rich*
100. Lung Growth and Development, *edited by J. A. McDonald*
101. Parasitic Lung Diseases, *edited by A. A. F. Mahmoud*
102. Lung Macrophages and Dendritic Cells in Health and Disease, *edited by M. F. Lipscomb and S. W. Russell*
103. Pulmonary and Cardiac Imaging, *edited by C. Chiles and C. E. Putman*
104. Gene Therapy for Diseases of the Lung, *edited by K. L. Brigham*
105. Oxygen, Gene Expression, and Cellular Function, *edited by L. Biadasz Clerch and D. J. Massaro*
106. $Beta_2$-Agonists in Asthma Treatment, *edited by R. Pauwels and P. M. O'Byrne*
107. Inhalation Delivery of Therapeutic Peptides and Proteins, *edited by A. L. Adjei and P. K. Gupta*
108. Asthma in the Elderly, *edited by R. A. Barbee and J. W. Bloom*
109. Treatment of the Hospitalized Cystic Fibrosis Patient, *edited by D. M. Orenstein and R. C. Stern*
110. Asthma and Immunological Diseases in Pregnancy and Early Infancy, *edited by M. Schatz, R. S. Zeiger, and H. N. Claman*
111. Dyspnea, *edited by D. A. Mahler*
112. Proinflammatory and Antiinflammatory Peptides, *edited by S. I. Said*
113. Self-Management of Asthma, *edited by H. Kotses and A. Harver*
114. Eicosanoids, Aspirin, and Asthma, *edited by A. Szczeklik, R. J. Gryglewski, and J. R. Vane*
115. Fatal Asthma, *edited by A. L. Sheffer*
116. Pulmonary Edema, *edited by M. A. Matthay and D. H. Ingbar*
117. Inflammatory Mechanisms in Asthma, *edited by S. T. Holgate and W. W. Busse*
118. Physiological Basis of Ventilatory Support, *edited by J. J. Marini and A. S. Slutsky*

119. Human Immunodeficiency Virus and the Lung, *edited by M. J. Rosen and J. M. Beck*
120. Five-Lipoxygenase Products in Asthma, *edited by J. M. Drazen, S.-E. Dahlén, and T. H. Lee*
121. Complexity in Structure and Function of the Lung, *edited by M. P. Hlastala and H. T. Robertson*
122. Biology of Lung Cancer, *edited by M. A. Kane and P. A. Bunn, Jr.*
123. Rhinitis: Mechanisms and Management, *edited by R. M. Naclerio, S. R. Durham, and N. Mygind*
124. Lung Tumors: Fundamental Biology and Clinical Management, *edited by C. Brambilla and E. Brambilla*
125. Interleukin-5: From Molecule to Drug Target for Asthma, *edited by C. J. Sanderson*
126. Pediatric Asthma, *edited by S. Murphy and H. W. Kelly*
127. Viral Infections of the Respiratory Tract, *edited by R. Dolin and P. F. Wright*
128. Air Pollutants and the Respiratory Tract, *edited by D. L. Swift and W. M. Foster*
129. Gastroesophageal Reflux Disease and Airway Disease, *edited by M. R. Stein*
130. Exercise-Induced Asthma, *edited by E. R. McFadden, Jr.*
131. LAM and Other Diseases Characterized by Smooth Muscle Proliferation, *edited by J. Moss*
132. The Lung at Depth, *edited by C. E. G. Lundgren and J. N. Miller*
133. Regulation of Sleep and Circadian Rhythms, *edited by F. W. Turek and P. C. Zee*
134. Anticholinergic Agents in the Upper and Lower Airways, *edited by S. L. Spector*
135. Control of Breathing in Health and Disease, *edited by M. D. Altose and Y. Kawakami*
136. Immunotherapy in Asthma, *edited by J. Bousquet and H. Yssel*
137. Chronic Lung Disease in Early Infancy, *edited by R. D. Bland and J. J. Coalson*
138. Asthma's Impact on Society: The Social and Economic Burden, *edited by K. B. Weiss, A. S. Buist, and S. D. Sullivan*
139. New and Exploratory Therapeutic Agents for Asthma, *edited by M. Yeadon and Z. Diamant*
140. Multimodality Treatment of Lung Cancer, *edited by A. T. Skarin*
141. Cytokines in Pulmonary Disease: Infection and Inflammation, *edited by S. Nelson and T. R. Martin*
142. Diagnostic Pulmonary Pathology, *edited by P. T. Cagle*
143. Particle–Lung Interactions, *edited by P. Gehr and J. Heyder*
144. Tuberculosis: A Comprehensive International Approach, Second Edition, Revised and Expanded, *edited by L. B. Reichman and E. S. Hershfield*
145. Combination Therapy for Asthma and Chronic Obstructive Pulmonary Disease, *edited by R. J. Martin and M. Kraft*
146. Sleep Apnea: Implications in Cardiovascular and Cerebrovascular Disease, *edited by T. D. Bradley and J. S. Floras*
147. Sleep and Breathing in Children: A Developmental Approach, *edited by G. M. Loughlin, J. L. Carroll, and C. L. Marcus*

148. Pulmonary and Peripheral Gas Exchange in Health and Disease, *edited by J. Roca, R. Rodriguez-Roisen, and P. D. Wagner*
149. Lung Surfactants: Basic Science and Clinical Applications, *R. H. Notter*
150. Nosocomial Pneumonia, *edited by W. R. Jarvis*
151. Fetal Origins of Cardiovascular and Lung Disease, *edited by David J. P. Barker*
152. Long-Term Mechanical Ventilation, *edited by N. S. Hill*
153. Environmental Asthma, *edited by R. K. Bush*
154. Asthma and Respiratory Infections, *edited by D. P. Skoner*
155. Airway Remodeling, *edited by P. H. Howarth, J. W. Wilson, J. Bousquet, S. Rak, and R. A. Pauwels*
156. Genetic Models in Cardiorespiratory Biology, *edited by G. G. Haddad and T. Xu*
157. Respiratory-Circulatory Interactions in Health and Disease, *edited by S. M. Scharf, M. R. Pinsky, and S. Magder*
158. Ventilator Management Strategies for Critical Care, *edited by N. S. Hill and M. M. Levy*
159. Severe Asthma: Pathogenesis and Clinical Management, Second Edition, Revised and Expanded, *edited by S. J. Szefler and D. Y. M. Leung*
160. Gravity and the Lung: Lessons from Microgravity, *edited by G. K. Prisk, M. Paiva, and J. B. West*
161. High Altitude: An Exploration of Human Adaptation, *edited by T. F. Hornbein and R. B. Schoene*
162. Drug Delivery to the Lung, *edited by H. Bisgaard, C. O'Callaghan, and G. C. Smaldone*
163. Inhaled Steroids in Asthma: Optimizing Effects in the Airways, *edited by R. P. Schleimer, P. M. O'Byrne, S. J. Szefler, and R. Brattsand*
164. IgE and Anti-IgE Therapy in Asthma and Allergic Disease, *edited by R. B. Fick, Jr., and P. M. Jardieu*
165. Clinical Management of Chronic Obstructive Pulmonary Disease, *edited by T. Similowski, W. A. Whitelaw, and J.-P. Derenne*
166. Sleep Apnea: Pathogenesis, Diagnosis, and Treatment, *edited by A. I. Pack*
167. Biotherapeutic Approaches to Asthma, *edited by J. Agosti and A. L. Sheffer*
168. Proteoglycans in Lung Disease, *edited by H. G. Garg, P. J. Roughley, and C. A. Hales*
169. Gene Therapy in Lung Disease, *edited by S. M. Albelda*
170. Disease Markers in Exhaled Breath, *edited by N. Marczin, S. A. Kharitonov, M. H. Yacoub, and P. J. Barnes*
171. Sleep-Related Breathing Disorders: Experimental Models and Therapeutic Potential, *edited by D. W. Carley and M. Radulovacki*
172. Chemokines in the Lung, *edited by R. M. Strieter, S. L. Kunkel, and T. J. Standiford*
173. Respiratory Control and Disorders in the Newborn, *edited by O. P. Mathew*
174. The Immunological Basis of Asthma, *edited by B. N. Lambrecht, H. C. Hoogsteden, and Z. Diamant*

175. Oxygen Sensing: Responses and Adaptation to Hypoxia, *edited by S. Lahiri, G. L. Semenza, and N. R. Prabhakar*
176. Non-Neoplastic Advanced Lung Disease, *edited by J. Maurer*

ADDITIONAL VOLUMES IN PREPARATION

Therapeutic Targets in Airway Inflammation, *edited by N. T. Eissa and D. Huston*

Respiratory Infections in Asthma and Allergy, *edited by S. Johnston and N. Papadopoulos*

Acute Respiratory Distress Syndrome, *edited by M. A. Matthay*

Upper and Lower Respiratory Disease, *edited by J. Corren, A. Togias, and J. Bousquet*

Venous Thromboembolism, *edited by J. E. Dalen*

Acute Exacerbations of Chronic Obstructive Pulmonary Disease, *edited by N. Siafakas, N. Anthonisen, and D. Georgopolous*

Lung Volume Reduction Surgery for Emphysema, *edited by H. E. Fessler, J. J. Reilly, Jr., and D. J. Sugarbaker*

OXYGEN SENSING

RESPONSES AND ADAPTATION TO HYPOXIA

Edited by

Sukhamay Lahiri

University of Pennsylvania School of Medicine
Philadelphia, Pennsylvania, U.S.A.

Gregg L. Semenza

Johns Hopkins University School of Medicine
Baltimore, Maryland, U.S.A.

Nanduri R. Prabhakar

Case Western Reserve University School of Medicine
Cleveland, Ohio, U.S.A.

MARCEL DEKKER

MARCEL DEKKER, INC. NEW YORK • BASEL

Cover art designed by Santhosh M. Baby.

Library of Congress Cataloging-in-Publication Data
A catalog record for this book is available from the Library of Congress.

ISBN: 0-8247-0960-8

This book is printed on acid-free paper.

Headquarters
Marcel Dekker, Inc.
270 Madison Avenue, New York, NY 10016
tel: 212-696-9000; fax: 212-685-4540

Eastern Hemisphere Distribution
Marcel Dekker AG
Hutgasse 4, Postfach 812, CH-4001 Basel, Switzerland
tel: 41-61-260-6300; fax: 41-61-260-6333

World Wide Web
http://www.dekker.com

The publisher offers discounts on this book when ordered in bulk quantities. For more information, write to Special Sales/Professional Marketing at the headquarters address above.

Current printing (last digit):
10 9 8 7 6 5 4 3 2 1

PRINTED IN THE UNITED STATES OF AMERICA

INTRODUCTION

These days one can be virtually certain that all symposia, scientific meetings of professional societies, or even world congresses will include at least one session titled "From Gene to Bedside."

What happens at the bedside is not a focus of this volume, but if oxygen sensing is controlled by genes, then the responses and adaptation to hypoxia are de facto equivalent to translating genomic research to application.

The applications in this case are multifaceted, since hypoxia can be environmentally dependent—as at high altitude—or it can occur because of pathological conditions. Both situations represent models of biological processes that are critically important to understand.

Adaptation to oxygen by living forms has been an extraordinarily slow process. It took millions—if not billions—of years to change the atmosphere from anoxic to oxygenic. In fact, this change controlled evolution as we know it today. Oxygen sensing became critical, and the response and adaptation to hypoxia appears to have been very slow to develop. Oxygen is a product of photosynthesis, and if this process ceases, living forms may be endangered. Such was the destiny of the dinosaurs, which are said to have become extinct because photosynthesis was interrupted by dust clouds resulting from asteroids crashing into the Earth.

Dinosaurs no longer roam the Earth, but today millions of living beings are exposed to hypoxic conditions, either natural or pathological. Oxygen pathways and sensing are controlled by cellular and molecular processes, but also by . . . yes . . . genetic determinants.

Other monographs and chapters in other volumes of the Lung Biology in Health and Disease series have addressed the many aspects of oxygen and oxygenation at the cellular and whole-organ level in both health and disease, but no volume has dissected, analyzed, and presented oxygen sensing and the responses to hypoxia like this one does. No doubt, the journey from gene to "bedside" is completed in this monograph.

When the notion arose that this topic should be included in the series, only one potential editor stood out in my mind: Sukhamay Lahiri, a friend and colleague of several decades, and also a scientist whose expertise in oxygen sensing is recognized worldwide. The addition of Gregg Semenza and Nanduri Prabhakar brought even greater distinction to the editorial team. One need only examine the titles of all the chapters and the roster of authors to be assured of the uniqueness and excellence of this volume.

I am personally indebted to the editors and all the contributors for allowing me to present this volume to the readership of the series.

Claude Lenfant, M.D.
Bethesda, Maryland

FOREWORD

There has been a revolution in thinking in respiratory biology. Concepts of homeostatic control, which were based on a sensor with fixed properties feeding information back to a central controller also with fixed properties, have proved to be too simplistic. Instead it has been recognized that sensors and control systems are plastic, changing with maturation and with altered internal and environmental conditions. Moreover, in the new view, homeostatic systems frequently have multiple sensors, with different thresholds, sensitivities, and dynamics but with overlapping fields of reception and overlapping and often multiple targets. The properties of the system, the targets that are excited, and the pathways used to activate targets are commonly modulated by the intensity with which the sensors are stimulated and perhaps by the pattern of stimulation as well. Nowhere has this revolution in thinking been more profound than in the field of oxygen sensing and the control of oxygen homeostasis.

The carotid body remains the main but no longer the only attraction in oxygen sensing. It shares the spotlight with multiple sensors located seemingly everywhere in the body; changes in oxygen availability alter specific molecules, initiating chains and cascades of biochemical reactions that can open or close ion channels in membranes, alter the configuration of various molecule targets, and/or trigger gene expression and thus produce short- and long-term adaptations to rising or falling oxygen levels. These oxygen-sensing molecules, which can be found in all forms of life, are present in almost every cell in higher animals where they are more or less active. The fascinating thing is that the molecules that act as oxygen sensors in various cell types do not seem to be always the same. Oxygen exerts its molecular effects at different sites and in different ways. Biology seems to exploit many of them to sense oxygen; however, not every molecule that reacts with oxygen need be a sensor initiating some adaptive process.

It is frustrating that while the main features of oxygen sensing and the pathways used have been delineated in many cell types, the molecular and cellular processes involved in oxygen sensing in the carotid body remain controversial. Perhaps because of its major role in adaptations to hypoxia, more than one molecular mechanism is present in the carotid body (as has been suggested in other cell types) to allow for both short- and long-term responses.

Hypoxia has both excitatory and inhibitory effects on ventilation, and the carotid body seems to release both inhibitory and stimulating neurotransmitters. This may allow a finely graded response to hypoxia and prevent the hypoxic response from being excessive. Similar push-pull processes may exist at the subcellular level, allowing finer adjustments in other functions affected by oxygen level such as vascular resistance, red cell formation, and angiogenesis. While it may seem that an extremely sensitive oxygen-sensing system would be desirable, increased sensitivity has its dark side. For example, with respect to ventilatory regulation, too much sensitivity can provoke apneas and intermittent hypoxia, which recent studies show can in some respects be more harmful than chronic hypoxia. It may also be that too little as well as too much oxygen can lead to the formation of oxygen radicals, which have been labeled the villains in many adverse processes, including carcinogenesis and aging, but may mediate oxygen sensing in some cell types. It may be that it is not oxygen levels that are homeostatically regulated but rather the level of oxygen radicals.

This volume tells the story of the new discoveries that have revolutionized concepts and makes clear the complexity of the processes that influence oxygen sensing. It is an exciting story but raises many new questions. At first the new results may seem not to simplify but to suggest that biological processes are like nested sets of Russian dolls in which each smaller doll is as elaborate as the one a size larger. But the real message is that oxygen homeostasis and other regulations are not just an assortment of Rube Goldberg–like mechanisms that allow life to continue only because there is much redundancy. Rather, they are based on discoverable fundamental chemical and physical principles. As you read through this volume you can see them emerging.

Neil Cherniack, M.D.
Professor of Medicine and Physiology
New Jersey Medical School
University of Medicine and Dentistry of New Jersey
Newark, New Jersey, U.S.A.

PREFACE

Claude Lenfant wrote to me (S.L.) in June 2000 about his interest in functional genomics of oxygen sensing and the cellular mechanisms of the process and said that he would like to see a volume on this topic for the Lung Biology in Health and Disease series. Interestingly, at about the same time, I, with Nanduri Prabhakar and Gregg Semenza, finished working on a symposium volume on oxygen sensing and Claude's proposal fell into place for the three of us. We decided to take up the challenge, and this volume was created. Almost all the invited authors agreed to contribute a chapter. When the chapters were assigned, no one knew what the others were going to write. We did not give any direction nor did we try to control the minds of the contributors. We decided to let each chapter stand alone. Remarkably, there was little overlap between the chapters.

The first 10 chapters are about the genomics of oxygen sensing, which is time dependent. The next 18 chapters are about oxygen sensing by the carotid body, the primary instantaneous oxygen sensor in mammals. These contain a mix of acute hypoxia with chronic intermittent and sustained hypoxia. The remaining chapters deal with very interesting aspects of oxygen biology by neurons and other cells. Oxygen sensing takes place in the cell membrane, in the cytoplasm interaction between endoplasmic reticulum and mitochondria, and in the nucleus. Our emphasis was not just on what happens but how it happens.

There is very little explicit discussion of the clinical aspects of oxygen deprivation, but it is implicit throughout the volume.

This book would not exist without the incessant work of Mary Pili to whom we, particularly S.L., owe our gratitude.

Sukhamay Lahiri
Gregg L. Semenza
Nanduri R. Prabhakar

CONTRIBUTORS

Verónica Abudara, Ph.D. Adjunct Professor, Department of Physiology, University of Montevideo Medical School, Montevideo, Uruguay

Helmut Acker, Prof.Dr.med. Laboratory of Optical Physiology, Max Planck Institute of Molecular Physiology, University of Essen, Dortmund, Germany

Faton H. Agani, M.D., Ph.D. Assistant Professor, Department of Anatomy, Case Western Reserve University School of Medicine, Cleveland, Ohio, U.S.A.

Maria Teresa Agapito, Ph.D. Associate Professor, Department of Biochemistry, Molecular Biology, and Physiology, Faculty of Medicine, University of Valladolid, Valladolid, Spain

Julio Alcayaga, Ph.D. Associate Professor, Department of Biology, Faculty of Science, University of Chile, Santiago, Chile

Stephen L. Archer, M.D., F.R.C.P.(C) Director, Division of Cardiology, Department of Medicine, University of Alberta, Edmonton, Alberta, Canada

Santhosh M. Baby, Ph.D. Fellow, Department of Physiology, University of Pennsylvania School of Medicine, Philadelphia, Pennsylvania, U.S.A.

Amy L. Bauer, B.S. Department of Molecular and Cellular Physiology, University of Cincinnati College of Medicine, Cincinnati, Ohio, U.S.A.

Utta Berchner-Pfannschmidt, Ph.D. Max Planck Institute of Molecular Physiology, University of Essen, Dortmund, Germany

Donald G. Buerk, Ph.D. Research Associate Professor, Department of Physiology, University of Pennsylvania School of Medicine, Philadelphia, Pennsylvania, U.S.A.

H. Franklin Bunn, M.D. Professor and Director of Hematology Research, Department of Medicine, Harvard Medical School, and Brigham & Women's Hospital, Boston, Massachusetts, U.S.A.

Mark L. Burleson, Ph.D. Assistant Professor, Department of Biology, The University of Texas at Arlington, Arlington, Texas, U.S.A.

Joshua D. Cahn, B.S. Technical Research Assistant, Department of Hematology, Harvard Medical School, and Brigham & Women's Hospital, Boston, Massachusetts, U.S.A.

John L. Carroll, M.D. Professor, Departments of Pediatrics and Physiology, University of Arkansas for Medical Sciences, and Arkansas Children's Hospital, Little Rock, Arkansas, U.S.A.

Juan Carlos Chávez, B.Sc. Department of Anatomy, Case Western Reserve University School of Medicine, Cleveland, Ohio, U.S.A.

Jia Chen, M.D. Research Associate, Department of Physiology, University of Utah School of Medicine, Salt Lake City, Utah, U.S.A.

Ernest Cutz, M.D., F.R.C.P.C. Professor, Department of Laboratory Medicine and Pathobiology, University of Toronto, and The Hospital for Sick Children, Toronto, Ontario, Canada

Maria F. Czyzyk-Krzeska, M.D., Ph.D. Associate Professor, Department of Molecular and Cellular Physiology, University of Cincinnati College of Medicine, Cincinnati, Ohio, U.S.A.

Dominic D'Agostino, Ph.D. Research Fellow, Department of Medicine, UMDNJ–Robert Wood Johnson Medical School, New Brunswick, New Jersey, U.S.A.

Peter A. Daudu, Ph.D. Research Associate, Department of Physiology, University of Pennsylvania School of Medicine, Philadelphia, Pennsylvania, U.S.A.

Bruce Dinger, Ph.D. Research Associate, Department of Physiology, University of Utah School of Medicine, Salt Lake City, Utah, U.S.A.

Reinhard P. Dirmeier, Ph.D. Department of Molecular, Cellular, and Developmental Biology, University of Colorado, Boulder, Colorado, U.S.A.

Norman H. Edelman, M.D. Professor, Departments of Medicine and of Physiology and Biophysics, and Dean, State University of New York at Stony Brook, Stony Brook, New York, U.S.A.

Carlos Eyzaguirre, M.D. Professor, Department of Physiology, University of Utah School of Medicine, Salt Lake City, Utah, U.S.A.

Joachim Fandrey, M.D. Professor, Department of Physiology, University of Essen, Essen, Germany

Ian M. Fearon, Ph.D. Assistant Professor, Department of Biology, McMaster University, Hamilton, Ontario, Canada

Salvatore J. Fidone, Ph.D. Professor and Chairman, Department of Physiology, University of Utah School of Medicine, Salt Lake City, Utah, U.S.A.

Robert S. Fitzgerald, Litt.B., S.T.B., S.T.M., Ph.D. Professor, Departments of Environmental Health Sciences, Physiology, and Medicine, Johns Hopkins University, Baltimore, Maryland, U.S.A.

Xiao Wen Fu, Ph.D. Research Associate, Department of Pediatric Laboratory Medicine, University of Toronto, and The Hospital for Sick Children, Toronto, Ontario, Canada

Marie-Alda Gilles-Gonzalez, Ph.D. Associate Professor, Department of Biochemistry, University of Texas Southwestern Medical Center, Dallas, Texas, U.S.A.

Constancio Gonzalez, M.D., Ph.D. Professor and Chairman, Department of Biochemistry, Molecular Biology, and Physiology, Faculty of Medicine, University of Valladolid, Valladolid, Spain

Thomas A. Gorr, Ph.D. Research Fellow, Department of Hematology, Harvard Medical School, and Brigham & Women's Hospital, Boston, Massachusetts, U.S.A.

Lucy R. Green, B.Sc., Ph.D. Centre for Fetal Origins of Adult Disease, University of Southampton, and Princess Anne Hospital, Southampton, England

Sachin A. Gupte, M.D., Ph.D. Research Associate, Department of Physiology, New York Medical College, Valhalla, New York, U.S.A.

Gabriel G. Haddad, M.D. Professor of Pediatrics and Neuroscience and Chairman, Department of Pediatrics, Albert Einstein College of Medicine, and Pediatrician-in-Chief, Children's Hospital at Montefiore, New York, New York, U.S.A.

Mark A. Hanson, D.Phil., Cert.Ed., F.R.C.O.G. BHF Professor of Cardiovascular Science and Director, Centre for Fetal Origins of Adult Disease, University of Southampton, and Princess Anne Hospital, Southampton, England

Liang He, M.D. Research Associate, Department of Physiology, University of Utah School of Medicine, Salt Lake City, Utah, U.S.A.

Thomas Hellwig-Bürgel, Dr.rer.nat. Postdoctoral Research Fellow, Institute of Physiology, University of Lübeck, Lübeck, Germany

Tomoko Higashi, M.D. Postdoctoral Fellow, Department of Environmental Health Sciences, Johns Hopkins University, Baltimore, Maryland, U.S.A.

Serabi Hirasawa, M.D. Postdoctoral Fellow, Department of Environmental Health Sciences, Johns Hopkins University, Baltimore, Maryland, U.S.A.

Pavel Hradecky, M.D. Research Associate, Department of Molecular and Cellular Biology, Harvard Medical School, Boston, Massachusetts, U.S.A.

Christine Huckstorf, Dr.med. Physiology Institute, University of Rostock, Rostock, Germany

Anna S. Hui, M.S. Research Associate, Department of Molecular and Cellular Physiology, University of Cincinnati College of Medicine, Cincinnati, Ohio, U.S.A.

Rodrigo Iturriaga, Ph.D. Associate Professor, Neurobiology Laboratory, Department of Physiology, Catholic University of Chile, Santiago, Chile

Adele Jackson, Ph.D. Postdoctoral Fellow, Department of Neuroscience, Ottawa Health Research Institute, Ottawa, Ontario, Canada

Wolfgang Jelkmann, M.D. Professor, Institute of Physiology, University of Lübeck, Lübeck, Germany

Rugang Jiang, M.D. Professor, Department of Physiology, Tianjin Medical University, Tianjin, People's Republic of China

Vincent Joseph, Ph.D. Department of Pediatrics, Laval University School of Medicine, Quebec, Ontario, Canada

Paul J. Kemp, D.Phil.(Oxon) School of Biomedical Sciences, University of Leeds, Leeds, England

Ganesh K. Kumar, Ph.D. Associate Professor, Department of Biochemistry, Case Western Reserve University School of Medicine, Cleveland, Ohio, U.S.A.

Prem Kumar, D.Phil. Department of Physiology, The Medical School, University of Birmingham, Birmingham, England

Anna V. Kuznetsova, Ph.D. Visiting Scientist, Department of Molecular and Cellular Physiology, University of Cincinnati College of Medicine, Cincinnati, Ohio, U.S.A.

Hugo Lagercrantz, M.D., Ph.D. Professor, Neonatal Research Unit, Karolinska Institute, Astrid Lindgren Children's Hospital, Stockholm, Sweden

Sukhamay Lahiri, D.Phil. Professor, Department of Physiology, University of Pennsylvania School of Medicine, Philadelphia, Pennsylvania, U.S.A.

Joseph C. LaManna, Ph.D. Professor, Department of Neurology, Case Western Reserve University School of Medicine, Cleveland, Ohio, U.S.A.

Boris Lande, Ph.D. Research Assistant, Department of Environmental Health Sciences, Johns Hopkins University, Baltimore, Maryland, U.S.A.

Anthony Lewis, B.Sc., Ph.D. School of Biomedical Sciences, University of Leeds, Leeds, England

Jinquing Li, M.D., Ph.D. Senior Research Investigator, Department of Physiology, University of Pennsylvania School of Medicine, Philadelphia, Pennsylvania, U.S.A.

José López-Barneo, M.D., Ph.D. Professor, Department of Physiology, University of Seville, Seville, Spain

Gang Lu, Ph.D. Postdoctoral Fellow, Department of Genome Science, and Genome Research Institute, University of Cincinnati College of Medicine, Cincinnati, Ohio, U.S.A.

Patrick H. Maxwell, M.B.B.S., D.Phil., F.R.C.P. Professor, Renal Section, Hammersmith Campus, Imperial College, London, England

Evangelos D. Michelakis, M.D., F.A.C.C. Assistant Professor, Division of Cardiology, Department of Medicine, University of Alberta, Edmonton, Alberta, Canada

David E. Millhorn, Ph.D. Director and Chairman, Department of Genome Science, and Genome Research Institute, University of Cincinnati College of Medicine, Cincinnati, Ohio, U.S.A.

William K. Milsom, Ph.D. Professor, Department of Zoology, University of British Columbia, Vancouver, British Columbia, Canada

Anil Mokashi, M.Sc. Research Specialist, Department of Physiology, University of Pennsylvania School of Medicine, Philadelphia, Pennsylvania, U.S.A.

James A. Nash, O.D. Optometrist, Department of Molecular and Cellular Physiology, University of Cincinnati College of Medicine, Cincinnati, Ohio, U.S.A.

Judith A. Neubauer, Ph.D. Professor, Department of Medicine, UMDNJ–Robert Wood Johnson Medical School, New Brunswick, New Jersey, U.S.A.

James P. Newman, Ph.D. Centre for Fetal Origins of Adult Disease, University of Southampton, and Princess Anne Hospital, Southampton, England

Colin A. Nurse, Ph.D. Professor, Department of Biology, McMaster University, Hamilton, Ontario, Canada

Ana Obeso, M.D., Ph.D. Associate Professor of Human Physiology, Department of Biochemistry, Molecular Biology, and Physiology, Faculty of Medicine, University of Valladolid, Valladolid, Spain

Kristin M. O'Brien, Ph.D. Research Associate, Department of Molecular, Cellular, and Developmental Biology, University of Colorado, Boulder, Colorado, U.S.A.

Richard A. Oeckler, M.S. Department of Physiology, New York Medical College, Valhalla, New York, U.S.A.

Jeffery L. Overholt, Ph.D. Assistant Professor, Department of Physiology and Biophysics, Case Western Reserve University School of Medicine, Cleveland, Ohio, U.S.A.

Ricardo Pardal, Ph.D. Postdoctoral Fellow, Department of Physiology, University of Seville, Seville, Spain

Hugh A. Pearson, B.Sc. School of Biomedical Sciences, University of Leeds, Leeds, England

Chris Peers, B.Sc., Ph.D. Professor, Institute for Cardiovascular Research, University of Leeds, Leeds, England

Jean-Marc Pequignot, Ph.D., D.Sc. Professor, Department of Physiology, Centre National de la Recherche Scientifique, Université Claude-Bernard Lyon I, Lyon, France

Julie Peyronnet, Ph.D. Neonatal Research Unit, Karolinska Institute, Astrid Lindgren Children's Hospital, Stockholm, Sweden

Paola Pichiule, B.Sc. Department of Anatomy, Case Western Reserve University School of Medicine, Cleveland, Ohio, U.S.A.

Leigh D. Plant, B.Sc. School of Biomedical Sciences, University of Leeds, Leeds, England

Robert O. Poyton, Ph.D. Professor, Department of Molecular, Cellular, and Developmental Biology, University of Colorado, Boulder, Colorado, U.S.A.

Zlatko I. Pozeg, M.D. Division of Cardiac Surgery, Department of Surgery, University of Alberta, Edmonton, Alberta, Canada

Nanduri R. Prabhakar, Ph.D., D.Sc. Professor and Vice-Chairman, Department of Physiology and Biophysics, Case Western Reserve University School of Medicine, Cleveland, Ohio, U.S.A.

Peter J. Ratcliffe, M.D., F.R.C.P. Professor, Department of Medicine, University of Oxford, Oxford, England

Nicola Ritucci, Ph.D. Postdoctoral Fellow, Division of Pulmonary and Critical Care Medicine, Department of Medicine, UMDNJ–Robert Wood Johnson Medical School, New Brunswick, New Jersey, U.S.A.

Asunción Rocher, Ph.D. Lecturer, Department of Biochemistry, Molecular Biology, and Physiology, Faculty of Medicine, University of Valladolid, Valladolid, Spain

Jean-Christophe Roux, Ph.D. Neonatal Research Unit, Karolinska Institute, Astrid Lindgren Children's Hospital, Stockholm, Sweden

Arijit Roy, Ph.D. Senior Research Investigator, Department of Physiology, University of Pennsylvania School of Medicine, Philadelphia, Pennsylvania, U.S.A.

Gloria Sanz Alfayate, B.S. Research Fellow, Department of Biochemistry, Molecular Biology, and Physiology, Faculty of Medicine, University of Valladolid, Valladolid, Spain

Phillip O. Schnell, B.S., M.S. Research Associate, Department of Molecular and Cellular Physiology, University of Cincinnati College of Medicine, Cincinnati, Ohio, U.S.A.

Gregg L. Semenza, M.D., Ph.D. Professor, Department of Pediatrics, and McKusick-Nathans Institute of Genetic Medicine, Johns Hopkins University School of Medicine, Baltimore, Maryland, U.S.A.

Karen A. Seta, Ph.D. Research Scientist, Department of Genome Science, and Genome Research Institute, University of Cincinnati College of Medicine, Cincinnati, Ohio, U.S.A.

Machiko Shirahata, M.D., D.M.Sc. Associate Research Professor, Department of Environmental Health Sciences, Johns Hopkins University, Baltimore, Maryland, U.S.A.

Irene C. Solomon, Ph.D. Assistant Professor, Department of Physiology and Biophysics, State University of New York at Stony Brook, Stony Brook, New York, U.S.A.

Erick Spears, M.A. Project Manager, Pharmatech Inc., Denver, Colorado, U.S.A.

Zachary Spicer, Ph.D. Postdoctoral Fellow, Department of Genome Science, and Genome Research Institute, University of Cincinnati College of Medicine, Cincinnati, Ohio, U.S.A.

L. J. Stensaas, Ph.D. Professor, Department of Physiology, University of Utah School of Medicine, Salt Lake City, Utah, U.S.A.

Daniel Phillip Stiehl, Cand.Med. Institute of Physiology, University of Lübeck, Lübeck, Germany

Tino Streller, Dr.rer.nat. Physiology Institute, University of Rostock, Rostock, Germany

Justin B. Striet, B.A., B.S. Research Assistant, Department of Molecular and Cellular Physiology, University of Cincinnati College of Medicine, Cincinnati, Ohio, U.S.A.

Jagadeeshan Sunderram, M.D. Assistant Professor, Department of Medicine, UMDNJ–Robert Wood Johnson Medical School, New Brunswick, New Jersey, U.S.A.

Bernard Thébaud, M.D., Ph.D. Assistant Professor, Department of Pediatrics, Vascular Biology Group, University of Alberta, Edmonton, Alberta, Canada

Roger J. Thompson, Ph.D. Postdoctoral Fellow, Department of Cellular and Structural Biology, University of Colorado Health Sciences Center, Denver, Colorado, U.S.A.

Benjamin D. Tyrrell, M.D. Division of Cardiology, Department of Medicine, Vascular Biology Group, University of Alberta, Edmonton, Alberta, Canada

Rodrigo Varas, M.S. Neurobiology Laboratory, Department of Physiology, Catholic University of Chile, Santiago, Chile

Hay-Yan Wang, Ph.D. Postdoctoral Fellow, Department of Brain Chemistry, National Institute on Drug Abuse, National Institutes of Health, Baltimore, Maryland, U.S.A.

Roland H. Wenger, Ph.D. Professor, Carl-Ludwig-Institute of Physiology, University of Leipzig, Leipzig, Germany

Michael S. Wolin, Ph.D. Professor, Department of Physiology, New York Medical College, Valhalla, New York, U.S.A.

Christoph Wotzlaw, Dipl.Ing.Biotech. Microscopy Specialist, Max Planck Institute of Molecular Physiology, University of Essen, Dortmund, Germany

Shigeki Yamaguchi, M.D., Ph.D. Postdoctoral Fellow, Department of Environmental Health Sciences, Johns Hopkins University, Baltimore, Maryland, U.S.A.

Herman Yeger, Ph.D. Senior Scientist, Department of Pediatric Laboratory Medicine, University of Toronto, and The Hospital for Sick Children, Toronto, Ontario, Canada

Yong Yuan, Ph.D. Research Associate, Department of Genome Science, and Genome Research Institute, University of Cincinnati College of Medicine, Cincinnati, Ohio, U.S.A.

Patricio Zapata, M.D. Professor, Department of Physiological Sciences, Laboratory of Neurobiology, Catholic University of Chile, Santiago, Chile

CONTENTS

OXYGEN SENSING

Part One

Genomics of Oxygen Sensing

GREGG L. SEMENZA

Intracellular O_2 concentrations in humans and other mammals are maintained within a relatively narrow range, with partial pressures of O_2 varying between a high of approximately 110 mmHg (16% O_2 at sea level) in the pulmonary alveoli to a low of less than 20 mmHg (3% O_2) in some areas of the heart, kidney, and brain. The physiological O_2 concentration within any cell reflects the net effect of O_2 delivery and O_2 consumption and is associated with adaptive intracellular responses to protect against the deleterious effects of O_2 deprivation (hypoxia) or O_2 excess (hyperoxia). Oxygen homeostasis thus reflects the requirement for O_2 as a substrate for essential biochemical reactions, most notably oxidative phosphorylation, which is balanced by the risk of oxidative damage to cellular macromolecules. The molecular mechanisms underlying homeostatic responses to hypoxia represent the central focus of this volume.

Stimulus-response pathways induced by hypoxia can be categorized as either acute or chronic. Acute responses are of rapid onset and short-term duration, whereas chronic responses are of delayed onset and long-term duration. These kinetics reflect the underlying molecular mechanisms: acute responses involve posttranslational modifications of existing proteins that alter their activity, whereas chronic responses involve changes in gene expression that result in the synthesis of novel proteins or increased synthesis of proteins already present within the cell.

In general, the first part of this volume will describe mechanisms for the regulation of gene expression in response to chronic hypoxia, whereas the second part will describe the mechanisms regulating electrophysiological responses to hypoxia by specialized oxygen-sensing cells that involve O_2-dependent regulation of ion channel activity.

Changes in O_2 concentration are sensed and elicit biological responses by the cells of all organisms on Earth, from bacteria to yeast to multicellular animals. In Chapter 1, Marie Gilles-Gonzalez describes O_2-sensing systems in Bacteria and Archaea. Among these are several heme-containing proteins whose biochemical activity is regulated by the presence or absence of O_2 bound to the heme group. These proteins bind heme through a PAS domain. PAS is an acronym referring to the eukaryotic proteins PER, ARNT, and SIM, which were first recognized to contain this domain. A limited number of PAS domain proteins have been identified in eukaryotes and in most of these the PAS domain appears to mediate protein dimerization. In contrast, a much larger number of PAS domain proteins have been identified in bacteria that function as sensors of stimuli as varied as light, O_2, and redox potential (1).

Despite the presence of PAS domain proteins in bacteria and higher animals, few, if any, of these proteins appear to be present in yeast. These single-celled eukaryotes, like their multicellular counterparts, can generate ATP either anaerobically via glycolysis or aerobically via oxidative phosphorylation. Yet unlike higher organisms, which oxidize glucose in the presence of adequate O_2 concentrations, yeast utilize fermentation to generate ATP in the presence of adequate glucose concentrations. It is perhaps not surprising, therefore, that yeast utilize mechanisms of O_2 sensing that differ both from prokaryotes and from higher eukaryotes. In Chapter 2, Robert Poyton and his colleagues describe a variety of mechanisms by which yeast gene expression is regulated in response to changes in environmental O_2 concentrations. Although the electron transport chain in the mitochondria utilizes O_2 as the final electron acceptor in the process of oxidative phosphorylation, some electron transfer to O_2 occurs at earlier steps in the electron transport chain, generating reactive oxygen species (ROS), which at low concentrations may function as signal transduction molecules and at high concentrations can lead to cell death. Poyton and colleagues discuss the potential role of ROS generated by the electron transport chain in modulating yeast gene expression. The O_2-dependent generation of ROS and its effect on O_2-regulated gene expression in mammalian cells is discussed by Joachim Fandrey in Chapter 4.

Whereas O_2 delivery to a single cell is a simple matter of diffusion, the anatomical complexity of higher multicellular animals presents a requirement for specialized physiological systems to ensure that all cells receive adequate oxygenation. In mammals, two major systems have evolved. The respiratory system includes the lungs, which provide a surface area for transfer of O_2 from the atmosphere to the hemoglobin present in erythrocytes, the diaphragm and accessory muscles of respiration, and the neural respiratory control centers. The circulatory system includes the O_2 transport vehicles (erythrocytes), the transportation engine (the heart), and transportation infrastructure (the vasculature). The development of these systems and their physiological maintenance require the coordinated expression of thousands of genes under the control of hundreds of transcription factors. Among these, hypoxia-inducible factor 1 (HIF-1) appears to play the role of master regulator of O_2 homeostasis. HIF-1 is required for the proper development of the heart, blood, blood vessels, and respiratory control centers (2–5). HIF-1 also controls physiological responses to hypoxia and ischemia in adult organisms, stimulating increased O_2 delivery via erythropoiesis and angiogenesis, and metabolic adaptation to O_2 deprivation via increased glucose transport and glycolysis. In Chapter 7, Joseph LaManna and colleagues present a molecular analysis of physiological adaptations that occur in the brain in response to prolonged mild hypoxia. Over 50 HIF-1 target genes are now known (6) and it is likely that 1–5% of all human genes are regulated by HIF-1. In Chapter 10, Franklin Bunn and colleagues describe a bioinformatic approach that may be useful for identifying candidate HIF-1 target genes in organisms such as *Drosophila melanogaster* and *Caenorhabditis elegans* for which the entire genome sequence is available to search for HIF-1-binding sites. However, studies have demonstrated that the presence of a HIF-1-binding site is necessary but not sufficient to mediate hypoxia-inducible gene expression (7), indicating the need for additional levels of analysis to determine whether genes containing putative HIF-1-binding sites are in fact regulated by HIF-1.

HIF-1 is a heterodimeric protein consisting of a constitutively expressed HIF-1β-subunit and an O_2-regulated HIF-1α-subunit (8,9). In Chapter 5, Roland Wenger describes the organization and regulation of the mouse gene encoding HIF-1α. The HIF-1α and HIF-1β subunits dimerize by interaction between helix-loop-helix (HLH) and PAS domains present within each protein (10). Dimerization juxtaposes basic domains, which, located immediately amino-terminal to the HLH domains, form a functional DNA-binding domain that recognizes the consensus binding-site sequence 5′-RCGTG-3′ (R = A or G). Both the expression of HIF-1α and its ability to activate transcription are subject to negative regulation under normoxic conditions. The presence of a PAS domain suggested the possibility for binding of O_2 via a heme prosthetic group, but no evidence in support of this hypothesis was ever obtained. In Chapter 3 Patrick Maxwell and Peter Ratcliffe describe the novel molecular mechanism by which HIF-1α protein expression is regulated via O_2-dependent prolyl hydroxylation of a regulatory domain located carboxyl-terminal to the PAS domain (11–14). Recently, HIF-1α transactivation domain function has been shown to be regulated by O_2-dependent asparaginyl hydroxylation (15). Three HIF-1α prolyl hydroxylases have been identified (11,14), whereas the molecular characterization of the asparaginyl hydroxylase(s) has not yet been reported. These discoveries have revealed a novel posttranslational modification which, like phosphorylation, regulates protein-protein interactions but which is uniquely regulated by the cellular O_2 concentration. O_2-dependent protein hydroxylation provides a potential mechanism linking acute and chronic responses to hypoxia and it will be interesting to determine how many other proteins are subject to this posttranslational modification.

Whereas HIF-1 activity is induced by hypoxia in all nucleated cells, the activity of several other transcription factors is also affected by hypoxia in a cell-type-specific manner via distinct signal transduction pathways. Included among these is HIF-2α (also known as EPAS-1), which is structurally similar to HIF-1α (16) and regulated in an O_2-dependent manner by prolyl and asparaginyl hydroxylases (11–15) but is expressed in a tissue-specific manner, including the vascular endothelium and carotid body (16,17). In Chapter 8, David Millhorn and colleagues describe the effects of hypoxia on signal transduction and gene expression in PC12 cells, a depolarizable cell line derived from a rat pheochromocytoma, which has many properties of carotid body glomus cells including the expression of tyrosine hydroxylase. These results also suggest connections between the acute responses of specialized O_2-sensing cells, such as the carotid body glomus cells, which depolarize in response to hypoxia, and the chronic responses of nondepolarizable cells. This connection has also been established by the demonstration that in mice partially deficient for HIF-1α expression, the carotid body does not depolarize in response to hypoxia (3). In addition to effects on protein hydroxylation and gene expression, hypoxia induces changes in tyrosine hydroxylase mRNA stability, as described by Maria Czyzyk-Krzeska and colleagues in Chapter 9. The complexity of responses to hypoxia provides evidence for the fundamental nature of this physiological stimulus and the requirement to respond on many biological levels.

For oxygen homeostasis to be maintained, O_2 delivery must match O_2 consumption. Since cell proliferation results in an inevitable increase in O_2 consumption, it is perhaps not surprising that stimulation of cells by growth factors also results in increased HIF-1α expression. As described in Chapter 6 by Wolfgang Jelkmann and colleagues, a large number of growth factors and cytokines share this property. Binding of these ligands to their cognate receptors activates signal transduction pathways within the cell that mediate increased HIF-1α expression. The increased HIF-1α expression in response to growth factor stimulation was tacitly assumed to be due to increased protein stability as in response to hypoxia. However, receptor tyrosine kinase signaling via the phosphatidylinositol-3-kinase pathway stimulates an increase in the translation of HIF-1α protein, an effect that is dependent upon the 5′-untranslated region of HIF-1α mRNA and activity of the downstream kinase FRAP, also known as the mammalian target of rapamycin (18). In contrast, signaling via ERK and p38 MAP kinases has been shown to increase HIF-1α transactivation domain function (19–21). Thus, like hypoxia, growth factor stimulation of signal transduction pathways can result in both increased HIF-1α protein expression and transcriptional activity.

Taken together, the papers in Part One demonstrate the complex interrelationships between cell proliferation, energy metabolism, oxygen homeostasis, redox state, and signal transduction in humans and other organisms. Alterations in oxygen homeostasis play critical roles in the pathophysiology of heart disease, cancer, stroke, and chronic lung disease, which are the most common diseases in the U.S. population, accounting for two-thirds of all deaths annually. The elucidation of the molecular mechanisms underlying adaptive and maladaptive responses to hypoxia will provide a better understanding of these disorders and may ultimately lead to new approaches to their treatment.

References

1. Taylor BL, Zhulin IB. PAS domains: internal sensors of oxygen, redox potential and light. Microbiol Mol Biol Rev 1999; 63:479–506.
2. Iyer NV, Kotch LE, Agani F, Leung SW, Laughner E, Wenger RH, Gassmann M, Gearhart JD, Lawler AM, Yu AY, Semenza GL. Cellular and developmental control of O_2 homeostasis by hypoxia-inducible factor 1α. Genes Dev 1998; 12:149–162.
3. Kline DD, Peng YJ, Manalo DJ, Semenza GL, Prabhakar NR. Defective carotid body function and impaired ventilatory responses to chronic hypoxia in mice partially deficient for hypoxia-inducible factor 1α. Proc Natl Acad Sci USA 2002; 99:821–826.
4. Ryan HE, Lo J, Johnson RS. HIF-1α is required for solid tumor formation and embryonic vascularization. EMBO J 1998; 17:3005–3015.
5. Yu AY, Shimoda LA, Iyer NV, Huso DL, Sun X, McWilliams R, Beaty T, Sham JSK, Wiener CM, Sylvester JT, Semenza GL. Impaired physiological responses to chronic hypoxia in mice partially deficient for hypoxia-inducible factor 1α. J Clin Invest 1999; 103:691–696.
6. Semenza GL. Hypoxia-inducible factor 1: oxygen homeostasis and disease pathophysiology. Trends Mol Med 2001; 7:345–350.
7. Semenza GL, Jiang BH, Leung SW, Passantino R, Concordet JP, Maire P, Giallongo A. Hypoxia response elements in the aldolase A, enolase 1, and lactate dehydrogenase A gene promoters contain essential binding sites for hypoxia-inducible factor 1. J Biol Chem 1996; 271:32529–32537.
8. Wang GL, Jiang BH, Rue EA, Semenza GL. Hypoxia-inducible factor 1 is a basic-helix-loop-helix-PAS heterodimer regulated by cellular O_2 tension. Proc Natl Acad Sci USA 1995; 92:5510–5514.
9. Wang GL, Semenza GL. Purification and characterization of hypoxia-inducible factor 1. J Biol Chem 1995; 270:1230–1237.
10. Wang BH, Rue E, Wang GL, Roe R, Semenza GL. Dimerization, DNA binding, and transactivation properties of hypoxia-inducible factor 1. J Biol Chem 1996; 271:17771–17778.
11. Bruick RK, McKnight SL. A conserved family of prolyl-4-hydroxylases that modify HIF. Science 2001; 294:1337–1340.
12. Ivan M, Kondo K, Yang H, Kim W, Valiando J, Ohh M, Salic A, Asara JM, Lane WS, Kaelin WG Jr. HIFα targeted for VHL-mediated destruction by proline hydroxylation: implications for O_2 sensing. Science 2001; 292:464–468.
13. Jaakkola P, Mole DR, Tian YM, Wilson MI, Gielbert J, Gaskell SJ, Kriegsheim AV, Hebestreit HF, Mukherji M, Schofield CJ, Maxwell PH, Pugh CW, Ratcliffe PJ. Targeting of HIF-α to the von Hippel–Lindau ubiquitylation complex by O_2-regulated prolyl hydroxylation. Science 2001; 292:468–472.
14. Epstein AC, Gleadle JM, McNeill LA, Hewitson KS, O'Rourke J, Mole DR, Mukherji M, Metzen E, Wilson MI, Dhanda A, Tian YM, Masson N, Hamilton DL, Jaakkola P, Barstead R, Hodgkin J, Maxwell PH, Pugh CW, Schofield CJ, Ratcliffe PJ. *C. elegans* EGL-9 and mammalian homologs define a family of dioxygenases that regulate HIF by prolyl hydroxylation. Cell 2001; 107:43–54.
15. Lando D, Peet DJ, Whelan DA, Gorman JJ, Whitelaw ML. Asparagine hydroxylation of the HIF transactivation domain: a hypoxic switch. Science 2002; 295:858–861.
16. Tian H, McKnight SL, Russell DW. Endothelial PAS domain protein 1 (EPAS1), a transcription factor selectively expressed in endothelial cells. Genes Dev 1997; 11:72–82.

17. Tian H, Hammer RE, Matsumoto AM, Russell DW, McKnight SL. The hypoxia-responsive transcription factor EPAS1 is essential for catecholamine homeostasis and protection against heart failure during embryonic development. Genes Dev 1998; 12:3320–3324.
18. Laughner E, Taghavi P, Chiles K. Mahon PC, Semenza GL. HER2 (neu) signaling increases the rate of hypoxia-inducible factor 1 alpha (HIF-1alpha) synthesis: novel mechanism for HIF-1-mediated vascular endothelial growth factor expression. Mol Cell Biol 2001; 21:3995–4004.
19. Hirota K. Semenza GL. Rac1 activity is required for the activation of hypoxia-inducible factor 1. J Biol Chem 2001; 276:21166–21172.
20. Richard DE, Berra E, Gothie E, Roux D, Pouyssegur J. p42/p44 mitogen-activated protein kinases phosphorylate hypoxia-inducible factor 1 alpha (HIF-1alpha) and enhance the transcriptional activity of HIF-1. J Biol Chem 1999; 274:32631–32637.
21. Sodhi A, Montaner S, Patel V, Zohar M, Bais C, Mesri EA, Gutkind JS. The Kaposi's sarcoma-associated herpes virus G protein-coupled receptor up-regulates vascular endothelial growth factor expression and secretion through mitogen-activated protein kinase and p38 pathways acting on hypoxia-inducible factor 1α. Cancer Res 2000; 60:4873–4880.

1

Biochemistry and Physiological Importance of Heme Proteins as Oxygen Sensors

MARIE-ALDA GILLES-GONZALEZ

University of Texas Southwestern
Medical Center
Dallas, Texas, U.S.A.

I. Introduction to Heme-Based Sensors

"Heme-based sensors" are increasingly appreciated as a broad and important class of heme proteins that detect and transduce physiological O_2, CO, or NO signals (1,2). Reversible binding of those ligands to heme iron represents one of the simplest mechanisms for their detection. This strategy for sensing of O_2, although not yet demonstrated for Eukarya, is widely used by Bacteria and Archaea. This review presents the current state of knowledge about heme-based sensors, emphasizing the physiology and biochemistry of O_2-sensing systems.

In heme-based sensors, a ligand-induced conformational change, initiated at a heme-binding domain, controls the function of a separate transducing domain. For such sensors, there is a considerable range in both the types of heme-binding domains that are possible and the activities to which they can couple. For example, in mammalian soluble guanylyl cyclases (sGC), coupling of a sensory heme-binding domain to a guanylyl cyclase domain permits NO regulation of cGMP levels (3). A three-dimensional structure is not yet available for the heme-binding domain of sGC, but its sequence suggests that it will not resemble any known heme-protein fold. In the CooA protein of *Rhodospirullum rubrum*, coupling of a different type of heme-binding domain to a DNA-binding domain allows CO-regulated induction of genes for CO catabolism (4,5). The CooA protein is unique in that it structurally resembles a catabolite-activator protein (CAP) with heme replacing cAMP in the regulatory domain (6). So far, O_2 signal transduction has offered the greatest variety in the processes controlled, the sensory domains used, and the transducer domains coupled (7). The processes regulated by O_2 include nitrogen fixation, cellulose production, and aerotaxis (8–13). At least two types of O_2-sensing domains have been identified, one that has a heme-PAS fold and another that is likely to have a myoglobin fold (12,14). The transducing domains are even more varied. They include chemotaxis domains, phosphodiesterase, and histidine-protein kinase domains (2,10,11,15).

II. The HemAT O_2 Sensors Governing Aerotaxis

In a variety of Bacteria and Archaea as different as *Bacillus subtilus*, *Bacillus radiodurans*, and *Halobacterium salinarum*, O_2 directly regulates aerotaxis by way of the HemAT proteins (11–13). Regulation is accomplished by coupling of a C-terminal chemotaxis domain in HemAT to a sensory N-terminal heme-binding domain. The chemotaxis domain of HemAT proteins is about 30% identical to the cytoplasmic domain of the *Escherichia coli* Tsr methyl-carrier protein (MCP). In contrast, the overall homology of the heme-binding domain to myoglobins is poor. Nevertheless, conservation of key residues, such as the proximal or F8 histidine of heme attachment, the CD1 phenylalanine, and the distal E7 histidine, suggests that the heme-binding domain in HemAT has a myoglobin fold. Alam and colleagues have shown for several organisms that a *hemAT* gene is required for a physiological aerophilic response. They have also unambiguously demonstrated binding of heme within the N-terminal domain of HemAT proteins. Mutagenesis experiments and resonance Raman studies support attachment of the heme to the F8 histidine, as suggested by the sequence (12,16). Based on the kinetic parameters for binding of O_2, the *B. subtilis* HemAT protein is estimated to have an O_2 equilibrium dissociation constant of $\sim 0.7\,\mu M$ (16).

III. Overview of Heme-Based O_2 Sensors Having a Heme-Binding PAS Domain

The broadest category of O_2 sensors are those proteins that detect O_2 with a heme-binding PAS domain (7). Heme-PAS domains are widespread in Bacteria and Archaea. The great evolutionary adaptability of those domains is further reflected by their ability to couple to at least three different types of activities: a histidine-protein kinase activity for signal transduction by phosphoryl transfer, a phosphodiesterase activity for signal transduction by hydrolysis of cyclic nucleotides, and a helix-loop-helix DNA-binding domain for signal transduction by direct regulation of transcription.

A. O_2 Regulation of Cyclic Nucleotide Second Messengers

Regulation of Cellulose Production by Acetobacter xylinum

In the cellulose-producing bacterium *Acetobacter xylinum*, the phosphodiesterase A1 protein (*Ax*PDEA1) functions to hydrolyze a potent allosteric activator of the bacterial cellulose synthase, the cyclic nucleotide [cyclic bis(3′ → 5′)diguanylic acid, or c-di-GMP] (17). *Ax*PDEA1 was recently shown to contain an N-terminal regulatory heme-PAS domain that, when bound to O_2, inhibits the C-terminal phosphodiesterase activity (10). Thus in cultures of this obligate aerobe, O_2 is now thought to enhance the construction of a buoyant cellulose pellicle by the cells nearest to the media surface for possible use as a biofilm. The type of phosphodiesterase domain found in *Ax*PDEA1, though not extensively studied, is widespread in Bacteria.

The Dos Protein of Escherichia coli

The Dos protein of *E. coli*, *Ec*Dos, provides another example of the occurrence of a phosphodiesterase domain jointly with a heme-PAS (15). The physiological role of *Ec*Dos is not yet known. Although *E. coli* has a variety of mechanisms for sensing O_2, all of those systems involve new gene expression and likely deal with long-term adaptation to changes in O_2 tension (7). The *Ec*Dos system is unique in that it provides a possible mechanism for rapid and global adaptation involving changes in the activities of preexisting proteins. Despite the strong homology of *Ec*Dos to *Ax*PDEA1 (~40% identity), and their nearly identical O_2 affinity ($K_d \sim 10\,\mu M$), the association and dissociation rates constants for binding of O_2 to *Ec*Dos were 2000-fold lower. A unique coordination of the heme iron in *Ec*Dos, with both axial ligands to the iron atom supplied by side chains in the protein, is likely responsible for the slow kinetics of ligand binding (15,18). Thus, unlike all other known heme-PAS domains where a ligand binds to an empty coordination site of the heme iron, in *Ec*Dos the ligand must displace a resident side chain of the protein before binding to iron atom. The disruption of a coordinate bond and displacement of a distal residue suggest a very direct mechanism involving repositioning that residue for driving the regulatory conformational change in *Ec*Dos.

B. O_2 Regulation of Phosphoryl Transfer

Physiological Importance of the FixL/FixJ System

The best-studied group of heme-PAS containing proteins are the O_2-sensing FixL histidine-protein kinases of Rhizobia (7,19). Under the hypoxic conditions that prevail during the symbiosis of Rhizobia with leguminous plants, FixL works jointly with a rhizobial transcription factor, the FixJ protein, to induce the expression of N_2 fixation genes (8,9,20–24). In *Sinorhizobium meliloti*, the facultative symbiont of alfalfa, more than 21 N_2-fixation genes, including those encoding the subunits of the dinitrogenase and dinitrogenase reductase enzymes, are under FixL/FixJ control (8,9). In *Bradyrhizobium japonicum*, the facultative symbiont of soybean, the

FixL/FixJ system regulates the expression of fewer N_2-fixation genes, but these include the genes encoding alternative terminal oxidases (20,23,24). Those high-affinity oxidases, which permit respiration with O_2 in the nearly anaerobic root nodule, are essential for survival of the Rhizobia during symbiosis.

A Classic Two-Component Regulatory System

The FixL and FixJ proteins constitute a classic two-component regulatory system (Fig. 1) (25–29). Such modular systems employ two conserved domains (transmitter and receiver) and two nonconserved domains (input and output) to detect various signals and bring about appropriate adaptive responses with a common mechanism of phosphoryl transfer. The phosphoryl-transfer reactions result in a net transfer of a γ-phosphoryl group from ATP to an aspartate residue in the receiver domain of the response regulator. An apparent intermediate in this transfer is a relatively stable phosphorylated sensor, with the ATP-derived phosphoryl group attached to a conserved histidine of its transmitter (or kinase) domain. Input signals vary greatly in their complexity. Binding of O_2 to a heme center represents one of the most readily measurable inputs for such systems (2). Transcriptional regulation, such as that accomplished by phosphorylated FixJ, is a common output. The sensor proteins of two-component systems are usually dimeric under physiological conditions, regardless of their phosphorylation state. For the response regulators, phosphorylation may control DNA binding by controlling dimerization. In at least the case of the *S. meliloti* FixJ, phosphorylation of the receiver domain clearly shifts the monomer-dimer equilibrium toward the dimeric form (30).

Sequences are available for FixL and FixJ proteins from several rhizobial species, including *S. meliloti*, *B. japonicum*, *Azorhizobium caulinodans*, and *Rhizobium leguminosarum* (9,20,21,31). As a rule, FixL proteins contain three

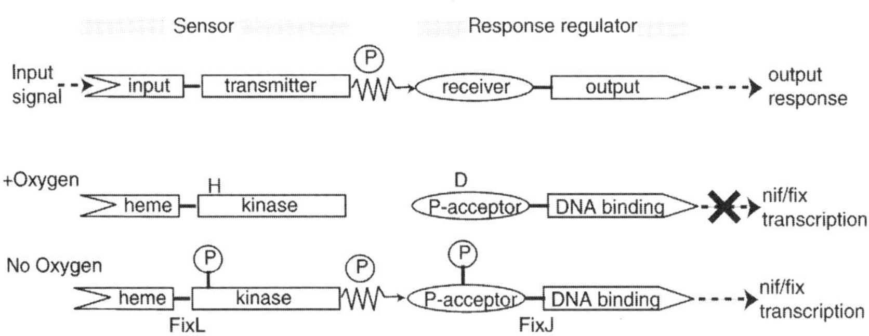

Figure 1 The FixL/FixJ two-component regulatory system. In two-component regulatory systems, the inputs and outputs are variable, but the transmitter and receiver domains from unrelated systems share ~20% identity. The input of the sensor FixL is a heme-binding domain that inhibits phosphorylation when O_2 binds and that permits enzymatic activity when O_2 leaves. The output response is transcriptional activation at nitrogen fixation promoters for only hypoxic conditions. (From Ref. 36.)

identifiable domains spanning about 500 residues. The N-terminal domain ($\sim$125 residues) is not conserved. Though this domain is usually hydrophobic and membrane-attached, in *B. japonicum* FixL it is soluble (1,32). The central heme-binding domain ($\sim$130 residues) and the C-terminal kinase domain ($\sim$250 residues) of FixL are well conserved ($\sim$55% sequence identity). FixJ proteins are equally well conserved over their entire lengths ($\sim$200 residues).

Enzymatic Studies of the Phosphoryl-Transfer Reactions

The Component Reactions

The overall reaction catalyzed by the FixL enzyme is the transfer of a γ-phosphoryl group from ATP to an asparate residue of FixJ. The overall *turnover* reaction may be written simply as follows, although this reaction has been treated as a "ping-pong, bi-bi" scheme.

$$2\text{FixJ} + 2\text{ATP} \overset{\text{FixL}_2}{\longleftrightarrow} \text{P-FixJ}_2 + 2\text{ADP}$$

In a classic ping-pong, bi-bi enzymatic reaction mechanism, an enzyme produces two products from two substrates, but those substrates are never simultaneously associated with the enzyme or allowed to influence each other's reaction. For the *S. meliloti* proteins, the *Rm*FixL enzyme can clearly accomplish a net transfer of a phosphoryl group from ATP to *Rm*FixJ by this sequential scheme (33–35). Specifically, under those conditions, dimeric *Rm*FixL reacts with ATP in a slow regulated step, forming a phosphorylated enzyme intermediate. The phospho-*Rm*FixL thus formed reacts with *Rm*FixJ, in a faster unregulated step, forming phospho-*Rm*FixJ. Nevertheless, three lines of evidence now indicate that Scheme 1 makes no significant contribution to the production of phospho-*Rm*FixJ in vivo (36).

Scheme 1: *"Ping-pong, Bi-bi" or Sequential Introduction of Substrates*

$$\text{Step 1, autophosphorylation}: Rm\text{FixL}_2 + 2\text{ATP} \leftrightarrow \text{P-}Rm\text{FixL}_2 + 2\text{ADP}$$
$$\text{Step 2, phosphotransfer}: \text{P-}Rm\text{FixL}_2 + 2Rm\text{FixJ} \leftrightarrow Rm\text{FixL}_2 + \text{P-}Rm\text{FixJ}_2$$

In Scheme 1, the rate of production of phospho-FixJ could never exceed the maximum rate of FixL autophosphorylation (Step 1). Nevertheless, simultaneous presentation of ATP and *Rm*FixJ to the enzyme increases the turnover number for production of phospho-*Rm*FixJ 30-fold, compared to the maximum autophosphorylation rate measured for Scheme 1, Step 1.

For Scheme 1, with all reactions at maximum velocity, the ratio of phosphorylated FixL to unphosphorylated FixL at equilibrium will be k_1/k_{-1} for autophosphorylation and $k_1/(k_{-1}+k_2)$ for turnover, where k_1 and k_{-1} are the forward and reverse rate constants for Step 1, and k_2 is the forward rate for Step 2. Clearly, with k_2 much faster than k_1, we should expect much lower concentrations of phospho-FixL at equilibrium during turnover than during autophosphorylation if Scheme 1 is operating. Instead, during turnover the equilibrium levels of phospho-FixL are over 17%, but during reaction of FixL with ATP in the absence of FixJ, the

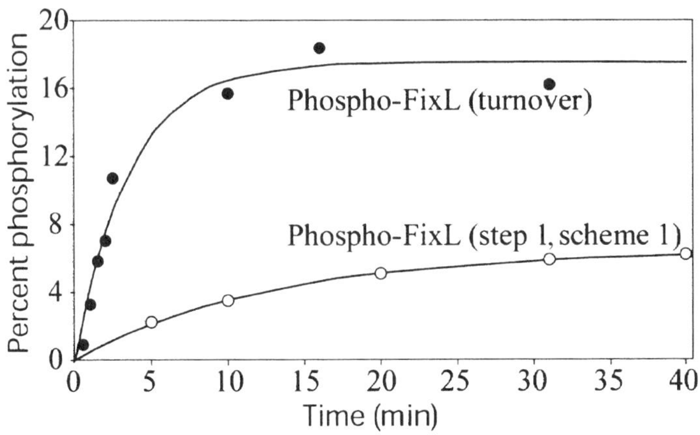

Figure 2 Enhanced phosphorylation of the FixLJ complex. Autophosphorylation of 5 µM *S. meliloti* deoxy-FixL without FixJ (open circles), phosphorylation following a 5-min preincubation of 2.5 µM *S. meliloti* deoxy-FixL with 2.5 µM *S. meliloti* FixJ (closed circles) at 23°C and pH 8.0. (From Ref. 36.)

levels of phospho-FixL reach only 4% (Fig. 2) (33,36). The implication of these results is that FixJ either increases k_1 or reduces k_{-1}. This would require complexation with FixJ to precede all phosphoryl transfer steps (36).

The uncomplexed P-*Rm*FixL$_2$ produced in Step 1 doubles as a powerful phosphatase that hydrolyzes P-*Rm*FixJ$_2$ almost as rapidly as it is made (33). However, when all the reactions are allowed to proceed in the presence of both ATP and FixJ, no detectable free phosphate is generated during a reaction time course of up to 2 hr (Fig. 3) (36).

A Revised Phosphoryl-Transfer Scheme Involving Simultaneous Introduction of Both Substrates

The ping-pong, bi-bi scheme (Scheme 1) is probably permissible only in vitro, where FixL can be made to phosphorylate in the absence of FixJ. In vivo, of course, FixL and FixJ are always present simultaneously. Indeed, the slightly overlapping coding regions of *RmfixL* and *RmfixJ* lead to coupling of their translation and synthesis of comparable levels of the *Rm*FixL and *Rm*FixJ proteins (9). Under those conditions, a far more efficient scheme for the *Rm*FixL-catalyzed turnover of *Rm*FixJ is available, in which complexation of *Rm*FixL with *Rm*FixJ precedes all the phosphoryl transfers.

Scheme 2: Simultaneous Introduction of Substrates

$$Rm\text{FixL}_2 + Rm\text{FixJ}_2 + 2\text{ATP} \leftrightarrow Rm\text{FixL}_2 \cdot Rm\text{FixJ}_2 \cdot 2\text{ATP}$$

$$Rm\text{FixL}_2 \cdot Rm\text{FixJ}_2 \cdot 2\text{ATP} \leftrightarrow (\text{P-}Rm\text{FixL} \cdot Rm\text{FixJ})_2 + 2\text{ADP}$$

$$(\text{P-}Rm\text{FixL} \cdot Rm\text{FixJ})_2 \leftrightarrow Rm\text{FixL}_2 + \text{P-}Rm\text{FixJ}_2$$

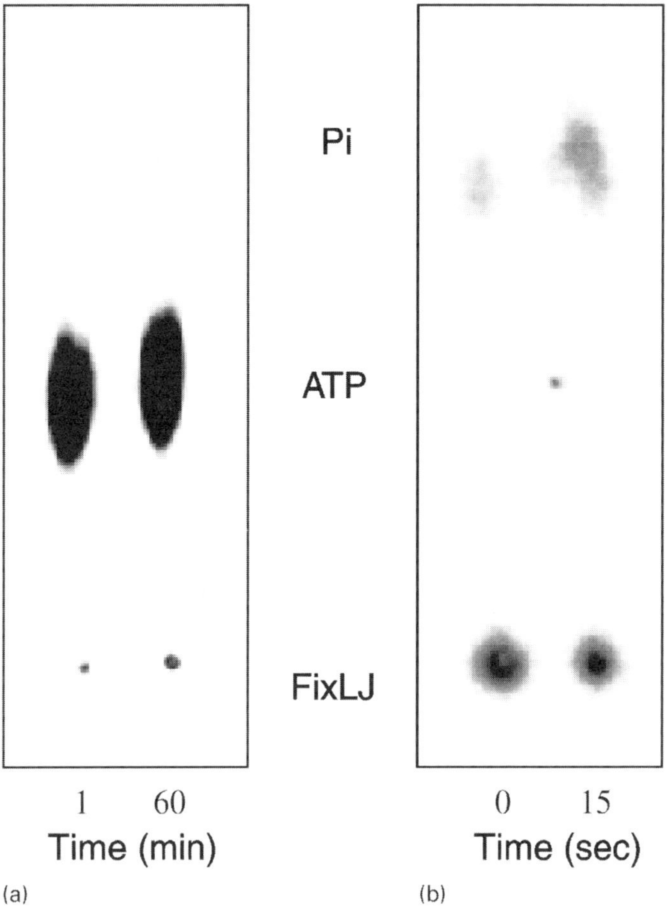

Figure 3 Presence of a phosphatase activity during phosphotransfer and its absence during turnover. The accumulation of inorganic phosphate during a 60-minute phosphorylation of a *S. meliloti* FixLJ complex (a) is contrasted with the accumulation of free phosphate in only 15 sec of Scheme 1 phosphotransfer (b), in which FixJ reacts with phospho-FixL prepared in the absence of FixJ. (From Ref. 36.)

Here *Rm*FixL is permitted to interact simultaneously with ATP and *Rm*FixJ. Interestingly, under those conditions turnover of *Rm*FixJ is more stringently and selectively regulated by heme ligands (see the next section). The phosphorylation of *Rm*FixL is much faster than the reaction of isolated *Rm*FixL with ATP in Scheme 1, despite the concurrent phosphorylation of *Rm*FixJ. The absence of free P-*Rm*FixL$_2$, uncomplexed with *Rm*FixJ, is an important feature of Scheme 2. The absence of this phospho-*Rm*FixJ phosphatase is supported by the failure of ^{32}Pi to accumulate in reactions containing γ-[^{32}P]-ATP, as measured by autoradiography of low-molecular-weight species following their separation by anion-exchange thin-layer chromatography (Fig. 3) (36).

Regulation of Phosphoryl Transfer by Heme Status

In reactions carried out according to Scheme 1, heme ligands were found to regulate the autophosphorylation step (Step 1) but not the phosphotransfer step (Step 2) (33,35). Neither reaction step in this scheme was affected by the oxidation state of the heme iron, with unliganded Fe^{II} or Fe^{III} *Rm*FixL being equally active (35). Binding of O_2 to Fe^{II} *Rm*FixL had the identical effect as binding of CN^- to Fe^{III} *Rm*FixL, with a 15-fold inhibition of the autophosphorylation resulting in either case (Table 1).

To explain the apparent insensitivity of the regulatory mechanism to oxidation state and its lack of heme-ligand specificity, the initial event in sensing by *Rm*FixL was hypothesized to be the switch of the iron atom from high to low spin on binding of any strong-field ligand (35). This event would be analogous to the conformational change thought to be triggered by a high- to low-spin transition of heme iron in human hemoglobin (37,38).

If the overall turnover of FixJ really represents a combination of the two reaction steps in Scheme 1, then the overall process should be:

- Blind to heme-iron oxidation state.
- Inhibited by heme ligands identically to the autophosphorylation step in Scheme 1, Step 1.

Table 1 Inhibition of the Autophosphorylation and Turnover Reactions by Heme Ligands

	Autophosphorylation[a]		Turnover[b]	
*Rm*FixL heme status	k_{ph} (% h^{-1})	I_{ph}	k_j (h^{-1})	I_j
Deoxy (Fe^{II})	33	1	12	1
Oxy ($Fe^{II}O_2$)[c]	2.2	15	<0.1	>100
Carbonmonoxy ($Fe^{II}CO$)	6.6	5	4.5	2.7
Nitrosyl ($Fe^{II}NO$)	15.7	2.1	5.2	2.3
Met (Fe^{III})	33	1	0.12	1
Cyanomet ($Fe^{III}CN^-$)	2.2	15	0.06	2
Imidazolemet (Fe^{III}Imid)	<0.4	>75	0	
Fluoromet ($Fe^{III}F^-$)	33	1	0.12	1

[a]The rate constant k_{ph} is the initial rate of autophosphorylation. I_{ph} is the inhibition factor for *Rm*FixL autophosphorylation, obtained by dividing the activity of the unliganded form by the activity of the liganded form of *Rm*FixL, for the same oxidation state.

[b]The rate constant k_j is the number of *Rm*FixJ molecules phosphorylated by one molecule of *Rm*FixJ per hour. I_j is the inhibition factor for *Rm*FixJ turnover, obtained by dividing the activity of the unliganded form by the activity of the liganded form of *Rm*FixL, for the same oxidation state.

[c]In air, *Rm*FixL is 90% oxy and 10% deoxy. For oxy-*Rm*FixL, the value of k_{ph} was calculated by subtracting the contribution of deoxy-*Rm*FixL (3.6% hr^{-1}) from the observed phosphorylation rate of *Rm*FixL in air (5.9% hr^{-1}) (39).

Instead the turnover of *Rm*FixJ to phospho-*Rm*FixJ was over 100 times faster for Fe^{II} *Rm*FixL than for Fe^{III} *Rm*FixL (Table 1) (39). The inhibition of turnover by O_2 was at least five times stronger than the inhibition of FixL autophosphorylation. Finally, heme ligands other than O_2 were much less inhibitory to the turnover reaction. Those effects of ligands and oxidation state on turnover were fully reversible.

C. Structure of an O_2-Sensing Module

The Heme-PAS Structural Fold

From genome sequence analyses, over 1100 proteins spanning all three major kingdoms of life are expected to contain PAS domains. This domain was initially noted for the eukaryotic proteins **P**eriod clock, **A**ryl hydrocarbon receptor nuclear translocator, and **S**imple-minded (40,41). Though relatively few PAS-domain proteins have been studied at the molecular biological or biochemical level, the known functions of PAS domains suggest that they represent a spectacularly adaptable signal-transduction module. The first PAS-domain structure to be solved was that of the light-detecting photoactive yellow protein (PYP) (42,43). Pellequer and colleagues hypothesized that PAS domains, despite their relatively poor sequence homology, might share a common structural fold (44). The crystal structures reported for the O_2-sensing heme-PAS domain from *B. japonicum* FixL and for the voltage-detecting PAS domain from the human HERG protein have proved this hypothesis correct (14,45). In addition, the structure of the *B. japonicum* heme-PAS has shown that heme can bind to an entirely novel alpha-beta fold having no resemblance to hemoglobins (Fig. 4) (14).

Interestingly, PAS domains exist that:

- Accomplish their sensing function without any cofactor
- Require a cofactor for their sensing functions
- Use one type of cofactor to detect more than one signal
- Use two different cofactors to detect a similar type of signal

For example, no cofactor has been found in the voltage sensor HERG (45). Flavin cofactors are bound to the NifL and Aer redox sensors (46–48). In plants, it is a flavin-mononucleotide-bound PAS in the NPH1 protein that serves as a detector of blue light, guiding phototropism (49,50).

PAS as a Putative Dimerization Domain

Some researchers have suggested that PAS domains are "dimerization domains" that regulate the activities of signal-transducing proteins by controlling their association. This is very unlikely for the photoactive yellow protein, and for the PAS domains from *B. japonicum* FixL, the NPH1 protein, and the HERG protein, all of which crystallize as monomers (14,42,45,50). In sensor kinases of the two-component

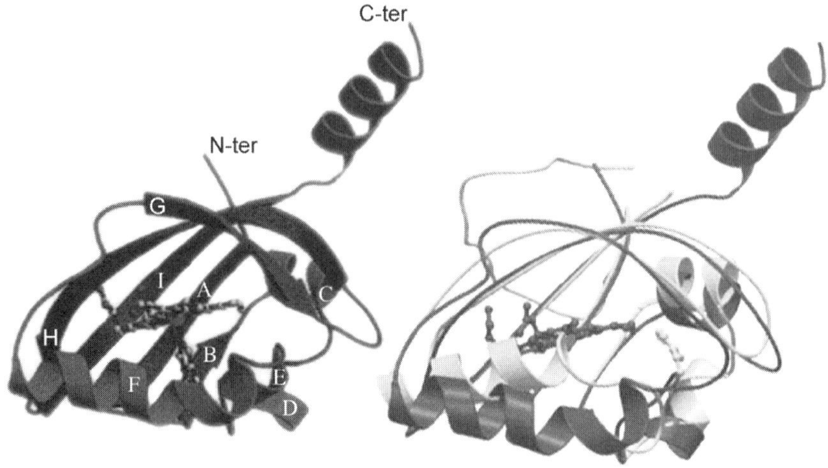

Figure 4 Structure of the *Bradyrhizobium japonicum* FixL heme-binding PAS domain (*Bj*FixLH) and its close resemblance to the structure of photoactive yellow protein (PYP). On the left is a ribbons diagram showing secondary-structure elements in *Bj*FixLH; on the right is a comparison of the structures of *Bj*FixLH (dark gray) and PYP (light gray), together with their heme and hydroxycinnamate cofactors, respectively. Consecutive letters denote consecutive regions of secondary structure. (From Ref. 14.)

class, of which FixL is a member, the major dimerization determinant is known to reside in the histidine-kinase domain (51). Based on these findings, it seems more reasonable to suppose that when PAS domains occur in multimeric proteins, they interact more strongly with their regulatory targets than with each other.

B. japonicum FixL and *S. meliloti* FixL heme-binding PAS domains have been shown by gel filtration and nuclear magnetic resonance to be monomeric in both active and inactive conformations (1,52). Miyatake and colleagues, who prepared a somewhat longer version of the *S. meliloti* FixL heme-PAS for their crystallographic studies, have reported a "dimerization surface" near the N-terminal end of the domain (53). However, biochemical measurements showed that deletion of the entire alpha-helix of presumed contact from full-length protein failed to affect activity (53). Rodgers and colleagues recently reported an intriguing O_2-governed dimerization of the *S. meliloti* heme-binding domain, possibly involving the same surface (54). These observations were not extended to the full-length protein, which aggregated into tetramers and nonamers for the deoxy form. Suggestions that this oligomerization contributes to regulation are unsupported by measurements of enzymatic activity or regulation (54). In fact, the FixL concentration dependence of the autophosphorylation reaction (Scheme 1, Step 1) is inconsistent with an activation by oligomerization but is readily explained by an equilibrium between an inactive monomeric form and an active dimeric form having an equilibrium dissociation constant (K_d) of 0.1 μM (39). This value is in excellent agreement with the K_d values (0.05–0.4 μM) reported so far for histidine-kinase dimerizations (55,56).

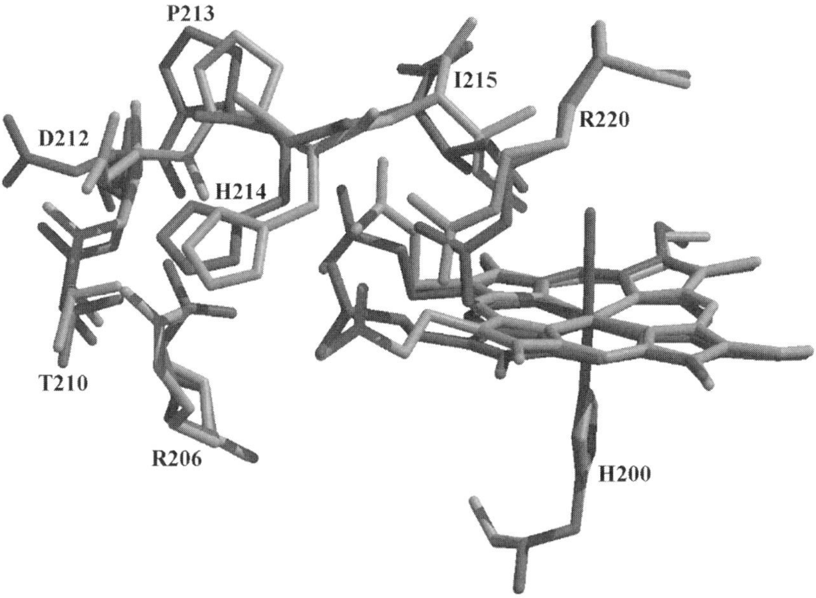

Figure 5 Structures of the regulatory FG loop of the *B. japonicum* FixL heme-binding domain (*Bj*FixLH) in the "on" and "off" states, as defined by measurements of autophosphorylation (enzymatic Scheme 1, Step 1). The figure shows an overlap of the refined models for both the unliganded "on" state (light gray), and the cyanide-bound "off" state (dark gray), based on the crystal structures of met- and cyanomet-*Bj*FixLH determined to 2.4-Å and 2.7-Å resolution, respectively. (From Ref. 14.)

The Ligand-Induced Conformational Change Proposed for FixL

A variety of structures have been solved for FixL heme-binding PAS domains (14,53,57). They include the initial crystal structure of the *B. japonicum* FixL heme-PAS domain in both "on" (unliganded) and "off" (cyanide-bound) conformations, as defined by measurements of autophosphorylation (enzymatic Scheme 1, Step 1) (14). The structures of the "on" Fe^{II} and Fe^{III} forms of the *S. meliloti* FixL heme-binding domain and additional "off" $Fe^{II}O_2$ and Fe^{III}imidazole forms of the *B. japonicum* FixL heme-binding domain are also available (53,57). Based on the structures of the ferric unliganded (Fe^{III}, or met) and cyanide-bound ($Fe^{III}CN^-$, or cyanomet) forms of the heme-PAS domains, a small ligand-induced conformational change was postulated to cause the kinase inhibition (14) (Fig. 5).

On binding of cyanide, the out-of-plane distortions of the porphyrin were significantly reduced, yielding a structure less "ruffled" by over 0.4 Å, according to normal-coordinate structural decomposition (NSD) analysis. In addition, three salt bridges between protein residues and the heme propionates were repositioned, causing displacement of a loop, i.e., the FG regulatory loop, by about 1.6 Å. There is considerable controversy over how binding of a ligand causes rearrangement of the salt bridges and displacement of the FG loop. Gong and colleagues originally proposed that alteration of the shape and reactivity of the heme itself, brought about by the coordination of a strong-field ligand, drove the conformational change (14). This hypothesis was largely inspired by the "spin state hypothesis" of FixL action, which had its basis in the apparently equal inhibition of FixL enzymatic activity by all strong-field ligands of heme (35). Perutz and colleagues have argued that the conformational change is driven by a simple displacement of distal residues by the bound ligand (19). Specifically, they have hypothesized that a slight displacement of two isoleucine residues (I209 and I210 in *Rm*FixL) from the heme pocket favors the second set of salt bridges over the first, triggering a long-range movement of the FG regulatory loop. This displacement hypothesis would predict that substitution of the isoleucines with bulkier residues would favor the sterically displaced "off" state even in the absence of ligands, whereas their replacement with smaller residues would favor the "on" state, making the ligand-bound forms more active. In fact, mutation of those residues had little effect on ligand inhibition, and all substitutions, whether to bulkier or smaller residues, raised the activity of liganded forms (58).

D. Toward an Integrated View of Structure and Function

So far, most of the research on O_2 signal transduction by FixL and FixJ has focused on analyzing the components of this system, such as the structures of isolated domains and the presumed component steps in phosphoryl transfer. It is now clear that the sensing unit is a FixLJ *complex* that operates differently from its constituent parts (36). Support for this view includes the observation that phosphoryl-transfer reactions done in the presence of FixL and FixJ are quite different, in both rate and regulation, from the reactions done in isolation (Table 1). Another line of support comes from the structures of isolated domains. Those structures fail to provide a consistent correlation between a presumed "off" state of the heme-PAS and inhibition of the kinase. For example, the structures of Fe^{II} and Fe^{III} *Rm*FixL heme-PAS are essentially the same (53). Nevertheless, the Fe^{II} *Rm*FixL enzyme is 100 times faster than Fe^{III} *Rm*FixL in the turnover of *Rm*FixJ (Table 1). Given the strong synergy observed for the FixL/FixJ system, the challenge of the next few years will be to devise approaches for observing the reaction steps and conformational changes as they occur in the intact O_2-sensing complex.

References

1. Gilles-Gonzalez MA, Gonzalez G, Perutz MF, Kiger L, Marden MC, Poyart C. Heme-based sensors, exemplified by the kinase FixL, are a new class of heme protein with distinctive ligand binding and autoxidation. Biochemistry 1994; 33:8067–8073.
2. Gilles-Gonzalez MA, Ditta GS, Helinski DR. A haemoprotein with kinase activity encoded by the oxygen sensor of *Rhizobium meliloti.* Nature 1991; 350:170–172.
3. Denninger JW, Marletta MA. Guanylate cyclase and the NO/cGMP signaling pathway. Biochim Biophys Acta 1999; 1411:334–350.

4. Shelver D, Kerby RL, He Y, Roberts GP. CooA, a CO-sensing transcription factor from *Rhodospirillum rubrum*, is a CO-binding heme protein. Proc Natl Acad Sci USA 1997; 94:11216–11220.
5. Roberts GP, Thorsteinsson MV, Kerby RL, Lanzilotta WN, Poulos T. CooA: a heme-containing regulatory protein that serves as a specific sensor of both carbon monoxide and redox state. Prog Nucleic Acid Res Mol Biol 2001; 67:35–63.
6. Lanzilotta WN, Schuller DJ, Thorsteinsson MV, Kerby RL, Roberts GP, Poulos TL. Structure of the CO sensing transcription activator CooA. Nat Struct Biol 2000; 7:876–880.
7. Gilles-Gonzalez MA. Oxygen signal transduction. IUBMB Life 2001; 51:165–173.
8. Ditta G, Virts E, Palomares A, Kim CH. The *nifA* gene of *Rhizobium meliloti* is oxygen regulated. J Bacteriol 1987; 169:3217–3223.
9. David M, Daveran ML, Batut J, Dedieu A, Domergue O, Ghai J, Hertig C, Boistard P. Kahn D. Cascade regulation of *nif* gene expression in *Rhizobium meliloti.* Cell 1988; 54:671–683.
10. Chang AL, Tuckerman JR, Gonzalez G, Mayer R, Weinhouse H, Volman G, Amikam D, Benziman M, Gilles-Gonzalez MA. Phosphodiesterase A1, a regulator of cellulose synthesis in *Acetobacter xylinum*, is a heme-based sensor. Biochemistry 2001; 40:3420–3426.
11. Hou S, Larsen RW, Boudko D, Riley CW, Karatan E, Zimmer M, Ordal GW, and Alam M. Myoglobin-like aerotaxis transducers in Archaea and Bacteria. Nature 2000; 403:540–544.
12. Hou S, Freitas T, Larsen RW, Piatibratov M, Sivozhelezov V, Yamamoto A, Meleshkevitch EA, Zimmer M, Ordal GW, Alam M. Globin-coupled sensors: a class of heme-containing sensors in Archaea and Bacteria. Proc Natl Acad Sci USA 2001; 98:9353–9358.
13. Hou S, Belisle C, Lam S, Piatibratov M, Sivozhelezov V, Takami H, Alam M. A globin-coupled oxygen sensor from the facultatively alkaliphilic *Bacillus halodurans C-125.* Extremophiles 2001; 5:351–354.
14. Gong W, Hao B, Mansy SS, Gonzalez G, Gilles-Gonzalez MA, Chan, MK. Structure of a biological oxygen sensor: A new mechanism for heme-driven signal transduction. Proc Natl Acad Sci USA 1998; 95:15177–15182.
15. Delgado-Nixon VM, Gonzalez G, Gilles-Gonzalez MA. Dos, a heme-binding PAS protein from *Escherichia coli*, is a direct oxygen sensor. Biochemistry 2000; 39:2685–2691.
16. Aono S, Kato T, Matsuki M, Nakajima H, Ohta T, Uchida T, Kitagawa T. Resonance Raman and ligand binding studies of the oxygen sensing signal transducer protein HemAT from *Bacillus subtilis.* J Biol Chem 2002; 277:13528–13538.
17. Ross P, Mayer R, Benziman M. Cellulose biosynthesis and function in bacteria. Microbiol Rev 1991; 55:35–58.
18. Tomita T, Gonzalez G, Chang AL, Ikeda-Saito M, Gilles-Gonzalez MA. A comparative resonance Raman analysis of heme-binding PAS domains: heme-iron coordination structures of the BjFixL, AxPDEA1, EcDos, and MtDos proteins. Biochemistry 2002; 41:4819–4826.
19. Perutz MF, Paoli M, Lesk AM. FixL, a haemoglobin that acts as an oxygen sensor: signalling mechanism and structural basis of its homology with PAS domains. Chem Biol 1999; 6:291–297.

20. Anthamatten D, Hennecke H. The regulatory status of the *fixL*- and *fixJ*-like genes in *Bradyrhizobium japonicum* may be different from that in *Rhizobium meliloti*. Mol Gen Genet 1991; 225:38–48.
21. Pawlowski K, Klosse U, de Bruijn FJ. Characterization of a novel *Azorhizobium caulinodans ORS571* two-component regulatory system, NtrY/NtrX, involved in nitrogen fixation and metabolism. Mol Gen Genet 1991; 231:124–138.
22. Soupene E, Foussard M, Boistard P, Truchet G, Batut J. Oxygen as a key developmental regulator of *Rhizobium meliloti* N_2 fixation gene expression within the alfalfa root nodule. Proc Natl Acad Sci USA 1995; 92:3759–3763.
23. Nellen-Anthamatten D, Rossi P, Preisig O, Kullik I, Babst M, Fischer HM, Hennecke H. *Bradyrhizobium japonicum* FixK2, a crucial distributor in the FixLJ-dependent regulatory cascade for control of genes inducible by low oxygen levels. J Bacteriol 1998; 180:5251–5255.
24. Preisig O, Zufferey R, Thony-Meyer L, Appleby CA, Hennecke H. A high-affinity cbb3-type cytochrome oxidase terminates the symbiosis-specific respiratory chain of *Bradyrhizobium japonicum*. J Bacteriol 1996; 178:1532–1538.
25. Stock JB, Ninfa AJ, Stock AM. Protein phosphorylation and regulation of adaptive responses in bacteria. Microbiol Rev 1989; 53:450–490.
26. Parkinson JS, Kofoid EC. Communication modules in bacterial signaling proteins. Annu Rev Genet 1992; 26:71–112.
27. Swanson RV, Alex LA, Simon MI. Histidine and aspartate phosphorylation: two-component systems and the limits of homology. Trends Biol Sci 1994; 19: 485–490.
28. Chang C, Stewart RC. The two-component system. Regulation of diverse signaling pathways in prokaryotes and eukaryotes. Plant Physiol 1998; 117:723–731.
29. Stock AM, Robinson VL, Goudreau PN. Two-component signal transduction. Annu Rev Biochem 2000; 69:183–215.
30. Da Re S, Schumacher J, Rousseau P, Fourment J, Ebel C, Kahn D. Phosphorylation-induced dimerization of the FixJ receiver domain. Mol Microbiol 1999; 34:504–511.
31. Patschkowski T, Schluter A, Priefer UB. *Rhizobium leguminosarum bv. viciae* contains a second *fnr/fixK*-like gene and an unusual *fixL* homologue. Mol Microbiol 1996; 21:267–280.
32. Lois AF, Ditta GS, Helinski DR. The oxygen sensor FixL of *Rhizobium meliloti* is a membrane protein containing four possible transmembrane segments. J Bacteriol 1993; 175:1103–1109.
33. Gilles-Gonzalez MA, Gonzalez G. Regulation of the kinase activity of heme protein FixL from the two-component system FixL/FixJ of *Rhizobium meliloti*. J Biol Chem 1993; 268:16293–16297.
34. Lois AF, Weinstein M, Ditta GS, Helinski DR. Autophosphorylation and phosphatase activities of the oxygen-sensing protein FixL of *Rhizobium meliloti* are coordinately regulated by oxygen. J Biol Chem 1993; 268:4370–4375.
35. Gilles-Gonzalez MA, Gonzalez G, Perutz MF. Kinase activity of oxygen sensor FixL depends on the spin state of its heme iron. Biochemistry 1995; 34:232–236.
36. Tuckerman JR, Gonzalez G, Gilles-Gonzalez MA. Complexation precedes phosphorylation for two-component regulatory system FixL/FixJ of *Sinorhizobium meliloti*. J Mol Biol 2001; 308:449–455.
37. Perutz MF, Sanders JKM, Chenery DH, Noble RW, Pennelly RR, Fung LWM, Ho C, Giannini I, Porschke D, Winkler H. Interactions between the quaternary structure of the

globin and the spin state of the heme in ferric mixed spin derivatives of hemoglobin. Biochemistry 1978; 17:3640–3651.
38. Perutz MF. Mechanisms of Cooperativity and Allosteric Regulation in Proteins. Cambridge: Cambridge University Press, 1989.
39. Tuckerman JR, Gonzalez G, Dioum EM, Gilles-Gonzalez MA. Ligand and oxidation-state specific regulation of the heme-based oxygen sensor FixL from *Sinorhizobium meliloti.* Biochemistry. Biochemistry 2002; 41:6170–6177.
40. Zhulin IB, Taylor BL, Dixon R. PAS domain S-boxes in Archaea, Bacteria and sensors for oxygen and redox. Trends Biochem Sci 1997; 22:331–333.
41. Taylor BL, Zhulin IB. PAS domains: internal sensors of oxygen, redox potential, and light. Microbiol Mol Biol Rev 1999; 63:479–506.
42. Genick UK, Borgstahl GE, Ng K, Ren Z, Pradervand C, Burke PM, Srajer V, Teng TY, Schildkamp W, McRee DE, Moffat K, Getzoff ED. Structure of a protein photocycle intermediate by millisecond time-resolved crystallography. Science 1997; 275:1471–1475.
43. Genick UK, Soltis SM, Kuhn P, Canestrelli IL, Getzoff ED. Structure at 0.85 A resolution of an early protein photocycle intermediate. Nature 1998; 392:206–209.
44. Pellequer JL, Wager-Smith KA, Kay SA, Getzoff ED. Photoactive yellow protein: a structural prototype for the three-dimensional fold of the PAS domain superfamily. Proc Natl Acad Sci USA 1998; 95:5884–5890.
45. Morais-Cabral JH, Lee A, Cohen SL, Chait BT, Li M, Mackinnon R. Crystal structure and functional analysis of the HERG potassium channel N Terminus: a eukaryotic PAS domain. Cell 1998; 95:649–655.
46. Hill S, Austin S, Eydmann T, Jones T, Dixon R. *Azotobacter vinelandii* NifL is a flavoprotein that modulates transcriptional activation of nitrogen-fixation genes via a redox-sensitive switch. Proc Natl Acad Sci USA 1996; 93:2143–2148.
47. Rebbapragada A, Johnson MS, Harding GP, Zuccarelli AJ, Fletcher HM, Zhulin IB, Taylor BL. The Aer protein and the serine chemoreceptor Tsr independently sense intracellular energy levels and transduce oxygen, redox, and energy signals for *Escherichia coli* behavior. Proc Natl Acad Sci USA 1997; 94:10541–10546.
48. Hefti M, Hendle J, Enroth C, Vervoort J, Tucker PA. Crystallization and preliminary crystallographic data of the PAS domain of the NifL protein from *Azotobacter vinelandii.* Acta Crystallogr D Biol Crystallogr 2001; 57:1895–1896.
49. Christie JM, Salomon M, Nozue K, Wada M, Briggs WR. LOV light, oxygen, or voltage domains of the blue-light photoreceptor phototropin nph1: binding sites for the chromophore flavin mononucleotide. Proc Natl Acad Sci USA 1999; 96:8779–8783.
50. Crosson S, Moffat K. Structure of a flavin-binding plant photoreceptor domain: insights into light-mediated signal transduction. Proc Natl Acad Sci USA 2001; 98:2995–3000.
51. Park H, Saha SK, Inouye M. Two-domain reconstitution of a functional protein histidine kinase. Proc Natl Acad Sci USA 1998; 95:6728–6732.
52. Bertolucci C, Ming LJ, Gonzalez G, Gilles-Gonzalez MA. Assignment of the hyperfine-shifted ^{1}H-NMR signals of the heme in the oxygen sensor FixL from *Rhizobium meliloti.* Chem Biol 1996; 3:561–566.
53. Miyatake H, Masahiro M, Park SY, Adachi SI, Tamura K, Nakamura H, Nakamura K, Tsuchiya T, Iizuka T, Shiro Y. Sensory mechanism of oxygen sensor FixL from *Rhizobium meliloti*: crystallographic, mutagenesis and resonance Raman spectroscopic studies. J Mol Biol 2000; 301:415–431.

54. Rodgers KR, Tang L, Lukat-Rodgers GS, Wengenack NL. Insights into the signal transduction mechanism of RmFixL provided by carbon monoxide recombination kinetics. Biochemistry 2001; 40:12932–12942.
55. Surette MG, Levit M, Liu Y, Lukat G, Ninfa EG, Ninfa A, Stock JB. Dimerization is required for the activity of the protein histidine kinase CheA that mediates signal transduction in bacterial chemotaxis. J Biol Chem 1996; 271:939–945.
56. Stewart RC, VanBruggen R, Ellefson DD, Wolfe AJ. TNP-ATP and TNP-ADP as probes of the nucleotide binding site of CheA, the histidine protein kinase in the chemotaxis signal transduction pathway of *Escherichia coli.* Biochemistry 1998; 37:12269–12279.
57. Gong W, Hao B, Chan MK. New mechanistic insights from structural studies of the oxygen-sensing domain of *Bradyrhizobium japonicum* FixL. Biochemistry 2000; 39:3955–3962.
58. Mukai M, Nakamura K, Nakamura H, Iizuka T, Shiro Y. Roles of Ile209 and Ile210 on the heme pocket structure and regulation of histidine kinase activity of oxygen sensor FixL from *Rhizobium meliloti.* Biochemistry 2000; 39:13810–13816.

2

A Role for the Mitochondrion and Reactive Oxygen Species in Oxygen Sensing and Adaptation to Hypoxia in Yeast

ROBERT O. POYTON, REINHARD P. DIRMEIER, and KRISTIN M. O'BRIEN

University of Colorado
Boulder, Colorado, U.S.A.

ERICK SPEARS

Pharmatech Inc.
Denver, Colorado, U.S.A.

I. Introduction

Oxygen has a profound effect on the survival, growth, and metabolism of most organisms. The presence or absence of oxygen, as well as oxygen tension itself, determines whether ATP will be produced primarily by respiration-linked oxidative phosphorylation, anaerobic respiration, or fermentation-linked substrate-level phosphorylation (1). Because oxygen tension affects the ratio of ATP produced by glycolysis versus oxidative phosphorylation it has important effects on the amount of biomass that accumulates, the amount of heat released, and the types and amounts of metabolic end products produced. Consequently, the ability of cells, tissues, and organisms to sense and respond to changes in oxygen levels is often crucial to their survival. Cellular adaptation to hypoxia is a good example of this. Intermittent episodes of hypoxia are associated with a variety of human pathophysiological states (e.g., sleep apnea, central hypoventilation syndrome, and vascular occlusion); these can alter metabolism, induce angiogenesis, and affect inflammatory responses. And the failure of cells to respond properly to hypoxia can lead to anemia, myocardial infarction, retinopathy, and the growth of tumors (2–4).

In many organisms, adaptation to changing oxygen concentrations is achieved both by the short-term effects of oxygen on energy metabolism and the long-term effects of oxygen on gene expression (5,6). Short-term regulation is immediate and can be affected by allosteric regulation of metabolic enzymes. It does not require the expression of new proteins. Long-term effects of oxygen on cell metabolism are mediated at the level of transcription, via oxygen-responsive genes in the nucleus. These oxygen-regulated genes can be placed into one of two groups: *aerobic* genes, which are transcribed optimally in the presence of air, and *hypoxic* genes, which are transcribed optimally under anoxic or microaerophilic conditions. Some aerobic and some hypoxic genes are members of families in which one member is an aerobic gene and another is a hypoxic gene. Often, these gene pairs encode protein isoforms that are interchangeable but functionally different (e.g., 7,8). Among those nuclear genes whose expression is sensitive to oxygen are genes for respiratory chain proteins, enzymes involved in the oxidative stress response, and reductases and desaturases involved in oxygen-dependent biosynthetic pathways (e.g., in the synthesis of heme, sterols, and unsaturated fatty acids). Changes in the levels of some of these proteins serves to redirect energy flow through metabolic pathways that are better able to match cellular energy production to need. When considering how cells sense and respond to oxygen it is important to keep in mind that some organisms can grow over a much wider range of oxygen concentrations than others. Facultative anaerobes, like *Escherichia coli* or *Saccharomyces cerevisiae*, can grow at oxygen concentrations anywhere between 0 μM (anoxia) and 220 μM (normoxia = 159 torr, or 21% O_2), and some grow well even under hyperoxic conditions (i.e., 250 μM O_2 and above). In contrast, a given mammalian cell or tissue generally experiences a more restricted range of oxygen concentrations (Fig. 1). Given the wide range of oxygen concentrations that cells can experience, it is likely that multiple pathways for sensing and responding to changes in oxygen have been developed and that different pathways are operative at different oxygen tensions and in different cells and tissues.

In recent years, considerable progress has been made in our understanding of the molecular components that are involved in regulating the oxygen-responsive transcription of nuclear genes (9–12). Numerous oxygen-responsive transcription factors have been identified in a number of different organisms, including mammals, yeast, and bacteria. Despite a great deal of progress in understanding how these transcription factors function, the nature of the more proximal events involved in oxygen sensing has remained elusive. It is possible to envision two fundamentally different types of model. In the first model, oxygen itself has a direct effect on transcription either by acting on a transcription factor itself or by affecting a component that interacts with a transcription factor. In a second model, oxygen is sensed more distally by a proximal oxygen sensor, a signal is produced, and the signal initiates a signal transduction cascade that affects distal transcription factors. Support for the first type of model was initially provided by studies on the *E. coli* FNR protein (13), iron-sulfur clusters in central neurons (14), mammalian iron-responsive elements (15), and more recently, from studies on the HIF-1α protein in the nematode *C. elegans* (12). Evidence for the second type of model has come from an analysis of a two-component oxygen-sensing system in the nitrogen-fixing bacterium *Rhizobium meliloti*. Here, the hemokinase FixL functions as a proximal

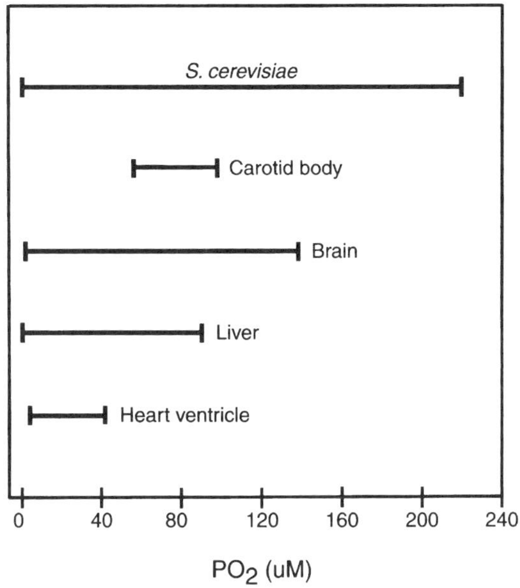

Figure 1 Tissue oxygen concentrations experienced by some mammalian cells. Values for the carotid body are from measurements of the carotid body microvasculature with an arterial PO_2 of 145 μM, taken from Lahiri et al. (110). Values for brain are from measurements taken within many regions of the brain in rat, cat, rabbit, and piglet, reviewed in Erecinska and Silver (111). Liver PO_2 measurements are from rat, taken from deGroot and Noll (112) and Vollmar and Menger (113). Measurements of heart ventricle are from the epicardium of ventricles in cat, dog, and piglet, taken from Rumsey et al. (114,115) and Honig and Gayeski (116). Tissue PO_2 measurements reported in the literature were converted from units of torr to μM O_2 using the conversion factor of 1.4 μM O_2 per unit torr.

oxygen sensor, which starts a phosphorylation cascade that ultimately leads to the up-regulation of genes involved in nitrogen fixation and hydrogen utilization (1,11). FixL functions to turn on genes only when the oxygen concentration reaches extremely low levels (below ~10 nM). Evidence for the involvement of hemoprotein oxygen sensors has also come from studies with other bacteria, yeast, and mammalian cells (for review see Ref. 1). Although it is not entirely clear how these hemoproteins function in oxygen sensing, there is growing evidence that reactive oxygen species (ROS) are involved (16–19). In eukaryotic cells, ROS are the by-products of mitochondrial respiration and, to a lesser extent, of oxygen-consuming reactions in the cytosol. These highly unstable molecules are Janus-faced. If their intracellular levels are abnormally elevated, ROS can lead to protein oxidation and degradation, the accumulation of oxidized nucleosides in DNA, and the production and release of lipid peroxides. This oxidative stress can have extremely deleterious effects on cells and their components. It can also promote apoptosis, premature aging, and a variety of degenerative human diseases (20–22).

On the other hand, low levels of ROS, especially those produced by the mitochondrion, appear to play essential roles as physiological signals (23–25).

One common element that seems to link the ability of cells to respond to oxygen or ROS with disease and aging is the mitochondrion. Mitochondrial dysfunction and/or the cross-talk that takes place between the mitochondrion and nucleus have been implicated in programmed cell death, aging, hypoxic gene induction, and the expression of aerobic genes in a diverse array of organisms (e.g., yeast and other fungi, nematodes, and mammals) (16,26–30). In this chapter, we focus on mitochondrial involvement in the adaptation of yeast cells to hypoxia. First, we consider oxygen-regulated gene expression. Then, we focus on adaptation to hypoxia as a form of oxidative stress. Finally, we discuss mitochondrial adaptation to hypoxia. Because mitochondrial involvement in oxygen sensing, aging, and disease appears to be ubiquitous, it is likely that the molecular mechanisms involved have been conserved during the course of eukaryotic evolution and that lessons learned from studies with the yeast *Saccharomyces cerevisiae* will be applicable to higher eukaryotes as well. Moreover, this yeast offers biochemical, physiological, and genetic approaches and is uniquely suited for assessing the involvement of *both* nuclear and mitochondrial genes in oxygen-sensing pathways.

II. Yeast Cells Adapt to Growth at Different Oxygen Concentrations via Differential Gene Expression

S. cerevisiae is a facultative anaerobe that can grow under normoxic or anoxic conditions and at any oxygen concentration in between. When grown in the presence of air *S. cerevisiae* respires, but under anaerobic conditions it supports its energy needs by fermentation. Because the expression of many respiratory protein genes is regulated by oxygen, the presence or absence of oxygen determines whether yeast cells will develop a functional respiratory chain. Interestingly, full respiratory function is achieved in cells grown at oxygen concentrations ranging from 220 μM O_2 (normoxia) to 5 μM O_2 (hypoxia). It is only at very low levels of oxygen (between 0 and 3.5 μM) that the level of respiration is related to oxygen concentration (31). This somewhat surprising finding is most likely attributable to the effects of oxygen tension per se on the expression of respiratory protein genes. Recent studies have demonstrated that the expression of many of these genes is determined by the actual concentration of oxygen and not merely by its presence or absence (9,32,33). This has been called oxygen-concentration-dependent gene expression. By examining the oxygen-concentration-dependent expression of a number of genes these studies have also demonstrated that there are at least four classes of oxygen-dependent genes in yeast: (1) aerobic genes with hypoxic isoform counterparts, (2) aerobic genes without hypoxic isoform counterparts, (3) hypoxic genes with aerobic isoform counterparts, and (4) hypoxic genes without aerobic isoform counterparts. In general, those genes with counterparts that are oppositely regulated by oxygen are much more stringently regulated by oxygen concentration than those genes that lack them. The finding that these four classes of genes are regulated differently in response to oxygen concentration lends further support to the idea that eukaryotic cells, even those as simple as *S. cerevisiae*, possess multiple mechanisms for sensing oxygen.

A. Effects of Hypoxia on Expression of Oxygen-Dependent Genes

Microarray analysis suggests that there are some 140 aerobic and 290 hypoxic nuclear genes in yeast (34). Together, these genes represent about 7% of all of the genes in the yeast nuclear genome. The remainder of the 6200 genes present in the yeast nuclear genome are unaffected by the presence or absence of oxygen. When yeast cells are shifted from normoxic to anoxic conditions the hypoxic genes are induced with remarkably divergent kinetics. Some hypoxic genes (e.g., *COX5b*, *ANB1*, *AAC3*) are induced immediately and their expression increases slowly until it reaches its anoxic levels, others (e.g., *CYC7*) exhibit a long lag and then their expression increases rapidly to their anoxic level, while the expression of others (e.g., *HEM13*, *OLE1*, *CPR1*, and *ERG11*) rises rapidly to a peak that is 1.5- to 2.5-fold higher than their anoxic levels and then falls back to their anoxic level. At the same time that transcript levels from hypoxic genes are increasing, transcript levels from aerobic genes are decreasing. This decay in transcript level is too rapid to result from simple dilution due to cell growth after a shift to anoxia and, therefore, is due to active degradation. So, in considering how yeast cells adapt to exposure to hypoxia or anoxia it is important to consider both the induction of hypoxic gene transcription and the degradation of transcripts from aerobic genes. At present, little is known about how oxygen concentration selectively affects the stability of transcripts in yeast or any other organism. However, our understanding of hypoxic gene induction is progressing rapidly.

B. Oxygen-Regulated Transcription

Several studies have revealed that heme plays a pivotal role in regulating the expression of many oxygen-regulated yeast genes, acting as a positive modulator for transcription of the aerobic genes and a negative modulator for the transcription of the hypoxic genes (1,33,35–38). The influence of heme on the expression of aerobic genes is at least partially mediated by the transcription factors Hap1p and Hap2/3/4/5p. In addition to its effects on aerobic genes, Hap1p regulates the expression of some hypoxic genes. Heme affects Hap1p by modulating its activity and Hap2/3/4/5p by regulating its intracellular level (c.f., 1,33,39,40). The influence of heme on the expression of hypoxic genes appears to be more complex. There are two groups of heme-dependent hypoxic genes. The best-understood group includes those hypoxic genes that are repressed in aerobic cells by Rox1p in conjunction with Tup1/Ssn6p (41–43). Genes in this group encode subunits of cytochrome complexes, enzymes involved in sterol, heme, and fatty acid biosynthesis, and at least one translation factor. Although heme does not influence the activity of Rox1p (44), *ROX1* is an aerobic gene whose expression is activated by Hap1p. Hence under conditions of low oxygen, heme and Rox1p levels drop (45) and Rox1p-regulated hypoxic genes are derepressed. In addition, Hap1p actively represses transcription of *ROX1* in oxygen-limiting conditions (46). Recently, two groups have identified a new *cis* element, LORE (Low Oxygen Response Element), that is involved in the induction of *OLE1*, the gene for Δ9-fatty acid desaturase, under low oxygen conditions (47–49). LORE functions positively in the induction of *OLE1* in response to hypoxia and negatively to repress *OLE1* in the presence of unsaturated fatty acids. Both the induction of *OLE1* by hypoxia and its repression by unsaturated fatty acids is mediated by Mga2p, which binds LORE. Thus, *OLE1* is oxygen-regulated by two different types of transcription factor; ROX1p, which represses it in air, and Mga2p, which induces it in the absence of air. The second group of heme-regulated hypoxic genes shows weak or no dependence on Rox1p (50,51). This group includes genes (*DAN1*, *DAN2*, *DAN3*, *DAN4*, *TIR1*, *TIR2*, *TIR3*, and *TIR4*) that encode cell wall mannoproteins. They are regulated by at least one transcriptional activator (Mox4/Upc2p) and three repression factors, Mox1p, Mox2p, and Rox7/Mot3p (52). *MOX4* is a hypoxic gene and is repressed under normoxic conditions by heme. In contrast, *MOX1*, *MOX2*, and *MOT3* are heme-dependent aerobic genes.

The precise role of heme in regulating the activity of Hap1p or in inducing *MOX1*, *MOX2*, and *MOT3* expression is unclear. Although it has been argued that intracellular heme concentration controls the activity of Hap1p (40,53,54), there is no direct evidence relating intracellular heme levels to Hap1p activity and the possibility that heme works as a redox-sensitive prosthetic group in Hap1p or in a hemoprotein oxygen sensor that functions upstream of Hap1p or *ROX1* has not been ruled out (c.f., 9). The recent finding that mutations in the respiratory chain and drugs that inhibit respiration prevent the induction of some ROX1-dependent hypoxic genes raises the distinct possibility that the redox state of one or more of the hemoprotein electron carriers in the respiratory chain functions to sense a change in oxygen concentration and translate it into downstream changes in gene expression.

III. Mitochondrial–Nuclear Cross-Talk Is Involved in Oxygen-Regulated Nuclear Gene Expression

It has been known for some time that mutations in mitochondrial DNA accompany aging and degenerative disease in a number of organisms (55). This can have profound effects on mitochondrial respiratory function, on cellular adaptation to hypoxia, and on the cross-talk between the mitochondrion and the nucleus. The latter is especially interesting in the context of oxygen-regulated gene expression because it has also been implicated in hypoxic gene induction and in the regulation of aerobic gene expression. This cross-talk involves signaling pathways that connect *either* mitochondrial respiration *or* the mitochondrial genome (independently of its respiratory function) to the expression of specific nuclear genes.

A. Cross-Talk Between the Mitochondrion and Nucleus

Although cross-talk between the mitochondrion and nucleus occurs in all eukaryotes it is best understood in *S. cerevisiae* (56). In this organism, the mitochondrion can affect the expression of specific nuclear genes by two fundamentally different pathways. The first pathway, *retrograde regulation*, is used by aerobically growing cells to sense the energy state of their mitochondria (56–58). It functions to upregulate some aerobic genes in response to the lack of mitochondrial respiration. Components of this pathway include: two subunits (Rtg1p and Rtg3p) of a heterodimeric transcription factor, Rtg2p, a cytoplasmic protein that contains an hsp-70-like ATP binding site (59), and the Tup1-Cyc8 protein complex, which interacts with the Rtg1-Rtg3 heterodimer and which can either activate or repress transcription (60). The second pathway, *intergenomic signaling*, downregulates aerobic genes in response to the lack of a mitochondrial genome or the reduced expression of one of its genes, and not in response to a lack of respiration (61,62). It functions to regulate the expression of a number of respiratory protein genes (62,63) and in the survival of *S. cerevisiae* in a mammalian host environment (64). This pathway works via the differential phosphorylation of the transcription factor ABF1 (65).

Intergenomic signaling is distinguishable from retrograde regulation in three ways. First, mitochondrial respiration is important for retrograde regulation but not for intergenomic signaling. Second, those genes that are subject to retrograde regulation are upregulated in the absence of mitochondrial respiration while those genes that are subject to intergenomic signaling are downregulated in the absence of a mitochondrial genome. Third, intergenomic signaling affects expression of genes that encode proteins involved in energy production while retrograde regulation does not.

B. Role of Cytochrome c Oxidase and the Respiratory Chain in the Induction of Hypoxic Genes

Recent studies with both yeast and mammalian cells in culture have suggested that cytochrome c oxidase and the mitochondrial respiratory chain may function as oxygen sensors in the induction of some hypoxic genes (16,26). Evidence for the involvement of cytochrome c oxidase and the respiratory chain in yeast comes from the finding that carbon monoxide and other respiratory inhibitors as well as mutations in respiratory chain proteins prevent induction of some hypoxic genes. These findings directly implicate the respiratory chain in the induction of some hypoxic genes. Are either the retrograde regulation or intergenomic signaling pathways involved? It is unlikely that the intergenomic signaling pathway is involved because it is the inhibition of respiration and not the lack of a mitochondrial genome that prevents the induction of hypoxic yeast nuclear genes. And it is unlikely that the retrograde regulation pathway is involved because hypoxic yeast genes whose induction requires the respiratory chain lack the R-box promoter elements that have been shown to function as binding sites for Rtg1-Rtg3. The considerations imply that other, as yet undiscovered, pathways or mechanisms are involved in mitochondrial–nuclear cross-talk.

C. The Mitochondrial Genome Itself Is Required for Optimal Expression of Some Aerobic Genes but Not for Their Downregulation in Hypoxic Cells

The mechanisms by which aerobic nuclear genes are downregulated during a shift to anoxia have been largely ignored in most organisms. Recently, this question has been addressed with yeast by looking at the downregulation of nuclear *COX* genes after a shift to anoxia (61,62). These studies have provided several new insights concerning the regulation of aerobic yeast genes. First, they demonstrated that both oxygen and the mitochondrial genome exert a positive effect on the expression of these genes but that mitochondrial respiration per se has no effect. Second, they showed that a mitochondrial genome is required for optimal expression of these genes under both normoxic and anoxic conditions. And third, they showed that the downregulation of these genes brought about by exposure to low oxygen concentrations and the down-regulation brought about by the absence of a mitochondrial genome are independent of one another, ruling out the possibility that the downregulation of aerobic genes in yeast cells exposed to hypoxia is mediated by the mitochondrion. Taken together, these findings support a role for intergenomic signaling in the expression of aerobic nuclear genes and imply that at least one mitochondrial gene is required for the optimal expression of these genes under both normoxic and anoxic conditions.

IV. Adaptation to Hypoxia as a Form of Oxidative Stress

Insofar as neither of the known signaling pathways that facilitate mitochondrial–nuclear cross-talk in yeast function in hypoxic gene induction, it is likely that a hitherto undescribed pathway or mechanism is involved. In trying to identify this pathway several questions are of interest. First, how do cytochrome c oxidase and the respiratory chain "sense" oxygen when cells are shifted from normoxic to anoxic conditions? Second, what signal(s) is (are) released from the mitochondrion during this process? And third, what pathway(s) are involved in transducing this signal into an effect on nuclear gene expression?

Because the respiratory chain can produce ROS and because ROS are known to mediate some intracellular signal cascades, we have recently addressed the possibility that ROS are involved in hypoxic gene induction, by first asking if a shift to anoxia is a form of oxidative stress. Previous studies with mammalian cells in culture have reported that when cells are shifted from normoxic to hypoxic conditions ROS levels increase and that it is these increased ROS levels that signal the induction of nuclear hypoxic genes (16,66,67). In contrast, other studies have reported that ROS levels decrease when mammalian cells in culture are exposed to hypoxia (c.f., 68). Both sets of studies used fluorescent dyes (either 2′,7′-dichlorofluorescein diacetate or dihydrorhodamine 123) to measure H_2O_2 levels in cells exposed to hypoxia. Because of this discrepancy it has been suggested that fluorescent dyes lack the necessary precision to monitor ROS levels in mammalian cells exposed to hypoxia (69). Alternatively, because ROS are transient and highly reactive it is possible that cells shifted to hypoxia do experience increased oxidative stress but that free ROS levels do not reflect the increase because they rapidly oxidize macromolecules (proteins, lipids, and nucleic acids) in their immediate vicinity. In this case, a more reliable method for measuring changes in ROS levels would be to measure their damage (70) by measuring the accumulation of lipid peroxides (e.g., malondialdehyde and hydroxyalkenals), oxidized nucleosides [8-hydroxy-2′-deoxyguanosine (8OH2dG)], or oxidized amino acid side chains on proteins (e.g., o-tyrosine, m-tyrosine, dityrosine, and carbonyl derivatives).

To determine whether ROS levels change in yeast cells exposed to anoxia, to ask if the respiratory chain is primarily responsible for ROS production in yeast, and to assess the possibility that ROS are involved in the induction of hypoxic yeast genes we have recently monitored ROS levels both by a fluorescent dye (carboxy-H2-dichloro-dihydrofluorescein-diacetate) and by assessing levels of protein and DNA damage (19) during a shift from normoxia to anoxia. These studies revealed that whereas carboxy-H2-dichloro-dihydrofluorescein-diacetate is useful for assessing oxidative stress in yeast cells grown at different steady-state oxygen concentrations, it is not useful for assessing ROS levels in cells shifted from one oxygen concentration to another. Indeed, during a shift from normoxic to anoxic conditions the intracellular concentration of the carboxy-H2-dichloro-dihydrofluorescein-diacetate decreases (19), probably reflecting dye efflux out of cells (71). This decrease in intracellular dye concentration makes it virtually impossible to use this dye as a reporter for intracellular oxidant levels. When considered with the results from similar studies with mammalian cells these findings support the view (69) that carboxy-H2-dichloro-dihydrofluorescein-diacetate and similar fluorescent dyes are not reliable for assessing ROS levels in cells shifted from one oxygen concentration to another.

It is generally accepted that the most reliable way to assess oxidative stress in cells is to measure oxidative damage (70). One useful indicator of oxidative stress is the level of the oxidized nucleoside 8-hydroxy-2′-deoxyguanosine (8OH2dG) in DNA. The levels of 8OH2dG are most easily measured using HPLC in conjunction with electrochemical detection (Fig. 2). Applying this assay to cells shifted from normoxia to anoxia it has been found that the levels of 8OH2dG in mitochondrial DNA are substantially higher than those in nuclear DNA, and that in normoxic cells the levels of 8OH2dG in both nuclear and mitochondrial DNA increase transiently, after a shift to anoxia (19). This increase suggests a transient rise in ROS in cells exposed to anoxia. To confirm this we used a second type of assay for oxidative damage: the accumulation of protein carbonyls. Protein carbonyls are easily quantitated after derivatization with 2,4-dinitrophenylhydrazine (2,4-DNP). The 2,4-dinitrophenylhydrazine is converted to 2,4-dinitrophenylhydrazone by interaction with carbonyl groups and the DNP-proteins are then subjected to size fractionation on an HPLC linked to a photo detector (72,73). DNP-proteins are also easily identified by immunoblotting, using anti-DNP antisera (74,75). In yeast, most of the protein carbonylation of both mitochondrial and cytosolic proteins results from ROS released by mitochondrial respiration (19). By analyzing mitochondrial and cytosolic fractions taken from yeast cells after a shift from normoxia to anoxia

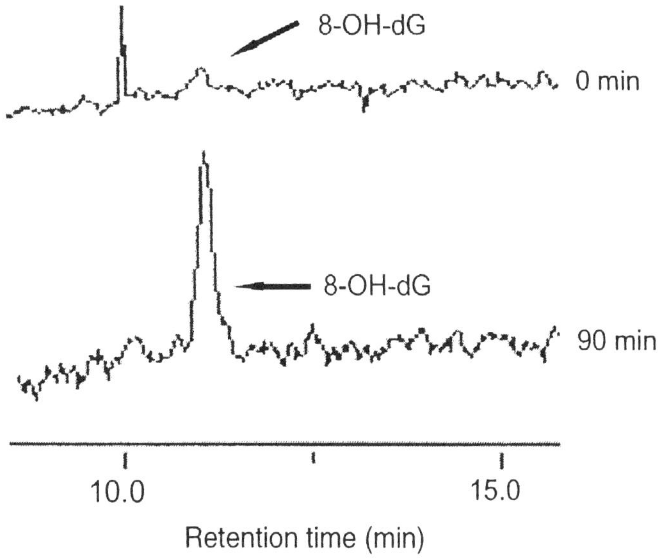

Figure 2 Detection of 8OHdG in mitochondrial DNA after a shift to anoxia. Mitochondrial DNA was isolated, hydrolyzed, and analyzed for 8-OHdG from cells before a shift (top trace) and 90 min after a shift from normoxia to anoxia (bottom trace). 8-OHdG was detected with an electrochemical detector.

we have found that mitochondrial protein carbonylation initially drops, then increases dramatically between 120 and 180 min after a shift, and finally drops back to its anoxic level after 12 h. An analysis of the carbonylated proteins by two-dimensional electrophoresis followed by immunoblotting with anti-DNP antibodies has demonstrated that specific proteins become carbonylated between 120 and 180 min after the shift. Carbonylation of these proteins decreases after 180 min. The level of carbonylation of cytosolic proteins increases slightly between 50 and 100 min after the shift and then declines. Taken together, these findings clearly indicate that ROS levels increase transiently when cells are shifted from normoxia to anoxia.

These observations are interesting in the context of studies on hypoxic gene induction in mammalian cells, where it has been proposed that ROS participate as signals in a signaling pathway that mediates hypoxic stabilization of the α subunit of the HIF-1 transcription factor (67). It is not clear from these studies if, or how, ROS are released from mitochondria in vivo. Also unclear is whether ROS stabilize HIF-1α by interacting with it directly or affect it indirectly, by pathways that result in posttranslational modification of HIF-1α (e.g., proline or asparagine hydroxylation) (12,76). In this regard, it is interesting that the prolyl hydroxylase that modifes HIF-1α in *C. elegans* (12) is homologous to the SM-20 gene product, a mammalian prolyl hydroxylase that resides in the mitochondrion (77).

It is not yet known if mammalian mitochondria experience transient changes in protein and MtDNA oxidation, as they do in yeast. Nonetheless, there is mounting evidence that mitochondrially generated ROS in both yeast and mammalian cells play a pivotal role in the induction of some hypoxic genes. From the finding that protein carbonylation and 8OH2dG levels in mitochondrial DNA increase after yeast cells are shifted to anoxia or hypoxia it is possible to envision at least three pathways by which mitochondrially produced ROS could participate in hypoxic gene induction (Fig. 3). In the first pathway, ROS oxidize a mitochondrial protein, which initiates a signaling pathway to the nucleus. In the second pathway, free ROS are released from mitochondria and initiate a signaling pathway to the nucleus. This is the pathway that has been proposed for mammalian cells (67). And in the third pathway, ROS modify mitochondrial gene expression via oxidative damage to MtDNA, which initiates a signaling pathway to the nucleus. In the first and third pathways the ROS signal originates in the mitochondrion and starts a signaling pathway while still in the mitochondrion, but in the second pathway, mitochondrially produced ROS leave the mitochondrion and start a signaling pathway in the cytosol or nucleus. At present, it is not known which of these pathway(s) is (are) operative. However, microarray analysis of gene expression data suggests that some genes involved in the oxidative stress response are induced by hypoxia (34) but that hypoxic genes are not induced by exogenously added oxidants like hydrogen peroxide (78,79). These microarray data are consistent with the hypothesis that the

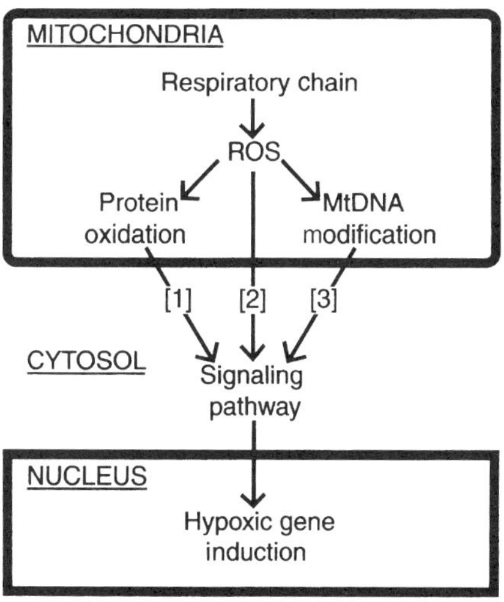

Figure 3 Three ways that ROS generated by the mitochondrial respiratory chain may be involved in hypoxic gene induction in yeast. In the first, ROS oxidize a mitochondrial protein, which initiates a signaling pathway to the nucleus. In the second, free ROS are released from mitochondria and initiate a signaling pathway to the nucleus. In the third, ROS modify mitochondrial gene expression via oxidative damage to MtDNA, which initiates a signaling pathway to the nucleus.

stress response that yeast cells experience during a shift to anoxia comes from within the cell and cannot be mimicked by exogenous oxidants.

The finding that normoxic cells undergo transient oxidative stress upon exposure to anoxia is surprising. Indeed, it raises several questions. First, does the respiratory chain release elevated levels of ROS during a shift to anoxia, and if so, why? Second, does the enhanced protein carbonylation observed during a shift to anoxia result from increased levels of ROS and where does the oxygen that is present in these ROS come from—free oxygen, lipids, or carbohydrates? Third, are specific genes affected by the increase in 8OH2dG levels or is the distribution of this oxidized base random? If specific genes are oxidized does this affect their function or expression, and if so, is this important for hypoxic gene induction? Fourth, does nitric oxide, a known inhibitor of cytochrome c oxidase, function in hypoxic gene induction? And fifth, do the proteins whose levels of carbonylation increase after a shift to hypoxia play a role in the induction of hypoxic nuclear genes, and are other types of oxidative protein modification part of signaling pathways from the mitochondrion to the nucleus?

V. Mitochondrial Adaptation to Hypoxia

In view of the growing importance of the mitochondrion in oxygen sensing and adaptation to hypoxia it is relevant to consider what is known about mitochondrial structure and function in cells grown at low oxygen concentrations.

A. Functional Studies with Yeast

Early studies by Schatz and co-workers examined the properties of mitochondria from anaerobic cells (80–83). These organelles, called promitochondria, have all of the compartments that are present in aerobic mitochondria but differ from them in several ways. Promitochondria lack, or have greatly diminished, respiratory chain enzyme activities (31,82) and extremely low levels of cytochromes aa_3, b, c, and c_1 (31,82). In addition, promitochondria have significantly lower levels of unsaturated fatty acids, especially oleic and palmitoleic acid, and high levels of short-chain saturated fatty acids (81,84). Even when cells are supplemented with ergosterol and Tween 80, a source of oleic acid, the lipid composition of promitochondria is somewhat different from that of mitochondria. In addition, the relative amounts of phospholipids vary between promitochondria and mitochondria. The most notable difference is the level of cardiolipin; it is lower in promitochondria than in mitochondria. This is interesting because in normoxic cells cardiolipin is not required for respiratory growth on a nonfermentable carbon source (85). Mitochondrial gross morphology and ultrastructure are also affected by the presence or absence of oxygen (86,87). The volume density of promitochondria is only 0.7% of the total cell volume compared to 10.5% for mitochondria in normoxic cells (86) and the surface area of cristae per mitochondrion is reduced by half in promitochondria compared to mitochondria (87).

Even though promitochondria lack a functional aerobic respiratory chain and even though mitochondria and promitochondria have different lipid compositions, the proteomes of these two organelles are remarkably similar (Fig. 4). This is consistent with the finding that many of the 450 or so nuclear genes that encode mitochondrial proteins are only marginally affected by oxygen (34). This finding suggests that the functional changes that distinguish mitochondria and promitochondria result from the effects of oxygen on relatively few genes and gene products, or on posttranslational events that affect the assembly of multimeric respiratory proteins or the insertion of essential prosthetic groups (e.g., heme, Fe.S centers, Cu) into them. Oxygen can affect these later processes by altering the expression or activity of one of the many "assembly facilitators" that are required for respiratory enzyme assembly (e.g., 34,88,89), by affecting rates of polypeptide subunit degradation (90), or by reducing the level of import of iron or copper into mitochondria. Interestingly, an oligomycin-sensitive F_1-ATPase is present and functional in promitochondria, suggesting some electron transport may be occurring, even in the absence of air (83). It is also interesting that nearly all the subunits of cytochrome c oxidase are present in promitochondria (62). Only subunit VIII, which is not required to form a functional enzyme, is undetectable (62). In addition, many proteins required for the assembly of respiratory protein complexes (56) are expressed in promitochondria (34). These findings raise the possibility that the low levels of respiratory protein subunits that are present in promitochondria assemble into partial or complete respiratory complexes and that these subunits or complexes have previously unidentified functions in anoxic cells. These functions may relate to the nonrespiratory functions (e.g., synthesis of amino acids, pyrimidines, heme, and sterols, and in mitochondrial gene expression pathways) carried out by mitochondria and promitochondria . The importance of these energy-requiring processes, even under anoxia, is evident from the 28-fold induction of *AAC3*, the gene encoding a hypoxic isoform of the mitochondrial ATP/ADP translocator (34), and from the finding that inhibition of the translocator by bongkrekic acid inhibits cell growth under both aerobic and anaerobic conditions (92). Alternatively, these respiratory chain subunits may have entirely new functions, which remain to be discovered, in anoxic cells.

A comparison of mitochondria and promitochondria, per se, may have little relevance to mammalian systems, which cannot survive anoxia. However, oxygen concentrations that lie between normoxia and hypoxia have substantial effects on mitochondrial function in yeast. It is mitochondrial response to these low oxygen concentrations that may be useful for understanding mitochondrial adaptation to hypoxia in mammalian cells. As mentioned above, respiration rate and inner mitochondrial membrane potential in yeast increases linearly with oxygen tension when cells are grown at oxygen tensions between 0 μM and 3.5 μM (31). Despite the linearity of these changes, the development of respiratory enzymes is not a coordinated response. Respiratory enzymes and cytochromes have different sensitivities to oxygen; their levels differ at varying oxygen concentrations. Only cytochrome c and ubiquinone levels parallel the development of maximum respiration rate. All other components develop maximal activity at oxygen levels

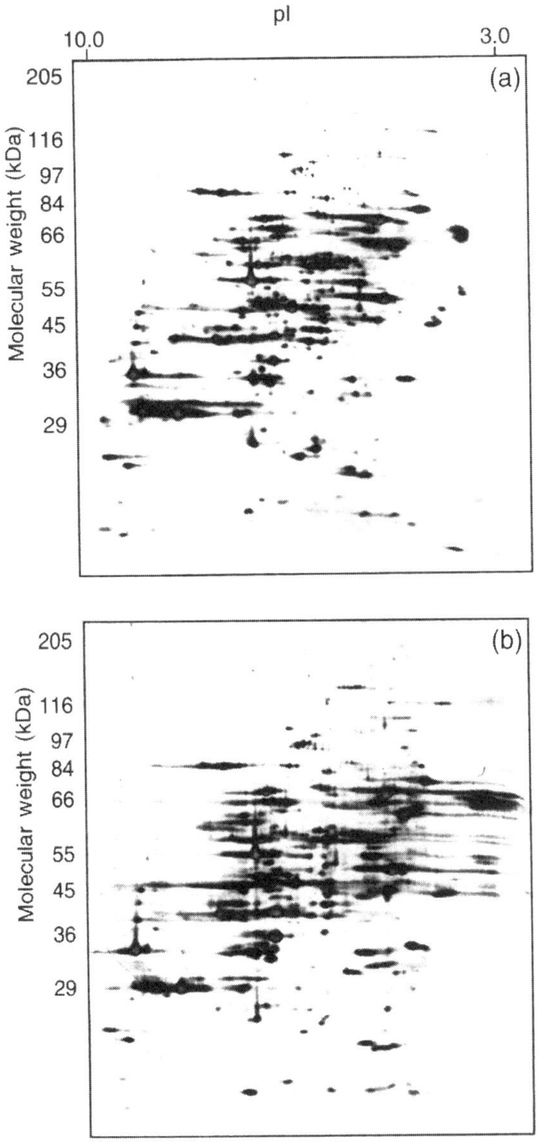

Figure 4 The proteomes of mitochondria (a) and promitochondria (b). Isolated mitochondria and promitochondria were solubilized with Triton X-100 and CHAPS detergents and separated by isolelectic focusing followed by SDS-PAGE.

lower than what is required to achieve maximal respiration rate. Also notable is that the activities of two antioxidants, cytochrome c peroxidase and catalase, do not decrease substantially until oxygen levels fall below 0.5 μM. These data indicate that mitochondria alter their function in response to low oxygen (hypoxia) and not just in response to anoxia. We also know that yeast cytochrome c oxidase adapts to hypoxia by expressing a hypoxic isoform, Vb, of subunit V. Subunit Vb is expressed only at oxygen levels below 1 μM (32). The substitution of Vb for Va, the isoform that is present in normoxic cells, increases the turnover rate of holo-cytochrome c oxidase and thus enhances its activity under low oxygen (91).

B. Functional Studies with Mammals

Schumacker and colleagues have reported that rat hepatocytes exposed to hypoxia (25–31 μM O_2) for 6 hr exhibit reduced rates of oxygen consumption (93) and that a gradual reduction in oxygen tension (from 140 to 0 μM O_2) leads to reduced respiration rates, at oxygen levels as high as 98 μM O_2. This response was not observed in cells that were rapidly shifted from 140 to 0 μM O_2 and was, therefore, proposed to be an adaptive response to hypoxia. Moreover, the effects of preconditioning to hypoxia were found to be reversible. This temporary decrease in oxygen consumption by cells at oxygen concentrations above the oxygen concentration that limits respiration has been referred to as reversible oxygen conformance. The same group has reported that isolated mitochondria from rat hepatocytes also exhibit reversible oxygen conformance. Mitochondria exposed to 3 μM O_2 for 1–2 hr have respiration rates that are 40% lower when measured at 28 μM O_2 compared to 140 μM O_2 (94). The inhibition of respiration parallels a decrease in the V_{max} of cytochrome c oxidase. A similar phenomenon has been observed with cardiac myocytes, except that these cells requires a much shorter exposure time to hypoxia to induce reversible oxygen conformance. The results of Schumacker and colleagues (93,94) contradict the findings of other groups, which have shown that hepatocyte whole cell respiration rates do not decrease until oxygen tension falls below ~7 μM O_2 (95), that oxygen consumption rates of isolated hepatocyte mitochondria are maintained at a maximum level at oxygen levels above ~2 μM O_2 (96), and that prolonged exposure to hypoxia has no significant effect on oxygen-concentration-dependent respiration in isolated rat hepatocyte mitochondria (97). It is not clear why these different sets of studies are in conflict.

Collectively, the findings of Schumacker and colleagues (16,66,67) have led to the hypothesis that preexposure to hypoxia limits the rates of mitochondrial respiration by reducing the V_{max} of cytochrome c oxidase, that the slowing of electron flux through cytochrome c oxidase leads to the release of ROS from upstream electron carriers (primarily ubisemiquinone), and that the rise in ROS levels starts a signal transduction pathway involved in the up-regulation of nuclear hypoxic genes. Although interesting, this hypothesis is controversial and suffers from being pieced together from partial data derived from several different systems [e.g., rat hepatocytes, chick cardiomyocytes, rat liver mitochondria, human hepatoma (Hep3B) cells, and bovine cytochrome c oxidase]. Also, this hypothesis leaves many questions unanswered. For example, does a change in the V_{max} of cytochrome c oxidase precede an increase in ROS in a given type of cell? While it has been proposed that a decrease in the V_{max} of cytochrome c oxidase is responsible for the increase in ROS levels, it is equally plausible that an increase in ROS levels is responsible for a decrease in the V_{max} of cytochrome c oxidase. This hypothesis also fails to explain how the reduction in respiration is reversed when cells, mitochondria, or cytochrome c oxidase is returned to normoxia. Is it merely an effect of oxygen itself on the oxygen binding site of the enzyme, as proposed (94), or are other factors involved? Finally, this hypothesis fails to recognize that the increased levels of reduced pyridine nucleotides that are observed in cells preconditioned to hypoxia can contribute to oxidative stress by facilitating Fenton chemistry (109) and may, themselves, be involved in a signaling pathway that affects induction of hypoxic nuclear genes. Studies on the role of the mitochondrion in hypoxic gene induction in well-characterized, genetically marked isogenic yeast strains (both wild type and mutants) may help resolve some of these questions.

C. Structural Considerations

As in yeast, hypoxia induces changes in morphology of mammalian cell mitochondria. The degree of alteration is dependent on the severity of the hypoxic treatment. In hypoxia induced by hypobaric conditions, the number of mitochondria in rat liver and muscle cells increases, and size decreases (98,99). These features may be important for increasing the surface area of mitochondria and thus enhancing the diffusion of oxygen and/or other metabolites. Under conditions of more severe hypobaric hypoxia, the percent cell volume occupied by mitochondria decreases (98). In liver cells, the distribution of mitochondria within the cell changes such that mitochondria are more homogeneously distributed throughout the cell (99). This is believed to decrease the diffusion distance for oxygen, and thus facilitate oxygen delivery to the mitochondrion. Mammalian heart cells exposed to anoxia show changes in mitochondrial structure more similar to those seen in anaerobically grown yeast. In this case, mitochondria fuse to form a "gigantic" mitochondrion with more diffuse cristae membranes (100).

Although the processes that regulate changes in mitochondrial structure in response to oxygen are unknown, many proteins that are important for maintaining the gross morphology of mitochondria have been identified (reviewed in Refs. 101 and 102). Several of these proteins are located on the outer mitochondrial membrane and are involved in tethering the mitochondrion to the cytoskeleton (103–105). So, it may be that changes in the function of these proteins regulate changes in morphology in response to oxygen. Considerably less is known about the processes that control mitochondrial ultrastructure. *S. cerevisiae* rho° cells, which lack a mitochondrial genome, have normal-shaped mitochondria with both an outer and inner membrane, but are completely devoid of cristae membrane; a structure remarkably similar to promitochondria (106). Similarly, mammalian rho° cells have a reduced cristae surface area compared to wild-type cells (107). It is not simply a lack of respiration that results in a decrease in cristae surface area, as respiration-deficient mutants have normal mitochondria (108). These studies intriguingly suggest that one, or several, mitochondrial genes may regulate changes in the surface area of cristae in response to hypoxia.

VI. Conclusions and Future Questions

It has been known for several years that the mitochondrion plays an important role in cellular energy generation and in metabolic pathways involved in amino acid and tetrapyrrole synthesis. It has also been known for several years that although the mitochondrion contains a genome it is small and that most of the 450 or so proteins that reside in the mitochondrion are encoded in the nucleus. These findings have tended to minimize and circumscribe mitochondrial contributions to cell function, cell growth, and cell division. Recently, however, it has become clear that the mitochondrion has a much wider role in cell function than previously thought. Indeed, during the past few years, cross-talk between the mitochondrion and nucleus has taken on increased importance in models of aging, degenerative disease, and apoptosis. It now appears that the mitochondrion is also important for the oxygen-regulated expression of nuclear genes in yeast and other organisms. Although our understanding of this at the molecular level is somewhat limited, a great deal of work is focused on mitochondrially produced signals and the signaling pathways that they affect. One class of signals that has emerged as important in yeast and mammalian cells are reactive oxygen species. These are produced by the respiratory chain and probably initiate a signaling pathway while still resident in the mitochondrion. This pathway is important for the induction of some hypoxic genes. It has been suggested that electron transport through cytochrome c oxidase slows during exposure to hypoxia and that this reduced rate of cytochrome c oxidase leads to the liberation of ROS further upstream at cytochrome b (16). It is also possible however, that reduced pyridine nucleotide levels increase during exposure to hypoxia and that this increase drives Fenton reactions that oxidize macromolecules (109), or that intramitochondrial iron stores are increased giving way to enhanced opportunity for Fenton chemistry. A second class of signals affects the phosphorylation of one or more nuclear transcription factors. This type of signaling is essential for the optimal expression of aerobic genes, which may in turn impact cellular aging and the ability of yeast cells to survive in a mammalian host. The identity of these signals is unknown but their production is dependent on one or more mitochondrial genes and not on respiration. In sum, it is now clear that the mitochondrion and its genome play a far greater role in cellular growth and division than previously thought. Future studies on the role of mitochondrial–nuclear cross-talk in oxygen-regulated gene expression should unveil some of the mechanisms involved.

Acknowledgments

The authors wish to thank their colleagues Dr. Pamela David, Marcella Li, and Athena Dodd for helpful discussions. Some of the work mentioned in this paper was supported by research grants GM30228 and HL63324 from the National Institutes of Health and a postdoctoral fellowship HL10449 to K.O. from the National Institutes of Health.

References

1. Bunn FH, Poyton RO. Oxygen sensing and molecular adaptation to hypoxia. Physiol Rev 1996; 76:839–885.
2. Guillemin K, Krasnow MA. The hypoxic response: huffing and hifing. Cell 1997; 89:9–12.
3. Dang CV, Semenza GL. Oncogenic alterations of metabolism. T.I.B.S. 1999; 24:68–72.
4. Semenza GL. HIF-1 and human disease: one highly involved factor. Genes Dev 2000; 14:1983–1991.
5. Hochachka PW, Buck LT, Doll C, Land SC. Unifying theory of hypoxia tolerance: molecular/metabolic defense and rescue mechanisms for surviving oxygen lack. Proc Natl Acad Sci USA 1996; 93:9493–9498.
6. Hochachka PW. Mechanism and evolution of hypoxia tolerance in humans. J Exp Biol 1998; 201:1243–1254.
7. Allen LA, Zhao X-J, Caughey W, Poyton RO. Isoforms of yeast cytochrome c oxidase subunit V affect the binuclear reaction center and alter the kinetics of interaction with the isoforms of yeast cytochrome c. J Biol Chem 1995; 270:110–118.
8. Burke PV, Poyton RO. Structure/function of oxygen-regulated isoforms of cytochrome c oxidase. J Exp Biol 1998; 201:1163–1175.
9. Poyton RO. Models for oxygen sensing in yeast: implications for oxygen-regulated gene expression in higher eucaryotes. Respir Physiol 1999; 115:119–133.
10. Taylor BL, Zhulin IB. Pas domains: internal sensors of oxygen, redox potential, and light. Microbiol Mol Bio Rev 1999; 63:479–506.
11. Gilles-Gonzalez MA. Oxygen signal transduction. IUMB Life 2001; 51:165–173.
12. Epstein AC, Gleadle JM, McNeill LA, Hewitson KS, O'Rourke J, Mole DR, et al. *C. elegans* EGL-9 and mammalian homologs define a family of dioxygenases that regulate HIF by prolyl hydroxylation. Cell 2001; 107:43–54.
13. Spiro S, Guest JR. Adaptive responses to oxygen limitation in *Escherichia coli.* T.I.B.S. 1991; 16:310–314.
14. Jiang C, Haddad GG. A direct mechanism for sensing low oxygen levels by central neurons. Proc Natl Acad Sci USA 1994; 91:7198–7201.
15. Rouault TA, Klausner RD. Iron-sulfur clusters as biosensors of oxidants and iron. T.I.B.S. 1996; 21:174–177.
16. Chandel NS, Maltepe E, Goldwasser E, Mathieu CE, Simon MC, Schumacker PT. Mitochondrial reactive oxygen species trigger hypoxia-induced transcription. Proc Natl Acad Sci USA 1998; 95:11715–11720.
17. Oh J-I, Kaplan S. Redox signaling: globalization of gene expression. EMBO J 2000; 19:4237–4247.
18. Oh J-I, Kaplan S. Generalized approach to the regulation and integration of gene expression. Mol Microbiol 2001; 39:1116–1123.
19. Dirmeier R, O'Brien KM, Li M, Dodd E, Spears E, Poyton RO. Exposure of yeast cells to anoxia induces transient oxidative stress: implications for the induction of hypoxic genes. J Biol Chem 2002; 277: 34773–34784

20. Ames BN, Shigenaga TM, Hagen TM. Oxidants, antioxidants, and the degenerative diseases of aging. Proc Natl Acad Sci USA 1993; 90:7915–7922.
21. Shigenaga MK, Hagen TM, Ames BN. Oxidative damage and mitochondria decay in aging. Proc Natl Acad Sci USA 1994; 91:10771–10778.
22. Wickens A. Ageing and the free radical theory. Respir Physiol 2001; 128:379–391.
23. Lander HM. An essential role for free radicals and derived species in signal transduction. FASEB J 1997; 11:118–124.
24. Finkel T. Oxygen radicals and signaling. Curr Opin Cell Biol 1998; 10:248–253.
25. Nemeto S, Takeda K, Yu Z-X, Ferrans VJ, Finkel T. Role for mitochondrial oxidants as regulators of cellular metabolism. Mol Cell Biol 2000; 20:7311–7318.
26. Kwast KE, Burke PV, Staahl BT, Poyton RO. Oxygen sensing in yeast: evidence for the involvement of the respiratory chain in regulating the transcription of a subset of hypoxic genes. Proc Natl Acad Sci USA 1999; 96:5446–5451.
27. Kirchman PA, Kim S, Lai CY, Jazwinski SM. Interorganelle signaling is a determinant of longevity in *Saccharomyces cerevisiae.* Genetics 1999; 152:179–190.
28. Dufour E, Boulay J, Rincheval V, Sainsard-Chanet A. A causal link between respiration and senescence in *Podospora anserina.* Proc Natl Acad Sci USA 2000; 97:4138–4143.
29. Branicky RC, Benard C, Hekimi S. clk-1, mitochondria, and physiological rates. BioEssays 2000; 22:48–56.
30. Salvioli S, Bonafe M, Capri M, Monti D, Franceschi C. Mitochondria, aging and longevity—a new perspective. FEBS Lett 2001; 492:9–13.
31. Rogers PJ, Stewart PR. Respiratory development in *Saccharomyces cerevisiae* grown at controlled oxygen tension. J Bacteriol 1973; 115:88–97.
32. Burke PV, Raitt DC, Allen LA, Kellogg EA, Poyton RO. Effects of oxygen concentration on the expression of cytochrome c and cytochrome c oxidase genes in yeast. J Biol Chem 1997; 272:14705–14712.
33. Kwast KE, Burke PV, Poyton RO. Oxygen sensing and the transcriptional regulation of oxygen responsive genes in yeast. J Exp Biol 1998; 201:1177–1195.
34. Ter Linde JJM, Liang H, Davis RW, Steensma HY, van Dilken JP, Pronk JT. Genome-wide transcriptional analysis of aerobic and anaerobic chemostat cultures of *Saccharomyces cerevisiae.* J Bacteriol 1999; 181:7409–7413.
35. Lowry CV, Lieber RH. Negative regulation of the *Saccharomyces ANB1* gene by heme, is mediated by the *ROX1* gene product. Mol Cell Biol 1986; 9:4145–4148.
36. Hodge MR, Kim G, Singh K, Cumsky MG. Inverse regulation of yeast COX5b genes by oxygen and heme. Mol Cell Biol 1989; 9:1958–1964.
37. Poyton RO, Burke PV. Oxygen regulated transcription of cytochrome c and cytochrome c oxidase genes in yeast. Biochim Biophys Acta 1992; 1101:252–256.
38. Pinkham JL, Keng T. Heme-mediated gene regulation in *Saccharomyces cerevisiae.* In: Winkelman G, Winge D, eds. Metal Ions in Fungi. New York: Marcel Dekker, 1994:455–501.
39. Zitomer RS, Lowry CV. Regulation of gene expression by oxygen in *Saccharomyces cerevisiae.* Microbiol Rev 1992; 56:1–11.
40. Zhang L, Hach A. Molecular mechanism of heme signaling in yeast: the transcriptional activator Hap1 serves as a key mediator. Cell Mol Life Sci 1999; 56:415–426.
41. Balasubramanian B, Lowry CV, Zitomer RS. The Rox1 repressor of the *Saccharomyces cerevisiae* hypoxic genes is a specific DNA-binding protein with a high mobility-group motif. Mol Cell Biol 1993; 13:6071–6078.

42. Deckert J, Perini R, Balasubramanian B, Zitomer RS. Multiple elements and autorepression regulate Rox1, a repressor of hypoxic genes in *Saccharomyces cerevisiae*. Genetics 1995; 139:1149–1158.
43. Kastaniotis AJ, Zitomer RS. *ROX1* mediated repression. Adv Exp Med Biol 2000; 475:185–195.
44. Keng T. *HAP1* and *ROX1* form a regulatory pathway in the repression of *HEM13* transcription in *Saccharomyces cerevisiae*. Mol Cell Biol 1992; 12:2616–2632.
45. Zitomer RS, Limbach MP, Rodriguez-Torres AM, Balasubramanian B, Deckert J, Snow PM. Approaches to study Rox1 repression of the hypoxic genes in yeast *Saccharomyces cerevisiae*. Methods 1997; 11:279–288.
46. Chandrel Y, Gaisne C, Lions C, Verdiere J. The transcriptional regulator Hap1p (Cyp1p) is essential for anaerobic or heme-deficient growth of *Saccharomyces cerevisiae*: genetic and molecular characterization of an extragenic suppressor that encodes a WD repeat protein. Genetics 1998; 148:559–569.
47. Vasconcelles MJ, Jiang Y, McDaid K, Gilooly L, Wretzel S, Porter DL. Identification and characterization of a low oxygen response element involved in the hypoxic induction of a family of *Saccharomyces cerevisiae* genes. J Biol Chem 2001; 276:14374–14384.
48. Jiang Y, Vasconcelles MJ, Wretzel S, Light A, Martin CE, Goldberg MA. MGA2 is involved in the low oxygen response element-dependent hypoxic induction of genes in *Saccharomyces cerevisiae*. Mol Cell Biol 2001; 21:6161–6169.
49. Nakagawa Y, Sugioka S, Kaneko Y, Harashima S. O2R, a novel regulatory element mediating Rox1p-independent O_2 and unsaturated fatty acid repression of OLE1 in *Saccharomyces cerevisiae*. J Bacteriol 2001; 183:745–751.
50. Donzeau M, Bourdineaud JP, Lauquin GJ. Regulation by low temperature and anaerobiosis of a yeast gene specifying a putative GPI-anchored plasma membrane protein. Mol Microbiol 1996; 20:449–459.
51. Sertil O, Cohen B, Davies KJD, Lowry CV. The *DAN1* gene of *S. cerevisiae* is regulated in parallel with the hypoxic genes, but by a different mechanism. Gene 1997; 192:199–205.
52. Cohen BD, Sertil O, Abramova NE, Mehta S, Davies KJA, Lowry CV. Induction and repression of *DAN1* and the family of anaerobic mannoprotein genes in *S. cerevisiae* occurs through a complex array of regulatory sites. Nucleic Acids Res 2001; 29:798–808.
53. Fytlovich S, Gervais M, Agrimonti C, Guiard B. Evidence for an interaction between the *CYP1* (*HAP1*) activator and a cellular factor during heme-dependent transcriptional regulation in the yeast *Saccharomyces cerevisiae*. EMBO J 1993; 62:1209–1218.
54. Zhang L, Hach A, Wang C. Molecular mechanisms governing heme signaling in yeast: a higher order complex mediates heme regulation of the transcriptional activator Hap1. Mol Cell Biol 1998; 18:3819–3828.
55. Wallace DC. Mitochondrial diseases in man and mouse. Science 1999; 1482–1488.
56. Poyton RO, McEwen JE. Crosstalk between nuclear and mitochondrial genomes. Annu Rev Biochem 1996; 65:563–607.
57. Liao X, Small WC, Srere PA, Butow RA. Intramitochondrial functions regulate nonmitochondrial citrate synthase (*CIT2*) expression in *Saccharomyces cerevisiae*. Mol Cell Biol 1991; 11:38–46.
58. Liao X, Butow RA. RTG1 and RTG2: two yeast genes required for a novel path of communication from mitochondria to nucleus. Cell 1993; 72:61–71.

59. Sekito T, Thornton J, Butow RA. Mitochondria-to-nuclear signaling is regulated by the subcellular localization of the transcription factors Rtg1p and Rtg3p. Mol Biol Cell 2000; 11:2103–2155.
60. Conlan RS, Gounalaki N, Hatzis P, Tzamarias D. The Tup1-Cyc8 protein complex can shift from transcriptional co-repressor to a transcriptional co-activator. J Biol Chem 1999; 274:205–210.
61. Poyton RO, Dagsgaard CJ. Mitochondrial-nuclear crosstalk is involved in oxygen-regulated gene expression in yeast. Adv Exp Med Biol 2000; 475:177–184.
62. Dagsgaard C, Taylor LE, O'Brien KM, Poyton RO. Effects of anoxia and the mitochondrion on expression of aerobic nuclear *COX* genes in yeast: evidence for a signaling pathway from the mitochondrial genome to the nucleus. J Biol Chem 2001; 276:7593–7601.
63. Epstein CB, Waddle JA, Hale W, Dave V, Thornton TL, Garner HR, et al. Genome-wide responses to mitochondrial dysfunction. Mol Biol Cell 2001; 12:297–308.
64. Goldstein AL, McCusker JH. Development of *Saccharomyces cerevisiae* as a model pathogen: a system for the genetic identification of gene products required for survival in the mammalian host environment. Genetics 2001; 159:499–513.
65. Trawick TJ, Kraut N, Poyton RO. Involvement of Abf1 in intergenomic signaling from mitochondrion to nucleus in yeast. J Biol Chem 2002; submitted.
66. Van deHoeck TL, Becker LB, Shao Z, Li C, Schumacker PT. Reactive oxygen species released from mitochondria during brief hypoxia induce preconditioning in cardiomyocytes. J Biol Chem 1998; 273:18092–18098.
67. Chandel NS, McClintock DS, Feliciano CE, Wood TM, Melendez JA, Rodriguez AM, et al. Reactive oxygen species generated at mitochondrial complex III stabilize hypoxia-inducible factor-1α during hypoxia. J Biol Chem 2000; 275:25130–25138.
68. Kietzmann T, Fandrey J, Acker H. Oxygen radicals as messengers in oxygen-dependent gene expression. News Physiol Sci 2000; 15:202–208.
69. Semenza GL. Perspectives on oxygen sensing. Cell 1999; 98:281–284.
70. Sohal RS, Weindruch R. Oxidative stress, caloric restriction, and aging. Science 1996; 273:59–63.
71. Jakubowski W, Bartosz G. Estimation of oxidative stress in *Saccharomyces cerevisiae* with fluorescent probes. Int J Biochem Cell Biol. 1997; 29:1297–1301.
72. Levine RL, Garland D, Oliver CN, Amici A, Climent I, Lenz A-G, et al. Determination of carbonyl content in oxidatively modified proteins. Methods Enzymol 1990; 186:464–478.
73. Agarwal S, Sohal RS. Differential oxidative damage to mitochondrial proteins during aging. Mech Ageing Dev 1995; 85:55–63.
74. Tamarit J, Cabiscol E, Ros J. Identification of the major oxidatively damaged proteins in *Escherichia coli* cells to oxidative stress. J Biol Chem 1998; 273:3027–3032.
75. Cabiscol E, Piulats E, Echave P, Herrero E, Ros J. Oxidative stress promotes specific protein damage in *Saccharomyces cerevisiae.* J Biol Chem 2000; 275:27393–27398.
76. Lando P, Peet DJ, Whelan DA, Gorman JJ, Whitelaw ML. Asparagine hydroxylation of the HIF transactivation domain a hypoxic switch. Science 2002; 295:858–861.
77. Lipscomb EA, Sarmieres PD, Freeman RS. SM-20 is a novel mitochondrial protein that causes caspase-dependent cell death in nerve growth factor-dependent neurons. J Biol Chem 2001; 276:5085–5092.

78. Gasch AP, Spellman PT, Kao CM, Carmel-Harel O, Eisen MB, Storz G, et al. Genomic expression programs in the response of yeast cells to environmental changes. Mol Biol Cell 2000; 11:4241–4257.
79. Causton HC, Ren B, Koh SS, Harbison CT, Kanin E, Jennings EG, et al. Remodeling of yeast genome expression in response to environmental changes. Mol Biol Cell 2001; 12:337.
80. Plattner H, Schatz G. Promitochondria of anaerobically grown yeast. III. Morphology. Biochemistry 1969; 8:339–343.
81. Paltauf F, Schatz G. Promitochondria of anaerobically grown yeast. II. Lipid composition. Biochemistry 1969; 8:335–339.
82. Criddle RS, Schatz G. Promitochondria of anaerobically grown yeast. I. Isolation and biochemical properties. Biochemistry 1969; 8:322–334.
83. Groot GSP, Kovac L, Schatz G. Promitochondria of anaerobically grown yeast. V. Energy transfer in the absence of an electron transfer chain. Proc Natl Acad Sci USA 1971; 68:308–311.
84. Austin K, Brown LR, Stewart PR. Proton magnetic resonance spectroscopy of mitochondrial membranes from yeast grown under different regimes of lipid supplementation. J Membrane Biol 1975; 24:55–69.
85. Jiang F, Rizavi HS, Greenberg ML. Cardiolipin is not essential for the growth of *Saccharomyces cerevisiae* on fermentable or non-fermentable carbon sources. Mol Microbiol 1997; 26:481–491.
86. Damsky CH. Environmentally induced changes in mitochondria and endoplasmic reticulum of *Saccharomyces cerevisiae.* J Cell Biol 1976; 71:123–135.
87. Visser W, van Spronsen EA, Nannings N, Pronk JT, Kuenen JG, van Dijken JP. Effects of growth conditions on mitochondrial morphology in *Saccharomyces cerevisiae.* Antonie von Leeuwenhoek 1995; 67:243–253.
88. Poyton RO. Assembling a time bomb cytochrome c oxidase and diseases. Nature Genet 1998; 20:316–317.
89. Forsha DT, Church C, Wazny P, Poyton RO. Structure and function of pet100p, a molecular chaperone required for the assembly of cytochrome c oxidase in *Saccharomyces cerevisiae.* Biochem Soc Trans 2001; 29:436–441.
90. Rep M, Grivell LA. The role of protein degradation in mitochondrial function and biogenesis. Curr Genet 1996; 30:367–380.
91. Waterland RA, Basu A, Chance B, Poyton RO. The isoforms of yeast cytochrome *c* oxidase subunit V alter the in vivo kinetic properties of the holoenzyme. J Biol Chem 1991; 266:4180–4186.
92. Visser W, Van der Baan AA, Batenburg-van der Vegte W, Scheffers WA, Kramer R, van Dijken JP. Involvement of mitochondria in the assimilatory metabolism of anaerobic *Saccharomyces cerevisiae* cultures. Microbiology 1994; 140:3039–3046.
93. Schumacker PT, Chandel NS, Agusti AG. Oxygen conformance of cellular respiration in hepatocytes. Am J Physiol 1993; 265:L395–L402
94. Chandel NS, Budinger GR, Kemp RA, Schumacker PT. Inhibition of cytochrome-c oxidase activity during prolonged hypoxia. Am J Physiol 1995; 268:L918–L925.
95. Jones DP, Kennedy FG. Intracellular oxygen supply during hypoxia. Am J Physiol 1982; 243:C247–C253.
96. Wilson DF, Rumsey WL, Green TJ, Vanderkooi JM. The oxygen dependence of mitochondrial oxidative phosphorylation measured by a new optical method for measuring oxygen concentration. J Biol Chem 1988; 263:2712–2718.

97. Taylor DE, Kantrow SP, Piantadosi CA. Mitochondrial respiration after sepsis and prolonged hypoxia. Am J Physiol 1998; 275:L139–L144
98. Lund DD, Tomanek RJ. The effects of chronic hypoxia on the mycardial cell of normotensive and hypertensive rats. Anat Rec 1980; 196:421–430.
99. Costa LE, Boveris A, Koch OR, Taquini AC. Liver and heart mitochondria in rats submitted to chronic hypobaric hypoxia. Am J Physiol 1998; 255:C123–C129
100. Sun CN, Dhalla NS, Olsen RE. Formation of gigantic mitochondria in hypoxic isolated perfused rat hearts. Experientia 1969; 25:763–764.
101. Herman GJ, Shaw JM. Mitochondrial dynamics in yeast. Annu Rev Cell Dev Biol 1998; 14:256–303.
102. Yaffe MP. The machinery of mitochondrial inheritance and behavior. Science 1999; 283:1493–1497.
103. Burgess SM, Delannoy M, Jensen RE. MMM1 encodes a mitochondrial outer membrane protein essential for establishing and maintaining the structure of yeast mitochondria. J Cell Biol 1994; 126:1375–1391.
104. Sogo LF, Yaffe MP. Regulation of mitochondrial morphology and inheritance by Mdm10p, a protein of the mitochondrial outer membrane. J Cell Biol 1994; 126:1361–1373.
105. Berger KH, Sogo LF, Yaffe MP. Mdm12p, a component required for mitochondrial inheritance that is conserved between budding and fission yeast. J Cell Biol 1997; 136:545–553.
106. Montisano DF, James TW. Mitochondrial morphology in yeast with and without mitochondrial DNA. Ultrastructure Res 1979; 67:288–296.
107. Gilkerson RW, Margineantu DH, Capaldi RA, Selker JML. Mitochondrial DNA depletion causes morphological changes in the mitochondrial reticulum of cultured human cells. FEBS Lett 2000; 474:1–4.
108. Church C, Poyton RO. Neither respiration nor cytochrome c oxidase affects mitochondrial morphology in *Saccharomyces cerevisiae.* J Exp Biol 1998; 201:1729–1737.
109. Linn S. DNA damage by iron and hydrogen peroxide in vitro and in vivo. Drug Metab Rev 1998; 30:313–326.
110. Lahiri S, Rumsey WL, Wilson DF, Iturriaga R. Contribution of in vivo microvascular PO_2 in the cat carotid body chemotransduction. J Appl Physiol 1993; 75:1035–1043.
111. Erecinska M, Silver IA. Tissue oxygen tension and brain sensitivity to hypoxia. Respir Physiol 2001; 128:263–276.
112. deGroot H, Littauer A, Noll T. Metabolic and pathological aspects of hypoxia in liver cells. In: Acker H, ed. Oxygen Sensing in Tissues. Berlin: Springer-Verlag, 1988:49–64.
113. Vollmar B, Menger MD. Assessment of microvascular oxygen supply and tissue oxygenation in hepatic ischemia/reperfusion. Adv Exp Med Biol 1997; 428:403–408.
114. Rumsey WL, Pawlowski M, Lejavardi N, Wilson DF. Measurement of oxygen pressure in the heart in vivo using phosphorescence quenching. Adv Exp Med Biol 1994a; 361:93–97.
115. Rumsey WL, Kuczynski B, Patel B. Detecting hypoxia in heart using phosphorescence quenching and ^{99M}technetium-nitroimidazoles. Adv Exp Med Biol 1994b; 361:99–104.

116. Honig CR, Gayeski TEJ. Comparison of intracellular PO_2 and conditions for blood-tissue O_2 transport in heart and working red skeletal muscle. Adv Exp Med Biol 1987; 215:309–321.

3

Regulation of HIF-1 by Oxygen

The Role of Prolyl Hydroxylase and the VHL Tumor Suppressor

PATRICK H. MAXWELL

Hammersmith Campus
Imperial College
London, England

PETER J. RATCLIFFE

University of Oxford
Oxford, England

I. Introduction

The commitment to oxygen as a terminal electron acceptor for mitochondrial respiration is a central aspect of the biology of more complicated multicellular organisms; as a consequence oxygen is an absolute requirement for the viability of cells and tissues. Many aspects of anatomy and physiology could be regarded as primarily designed to achieve the basic aim of distributing oxygen reliably to all the constituent cells in an organism. Meeting this challenge requires continuous adjustments to the oxygen acquisition and transport mechanisms, including respiration and the circulation, and to cellular metabolism.

A broad range of control mechanisms, such as regulation of arterial blood pressure, are important in achieving reliable oxygen distribution, although the extent to which they are directly regulated by oxygen is unclear. On the other hand, it is clear that many key responses are governed by cellular responses to the local level of oxygen itself. The most notable examples are provided by the respiratory control system centered on the carotid body, and red blood cell production, which is regulated by erythropoietin production from the kidney (1,2). The carotid body increases its discharge rate when the pO_2 of arterial blood is reduced, leading to an increase in the rate of respiration. The renal fibroblasts increase erythropoietin production when oxygen delivery is reduced due to anemia or hypoxemia (3,4). A long-standing question has been the nature of the oxygen sensors underlying these important homeostatic responses. However, the nature of the sensors, or even their existence as discrete entities, has not been clear.

This chapter will consider recent advances in understanding how the HIF-1 transcription control system responds to oxygen. HIF-1 was originally identified as a transcription factor in erythropoietin-producing hepatoma cells that binds to an oxygen-responsive enhancer element lying 3′ to the erythropoietin gene (5,6). Despite the regulated tissue expression of erythropoietin, the underlying transcriptional control system actually operates much more widely (7), being present in all mammalian cell types examined to date. HIF-1 regulates the expression of a very broad range of genes, including erythropoietin, angiogenic growth factors, and enzymes involved in glucose transport and glycolysis (8). A HIF-1 transcription complex consists of an alpha and a beta subunit, both of which are basic-helix-loop-helix PAS domain transcription factors and are members of multigene families (6). The HIF system is highly conserved in evolution of multicellular animals, being present in *Caernorhabditis elegans* (9,10) and in *Drosophila melanogaster* (11,12). The regulatory alpha subunit is subjected to oxygen-dependent destruction by the proteasome (13–16). Importantly, although this appears to be a dominant regulatory mechanism, the HIF-1 complex is also oxygen-responsive in other ways. Thus translocation of the alpha subunit to the nucleus can be oxygen-regulated, as can transactivator recruitment by the C terminus of HIF alpha (17–19).

Over time substantial effort has focused on understanding how the erythropoietin system, and subsequently how HIF itself, is regulated by oxygen. Many possibilities have been proposed for the oxygen-sensing process. Important in shaping these ideas were the observations that iron chelation or exposure to transition metals such as cobalt, manganese, and nickel mimic the hypoxia response, stabilizing HIF-α and up-regulating the expression of HIF target genes (20–22). Moreover, the efficacy of cobalt was noted to be inversely related to the availability of iron, suggesting that metal might be competing at an oxygen-sensing iron center (23). Two broad categories of process have been proposed.

In the first type of model a specific oxygen-sensing protein has been postulated. Several examples of such sensing proteins have now been defined in prokaryotic systems, including the dioxygen-liganding haemoprotein *fixL* in *Rhizobia* (24), and oxygen-radical sensitive iron-sulfur clusters such as *sox R/S* in *Escherichia coli* (25). In such models it has been proposed that cobalt might substitute for iron and inhibit activation by oxygen. One potential problem with such a model for the oxygen-sensing mechanism underlying HIF regulation is that many iron centers (for instance heme iron) do not exchange in this way, so that it was necessary to propose that the sensing molecule was itself turning over rapidly.

The second type of model followed from the recognition that exposure of cells to hydrogen peroxide inhibits activation of the HIF system by hypoxia (26), and proposed that the iron-and-oxygen-sensitive degradation of HIF derived from Fenton chemistry. In the majority of such models it has been proposed that increases in the generation of reactive oxygen species (ROS) at higher dioxygen concentrations were responsible for activating pathways that promote HIF-α degradation. Several sources of such ROS have been proposed including different isoforms of NADPH oxidase (27) and specific cytochrome/cytochrome reductase systems (28). In contrast, one group has proposed that an increase in mitochondrial ROS production in hypoxia provides a signal that stabilizes HIF-α (29,30). In considering the problem of cellular oxygen sensing, postulated ROS signaling mechanisms do have the potential to provide responses over a broad range of cellular oxygen levels. However, there are difficulties in understanding how precise physiological responses to oxygen availability (for instance, feedback control of haematocrit by erythropoietin) could be regulated by such broadly reactive species. To address this, two proposals have been put forward. First, a sensing/signaling molecule might contain particularly sensitive residues, such as the reactive cysteine residues in the bacterial ROS sensor *oxyR/S* (31). Second, specific local interactions of iron with metal-binding sites in proteins might generate localized ROS by Fenton chemistry, resulting in modification of adjacent amino acids.

However, a weakness of these experimental approaches is that they cannot clearly define the point of molecular interaction with the HIF system. For example, redox-active agents that are proposed to affect the oxygen-sensing mechanism might do so directly, or very indirectly in a way that does not mimic physiological regulation by oxygen.

Consequently many groups have sought to move proximally from an understanding of the regulation of HIF to the sensing/transduction process by molecular biochemical and genetic approaches. This has yielded insights into how proteasomal destruction of HIF alpha is regulated by oxygen and has recently led to the definition of an underlying mechanism that can be regarded as a molecular oxygen sensor. As outlined below, the response system is based on an enzymatic prolyl hydroxylation reaction, which uses molecular oxygen as a cosubstrate (Fig. 1). The enzyme itself has a loosely coordinated iron at the catalytic center; removal or substitution of this iron atom therefore accounts for the action of iron chelators and cobaltous ions. Following conversion, the hydroxyproline-containing HIF alpha chains are captured efficiently by the von Hippel–Lindau (VHL) tumor suppressor protein. This leads to their ubiquitylation and subsequent destruction. These insights into HIF regulation are of interest for several reasons including the following. First, prolyl hydroxylation is a novel signal regulating ubiquitylation and proteolytic destruction, and raises the question that it may be important in regulating destruction of other targets. Second, inhibition of the prolyl hydroxylation step provides a potentially exciting route to activating the HIF pathway in normoxic cells. Third, the identification of a molecular mechanism for sensing oxygen in mammalian cells raises the possibility that similar mechanisms may regulate other oxygen-response systems besides HIF.

II. The Role of the von Hippel–Lindau Tumor Suppressor in HIF Regulation

A significant step toward understanding how HIF destruction is regulated was the recognition of the critical role of the von Hippel–Lindau (VHL) tumor suppressor protein (32). The VHL gene is situated at 3p25 and was originally cloned in 1993 as

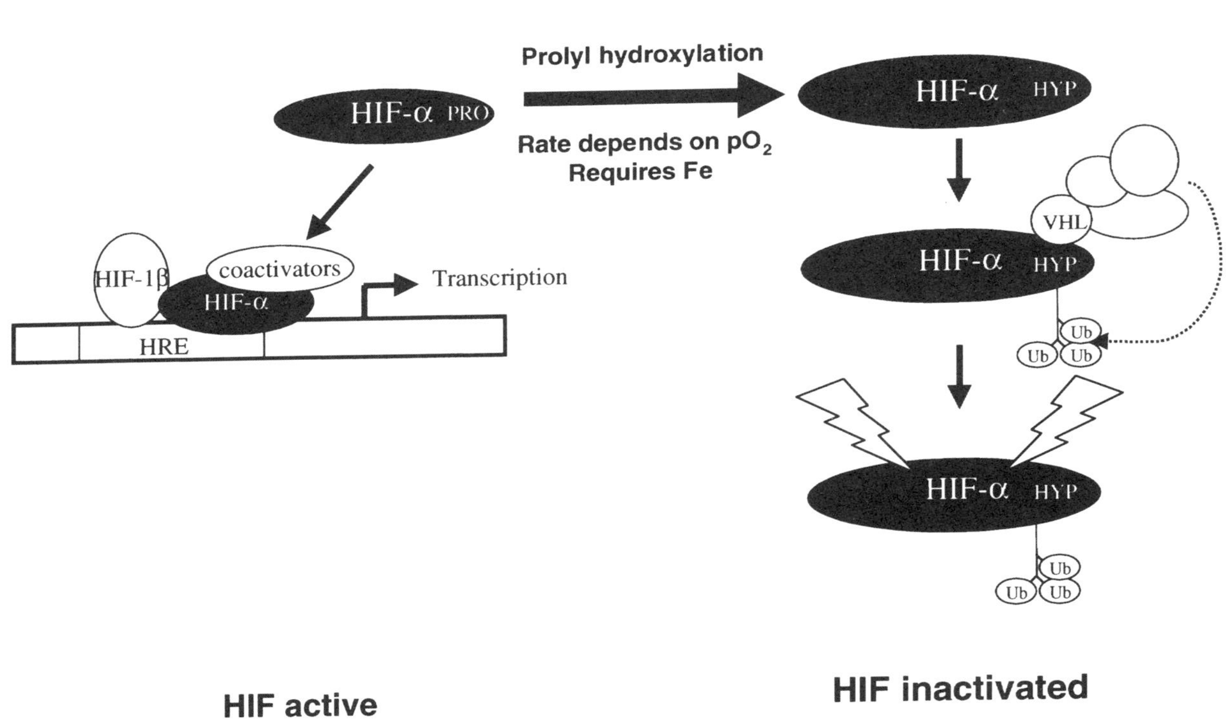

Figure 1 Role of prolyl hydroxylases as an oxygen-sensor regulating HIF activation. In the presence of oxygen, PHD enzymes convert specific prolyl residues in HIF-α chains to hydroxyproline. The enzymes require iron as a cofactor. Modified HIF-α chains are captured by the pVHL ubiquitin E3 ligase complex, ubiquitylated, and destroyed by the proteasome. Under reduced oxygenation, HIF-α rapidly accumulates, enters the nucleus, dimerizes with a β subunit, recruits coactivators, and activates transcription of genes with hypoxic response elements (HRE). HYP, hydroxyproline.

the gene underlying a hereditary cancer syndrome, VHL disease (33). Transmission is autosomal dominant, and affected individuals have a high lifetime risk of retinal and central nervous system hemangioblastoma and clear cell renal cell carcinoma (CCRCC). Other manifestations include phaeochromocytoma, pancreatic islet cell tumors, benign cystic lesions in the kidney and pancreas, endolymphatic sac tumors, and papillary cystadenomas of the epidydimis (in males) or broad ligament (in females). Individuals with VHL disease inherit one defective allele in the germline, and in the tumors that they develop the second allele is inactivated. Importantly, mutations in both VHL alleles occur in the majority of sporadic cases of CCRCC and hemangioblastoma (34,35). VHL therefore conforms to Knudson's two-hit hypothesis of tumor suppressor inactivation and appears to have a pivotal role in the genesis of the most common form of kidney cancer.

The VHL gene encodes a protein of 213 amino acids, which did not show obvious similarities to other proteins. Subsequently it was recognized that a second isoform consisting of amino acids 54–213 is also produced by translation initiation from an internal ATG codon (36,37). This isoform is also a functional tumor suppressor protein and both isoforms are expressed in a range of cultured cell lines. Since the first 53 residues of the higher-molecular-weight pVHL are not well conserved, and mutations in these residues are not generally associated with VHL disease, the functional significance of this part of the protein is uncertain; it is not required for HIF regulation.

Since the majority of CCRCCs have inactivation of both copies of the VHL gene, many cell lines derived from these tumors are available that are defective for VHL function. Comparison of sublines stably transfected with wild-type VHL genes, mutant VHL genes, and empty expression vector has provided a powerful means of investigating pVHL function. Such studies in renal carcinoma cell lines established that the product of the VHL tumor suppressor gene was required for oxygen-dependent proteolysis of HIF alpha subunits (32). Whereas in other mammalian cell types HIF alpha subunits are rapidly destroyed, in CCRCC cell lines lacking pVHL HIF alpha subunits are constitutively stable. This is corrected by reintroduction of pVHL. pVHL acts as the recognition component of a ubiquitin E3 ligase complex, which captures and ubiquitylates HIF alpha subunits in the presence of oxygen (38–40). pVHL is part of a multiprotein complex with elongin B, elongin C, Rbx1, and Cul2, a complex that is extensively similar to the SCF class of ubiquitin ligases (for Skp1/Cdc53/F-box protein) (41–45). The crystal structure of the VHL-elongin C-elongin B complex was recently solved (46), and mutational and protein interaction studies demonstrated that an exposed subdomain of pVHL, termed the beta domain, interacts directly with HIF alpha subunits (39,40). The interaction surface of pVHL is predicted to include the region around Tyr98, since single amino acid substitutions here that are not predicted to have structural effects prevent HIF alpha capture (39).

As yet it is not clear whether pVHL is essential for HIF alpha destruction in all cells, but there are now several settings apart from CCRCC where it clearly is. These include *C. elegans*, in which worms with mutations in the *vhl* gene exhibit constitutive stabilization of HIF1A protein and activation of the HIF system (10).

Similarly, in mice bearing a conditionally targeted *vhl* allele, Cre-recombinase-mediated excision results in activation of the HIF pathway in hepatocytes (47). In a further example, some sublines of Chinese hamster ovary-K1 cells selected for activating mutations in the HIF pathway were found to have *vhl* mutations, and expression of a wild-type human *VHL* gene restored HIF regulation to normal (48).

III. The Interaction Between HIF Alpha and pVHL

The VHL-HIF alpha interaction is robust in protein extracts prepared from cells in the presence of proteasomal inhibitors and oxygen and can be demonstrated using supershift assays and coimmunoprecipitation (32). Importantly, in extracts of cells exposed to iron chelators or cobaltous ions VHL and HIF alpha are not associated with each other, establishing that the interaction between these two proteins was regulated by stimuli relevant to the oxygen sensor (32). Furthermore, hypoxia was also found to prevent the VHL-HIF alpha interaction provided care was taken to prevent reoxygenation effects during cell harvesting, lysis, and analysis (49,50).

Regulation of the interaction between a protein and its cognate ubiquitin E3 ligase complex is in keeping with the best understood examples of regulated ubiquitin-mediated proteolysis (51). In these examples phosphorylation of the target, or less commonly the ubiquitin E3 ligase, is necessary for the interaction to occur and allows ubiquitylation and destruction of the target to proceed. In the case of HIF-alpha, destruction also depends on a posttranslational modification, but rather than phosphorylation the modification is hydroxylation of specific prolyl residues. At least two prolyl residues in HIF-1 alpha (Pro402 and Pro564) are subject to prolyl hydroxylation, and polypeptides containing each hydroxyproline substitution are captured efficiently by the pVHL ubiquitylation complex (49,50,52). The two prolyl hydroxylation sites in HIF-1 alpha and HIF-2 alpha, and the single site in *C. elegans* HIF1A conform to the sequence LXXLAP (10,52). The conversion to hydroxyproline is enzymatic and requires 2-oxoglutarate. It was therefore predicted that the enzyme (or enzymes) responsible would belong to the overall class of 2-oxoglutarate-dependent dioxygenases. Members of this family of enzymes belong to a class of iron-dependent oxygenases that have a highly conserved beta-barrel jelly-roll conformation aligning particular residues involving iron coordination at the catalytic site (53). Interestingly, the structurally defined members all employ a 2-histidine-1-carboxylate triad that forms one face of an octahedral iron coordination system (54). This allows the other sites to be occupied by substrate, cosubstrate, or other endogenous ligands before interaction of oxygen at the sixth site. It is probable that such an arrangement permits greater flexibility than is possible with the use of the porphyrin in heme-containing enzymes, where only one site is available for coordination of exogenous ligands such as oxygen. The overall class of 2-oxoglutarate-dependent dioxygenases is emerging as an important class with diverse biological roles; pertinent to HIF regulation, the class includes the previously recognized prolyl-4-hydroxylases, which are responsible for modifying procollagen (55).

IV. Identification of the HIF Prolyl Hydroxylase Enzymes

An important aim was to define the molecular identity of the prolyl hydroxylase enzyme(s) responsible for modifying HIF alpha subunits. Using the structure of several 2-oxoglutarate-dependent dioxygenases and related enzymes together with sequence from genome programs in *C. elegans* and *Homo sapiens*, candidate genes encoding proteins with similarities to known members of the class were identified and screened using genetic and/or biochemical assays (10,56). In *C. elegans* a single HIF prolyl hydroxylase gene was identified—EGL9 (10). This gene was originally identified in a mutant screen with selection for worms with egg-laying defects (57). Interestingly, mutations in this gene were subsequently found to protect worms from cyanide-mediated killing by *Pseudomonas aeruginosa* (58). As predicted for a HIF prolyl hydroxylase gene, *egl9* mutant worms have a constitutively activated HIF system. This is because without the enzyme HIF-1 alpha does not undergo modification of the key prolyl residue despite the presence of iron and oxygen. In vitro experiments established that when EGL9 and HIF1A are co-expressed in *E. coli* the HIF1A undergoes prolyl hydroxylation, confirming that EGL9 is indeed a HIF prolyl hydroxylase (10).

Three different, but closely related, genes encoding mammalian HIF prolyl hydroxylase enzymes have been identified in the human genome, which are also present in *Mus musculus*, referred to as PHD1, PHD2, and PHD3 (Table 1) (10,56). In vitro experiments have established that each functions as a prolyl hydroxylase for HIF-1 alpha and have demonstrated some selectivity for the two different hydroxylation sites within HIF-1 alpha (10). As yet, mammalian cells with defects in these enzymes have not been identified, and it is not clear how much redundancy there may be in the system. What is clear is that the rate of the enzymatic reaction in cell extracts is sensitive to the amount of oxygen (10).

A prolyl hydroxylase–based oxygen-sensing system fits well with key attributes of the HIF system—notably that iron chelators or cobaltous ions mimic the effect of hypoxia. In particular, iron chelators remove the ferrous iron atom from the active site of the enzyme, preventing the conversion of HIF alpha subunits and thereby mimicking hypoxia. Exposure to cobaltous ions results in substitution of cobalt for the iron at the active center, also disabling the enzyme. Thus, the prolyl hydroxylase sensor accommodates the effects of cobalt and iron chelators without requiring the sensor itself to be turning over rapidly.

Table 1 The Three Mammalian Prolyl Hydroxylases That Have Been Shown to Modify HIF-1 Alpha

	Chr (human)	Size	Other terms	References	Induced by hypoxia
PHD1	19q13.2	407 aa	EGLN2, Falkor	10	No
PHD2	1q42–43	426 aa	EGLN1, C1orf12	10,94–95	Yes
PHD3	14q12	239 aa	EGLN3, SM-20	10,60–64	Yes

PHD1-3 have also been referred to as HIF-PH1-3 (56).

V. Level of Expression of the Prolyl Hydroxylase Enzymes

As the reaction is clearly not at equilibrium, an increased level of expression of the enzymes is predicted to increase the rate of conversion of HIF substrate at a given oxygen concentration, provided the concentrations of 2-oxoglutarate, iron, and HIF substrate are not limiting. Although as yet our knowledge of the expression patterns of the enzymes is limited, it is clear that expression of PHD2 and PHD3 can be increased by hypoxia in some cell types, which would be predicted to provide a negative feedback mechanism tending to decrease the magnitude of the HIF response (10). This may well account for the previously observed decrease in the half-life of HIF alpha subunits on prolonged hypoxic exposure (59).

Of the three PHD genes, most is known about PHD3, the rat ortholog of which is SM-20, a gene first identified on the basis of transiently induced expression in rat vascular smooth muscle cells on exposure of cultured fibroblasts to PDGF (60). Working independently, a second group identified this gene on the basis of increased expression when rat embryonic fibroblasts with a temperature-sensitive mutation in p53 were moved to 32°C, the permissive temperature for expression of functional p53 protein (61). A third group identified SM-20 as a gene whose expression was increased in sympathetic neurons after NGF withdrawal (62). Subsequently this group have shown that overexpression of SM-20 could induce apoptosis in sympathetic neurons (63). Finally, this gene was also identified as a gene whose expression was induced during epithelial conversion of cultured renal mesenchymal cells (64). At present it is not clear how these observations relate to the role of PHD3 in HIF regulation—but the implication is that the level of prolyl hydroxylase may be significantly modulated by a range of signals involved in cell proliferation, differentiation, and apoptosis. This might provide a way of auto-matically adjusting oxygen provision and consumption when a cell embarks on a course that will alter its energy requirements or relationship to the vasculature.

VI. Availability of Cosubstrates and Cofactors

Besides the potential influence of altered enzyme concentrations on the rate of the prolyl hydroxylase reaction, physiologically relevant changes in the level of 2-oxoglutarate, or of available intracellular iron, could have significant effects and thereby influence the oxygen-sensitive signal. In support of this, previous experiments in cultured cells suggested that changes in iron availability alter HIF activation (23). Understanding whether fluctuations in 2-oxoglutarate and intracellular iron do in fact modulate the prolyl hydroxylase reaction will require detailed knowledge of the kinetic properties of the different enzymes and the compartments that they operate in. Of interest, several downstream targets of the HIF system are important in cellular iron homeostasis, suggesting that under some circumstances the system could be regarded as a physiological iron sensor.

Ascorbate has an important role in the procollagen prolyl hydroxylase reaction, most probably allowing recovery of the enzyme following abortive reactions in which the enzyme rather than substrate is oxidized. Clearly it will be of interest to understand whether ascorbate plays a similar role in the HIF prolyl hydroxylase reactions, and whether manipulating ascorbate could alter activity of the system. Preliminary observations show that addition of ascorbate can increase the level of prolyl hydroxylase activity.

VII. Does the Expression Level and Intracellular Localization of pVHL Modulate the HIF System?

An important question is whether alterations in the amount and/or cellular localization of pVHL could also regulate the HIF system. pVHL is predominantly found in the cytoplasmic and membrane compartments (36,65–67). Since the role of HIF alpha chains is in the nucleus there is an apparent paradox. However, pVHL continuously moves through the nuclear compartment, from which it is actively transported (68,69). This continuous flux through the nuclear compartment may account for apparently differing reports concerning intracellular localization of pVHL (65,70). Clearly shuttling of pVHL through the nuclear compartment provides a potential mechanism by which hydroxylated HIF alpha chains in the nucleus can be captured and ubiquitylated.

Experimental observations in renal carcinoma cell lines suggest that reexpression of any detectable amount of pVHL is sufficient to restore HIF regulation, and that expression of higher levels of pVHL does not significantly influence activity of the HIF system in normoxia or in hypoxia. At least in this system it therefore appears that there is a low threshold level of pVHL that is necessary for HIF alpha capture, and that above this threshold other steps in the pathway are limiting. In contrast, in the context of forced HIF alpha expression at high levels in COS cells, it is clear that normal cellular levels of pVHL can be limiting, showing that the level of pVHL can influence the HIF response, albeit in an artificial setting (71). In this context, two recent reports describe circumstances in which the amount of native pVHL is altered in a potentially relevant way. First, in explants of early gestation placentas exposure to 2% oxygen for 48 h was found to result in a marked increase in pVHL (72). Second, in cultured normal renal proximal tubular epithelial cells the level of pVHL was reported to be 100-fold higher in dense cultures compared to sparse cultures (72). Whether these alterations in the amount of pVHL have significant effects on the function of the HIF pathway remains to be determined.

VIII. What Other Processes Are Mediated by pVHL and Do These Involve Prolyl Hydroxylation?

VHL's role in recognizing the oxygen-dependent destruction domain of HIF alpha after conversion to the hydroxyproline-containing form appears not to be the only role of this tumor suppressor protein. pVHL's involvement in other processes can be considered in three categories; regulation of aspects of HIF other than ODDD-mediated proteolysis, regulation of oxygen-responsive pathways via routes other than HIF-1, and effects on other cellular processes. As discussed above, activity of HIF-1 alpha is regulated by oxygen not only through proteasomal destruction, but also by regulated transactivator recruitment and nuclear translocation. A relevant observation is that in cells lacking pVHL the HIF system appears to be fully activated, i.e., to a level equivalent to maximal induction by hypoxia or iron chelation. This strongly suggests that pVHL may be involved in aspects of HIF regulation by oxygen other than ODDD-mediated destruction. Very recently direct evidence for such a link has come from the identification of a protein termed "factor inhibiting HIF-1" (FIH-1), which interacts with both pVHL and the C terminal portion of HIF-1 alpha and represses transactivation by the C terminus probably by recruiting histone deacetylases (73). An attractive possibility would be that regulation of other aspects of HIF alpha by oxygen besides ODDD ubiquitylation involves oxygen-regulated hydroxylation of prolyl residues in HIF alpha or another protein.

The second category of potential pVHL targets are oxygen-response processes other than HIF-1. A striking feature of renal carcinoma cell lines lacking pVHL is that the level of expression of VEGF and GLUT-1 is constitutively high (74). In normal cells expression of these genes in hypoxia is increased not only by increased transcription, but also by stabilization of their mRNA. In cells lacking pVHL, not only is transcription increased by activation of HIF-1, but the mRNA is constitutively stable. One theoretical possibility, which has not yet been excluded, is that the mRNA stabilization is mediated either directly or indirectly by HIF-1. However, recent data suggest that a ribonucleoprotein, hnRNP A2, is a target of pVHL and that it regulates stability of these genes (75). It is attractive to speculate that a number of cellular proteins could be subjected to oxygen-regulated hydroxylation of specific prolyl residues, which could then alter the function of the protein directly, or alter its interaction with another protein, or permit capture by pVHL and ubiquitylation. Candidates for this mode of regulation by oxygen include oxygen-responsive ion channels, and other oxygen-responsive-transcription factors such as p53. In considering the possibility that prolyl hydroxylases may have a more general role in cell signaling it is important to recognize that enzymes of this class are not equilibrium enzymes that catalyze reversible reactions. Therefore, rapid modulation of a signal pathway in both directions requires either that the enzyme target can be rapidly destroyed and resynthesized in a nonhydroxylated form (as is the case for HIF alpha subunits) or that there is a second enzyme system that provides a reversal mechanism.

While it is attractive to speculate in this fashion that there may be other oxygen-regulated targets of pVHL or prolyl hydroxylases, there is considerable evidence that pVHL is involved in cellular processes that appear quite distinct from this. These putative functions of pVHL have come from two broad approaches. One is to identify interacting proteins, and a second is to examine the effect of VHL reexpression on candidate cellular processes. As an example, there is strong evidence for a role of pVHL in fibronectin assembly (76). This followed the identification of fibronectin as a protein interacting with pVHL in transfected renal carcinoma cells. Both in renal carcinoma cells and in murine embryonic stem cells, lack of pVHL was shown to result in failure to assemble an extracellular fibronectin matrix, with accumulation of fibronectin in the endoplasmic reticulum. The precise role of pVHL in fibronectin assembly is not yet completely understood, and to date the ubiquitin-proteasome pathway has not been implicated. Besides fibronectin and the elements of the E3 ubiquitin ligase complex discussed above (elongin B, elongin C, Rbx1, and Cul2), there are other cellular proteins that have been reported to interact specifically with pVHL. These include the transcription factor Sp-1 (77), a protein termed VBP-1 (for VHL binding protein-1) (78), and more recently VDU1 (for VHL-interacting deubiquitinating enzyme 1) (79). Whether these interactions involve epitopes containing hydroxyproline is not yet known.

Overall, at the current time it is unclear whether a unifying view of pVHL's functions could reasonably be taken, although this has not yet been excluded. Currently it remains possible that all the functions of pVHL are relevant to cellular oxygen responses or are concerned with hydroxyproline-specific recognition.

IX. What Is the Role of HIF Dysregulation in VHL-Associated Tumors?

There has been much recent interest in the role of HIF activation in tumor biology (for reviews see Refs. 80,81). The recognition that defects in the VHL gene give rise to cancer suggests that a genetic event that activates the HIF pathway may be a sufficient event to lead to cancer development in some cells. Before considering this issue in cells with VHL inactivation it is appropriate to consider the relevance of HIF to tumor biology in general. Extensive regions of solid tumors are hypoxic, and it is clear that HIF is activated in these regions (82,83). This results in high levels of expression of target genes including glycolytic enzymes, glucose transporters, and angiogenic growth factors in focal regions of the tumor (84). In several studies, comparison of model tumors with and without defects in the HIF pathway has shown that disabling the HIF pathway generally decreases tumor angiogenesis and tumor growth (84–86). Using cultured cell lines, a number of genetic events associated with cancer have been shown to modulate the HIF response, leading to a higher level of HIF activity both under standard culture conditions and in response to hypoxia (for review see Ref. 81). In contrast to the effect of mutations in the VHL gene, the HIF pathway is up-regulated but remains oxygen-responsive. We have previously argued that it would be teleologically attractive if increased proliferation itself resulted in up-regulation of the HIF pathway (81). This would provide a way in which the increased energy and synthetic requirements associated with proliferation could be anticipated. An important addition to the ways that HIF activity may be modulated in evolving cancer cells is that prolyl hydroxylase activity could be altered. This could be achieved by altering expression or activity of the PHD enzymes themselves, or altered availability of iron or 2-oxoglutarate. Interestingly, as outlined above, PHD3 mRNA expression is increased in fibroblasts exposed to several mitogenic stimuli, which would be predicted to decrease, rather than increase, HIF activation.

A rather striking observation is that aside from CCRCC and hemangioblastoma, VHL mutations are very uncommon in tumors and cell lines. This suggests that in other cell types constitutive HIF activation would not offer a net selective advantage to a cancer cell. Relevant to this, wild-type embryonic stem cells were more likely to undergo apoptosis in some settings than HIF-1alpha −/− or HIF-2 alpha −/− cells (87,88), and HIF activation has been shown to increase the transcription of genes that are proapoptotic (89,90). Since HIF activation regulates the expression of a very broad range of target genes, which vary from one cell type to another, it would be anticipated that the balance of effects is different from one cell type to another. That the HIF oxygen response system is intact in all cell lines except for CCRCC suggests that there is a net benefit from retaining an intact HIF oxygen-response system in the great majority of tumor types.

Returning to VHL-associated tumors, it seems reasonable to assume that constitutive activation of the HIF pathway accounts for the angiogenic phenotype. But is it HIF activation or another effect of VHL loss that initiates tumor formation when VHL function is lost? Looked at in another way, is constitutive activation of the oxygen-response pathway the primary defect in VHL-associated tumors? At the moment insights into this question are limited. One route of investigation has been to test different disease-associated VHL mutations and examine whether the clinical consequences correlate with effects on the HIF system. Clinically VHL disease has been categorized, as shown in Table 2, on the basis of different risks of phaeochromocytoma, hemangioblastoma, and renal cell carcinoma. Individual mutations associated with type 1, type 2A, and type 2B were all found to result in failure to regulate HIF-1alpha and HIF-2alpha when stably transfected into VHL-defective CCRCC cells (91,92). Thus all mutations that are associated with CCRCC appear to interfere with pVHL's ability to regulate HIF. This offers some support for the possibility that the primary defect leading to CCRCC in renal epithelial cells is activation of the HIF pathway.

Table 2 Clinical Categories of VHL Disease, Associated Genetic Mutations and the Effect on Regulation of the HIF System

Category	HAB	RCC	Phaeo	Genotype	HIF regulation in RCC
Type I	✓	HIGH	X	Often deletions/ truncations	Disabled
Type II				Usually missense	
IIA	✓	LOW	✓	e.g., Tyr98His	Disabled
IIB	✓	HIGH	✓	e.g., Arg167Gln	Disabled
IIC	X	X	✓	e.g., Leu188Val	Not affected

Clinical categories are defined on whether or not hemangioblastoma (HAB) and pheochromocytoma are found in affected kindreds and according to the risk of renal cell carcinoma (RCC). For each clinical category the most frequently associated mutation is indicated. The effect of the mutations on the ability to restore HIF regulation in stably transfected, VHL-defective renal carcinoma cells is also indicated.

X. HIF Prolyl Hydroxylase and the VHL-HIF Interaction as Therapeutic Targets

Inhibition of HIF prolyl hydroxylase activity, or preventing the capture of modified HIF alpha chains by pVHL, should be highly effective methods of activating the HIF pathway. Indeed we have recently shown that the cell-penetrant 2-oxoglutarate analog dimethyloxalylglycine activates the HIF pathway in cultured cells (49). Activation of the HIF pathway has already been shown to be an effective way of promoting angiogenesis in in vivo experiments. Furthermore, vessels produced by expression of a constitutively active HIF molecule appear to be less leaky than those produced in response to vascular endothelial growth factor alone (93).

Importantly, HIF prolyl hydroxylase inhibitors might antagonize other enzymes in the overall class of 2-oxoglutarate-dependent dioxygenases. Conversely, it is possible that compounds developed as inhibitors of other enzymes in the class, such as collagen prolyl 4-hydroxylase, may have significant activity as inhibitors of PHD enzymes. To target the HIF system with precision it will clearly be important to develop inhibitors that are specific for HIF prolyl hydroxylase.

References

1. Jelkmann W. Erythropoietin: Structure, control of production, and function. Physiol Rev 1992; 72:449–489.
2. Lahiri S. Historical perspectives of cellular oxygen sensing and responses to hypoxia. J Appl Physiol 2000; 88:1467–1473.
3. Maxwell PH, Osmond MK, Pugh CW, Heryet A, Nicholls LG, Tan CC, Doe BG, Ferguson DJP, Johnson MH, Ratcliffe PJ. Identification of the renal erythropoietin-producing cells using transgenic mice. Kidney Int 1993; 44:1149–1162.
4. Bachmann S, Le Hir M, Eckardt K-U. Co-localization of erythropoietin messenger RNA and ecto-5′-nucleotidase immunoreactivity in peritubular cells of rat renal cortex indicates that fibroblasts produce erythropoietin. J Histochem Cytochem 1993; 41:335–341.
5. Wang GL, Semenza GL. Characterization of hypoxia-inducible factor 1 and regulation of DNA binding activity by hypoxia. J Biol Chem 1993; 268:21513–21518.
6. Wang GL, Jiang B-H, Rue EA, Semenza GL. Hypoxia-inducible factor 1 is a basic-helix-loop-helix-PAS heterodimer regulated by cellular O_2 tension. Proc Natl Acad Sci USA 1995; 92:5510–5514.
7. Maxwell PH, Pugh CW, Ratcliffe PJ. Inducible operation of the erythropoietin 3′ enhancer in multiple cell lines: evidence for a widespread oxygen sensing mechanism. Proc Natl Acad Sci USA 1993; 90:2423–2427.
8. Semenza GL. HIF-1: mediator of physiological and pathophysiological responses to hypoxia. J Appl Physiol 2000; 88:1474–1480.
9. Jiang H, Guo R, Powell-Coffman JA. The *Caernorhabditis elegans hif-1* gene encodes a bHLH-PAS protein that is required for adpatation to hypoxia. Proc Natl Acad Sci USA 2001; 98:7916–7921.
10. Epstein ACR, Gleadle JM, McNeill LA, Hewitson KS, O'Rourke J, Mole DR, Mukherji M, Metzen E, Wilson MI, Dhanda A, Tian Y-M, Masson N, Hamilton DL,

Jaakkola P, Barstead R, Hodgkin J, Maxwell PH, Pugh CW, Schofield CJ, Ratcliffe PJ. *C. elegans* EGL-9 and mammalian homologues define a family of dioxygenases that regulate HIF by prolyl hydroxylation. Cell 2001; 107:43–45.
11. Nagao M, Ebert BL, Ratcliffe PJ, Pugh CW. *Drosophila melanogaster* SL2 cells contain a hypoxically inducible DNA binding complex which recognises mammalian HIF-1 binding sites. FEBS Lett 1996; 387:161–166.
12. Bacon NC, Wappner P, O'Rourke JF, Bartlett SM, Shilo B, Pugh CW, Ratcliffe PJ. Regulation of the *Drosophila* basic helix-loop-helix PAS protein Sima by hypoxia: functional evidence for homology with mammalian HIF-1 alpha. Biochem Biophy Res Commun 1998; 249:811–816.
13. Salceda S, Caro J. Hypoxia-inducible factor 1α (HIF-1α) protein is rapidly degraded by the ubiquitin-proteasome system under normoxic conditions. J Biol Chem 1997; 272:22642–22647.
14. Pugh CW, O'Rourke JF, Nagao M, Gleadle JM, Ratcliffe PJ. Activation of hypoxia inducible factor-1; definition of regulatory domains within the α subunit. J Biol Chem 1997; 272:11205–11214.
15. Huang LE, Gu J, Schau M, Bunn HF. Regulation of hypoxia-inducible factor 1α is mediated by an oxygen-dependent domain via the ubiquitin-proteasome pathway. Proc Natl Acad Sci USA 1998; 95:7987–7992.
16. Wiesener MS, Turley H, Allen WE, William C, Eckardt K-U, Talks KL, Wood SM, Gatter KC, Harris AL, Pugh CW, Ratcliffe PJ, Maxwell PH. Induction of endothelial PAS domain protein-1 by hypoxia: characterization and comparison with hypoxia-inducible factor-1α. Blood 1998; 92:2260–2268.
17. Kallio PJ, Okamoto K, O'Brien S, Carrero P, Makino Y, Tanaka H, Poellinger L. Signal transduction in hypoxic cells: inducible nuclear translocation and recruitment of the CBP/p300 coactivator by the hypoxia-inducible factor-1α. EMBO J 1998; 17:6573–6586.
18. Ema M, Hirota K, Mimura J, Abe H, Yodoi J, Sogawa K, Poellinger L, Fujii-Kuriyama Y. Molecular mechanisms of transcription activation by HLF and HIF1alpha in response to hypoxia: their stabilization and redox signal-induced interaction with CBP/p300. EMBO J 1999; 18:1905–1914.
19. Bhattacharya S, Michels CL, Leung M-K, Arany ZP, Kung AL, Livingston DM. Functional role of p35srj, a novel p300/CBP binding protein, during transactivation by HIF-1. Genes Dev 1999; 13:64–75.
20. Goldwasser E, Jacobson LO, Fried W, Plazk LF. Studies on erythropoiesis V: The effect of cobalt on the production of erythropoietin. Blood 1958; 13:55–60.
21. Goldberg MA, Dunning SP, Bunn HF. Regulation of the erythropoietin gene: evidence that the oxygen sensor is a heme protein. Science 1988; 242:1412–1415.
22. Wang GL, Semenza GL. Desferrioxamine induces erythropoietin gene expression and hypoxia-inducible factor 1 DNA-binding activity: implications for models of hypoxia signal transduction. Blood 1993; 82:3610–3615.
23. Ho VT, Bunn HF. Effects of transition metals on the expression of the erythropoietin gene: further evidence that the oxygen sensor is a heme protein. Biochem Biophy Res Commun 1996; 223:175–180.
24. Gilles-Gonzalez MA, Gonzalez G, Perutz MF. Heme-based sensors, exemplified by the kinas FixL, are a new class of heme protein with distinctive ligand binding and autoxidation. Biochemistry 1994; 33:8067–8073.
25. Hidalgo E, Ding H, Demple B. Redox signal transduction: mutations shifting [2Fe-2S] centers of the SoxR sensor-regulator to the oxidized form. Cell 1997; 88:121–129.

26. Fandrey J, Frede S, Jelkmann W. Role of hydrogen peroxide in hypoxia-induced erythropoietin production. Biochem J 1994; 303:507–510.
27. Acker H. Mechanisms and meaning of cellular oxygen sensing in the organism. Respir Physiol 1994; 95:1–10.
28. Zhu H, Qiu H, Yoon H-W, Huang S, Bunn H. Identification of a cytochrome b–type NAD(P)H oxidoreductase ubiquitously expressed in human cells. Proc Natl Acad Sci USA 1999; 96:14742–14747.
29. Chandel NS, McClintock DS, Feliciano CE, Wood TM, Melendez JA, Rodriguez AM, Schumacker PT. Reactive oxygen species generated at mitochondrial complex III stabilize hypoxia-inducible factor-l alpha during hypoxia: a mechanism of oxygen sensing. J Biol Chem 2000:1–37.
30. Chandel NS, Maltepe E, Goldwasser E, Mathieu CE, Simon MC, Schumacker PT. Mitochondrial reactive oxygen species trigger hypoxia-induced transcription. Proc Natl Acad Sci USA 1998; 95:11715–11720.
31. Storz G, Imlay J. Oxidative stress. Curr Opin Microbiol 1999; 2:188–194.
32. Maxwell PH, Wiesener MS, Chang G-W, Clifford SC, Vaux EC, Cockman ME, Wykoff CC, Pugh CW, Maher ER, Ratcliffe PJ. The tumour suppressor protein VHL targets hypoxia-inducible factors for oxygen-dependent proteolysis. Nature 1999; 399:271–275.
33. Latif F, Tory K, Gnarra J, Yao M, Duh F-M, Orcutt ML, Stackhouse T, Kuzmin I, Modi W, Geil L, Schmidt L, Zhou F, Li H, Wei MH, Chen F, Glenn G, Choyke P, Walther MM, Weng Y, Duan D-SR, Dean M, Glavac D, Richards FM, Crossey PA, Ferguson-Smith MA, Le Paslier D, Chumakov I, Cohen D, Chinault AC, Maher ER, Linehan WM, Zbar B, Lerman MI. Identification of the von Hippel–Lindau disease tumor suppressor gene. Science 1993; 260:1317–1320.
34. Gnarra JR, Tory K, Weng Y, Schmidt L, Wei MH, Li H, Latif F, Liu S, Chen F, Duh F-M, Lubensky I, Duan DR, Florence C, Pozzatti R, Walther MM, Bander NH, Grossman HB, Brauch H, Pomer S, Brooks JD, Isaacs WB, Lerman MI, Zbar B, Linehan WM. Mutations of the *VHL* tumour suppressor gene in renal carcinoma. Nature Genet 1994; 7:85–90.
35. Kanno H, Kondo K, Ito S, Yamamoto I, Fujii S, Torigoe S, Sakai N, Hosaka M, Shuin T, Yao M. Somatic mutations of the Von Hippel–Lindau tumor suppressor gene in sporadic central nervous system hemangioblastomas. Cancer Res 1994; 54:4845–4847.
36. Iliopoulos O, Ohh M, Kaelin WG, Jr. $pVHL_{19}$ is a biologically active product of the von Hippel–Lindau gene arising from internal translation initiation. Proc Natl Acad Sci USA 1998; 95:11661–11666.
37. Schoenfeld A, Davidowitz EJ, Burk RD. A second major native von Hippel–Lindau gene product, initiated from an internal translation start site, functions as a tumor suppressor. Proc Natl Acad Sci USA 1998; 95:8817–8822.
38. Kamura T, Sato S, Iwai K, Czyzyk-Krzeska M, Conaway RC, Conaway JW. Activation of HIF1α ubiquitination by a reconstituted von Hippel–Lindau (VHL) tumor suppressor complex. Proc Natl Acad Sci USA 2000; 97:10430–10435.
39. Cockman ME, Masson N, Mole DR, Jaakkola P, Chang GW, Clifford SC, Maher ER, Pugh CW, Ratcliffe PJ, Maxwell PH. Hypoxia inducible factor-α binding and ubiquitylation by the von Hippel–Lindau tumor suppressor protein. J Biol Chem 2000; 275:25733–25741.
40. Ohh M, Park CW, Ivan M, Hoffman MA, Kim TY, Huang LE, Pavletich N, Chau V, Kaelin WG. Ubiquitination of hypoxia-inducible factor requires direct binding to the beta-domain of the von Hippel–Lindau protein. Nat Cell Biol 2000; 2:423–427.

41. Duan DR, Pause A, Burgess WH, Aso T, Chen DYT, Garrett KP, Conaway RC, Conaway JW, Linehan WM, Klausner RD. Inhibition of transcription elongation by the VHL tumor suppressor protein. Science 1995; 269:1402–1406.
42. Kibel A, Iliopoulos O, DeCaprio JA, Kaelin WG, Jr. Binding of the von Hippel–Lindau tumor suppressor protein to elongin B and C. Science 1995; 269:1444–1446.
43. Pause A, Lee S, Worrell RA, Chen DYT, Burgess WH, Linehan WM, Klausner RD. The von Hippel–Lindau tumor-suppressor gene product forms a stable complex with human CUL-2, a member of the Cdc53 family of proteins. Proc Nat Acad Sci USA 1997; 94:2156–2161.
44. Lonergan KM, Iliopoulos O, Ohh M, Kamura T, Conaway RC, Weliky Conaway J, Kaelin WG, Jr. Regulation of hypoxia-inducible mRNAs by the von Hippel–Lindau tumor suppressor protein requires binding to complexes containing elongins B/C and Cul2. Mol Cell Biol 1998; 18:732–741.
45. Kamura T, Koepp DM, Conrad MN, Skowyra D, Moreland RJ, Iliopoulos O, Lane WS, Kaelin WG, Jr., Elledge SJ, Conaway RC, Harper JW, Conaway JW. Rbx1, a component of the VHL tumor suppressor complex and SCF ubiquitin ligase. Science 1999; 284:657–661.
46. Stebbins CE, Kaelin WG, Jr., Pavletich NP. Structure of the VHL-ElonginC-ElonginB complex: implications for VHL tumor suppressor function. Science 1999; 284:455–461.
47. Haase VH, Glickman, JN, Socolovsky M, Jaenisch, R. Vascular tumors in livers with targeted inactivation of the von Hippel–Lindau tumor suppressor. Proc Nat Acad Sci USA 2001; 98:1583–1588.
48. Vaux EC, Wood SM, Cockman ME, Nicholls LG, Yeates KM, Pugh CW, Maxwell PH, Ratcliffe PJ. Selection of mutant CHO cells with constitutive activation of the HIF system and inactivation of the von Hippel–Lindau tumor suppressor. J Biol Chem 2001; 276:44323–44330.
49. Jaakkola P, Mole DR, Tian Y-M, Wilson MI, Gielbert J, Gaskell SJ, von Kriegsheim A, Hebestreit HF, Mukherji M, Schofield CJ, Maxwell PH, Pugh CW, Ratcliffe PJ. Targeting of HIF-α to the von Hippel–Lindau ubiquitylation complex by O_2-regulated prolyl hydroxylation. Science 2001; 292:468–472.
50. Ivan M, Kondo K, Yang H, Kim W, Valiando J, Ohh M, Salic A, Asara JM, Lane WS, Kaelin WG Jr. HIFα targeted for VHL-mediated destruction by proline hydroxylation: implications for O_2 sensing. Science 2001; 292:464–468.
51. Kornitzer D, Ciechanover A. Modes of regulation of ubiquitin-mediated protein degradation. J Cell Physiol 2000; 182:1–11.
52. Masson N, Willam C, Maxwell PH, Pugh CW, Ratcliffe PJ. Independent function of two destruction domains in hypoxia-inducible factor-α chains activated by prolyl hydroxylation. EMBO J 2001; 20:5197–5206.
53. Schofield CJ, Zhang Z. Structural and mechanistic studies on 2-oxoglutarate-dependent oxygenases and related enzymes. Curr Opin Struct Biol 1999; 9:722–731.
54. Hegg EL, Que L. The 2-His-1-carboxylate facial triad—an emerging structural motif in mononuclear non-heme iron (II) enzymes. Eur J Biochem 1997; 250:625–629.
55. Kivirikko KI, Myllyharju J. Prolyl 4-hydroxylases and their protein disulfide isomerase subunit. Matrix Biol 1998; 16:357–368.
56. Bruick RK, McKnight SL. A conserved family of prolyl-4-hydroxylases that modify HIF. Science 2001; 11:11.
57. Trent C, Tsung N, Horvitz HR. Egg-laying defective mutants of the nematode *Caenorhabditis elegans*. Genetics 1983; 104:619–647.

58. Gallagher LA, Manoil C. *Pseudomonas aeruginosa* PAO1 kills *Caenorhabditis elegans* by cyanide poisoning. J Bacteriol 2001; 183:6207–6214.
59. Berra E, Richard DE, Gothie E, Pouyssegur J. HIF-1-dependent transcriptional activity is required for oxygen-mediated HIF-1α degradation. FEBS Lett 2001; 491:85–90.
60. Wax SD, Rosenfield CL, Taubman MB. Identification of a novel growth factor–responsive gene in vascular smooth muscle cells. J Biol Chem 1994; 269:13041–13047.
61. Madden SL, Galella EA, Riley D, Bertelsen AH, Beaudry GA. Induction of cell growth regulatory genes by *p53*. Cancer Res 1996; 56:5384–5390.
62. Lipscomb EA, Sarmiere PD, Crowder RJ, Freeman RS. Expression of the SM-20 gene promotes death in nerve growth factor-dependent sympathetic neurons. J Neurochemi 1999; 73:429–432.
63. Lipscomb EA, Sarmiere PD, Freeman RS. SM-20 is a novel mitochondrial protein that causes caspase-dependent cell death in nerve growth factor-dependent neurons. J Biol Chem 2001; 276:5085–5092.
64. Plisov SY, Ivanov SV, Yoshino K, Dove LF, Plisova TM, Higinbotham KG, Karavanova I, Lerman M, Perantoni AO. Mesenchymal-epithelial transition in the developing metanephric kidney: gene expression study by differential display. Genesis 2000; 27:22–31.
65. Lee S, Chen DYT, Humphrey JS, Gnarra JR, Linehan WM, Klausner RD. Nuclear/cytoplasmic localization of the von Hippel–Lindau tumor supressor gene product is determined by cell density. Proc Nat Acad Sci USA 1996; 93:1770–1775.
66. Los M, Jansen GH, Kaelin WG, Lips CJM, Blijham GH, Voest EE. Expression pattern of the von Hippel–Lindau protein in human tissues. Lab Invest 1996; 75:231–238.
67. Corless CL, Kibel AS, Iliopoulos O, Kaelin WG, Jr. Immunostaining of the von Hippel–Lindau gene product in normal and neoplastic human tissues. Hum Pathol 1997; 28:459–464.
68. Lee S, Neumann M, Stearman R, Stauber R, Pause A, Pavlakis GN, Klausner RD. Transcription-dependent nuclear-cytoplasmic trafficking is required for the function of the von Hippel–Lindau tumor suppressor protein. Mol Cell Biol 1999; 19:1486–1497.
69. Groulx I, Bonicalzi ME, Lee S. Ran-mediated nuclear export of the von Hippel–Lindau tumor suppressor protein occurs independently of its assembly with cullin-2. J Biol Chem 2000; 275:8991–9000.
70. Ye Y, Vasavada S, Kuzmin I, Stackhouse T, Zbar B, Williams BRG. Subcellular localization of the von Hippel–Lindau disease gene product is cell cycle-dependent. Int J Cancer 1998; 78:62–69.
71. Tanimoto K, Makino Y, Pereira T, Poellinger L. Mechanism of regulation of the hypoxia-inducible factor-1α by the von Hippel–Lindau tumor suppressor protein. EMBO J 2000; 19:4298–4309.
72. Genbacev O, Krtolica A, Kaelin W, Fisher SJ. Human cytotrophoblast expression of the von Hippel–Lindau protein is downregulated during uterine invasion in situ and upregulated by hypoxia in vitro. Dev Biol 2001; 233:526–536.
73. Mahon PC, Hirota K, Semenza GL. FIH-1: a novel protein that interacts with HIF-1alpha and VHL to mediate repression of HIF-1 transcriptional activity. Genes Dev 2001; 15:2675–2686.
74. Iliopoulos O, Levy AP, Jiang C, Kaelin WG Jr, Goldberg MA. Negative regulation of hypoxia-inducible genes by the von Hippel–Lindau protein. Proc Nat Acad Sci USA 1996; 93:10595–10599.

75. Pioli PA, Rigby WFC. The von Hippel–Lindau protein interacts with hnRNP A2 and regulates its expression. J Biol Chem 2001; 276:40346–40352.
76. Ohh M, Yauch RL, Lonergan KM, Whaley JM, Stemmer-Rachamimov AO, Louis DN, Gavin BJ, Kley N, Kaelin WG, Jr, Iliopoulos O. The von Hippel–Lindau tumor suppressor protein is required for proper assembly of an extracellular fibronectin matrix. Mol Cell 1998; 1:959–968.
77. Mukhopadhyay D, Knebelmann B, Cohen HT, Ananth S, Sukhatme VP. The von Hippel–Lindau tumor suppressor gene product interacts with Sp1 to repress vascular endothelial growth factor promoter activity. Mol Cell Biol 1997; 17:5629–5639.
78. Tsuchiya H, Iseda T, Hino O. Identification of a novel protein (VBP-1) binding to the von Hippel–Lindau (*VHL*) tumor suppressor gene product. Cancer Res 1996; 56:2881–2885.
79. Li Z, Na X, Wang D, Schoen SR, Messing EM, Wu G. Ubiquitination of a novel deubiquitinating enzyme requires direct binding to von Hippel–Lindau tumor suppressor protein. J Biol Chem 2001; 5:5.
80. Semenza GL. Hypoxia, clonal selection, and the role of HIF-1 in tumor progression. Crit Rev Biochem Mol Biol 2000; 35:71–103.
81. Maxwell PH, Pugh CW, Ratcliffe PJ. Activation of the HIF pathway in cancer. Curr Opin Genet Dev 2001; 11:293–299.
82. Zhong H, De Marzo AM, Laughner E, Lim M, Hilton DA, Zagzag D, Buechler P, Isaacs WB, Semenza GL, Simons JW. Overexpression of hypoxia-inducible factor 1α in common human cancers and their metastases. Cancer Res 1999; 59:5830–5835.
83. Talks K, Turley H, Gatter KC, Maxwell PH, Pugh CW, Ratcliffe PJ, Harris AL. The expression and distribution of the hypoxia-inducible factors HIF-1alpha and HIF-2alpha in normal human tissues, cancers, and tumor-associated macrophages. Am J Pathol 2000; 157:411–421.
84. Maxwell PH, Dachs GU, Gleadle JM, Nicholls LG, Harris AL, Stratford IJ, Hankinson O, Pugh CW, Ratcliffe PJ. Hypoxia inducible factor-1 modulates gene expression in solid tumors and influences both angiogenesis and tumor growth. Proc Nat Acad Sci USA 1997; 94:8104–8109.
85. Ryan HE, Poloni M, McNulty W, Elson D, Gassmann M, Arbeit JM, Johnson RS. Hypoxia-inducible factor-1α is a positive factor in solid tumor growth. Cancer Res 2000; 60:4010–4015.
86. Ryan HE, Lo J, Johnson RS. HIF-1a is required for solid tumor formation and embryonic vascularization. EMBO J 1998; 17:3005–3015.
87. Carmeliet P, Dor Y, Herbert J-M, Fukumura D, Brusselmans K, Dewerchin M, Neeman M, Bono F, Abramovitch R, Maxwell PH, Koch CJ, Ratcliffe PJ, Moons L, Jain RK, Collen D, Keshert E. Role of HIF-1alpha in hypoxia-mediated apoptosis, cell proliferation and tumour angiogenesis. Nature 1998; 394:485–490.
88. Brusselmans K, Bono F, Maxwell P, Dor Y, Dewerchin M, Collen D, Herbert JM, Carmeliet P. Hypoxia-inducible factor-2alpha (HIF-2alpha) is involved in the apoptotic response to hypoglycemia but not to hypoxia. J Biol Chem 2001; 276:39192–39196.
89. Bruick RK. Expression of the gene encoding the proapoptotic Nip3 protein is induced by hypoxia. Proc Nat Acad Sci USA 2000; 97:9082–9087.
90. Sowter HM, Ratcliffe PJ, Watson P, Greenberg AH, Harris AL. HIF-1-dependent regulation of hypoxic induction of the cell death factors BNIP3 and NIX in human tumors. Cancer Res 2001; 61:6669–6673.

91. Hoffman MA, Ohh M, Yang H, Klco JM, Ivan M, Kaelin WGJ. von Hippel–Lindau protein mutants linked to type 2C VHL disease preserve the ability to downregulate HIF. Hum Mol Genet 2001; 10:1019–1027.
92. Clifford SC, Cockman ME, Smallwood AC, Mole DR, Woodward ER, Maxwell PH, Ratcliffe PJ, Maher ER. Contrasting effects on HIF-1α regulation by disease-causing pVHL mutations correlate with patterns of tumourigenesis in von Hippel–Lindau disease. Hum Mol Genet 2001; 10:1029–1038.
93. Elson DA, Thurston G, Huang LE, Ginzinger DG, McDonald DM, Johnson RS, Arbeit JM. Induction of hypervascularity without leakage or inflammation in transgenic mice overexpressing hypoxia-inducible factor-1alpha. Genes Dev 2001; 15:2520–2532.
94. Dupuy D, Aubert I, Duperat VG, Petit J, Taine L, Stef M, Bloch B, Arveiler B. Mapping, characterization, and expression analysis of the SM-20 human homologue, c1orf12, and identification of a novel related gene, SCAND2. Genomics 2000; 69:348–354.
95. Taylor MS. Characterization and comparative analysis of the EGLN gene family. Gene 2001; 275:125–132.

4

Oxygen- or Redox-Dependent Regulation

The Role of Hydrogen Peroxide in the Regulation of Erythropoietin Gene Expression

JOACHIM FANDREY

University of Essen
Essen, Germany

I. Introduction

The hormone erythropoietin (EPO), a 30.4-kDa glycoprotein, is the key regulator in erythropoiesis (1). Either the decrease of the oxygen-carrying capacity of the blood as in anemia or a reduction in the inspiratory PO_2 can lead to an up to 1000-fold increase in the production of the hormone. In the adult, EPO is sufficiently produced only by the kidneys. Earlier animal experiments by Jacobson et al. (2) revealed this production site by a classic series of organ ablation experiments. Biologically active EPO was found to be produced by the kidneys upon hypoxic stimulation without the requirement of other organs or plasma protein precursors [excellently reviewed by Jelkmann (1)]. Still, it took some decades to identify interstitial fibroblasts of the kidney as the EPO-expressing cells (3). In humans, the role of the kidneys becomes obvious when patients who suffer from terminal renal insufficiency develop an anemia due to the lack of the hormone. Fortunately, due to progress in recombinant DNA technology today the recombinant human EPO is successfully used to compensate for the lack of endogenous production and to cure the anemia of chronic renal insufficiency. Interestingly, in animal (4) and human fetuses (5), the liver is the predominant organ for EPO production and obviously—as in the anemia of chronic renal insufficiency—cannot take over this function again after birth and substitute for the lack of production by the kidneys. During ontogeny the production of EPO starts in the liver and, at least in humans, makes up to more than 90% of the total mRNA until birth. There is minor expression in the kidneys, the spleen, and the bone marrow, and, in addition, in the central nervous system (5,6). One important, but yet unanswered question of EPO research is the switch of the production site from liver to kidney after the first month of life.

II. How to Study Oxygen-Dependent EPO Expression

Studies on the mechanism of hypoxic induction of EPO had always been hampered by models that sufficiently reflect the in vivo production. Until today there are no reports on successful attempts to isolate the EPO-producing cells from the kidney and bring them into culture. No permanent renal tumor cell line is available that shows oxygen-regulated EPO production. Goldberg and colleagues (7) reported two hepatoma cell lines, HepG2 and Hep3B, that produce EPO in an oxygen-dependent manner. It was found that the EPO gene was expressed in exact dependency of the pericellular PO_2 (8,9). For these studies it was important to establish cell culture systems where the pericellular oxygen tension was tightly controlled and actually reflected the oxygen tension in the incubation gas (8,10). From these in vitro studies in hepatoma cells it became evident that EPO production depended on EPO gene expression, which was strictly regulated by the PO_2. Therefore, EPO gene regulation in hepatoma provided an ideal system to study oxygen-dependent gene expression in general. Most of the findings obtained with these tumor cell lines—which may, in fact, partly reflect the fetal EPO expression in hepatocytes—are relevant to EPO expression in the kidneys as well. Due to the lack of suitable kidney cell lines the isolated perfused rat kidney was used as a model system to study EPO expression (11–13). Many results from initial studies with HepG2 hepatoma cells were confirmed in this model like the inhibition of EPO expression by proinflammatory cytokines (14) or by reactive O_2 species (see below) (15). However, studies with the isolated perfused rat kidneys, like whole animal experiments, cannot exclude that changes in the PO_2 profile in the kidneys may affect expression of the EPO gene in interstitial fibroblasts although the PO_2 in the perfusion medium (reflecting the blood in vivo) remained unchanged.

III. Regulatory DNA Sequences in the Human EPO Gene

Hypoxia induces EPO mRNA levels several 100-fold. In an initial attempt to identify regulatory DNA sequences that are responsible for the increased transcription of the EPO gene under hypoxia Semenza et al. generated mice that were transgenic for the human EPO gene (16). Fragments of human DNA were used that contained the full EPO gene with all its exons and introns but differed from 4 to 22 kb depending on the size of the 5′ and 3′ flanking sequences that were included (16). These studies provided important data about potential regulatory DNA sequences surrounding the human EPO gene. It was found that a fragment consisting of 400 base pairs of 5′

flanking sequence and 700 base pairs of 3′ flanking sequence was sufficient for hypoxia-inducible expression in hepatic cells but that the expression of the EPO gene was not restricted to the liver or the kidneys. Further extension of the 5′ fragment to 6 kb, however, prevented aberrant expression in organs other than the liver indicating that organ specificity of EPO expression was determined by negative regulatory sequences. Nevertheless, sequences lying 6–14 kb 5′ to the gene were necessary for oxygen-regulated expression in the kidney. Still, in these transgenic mice EPO expression was prominently found in the liver indicating that sequences that suppress hepatic expression in the adult may lie more than 16.5 kb upstream or more than 2.2 kb downstream of the human EPO gene. In addition, and much more important for the following years of work, was the identification of the regulatory DNA sequences for oxygen-dependent expression of the gene. It was of major interest that these regulatory elements were lying 5′ of the EPO gene for the kidney but 3′ for the liver. However, in both organs EPO expression is hypoxia-inducible. This phenomenon—increase in production of a particular protein under hypoxic conditions when general protein synthesis is reduced to up to 70%—has initiated and promoted research on hypoxia-inducible gene expression in general.

IV. Hypoxia-Inducible EPO Expression In Vitro: The Identification of HIF-1

Further studies that made use of the Hep3B hepatoma cell as an in vitro model focused on the liver-inducible element that was responsible for hypoxia-induced expression of the EPO gene. Experiments were therefore aimed at elucidating the exact size and sequence of the 3′ flanking enhancer of the EPO gene. Only 50 nucleotides in length was finally the DNA element that provided full enhancer function in a way that a reporter gene under the control of this enhancer was induced 10- to 50-fold under hypoxic conditions (17,18). Wang et al. purified and isolated the proteins that bound to the regulatory DNA element and named them hypoxia-inducible factor 1 (HIF-1) (17). HIF-1 was found to be a dimer consisting of a 120-kDa α-subunit and a β-subunit that was about 93 kDa in size. Sequencing of both subunits revealed that the α- and β-subunit belong to a family of transcription factors that are characterized through a basic helix-loop-helix-DNA-binding domain and two regions of high homology that group them into a family of PAS (for Period-Arylhydrocarbon(=dioxin)-Similar, which are other transcription factors of this family) transcription factors. It was found that the α-subunit was an entirely new and so far unrecognized protein, whereas the β-subunit was identical to a previously isolated member of the PAS family called ARNT (=Arylhydrocarbon Receptor Nuclear Translocator).

Subsequent studies on the expression of HIF-1α and HIF-1β mRNA revealed that both transcription factor subunits are expressed in all adult and embryonic mouse and human tissues (19,20). HIF-1α is absolutely required for proper embryonic development since HIF-1$\alpha^{-/-}$ mice die around day 9 of gestation (21). Two further members of the PAS transcription factor family, HIF-2α and ARNT-2, share high degrees of homology with HIF-1α and HIF-1β, respectively. Interestingly, ARNT-2 is primarily expressed in the brain and kidney, whereas HIF-2α is found in endothelial cells, catecholamine-producing cells in the kidney and lung during embryogenesis. HIF-1α and HIF-2α are highly homologous within their DNA binding and transactivating sequence. Both HIF-1α and HIF-2α can dimerize with ARNT (=HIF-1β) or ARNT-2 and bind to DNA to activate the expression of HIF-1-regulated genes. With respect to hypoxia-induced EPO gene expression it is not clear which combination of PAS transcription factors, e.g., HIF-1α/HIF-1β or HIF-2α/HIF-1β or HIF-1α/ARNT-2, is responsible for hypoxia-inducible expression. Since the HIF-1α/HIF-1β dimer was isolated from the DNA element that regulates expression in hepatoma cells, it is most likely that HIF-1α and HIF-1β regulate expression in liver cells. However, recent preliminary reports suggest that HIF-2α may be responsible for hypoxia-inducible expression of the EPO gene in the kidney (22). With respect to what will be pointed out below, it is of interest that despite the high degree of homology a cysteine residue present in the basic domain of HIF-2α is not found in HIF-1α. This redox-sensitive cysteine residue may provide a target for specific redox regulation of DNA binding by HIF-2α/HIF-1β heterodimers (23).

V. Oxygen-Dependent Prolyl Hydroxylation Determines O_2 Lability of HIF-1α

However, the regulation of HIF-1α is primarily determined by the intracellular oxygen tension. Under conditions of a high PO_2—probably a PO_2 of 50–70 mmHg is sufficient—HIF-1α is rapidly degraded by the ubiquitin proteasome pathway, while hypoxic exposure prevents its degradation (24,25). Central to this normoxic degradation is the enzymatic hydroxylation of proline 402 and proline 564 of HIF-1α, which tags the protein for interaction with the E3-ligase of the von Hippel–Lindau multiprotein complex (26–28). The multiprotein complex formed by VHL protein contains at least the elongins B and C, Cul2, and RBX and ubiquitinates HIF-1α that had been hydroxylated at its proline residues. Interestingly, transition metals like cobalt, iron chelators, or inhibitors of the prolyl hydroxylase enzymes can mimic the hypoxic effect on this enzyme. Currently, the understanding of the function of the oxygen-sensing process is that oxygen itself is used by the prolyl hydroxylase (PHD) enzymes and becomes limiting under hypoxic conditions (29). Thus, in the absence of oxygen, proline 564 (and proline 402) are no longer hydroxylated, HIF-1α is not recognized by the VHL complex and will not be degraded. Thus, when PO_2-dependent changes of HIF-1α protein levels are achieved through altered enzymatic activity, PHDs fulfill the requirements for an oxygen sensor.

VI. Reactive Oxygen Species and EPO Expression

Earlier work from Goldberg and colleagues (30) and the author (31) gave reason to propose a heme protein as part of the cellular oxygen-sensing process. Since heme proteins can easily react with oxygen as in hemoglobin or cytochromes, pharmacological modification of type b-cytochromes of the P450 system was taken as an argument of the potential involvement of b-cytochromes in the oxygen-sensing pathway. Moreover, spectrophotometric analysis of hepatoma cells (line HepG2) that express EPO in an oxygen-dependent manner revealed that these cells contain a particular nonmitochondrial heme protein, a b-cytochrome, that changed its light-absorbing characteristics depending on the oxygenation state of the cells (32,33). Within the family of b-cytochromes are oxidases like the P450 oxidase-reductase system or the NADPH-dependent oxidase from phagocytes that generate reactive O_2 species. They catalyze the production of reactive oxygen species, e.g., superoxide anions (O_2^-), either as a by-product or directly. O_2^- is then dismutated, either spontaneously or by the enzyme superoxide dismutase, to hydrogen peroxide. With use of redox-sensitive fluorescent dyes like 123-dihydrorhodamine, it was found that hepatoma cells produce reactive oxygen species (ROS) (33). More interestingly, Western blots and immunohistochemical studies using antibodies generated against the different subunits of the NADPH oxidase from phagocytes revealed that proteins of similar size like the phagocyte enzyme subunits were present in extracts and intact HepG2 cells. Since the NADPH oxidase from phagocytes is composed of at least constitutive membrane-bound subunits, a $p22_{phox}$ and a large glycoprotein $gp91_{phox}$ subunit, to which a $p40_{phox}$, a $p47_{phox}$, and a $p67_{phox}$ subunit colocalize upon activation of the enzyme, the expression of these subunits was also tested in HepG2 cells. Surprisingly, only the mRNA for the $p22_{phox}$ subunit, which is ubiquitously expressed, was found whereas all other mRNAs are not expressed in hepatoma cells (own unpublished results). Thus, the finding that proteins with similar epitopes were recognized by antibodies against the NADPH oxidase subunits from phagocytes gave reason to suggest an enzyme similar to the phagocyte oxidase in HepG2 cells. However, this oxidase would clearly not be identical to the burst oxidase because the mRNAs of the burst oxidase—with the exception of $p22_{phox}$—are absent in HepG2.

Subsequent studies on hydrogen peroxide production rates by hepatoma cells revealed that production of ROS was continuous and closely related to the pericellular PO_2 (34). Production rates for H_2O_2 were highest at PO_2 values above 90 mmHg and decreased with lower PO_2 values (Fig. 1). Interestingly, the greatest changes in the hydrogen peroxide production rate were found in the PO_2 range that appears to be most physiological with respect to regulation of the cellular function, i.e., between 70 and 25 mmHg. Correspondingly, in the extracts of cells that produced hydrogen peroxide, EPO production was measured (34). It was found that highest EPO expression at low PO_2 values correlated with low ROS production rates. In contrast, high oxygen tension and high ROS production were correlated with low EPO gene expression. In addition, EPO secreted into the culture supernatant by HepG2 cells was measured. At high pericellular PO_2 values of about 140 mmHg EPO production was low, but could be increased by scavengers of hydrogen peroxide like catalase or 123-dihydrorhodamine. In a way, by scavenging endogenously produced hydrogen peroxide it appeared that cells were brought to conditions of hypoxia. In this situation EPO gene expression and protein production

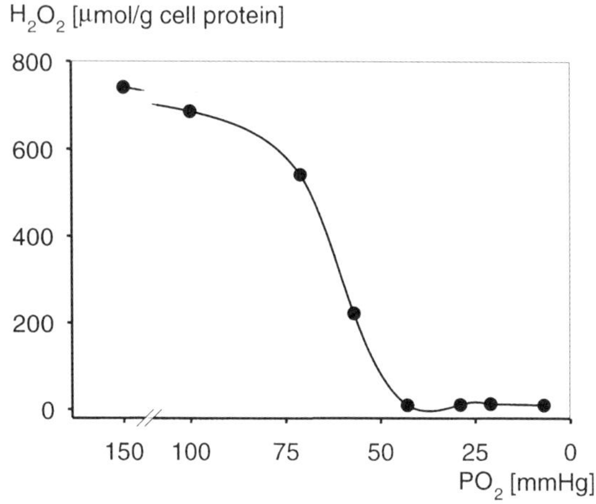

Figure 1 Hydrogen peroxide (H_2O_2) production by HepG2 cells depends on the pericellular PO_2. HepG2 cells were cultivated on gas-permeable cell culture dishes to ensure equilibration of the PO_2 in the incubation chamber with the pericellular PO_2 (8). Cells were treated with menadione (600 μM) for 24 hr before the start of the experiment. Cells were then incubated for 2 hr at different PO_2 values, and H_2O_2 was measured in the cell-free supernatant as described (34).

were increased. Conversely at PO_2 values of about 20 mmHg (hypoxia) EPO concentration in the supernatant was high. The addition of exogenous hydrogen peroxide to these hypoxic cells dramatically reduced the production of EPO. Thus, in this case low endogenous production of ROS because of hypoxic incubation conditions was changed to a situation similar to normoxia where high hydrogen peroxide concentration is found in the culture supernatant. Subsequently, EPO gene expression was lowered. This first evidence from in vitro experiments led to the conclusion that reactive oxygen species may be mediating the effect of oxygen on the expression of the EPO gene.

Thus, evidence from pharmacological manipulation of the P450 system, a type b-cytochrome (31), spectroscopic alterations upon changes in the oxygenation state (32,33,35), and immunological similarity with the NADPH oxidase were combined to conclude that a b-cytochrome-producing hydrogen peroxide in dependence of the PO_2 could serve as an oxygen sensor and ROS as intracellular messenger molecules within the cell. These in vitro studies in human hepatoma cell lines were subsequently extended to test for the role of hydrogen peroxide for hypoxia-induced EPO gene expression in the isolated perfused rat kidney. Since the kidney is the major organ for EPO production, it was of considerable interest that treatment with hydrogen peroxide of the hypoxically perfused isolated rat kidney reduced EPO gene expression by 40% and EPO protein production by about 30% (15). To test whether the inhibition of EPO was specific and not due to overall impairment of kidney function and gene expression, filtration and reabsorption functions of the perfused organs were measured and remained unchanged by hydrogen peroxide treatment (15). In addition, the redox cycler pyrrogallol that generates O_2^- also reduced the production of erythropoietin by the kidney without significantly affecting overall kidney function (such as sodium reabsorption or water clearance) (15). Expression of VEGF, which was also measured by quantitative RT-PCR, was in fact increased by oxidative treatment ruling out the possibility that the reduced EPO expression was simply the result of an overall reduced gene expression rate (Fig. 2). Moreover, iron chelators or scavengers for hydroxyl radicals were able to fully antagonize the inhibitory effect of reactive oxygen species on EPO expression in the kidney. This result indirectly suggested that hydrogen peroxide may react in Fenton-type reaction in which hydrogen peroxide and iron ions give rise to highly reactive hydroxyl radicals (15).

VII. The Potential Mode of Action of Hydrogen Peroxide

As just pointed out, the inhibitory effect of hydrogen peroxide on EPO expression could be fully antagonized by iron chelation. This was also found in hepatoma cells and suggested that reactive hydrogen peroxide may react in the presence of ferrous iron to form hydroxyl radicals (36). These highly reactive molecules would then be the immediate compound to inhibit gene expression. With respect to the hypothesis of hydrogen peroxide as a signaling molecule that controls the hypoxia-inducible expression of erythropoietin, the following reaction cascade was appealing: Since the half-life of hydrogen peroxide is sufficiently long to allow passage through the cells,

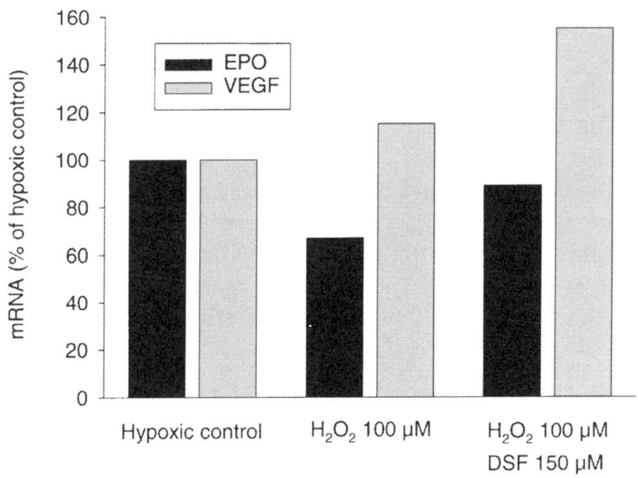

Figure 2 Hydrogen peroxide (H_2O_2) and the iron chelator desferrioxamine (DSF) affect the expression of EPO and VEGF in the isolated perfused rat kidney. Rat kidneys were hypoxically (PO_2 26 ± 1 mmHg) perfused for 3 hr as described (15). EPO and VEGF mRNA levels were measured by competitive RT-PCR.

the Fenton reaction could take place locally when ferrous iron is in the ultimate vicinity of transcription factors or other regulatory proteins that are affected by the emerging highly reactive hydroxyl radicals. Evidence for such a locally defined reaction was obtained by using the hydrogen peroxide–sensitive dye 123-dihydrorhodamine, which also needs ferrous iron to generate the fluorescence signal (37). Fluorescence of this dye was very clearly localized around the nucleus of HepG2 cells and therefore suggested that the reaction with hydrogen peroxide and the dye in the presence of iron is localized to this compartment. This localized fluorescence signal can in principle have two reasons: Either the ferrous iron is unequally distributed throughout the cell or the ROS are produced focally. Studies on iron distribution within the cells using the fluorescent dye Phengreen revealed that iron is not localized in particularly higher concentrations around the nucleus (38). Therefore, we believe that within the cell ROS production and the respective Fenton chemistry are focally localized. Moreover, we could obtain clear evidence that the localized fluorescence signal was not derived from 123-rhodamine that accumulated in mitochondria (39). This was of particular importance since a role for mitochondria in hypoxic signaling had been proposed (40).

Certainly, mitochondria are a source of ROS and have always been considered to be involved in O_2 sensing. However, recent work (41,42) clearly showed that hypoxia-inducible gene expression is unimpaired in cells deficient of biologically active mitochondria and that hydrogen peroxide production is dependent on the PO_2 in cells without functional mitochondria (42). Thus, the perinuclear localization of Fenton-type reaction may be the ultimate localization at which redox control is exerted on factors that affect EPO gene expression. More recently, a so-far-unknown type b-cytochrome NAD(P)H-dependent oxidoreductase has been identified that produces ROS in an oxygen-sensitive manner (43), but it is not yet known whether cells lacking this enzyme show defects in oxygen-dependent gene expression.

VIII. The Effect of ROS on HIF-1

So far, these experiments had indicated that a decrease in the levels of ROS under hypoxic conditions had led to increased EPO gene expression. Since EPO is the predominant example of a HIF-1-regulated gene, subsequent studies were performed to study effects of ROS on HIF-1 mRNA expression or, more importantly, HIF-1α protein levels. From our own studies in hepatoma HepG2 cells we knew that at least in these cells only a very moderate increase in HIF-1α mRNA levels upon hypoxic stimulation is observed. However, even this small increase was abolished by the addition of exogenous hydrogen peroxide. Studies performed by Huang and co-workers (44) indicated, however, that the far more important regulation of HIF-1α protein levels also depended upon redox-sensitive stabilization of the α-subunit. As was pointed out earlier, the addition of hydrogen peroxide inhibited the accumulation of HIF-1α protein and this process was also critically dependent on free ferrous iron in the cell, because desferroxamin as an iron chelator was able to antagonize the effect of hydrogen peroxide. Our own unpublished studies using the redox cycler DMNQ confirmed these results. Under moderate hypoxia where HIF-1α was stabilized and accumulated in the cell, the addition of the redox cycler and thus generation of O_2^- molecules led to a degradation of HIF-1α (Fig. 3). Interestingly, the effect of DMNQ was lost in severe hypoxia (or almost anoxia with PO_2 values < 1 mmHg). Under these conditions HIF-1α was still stabilized and could still be activated in the cell (unpublished data from reporter gene assays with luciferase driven by three copies of a HIF-1 binding site). The most likely explanation for this effect is that redox cyclers need oxygen to generate O_2^- molecules. In addition, unchanged levels of HIF-1α under severe hypoxia in the presence of DMNQ rule out a simple toxic degradation of HIF-1α or even a general inhibition of translation under conditions of moderate hypoxia (20 mmHg PO_2). HIF-1α mRNA levels were unchanged under all conditions. However, conflicting data from Chandel et al. (45) arose from observations that application of hydrogen peroxide to cells might also activate HIF-1 and HIF-1 target genes. This was in support of their previous proposed model that increased mitochondrial ROS under hypoxic conditions might be a central part of the oxygen sensor. Again, as pointed out above, Vaux et al. (42) clearly found hypoxic regulation in the absence of functional mitochondria, which made these organelles unlikely the central oxygen sensor.

One important difference between the experimental setting of the exogenous addition of hydrogen peroxide and endogenous production has to be considered: Whereas the addition of 100–250 μM of exogenous hydrogen peroxide leads to an enormously rapid increase of reactive species, this "oxidative stress" lasts only a couple of minutes. As measured for HepG2 cells (34) and known for many other cells, highly active catalase and glutathione peroxidase very rapidly degrade

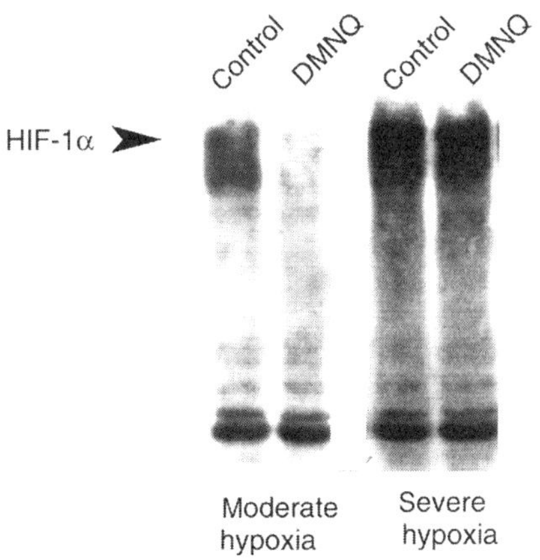

Figure 3 The redox cycler and O_2^- generator DMNQ inhibits hypoxic HIF-1α accumulation only under moderate (PO_2~20 mmHg) but not severe (PO_2 < 1 mmHg) hypoxia. HepG2 cells were incubated with 50 μM for 4 hr under moderate or severe hypoxia and HIF-1α protein was detected by Western blot in total cell extracts (53).

hydrogen peroxide. Hydrogen peroxide added to HepG2 cells in culture has a half-life of only a few minutes. In contrast, hydrogen peroxide production by these hepatoma cells was found to be continuous over time and depending on the PO_2. Thus, additional experiments were performed, in which glucose oxidase and glucose were used as a system that continuously generates hydrogen peroxide. As shown in Figure 4, under these conditions hypoxia-induced EPO expression was significantly reduced. The calculated concentration of hydrogen peroxide in these experiments (calculated from glucose concentration and enzyme activity) was in the range of several 100 nM of hydrogen peroxide—concentrations that can certainly be reached by cells in vivo.

A very recent study by Srinivas et al. (41) supported the notion that an active mitochondrial respiratory chain was not necessary for HIF-1 activation. Moreover, the authors performed interesting experiments in which catalase, the main hydrogen peroxide–degrading enzyme, was overexpressed in mitochondria and the cytosol. However, exposure to normoxia or hypoxia or the iron chelator desferrioxamine of these catalase-overexpressing cells did not affect the response to hypoxia or change the basal levels of HIF-1α protein expression. Therefore, an increase in hydrogen peroxide degradation in these two cellular compartments was not able to increase HIF-1α protein levels and explain the higher EPO expression seen in cells that were treated with hydrogen peroxide scavengers.

From the findings reported by Vaux et al. (42) and Srinivas et al. (41) three important conclusions may be drawn: (1) An active mitochondrial respiratory chain

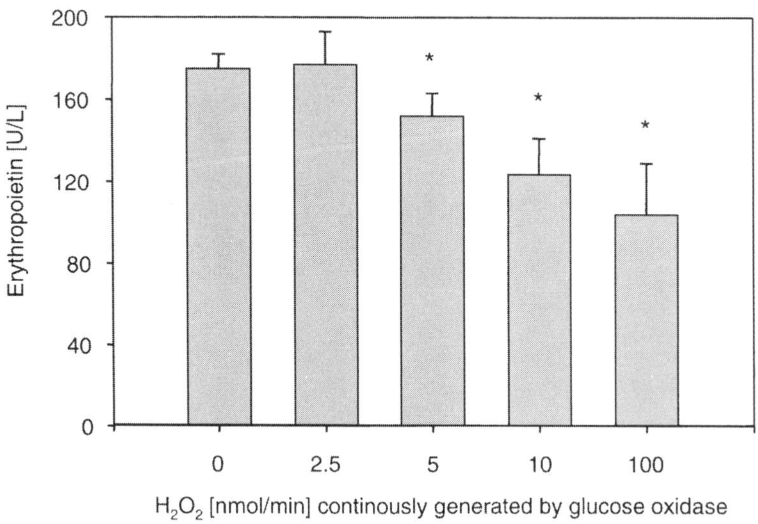

Figure 4 Continuously generated H_2O_2 from glucose by glucose oxidase inhibits hypoxia-induced EPO production in HepG2 cells. H_2O_2 concentrations were calculated from the glucose concentration (5.5 mmol/L) and the enzyme activity of the glucose oxidase. EPO protein was measured in the culture supernatant after a 24-hr hypoxic incubation.

was not necessary for oxygen sensing. (2) Despite nonactive mitochondria the levels of ROS were modulated by the pericellular oxygen tension resulting in higher levels of ROS under high oxygen tension and less hydrogen peroxide under hypoxia. (3) Modulation of HIF-1α protein levels by the addition of exogenous hydrogen peroxide—mostly in higher concentrations that are found in the vicinity of hepatoma cells that were used in these studies—led to controversial results with respect to HIF-1α protein levels. On the other hand, introduction of hydrogen peroxide–degrading enzymes like catalase was not sufficient to increase HIF-1α protein levels under normoxic conditions.

In view of the very recent and elegant work from the groups of Peter Ratcliffe at Oxford (27,29), Steven McKnight in Dallas (46), and William Kaelin in Boston (26), a prominent role for oxygen sensing by b-cytochromes or other ROS-generating systems is unlikely since the principal oxygen-sensing system appears to be a group of proline hydroxylases. Although redox chemistry is involved in proline hydroxylases that depend on ferrous iron that is redox-sensitive, no definite experiments have been reported whether changes in the cellular redox state affect the rate of proline hydroxylation and might therefore interfere with HIF-1α degradation.

IX. The Role of NO and ROS in HIF-1α-Dependent Gene Expression

Earlier studies on the role of nitric oxide (NO) on EPO production had revealed that NO at concentrations that may be produced by macrophages in the ultimate vicinity of EPO-producing cells inhibits EPO gene expression and EPO protein production. Studies that focused on the role of NO on HIF activation again led to controversial results: Initial studies (47,48) claimed that NO inhibits HIF-1α protein accumulation and might therefore be responsible for the inhibition of EPO gene expression. However, subsequent studies by several groups including our own revealed that low concentrations of NO released by certain NO donors that release only NO and no other oxidative products provided clear evidence of NO-dependent stabilization of HIF-1α protein (49). In a study from the author's laboratory that focused on the interaction of NO and ROS with respect to activation of a HIF-1-driven reporter gene and the endogenous EPO gene it was found that NO was able to transiently reduce ROS levels in hepatoma cells. However, this effect was observed only during an initial incubation for up to 4 hr. During this time NO was able to activate HIF-1 reporter gene activity and to moderately increase the expression of the HIF-1 target gene EPO. During this short-term incubation NO donors inhibited the production of ROS after stimulation with NADH or NADPH (50). Interestingly, extended incubation for up to 48 hr led to a complete inversion of ROS production and also reporter gene activity. Now ROS levels were increased and HIF-1 activity was inhibited. A most likely explanation for the late phase was that NO reduced catalase activity in these cells and by that means disturbed the balance of ROS generation and degradation. This was confirmed by determination of the total antioxidative capacity (sum of catalase and glutathion peroxidase activity) of the HepG2 cells in the virtual absence of any glutathion peroxidase activity (50). It is, however, unclear by which means the initial activation of HIF-1α by NO was achieved. Interestingly, this biphasic effect of NO on HIF-1α also showed a peculiar dose dependency as reported by Sandau et al. (49). Low concentrations of NO (100 μM) led to a rapid increase in HIF-1α protein levels between 2 and 4 hr, whereas higher concentrations (1 mM) increased HIF-1α protein only after more than 4 hr. Interestingly, hypoxic accumulation of HIF-1α was also inhibited after more than 4 hr supporting the data from Genius and Fandrey (50) that a biphasic effect of NO is responsible for the discrepant results of this substance. In addition, differences in the NO donors used and the kinetics by which NO is released may account for some of the conflicting data. With respect to EPO expression, NO donors only moderately increased EPO mRNA during the short-term treatment with NO. After 24 hr of incubation, however, only the hypoxic response of the EPO gene is reduced in a dose-dependent manner, whereas basal expression is not affected (50). Since the redox chemistry that is involved in ROS and NO is at least as complicated as the above-mentioned Fenton chemistry, further studies on this issue are clearly needed. Moreover, so far an effect of NO on the oxygen-dependent proline hydroxylation has not been reported, but several studies of this kind are currently in progress.

X. Redox or Oxygen-Dependent Regulation of the EPO Gene

From what has been said before, it appears as if no clear-cut picture can be drawn with respect to the effect of ROS on HIF-1-dependent gene expression. It appears clear that oxygen-dependent expression of the EPO gene depends on activation of the HIF-1 complex. In addition, EPO expression is sensitive to redox changes. Therefore, one has to consider that the EPO gene, like many other genes, might be regulated by hypoxia and redox changes as well. Studies by Tabata and colleagues (51) have pointed out that negative regulatory factors like GATA-2 are able to bind to the promoter of the EPO gene and repress this gene. In search of a target for hydrogen peroxide Tabata and colleagues reported that hydrogen peroxide increased GATA-2 levels in Hep3B cells (51). Increased GATA-2 then bound to the promoter and repressed EPO gene expression. That mechanism, which is dependent on redox changes but independent of HIF-1 activation, might control the production of EPO and make this gene a redox-sensitive gene. A model for this mode of action is presented in Figure 5. Since a similar mechanism has been reported for the NO-dependent inhibition of EPO gene expression (52), one might explain at least the inhibition of EPO production by a long-term treatment with NO donors.

In conclusion, EPO expression is induced by hypoxia via activation of HIF-1. The PO_2-dependent modulation of ROS levels within the physiological range of PO_2 values can modulate EPO gene expression. HIF-1 regulation, however, depends on oxygen-sensitive hydroxylation of proline residues of HIF-1α by prolyl hydroxylases

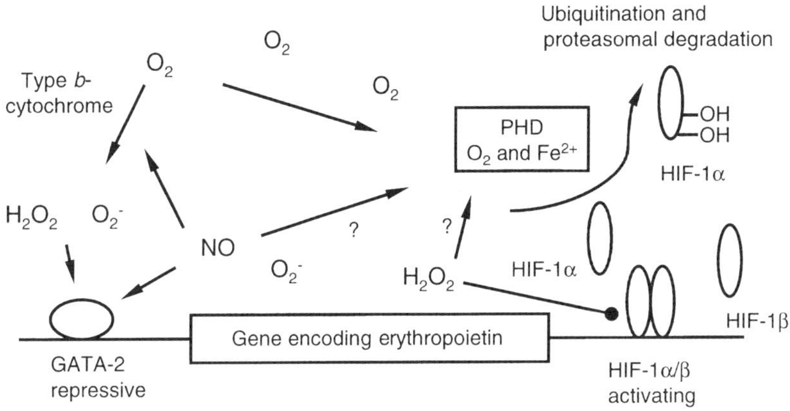

Figure 5 Schematic drawing of the potential oxygen- and redox-sensitive regulation of the EPO gene. Oxygen sensing is most likely achieved through a group of prolyl hydroxylases (PHD) that tag HIF-1α for degradation under high PO_2 conditions. Additional modulation of EPO expression by H_2O_2 and derived species could act through repressive transcription factors like GATA-2 (51) or by interaction with PHD.

and not on oxidative modification by ROS. Oxygen- and redox-dependent regulation therefore most likely merge at the level of interaction between different activators and repressors of the EPO gene.

Acknowledgments

I gratefully acknowledge the continuous support and friendship of Helmut Acker, Frank Bunn, and Wolfgang Jelkmann, as well as my co-workers Stilla Frede, Patricia Freitag, Just Genius, and Horst Pagel.

References

1. Jelkmann W. Erythropoietin: structure, control of production, and function. Physiol Rev 1992; 72(2):449–489.
2. Jacobson LO, Goldwasser E, Fried W, Plzak L. Role of the kidney in erythropoiesis. Nature 1957; 179:633–634.
3. Bachmann S, Hir ML, Eckardt KU. Co-localization of erythropoietin mRNA and ecto-5-nucleotidase immunoreactivity in peritubular cells of rat renal cortex indicates that fibroblasts produce erythropoietin. J Histochem Cytochem 1993; 41:335–340.
4. Moritz KM, Lim GB, Wintour EM. Developmental regulation of erythropoietin and erythropoiesis. Am J Physiol 1997; 273(6 Pt 2):R1829–R1844.
5. Dame C, Fahnenstich H, Freitag P, Hofmann D, Abdul-Nour T, Bartmann P, et al. Erythropoietin mRNA expression in human fetal and neonatal tissue. Blood 1998; 92(9):3218–3225.

6. Dame C, Bartmann P, Wolber E, Fahnenstich H, Hofmann D, Fandrey J. Erythropoietin gene expression in different areas of the developing human central nervous system. Brain Res Dev Brain Res 2000; 125(1–2):69–74.
7. Goldberg MA, Glass GA, Cunningham JM, Bunn HF. The regulated expression of erythropoietin by two human hepatoma cell lines. Proc Natl Acad Sci USA 1987; 84(22):7972–7976.
8. Wolff M, Fandrey J, Jelkmann W. Microelectrode measurements of pericellular PO_2 in erythropoietin-producing human hepatoma cell cultures. Am J Physiol 1993; 265(5 Pt 1):C1266–C1270.
9. Fandrey J, Bunn HF. In vivo and in vitro regulation of erythropoietin mRNA: measurement by competitive polymerase chain reaction. Blood 1993; 81(3):617–623.
10. Metzen E, Wolff M, Fandrey J, Jelkmann W. Pericellular PO_2 and O_2 consumption in monolayer cell cultures. Respir Physiol 1995; 100(2):101–106.
11. Fisher JW, Langston JW. The influence of hypoxemia and cobalt on erythropoietin production in the isolated perfused dog kidney. Blood 1967; 29(1):114–125.
12. Ratcliffe PJ, Jones RW, Phillips RE, Nicholls LG, Bell JI. Oxygen-dependent modulation of erythropoietin mRNA levels in isolated rat kidneys studied by RNase protection. J Exp Med 1990; 172(2):657–660.
13. Pagel H, Jelkmann W, Weiss C. Isolated serum-free perfused rat kidneys release immunoreactive erythropoietin in response to hypoxia. Endocrinology 1991; 128:2633–2638.
14. Jelkmann W, Pagel H, Wolff M, Fandrey J. Monokines inhibiting erythropoietin production in human hepatoma cultures and in isolated perfused rat kidneys. Life Sci 1992; 50(4):301–308.
15. Neumcke I, Schneider B, Fandrey J, Pagel H. Effects of pro- and antioxidative compounds on renal production of erythropoietin. Endocrinology 1999; 140(2): 641–645.
16. Semenza GL, Koury ST, Nejfelt MK, Gearhart JD, Antonarakis SE. Cell-type-specific and hypoxia-inducible expression of the human erythropoietin gene in transgenic mice. Proc Natl Acad Sci USA 1991; 88(19):8725–8729.
17. Wang GL, Jiang BH, Rue EA, Semenza GL. Hypoxia-inducible factor 1 is a basic-helix-loop-helix-PAS heterodimer regulated by cellular O_2 tension. Proc Natl Acad Sci USA 1995; 92(12):5510–5514.
18. Blanchard KL, Acquaviva AM, Galson DL, Bunn HF. Hypoxic induction of the human erythropoietin gene: cooperation between the promoter and enhancer, each of which contains steroid receptor response elements. Mol Cell Biol 1992; 12(12):5373–5385.
19. Wiener CM, Booth G, Semenza GL. In vivo expression of mRNAs encoding hypoxia-inducible factor 1. Biochem Biophys Res Commun 1996; 225(2):485–488.
20. Jain S, Maltepe E, Lu MM, Simon C, Bradfield CA. Expression of ARNT, ARNT2, HIF1 alpha, HIF2 alpha and Ah receptor mRNAs in the developing mouse. Mech Dev 1998; 73(1):117–123.
21. Iyer NV, Kotch LE, Agani F, Leung SW, Laughner E, Wenger RH, et al. Cellular and developmental control of O_2 homeostasis by hypoxia-inducible factor 1 alpha. Genes Dev 1998; 12(2):149–162.
22. Rosenberger C, Mandriota S, Jürgensen JS, Wiesener MS, Ratcliffe PJ, Frei U, et al. Expression of HIF-1 and -2α in hypoxic and ischemic rat kidneys (abstract). Nieren- und Hochdruckkrankheiten 2001.

23. Lando D, Pongratz I, Poellinger L, Whitelaw ML. A redox mechanism controls differential DNA binding activities of hypoxia-inducible factor (HIF) 1alpha and the HIF-like factor. J Biol Chem 2000; 275(7):4618–4627.
24. Salceda S, Caro J. Hypoxia-inducible factor 1alpha (HIF-1alpha) protein is rapidly degraded by the ubiquitin-proteasome system under normoxic conditions. Its stabilization by hypoxia depends on redox-induced changes. J Biol Chem 1997; 272(36):22642–22647.
25. Huang LE, Gu J, Schau M, Bunn HF. Regulation of hypoxia-inducible factor 1alpha is mediated by an O_2-dependent degradation domain via the ubiquitin-proteasome pathway. Proc Natl Acad Sci USA 1998; 95(14):7987–7992.
26. Ivan M, Kondo K, Yang H, Kim W, Valiando J, Ohh M, et al. HIF-1alpha targeted for VHL-mediated destruction by proline hydroxylation: implications for O_2 sensing. Science 2001; 292(5516):464–468.
27. Jaakkola P, Mole DR, Tian YM, Wilson MI, Gielbert J, Gaskell SJ, et al. Targeting of HIF-1alpha to the von Hippel-Lindau ubiquitylation complex by O_2-regulated prolyl hydroxylation. Science 2001; 292(5516):468–472.
28. Masson N, Willam C, Maxwell PH, Pugh CW, Ratcliffe PJ. Independent function of two destruction domains in hypoxia-inducible factor-alpha chains activated by prolyl hydroxylation. EMBO J 2001; 20(18):5197–5206.
29. Epstein AC, Gleadle JM, McNeill LA, Hewitson KS, O'Rourke J, Mole DR, et al. *C. elegans* EGL-9 and mammalian homologs define a family of dioxygenases that regulate HIF by prolyl hydroxylation. Cell 2001; 107(1):43–54.
30. Goldberg MA, Dunning SP, Bunn HF. Regulation of the erythropoietin gene: evidence that the oxygen sensor is a heme protein. Science 1988; 242(4884):1412–1415.
31. Fandrey J, Seydel FP, Siegers CP, Jelkmann W. Role of cytochrome P450 in the control of the production of erythropoietin. Life Sci 1990; 47(2):127–134.
32. Görlach A, Holtermann G, Jelkmann W, Hancock JT, Jones SA, Jones OTG, et al. Photometric characteristics of haem proteins in erythropoietin-producing hepatoma cells (HepG2). Biochem J 1993; 290:771–776.
33. Görlach A, Fandrey J, Holtermann G, Acker H. Effects of cobalt on haem proteins of erythropoietin-producing HepG2 cells in multicellular spheroid culture. FEBS Lett 1994; 348(2):216–218.
34. Fandrey J, Frede S, Jelkmann W. Role of hydrogen peroxide in hypoxia-induced erythropoietin production. Biochem J 1994; 303(Pt 2):507–510.
35. Porwol T, Ehleben W, Zierold K, Fandrey J, Acker H. The influence of nickel and cobalt on putative members of the oxygen-sensing pathway of erythropoietin-producing HepG2 cells. Eur J Biochem 1998; 256(1):16–23.
36. Fandrey J, Frede S, Ehleben W, Porwol T, Acker H, Jelkmann W. Cobalt chloride and desferrioxamine antagonize the inhibition of erythropoietin production by reactive oxygen species. Kidney Int 1997; 51(2):492–496.
37. Ehleben W, Porwol T, Fandrey J, Kummer W, Acker H. Cobalt and desferrioxamine reveal crucial members of the oxygen sensing pathway in HepG2 cells. Kidney Int 1997; 51(2):483–491.
38. Fandrey J, Petrat F, Rauen U, de Groot H, Porwol T, Acker H, et al. The role of intracellular distribution of reactive oxygen species and chelatable iron for the activation process of hypoxia inducible factor 1 (abstract). FASEB J 2001; 15:690.5.
39. Kietzmann T, Fandrey J, Acker H. Oxygen radicals as messengers in oxygen-dependent gene expression. News Physiol Sci 2000; 15:202–208.

40. Chandel NS, Maltepe E, Goldwasser E, Mathieu CE, Simon MC, Schumacker PT. Mitochondrial reactive oxygen species trigger hypoxia-induced transcription. Proc Natl Acad Sci USA 1998; 95(20):11715–11720.
41. Srinivas V, Leshchinsky I, Sang N, King MP, Minchenko A, Caro J. Oxygen sensing and HIF-1 activation does not require an active mitochondrial respiratory chain electron-transfer pathway. J Biol Chem 2001; 276(25):21995–21998.
42. Vaux EC, Metzen E, Yeates KM, Ratcliffe PJ. Regulation of hypoxia-inducible factor is preserved in the absence of a functioning mitochondrial respiratory chain. Blood 2001; 98(2):296–302.
43. Zhu H, Qiu H, Yoon HW, Huang S, Bunn HF. Identification of a cytochrome b-type NAD(P)H oxidoreductase ubiquitously expressed in human cells. Proc Natl Acad Sci USA 1999; 96(26):14742–14747.
44. Huang LE, Arany Z, Livingston DM, Bunn HF. Activation of hypoxia-inducible transcription factor depends primarily upon redox-sensitive stabilization of its alpha subunit. J Biol Chem 1996; 271(50):32253–32259.
45. Chandel NS, McClintock DS, Feliciano CE, Wood TM, Melendez JA, Rodriguez AM, et al. Reactive oxygen species generated at mitochondrial complex III stabilize hypoxia-inducible factor-1alpha during hypoxia: a mechanism of O_2 sensing. J Biol Chem 2000; 275(33):25130–25138.
46. Bruick RK, McKnight SL. A conserved family of prolyl-4-hydroxylases that modify HIF. Science 2001; 294(5545):1337–1340.
47. Sogawa K, Numayama-Tsuruta K, Ema M, Abe M, Abe H, Fujii-Kuriyama Y. Inhibition of hypoxia-inducible factor 1 activity by nitric oxide donors in hypoxia. Proc Natl Acad Sci USA 1998; 95(13):7368–7373.
48. Huang LE, Willmore WG, Gu J, Goldberg MA, Bunn HF. Inhibition of hypoxia-inducible factor 1 activation by carbon monoxide and nitric oxide. Implications for oxygen sensing and signaling. J Biol Chem 1999; 274(13):9038–9044.
49. Sandau KB, Fandrey J, Brüne B. Accumulation of HIF-1 under the influence of nitric oxide. Blood 2001; 97:1009–1015.
50. Genius J, Fandrey J. Nitric oxide affects the production of reactive oxygen species in hepatoma cells: implications for the process of oxygen sensing. Free Radic Biol Med 2000; 29(6):515–521.
51. Tabata M, Tarumoto T, Ohmine K, Furukawa Y, Hatake K, Ozawa K, et al. Stimulation of GATA-2 as a mechanism of hydrogen peroxide suppression in hypoxia-induced erythropoietin gene expression. J Cell Physiol 2001; 186(2):260–267.
52. Tarumoto T, Imagawa S, Ohmine K, Nagai T, Higuchi M, Imai N, et al. N(G)-monomethyl-L-arginine inhibits erythropoietin gene expression by stimulating GATA-2. Blood 2000; 96(5):1716–1722.
53. Metzen E, Fandrey J, Jelkmann W. Evidence against a major role for Ca^{2+} in hypoxia-induced gene expression in human hepatoma cells (Hep3B). J Physiol (Lond) 1999; 517(Pt 3):651–657.

5

Structure and Regulation of the Mouse Hypoxia-Inducible Factor-1α Gene

ROLAND H. WENGER

University of Leipzig
Leipzig, Germany

I. Introduction

The hypoxia-inducible factor-1 (HIF-1) is a ubiquitously expressed transcriptional master regulator of a number of genes involved in mammalian oxygen homeostasis and glycolytic energy metabolism (reviewed in Refs. 1,2). The rate-limiting HIF-1α subunit is expressed in an inverse relation to the oxygen partial pressure (3,4). To ensure the cell's very rapid adaptation to hypoxic conditions, HIF-1α is regulated by the unusual mechanism of protein stability. Oxygen-dependent prolyl hydroxylases modify HIF-α, enabling the binding of the von Hippel–Lindau tumor suppressor protein (pVHL), ubiquitinylation, and subsequent proteasomal degradation (5–9). Because of the ubiquitous nature of mRNA expression and the conditional regulation at the protein level, so far, transcriptional regulation of HIF-1α expression did not seem to be of prime importance for its function. Nevertheless, recent findings on tissue-specific transcription of distinct HIF-1α mRNA isoforms in the mouse shed new lights on the physiological functions of HIF-1α.

II. Cloning of the *Hif1a* Gene Encoding Mouse HIF-1α

Cloning of the human (10), mouse (11,12), and rat (13) HIF-1α cDNAs revealed a conserved predicted protein (overall similarity is around 90%) of 826 amino acids (for human HIF-1α) and a relatively long 3′ untranslated region (UTR) containing several mRNA destabilization elements. While the mouse HIF-1α cDNA reported by Li et al. (12) corresponded to the human cDNA, the mouse mRNA isoform that we reported differed at its 5′ end by an alternative UTR and the lack of the first 12 codons (11). Cloning and characterization of the mouse *Hif1a* gene provided the explanation for this discrepancy. Mouse *Hif1a* is a single-copy gene (11) containing 15 exons spread over 47 kb (14). As indicated in Figure 1, *Hif1a* contains two alternative first exons (designated exon I.1 and exon I.2) giving rise to two different mRNA isoforms (termed mHIF-1αI.1 and mHIF-1αI.2). We demonstrated that expression of exons I.1 and I.2 is driven by their own promoters (14,15). No mRNA isoform could be cloned that contained both first exons, suggesting that exon I.2 is spliced out of the pre-mRNA transcribed from the exon I.1 promoter. The structure of the *Hif1a* locus has also been reported by Luo et al. (16). Closer inspection of their sequence revealed a third mouse HIF-1α mRNA isoform that contains 14-amino-acid insertion due to alternative splicing at the 5′ end of exon XI (Fig. 1). However, the functional significance of this mRNA isoform still needs to be defined.

The human *HIF1A* gene has been assigned to chromosome 14q21–q24, and the mouse *Hif1a* gene has been assigned to a region on chromosome 12 that shows conserved synteny with the corresponding region bearing *HIF1A* on the human chromosome (11,17). Neither the human *HIF2A* gene (human chromosome 2) nor the human *HIF3A* gene (human chromosome 19) is located on the same chromosome as the *HIF1A* gene.

III. Differential Structures of the Mouse *Hif1a* I.1 and I.2 Promoters

Whereas the exon I.1 promoter exhibits tissue-specific features, exon I.2 is associated with a so-called CpG island as it is typically found in housekeeping-type promoters of ubiquitously expressed genes (15). CpG methylation plays an important regulatory role in mammalian gene expression, contributing to X-chromosome inactivation and genomic imprinting as well as tissue- and developmental-stage-specific transcriptional regulation. CpG dinucleotides are underrepresented in the mammalian genome and are often methylated if located outside of the CpG islands. The human genome contains around 45,000 CpG islands approximately 1 kb in length. In contrast to the bulk genome, these regions are G + C rich and entirely methylation-free. CpG islands are associated with the promoters of all housekeeping genes and of 40% of the tissue-specific genes. Methylated CpG (containing 5-methylcytosine) interferes with transcription factor binding to DNA through both direct steric hindrance and the binding of repressor proteins. Also HIF-1 function is impaired by CpG methylation since the HIF-1 core

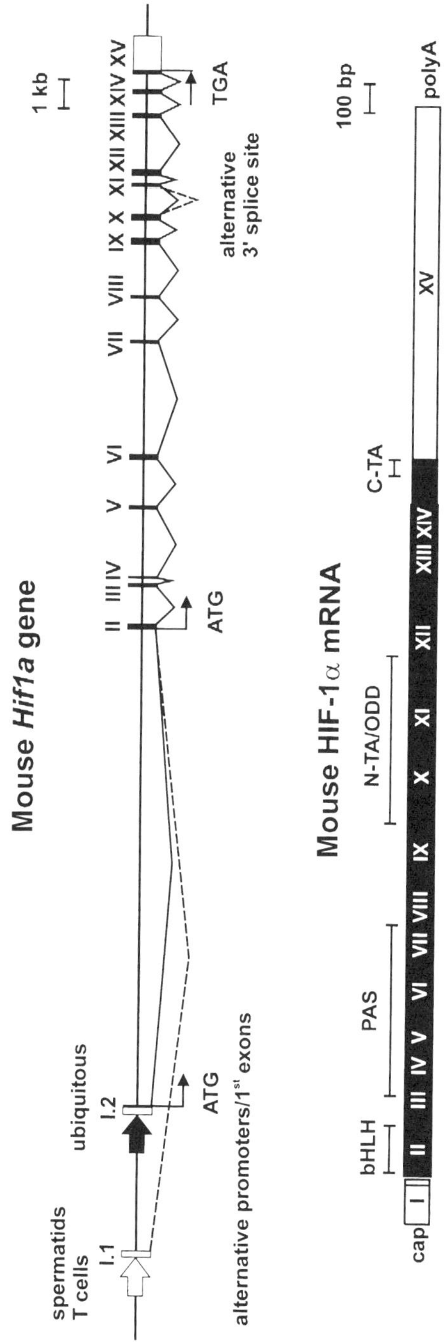

Figure 1 Structure of the mouse gene encoding HIF-1α (*Hif1a*) and the mature HIF-1α mRNA. Domains: bHLH, basic-helix-loop-helix; PAS, Per-Arnt-Sim; N-TA, N-terminal *trans*-activation; ODD, oxygen-dependent degradation; C-TA, C-terminal TA.

DNA-binding site (ACGTG) contains a CpG dinucleotide (18,19). Thus, CpG islands are important structural elements that keep the chromatin in a constitutively "open" conformation, allowing access of transcription factors to promoter regions.

The exon I.1 promoter contains a region of 212 bp sharing 74% homology with a LINE-1 repetitive element, starting 364 bp upstream of the transcription initiation site (14). This LINE-1 element was found to be most homologous to the 5′ flanking region of the TNF-α locus. This element is also 67% homologous to a 146-bp stretch in the 3′ UTR of the gene encoding the aryl hydrocarbon receptor. The biological significance of this finding, however, awaits further investigations. The *Hif1a* exon I.1 flanking regions are enriched in low-complexity repetitive elements such as $(A)_n$, $(T)_n$, $(GT)_n$, $(GA)_n$, and $(ATCN)_n$ repeats. Interestingly, the $(ATCN)_n$ repeat is located within an open reading frame (ORF) that potentially encodes a 14.1-kDa peptide. These repeats are concentrated in the sex-determining region of the mouse Y chromosome and have been proposed to encode primordial proteins (20,21). mRNAs containing such 4-bp repeats (also termed GATA repeats for the complementary strand) are sometimes transcribed (20,21), but whether the $(ATCN)_n$ element in the *Hif1a* I.1 promoter is also transcribed is currently unknown.

IV. 5′ End Heterogeneity of HIF-1α

The mHIF-1αI.1 mRNA isoform encodes for a predicted protein product that is 12 amino acids shorter than the predicted mHIF-1αI.2 protein. This prediction is based on the fact that no ATG translation initiation codon could be identified in *Hif1a* exon I.1. Rather the ATG codon at the very 5′ end of exon II might serve as the translation initiation codon of the mHIF-1αI.1 mRNA. In contrast, *Hif1a* exon I.2 contains an ATG start codon that is in frame with the ATG codon on exon II, resulting in a protein that contains 12 additional amino acids (Fig. 1). Despite its vicinity to the basic-helix-loop-helix DNA-binding domain (Table 1), this N-terminal deletion in the mHIF-1αI.1 protein does not affect the DNA-binding efficiency when compared with the mHIF-1αI.2 protein (22). Recently, Lukashev et al. confirmed these results by demonstrating that two HIF-1α isoforms displayed similar *trans*-activation activities (23).

As shown in Table 1, amino acid sequence heterogeneity has also been found at the 5′ ends of the rat and human HIF-1α proteins. It should be noted that these sequences are only predictions based on the cDNA sequences. The actual N-termini of the HIF-1α proteins have not been determined so far, and they might differ from the predictions due to posttranslational processing. Apart from the two different promoters/first exon combinations, differential 5′ termini are derived from alternative splicing, or they are due to polymorphisms. The comparison with HIF-1α derived from other species revealed a high degree of conservation within the basic domains (responsible for DNA-binding). Interestingly, Drutel et al. reported a KKMSS protein kinase A (PKA) consensus phosphorylation site in the rat HIF-1α protein (24). The presence or absence of this site is dependent on alternative splicing of the pre-mRNA. Currently, it is not known whether this site represents a functional target of PKA. However, it is noteworthy that in contrast to mHIF-1αI.2, the short

Table 1 5′ End Heterogeneity of HIF-1α

	Basic domain	References
Mouse HIF-1αI.1	MSS*ERRKEKSRDAARSRR*...	(11,14)
Mouse HIF-1αI.2	MEGAGGENEKKKMSS*ERRKEKSRDAARSRR*...	(12,15,16)
Rat HIF-1α	MSS*ERRKEKSRDAARSRR*...	(13)
Rat HIF-1α	MEGAGGENEKKKMSS*ERRKEKSRDAARSRR*...	(46)
Rat HIF-1α	MEGAGGENEKKNRMSS*ERRKEKSRDAARSRR*...	(47)
Rat HIF-1α(S)	...NEKKMSS...	(24)
Rat HIF-1α(L)	...NEKNRMSS...	(24)
Human HIF-1α	MEGAGGANDKKKISS*ERRKEKSRDAARSRR*...	(10,48–50)
Human HIF-1α	MEGAGGANDKKNRISS*ERRKEKSRDAARSRR*...	(50)
Human HIF-1α	MEGAGGANDKKKISS*ERRKEKSRDAARSRR*...	Unpublished[a]
Bovine HIF-1α	MEGAGGANDKKKISS*ERRKEKSRDAARSRR*...	(51)
Chicken HIF-1α	MDSPGGVTDKKRISS*ERRKEKSRDAARCRR*...	(52)
Xenopus HIF-1α	MEGSVVVSEKKRISS*ERRKEKSRDAARCRR*...	Unpublished[b]
Trout HIF-1α	MDTGVVPEKKSRVSS*DRRKEKSRDAARCRR*...	(53)
Drosophila HIF-1α	61-GKPKEKRRNN*EKRKEKSRDAARCRR*...	(54)
C. elegans HIF-1α	MEDNRKRNM*ERRRETSRHAARDRR*...	(55)

Alignment of the N-terminal amino acid sequences derived from various species. Amino acid residues constituting the basic domain are shown in italics. Accession numbers of unpublished sequences: [a]BAB70608; [b]CAB96628.

mHIF-1αI.1 isoform is devoid of this site, suggesting that in both rat and mouse functionally similar protein isoforms might be the result of two different molecular biological regulatory pathways.

V. The *Hif1a* Exon I.1 Promoter Is Only Weakly Active in Cultured Cell Lines

By exon-specific RT-PCR, we detected both mHIF-1α mRNA isoforms in all organs analyzed (with the exception of the mHIF-1αI.1 mRNA, which was not detectable in the mouse liver). However, by quantitative RNase protection analysis, we found that the mHIF-1αI.1 mRNA isoform is differentially expressed in adult mouse tissues, being highest in kidney, tongue, stomach, and testis but undetectable in liver (15). Compared with the ubiquitously expressed mHIF-1αI.2 mRNA isoform, mHIF-1αI.1 mRNA expression levels were markedly lower in these tissues. Interestingly, the predominant mHIF-1αI.1 mRNA isoform in testis as well as in embryonic tissue, but not in kidney, tongue, and stomach, seemed to be extended at its 5′ terminus when compared to the transcription initiation site mapped in a cultured cell line (15).

Reporter gene studies using various exon I.1 and I.2 5′ flanking fragments revealed a strong and ubiquitous promoter activity of the *Hif1a* exon I.2 promoter, but only a weak activity of the exon I.2 promoter in all cell lines analyzed (14,15). Despite the presence of multiple potential HIF-1 DNA-binding consensus sequences, both promoters were nonresponsive to hypoxic conditions, confirming our earlier observations that the HIF-1α mRNA is not up-regulated by hypoxia (11,25). Thus, to define the cell type(s) expressing the tissue-specific mHIF-1αI.1 mRNA isoform, in situ hybridizations of mouse tissues were performed using antisense probes derived from *Hif1a* exon I.1 and exon I.2.

VI. Isoform-Specific Expression of HIF-1α in Mouse Testis

Surprisingly, in situ hybridization revealed a highly specific and abundant expression pattern of mHIF-1αI.1 mRNA in mouse testis (26). The signal with the *Hif1a* exon I.1 antisense probe was confined to the elongated nucleus of spermatids, corresponding to the late, postmeiotic, haploid stages of mouse spermatogenesis. In contrast, the *Hif1a* exon I.2 antisense probe hybridized to a region overlapping with spermatocytes and Sertoli cells, but did not hybridize with postmeiotic cells. Even more surprising, immunohistochemistry of HIF-1α in mouse testis and epididymis (26) as well as immunofluorescence of isolated mouse sperm revealed a specific signal localized to the midpiece of the sperm tail (Fig. 2). Thus, it seems likely that the mHIF-1αI.1 isoform is specifically expressed in mouse sperm, whereas mHIF-1αI.2 is ubiquitously expressed. However, because there are no isoform-specific antibodies existing, we cannot formally exclude the possibility that both mHIF-1α protein isoforms might be expressed in sperm. *Hif1a* exon I.1 promoter-driven reporter gene constructs were again only weakly active in several cell lines corresponding to testis-specific cell types (Leydig, Sertoli, spermatogonia, spermatocytes), confirming the postmeiotic nature of mHIF-1αI.1 mRNA expression (26). Thus, these results explain our inability to detect *Hif1a* exon I.1 promoter activity in cultured (diploid) cell lines.

VII. What Is the Biological Significance of HIF-1α Expression in the Testis?

In the testis, spermatogenesis is a highly ordered differentiation process, involving profound transcriptional and morphological changes. Spermatogenesis is initiated by the differentiation of self-replicating spermatogonia at the basal layers of the seminiferous tubuli to become committed intermediate spermatogonia. Following two further mitotic divisions, primary spermatocytes are formed that undergo the two rounds of meiotic divisions resulting first in secondary spermatocytes and later in haploid round spermatids. In the mouse, in a process called spermiogenesis, these round spermatids mature in 16 distinct steps via elongated spermatids to mature spermatozoa. After about 13.5 days of mouse spermiogenesis, spermatozoa are finally released into the lumen of the seminiferous tubuli. Spermatogenesis appears in synchronized waves of layers of differentiating germ cells that sequentially migrate from the basal toward the luminal regions of the seminiferous tubuli, making it possible to distinguish stages I–XII of the cycle of spermatogenesis (reviewed in Ref. 27). During nuclear elongation of haploid spermatids, the rate of

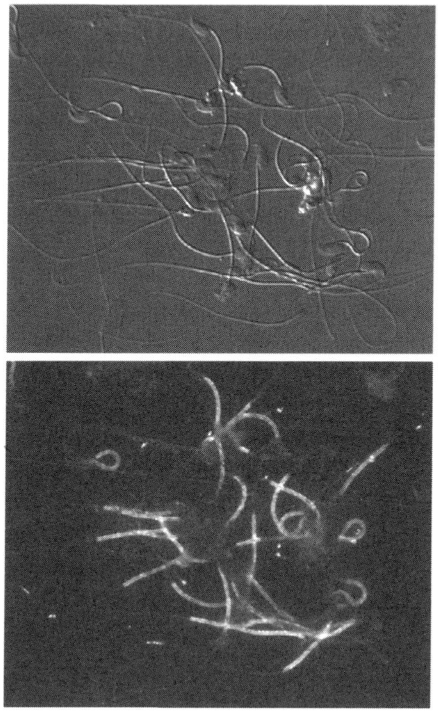

Figure 2 Localization of HIF-1α in mouse spermatozoa. Immunofluorescence analysis of spermatozoa isolated from epididymis and vas deferens using a monoclonal anti-HIF-1α antibody. Original magnification, 630×; scale bar length, 50 μm; (top) transmission; (bottom) immunofluorescence.

transcription declines and becomes undetectable in elongated spermatids. Nevertheless, ongoing translation of a number of sperm-cell-specific structural proteins and isoforms of metabolic enzymes is required for the production of spermatozoa. Therefore, specific mRNA isoforms containing long poly(A) tails are stored as translationally inactive messenger ribonucleoprotein (mRNP) particles. In transcriptionally inactive states, these mRNA isoforms are recruited into translationally highly active polysomes to ensure ongoing protein synthesis (reviewed in Refs. 28–30). Prominent examples include the nuclear transition proteins (31,32) and later in spermiogenesis the protamines (33–35), which replace the histones and lead to compaction of the chromatin.

There are also various testis-specific isoforms of glycolytic enzymes, which are expressed in the haploid stages of spermatogenesis and which are still active in mature spermatozoa. These include phosphoglycerate kinase 2 (36,37), glyceraldehyde 3-phosphate dehydrogenase-2 (38), and lactate dehydrogenase C (39). Because of the greater oxygen diffusion distance, the luminal regions of the seminiferous tubuli as well as the epididymis are likely to be hypoxic when compared to the basal regions of spermatogonial self-renewal. In addition, the high proliferative capacity of the germinal epithelium suggests a pronounced oxygen consumption, thereby further decreasing the oxygen concentration. In fact, oxygen partial pressures as low as 2 mmHg have been reported, which are among the lowest values found in the body and otherwise occur only in the vicinity of mitochondria (40). Thus, specific glycolytic enzyme isoform expression might be related to hypoxic adaptation by altered anaerobic energy metabolism. Indeed, sperm capacitation, motility changes, acrosome reaction, and fertilization are exclusively dependent on anaerobic glycolysis and can occur under strictly anaerobic conditions (41). However, the influence of oxygen partial pressure, consumption, and metabolism on the molecular events during spermatogenesis, spermatozoa release, in ejaculated sperm, and during fertilization is largely unknown.

Testis-specific transcription factor isoform expression, such as we found for HIF-1α, is not an uncommon finding. For example, expression of testis-specific promoter/first exon combinations of the TATA-binding protein (TBP) (42) and GATA-1 (43) or testis-specific alternative splicing of CREMτ (44) has been reported. TBP mRNA is stored as mRNP for up to 1 week, after which it is required for the transcription of sperm-specific proteins (e.g., protamines) under conditions when general transcription ceases (28). mRNA expression of the testis-specific TBP and CREMτ isoforms begins in early-round spermatids and the corresponding proteins are no longer detectable in mature spermatozoa.

In marked contrast, the detection of the mHIF-1αI.1 mRNA isoform in the head of mature haploid spermatids and of the HIF-1α protein in the principal piece of the flagellum of spermatozoa represents to our knowledge the first example of testis-specific transcription factor expression beyond the release of spermatozoa from the seminiferous tubuli. To date, the function of HIF-1α at this specific site is unclear. As outlined above, many HIF-1 target genes, especially the glycolytic enzymes, are also expressed in the testis as specific isoforms (36–39). However, because the spermatozoal nuclei are thought to be transcriptionally inactive, it seems unlikely that HIF-1 functions as a transcription factor in sperm. Rather, the high levels of HIF-1α might be required in oocytes after fertilization that occurs in a hypoxic environment. Because mice containing a null mutation in the second exon of the *Hif1a* gene are nonviable (45), an exonI.1-specific mouse knockout model will be required to elucidate the functional significance of specific HIF-1α isoform expression during mouse spermiogenesis.

Another intriguing question is the exact mechanism of the mHIF-1αI.1 mRNA accumulation in the differentiated spermatids. Our results with in vitro cultured spermatogonia and spermatocyte cell lines indicate that the exon I.1 promoter does not become activated during spermatogenesis in these premeiotic stages. Interestingly, ORFs for premordial peptides, similar to the ORF located in the proximal exon I.1 promoter (see above), have been reported to be transcribed in the sex-determining region of the Y chromosome (21). While the mouse *Hif1a* gene is not located on the Y chromosome, this sequence might be responsible for activating the exon I.1 promoter during male postmeiotic spermiogenesis. Clearly, in vitro transcription assays and transgenic mouse models are needed to clarify this point.

While exposure to hypoxia induced mHIF-1αI.2 protein levels in the testis, it did not further induce mHIF-1αI.1 protein in epididymal spermatozoa or following in vitro exposure of isolated spermatozoa (26). The most obvious explanation for this effect might be the lack of proteasomal activity in mature spermatozoa, rendering mHIF-1αI.1 protein stability already under normoxic conditions. Although we cannot formally exclude that the mHIF-1αI.2 protein corresponds to the isoform detected in spermatozoa, the switch from mHIF-1αI.2 to mHIF-1αI.1 mRNA isoform expression during spermatogenesis and the complete lack of detectable mHIF-1αI.2 mRNA in spermatids strongly suggest that the mHIF-1αI.1 mRNA is translated in postmeiotic spermatids. The definitive proof would require antibodies specific for the two different predicted N-termini. However, as we did not find any functional differences of the two isoforms in vitro (22), the biological significance of our findings probably lies in the fact that an alternative promoter is active to ensure the expression of HIF-1α at unusually late stages of spermiogenesis, rather than the expression of a structurally and functionally different protein isoform.

VIII. Isoform-Specific Expression of HIF-1α in Mouse T Cells

Recently, Lukashev et al. reported that the mHIF-1αI.1 mRNA isoform is specifically up-regulated in T cells following in vitro activation by T-cell-receptor cross-linking, phorbol ester stimulation, or calcium ionophore treatment, but not by hypoxia (23). In vivo, concanavalin A activation also induced mHIF-1αI.1 mRNA levels in T cells. As we reported before (22), no functional differences between the two protein isoforms could be detected by Lukashev et al. (23). While the biological significance of isoform-specific HIF-1α induction in mouse T cells is unclear, these results explain our previous finding that low levels of the mHIF-1αI.1 mRNA isoform can be detected in almost all mouse organs by RT-PCR but not by RNase protection (15) or in situ hybridization (26).

Clearly, to unravel the physiological significance of these novel results on tissue-specific expression of a distinct isoform of HIF-1α, genetically altered mouse models will be required. Attempts to create such mouse strains are currently being performed.

Acknowledgments

I thank D. M. Katschinski and K. F. Wagner for critically reading the manuscript. This work was supported by grants from the Medical University of Lübeck and the Deutsche Forschungsgemeinschaft (We2672/1-1).

References

1. Semenza GL. HIF-1 and mechanisms of hypoxia sensing. Curr Opin Cell Biol 2001; 13:167–171.

2. Wenger RH. Mammalian oxygen sensing, signalling and gene regulation. J Exp Biol 2000; 203:1253–1263.
3. Jiang BH, Semenza GL, Bauer C, Marti HH. Hypoxia-inducible factor 1 levels vary exponentially over a physiologically relevant range of O_2 tension. Am J Physiol 1996; 271:C1172–1180.
4. Jewell UR, Kvietikova I, Scheid A, Bauer C, Wenger RH, Gassmann M. Induction of HIF-1α in response to hypoxia is instantaneous. FASEB J 2001; 15: 1312–1314.
5. Maxwell PH, Wiesener MS, Chang GW, et al. The tumour suppressor protein VHL targets hypoxia-inducible factors for oxygen-dependent proteolysis. Nature 1999; 399:271–275.
6. Jaakkola P, Mole DR, Tian YM, et al. Targeting of HIF-α to the von Hippel–Lindau ubiquitylation complex by O_2-regulated prolyl hydroxylation. Science 2001; 292:468–472.
7. Ivan M, Kondo K, Yang H, et al. HIFα targeted for VHL-mediated destruction by proline hydroxylation: implications for O_2 sensing. Science 2001; 292:464–468.
8. Epstein AC, Gleadle JM, McNeill LA, et al. *C. elegans* EGL-9 and mammalian homologs define a family of dioxygenases that regulate HIF by prolyl hydroxylation. Cell 2001; 107:43–54.
9. Bruick RK, McKnight SL. A conserved family of prolyl-4-hydroxylases that modify HIF. Science 2001; 294:1337–1340.
10. Wang GL, Jiang BH, Rue EA, Semenza GL. Hypoxia-inducible factor 1 is a basic-helix-loop-helix-PAS heterodimer regulated by cellular O_2 tension. Proc Natl Acad Sci USA 1995; 92:5510–5514.
11. Wenger RH, Rolfs A, Marti HH, Guénet JL, Gassmann M. Nucleotide sequence, chromosomal assignment and mRNA expression of mouse hypoxia-inducible factor-1α. Biochem Biophys Res Commun 1996; 223:54–59.
12. Li H, Ko HP, Whitlock JP. Induction of phosphoglycerate kinase 1 gene expression by hypoxia: roles of ARNT and HIF1α. J Biol Chem 1996; 271:21262–21267.
13. Ladoux A, Frelin C. Cardiac expressions of HIF-1α and HLF/EPAS, two basic loop helix/PAS domain transcription factors involved in adaptative responses to hypoxic stresses. Biochem Biophys Res Commun 1997; 240:552–556.
14. Wenger RH, Rolfs A, Kvietikova I, Spielmann P, Zimmermann DR, Gassmann M. The mouse gene for hypoxia-inducible factor-1α: genomic organization, expression and characterization of an alternative first exon and 5′ flanking sequence. Eur J Biochem 1997; 246:155–165.
15. Wenger RH, Rolfs A, Spielmann P, Zimmermann DR, Gassmann M. Mouse hypoxia-inducible factor-1α is encoded by two different mRNA isoforms: expression from a tissue-specific and a housekeeping-type promoter. Blood 1998; 91:3471–3480.
16. Luo G, Gu Y-Z, Jain S, et al. Molecular characterization of the murine *HIF-1α* locus. Gene Exp 1997; 6:287–299.
17. Semenza GL, Rue EA, Iyer NV, Pang MG, Kearns WG. Assignment of the hypoxia-inducible factor 1α gene to a region of conserved synteny on mouse chromosome 12 and human chromosome 14q. Genomics 1996; 34:437–439.
18. Wenger RH, Kvietikova I, Rolfs A, Camenisch G, Gassmann M. Oxygen-regulated erythropoietin gene expression is dependent on a CpG methylation-free hypoxia-inducible factor-1 DNA-binding site. Eur J Biochem 1998; 253:771–777.

19. Wenger RH, Gassmann M. Molecular biology of hypoxia-inducible factor-1. In: Abraham N, ed. Molecular Biology of Hematopoiesis. Vol. 6. New York: Kluwer Academic/Plenum Publishers, 1999:269–276.
20. Ohno S, Epplen JT. The primitive code and repeats of base oligomers as the primordial protein-encoding sequence. Proc Natl Acad Sci USA 1983; 80:3391–3395.
21. Singh L, Phillips C, Jones KW. The conserved nucleotide sequences of Bkm, which define Sxr in the mouse, are transcribed. Cell 1984; 36:111–120.
22. Görlach A, Camenisch G, Kvietikova I, Vogt L, Wenger RH, Gassmann M. Efficient translation of mouse hypoxia-inducible factor-1α under normoxic and hypoxic conditions. Biochim Biophys Acta 2000; 1493:125–134.
23. Lukashev D, Caldwell C, Ohta A, Chen P, Sitkovsky M. Differential regulation of two alternatively spliced isoforms of hypoxia-inducible factor-1α in activated T lymphocytes. J Biol Chem 2001:10.1074/jbc.M104782200.
24. Drutel G, Kathmann M, Héron A, et al. Two splice variants of the hypoxia-inducible factor HIF-1α as potential dimerization partners of ARNT2 in neurons. Eur J Neurosci 2000; 12:3701–3708.
25. Wenger RH, Kvietikova I, Rolfs A, Gassmann M, Marti HH. Hypoxia-inducible factor-1α is regulated at the post-mRNA level. Kidney Int 1997; 51:560–563.
26. Marti HH, Katschinski DM, Wagner KF, Schäffer L, Stier B, Wenger RH. Isoform-specific expression of hypoxia-inducible factor-1α during the late stages of mouse spermiogenesis. Mol Endocrinol 2002; 16:234–243.
27. Russell LD, Ettlin RA, Sinha Hikim AP, Clegg ED. Histological and Histopathological Evaluation of the Testis. Clearwater, FL: Cache River Press, 1990.
28. Steger K. Transcriptional and translational regulation of gene expression in haploid spermatids. Anat Embryol 1999; 199:471–487.
29. Kleene KC. Patterns of translational regulation in the mammalian testis. Mol Reprod Dev 1996; 43:268–281.
30. Goldberg E. Transcriptional regulatory strategies in male germ cells. J Androl 1996; 17:628–632.
31. Heidaran MA, Kistler WS. Transcriptional and translational control of the message for transition protein 1, a major chromosomal protein of mammalian spermatids. J Biol Chem 1987; 262:13309–13315.
32. Heidaran MA, Showman RM, Kistler WS. A cytochemical study of the transcriptional and translational regulation of nuclear transition protein 1 (TP1), a major chromosomal protein of mammalian spermatids. J Cell Biol 1988; 106:1427–1433.
33. Kleene KC, Distel RJ, Hecht NB. Translational regulation and deadenylation of a protamine mRNA during spermiogenesis in the mouse. Dev Biol 1984; 105:71–79.
34. Kleene KC. Poly(A) shortening accompanies the activation of translation of five mRNAs during spermiogenesis in the mouse. Development 1989; 106:367–373.
35. Bunick D, Johnson PA, Johnson TR, Hecht NB. Transcription of the testis-specific mouse protamine 2 gene in a homologous in vitro transcription system. Proc Natl Acad Sci USA 1990; 87:891–895.
36. Gold B, Fujimoto H, Kramer JM, Erickson RP, Hecht NB. Haploid accumulation and translational control of phosphoglycerate kinase-2 messenger RNA during mouse spermatogenesis. Dev Biol 1983; 98:392–399.
37. McCarrey JR, Kumari M, Aivaliotis MJ, et al. Analysis of the cDNA and encoded protein of the human testis-specific PGK-2 gene. Dev Genet 1996; 19:321–332.

38. Welch JE, Brown PL, O'Brien DA, et al. Human glyceraldehyde 3-phosphate dehydrogenase-2 gene is expressed specifically in spermatogenic cells. J Androl 2000; 21:328–338.
39. Li S, Zhou W, Doglio L, Goldberg E. Transgenic mice demonstrate a testis-specific promoter for lactate dehydrogenase, LDHC. J Biol Chem 1998; 273:31191–31194.
40. Max B. This and that: hair pigments, the hypoxic basis of life and the Virgilian journey of the spermatozoon. Trends Pharmacol Sci 1992; 13:272–276.
41. Fraser LR, Quinn PJ. A glycolytic product is obligatory for initiation of the sperm acrosome reaction and whiplash motility required for fertilization in the mouse. J Reprod Fertil 1981; 61:25–35.
42. Schmidt EE, Ohbayashi T, Makino Y, Tamura T, Schibler U. Spermatid-specific overexpression of the TATA-binding protein gene involves recruitment of two potent testis-specific promoters. J Biol Chem 1997; 272:5326–5334.
43. Ito E, Toki T, Ishihara H, et al. Erythroid transcription factor GATA-1 is abundantly transcribed in mouse testis. Nature 1993; 362:466–468.
44. Delmas V, van der Hoorn F, Mellstrom B, Jegou B, Sassone-Corsi P. Induction of CREM activator proteins in spermatids: down-stream targets and implications for haploid germ cell differentiation. Mol Endocrinol 1993; 7:1502–1514.
45. Iyer NV, Kotch LE, Agani F, et al. Cellular and developmental control of O_2 homeostasis by hypoxia-inducible factor 1α. Genes Dev 1998; 12:149–162.
46. Kietzmann T, Cornesse Y, Brechtel K, Modaressi S, Jungermann K. Perivenous expression of the mRNA of the three hypoxia-inducible factor α-subunits, HIF1α, HIF2α and HIF3α, in rat liver. Biochem J 2001; 354:531–537.
47. Zou AP, Yang ZZ, Li PL, Cowley AJ. Oxygen-dependent expression of hypoxia-inducible factor-1α in renal medullary cells of rats. Physiol Genom 2001; 6:159–168.
48. Hogenesch JB, Chan WK, Jackiw VH, et al. Characterization of a subset of the basic-helix-loop-helix-PAS superfamily that interacts with components of the dioxin signaling pathway. J Biol Chem 1997; 272:8581–8593.
49. Iyer NV, Leung SW, Semenza GL. The human hypoxia-inducible factor 1α gene: *HIF1A* structure and evolutionary conservation. Genomics 1998; 52:159–165.
50. Gothié E, Richard DE, Berra E, Pagès G, Pouysségur J. Identification of alternative spliced variants of human hypoxia-inducible factor-1α. J Biol Chem 2000; 275: 6922–6927.
51. Hara S, Kobayashi C, Imura N. Molecular cloning of cDNAs encoding hypoxia-inducible factor (HIF)-1α and -2α of bovine arterial endothelial cells. Biochim Biophys Acta 1999; 1445:237–243.
52. Takahashi T, Sugishita Y, Nojiri T, et al. Cloning of hypoxia-inducible factor 1α cDNA from chick embryonic ventricular myocytes. Biochem Biophys Res Commun 2001; 281:1057–1062.
53. Soitamo AJ, Rabergh CM, Gassmann M, Sistonen L, Nikinmaa M. Characterization of a hypoxia-inducible factor (HIF-1α) from rainbow trout: accumulation of protein occurs at normal venous oxygen tension. J Biol Chem 2001; 276:19699–19705.
54. Nambu JR, Chen W, Hu S, Crews ST. The *Drosophila melanogaster* similar bHLH-PAS gene encodes a protein related to human hypoxia-inducible factor 1α and *Drosophila* single-minded. Gene 1996; 172:249–254.
55. Jiang H, Guo R, Powell-Coffman JA. The *Caenorhabditis elegans hif-1* gene encodes a bHLH-PAS protein that is required for adaptation to hypoxia. Proc Natl Acad Sci USA 2001; 98:7916–7921.

6

Hypoxia-Inducible Factor-1

More Than a Hypoxia-Inducible Transcription Factor

THOMAS HELLWIG-BÜRGEL, DANIEL PHILLIP STIEHL, and WOLFGANG JELKMANN

University of Lübeck
Lübeck, Germany

I. Introduction

Transcription is an energy-consuming process. From an economical view it is reasonable that not all genes are transcribed simultaneously in one cell. Indeed, most genes are transcribed only when a specific gene product is needed. Regulation of gene expression is mainly achieved by the coordinate interplay of different transcription factors (TFs) and transcriptional coactivators with DNA-dependent RNA-polymerases. To gain spatial and temporal control in this process, TFs have to fulfill two prerequisites: sequence specificity and inducibility.

Hypoxia-inducible factor-1 (HIF-1) is the most important TF in mediating the physiological response to hypoxia, because in meeting the requirements, HIF-1 binds to a specific DNA sequence (consensus sequence 5′-RCGTG-3′) upon the onset of hypoxia (1). The importance of HIF-1 is emphasized by the fact that this TF is found in a wide variety of different multicellular organisms, ranging from nematodes to mammalias (2–8). Furthermore, HIF-1 plays a key role in tumor progression and metastasis (9). Besides hypoxia, a growing number of physiological and artificial stimuli as well as genetic alterations have proved to activate HIF-1 under high oxygen concentrations. This chapter will focus on the normoxic induction of HIF-1 and the intracellular signaling pathways involved in HIF-1 stabilization and activation.

II. Mechanism of HIF-1α Degradation in Normoxia

HIF-1 is a heterodimeric TF composed of α- and β-subunits (10). The α-subunit mediates oxygen susceptibility, whereas the β-subunit is no target of the oxygen dependent degradation (1). In the presence of oxygen the α-subunit is hydroxylated at Pro-564 (and/or Pro-402) by distinct prolyl hydroxylases (PHs) in a process that requires Fe^{2+} ions, 2-oxoglutarate, and molecular oxygen (11–13). Hydroxyproline in position 564 targets HIF-1α to the von Hippel–Lindau protein (pVHL), which is part of the E3 ubiquitin ligase complex (VHLE3 complex), and once ubiquitinated HIF-1α is rapidly degraded by the ubiquitin-proteasome pathway (14–20). The speed of this process is impressive, the in vivo half-life of HIF-1α under conditions of reoxygenation is < 1 min (21).

III. HIF-1α Stabilization

A. HIF-1α Stabilization in Hypoxia

During hypoxia, oxygen is limited as cosubstrate for PHs. The process of hydroxylation is dramatically reduced; consequently HIF-1α is not recognized by the VHLE3 complex and accordingly not subject to proteasomal degradation. The result is the accumulation of HIF-1α, which dimerizes with HIF-1β to form heterodimeric complexes (14–16). This process is extremely fast, too, and has been described to be instantaneous (22).

B. HIF-1α Stabilization by Growth Factors and Hormones in Normoxia

Besides hypoxia a variety of different stimuli were described to induce HIF-1 under high oxygen concentrations in a large number of different cell types (summarized in Table 1). Most of the stimuli exert their biological activities via receptor-coupled processes. With respect to the events following ligand binding to the receptor, the different stimuli can be separated into a few groups according to their intracellular signaling pathways.

The first group of normoxic inducers described (chemical compounds like iron chelating agents, $CoCl_2$, or Mersalyl are excluded) were insulin and insulin-like growth factor-I (IGF-I), which are able to induce HIF-1 DNA-binding in electrophoretic mobility shift assays (EMSAs), to enhance HIF-1 regulated expression of reporter genes, and to induce transcription of the glucose transporter 1 (GLUT-1) and vascular endothelial growth factor (VEGF) genes, two well-known HIF-1 targets (23). Interestingly, Bennett et al. (24) reported that insulin inhibits the ubiquitin-dependent degrading activity of the 26S proteasome. This may easily explain the increased amounts of HIF-1α in insulin-treated cells. Even if hydroxylated at

Table 1 Peptides Inducing HIF-1 in Normoxia

Normoxic stimulus	Cells	References
Insulin	HepG2, T47D (human ductal breast carcinoma), L8 murine myoblasts, Hepa 1c1c7	(23)
	Mouse embryo fibroblasts, human embryonic kidney 293 cells	(25)
	U2-OS (human osteosarcoma), ACHN (renal carcinoma derived), DU145, PC-3 (both are metastatic prostate carcinoma-derived cell lines)	(38)
	HepG2	(41)
IGF-1	HepG2	(23)
	Mouse embryo fibroblasts, human embryonic kidney 293 cells	(25)
IGF-2	Mouse embryo fibroblasts, human embryonic kidney 293 cells	(25)
FGF-2	Human embryonic kidney 293 cells	(25)
EGF	Human embryonic kidney 293 cells	(25)
	PC-3, DU145, TSU, PPC-1 (human prostate cancer cells)	(34)
	DU145, PC-3 (human prostate cancer cells)	(38)
HGF	HepG2	(28)
PDGF	Rat vascular smooth muscle cells	(26)
FGF-2	Rat vascular smooth muscle cells	(26)
TGF-β	HT-1080 fibrosarcoma cells	(29)
Interleukin-1β	HepG2	(41,52)
	Human proximal tubular epithelial cells	(53)
	Human gingival/synovial fibroblasts	(54)
TNF-α	HepG2	(52)
	Fetal rat alveolar type II epithelial cells	(71)
	Rat PMN and macrophages	(55)
	Proximal tubular LLC-PK_1 cells	(40)
Angiotensin II	Rat vascular smooth muscle cells	(26)
Thrombin	Human vascular smooth muscle cells	(27)
	Rat vascular smooth muscle cells	(26)
Heregulin	MCF-7	(42)

Pro-564 and subsequently polyubiquitinated, HIF-1α could not be degraded due to impaired proteasomal function. Feldser and coauthors (25) added IGF-II, epidermal growth factor (EGF), and fibroblast growth factor 2 (FGF 2) as further normoxic HIF-1 inducers to the menu. Moreover, the authors described an autocrine growth factor loop since IGF-II gene expression is HIF-1 regulated (25). Additionally, platelet-derived growth factor (PDGF), thrombin, and angiotensin II (Ang II) were identified to activate HIF-1α (26,27), which will be discussed below. Hepatocyte growth factor (HGF) and transforming growth factor-β (TGF-β) are the latest members of the growing family of HIF-1-inducing growth factors (28,29).

Biochemically, insulin-, IGF-, EGF-, FGF-, HGF-, and PDGF-receptors (IR, IGF-IR, EGFR, FGFR, HGFR, and PDGFR) belong to the group of receptor tyrosine kinases. Ligand binding to IR-kinase or IGF-IR-kinase, for instance, leads to receptor autophosphorylation and tyrosine phosphorylation of several target proteins such as Shc- and the family of insulin receptor substrate (IRS) proteins (IRS-1–IRS-4) (30–32). The phosphorylated tyrosine residues of these proteins serve as docking sites for downstream effector molecules. Finally, two major kinase cascades are triggered by these initial steps, the phosphatidylinositol 3-kinase (PI3K) and the mitogen-activated protein kinase (MAPK) pathways (Fig. 1). Growth factors and hormones other than insulin or IGFs initiate these two pathways by use of different adaptor proteins/kinases during the first steps of signaling.

Role of the PI3K Pathway

The involvement of the PI3K/AKT/FRAP pathway in the stabilization of HIF-1α has been demonstrated in several studies (33–41). In the present working model of the PI3K pathway (Fig. 1) the major steps of the cascade are depicted and the points at which a pharmacological blockade, a genetic alteration, or expression of dominant negative isoforms of pathway-specific protein kinases influence HIF-1 stabilization and/or activity are highlighted. The condensed information of the above-cited articles is that PI3K-activity is necessary for the stabilization of HIF-1α and DNA-binding activity of HIF-1. Thus, the enhanced transcription of HRE reporter genes or endogeneous HIF-1 target genes is mainly due to increased HIF-1α amounts and not (or to a lesser extent) to enhanced *trans*-activation properties of HIF-1α. Furthermore, a recent study has shown that HIF-1α synthesis is stimulated via the PIK3K/AKT/FRAP pathway in cancer cell cultures, which is of significance with regard to activated VEGF gene expression and tumor angiogenesis (42).

Role of the MAPK Pathway

In mammalian cells, at least six more or less independent MAPK signaling units appear to exist. Three of these have been characterized in detail [p42/p44 MAPK (ERK2/1), JNK/SAPK, and p38 MAPK] (Fig. 2). However, there are cross-talks between the individual pathways. For instance, looking inside HGFR signaling, it becomes obvious that HGF initializes the p42/p44 MAPK pathway as well as the JNK/SAPK and p38 MAPK pathways, and additionally the PI3K pathway. Sticking to HGFR signaling, Tacchini et al. (28) demonstrated that inhibition of the p42/p44 MAPK pathway with PD98059 (PD) by blocking the MEK-1 (MAPK kinase-1) has no effects on HGF-induced HIF-1 DNA binding and HIF-1α protein accumulation, whereas inhibition of PI3K by LY294002 (LY) lowers both. The exact role of JNK/SAPK in this model system has remained unclear, since this pathway is not affected by PD but inhibited to 50% by LY.

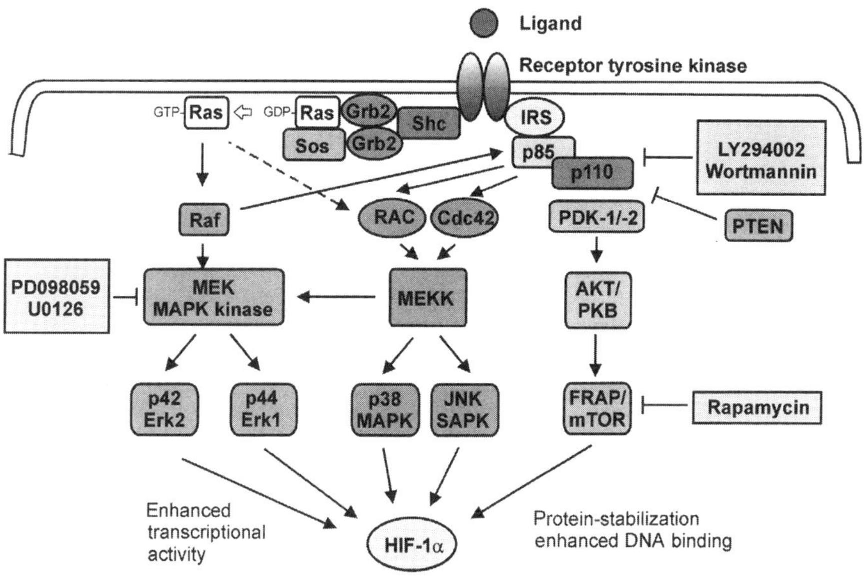

Figure 1 Scheme of signaling pathway from receptor tyrosine kinases to activated HIF-1α. *PI3K cascade:* The active heterodimer of PI3K consists of a regulatory p85 subunit and a catalytic p110 subunit. Once activated it catalyzes the conversion of PI-4-phosphate and PI-4,5-bisphosphate to PI-3,4-bisphosphate and PI-3,4,5-trisphosphate, respectively. PI-3,4,5-trisphosphate is an allosteric activator of PDK-1 and -2 (PI-dependent kinases-1 and -2). PDKs in turn phosphorylate AKT (protein kinase B). A major downstream target of AKT is FRAP (FKBP12/rapamycin-associated protein; also known as mTOR, mammalian target of rapamycin). Activation of the PI3K/AKT/FRAP pathway results in increased HIF-1α levels (39,42). Note that the activity of AKT is negatively regulated by several lipid phosphatases such as PTEN (phosphatase and tensin homolog deleted on chromosome ten) (74). *MAPK cascade:* Phosphorylated receptor tyrosine kinases (like the activated insulin-receptor) activate the Shc (Src homology collagen) protein, which interacts with the Grb2 adapter protein. Grb2 is bound constitutively to a proline-rich region of Sos (Son of sevenless), which is a GDP/GTP exchanger for Ras. Upon ligand stimulation Grb2/Sos binds to Shc, thus directing the relocalization of associated Sos to the membrane. Membrane-bound Sos in turn activates Ras by catalyzing the GDP/GTP exchange. Activated Ras initiates the downstream cascade including the sequential activation of Raf, MAPKK (MEK), p42/p44 MAPK. The MAPKs finally increase the transcriptional activity of HIF-1. *p38 MAPK and JNK/SAPK pathway:* Receptor tyrosine kinases are also able to activate several G-proteins such as RAC and Cdc42 (Cell division cycle protein 42) possibly by interference with the PI3K pathway. These initial events lead to an activation of MEKK, from which the p38 MAPK and JNK pathways are triggered.

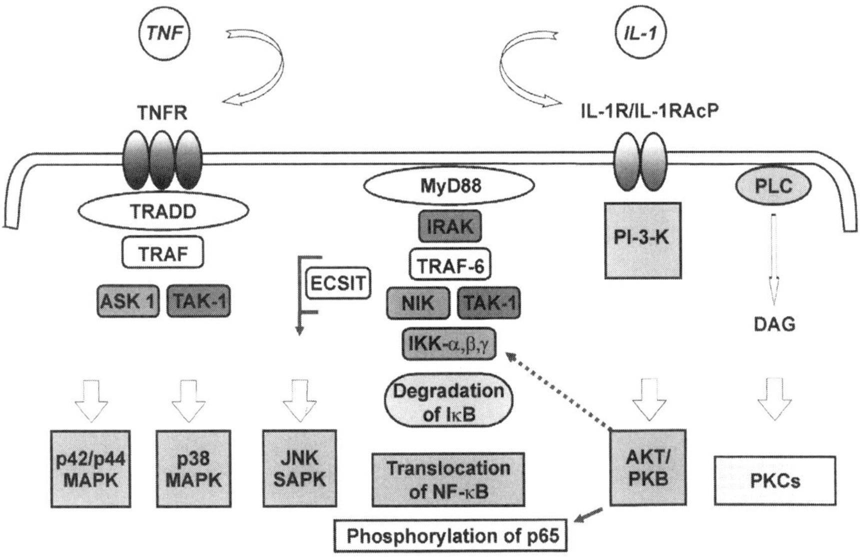

Figure 2 Cytokine signaling by TNF-α and IL-1β. *TNF-α* binding to trimeric TNFR leads to receptor oligomerization and subsequent binding of several adapter proteins/kinases. Binding of TRADD (TNFR-associated protein with a death domain) establishes a platform adapter that recruits several signaling molecules to the activated receptor. Coupling of TRAF (TNFR-associated factor) to TNFR/TRADD enables ASK1 (apoptosis signal-regulating kinase 1) or TAK-1 (MAPK complex involved in transforming growth factor-β signaling) binding and subsequent activation of JNK. Finally, the p38 MAPK, p42/p44 MAPK, JNK, and NF-κB pathways are activated by use of different intermediate molecular adapters. Upon *IL-1* binding to the IL-1R, the IL-1R/IL-1R AcP (IL-1R accessory protein) complex forms. IRAK (IL-1R-associating kinase) associates with IL-1R/IL-1R AcP through the adapter protein MyD88 (myeloid differentiation primary response gene product). Phosphorylated IRAK interacts with TRAF6 (TNFR-associated factor 6), which activates the downstream serine/threonine kinase NIK (NF-κB-inducing kinase). Alternatively, NIK can be activated through the action of TAK-1 (MAPK complex involved in transforming growth factor-β signaling). ECSIT (evolutionary conserved signaling intermediate in Toll pathways) interacts also with TRAF6, but this way does not lead to NIK activation but to MAPK-Erk kinase-kinase (MEKK-1)-dependent NF-κB and AP-1 activation. Moreover, PLC and PI3K signaling pathways are induced by IL-1R/IL-1R AcP but independently from MyD88/IRAK (75).

Using in vitro models, Richard et al. (43) have demonstrated that HIF-1α is phosphorylated by p42/p44 MAPK but not by p38 MAPK or JNK/SAPK and that this phosphorylated HIF-1 enhances the expression of an HRE (hypoxia responsive element)-driven reporter gene. Hur and co-workers (44) have reported that proteasomal inhibition alone does not suffice to induce HIF-1-dependent transcription. Moreover, they have shown that PD inhibition of MEK-1 has no significant influence on the stabilization or DNA-binding ability of HIF-1α/HIF-1, while the *trans*-activation ability of HIF-1 is reduced (44). In summary, p42/p44 MAPK appears to regulate the *trans*-activation properties of HIF-1α, but it is not (or to a smaller extent) involved in the regulation of protein stability or DNA-binding activity.

In an elegant series of experiments Miele et al. (45) have shown that VEGF mRNA induction by insulin and IGF-I in NIH3T3 fibroblasts transfected with either human IR or IGF-IR occurred by preferential use of the PI3K pathway in case of insulin/IR or the MAPK pathway in case of IGF-I/IGF-IR. Although the authors depicted no direct link to HIF-1, one can speculate that two different mechanisms led to the elevated VEGF mRNA levels. In the first case the mechanism was, in all likelihood, an increased HIF-1α protein amount rather than enhanced *trans*-activation ability, whereas in the second case it was an enhanced *trans*-activation ability rather than an elevated HIF-1α amount. However, in vivo phosphorylated and dephosphorylated HIF-1α seems to have different functions since phosphorylated HIF-1α is the major form that binds to HIF-1β and activates transcription, whereas the dephosphorylated form binds to and stabilizes p53, which in turn initiates cell cycle arrest and apoptosis (46).

Other Signaling Pathways

Some normoxic HIF-1α inducers are thought to act via other intracellular signaling pathways. With respect to the effect of Ang II it has been demonstrated that HIF-1α induction is mediated neither by PI3K nor by MAPK. Instead, a mechanism involving reactive oxygen species (ROS) has been proposed since HIF-1α induction was abrogated in the presence of diphenyleneiodonium (DPI) or catalase (26). Ang II receptors (AT_1 and AT_2 receptors) and thrombin receptors [protease-activated receptors (PAR)] 1, 3, and 4 belong to the group of G-protein-coupled receptors that activate phospholipase C (PLC). Activated PLC generates inositol trisphosphate (IP-3), which releases Ca^{2+} ions from intracellular stores, and diacylglycerol (DAG), which activates protein kinase C (PKC) (see below). Consequently, thrombin binding to PARs exerts intracellular effects like Ang II binding to AT-receptors. Görlach et al. (27) have shown that HIF-1 can be stimulated by thrombin and that p38 MAPK and PI3K are involved in the activation process. Additionally, thrombin signaling needs ROS (similar to Ang II) to induce HIF-1 since a functional $p22^{phox}$-containing NADPH oxidase is essential. Focused on DAG as an intracellular messenger it has been reported that its concentration is elevated during hypoxic incubation in Hep3B cells, HeLa cells, and neonatal rat ventricular myocytes (47–49). The direct link to HIF-1α is established, because pharmacological inhibition of DAG kinase (DGK) reduces hypoxia-induced HIF-1α protein accumulation in the nucleus as well as HRE reporter gene transcription (49). Nevertheless, the role of DAG during normoxic HIF-1α induction remains elusive.

The role of PKC in HIF-1 activation has been investigated by several groups. Kruger and co-workers (50) have reported that UNC-01, a PKC inhibitor, abrogates hypoxia-mediated *trans*-activation of a HRE reporter gene in PC3M cells. Baek and coauthors (51) have demonstrated that PKCδ is translocated to the membrane during hypoxia in RIF tumor cells and that pharmacological or genetic inhibition of PKCδ inhibits the transcriptional activation of HIF-1 by hypoxia. The initial regulatory unit in this process seems to be the PI3K, since PI3K inhibition abrogates PKC membrane translocation (51). But similar to DAG, the relevance of PKC during normoxic HIF-1α induction needs to be enlightened.

IV. Role of HIF-1 in Immune Reactions

Evidence is accumulating that HIF-1α plays an important role during inflammation. It has been reported that the proinflammatory cytokines interleukin 1β (IL-1β) and tumor necrosis factor-α (TNF-α) induce HIF-1α in various types of cells, including human hepatoma cells (HepG2), primary proximal tubular epithelial cells (PTEC), primary synovial and gingival fibroblasts, and polymorphonuclear leukocytes (PMN) and macrophages (52–55). On closer inspection some cell-type specificities become obvious. In HepG2 cells IL-1β increases both HIF-1 DNA binding and HIF-1α protein amount in nuclear extracts of normoxic cells, whereas TNF-α induces only DNA binding, leaving the HIF-1α protein concentration unaffected (52). The steady-state mRNA level of HIF-1α was not altered in this model. In contrast, Thornton and coauthors (54) described an autocrine positive feedback loop, since in cultured synovial fibroblasts IL-1β and TNF-α (and lipopolysaccharide) were found to increase steady-state levels of HIF-1α mRNA (54).

The intracellular events following binding of TNF-α or IL-1β to their receptors are complex and cell-specific. In particular, the amount of adapter proteins interacting with the TNF receptors type I and II (TNFR I and II) is difficult to survey and the intracellular signaling cascades activated by these two receptors differ (56). TNF-α and IL-1 are able to activate NF-κB, JNK, p38 MAPK, and p42/p44 MAPK pathways (via TNFR I and IL-1R) (for details see Fig. 2). IL-1β further activates the PI3K pathway. The contribution of the individual proteins involved in the first steps of cytokine signaling in regard to HIF-1α activation is not yet known. Considering the final activation of PI3K upon IL-1β binding, the different effects of TNF-α and IL-1β on HIF-1α stabilization and DNA binding reported by Hellwig-Bürgel et al. (52) are easily explained. The involvement of the PI3K pathway in the induction of HIF-1α by TNF-α (40) or IL-1β (41) has recently been demonstrated.

Inflammatory cytokines, microbial products, and microbes trigger the expression of inducible NO-synthase (iNOS) and, thus, NO synthesis (57). NO has been implicated as a modulator of HIF-1 DNA binding and transcriptional activity, but the observations are still controversial. Earlier studies showed NO donors, such as nitroprusside, to lower HIF-1 DNA binding and transcriptional activity in hypoxic Hep3B, Neuro 2A, HeLa, and vascular endothelial smooth muscle cells (58–60). More recent studies indicate that there is a bimodal action as NO was found to transiently increase HIF-1 activity in a reporter gene assay (61). In fact, NO can exert a number of effects following nitrosylation of proteins and peroxidation of lipids. NO donation has been shown to increase HIF-1α levels in human glioblastoma A172 (62) and porcine renal tubular cells LLC-PK1 (63) as well as HIF-1 DNA-binding activity in human hepatoma cells Hep3B (62,64), human prostate cells (65), HeLa cells (64), and green monkey kidney COS-1 cells (64). S-Nitrosoglutathione has been reported to support HIF-1α stabilization in LLC-PK1 cells (36) and in bovine pulmonary artery endothelial cells (66). In LLC-PK1 cells the induction of HIF-1 DNA binding by added S-nitrosoglutathione was lowered on pharmacological PI3K inhibition (36).

Another symptom of inflammation is local heat. Katschinski and coauthors (67) reported that elevated temperature (42°C) is able to induce the nonphosphorylated form of HIF-1α under normoxic conditions in HepG2 cells. Furthermore, in mice with elevated body temperature (rectal temperature 41°C) HIF-1α accumulated in kidney and liver and remained stable for several hours after recooling to normal body temperature. Thus implies a novel mechanism that stabilizes HIF-1α under elevated temperatures, possibly via interaction with HSP90 (67).

Further support of the hypothesis that HIF-1α plays a significant role during inflammation comes from the observation that HIF-1α protein is found in cells at the border of artificial skin incisions in adult sheep during the inflammatory phase of wound healing (68,69). Even in chronic inflammatory diseases like rheumatoid arthritis HIF-1 is participating. HIF-1α is abundantly present in macrophages of the rheumatoid synovium but not in macrophages from healthy synovia (70). The connection between HIF-1α and wound healing was further demonstrated by Albina and co-workers (55), who isolated inflammatory cells from a subcutaneously implanted polyvinyl alcohol spongue. In this model TNF-α—though not IL-1β—induced HIF-1α accumulation in PMN and macrophages. Possibly, TNF-α stimulates the formation of reactive oxygen species (ROS), which in turn activate HIF-1 (71).

In the context of wound healing angiogenesis is an important process to restore tissue integrity and function. TGF-β is known to induce VEGF gene expression and VEGF synthesis. Signaling by TGF-β receptors is mediated by Smad (artificial name derived from the vertebrate homolog *Sma* from *Caenorhabditis elegans* and *Mad* from *Drosophila melanogaster*) proteins that regulate gene transcription through association and functional cooperation with other transcription factors or transcriptional coactivators. Shih and Claffey (29) have shown that TGF-β induces HIF-1 and AP-1 DNA-binding activity and enhances the expression of HRE and AP-1 reporter genes in HT-1080 fibrosarcoma cells. Furthermore, the amount of HIF-1α protein and c-jun protein in nuclear extracts was increased in this cell type.

A synergistic cooperation between the TGF-β signaling pathway and hypoxia with respect to VEGF expression has been described by Sánchez-Elsner and coauthors (72). Functional analysis of the VEGF-promoter region revealed that intact HREs and binding sites for Smads are necessary for optimal VEGF promoter activity. HIF-1 dependent induction of the VEGF promoter was maximal in the presence of Smad3, and coimmunoprecipitation experiments revealed that HIF-1α physically interacts with Smad3 (72). On the other hand, it has been reported that hypoxia regulates TGF-β_3 in cytotrophoblasts and that a physiologically low oxygen concentration occurs in fetal skin leads to a HIF-1-mediated increase in TGF-β_3 expression (69,73). Thus, there is a positive feedback between HIF-1 and TGF-β.

Clearly, this review is far from providing a complete summary of all articles touching the field of normoxic HIF-1α induction. As mentioned at the beginning, all normoxic HIF-1α-inducing chemical compounds are omitted in the discussion as well as the HIF-1α-interacting proteins. But considering the complete menu presented here, it becomes evident that HIF-1 is much more than only a hypoxia-inducible transcription factor.

References

1. Jiang BH, Rue E, Wang GL, Roe R, Semenza GL. Dimerization, DNA binding, and transactivation properties of hypoxia-inducible factor 1. J Biol Chem 1996; 271:17771–17778.
2. Jiang H, Guo R, Powell-Coffman JA. The *Caenorhabditis elegans* hif-1 gene encodes a bHLH-PAS protein that is required for adaptation to hypoxia. Proc Natl Acad Sci USA 2001; 98:7916–7921.
3. Nambu JR, Chen W, Hu S, Crews ST. The Drosophila melanogaster similar bHLH-PAS gene encodes a protein related to human hypoxia-inducible factor 1α and *Drosophila* single-minded. Gene 1996; 172:249–254.
4. Bacon NC, Wappner P, O'Rourke JF, Bartlett SM, Shilo B, Pugh CW, Ratcliffe PJ. Regulation of the *Drosophila* bHLH-PAS protein Sima by hypoxia: functional evidence for homology with mammalian HIF-1α. Biochem Biophys Res Commun 1998; 249:811–816.
5. Soitamo AJ, Rabergh CM, Gassmann M, Sistonen L, Nikinmaa M. Characterization of a hypoxia-inducible factor (HIF-1α) from rainbow trout. accumulation of protein occurs at normal venous oxygen tension. J Biol Chem 2001; 276:19699–19705.
6. Catron T, Mendiola MA, Smith SM, Born J, Walker MK. Hypoxia regulates avian cardiac Arnt and HIF-1α mRNA expression. Biochem Biophys Res Commun 2001; 282:602–607.
7. Luo G, Gu YZ, Jain S, Chan WK, Carr KM, Hogenesch JB, Bradfield CA. Molecular characterization of the murine HIF-1α locus. Gene Exp 1997; 6:287–299.
8. Wang GL, Semenza GL. General involvement of hypoxia-inducible factor 1 in transcriptional response to hypoxia. Proc Natl Acad Sci USA 1993; 90:4304–4308.
9. Semenza GL. Hypoxia, clonal selection, and the role of HIF-1 in tumor progression. Crit Rev Biochem Mol Biol 2000; 35:71–103.
10. Wang GL, Jiang BH, Rue EA, Semenza GL. Hypoxia-inducible factor 1 is a basic-helix-loop-helix-PAS heterodimer regulated by cellular O_2 tension. Proc Natl Acad Sci USA 1995; 92:5510–5514.
11. Epstein ACR, Gleadle JM, McNeill LA, Hewitson KS, O'Rourke J, Mole DR, Mukherji M, Metzen E, Wilson MI, Dhanda A, Tian YM, Masson N, Hamilton DL, Jaakkola P, Barstead R, Hodgkin J, Maxwell PH, Pugh CW, Schofield CJ, Ratcliffe PJ. *C. elegans* EGL-9 and mammalian homologs define a family of dioxygenases that regulate HIF by prolyl hydroxylation. Cell 2001; 107:43–54.
12. Bruick RK, McKnight SL. A conserved family of prolyl-4-hydroxylases that modify HIF. Science 2001; 294:1337–1340.
13. Masson N, Willam C, Maxwell PH, Pugh CW, Ratcliffe PJ. Independent function of two destruction domains in hypoxia-inducible factor-α chains activated by prolyl hydroxylation. EMBO J 2001; 20:5197–5206.

14. Jaakkola P, Mole DR, Tian YM, Wilson MI, Gielbert J, Gaskell SJ, Kriegsheim A, Hebestreit HF, Mukherji M, Schofield CJ, Maxwell PH, Pugh CW, Ratcliffe PJ. Targeting of HIF-α to the von Hippel–Lindau ubiquitylation complex by O_2-regulated prolyl hydroxylation. Science 2001; 292:468–472.
15. Ivan M, Kondo K, Yang H, Kim W, Valiando J, Ohh M, Salic A, Asara JM, Lane WS, Kaelin Jr. WG. HIFα targeted for VHL-mediated destruction by proline hydroxylation: implications for O_2 sensing. Science 2001; 292:464–468.
16. Yu F, White SB, Zhao Q, Lee FS. HIF-1α binding to VHL is regulated by stimulus-sensitive proline hydroxylation. Proc Natl Acad Sci USA 2001; 98:9630–9635.
17. Ohh M, Park CW, Ivan M, Hoffman MA, Kim TY, Huang LE, Pavletich N, Chau V, Kaelin WG. Ubiquitination of hypoxia-inducible factor requires direct binding to the β-domain of the von Hippel–Lindau protein. Nat Cell Biol 2000; 2:423–427.
18. Maxwell PH, Wiesener MS, Chang GW, Clifford SC, Vaux EC, Cockman ME, Wykoff CC, Pugh CW, Maher ER, Ratcliffe PJ. The tumour suppressor protein VHL targets hypoxia-inducible factors for oxygen-dependent proteolysis. Nature 1999; 399:271–275.
19. Salceda S, Caro J. Hypoxia-inducible factor 1α (HIF-1α) protein is rapidly degraded by the ubiquitin-proteasome system under normoxic conditions: its stabilization by hypoxia depends on redox-induced changes. J Biol Chem 1997; 272:22642–22647.
20. Huang LE, Arany Z, Livingston DM, Bunn HF. Activation of hypoxia-inducible transcription factor depends primarily upon redox-sensitive stabilization of its α subunit. J Biol Chem 1996; 271:32253–32259.
21. Yu AY, Frid MG, Shimoda LA, Wiener CM, Stenmark K, Semenza GL. Temporal, spatial, and oxygen-regulated expression of hypoxia-inducible factor-1 in the lung. Am J Physiol 1998; 275:L818-L826.
22. Jewell UR, Kvietikova I, Scheid A, Bauer C, Wenger RH, Gassmann M. Induction of HIF-1α in response to hypoxia is instantaneous. FASEB J 2001; 15:1312–1314.
23. Zelzer E, Levy Y, Kahana C, Shilo BZ, Rubinstein M, Cohen B. Insulin induces transcription of target genes through the hypoxia-inducible factor HIF-1α/ARNT. EMBO J 1998; 17:5085–5094.
24. Bennett RG, Hamel FG, Duckworth WC. Insulin inhibits the ubiquitin-dependent degrading activity of the 26S proteasome. Endocrinology 2000; 141:2508–2517.
25. Feldser D, Agani F, Iyer NV, Pak B, Ferreira G, Semenza GL. Reciprocal positive regulation of hypoxia-inducible factor 1α and insulin-like growth factor 2. Cancer Res 1999; 59:3915–3918.
26. Richard DE, Berra E, Pouyssegur J. Nonhypoxic pathway mediates the induction of hypoxia-inducible factor 1α in vascular smooth muscle cells. J Biol Chem 2000; 275:26765–26771.
27. Görlach A, Diebold I, Schini-Kerth VB, Berchner-Pfannschmidt U, Roth U, Brandes RP, Kietzmann T, Busse R. Thrombin activates the hypoxia-inducible factor-1 signaling pathway in vascular smooth muscle cells: role of the $p22^{phox}$-containing NADPH oxidase. Circ Res 2001; 89:47–54.
28. Tacchini L, Dansi P, Matteucci E, Desiderio MA. Hepatocyte growth factor signalling stimulates hypoxia inducible factor-1 (HIF-1) activity in HepG2 hepatoma cells. Carcinogenesis 2001; 22:1363–1371.
29. Shih SC, Claffey KP. Role of AP-1 and HIF-1 transcription factors in TGF-β activation of VEGF expression. Growth Factors 2001; 19:19–34.

30. Zick Y. Insulin resistance: a phosphorylation-based uncoupling of insulin signaling. Trends Cell Biol 2002; 11:437–441.
31. Dupont J, LeRoith D. Insulin and insulin-like growth factor 1 receptors: similarities and differences in signal transduction. Horm Res 2002; 55:22–26.
32. Schaeffer HJ, Weber MJ. Mitogen-activated protein kinases: specific messages from ubiquitous messengers. Mol Cell Biol 1999; 19:2435–2444.
33. Mazure NM, Chen EY, Laderoute KR, Giaccia AJ. Induction of vascular endothelial growth factor by hypoxia is modulated by a phosphatidylinositol 3-kinase/Akt signaling pathway in Ha-ras-transformed cells through a hypoxia inducible factor-1 transcriptional element. Blood 1997; 90:3322–3331.
34. Zhong H, Chiles K, Feldser D, Laughner E, Hanrahan C, Georgescu MM, Simons JW, Semenza GL. Modulation of hypoxia-inducible factor 1α expression by the epidermal growth factor/phosphatidylinositol 3-kinase/PTEN/AKT/FRAP pathway in human prostate cancer cells: implications for tumor angiogenesis and therapeutics. Cancer Res 2000; 60:1541–1545.
35. Jiang BH, Zheng JZ, Aoki M, Vogt PK. Phosphatidylinositol 3-kinase signaling mediates angiogenesis and expression of vascular endothelial growth factor in endothelial cells. Proc Natl Acad Sci USA 2000; 97:1749–1753.
36. Sandau KB, Faus HG, Brune B. Induction of hypoxia-inducible-factor 1 by nitric oxide is mediated via the PI 3K pathway. Biochem Biophys Res Commun 2000; 278:263–267.
37. Zundel W, Schindler C, Haas KD, Koong A, Kaper F, Chen E, Gottschalk AR, Ryan HE, Johnson RS, Jefferson AB, Stokoe D, Giaccia AJ. Loss of PTEN facilitates HIF-1-mediated gene expression. Genes Dev 2000; 14:391–396.
38. Jiang BH, Jiang G, Zheng JZ, Lu Z, Hunter T, Vogt PK. Phosphatidylinositol 3-kinase signaling controls levels of hypoxia-inducible factor 1. Cell Growth Differ 2001; 12:363–369.
39. Sodhi A, Montaner S, Miyazaki H, Gutkind JS. MAPK and Akt act cooperatively but independently on hypoxia inducible factor-1α in *rasV12* upregulation of VEGF. Biochem Biophys Res Commun 2001; 287:292–300.
40. Sandau KB, Zhou J, Kietzmann T, Brune B. Regulation of the hypoxia-inducible factor 1alpha by the inflammatory mediators nitric oxide and tumor necrosis factor-alpha in contrast to desferroxamine and phenylarsine oxide. J Biol Chem 2001; 276:39805–39811.
41. Stiehl DP, Jelkmann W, Wenger RH, Hellwig-Bürgel T. Normoxic induction of the hypoxia-inducible factor-1α by insulin and interleukin-1β involves the phosphatidylinositol 3-kinase pathway. FEBS Lett. 2002; 512: 157–162.
42. Laughner E, Taghavi P, Chiles K, Mahon PC, Semenza GL. HER2 (neu) signaling increases the rate of hypoxia-inducible factor 1α (HIF-1α) synthesis: novel mechanism for HIF-1-mediated vascular endothelial growth factor expression. Mol Cell Biol 2001; 21:3995–4004.
43. Richard DE, Berra E, Gothie E, Roux D, Pouyssegur J. p42/p44 mitogen-activated protein kinases phosphorylate hypoxia-inducible factor 1α (HIF-1α) and enhance the transcriptional activity of HIF-1. J Biol Chem 1999; 274:32631–32637.
44. Hur E, Chang KY, Lee E, Lee S-K, Park H. Mitogen-activated protein kinase kinase inhibitor PD98059 blocks the trans-activation but not the stabilization or DNA binding ability of hypoxia-inducible factor-1α. Mol Pharmacol 2001; 59:1216–1224.

45. Miele C, Rochford JJ, Filippa N, Giorgetti-Peraldi S, Van Obberghen E. Insulin and insulin-like growth factor-I induce vascular endothelial growth factor mRNA expression via different signaling pathways. J Biol Chem 2000; 275:21695–21702.
46. Suzuki H, Tomida A, Tsuruo T. Dephosphorylated hypoxia-inducible factor 1α as a mediator of p53-dependent apoptosis during hypoxia. Oncogene 2001; 20:5779–5788.
47. Goldberg M, Zhang HL, Steinberg SF. Hypoxia alters the subcellular distribution of protein kinase C isoforms in neonatal rat ventricular myocytes. J Clin Invest 1997; 99:55–61.
48. Yoshioka K, Clejan S, Fisher JW. Activation of protein kinase C in human hepatocellular carcinoma (HEP3B) cells increases erythropoietin production. Life Sci 1998; 63:523–535.
49. Aragonés J, Jones DR, Martín S, San Juan MA, Alfranca A, Vida F, Vara A, Mérida I, Landázuri MO. Evidence for the involvement of diacylglycerol kinase in the activation of hypoxia-inducible transcription factor 1 by low oxygen tension. J Biol Chem 2001; 276:10548–10555.
50. Kruger EA, Blagosklonny MV, Dixon SC, Figg WD. UCN-01, a protein kinase C inhibitor, inhibits endothelial cell proliferation and angiogenic hypoxic response. Invasion Metast 1998; 18:209–218.
51. Baek SH, Lee UY, Park EM, Han MY, Lee YS, Park YM. Role of protein kinase Cδ in transmitting hypoxia signal to HSF and HIF-1. J Cell Physiol 2001; 188:223–235.
52. Hellwig-Bürgel T, Rutkowski K, Metzen E, Fandrey J, Jelkmann W. Interleukin-1β and tumor necrosis factor-α stimulate DNA binding of hypoxia-inducible factor-1. Blood 1999; 94:1561–1567.
53. El Awad B, Kreft B, Wolber E-M, Hellwig-Burgel T, Metzen E, Fandrey J, Jelkmann W. Hypoxia and interleukin-1β stimulate vascular endothelial growth factor production in human proximal tubular cells. Kidney Int 2000; 58:43–50.
54. Thornton RD, Lane P, Borghaei RC, Pease EA, Caro J, Mochan E. Interleukin 1 induces hypoxia-inducible factor 1 in human gingival and synovial fibroblasts. Biochem J 2000; 350:307–312.
55. Albina JE, Mastrofrancesco B, Vessella JA, Louis CA, Henry jr WL, Reichner JS. HIF-1 expression in healing wounds: HIF-1α induction in primary inflammatory cells by TNF-α. Am J Physiol Cell Physiol 2001; 281:C1971–C1977.
56. Jupp OJ, Mcfarlane SM, Anderson HM, Littlejohn AF, Mohamed AA, MacKay RH, Vandenabeele P, MacEwan, D.J. Type II tumour necrosis factor-alpha receptor (TNFR2) activates c-Jun N-terminal kinase (JNK) but not mitogen-activated protein kinase (MAPK) or p38 MAPK pathways. Biochem J 2001; 359:525–535.
57. Morris SM, Billiar TR. New insights into the regulation of inducible nitric oxide synthesis. Am J Physiol 1994; 266:E829–E839.
58. Sogawa K, Numayama TK, Ema M, Abe M, Abe H, Fujii KY. Inhibition of hypoxia-inducible factor 1 activity by nitric oxide donors in hypoxia. Proc Natl Acad Sci USA 1998; 95:7368–7373.
59. Liu Y, Christou H, Morita T, Laughner E, Semenza GL, Kourembanas S. Carbon monoxide and nitric oxide suppress the hypoxic induction of vascular endothelial growth factor gene via the 5′ enhancer. J Biol Chem 1998; 273:15257–15262.
60. Huang LE, Willmore WG, Gu J, Goldberg MA, Bunn HF. Inhibition of hypoxia-inducible factor 1 activation by carbon monoxide and nitric oxide. Implications for oxygen sensing and signaling. J Biol Chem 1999; 274:9038–9044.

61. Genius J, Fandrey J. Nitric oxide affects the production of reactive oxygen species in hepatoma cells: implications for the process of oxygen sensing. Free Radic Biol Med 2000; 29:515–521.
62. Kimura H, Weisz A, Kurashima Y, Hashimoto K, Ogura T, D'Acquisto F, Addeo R, Makuuchi M, Esumi H. Hypoxia response element of the human vascular endothelial growth factor gene mediates transcriptional regulation by nitric oxide: control of hypoxia-inducible factor-1 activity by nitric oxide. Blood 2000; 95:189–197.
63. Sandau K, Fandrey J, Brüne B. Accumulation of HIF-1α under the influence of nitric oxide. Blood 2001; 97:1009–1015.
64. Kimura H, Weisz A, Ogura T, Hitomi Y, Kurashima Y, Hashimoto K, D'Acquisto F, Makuuchi M, Esumi H. Identification of hypoxia-inducible factor 1 ancillary sequence and its function in vascular endothelial growth factor gene induction by hypoxia and nitric oxide. J Biol Chem 2001; 276:2292–2298.
65. Sheta EA, Trout H, Gildea JJ, Harding MA, Theodorescu D. Cell density mediated pericellular hypoxia leads to induction of HIF-1α via nitric oxide and Ras/MAP kinase mediated signaling pathways. Oncogene 2001; 20:7624–7634.
66. Palmer LA, Gaston B, Johns RA. Normoxic stabilization of hypoxia-inducible factor-1 expression and activity: redox-dependent effect of nitrogen oxides. Mol Pharmacol 2000; 58:1197–1203.
67. Katschinski DM, Le L, Heinrich D, Wagner KF, Hofer T, Schindler SG, Wenger RH. Heat induction of the unphosphorylated form of hypoxia-inducible factor-1α is dependent on heat shock protein-90 activity. J Biol Chem 2002; 4.
68. Scheid A, Wenger RH, Christina H, Camenisch I, Ferenc A, Stauffer UG, Gassmann M, Meuli M. Hypoxia-regulated gene expression in fetal wound regeneration and adult wound repair. Pediatr Surg Int 2000; 16:232–236.
69. Scheid A, Wenger R, Schaffer L, Camenisch I, Distler O, Ferenc A, Cristina H, Ryan H, Johnson R, Wagner K, Stauffer U, Bauer C, Gassmann M, Meuli M. Physiologically low oxygen concentrations determined in fetal skin regulate hypoxia-inducible factor 1 and transforming growth factor b3. FASEB J 2002; 14.
70. Hollander AP, Corke KP, Freemont AJ, Lewis CE. Expression of hypoxia-inducible factor 1alpha by macrophages in the rheumatoid synovium: implications for targeting of therapeutic genes to the inflamed joint. Arthritis Rheum 2001; 44:1540–1544.
71. Haddad JJ, Land SC. A non-hypoxic, ROS-sensitive pathway mediates TNF-α-dependent regulation of HIF-1α. FEBS Lett 2001; 505:269–274.
72. Sánchez-Elsner T, Botella LM, Velasco B, Corbi A, Attisano L, Bernabeu C. Synergistic cooperation between hypoxia and transforming growth factor-β pathways on human vascular endothelial growth factor gene expression. J Biol Chem 2001; 276:38527–38535.
73. Caniggia I, Mostachfi H, Winter J, Gassmann M, Lye SJ, Kuliszewski M, Post M. Hypoxia-inducible factor-1 mediates the biological effects of oxygen on human trophoblast differentiation through TGFβ_3. J Clin Invest 2000; 105:577–587.
74. Cantley LC, Neel BG. New insights into tumor suppression: PTEN suppresses tumor formation by restraining the phosphoinositide 3-kinase/AKT pathway. Proc Natl Acad Sci USA 1999; 96:4240–4245.
75. Daun JM, Fenton MJ. Interleukin-1/Toll receptor family members: receptor structure and signal transduction pathways. J Interferon Cytokine Res 2000; 20:843–855.

7

Brain Microvascular and Metabolic Adaptation to Prolonged Mild Hypoxia

FATON H. AGANI, JUAN CARLOS CHÁVEZ, PAOLA PICHIULE, and JOSEPH C. LAMANNA

Case Western Reserve University School of Medicine
Cleveland, Ohio, U.S.A.

I. Introduction

The need for adequate oxygen and glucose supply as well as removal of carbon dioxide is essential for tissue homeostasis. Thus, maintaining optimal blood flow is critical and complex mechanisms have evolved to match tissue demands under various pathophysiological conditions (1). The energy requirements of the brain are particularly large and immediate. In addition, brain tissue has some characteristic features that make it unique and different from other tissues. For example, an important constraint is that central nervous system neurons are very vulnerable to oxidative damage from reactive oxygen species (ROS). Thus, normal brain function appears to require low levels of oxygen in the brain regions that are not active. However, active areas will require an elevated blood flow while activity persists. Thus, the challenge for the cerebrovascular control mechanisms is to maintain low tissue oxygen during relative quiescence, but at the same time be able to rapidly provide a high oxygen supply during the period of neuronal activation. Although the major regulatory functions occur at the level of the larger brain vessels, the contribution of local regulation of brain flow may have an important impact in meeting the regional requirements.

Table 1 Summary of Effects of Acute Hypoxia on the Brain

1.	Hemoglobin disoxygenation
2.	Decreased tissue pO_2
3.	Increased blood volume
4.	Accelerated capillary mean transit time
5.	Capillary recruitment
6.	Increased blood flow
7.	Increased CMRglu
8.	Hypothermia

The blood flow adjustments necessary to keep this delicate balance of oxygen supply to neurons become even more complex in the conditions of reduced oxygen availability. Table 1 summarizes the acute cerebrovascular and metabolic responses to acute hypoxia.

Hypoxic exposure is manifested in the brain by a shift to a more disoxygenated capillary hemoglobin profile (2), despite increased blood flow (3,4), greater blood volume (2,5,6), and decreased capillary mean transit times (6). Brain parenchymal oxygen pressures fall (5,7,8), but local glucose utilization is increased (4,9,10).

In the central nervous system, capillary density is closely correlated with glucose and oxygen consumption—regions with higher rates of metabolism have higher capillary density (11). Consequently, states associated with chronic decrease in oxygen availability are also accompanied by adaptive brain capillary remodeling. For example, rats and mice exposed to prolonged hypoxia develop an increased brain capillary density (3,12–15). The increase in capillary density occurs as a result of angiogenesis (16,17). The structural changes in the capillaries begin within the first week of hypoxic exposure, and remodeling is completed before 3 weeks of continued exposure (18). Prolonged increases in neuronal activity and, thus, local metabolic demand, also appear to induce an angiogenetic response (19–21).

The process of developing a network of new blood vessels is initiated and controlled by complex signal transduction events. Some of these signals are recently discovered and they reveal a new picture of how activation of particular signaling pathways and components, as well as activation of transcription factors and their target genes, orchestrates tissue remodeling under conditions of reduced oxygen availability. For example, several transcription factors including hypoxia-inducible factor 1 (HIF-1) have been reported to be activated during hypoxia (reviewed in Refs. 22,23). HIF-1 has received the central attention among them as the result of studies conducted during the last few years that have indicated its crucial and dominant role in regulating many aspects of cellular responses to hypoxia. HIF-1 is a general hypoxia-inducible transcription factor (reviewed in Refs. 22–24) that activates multiple target genes whose products promote angiogenesis, erythropoiesis, glucose transport into cells, and glycolysis. Thus, HIF-1 can single-handedly orchestrate complex patterns of adaptational responses in oxygen-deprived cells by activating more than two dozen genes, and this prompted us to investigate its expression in the brain of rats during chronic hypoxia.

In this chapter we will present and attempt to summarize the body of data gathered in our laboratories from our recent investigations that broadly encompass two main directions. The first direction we took is to determine morphological and functional changes that occur during adaptation to hypoxia, at both systemic and local brain levels. Our second goal was to attempt to understand the molecular events that precede these structural and functional adaptations; thus we focused our efforts on the expression of HIF-1 and its target genes in rat brain during chronic hypobaric hypoxia, as well as signal transduction mechanisms that lead to activation of HIF-1 under conditions of reduced oxygen availability.

II. Adaptations to Hypoxia in the Rat

A. Systemic Adaptations

Adult Wistar rats were kept in hypobaric chambers maintained at a pressure of 380 Torr (0.5 ATM, equivalent to 10% normobaric hypoxia). Under these conditions arterial oxygen partial pressure initially falls to about 40–45 mmHg and this hypoxemia initiates short- and long-term physiological adaptation. Rats exposed to this level of hypoxemia survive at least several months and show successful adaptational capability reflected by hyperventilation, an increase in erythrocyte mass, and a decrease in weight gain. Freely breathing rats hyperventilate, which results in $PaCO_2$ falling below 20 mmHg (3), this acute respiratory alkalosis is compensated by bicarbonate excretion, which makes it possible to maintain normal arterial blood pH despite hypocarbia. An increase in hematocrit was gradual and progressive; it was not unusual for hematocrit to reach 60–70 or more after 3 weeks of hypoxia compared to the average hematocrit of control rats of $\sim$52 (3,16–18,25–27). Rats exposed to hypoxia stop gaining weight; in contrast, the control rats gain approximately 7–8 g/day (2% of body weight increment per day). Upon return to normoxia the exposed rats demonstrate weight gain similar to control rats and a return to normal hematocrit (18).

The systemic adaptation mechanisms in this rat model resemble those of humans adapting to high altitude (28–30), except that rat hematocrit is much more elevated than the human. The effect of hypoxia on body weight is also similar to human adaptations in high altitude (31), which is thought to be due to decreased body water content (32).

B. Brain Metabolic Adaptations

Prolonged hypoxia causes an increase in the concentration of the glucose in the brain (150% of control) and lactate (250% of control) and a decreased glycogen (75% of control) (25). The regional metabolic rate for glucose is also increased by 10–40%; however, ATP and phosphocreatine content, as well as intracellular pH, were unchanged consistent with the notion that adaptation was successful (25). A decrease of 16–27% in cytochrome oxidase activity (determined by quantitative immunohistochemistry) could be demonstrated (27), in agreement with reported enzymatic analyses (33,34). The moderate increase of glucose metabolism probably is not sufficient to result in significant increase of ATP production, but increased proton generation from glycolytic ATP hydrolysis could counterbalance hyperventilation-induced decreased tissue CO_2 and maintain acid-base balance (35,36). These findings appear consistent with the notion that brain hypometabolism follows exposure to chronic hypoxia (37,38), probably via hypoxia-associated hypothermia (39), which is also protective in acute hypoxia (40).

C. Brain Vascular Adaptations

Capillary density in the brain increases about 33% after 1 week of hypoxia (18) and continues to increase to about 70% at the end of 3 weeks (3,18). Regional increases in the number of capillaries, an increase in the length of capillary segments (41), as well as an increased capillary sprouting can be demonstrated (16). Similar results have been reported in mice exposed to hypoxia for 28 days (12). The new capillary growth appears to be at first hyperplastic and later hypertrophic (17). In addition, an increase of about 18% in capillary diameter was observed (42), without changes in the wall thickness, basement membrane thickness, or pericyte coverage after 3 weeks of hypoxia. The number of mitochondria per endothelial cell did not change; however, the density of mitochondria in the neuropil was decreased (42). All these changes result, after an initial increase in the cerebral cortical blood flow during the first few hours of hypoxia, in a return to normal tissue blood flow and oxygen tension by 3 weeks of continued environmental hypoxic exposure. The overall increased brain capillary density diminishes intercapillary distances thereby presumably making more oxygen available to the cells (3,12).

Table 2 summarizes the brain structural and functional adaptations to chronic hypoxia.

III. HIF-1 and Brain Adaptations to Hypoxia

Immunoblot assays demonstrated that the HIF-1α protein expression increased after 2 hr of exposure and remained elevated at other time points tested 6 hr, 12 hr, 1 day, and 4, 7, and 14 days after hypoxia. At the next time interval tested (21 days) HIF-1α protein levels had decreased to normoxic control levels. No apparent induction of

Table 2 Summary of Brain Adaptations to Prolonged Mild Hypoxia

1.	Vascular remodeling and restored tissue oxygen tension
2.	Increased glucose transport and metabolism
3.	Glycolytic contribution to tissue acid-base homeostasis
4.	Decreased mitochondrial oxidative energy metabolism

HIF-1β could be observed in brain cortex samples of hypoxic rats. These experiments, shown in Figure 1, demonstrated that HIF-1α protein expression during chronic hypobaric hypoxia returns to baseline (normoxic controls) between 14 and 21 days of exposure (43).

We also analyzed the expression of two HIF-1 downstream target genes: vascular endothelial growth factor (VEGF) and glucose transporter 1 (GLUT-1). It had been previously shown that prolonged hypoxia resulted in increased glucose

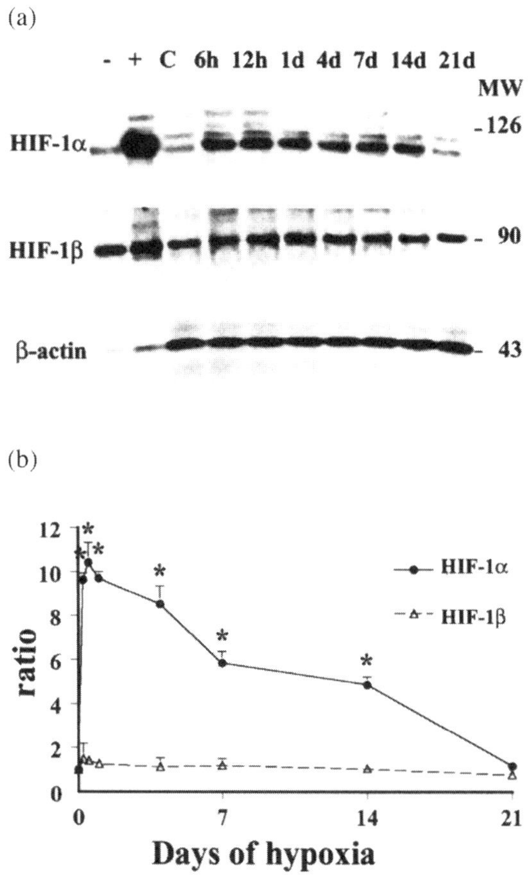

Figure 1 (a) Western blot analysis of HIF-1α, HIF-1β, and β-actin (loading control) in brain cortex of rats exposed to normoxic (C) and hypoxic (6 hr–21 days) conditions. Crude nuclear extracts of Hepal cells exposed to 20% (−) and 1% (+) oxygen were used as negative and positive controls, respectively, for the HIF-1α immunoblots. Indicated on the right are the positions of molecular weight standards. (b) The graph plots the ratio of HIF-1α (filled circle) and HIF-1β (open triangle) optical density relative to control values. Data represent mean $\pm$ SD from three experiments (*$p < 0.05$ vs. control).

influx through the blood-brain barrier consistent with an increase in microvessel GLUT-1 transporter density (18,26). Also, increased brain VEGF levels were associated with hypoxic adaptation in similar rat and mouse models of prolonged hypoxic exposure (44,45).

Northern blot analysis revealed a similar pattern of expression of these genes during chronic hypobaric hypoxia in the rat (43). Upregulation of transcripts for VEGF and GLUT-1 was detected at 6 hr and reached maximum levels at 24 hr of exposure. These expression levels were maintained through 4 days of exposure and were followed by gradual declines at day 7 and 14 and reached normoxic control levels by day 21. We also analyzed by immunoblotting VEGF protein levels and observed that they appropriately followed the pattern of VEGF mRNA levels. Taken together these results show that changes in HIF-1α protein expression during chronic exposure to hypoxia are closely paralleled by the expression of its target genes. Immunolocalization studies of HIF-1α in the rat brain revealed that in the normoxic forebrain there is little detectable HIF-1α however, HIF-1α immunostaining became detectable after exposure to hypoxia throughout the gray matter (43). The immunostaining was primarily nuclear, and involved almost all cells. Immunostaining was detected in small blood vessels and capillaries, suggesting that endothelial cells were also expressing HIF-1α. Nuclear immunostaining was observed in the pial layer, in cells lining the lateral and third ventricles, and in epithelial cells of the chorioid plexus. HIF-1α staining was colocalized with NeuN and GFAP, indicating that neurons and astrocytes express HIF-1α during hypoxia. As expected based on Western blotting data, no HIF-1α immunostaining could be detected in the forebrain after 21 days of hypoxia.

These studies show that HIF-1α protein accumulates in the brain during hypobaric hypoxia, The protein is expressed in different cell types, including neurons, astrocytes, endothelial, and ependymal cells, suggesting a generalized tissue response. This is in agreement with previous reports from our and other laboratories from cell culture data, which revealed that HIF-1α response is ubiquitous and not cell-type-restricted. HIF-1α levels progressively declined during prolonged hypoxia and returned to baseline levels sometime between 14 and 21 days of hypoxic exposure (see Fig. 1). To understand the basis of this return of HIF-1α to baseline levels we first attempted to exclude the possibility that the mechanisms that lead to HIF-1α accumulation were impaired during the prolonged exposure. To this end the rats exposed to hypobaric hypoxia for 21 days were transferred to normobaric hypoxic chambers and exposed to either 10% or 8% oxygen for a further 4 hr. This resulted in increased HIF-1α protein levels in rats exposed to 8% oxygen, but there was no detectable change in HIF-1α levels in rats exposed to 10% oxygen, the equivalent oxygen partial pressure as in the hypobaric exposure at 0.5 ATM. These results indicated that the mechanisms leading to HIF-1α accumulation are intact and that a further hypoxic stimulus was indeed capable of eliciting HIF-1α response. It is likely relevant that oxygen tension in the normal brain is not uniform but follows a log-normal distribution, with values ranging from below 1–2 Torr to above ~50 torr with a mean of ~10–20 torr (46,47), reflecting the expected diffusion field of the tissue capillary network (48). Hypoxia would cause a shift to the left (toward lower oxygen tensions) of the tissue oxygen histogram and increased HIF-1α accumulation throughout different brain regions. After 3 weeks of chronic hypoxic exposure the activated process of angiogenesis would result in increased blood vessel and capillary density thus decreasing diffusional distance from capillaries to the cells. Under these new circumstances presumably the tissue oxygen tension will return to the baseline levels, thus diminishing the signal for HIF-1α activation. Thus, we suggest that combined effects of angiogenesis, polycythemia, and metabolic adaptations during chronic hypoxia are able to restore oxygen tension in the tissue (35). Recent brain tissue oxygen tension measurements in hypoxic adapted rats are consistent with restored levels of tissue oxygen (49). Moreover, this indicates that HIF-1α levels in the brain are regulated by intrinsic signals from oxygen-deprived tissue, rather than by ambient decrease in oxygen availability per se.

IV. Mechanisms of HIF-1α Activation

HIF-1 was initially discovered through its ability to activate the erythropoietin gene during hypoxia via binding to the enhancer element located in the 3′-flanking region of the gene (50). Cloning of HIF-1 revealed a heterodimeric protein consisting of two subunits, HIF-1α and HIF-1β (51,52). HIF-1β, identical to the aryl hydrocarbon receptor nuclear translocator (ARNT), is a common binding subunit of many basic helix-loop-helix (bHLH) heterodimer transcription factors with Per-Arnt-Sim (PAS) domains, besides HIF-1α (53). HIF-1β is constitutively expressed, while HIF-1α is the oxygen-regulated subunit, rapidly accumulating in cells exposed to hypoxia (54). Under well-oxygenated conditions HIF-1α protein undergoes rapid ubiquitination and degradation by the proteasome system (55–57). HIF-1α is ubiquitinated by the von Hippel–Lindau protein (pVHL), which binds directly to the oxygen-dependent degradation domain of HIF-1α and targets it for proteasome degradation (58,59). VHL is associated with elongins B and C, cullin-2, and probably other factors that constitute part of a multiprotein complex (60). Interaction between pVHL and a specific domain of the HIF-1α subunit was reported to be regulated through hydroxylation of a proline residue by a specific prolyl hydroxylase (61,62). Prolyl hydroxylases have a requirement for dioxygen, iron, and 2-oxoglutarate (α-ketoglutarate) (63). In hypoxic conditions prolyl hydroxylation of the HIF-1α subunit is suppressed leading to stabilization of the protein. Models based on a putative hemeprotein with oxygen-sensing properties (64) and NADPH oxidase activity and reactive oxygen species formation (65) have been proposed earlier to explain the ability of cells to sense changes in the O_2 concentration. The process of HIF-1α activation is likely more complex and involves activation of multiple signal transduction pathways (reviewed in Ref. 23).

We (66) and others (67,68), have recently suggested that signaling from mitochondria might contribute to the activation of HIF-1α during hypoxia. Inhibition of complex I with rotenone (66–69), or the upstream part of complex III with myxothiazol (67,68), blocked hypoxic induction of HIF-1α in Hep3B cells. MPP+, a neurotoxin and different class of complex I inhibitor, prevented HIF-1α hypoxic induction in cells of neural origin PC12 and CATH.a (mouse neuroblastoma) (66). In contrast, inhibition at the downstream end of complex III with antimycin A did not prevent HIF-1α hypoxic stabilization (67,68). A mitochondria-based oxygen-sensing model has been suggested based on the observations that although different mitochondrial inhibitors all block electron transport, they exhibit differential effects on ROS generation by mitochondria (67,68). This depends on the site of their action relative to the ubisemiquinone generation site. Ubisemiquinone is a free radical that transfers electrons to molecular O_2, yielding superoxide. Ubisemiquinone is generated by oxidation of ubiquinol in the bc1 complex (complex III); this step is blocked by myxothiazol but not by antimycin A. Inhibitors that blocked electron transport chain upstream from the ubisemiquinone generation site (inhibitors of complex I and of the upstream part of complex III, such as myxothiazol) prevented both ROS formation and hypoxic induction of HIF-1α. In contrast, inhibitors that acted downstream of the ubisemiquinone generation site (inhibitors of downstream complex III, such as antimycin A, as well as inhibitors of complex IV) tended to augment ROS generation and did not prevent hypoxic induction of HIF-1α. Thus, the model suggested that ROS generated at complex III lead to stabilization of HIF-1α during hypoxia. The involvement of ROS in hypoxic stabilization of HIF-1α was further supported by findings that: (1) antioxidants prevented hypoxic induction of HIF-1α, and, (2) inhibitors of the voltage-dependent anion channel (VDAC) on the mitochondrial membrane also prevented hypoxic induction of HIF-1α presumably by blocking egress of superoxide from the mitochondria into the cytosol (67,68).

The notion that mitochondrial electron transport chain activity is necessary for HIF-1α stabilization during hypoxia was recently questioned by two studies showing that in Y0 cells (ethidium bromide-selected cells, which lack respiratory chain activity) HIF-1α hypoxic stabilization and transcriptional activation of HIF-1 target genes were preserved (70,71). In addition, when Y0 cells were treated with rotenone, hypoxic induction of HIF-1α was prevented, and this was taken as an argument that it is an unknown and nonspecific effect of rotenone, rather than its inhibitory effect on complex I, that suppresses HIF-1α stabilization in hypoxic cells (71). However, in both these studies (70,71), cells were exposed to near-anoxic conditions (0% and 0.5% O_2), rather than hypoxic conditions (1 and 1.5% O_2) as in previous studies (66–69). This is addressed in a recent report, which shows that hypoxic but not anoxic stabilization of HIF-1α requires mitochondrial respiratory chain activity (72). If prolyl hydroxylases require O_2 as a substrate, then severe oxygen deprivation close to anoxia will result in HIF-1α stabilization, because in the virtual absence of oxygen the proline hydroxylation would always be inhibited (72). However, the fact that anoxia (or near anoxia) stabilizes HIF-1α in cells does not exclude the possibility that mitochondrial respiratory chain activity might be required for HIF-1α response toward a less severe oxygen deprivation (hypoxia) (72). The exact contribution of mitochondria to oxygen sensing and HIF-1 activation remains to be established.

V. Nonhypoxic Mechanisms of HIF-1 Activation

In addition to hypoxia, HIF-1 can be activated by other factors (reviewed in Refs. 22,24). We were particularly intrigued by the findings that some growth factors, including insulin-like growth factor I (IGF-1), activate HIF-1 in normoxic cells via mechanisms distinct from hypoxia (73–75). IGF-1 is induced in brain after ischemia/reperfusion injury and is considered to have a major protective role in these conditions (reviewed in Ref. 76). Our recent studies show similar temporal patterns of IGF-1 and HIF-1α expression after global ischemia in the brain (77). Both IGF-1 and HIF-1α remained elevated for up to 7 days (77), suggesting that in the absence of hypoxia for prolonged periods after reperfusion, IGF-1 might be a likely candidate that maintains elevated HIF-1α levels in the brain.

VI. HIF-1 Activation in Hypoxia and Ischemia During Aging

We have recently reported a reduced HIF-1α accumulation after hypoxia and global ischemia in the cerebral cortex of old F334 rats compared to young rats (78). This suggests that HIF-1 activation in the brain in the response to hypoxia or ischemia is attenuated as a function of age. Our results are in agreement with those of Frenkel-Denkberg et al. (79). Attenuated HIF-1 response could be due to impaired oxygen-sensing mechanisms during aging or due to a decrease in HIF-1α transcription or a reduced rate of HIF-1α protein synthesis. In addition, it is tempting to speculate that aging-associated mitochondrial impairment may affect HIF-1 activation in senescent brain, which remains to be explored in future.

VII. Summary and Conclusions

Studies in our and other laboratories show that HIF-1α plays a major role in the physiological adaptation of the brain to environmental hypoxic exposure. It is likely that HIF-1α accumulation is responsible for the upregulation of EPO, VEGF, GLUT-1, and the glycolytic enzymes, supporting the brain metabolic adjustments and vascular remodeling of the tissue. In addition, activation of HIF-1 via nonhypoxic mechanisms in the brain opens new questions and promises new answers as to how HIF-1 is regulated in ischemia/reperfusion injury and how HIF-1 activation affects outcome in these conditions. Attenuated HIF-1 response in aging may explain, at least in part, why senescent organisms are much more susceptible to ischemia/reperfusion injury. Exploring the mechanisms of HIF-1 activation in brain adaptation to hypoxia and ischemia/reperfusion injury is important not only for understanding oxygen homeostasis but also for design of new therapies in these conditions.

References

1. Siesjö BK, Berntman L, Nilsson B. Regulation of microcirculation in the brain. Microvasc Res 1980; 19:158–170.
2. Rosenthal M, LaManna JC, Jöbsis FF, Levasseur JE, Kontos HA, Patterson JL Jr. Effects respiratory gases on cytochrome a in intact cerebral cortex: is there a critical PO_2? Brain Res 1976; 108:143–154.
3. LaManna JC, Ven del LM, Farrell RM. Brain adaptation to chronic hypobaric hypoxia in rats. J Appl Physiol 1992; 72:2238–2243.
4. Beck T, Krieglstein J. Cerebral circulation, metabolism, and blood-brain barrier in rats in hypocapnic hypoxia. Am J Physiol 1987; 252:H504–H512.
5. Kreisman NR, Sick TJ, La Manna JC, Rosenthal M. Local tissue oxygen tension–cytochrome a,a_3 redox relationships in rat cerebral cortex in vivo. Brain Res 1981; 218:161–174.
6. Shockley RP, LaManna JC. Determination of rat cerebral cortical blood volume changes by capillary mean transit time analysis during hypoxia, hypercapnia, and hyperventilation. Brain Res 1988; 454:170–178.
7. Leniger-Follert E, Lübbers DW, Wrabetz . Regulation of local tissue pO_2 of the brain cortex at different arterial O_2 pressures. Pflüg Arch 1975; 359:81–95.
8. Metzger H, Erdmann W, Thews G. Effect of short periods of hypoxia, hyperoxia, and hypercapnia on brain O_2 supply. J Appl Physiol 1971; 31:751–759.
9. Torbati D, Greenberg JH, Lambertsen CJ. Regional cerebral glucose metabolic rate during thirty minutes hypoxia of 7% oxygen in adult conscious rats. Neurosci Lett 1986; 65:253–258.
10. Pulsinelli WA, Duffy TE. Local cerebral glucose metabolism during controlled hypoxemia in rats. Science 1979; 204:626–629.
11. Kuschinsky W, Paulson OB. Capillary circulation in the brain. Cerebrovasc Brain Metab Rev 1992; 4:261–286.
12. Boero JA, Ascher J, Arregui A, Rovainen C, Woolsey TA. Increased brain capillaries in chronic hypoxia. J Appl Physiol 1999; 86(4):1211–1219.
13. Opitz E. Increased vascularization of the tissue due to acclimatization to high altitude and its significance for the oxygen transport. Exp Med Surg 1951; 9:389–403.
14. Miller AT Jr, Hale DM. Increased vascularity of brain, heart, and skeletal muscle of polycythemic rats. Am J Physiol 1970; 219:702–704.
15. Diemer K, Henn R. Kapillarvermehrung in der hirnrinde der ratte unter chronischem sauerstoffmangel. Die Natur 1965; 52:135–136.
16. Mironov V, Hritz MA, LaManna JC, Hudetz AG, Harik SI. Architectural alterations in rat cerebral microvessels after hypobaric hypoxia. Brain Res 1994; 660:73–80.
17. Harik SI, Hritz MA, LaManna JC. Hypoxia-induced brain angiogenesis in the adult rat. J Physiol (Lond) 1995; 485.2:525–530.
18. Harik N, Harik SI, Kuo N-T, Sakai K, Przybylski RJ, LaManna JC. Time course and reversibility of the hypoxia-induced alterations in cerebral vascularity and cerebral capillary glucose transporter density. Brain Res 1996; 737:335–338.
19. Black JE, Sirevaag AM, Greenough WT. Complex experience promotes capillary formation in young rat visual cortex. Neurosci Lett 1987; 83:351–355.
20. Black JE, Zelazny AM, Greenough WT. Capillary and mitochondrial support of neural plasticity in adult rat visual cortex. Exp Neurol 1991; 111:204–209.

21. Isaacs KR, Anderson BJ, Alcantara AA, Black JE, Greenough WT. Exercise and the brain: Angiogenesis in the adult rat cerebellum after vigorous physical activity and motor skill learning. J Cereb Blood Flow Metab 1992; 12:110–119.
22. Semenza GL. Regulation of mammalian O_2 homeostasis by hypoxia-inducible factor I. Annu Rev Cell Dev Biol 1999; 15:551–578.
23. Semenza GL. HIF-l and mechanisms of hypoxia sensing. Curr Opin Cell Biol 2001; 13(2):167–171.
24. Wenger RH. Mammalian oxygen sensing, signaling and gene regulation. J Exp Biol 2000; 203 Pt 8:1253–1263.
25. Harik SI, Lust WD, Jones SC, Lauro KL, Pundik S, LaManna JC. Brain glucose metabolism in hypobaric hypoxia. J Appl Physiol 1995; 79(1):136–140.
26. Harik SI, Behmand RA, LaManna JC. Hypoxia increases glucose transport at blood-brain barrier in rats. J Appl Physiol 1994; 77:896–901.
27. LaManna JC, Kutina-Nelson KL, Hritz MA, Huang Z, Wong-Riley MTT. Decreased rat brain cytochrome oxidase activity after prolonged hypoxia. Brain Res 1996; 720:1–6.
28. Lenfant C, Sullivan K. Adaptation to high altitude. N Engl J Med 1971; 284:1298–1309.
29. Dempsey JA, Forster HV. Mediation of ventilatory adaptations. Physiol Rev 1982; 62:262–346.
30. Monge C, León-Velarde F. Physiological adaptation to high altitude: Oxygen transport in mammals and birds. Physiol Rev 1991; 71:1135–1172.
31. Boyer SJ, Blume FD. Weight loss and changes in body composition at high altitude. J Appl Physiol 1984; 57:1580–1585.
32. Fregly MJ, Waters IW. Posthypoxic drinking response of rats. Fed Proc 1966; 25:1220–1226.
33. Chávez JC, Pichiule P, Boero J, Arregui A. Reduced mitochondrial respiration in mouse cerebral cortex during chronic hypoxia. Neurosci Lett 1995; 193:169–172.
34. Caceda R, Gamboa JL, Boero JA, Monge C, Arregui A. Energetic metabolism in mouse cerebral cortex during chronic hypoxia. Neurosci Lett 2001; 301(3):171–174.
35. Lauro KL, LaManna JC. Adequacy of cerebral vascular remodeling following three weeks of hypobaric hypoxia examined by an integrated composite analytical model. In: Nemoto EM, LaManna JC, eds. Oxygen Transport to Tissue XVIII (Advances in Experimental Medicine and Biology, Vol. 411). New York: Plenum Publishing Corp, 1997:369–376.
36. Musch TI, Dempsey JA, Smith CA, Mitchell GS, Bateman NT. Metabolic acids and $[H^+]$ regulation in brain tissue during acclimatization to chronic hypoxia. J Appl Physiol 1983; 55:1486–1495.
37. Hochachka PW, Clark CM, Brown WD, Stanley C, Stone CK, Nickles RJ, et al. The brain at high altitude: hypometabolism as a defense against chronic hypoxia? J Cereb Blood Flow Metab 1994; 14(4):671–679.
38. Mortola JP. Hypoxic hypometabolism in mammals. News in Physiological Sciences 1993; 8:79–82.
39. Wood SC. Interactions between hypoxia and hypothermia. Annu Rev Physiol 1991; 53:71–85.
40. Carlsson C, Hägerdal M, Siesjö BK. Protective effect of hypothermia in cerebral oxygen deficiency caused by arterial hypoxia. Anesthesiology 1976; 44(1):27–35.
41. LaManna JC, Cordisco BR, Kneuse DE, Hudetz AG. Increased capillary segment length in cerebral cortical microvessels of rats exposed to 3 weeks of hypobaric hypoxia. In: Vaupel P, Zander R, Bruley DF, eds. Oxygen Transport To Tissue XV (Advances in

Experimental Medicine and Biology, Vol. 345). New York: Plenum Publishing Corp, 1994:627–632.
42. Stewart PA, Isaacs H, LaManna JC, Harik SI. Ultrastructural concomitants of hypoxia-induced angiogenesis. Acta Neuropathol 1997; 93(6):579–584.
43. Chávez JC, Agani F, Pichiule P, LaManna JC. Expression of hypoxic inducible factor 1α in the brain of rats during chronic hypoxia. J Appl Physiol 2000; 89:1937–1942.
44. Xu F, Severinghaus JW. Rat brain VEGF expression in alveolar hypoxia: possible role in high-altitude cerebral edema. J Appl Physiol 1998; 85(l):53–57.
45. Kuo N-T, Benhayon D, Przybylski RJ, Martin RJ, LaManna JC. Prolonged hypoxia increases vascular endothelial growth factor mRNA and protein in adult mouse brain. J Appl Physiol 1999; 86(l):260–264.
46. Lübbers DW, Baumgärtl H. Heterogeneities and profiles of oxygen pressure in brain and kidney as examples of the pO_2 distribution in the living tissue. Kidney Int 1997; 51:372–380.
47. Sick TJ, Lutz PL, LaManna JC, Rosenthal M. Comparative brain oxygenation and mitochondrial redox activity in turtles and rats. J Appl Physiol 1982; 53:1354–1359.
48. Lübbers DW. Oxygen delivery and microcirculation in the brain. In: Manabe H, Zweifach BW, Messmer K, eds. Microcirculation in Circulatory Disorders. Tokyo: Springer-Verlag, 1988:33–50.
49. Dunn JF, Grinberg O, Roche M, Nwaigwe CI, Hou HG, Swartz HM. Noninvasive assessment of cerebral oxygenation during acclimation to hypobaric hypoxia. J Cereb Blood Flow Metab 2000; 20(12):1632–1635.
50. Semenza GL, Wang GL. A nuclear factor induced by hypoxia via de novo protein synthesis binds to the human erythropoietin gene enhancer at a site required for transcriptional activation. Mol Cell Biol 1992; 12(12):5447–5454.
51. Wang GL, Semenza GL. Purification and characterization of hypoxia-inducible factor 1. J Biol Chem 1995; 270(3):1230–1237.
52. Wang GL, Jiang BH. Rue EA, Semenza GL. Hypoxia-inducible factor 1 is a basic-helix-loop-helix-PAS heterodimer regulated by cellular O2 tension. Proc Natl Acad Sci USA 1995; 92(12):5510–5514.
53. Hoffman EC, Reyes H, Chu FF, Sander F, Conley LH, Brooks BA, Hankinson O. Cloning of a factor required for activity of the Ah (dioxin) receptor. Science 1991; 252(5008):954–958.
54. Huang LE, Arany Z, Livingston DM, Bunn HF. Activation of hypoxia-inducible transcription factor depends primarily upon redox-sensitive stabilization of its alpha subunit. J Biol Chem 1996; 271(50):32253–32259.
55. Salceda S, Caro J. Hypoxia-inducible factor 1 alpha (HIF-1α) protein is rapidly degraded by the ubiquitin-proteasome system under normoxic conditions—its stabilization by hypoxia depends on redox-induced changes. J Biol Chem 1997; 272(36):22642–22647.
56. Huang LE, Gu J, Schau M, Bunn HF. Regulation of hypoxia-inducible factor 1 alpha is mediated by an O-2-dependent degradation domain via the ubiquitin-proteasome pathway. Proc Natl Acad Sci USA 1998; 95(14):7987–7992.
57. Kallio PJ, Wilson WJ, O'Brien S. Makino Y, Poellinger L. Regulation of the hypoxia-inducible transcription factor 1α by the ubiquitin-proteasome pathway. J Biol Chem 1999; 274(10):6519–6525.
58. Maxwell PH, Wiesener MS, Chang GW, Clifford SC, Vaux EC, Cockman ME, Wycoff CC, Pugh CW, Maher ER, Ratcliffe PJ. The tumour suppressor protein VHL targets

hypoxia-inducible factors for oxygen-dependent proteolysis. Nature 1999; 399(6733):271–275.
59. Ohh M, Park CW, Ivan M, Hoffman MA, Kim TY, Huang LE, Pavletich N, Chau V, Kaelin WG. Ubiquitination of hypoxia-inducible factor requires direct binding to the beta-domain of the von Hippel–Lindau protein. Nat Cell Biol 2000; 2(7):423–427.
60. Tanimoto K, Makino Y, Pereira T, Poellinger L. Mechanism of regulation of the hypoxia-inducible factor-1 alpha by the von Hippel–Lindau tumor suppressor N protein. EMBO J 2000; 19(16):4298–4309.
61. Ivan M, Kondo K, Yang H, Kim W, Valiando J, Ohh M, Salic A, Asara J, Lane WS, Kaelin WG. HIF alpha targeted for VHL-mediated destruction by proline hydroxylation: implications for O_2 sensing. Science 2001; 292(5516):464–468.
62. Jaakkola P, Mole DR, Tian YM, Wilson MI, Gielbert J, Gaskell SJ, Kriegsheim A, Hebestreit HF, Mukherji M, Schofield CJ, Maxwell PH, Pugh CW, Ratcliffe PJ. Targeting of HIF-alpha to the von Hippel–Lindau ubiquitylation complex by O_2 regulated prolyl hydroxylation. Science 2001; 292(5516):468–472.
63. Kivirikko KI, Myllyharju J. Prolyl 4-hydroxylases and their protein disulfide isomerase subunit. Matrix Biol 1998; 16(7):357–368.
64. Goldberg MA, Dunning SP, Bunn HF. Regulation of the erythropoietin gene: evidence that the oxygen sensor is a heme protein. Science 1988; 242(4884):1412–1415.
65. Fandrey J, Frede S, Jelkmann W. Role of hydrogen peroxide in hypoxia-induced erythropoietin production. Biochem J 1994; 303 (Pt 2):507–510.
66. Agani F, Pichiule P, Chávez JC, LaManna J. The role of mitochondria in the regulation of hypoxia-inducible factor 1 expression during hypoxia. J Biol Chem 2000; 275:35863–35867.
67. Chandel NS, Maltepe E, Goldwasser E, Mathieu CE, Simon MC, Schumacker PT. Mitochondrial reactive oxygen species trigger hypoxia-induced transcription. Proc Natl Acad Sci USA 1998; 95(20):11715–11720.
68. Chandel NS, McClintock DS, Feliciano CE, Wood TM, Melendez JA, Rodriguez AM, Schumacker PT. Reactive oxygen species generated at mitochondrial complex III stabilize hypoxia-inducible factor-1alpha during hypoxia: a mechanism of O_2 sensing. J Biol Chem 2000; 275(33):25130–25138.
69. Hirota K, Semenza GL. Rac1 activity is required for the activation of hypoxia-inducible factor 1. J Biol Chem 2001; 276(24):21166–21172.
70. Srinivas V, Leshchinsky I, Sang N, King MP, Minchenko A, Caro J. Oxygen sensing and HIF-1 activation does not require an active mitochondrial respiratory chain electron-transfer pathway. J Biol Chem 2001; 276(25):21995–21998.
71. Vaux EC, Metzen E, Yeates KM, Ratcliffe PJ. Regulation of hypoxia-inducible factor is preserved in the absence of a functioning mitochondrial respiratory chain. Blood 2001; 98(2):296–302.
72. Schroedl C, McClintock D, Budinger GRS, Chandel N. Hypoxic but not anoxic stabilization of HIF-1α requires mitochondrial reactive oxygen species. Am J Physiol 2002; 283:L922–L931.
73. Agani F, and Semenza G. Mersalyl is a novel inducer of vascular endothelial growth factor gene expression and hypoxia-inducible factor 1 activity. Mol Pharmacol 1998; 54(5):749–754.
74. Feldser D, Agani F, Iyer NV, Pak B, Ferreira G, Semenza GL. Reciprocal positive regulation of hypoxia-inducible factor 1alpha and insulin-like growth factor 2. Cancer Res 1999(Aug 15); 59(16):3915–3918.

75. Zelzer E, Levy Y, Kahana C, Shilo BZ, Rubinstein M, Cohen B. Insulin induces transcription of target genes through the hypoxia-inducible factor HIF-1alpha/ARNT. EMBO J 1998; 17(17):5085–5094.
76. Lipton P. Ischemic cell death in brain neurons. Physiol Rev 1999; 79(4):1431–1568.
77. Chávez JC, LaManna J. IGF-1 and HIF-1α accumulation induced by transient global ischemia. J Cereb Blood Flow Metab 2001; 21(Suppl 1):S389.
78. Chávez JC, LaManna JC. HIF-1α accumulation in the rat brain in response to hypoxia and ischemia is attenuated during aging. ISOTT 2001; 21 Proceedings. In: Wilson DF, Biaglow J, Evans SM, Pastuszko A, eds. New York: Kluwer Academic/Plenum. In press.
79. Frenkel-Denkberg G, Gershon D, Levy AP. The function of hypoxia-inducible factor 1 (HIF-1) is impaired in senescent mice. FEBS Lett 1999; 462(3):341–344.

8

Molecular Adaptation to Hypoxia

KAREN A. SETA, YONG YUAN, ZACHARY SPICER, GANG LU, and DAVID E. MILLHORN

Genome Research Institute
University of Cincinnati College of Medicine
Cincinnati, Ohio, U.S.A.

I. Introduction

Mammalian cells are critically dependent upon oxygen for survival. Neurons in particular are extremely susceptible to hypoxic injury (1,2). In contrast, oxygen-sensing cells are able to survive and function even during sustained exposure to severe hypoxia. Oxygen sensing is a special phenotype that functions to detect changes in oxygen tension and transduce this signal into organ system functions that enhance the delivery of oxygen to tissue in a wide variety of different organisms. The mechanisms that mediate adaptation and tolerance to hypoxia in the oxygen-sensing phenotype are unknown, but most assuredly involve specific signaling events and the coordinated expression of numerous genes. Historically, progress in understanding adaptive mechanisms in O_2-sensing cells has been hampered by their relative scarcity and by their obscure locations in various tissues. These problems prevented rigorous molecular studies of O_2-sensing cells in situ.

Our approach to overcoming these problems was to identify a clonal cell line that exhibited functions consistent with those observed in O_2-sensing cells and to use these cells as a model system for studying the molecular basis of adaptation and tolerance to hypoxia in O_2-sensing cells. We established the dopaminergic pheochromocytoma (PC12) cell line as a reliable and valuable model for this purpose (3–7). In this chapter we shall describe and discuss outcomes from our studies on signal transduction and gene expression using this cell line and relate these findings to function in other O_2-sensing cells.

II. Hypoxia-Regulated Signal Transduction Pathways

The key event in most signaling pathways is the interaction between a chemical factor and a receptor molecule. In some cases the receptor is associated with the plasma membrane and in other cases it is located in the cytoplasm or nucleus. Regardless of receptor location, its activation by the chemical factor sets in motion a signaling cascade that ultimately regulates cell function including gene expression. Even the idea that oxygen can act as a chemical factor to specifically regulate cellular signaling is somewhat controversial. Nevertheless there is growing evidence that this is indeed the case.

For the purpose of this discussion, O_2-sensing cells will be divided into two groups: those that depolarize when exposed to reduced O_2 (e.g., excitable cells), and those that do not (e.g., nonexcitable cells). This is an important distinction because membrane depolarization can lead to changes in calcium flux across the plasma membrane, which in turn can lead to changes in intracellular free Ca^{2+} and activation of Ca^{2+}-dependent signaling pathways. Examples of excitable and nonexcitable O_2-sensing cells are the carotid body type I cells (8) and the erythropoietin-secreting cells of the kidney, respectively. Theoretically, the type I cells can have the same signaling capabilities as the erythropoietin-secreting cells, but because the erythropoietin-secreting cells do not depolarize they cannot have all the signaling capabilities as type I cells.

We have identified several critical signaling pathways in PC12 cells that regulate gene expression during hypoxia. These include the cAMP-protein kinase A (6), Ca^{2+}-calmodulin (6), p42/44 mitogen-activated protein kinase (MAPK) (9), stress-activated protein kinase (SAPK; p38 kinase) (10), and phosphatidylinositol 3-kinase-Akt (11) pathways as regulators of gene expression. We have also shown that several of these pathways regulate hypoxia-specific transcription factors that are members of the hypoxia-inducible factor (HIF) family.

A. Membrane Depolarization and Calcium Influx

Exposure of PC12 cells to reduced O_2 leads to membrane depolarization and an increase in intracellular free Ca^{2+} (4), which is similar to the response measured in carotid body type I cells (8–12). Whole-cell voltage-clamp experiments revealed that the membrane depolarization in PC12 cells during hypoxia is mediated by an O_2-sensitive potassium channel (K_{O2}) (13). We found that the outward K^+ current was inhibited in response to a reduction in O_2 (4) (Fig. 1). This reduction was more pronounced with increasing severity of hypoxia. The stepwise inhibition of the K_{O2} current during graded hypoxia was associated with a progressive membrane depolarization. Electrophysiological and pharmacological studies revealed that the hypoxia-sensitive K^+ channel in PC12 cells was most likely the Kv1.2 channel (13). For example, we found that Kv1.2 antibodies dialyzed through the patch recording

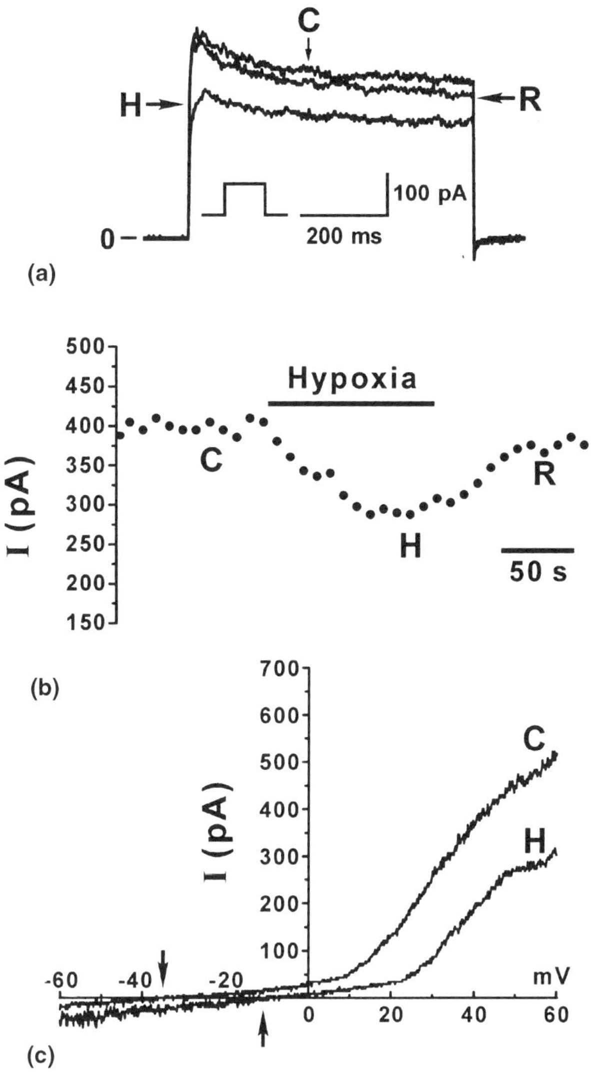

Figure 1 Effect of reduced O_2 on potassium (K^+) current (I_K). (a) Superimposed current traces recorded during control period (C, normoxia, 150 mmHg), after steady-state inhibition by hypoxia (H, 0 mmHg), and after return to normoxic conditions (R, 150 mmHg). Cells were depolarized to +50 mV for 800 msec from a holding potential of −90 mV. (b) Time course of the hypoxic inhibition of outward current (same cell as in panel a). Sampling rate was 0.1 Hz. Steady-state responses to changes in P_{O2} were attained <1 min after exposure. (c) Current-voltage relationships recorded during control period (C) and after steady-state inhibition by hypoxia (H). Cell voltage was ramped from −60 to +60 mV over 5 sec and repeated every 10 sec.

pipette completely blocked the K_{O2} current, while antibodies against another K^+ channel (Kv2.1) had no effect. We also found that introduction of the Kv1.2 channel gene into *Xenopus* oocytes led to expression of an O_2-sensitive K^+ current (13).

Depolarization and the resulting increase in intracellular Ca^{2+} levels that occur during hypoxia in PC12 cells suggest that a variety of Ca^{2+}-dependent protein kinases and phosphatases, including Ca^{2+}/calmodulin-dependent kinases, may also be regulated (14). Activation of these Ca^{2+} signaling enzymes can have profound effects on cell function including altered gene expression and altered cell phenotype. Depolarization of PC12 cells is known to activate Ca^{2+}/calmodulin-dependent protein kinases and to stimulate cyclic AMP response element (CRE)-dependent gene transcription (14,15). We examined the role of increased intracellular free Ca^{2+} on the expression of tyrosine hydroxylase (TH), the rate-limiting enzyme in the synthesis of dopamine and other catecholamine neurotransmitters (16). We had previously showed that TH is regulated by reduced O_2 in both carotid body type I cells and PC12 cells (3,17). We found that removal of extracellular Ca^{2+} or chelation of intracellular Ca^{2+} with BAPTA-AM inhibited the increase in cyosolic free Ca^{2+} and blocked the hypoxia-induced expression of TH (Fig. 2). In addition, we found that blockade of the L-type Ca^{2+} channel with nifedipine partially inhibited the increased expression of TH in PC12 cells during hypoxia. These findings support the possibility that an increase in intracellular free Ca^{2+} in the excitable PC12 and type I cells plays a role in regulating gene expression and the cellular response to hypoxia.

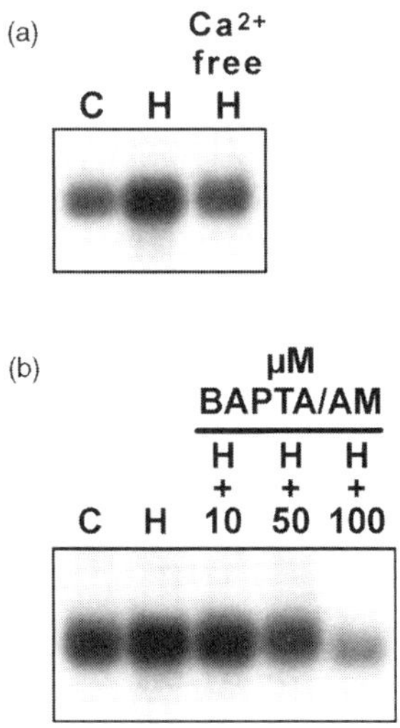

Figure 2 Effect of removal of intracellular and extracellular calcium on induction of tyrosine hydroxylase mRNA in PC12 cells by hypoxia. (a) Cells were exposed to normoxia (C, 21% O_2, 6 hr) or hypoxia (H, 5% O_2, 6 hr) in normal serum-free medium or to hypoxia in Ca^{2+}-free medium supplemented with 1 mM EGTA. (b) Same experiment as in panel a except, in addition, cells were preloaded with the indicated amount of BAPTA-AM for 40 min prior to hypoxia exposure to chelate intracellular calcium.

We have also investigated the Ca^{2+}-dependent signaling enzymes that might mediate changes in gene expression during hypoxia. Interestingly, we found that prolonged exposure (>6 hr) to 5% O_2 led an approximate 50% downregulation of Ca^{2+}/calmodulin-dependent kinase II and protein kinase A activities (6). This would seem to indicate that these signaling systems are not involved in regulation of increased gene expression during hypoxia. However, because these signaling systems mediate a wide variety of cellular functions that are unrelated to hypoxia, it is possible that these other functions are curtailed and that the regulation of hypoxia-related activities is spared. We tested this possibility by measuring the effect of calmodulin blockade on the expression of a reporter gene that was engineered to be responsive to hypoxia (9). We found that this reporter gene is stimulated by reduced O_2 in PC12 cells and, importantly, that pharmacological inhibition of calmodulin prevented its induction by hypoxia. This finding supports the notion that even though the overall activity of the Ca^{2+}/calmodulin system is inhibited by hypoxia, this system is still involved in regulation of hypoxia-responsive genes and the cellular adaptation to reduced O_2. The inhibition of the overall Ca^{2+}/calmodulin activity during hypoxia is probably related to an overall reduction in nonessential metabolic cellular activities, which is also an important adaptive mechanism (18). We have also examined the potential role of protein kinase C (PKC) in hypoxia-induced gene expression in PC12 cells (16). We found that blockade of calcium-dependent PKC inhibited expression of TH during hypoxia in PC12 cells. Taken together these data suggest that the increase in intracellular free Ca^{2+} that occurs in PC12 and type I cells during hypoxia is a key mechanism for hypoxia-induced gene expression and probably plays a major role in adaptation to hypoxia in excitable O_2-sensing cells.

B. The Stress- and Mitogen-Activated Pathways

The stress (SAPK)- and mitogen (MAPK)-activated pathways play critical roles in responding to cellular stress and promoting cell growth and survival (19,20). SAPKs and MAPKs are the downstream components of three-member protein kinase modules (21). Five homologous subfamilies of these kinases have been identified, and the three major families include p38/SAPK2, c-Jun N-terminal kinase (JNK), and p42/p44 MAPK (also referred to as extracellular signal-regulated kinase, ERK). In general, the stress-activated protein kinases (p38 and JNK) are activated primarily by noxious environmental stimuli such as ultraviolet light, osmotic stress, inflammatory cytokines, and inhibition of protein synthesis (22–26). Increasing evidence suggests that under certain conditions these pathways (p38 and JNK) can also be activated by mitogenic and neurotrophic factors (27,28). In contrast, p42/p44 MAP kinases are stimulated primarily by mitogenic and differentiative factors in a Ras-dependent manner (29,30). There is evidence that the p42/p44 ERKs can also be activated by environmental stimuli (19–21). We therefore embarked on a series of studies to determine if hypoxia, a prevalent physiological stressor in many disease states, regulates the activity of the SAPK and MAPK signaling pathways.

To investigate the effects of hypoxia on the stress- and mitogen-activated signaling pathways, PC12 cells were exposed to 5% O_2 for various times between 20 min and 6 hr. Western blot analysis revealed that exposure to hypoxia progressively induced phospho-p38 immunoreactivity in two closely migrating bands (10). These two bands represented p38α and p38γ phosphoproteins. When PC12 cells that were transfected with FLAG epitope-tagged versions of one of the different isoforms of p38 (α, β, β2, δ, or γ) were exposed to hypoxia, again only the p38α and p38γ isoforms were activated (Fig. 3) (10). Total p38 levels remained unchanged.

We next evaluated the effect of hypoxia on the p42/p44 MAPK pathway in PC12 cells (10). Western blot analysis showed that exposure to 5% O_2 for 6 hr induced a marked increase in tyrosine phosphorylation of both p42 and p44 MAPK, while the total MAPK was unchanged by hypoxia (Fig. 4). The phosphorylation of p42 and p44 by hypoxia was somewhat less than that measured when PC12 cells

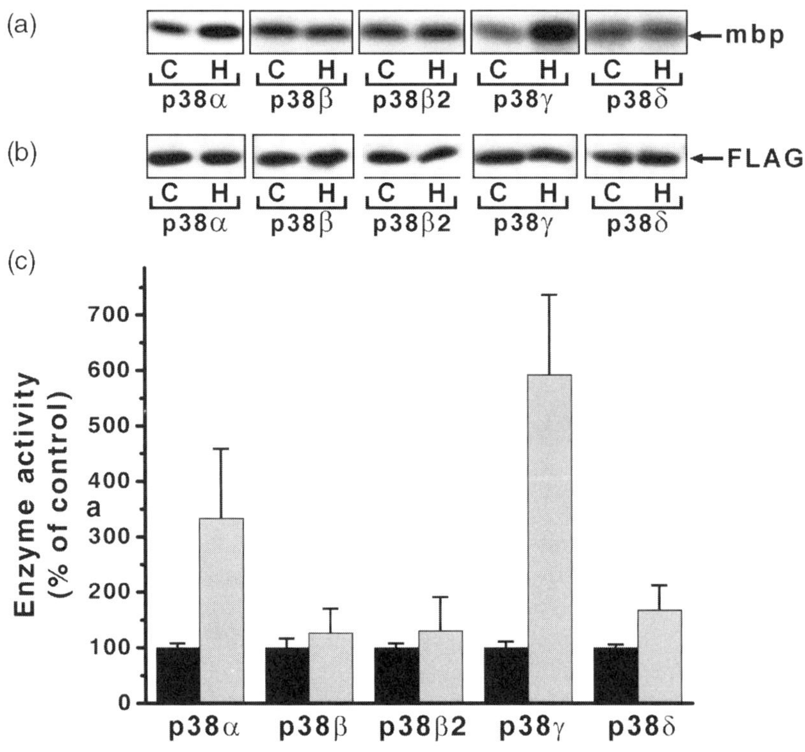

Figure 3 Effect of hypoxia on p38α and p38γ activity. PC12 cells were transfected with FLAG-tagged p38 isoforms and exposed to either normoxia (C, 21% O_2) or hypoxia (H, 5% O_2) for 6 hr. (a) Enzyme activity of various p38 isoforms was determined in immune complex kinase assays by determining the amount of 32p incorporated into myelin basic protein (mbp). (b) Whole-cell lysates were immunoblotted with antibody against the FLAG tag. Protein kinase activity of the various p38 isoforms after exposure to normoxia (black bars) or hypoxia (gray bars) is expressed as average percent of control $\pm$ SE ($n = 6$–9).

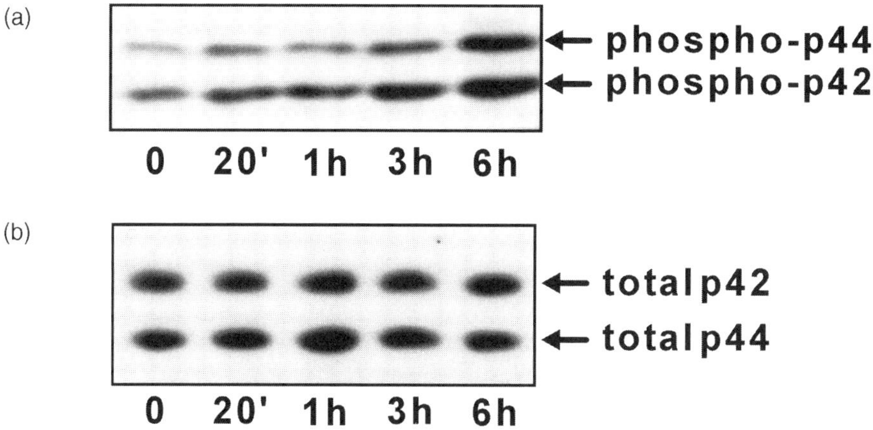

Figure 4 Increased phosphorylation of p42 and p44 MAPK during hypoxia in PC12 cells. PC12 cells were exposed to hypoxia (21% O_2) for the times indicated. Whole-cell lysates were immunoblotted with an antibody that recognizes phosphorylated p42/p44 (a) or total (phosphorylation state-independent) p42/p44 (b).

were exposed to prototypical activators of MAPK [e.g., nerve growth factor, NGF; ultraviolet (UV) light].

In contrast, hypoxia did not alter JNK enzyme activity significantly (10), whereas exposure of PC12 cells to UV light did increase JNK activity. Thus, this SAPK pathway does not appear to be involved in mediating the response to hypoxia in PC12 cells.

The cyclin D1 gene is one known target of p38 (31). It has been shown that activation of p38 kinase in CC139 cells leads to downregulation of cyclin D1 gene expression (31). We found that exposure to hypoxia (5% O_2) for periods ranging from 3 to 24 hr led to a progressive decrease in cyclin D1 gene expression in PC12 cells (10). This decrease was partially prevented in PC12 cells stably transfected with a kinase-inactive form of p38γ and by pharmacological blockade of p38α.

Cyclin D1 is a G_1 cyclin whose synthesis and associated cyclin-dependent kinase activity are required for progression through the G_1 phase of the cell cycle (32,33). Our finding that hypoxia induces a downregulation of cyclin D1 suggested that hypoxia might cause cells to accumulate in the G_0/G_1 phase of the cell cycle. We found that exposure to hypoxia for 24 hr did cause an increase (17%) in the number of cells in G_o/G_1. Moreover, treatment with a drug (SB203580) that blocks p38α was able to partially reverse the accumulation of cells in G_o/G_1. In addition, stable transfection of PC12 cells with the kinase-inactive form of p38γ partially prevented the accumulation of PC12 cells in G_o/G_1. Thus, activation of the p38 pathway (p38α and p38γ) and down-regulation of cyclin D1 by hypoxia can have a significant effect on progression of cells through the cell cycle, which might be an important adaptive mechanism in PC12 cells and in other cells that are undergoing cell division (e.g., embryonic development or tumors).

These studies show that hypoxia causes specific regulation of the SAPK and p42/p44 MAPK pathways, but not the JNK pathway in PC12 cells. Activation of these pathways may play an important role in hypoxia regulation of gene expression and in molecular adaptation to hypoxia. The downstream transcription factors and protein kinases that are targeted by these pathways are beginning to be elucidated. Nevertheless, little is known about the specific genes that are regulated by these pathways in response to extracellular stress. Further investigation is needed to gain insights into the specific mechanisms by which these pathways are activated by hypoxia, and to determine the downstream targets of these pathways during hypoxia.

C. The Phosphatidylinositol 3-Kinase (PI3K)-Akt Pathway

The cellular processes related to survival and apoptosis are mediated in part by the PI3K-Akt pathway. Akt (also known as protein kinase B) is a cytosolic serine/threonine kinase that has been shown to be critical for cell survival under adverse conditions (34,35). Akt has been shown to block apoptosis induced by a number of "death stimuli," and Akt is required for growth factor–mediated cell survival in neurons. In a number of cell types, withdrawal of growth factors triggers programmed cell death, and activation of Akt can block this process (34–36). There is growing evidence that certain stressors, including osmotic stress, H_2O_2, and sodium arsenite, can activate Akt (37,38). However, this is somewhat of a controversial area since other studies have failed to find effects of various stress stimuli on Akt (39).

The cellular mechanisms involved in Akt regulation are beginning to be understood (see Fig. 5). Akt has a pleckstrin homology (PH) domain that preferentially binds Ptdlns(3,4,5)P_3 and Ptdlns(3,4)P_2, phospholipids that are generated by PI3K. Akt translocates from the cytosol to the inner leaflet of the plasma membrane to bind Ptdlns(3,4,5)P_3 and Ptdlns(3,4)P_2 (40). At the lipid bilayer, Akt becomes phosphorylated by PDK, an upstream kinase (34,35). Akt becomes phosphorylated on Ser^{473} (in the regulatory domain) and Thr^{308} (in the catalytic domain), resulting in activation of the enzyme. Once activated, Akt detaches from the plasma membrane and translocates to the nucleus (40). In the nucleus Akt can phosphorylate a variety of transcription factors and other regulatory proteins (34,35), including glycogen synthase kinase, Bad, caspase 9, the Forkhead family of transcription factors, and IκB kinase. Some of these transcription factors are known to be involved in mediating cell death (34).

Given the role of Akt in promoting cell survival under adverse conditions, it was of interest to investigate the effects of hypoxia on Akt in PC12 cells (11). We measured a dramatic increase in phospho-Akt (Ser^{473}) that persisted after 24 hr of exposure to 5% O_2. Activation of Akt, and its associated antiapoptotic effects, is typically mediated through PI3K (34,35). Treatment of PC12 cells with wortmannin, a specific inhibitor of PI3K, prior to exposure to hypoxia, completely abolished the effect of hypoxia on phosphorylation of Akt (Fig. 6). Wortmannin did not alter total Akt protein levels (Fig. 6).

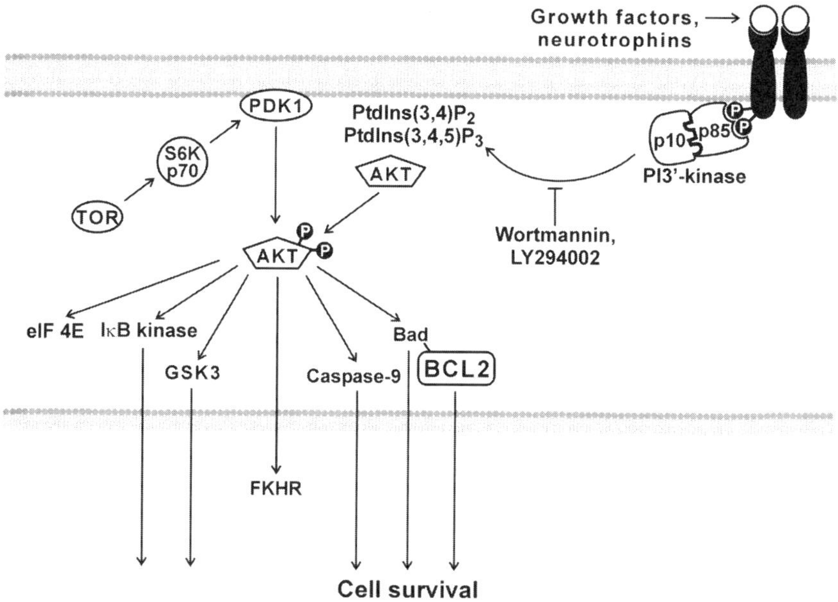

Figure 5 The PI3K-Akt signal transduction pathway.

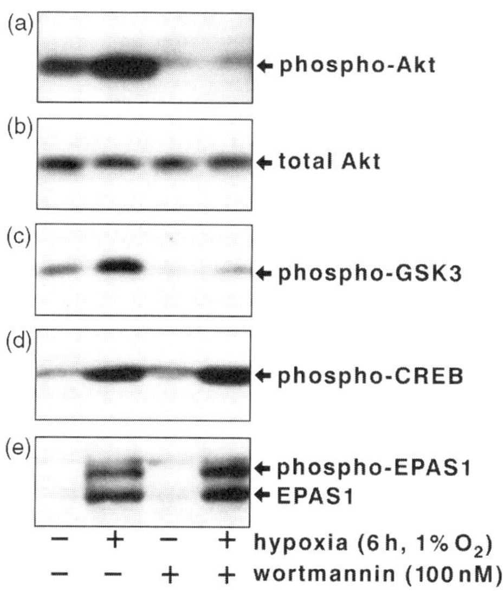

Figure 6 Effect of hypoxia on the phosphorylation of Akt. PC12 cells were pretreated with either wortmannin (100 nM), a specific inhibitor of PI3-Kinase, or vehicle (0.01% DMSO) for 1 hr prior to exposure to either normoxia (21% O_2) or hypoxia (5% O_2) for 6 hr. Whole-cell lysates were then immunoblotted for either phospo-Akt (a), total Akt (b), phospho-GSK3 (c), phospho-CREB (d), or EPAS1 (e). Note that CREB phosphorylation and EPAS1 phosphorylation and accumulation persist in the presence of wortmannin.

Glycogen synthase kinase-3 (GSK-3) is a well-characterized substrate of Akt (41). GSK-3 exists in two homologous isoforms, GSK-3α and GSK-3β. Western blot analysis revealed that hypoxia induced an increase in phospho-GSK-3 similar to that seen in Akt (Fig. 6). We also found that pretreatment with wortmannin blocked the hypoxia-induced phosphorylation of GSK-3 in a manner identical to that observed for Akt.

There is growing evidence that Akt and GSK-3 play an important role in mediating cell survival during hypoxia. It has been suggested that activation of Akt leads to stabilization of the hypoxia-inducible transcription factor HIF-1α (42). The HIF family of transcription factors binds to hypoxia response elements (HREs) localized with the regulatory region of hypoxia responsive genes (43). It is of interest to note that the HIF-1β subunit (also known as aryl hydrocarbon nuclear translocator, ARNT) contains an Akt consensus phosphorylation site (34). Mazure and coworkers (44) demonstrated that hypoxia-induced stimulation of an HRE-containing gene, VEGF, was attenuated by wortmannin.

Phosphorylation of GSK-3 results in its enzymatic inactivation (41). It is important to note that GSK-3 regulates a variety of events involved in cellular metabolism, including glycogen synthase (41,45). In addition to its metabolic functions, GSK-3 has been shown to regulate cell survival in PC12 cells (46). Transfection of constitutively active GSK-3 drives PC12 cells into apoptosis, whereas transfection with kinase-inactive GSK-3 blocks apoptosis (46). Thus, the activation of Akt (i.e., phosphorylated Akt) and the inactivation of GSK-3 (i.e., phosphorylated GSK-3) by hypoxia are consistent with promoting cell survival and tolerance to hypoxia in PC12 cells. Taken together, these studies suggest that the PI3K-Akt-GSK-3 signaling pathway may represent an important aspect of the cellular response to hypoxia.

D. Other Hypoxia-Regulated Pathways

A Novel CREB Kinase Pathway

Multiple signaling pathways converge at the level of cyclic AMP (cAMP) response element-binding protein (CREB), a transcription factor that regulates expression of genes that contain a specific DNA sequence called the cAMP response element (CRE) (47). CREB mediates cellular responses to a variety of physiological signals, including neurotransmitters, depolarization, synaptic activity, mitogenic and differentiative factors, and stressors (14,48–52). Upon phosphorylation of Ser^{133}, CREB can facilitate transcriptional activation of genes containing the CRE motif (53). Several protein kinases, including protein kinase A, calcium/calmodulin-dependent protein kinases, protein kinase C, RSK-2, and MAPKAP kinase-2, have been shown to mediate phosphorylation of CREB (15,48,51,54–56). Given the major role for CREB in regulating genes that mediate a wide variety of cellular functions to different stimuli, we investigated the potential role of CREB in regulating the cellular response to reduced O_2 (57).

Hypoxia induced a robust Ser^{133} phosphorylation of CREB (Fig. 7) that was greater than that produced by two prototypical stimuli used to activate CREB,

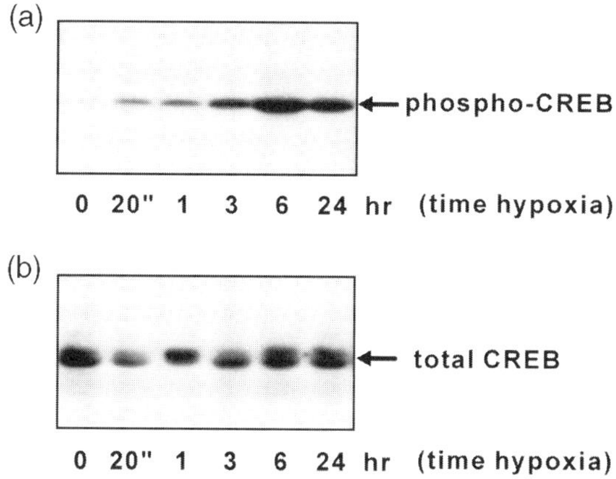

Figure 7 Effect of hypoxia on CREB phosphorylation. PC12 cells were exposed to either normoxia or hypoxia (5% O_2) for the indicated time. Whole-cell lysates were immunoblotted with an antibody to Ser^{133} phospho-CREB (a) or total (phosphorylation-state-independent) CREB (b).

forskolin (activation of PKA), and KCl-induced depolarization (57). To our knowledge, this is the first evidence for regulation of CREB by hypoxia. This suggests that the phosphorylation of CREB may be one mechanism by which PC12 cells adapt to hypoxia. An important finding was that the hypoxia-induced phosphorylation of CREB was not mediated by any of the previously known pathways that activate CREB (Fig. 8), including PKA- and Ca^{2+}-dependent protein kinases. This conclusion is based on our finding that phosphorylation of CREB by hypoxia persisted completely in PKA-deficient PC12 cells and in the absence of both extracellular and intracellular Ca^{2+}. Hypoxia-induced phosphorylation of CREB was not attenuated by blockade of the Ca^{2+}-dependent isoforms of PKC. We also found that the MAPK, p38 stress-activated protein kinase, and MAPKAP kinase pathways are not involved in the hypoxia-induced phosphorylation of CREB (57).

It has been established previously that TH gene expression is induced by hypoxia in both type I carotid body cells and PC12 cells (3,17). The TH gene contains a CRE that has been shown to be critical for cAMP- and Ca^{2+}-induced activation of gene expression (58–60). We found that mutation of the CRE significantly attenuated the hypoxia-induced activation of a TH-CAT reporter gene construct (Fig. 9). We previously showed that the TH gene contains several sequence motifs that participate in hypoxia-induced activation of gene expression, including AP1- and hypoxia response element (HRE)-like cis elements located between −284 and −190 nucleotides relative to transcription start site (3). Thus, one or more of these other upstream regulatory elements presumably mediates the residual activation of the TH gene in the absence of the CRE, and CREB is likely to be insufficient to mediate the entire induction by hypoxia.

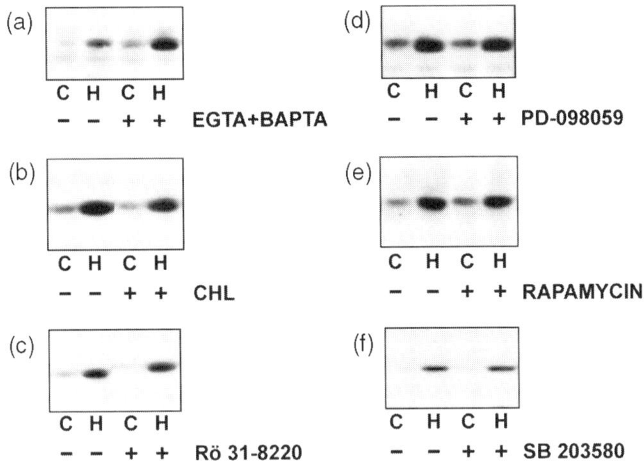

Figure 8 CREB phosphorylation by hypoxia does not require Ca^{2+}, PCK, RSK-2, MAPK, or p38. Cells were pretreated with various drugs or vehicle (0.1% dimethyl sulfoxide) as indicated. Cells were then exposed to either normoxia (C, 21% O_2) or hypoxia (H, 5% O_2) for 6 hr, and whole-cell lysates were immunoblotted with an antibody specific for Ser^{133} phospho-CREB. (a) Cells were preincubated for 40 min in serum-free medium in the presence of Ca^{2+} and vehicle (−) or in serum-free medium formulated without Ca^{2+} and supplemented with 1 mM EGTA + 100 μM BAPTA-AM (+). The medium was then replaced (minus drug or vehicle) and cells were exposed to either normoxia or hypoxia. (b) Cells were pretreated for 40 min in serum-free medium with either vehicle (−) or 20 μM chelerythrine chloride, an inhibitor of PKC (CHL; +), and exposed to either normoxia or hypoxia. (c) Cells were pretreated for 40 min in serum-free medium with either vehicle (−) or 0.3 μM Ro 31-8220, an inhibitor of RSK and p70 S6 kinase (+), and exposed to either normoxia or hypoxia. (d) Cells were pretreated for 40 min in serum-free medium with either vehicle (−) or 50 μM PD098059, an inhibitor of MEK1 and the downstream MAPKS (+), and exposed to either normoxia or hypoxia. (e) Cells were pretreated for 1 hr in serum-free medium with either vehicle (−) or 10 nM rapamycin, an inhibitor of p70 S6 kinase (+), and exposed to either normoxia or hypoxia. (f) Cells were pretreated for 1 hr in serum-free medium with either vehicle (−) or 20 μM SB203580, an inhibitor of p38α kinase and MAPKAP kinase (+), and then exposed to either normoxia or hypoxia. In all of these experiments, hypoxia did not alter the total levels of CREB.

The robust and persistent hypoxia-induced phosphorylation of CREB in PC12 cells (57) indicates that activation of CREB may be involved in functional activation of certain genes that are regulated by hypoxia, such as TH (3,17,57). We propose that phosphorylation of CREB and activation of CRE-containing hypoxia-responsive genes may mediate an adaptive cellular response to hypoxia. A surprising finding was that activation of CREB by hypoxia is not mediated by any of the known CREB activation pathways. Thus, hypoxic activation of this important transcription factor seems to involve a yet undescribed and novel pathway. Studies are underway to identify the signaling pathway by which CREB phosphorylation is induced during hypoxia.

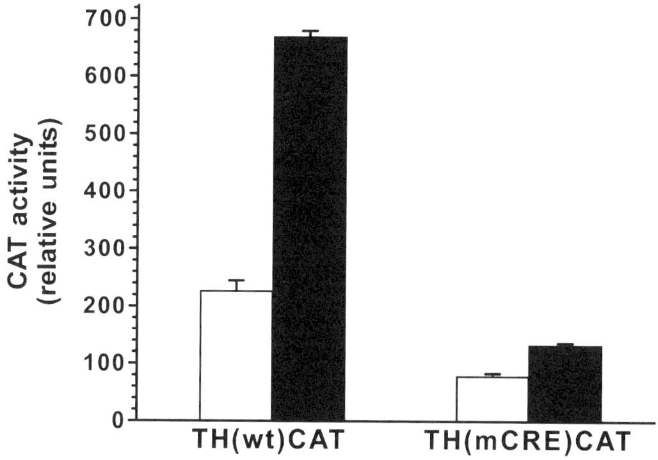

Figure 9 The CRE (cAMP response element) is involved in the activation of TH gene expression by hypoxia. PC12 cells were transfected with either a construct that consists of the wild-type −272 CRE [TH(wt)CAT] or a mutant −272 CRE [TH(mCRE)CAT] reporter gene construct. Cells were then exposed to normoxia (21% O_2, open bars), or hypoxia (5% O_2, closed bars) for 24 hr. CAT levels were analyzed by ELISA.

The PYK-2 Pathway

Pyk2 (also known as CADTK, CAKp, and RAFTK) is a proline-rich nonreceptor tyrosine kinase that is activated by an increase in intracellular calcium levels and is highly expressed in neural cell types and in PC12 cells (61–65). Pyk2 can also be activated by a variety of other signals, including activation of muscarinic acetylcholine (m1) receptors, protein kinase C, growth factors, fibronectin, reactive oxygen species, and various stress signals (61–67). Pyk2 is structurally related to the focal adhesion kinase (61–63,68). The proline-rich regions of Pyk2 provide binding sites for SH3 domain–containing proteins, such as $p130^{cas}$ and the GTPase-activating protein, Graf (69,70). Activation of Pyk2 has been associated with an activation of Src, JNK, and MAPK (61,65,71).

We have shown that PC12 cells respond very quickly to hypoxia with an increase in intracellular calcium levels (4,5). Furthermore, withdrawal of extracellular calcium blocks the hypoxia-induced increase in expression levels of certain hypoxia-regulated genes, including tyrosine hydroxylase and junB (16). We have examined the effect of hypoxia on phosphorylation of Pyk2 and characterized the role played by calcium in this regulation. We found that both hypoxia and depolarization by KCl induced a strong increase in phospho-tyrosine content of Pyk2 in the presence of Ca^{2+} (Fig. 10). However, in Ca^{2+}-free medium, the effects of both hypoxia and KCl on Pyk2 phosphorylation were completely abolished.

Downstream signaling pathways that are regulated by Pyk2 include MAPK, JNK, and c-src (61,64,65,72). We found hypoxia modestly activates MAPK, as indicated by an increase in its phosphorylation state (see Fig. 4). However, there was

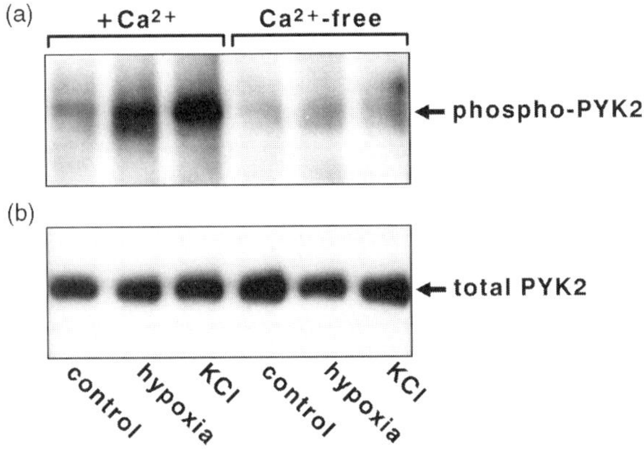

Figure 10 Hypoxia induces Ca^{2+}-dependent phosphorylation on Pyk2. PC12 cells were exposed to normoxia (21% O_2, control), hypoxia (5% O_2, 1 hr), or KCl (75 mM, 5 min) in the presence or absence of extracellular Ca^{2+}. Pyk2 immunoprecipitates were immunoblotted with an antibody to phosphotyrosine (a) or total Pyk2 (b).

no effect of hypoxia on the phosphorylation state of the stress-activated protein kinase JNK, although JNK was robustly stimulated by other stimuli, including EGF, NGF, and UV irradiation. Thus, there appears to be no link between activation of Pyk2 and the downstream stress-activated protein kinase JNK. We also observed no effect of hypoxia (5% O_2 or 1% O_2) on either the phosphorylation state or enzymatic activation of c-src. However, we were unable to observe an activation (phosphorylation) of c-src in response to any other stimuli we applied, including EGF, NGF, IGF-I, or KCI depolarization, although c-src immunoreactivity was readily detectable. We therefore cannot exclude the possibility that activation of Pyk2 by hypoxia is associated with activation of c-src in PC12 cells.

It is possible that the acute activation of Pyk2 by hypoxia specifically targets still other downstream substrates in addition to MAPK. For example, one of the known targets of Pyk2 is the voltage-dependent K^+ channel, Kv1.2 (61,66). Pyk2 has been shown to phosphorylate Kv1.2 on one or more tyrosine residues within the cytosolic carboxyl terminal portion of the channel (61,66). Interestingly, Kv1.2 is an O_2-sensitive K^+ channel (5,13). As stated above, one of the earliest known cellular events in response to hypoxia is a partial inhibition of the conductance of Kv1.2 channels (5,13). Furthermore, when oocytes expressing wild-type Pyk2 and Kv1.2 are treated with phorbol myristol acetate (PMA), Kv1.2 currents are markedly inhibited (61). However, this effect is absent when a mutant (kinase-inactive) form of Pyk2 is coexpressed with Kv1.2 (61). Our laboratory has clearly shown that the rapid inhibition of the Kv1.2 current by hypoxia is not a calcium-dependent process per se (4,5). Thus, the acute calcium-dependent activation of Pyk2 by hypoxia does not appear to be the mechanism by which hypoxia inhibits the Kv1.2 current. However, phosphorylation of Kv1.2 by Pyk2 may well have other, more long-term regulatory effects on the channel and its associated proteins. For example, long-term exposure to hypoxia selectively upregulates gene expression for the α subunit of Kv1.2 in PC12 cells (5). Further studies will be required to delineate what, if any, role Pyk2 plays in the long-term regulation of Kv1.2 function and protein : protein interactions.

Taken together, these studies show that activation of the Pyk2 protein tyrosine kinase is one of the early events in response to hypoxia in the O_2-responsive PC12 cell line. This effect occurs in a calcium-dependent manner and is therefore likely to be mediated by the rapid depolarization and calcium influx that occur in response to hypoxia in these excitable cells.

The MAP Kinase Phosphatase-1 (MKP-1) Pathway

MKP-1 (also known as CL100 and 3CH134) is a member of a family (MKPs) of dual-specificity phosphatases that oppose the effects of the mitogen- and stress-activated protein kinases (MAPKs and SAPKs) (73). Phosphorylation of MAPKs and SAPKs can be induced by a wide variety of cellular stimuli (see above). Upon phosphorylation of Thr-X-Tyr motifs, these signaling enzymes become activated and can translocate to the nucleus and phosphorylate various transcription factors, thereby regulating gene expression. The MKP family of enzymes is capable of dephosphorylating both phosphothreonine and phosphotyrosine in these Thr-X-Tyr motifs. Activation of MAPKs and SAPKs is frequently associated with activation of MKPs, suggesting that MKPs play a role in feedback control of MAPK signaling (74).

MKPs can be generally classified as being primarily localized either in the nucleus (MKP-1 and MKP-2) or in the cytosol (MKP-3, MKP-4, MKP-5, and M3/6) (75). The nuclear MKPs are highly inducible, and are considered to be immediate-early genes. Recently, it has been shown that the physical interaction of MAPKs and SAPKs with MKPs can stimulate the catalytic activity of both cytosolic and nuclear MKPs (76–79). It has been suggested that an increase in MKP gene expression represents another level of negative-feedback regulation on MAPK signaling pathways (74,80).

We previously showed that MAPKs and certain SAPKs are activated in PC12 cells by reduced O_2 (9,10). We found that MKP-1 mRNA (Fig. 11) and protein levels (Fig. 11) are strongly up-regulated by hypoxia (1% O_2) in PC12 cells (81). We found that this regulation was not unique to PC12 cells in that it also occurred in other cell lines (HepG2, Hep3B) that have been used to study O_2-responsive gene regulation, but not in cell lines (HEK 293, COS-7) that are not oxygen-sensing (Fig. 11). In addition, we demonstrated that cobalt chloride and deferoxamine caused an increase in MKP-1 gene expression to a similar extent as that measured during hypoxia. Both agents have been shown to mimic the effects of hypoxia, in part by increasing the binding activity and protein levels for the HIFs (hypoxia-inducible protein factors, HIF-1α and HIF-2α). This suggests that the hypoxia-induced increase in MKP-1 mRNA is dependent upon the activation of HIF.

There has been debate about the signaling mechanisms by which MKP-1 gene expression is regulated. Several studies have implicated calcium as playing a critical

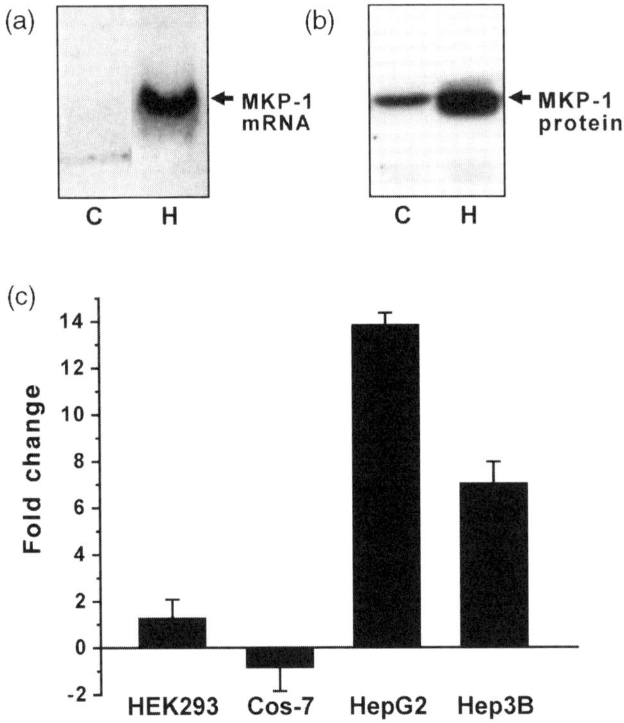

Figure 11 Hypoxia increases MKP-1 mRNA and protein levels in PC12 cells and in other O_2-sensing cells lines but not in non-O_2 sensing cell lines. (a and b) PC12 cells were exposed to normoxia (C, 21% O_2) or hypoxia (H, 1% O_2) for 4 hr. (a) MKP-1 RNA levels were determined by northern blot. (b) Whole-cell lysates were immunoblotted with an antibody to MKP-1. (c) O_2-sensing HepG2 and Hep3B and non-O_2-sensing HEK293 and Cos-7 cells were exposed to normoxia or hypoxia for 4 hr. MKP-1 mRNA levels were determined by real-time PCR and are expressed as fold change in hypoxia relative to normoxia (mean $\pm$ SE).

role in regulation of MKP-1 gene expression (80,82,83). However, other studies have indicated that the stress-activated protein kinases are involved in regulation of MKP-1 expression (84,85). It was therefore of considerable interest to determine which, if any, of these pathways induce MKP-1 gene expression in PC12 cells during hypoxia. Interestingly, the increase in MKP-1 mRNA induced by reduced O_2 was unaffected by the removal of extracellular and intracellular Ca^{2+} (81). This is in striking contrast to the effects of KCl-induced depolarization on MKP-1, which were abolished in the absence of Ca^{2+}.

It has been suggested that induction of MKP-1 gene expression occurs as a compensatory response to activation of MAPK (86,87). Although pharmacological blockade of MEK, the upstream activator of MAPK, prevented hypoxia-induced phosphorylation of MAPK (ERK1/2), it did not alter the effect of hypoxia on MKP-1 mRNA levels (81). Therefore, the hypoxia-induced activation of MAPK is not essential for the increase in MKP-1 gene expression. This finding is consistent with those from other studies in which inhibition of MEK was insufficient to prevent the induction of MKP-1 gene expression by other stimuli (88,89). We also investigated the possible role of the phosphoinositide 3′-kinase (PI3K)/Akt pathway in the regulation of MKP-1 during hypoxia. Blockade of PI3K had no effect on the induction of MKP-1 by hypoxia, indicating that the PI3K/Akt pathway is also not involved (81).

A significant finding was that the effects of hypoxia on MKP-1 were markedly attenuated, but not completely prevented, by a drug (SB203580) that inhibits the p38 family of SAPKs (81). SB203580 inhibits the p38α and p38β subtypes but not the p39γ and p38δ subtypes. As stated above, we found that the p38α and p38γ subtypes are regulated by hypoxia (10). The partial inhibition of MKP-1 gene expression in cells pretreated with SB203580 is therefore likely due to blockade of p38α. In addition, we propose that the hypoxia-induced expression of MKP-1 that was not inhibited by SB203580 might be associated with the hypoxic induction of the SB203580-insensitive p38γ. Alternatively, other as yet unidentified non-p38 kinase signaling pathways may also play a role in the regulation of MKP-1 by reduced O_2. Thus, while there may be cell-specific differences in the mechanism of regulation of key genes and proteins that mediate adaptation, HIF1α and 2α, the p38 protein kinases, and MKP-1 appear to be linked in a complex patter of coregulation.

III. Hypoxia-Regulated Gene Expression: The EPAS-1 Story

Regulation of gene expression is a primary response by which cells adapt to changes in the environment. The ability to adapt to reduced O_2 is essential for the viability of all organisms. The mechanisms involved in regulation of gene expression in response to hypoxia are beginning to be understood. Transcription factors that are activated by hypoxia include the hypoxia-inducible factor (HIF-1α), c-Fos, JunB, and CREB (16,57,90–92). HIF-1α has been shown to be critical for hypoxia-induced regulation of a number of genes in a variety of different cell types and tissues (for review see Refs. 93,94). Recently, another HIF-like protein called endothelial PAS domain protein 1 (EPAS-1, also known as HIF-2α) was identified in the type I O_2-sensing cells of the carotid body (95). Although most attention has been focused on HIF-1α, we found that EPAS-1 is more robustly activated by hypoxia in both PC12 cells and in carotid body cells (9).

EPAS-1 is a basic helix-loop-helix transcription factor that shares 48% sequence homology with HIF-1α (95). EPAS1 protein levels, like HIF-1α levels, are relatively low under basal conditions and accumulate upon exposure to reduced O_2 (9). EPAS1, and the other HIFs, translocate to the nucleus and *trans*-activate target genes containing the consensus sequence 5′-RCGTG-3′, which is commonly referred to as the hypoxia response element (HRE) (43). Once activated by hypoxia, EPAS1 (like HIF-1α) forms a heterodimer with the aryl hydrocarbon nuclear receptor translocator (ARNT) protein, which then binds to the HRE on hypoxia-responsive genes. Here, we shall briefly describe some of our recent work on regulation of EPAS1 activity in PC12 cells during hypoxia.

The specific signaling pathways involved in HIF-1α and EPAS1 activation are not well understood. We investigated the possibility that one or more of the hypoxia-activated signal transduction pathways discussed above might be involved in regulation of EPAS1 in PC12 cells. We found that the EPAS1 protein levels were robustly (12-fold) elevated during hypoxia (1% O_2 for 6 hr), and that a reporter gene that contained three repeats of the HRE was regulated in a dose-dependent manner with a progressive reduction in O_2 (9) (Fig. 12).

Because the MAPK pathway is known to regulate a number of transcription factors, including c-Fos, JunB, CREB, and Elk-1 (56,96,97), we hypothesized that the MAPK pathway might be important for EPAS1 activation during hypoxia. To test this hypothesis, PC12 cells were cotransfected with the HRE reporter gene and a plasmid encoding human EPAS1 cDNA and then exposed to normoxia or hypoxia (9). We found that transfection of EPAS1 stimulated expression of the HRE-luciferase reporter gene under both normoxic and hypoxic conditions (Fig. 13). Transfection of a constitutively active MEK1 (pFC-MEK1), a dual-specificity kinase that directly phosphorylates and activates MAPK (21), also enhanced HRE-luciferase activity during both normoxia and hypoxia (Fig. 13). When EPAS1 and pFC-MEK1 were coexpressed, they caused a much larger increase in the *trans*-activation of the HRE-luciferase reporter gene during hypoxia than transfection of either alone. We also found that pharmacological inhibition of MEK1 completely blocked the effect of hypoxia on both the basal and EPAS1-stimulated HRE-luciferase activity (Fig. 13). These results strongly indicate that the MEK1-MAPK signaling pathway is critical for mediating EPAS1 activation of HRE-dependent gene expression (9).

Ras is the initial step in the MAPK signaling pathway. To test whether Ras was involved in the EPAS1 *trans*-activation of the HRE-luciferase reporter gene, PC12 cells were cotransfected with the EPAS1 expression plasmid, the HRE-luciferase reporter gene, and increasing amounts of a dominant-negative Ras expression plasmid (RasN-17). We found that neutralization of Ras with increasing amounts of RasN-17 had no effect on the EPAS1 *trans*-activation of the HRE-luciferase gene. However, coexpression of the same amounts of RasN-17 did block activation of a c-fos-luciferase reporter gene by nerve growth factor (NGF), the prototypical activator of the Ras-Raf-MEK-MAPK pathway. These findings indicate that hypoxia activates MAPK and EPAS1 by a Ras-independent mechanism.

As described at the beginning of this chapter, hypoxia results in depolarization and Ca^{2+} influx into PC12 cells and carotid body type I cells during hypoxia. It was reported recently that depolarization of PC12 cells results in MAPK activation via a calmodulin-dependent mechanism (98,99). We therefore investigated the possibility that calmodulin could be involved in the activation of MAPK and EPAS1 during hypoxia. We found that pretreatment of PC12 cells with the calmodulin antagonists W13 and calmidazolium chloride caused a pronounced reduction in hypoxia-induced MAPK phosphorylation (9). Thus, MAPK activation of EPAS1 probably occurs via a calmodulin-dependent pathway.

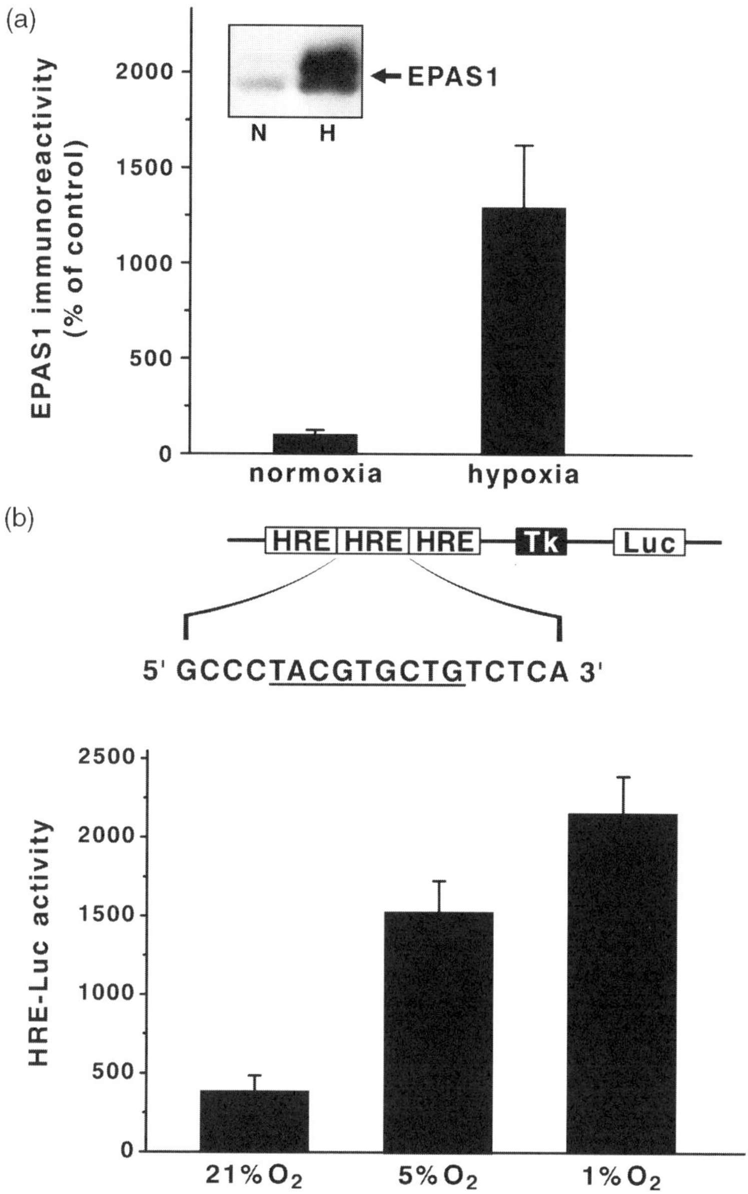

Figure 12 EPAS1 protein accumulates and is activated by hypoxia in PC12 cells. (a) PC12 cells were exposed to normoxia (21% O_2) or hypoxia (1% O_2, 6 hr) followed by SDS-PAGE and immunoblotting with an EPAS1 antibody (inset). EPAS1 immunoreactivity levels are expressed as percent of control. (b) PC12 cells were transfected with a reporter construct that contained three repeats of the HRE in front of the luciferase gene. Cells were then exposed to normoxia (21% O_2) or varying degrees of hypoxia as indicated for 6 hr, and luciferase activity was determined.

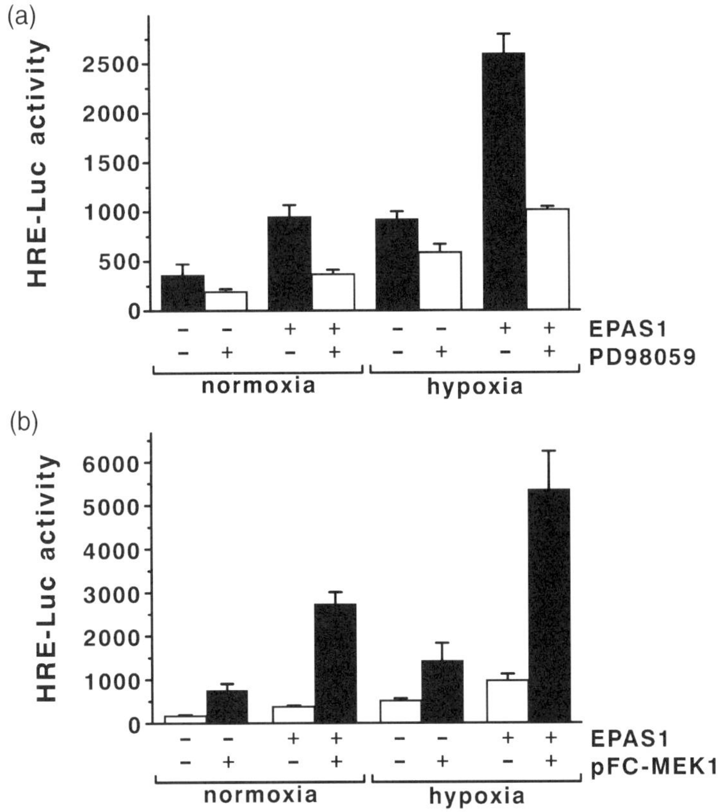

Figure 13 MAPK is critical for EPAS1 *trans*-activation of an HRE-luciferase reporter gene. PC12 cells were cotransfected with the EPAS1 and the HRE-Luc reporter construct described in Figure 12. (a) Cells were then exposed to normoxia (21% O_2) or hypoxia (1% O_2) for 6 hr in presence (+) or absence (−) of 50 μM PD98059. Cell lysates were assayed for luciferase activity. (b) Cells were additionally cotransfected with pFC-MEK1 (+) or empty vector (−) and then exposed to normoxia or hypoxia.

Although MAPK is critical for hypoxic regulation of EPAS1 function, it is not the kinase that phosphorylates EPAS1 during hypoxia. Neither hypoxia-induced phosphorylation nor accumulation of EPAS1 protein was inhibited by pharmacological inhibition of MEK1 with PD90859 (9). This is based on our finding that blockade of MEK1 failed to prevent incorporation of [^{32}P] orthophosphate into EPAS1. These results suggest that multiple MAPK-dependent and MAPK-independent signals are required for EPAS1 activation. We propose that a MAPK-independent signal leads to accumulation of EPAS1 protein, presumably by inhibition of ubiquitin-proteosome degradation (100), and that a second MAPK-independent signal leads to the phosphorylation of EPAS1.

Recent studies by others suggest that multiple signals are involved in regulation of EPAS1 and identify two domains of EPAS1 that are required for its activation by hypoxia (101–103). One of the critical EPAS1 domains is an internal domain that extends from amino acids 450–571 and shares homology with the oxygen-dependent domain of HIF-1α, which is critical for activation by reduced O_2. The second important EPAS1 regulatory domain is a C-terminal activation domain (amino acids 824–876), which is the site of posttranslational modification in EPAS1 during hypoxia (101). The mechanism of MAPK-dependent activation of EPAS1 is unknown. The fact that EPAS1 phosphorylation persists in the presence of MEK1 blockade suggests that the MAPK pathway does not directly target EPAS1, but instead targets other protein(s) that are critical for the formation of the EPAS1 DNA-binding complex. Others have shown that CREB-binding protein (CBP) interacts with HIF-1α and EPAS1 and potentiates the activation of these hypoxia transcription factors (101). It has also been shown that the C-terminal regions of CBP can be phosphorylated by MAPK (104). Thus, CBP might be a target of hypoxia-activated MAPK, which could then recruit EPAS1 to the DNA-binding complex. Finally, it has been proposed that several "general transcription factors" are present in the EPAS1 DNA-binding complex (101). These proteins are potential targets of MAPK regulation. Thus, it is likely that the MAPK-dependent activation of EPAS1 *trans*-activation involves the recruitment of proteins other than EPAS1 to the DNA-binding complex.

It is now well known that the α-subunit of the HIF transcription factors is degraded by proteasome pathways during normoxia, but stabilized under hypoxic conditions. The exact mechanisms by which hypoxia stabilizes HIFs remain unknown. The mRNA levels for HIF-1α and EPAS1 are unchanged during hypoxia. Thus, the stabilization of these important hypoxia-induced transcription factors occurs via regulation of their protein levels. Protein levels for both HIF-1α and EPAS1 accumulate in response to hypoxia, proteasomal inhibitors, transition metals (e.g., cobalt), iron chelators, and reducing agents (105). To gain insight into this mechanism we studied the effect of cobalt on EPAS1 protein level in PC12 cells. We found that EPAS1 binds to cobalt and that the binding activity is within the oxygen-dependent domain (ODD) (106). The ODD is necessary and sufficient for regulation of protein stability in response to hypoxia.

The von Hippel–Landau (VHL) tumor suppressor gene product has been shown to be involved in the regulation of HIF-1α and EPAS1 protein levels (107). There is a 17-amino-acid sequence within the ODD that is conserved among all HIFs (108). It has been reported that this sequence mediates the interaction between HIF and VHL (107). VHL mediates the degradation of HIF-1α through a ubiquitin-proteasome pathway. Recent findings suggest that hydroxylation of proline residues within the ODD regulates VHL binding to and eventual degradation of HIF (108–111). The interaction between VHL and HIF is disrupted by cobalt (107). We showed that the cobalt and VHL binding sites overlap (106). Mutation of the overlapping site prevents cobalt binding and leads to stabilization of EPAS1. Thus, a major component of the signaling mechanism that leads to accumulation of HIF-1α

or EPAS1 might be the activation of proteins that bind to this region of the ODD and prevent degradation by VHL.

Thus, we showed for the first time that EPAS1 is phosphorylated during hypoxia and that the MAPK pathway is critical for EPAS1 *trans*-activation during hypoxia in the O_2-sensing PC12 cell line. However, since EPAS1 is not directly phosphorylated by MAPK, the mechanism that regulates accumulation of EPAS appears to require the interaction of other signaling pathways and DNA-binding factors. Additional research is needed to elucidate the exact signal transduction mechanisms by which reduced O_2 leads to the accumulation of EPAS1 and the *trans*-activation of hypoxia responsive genes. It is important to realize that these mechanisms will probably differ greatly between excitable and nonexcitable O_2-sensing cells. In this regard, it is important to recognize that not all hypoxia-responsive genes contain an HRE and therefore are not targets for the HIFs. For example, the immediate early gene junB is robustly activated by hypoxia, yet we were unable to identify an HRE in the 5′ regulatory region of the junB gene. A primary mechanism for regulation of junB and other AP1 immediate early genes is Ca^{2+} and cAMP signaling pathways (60). Thus, a complete understanding of the gene expression pattern that is activated by hypoxia requires a comprehensive approach for identifying hypoxia-responsive genes in both excitable and non-excitable cells and genes whose expression is regulated by HIFs and non-HIF mechanisms.

IV. The Future: Identification of the Hypoxia Genome

The ongoing quest to understand the role of gene expression in mediating alterations in cell phenotype (e.g., a hypoxia-tolerant phenotype) has historically been restricted to the study of single genes and proteins. Yet, cellular adaptation to environmental stimuli such as hypoxia is a very complex physiological function that requires the simultaneous expression of hundreds of different genes and proteins. The Human Genome Project (and various other sequencing projects) is not only leading to identification of sequences for the entire genome, but has also sparked the development of new high-throughput approaches that allow identification of the genes involved in complex cellular functions. Gene expression (cDNA) microarray technology and the expanding gene sequence databases have made it possible to identify the complement of transcribed genes (the "transcriptome") within cells and tissues that are stimulated in response to an environmental stimulus or stress such as hypoxia (see Ref. 112). The gene expression profile that results from extracellular stimuli is analyzed by various computational approaches and then classified according to gene structure and function. Thus, functional genomics offers the potential to markedly expand our understanding of biological systems and how these systems adapt to environmental stimuli.

We have used a high-throughput functional genomics approach to identify the genes that regulated by hypoxia in PC12 cells. Our approach begins with the construction of a subtracted cDNA library, which isolates the known and unknown (novel) genes that are up- or downregulated by reduced O_2. The technique for making the subtracted cDNA library is called subtractive suppression hybridization (SSH) (81,113). SSH increases the probability of obtaining low-abundance, differentially expressed cDNAs. Each clone is sequenced and identified in either public or private databases. The clones are then used to print microarray slides.

We recently used this approach to identify approximately 200 unique known genes and 40 unknown sequences (novel genes) that are stimulated by hypoxia in PC12 cells. Theoretically, these genes represent the "hypoxia" genome, i.e., the genes that regulate the entire response to hypoxia in PC12 cells. The major challenge is to group these genes into clusters based on function and the mechanism of activation. This type of information will provide a more comprehensive picture concerning how O_2-sensing cells function and adapt to a persistent hypoxia environment. In addition, information from such studies will provide insights into the development of diagnostic and therapeutic products for ischemic/hypoxic disease. It is also important to determine if genes that regulate the cellular response to hypoxia stress also regulate the response to other cellular stresses.

Acknowledgments

This work was supported by NIH grants HL33831, DK58811, HL66312, HL59945, HL07571, and a grant from the U.S. Army.

References

1. Beal MF. Aging, energy, and oxidative stress in neurodegenerative diseases. Ann Neurol 1995; 38:357–366.
2. Choi DW. Ischemia-induced neuronal apoptosis. Curr Opin Neurobiol 1996; 6: 667–672.
3. Czyzyk-Krzeska MF, Furnari B, Lawson EL, Millhorn DE. Hypoxia increases rate of transcription and stability of tyrosine hydroxylase mRNA in pheochromocytoma (PC12) cells. J Biol Chem 1994; 269:760–764.
4. Zhu WH, Conforti L, Czyzyk-Krzeska MF, Millhorn DE. Membrane depolarization and dopamine secretion in PC12 cells during hypoxia are regulated by an O_2-sensitive K^+ current. Am J Physiol Cell 1996; 271:658–665.
5. Conforti L, Millhorn DE. Selective inhibition of a slow-inactivating voltage-dependent K^+ channel in rat PC12 cells by hypoxia. J Physiol (Lond) 1997; 502:293–305.
6. Beitner-Johnson D, Liebold J. Millhorn DE. Hypoxia regulates the cAMP and Ca^{2+}/calmodulin signaling systems in PC12 cells. Biochem and Biophys Res Commun 1998; 241:61–66.
7. Kobayashi S, Millhorn DE. Stimulation of expression for the adenosine A2A receptor gene by hypoxia in PC12 cells: a potential role in cell protection. J Biol Chem 1999; 274:20358–20365.
8. Lopez-Barneo J, Lopez-Lopez JR. Urena J, Gonzales C. Chemotransduction in the carotid body: K^+ current modulated by PO_2 in type I chemoreceptor cells. Science 1989; 241:580–582.

9. Conrad PW, Freeman TL, Beitner-Johnson D, Millhorn DE. EPAS1 transactivation during hypoxia requires p42/p44 MAPK. J Biol Chem 1999; 274:33709–33713.
10. Conrad PW, Rust RT, Han J, Millhorn DE, Beitner-Johnson D. Selective activation of p38α and p38γ by hypoxia: role in regulation of cyclin D1 by hypoxia in PC12 cells. J Biol Chem 1999; 274:23570–23576.
11. Beitner-Johnson D, Rust RT, Hsieh T, Millhorn DE. Hypoxia activates Akt and induces phosphorylation of GSK-3 in PC12 cells. Cell Signal 2001; 13:23–27.
12. Buckler KJ, Vaughan-Jones RD. Effects of hypoxia on membrane potential and intracellular calcium in rat neonatal carotid body type I cells. J Physiol 1994; 476:423–428.
13. Conforti L, Bodi I, Nisbet J, Millhorn DE. O_2-sensitive K^+ channels: role of the Kv1.2 α-subunit in mediating the hypoxic response. J Physiol 2000; 524:783–793.
14. Sheng ME, McFadden G, Greenberg ME. Membrane depolarization and calcium induce c-fos transcription via phosphorylation of transcription factor CREB. Neuron 1990; 4:571–582.
15. Sheng ME, Thompson MA, Greenberg ME. CREB: a Ca(2+)-regulated transcription factor phosphorylated by calmodulin-dependent kinases. Science 1991; 252:1427–1430.
16. Raymond R, Millhorn DE. Regulation of tyrosine hydroxylase gene expression during hypoxia: role of Ca^{2+} and PKC. Kidney Int 1997; 51:536–541.
17. Czyzyk-Krzeska MF, Bayliss D, Lawson E, Millhorn DE. Regulation of tyrosine hydroxylase gene expression in rat carotid body by hypoxia. J Neurochem 1992; 58:1538–1546.
18. Hochachka PW, Buck LT, Doll CJ, Land SC. Unifying theory of hypoxia tolerance: molecular/metabolic defense and rescue mechanisms for surviving oxygen lack. Proc Natl Acad Sci USA 1996; 93:9493–9498.
19. Widmann C, Gibson S, Jarpe B, Johnson GL. Mitogen-activated protein kinase: conservation of a three-kinase module from yeast to human. Physiol Rev 1999; 79:143–180.
20. Su B, Karin M. Mitogen-activated protein kinase cascades and regulation of gene expression. Curr Opin Immunol 1996; 8:402–411.
21. Garrington TP, Johnson GL. Organization and regulation of mitogen-activated protein kinase signaling pathways. Curr Opin Cell Biol 1999; 11:211–218.
22. Kyriakis JM, Avruch J. Sounding the alarm: protein kinase cascades activated by stress and inflammation. J Biol Chem 1996; 271:24313–24316.
23. Hibi M, Lin A, Smeal T, Minden A, Karin M. Identification of an oncoprotein- and UV-responsive protein kinase that binds and potentiates the c-Jun activation domain. Genes Dev 1993; 7:2135–2148.
24. Derijard B, Hibi M, Wu IH, Barrett G, Su B, Deng T, Karin M, Davis RJ. JNK1: a protein kinase stimulated by UV light and Ha-Ras that binds and phosphorylates the c-Jun activation domain. Cell 1994; 76:1025–1037.
25. Kyriakis JM, Banerjee P, Nikolakaki E, Dal T, Rubie EA, Ahmad MF, Avruch J, Woodgett JR. The stress-activated protein kinase subfamily of c-Jun kinases. Nature 1994; 369:156–160.
26. Han J, Lee JD, Bibbs L, Ulevitch RJ. A MAP kinase targeted by endotoxin and hyperosmolarity in mammalian cells. Science 1994; 265:808–811.
27. Logan SK, Falaska M, Hu P, Schlessinger J. Phosphatidylinositol 3-kinase mediates epidermal growth factor-induced activation of the c-Jun N-terminal kinase signaling pathway. Mol Cell Biol 1997; 17:5784–5790.

28. Xing J, Kornhauser JM, Xia Z, Thiele EA, Greenberg ME. Nerve growth factor activates extracellular signal-regulated kinase and p38 mitogen-activated protein kinase pathways to stimulate CREB serine 133 phosphorylation. Mol Cell Biol 1998; 18:1946–1955.
29. Raingeaud J, Gupta S, Rogers JS, Dickens M, Han J, Ulevitch RJ, Davis RJ. Proinflammatory cytokines and environmental stress cause p38 mitogen-activated protein kinase activation by dual phosphorylation on tyrosine and threonine. J Biol Chem 1995; 270:7420–7426.
30. Whitmarsh AJ, Davis RJ. Transcription factor AP-1 regulation by mitogen-activated protein kinase signal transduction pathways. J Mol Med 1994; 74:589–607.
31. Lavoie JN, L'Allemain G, Brunet A, Muller R, Pouyssegur J. Cyclin D1 expression is regulated positively by the p42/p44MAPK and negatively by the p38/HOGMAPK pathway. J Biol Chem 1996; 271:20608–20616.
32. Baldin V, Lucas J, Marcote MJ, Pagano M, Draetta G. Cyclin D1 is a nuclear protein required for cell cycle progression in G1. Genes Dev 1993; 7:812–821.
33. Quelle DE, Ashmum RA, Shurtleff SA, Kato J, Bar-Sagi D, Roussel MF, Sherr CJ. Overexpression of mouse D-type cyclins accelerates G1 phase in rodent fibroblasts. Genes Dev 1993; 7:1559–1571.
34. Datta SR, Brunet A, Greenberg ME. Cellular survival: a play in three Akts. Genes Dev 1999; 13:2905–2927.
35. Vanhaeselbroeck B, Alessi DR. The PI3K-PDK1 connection: more than just a road to PKB. Biochem J 2000; 346:561–576.
36. Dubek H, Datta SR, Franke TF, Birnbaum MJ, Yao R, Cooper GM, Segal RA, Kaplan DR, Greenberg ME. Regulation of neuronal survival by the serine-threonine protein kinase Akt. Science 1997; 275:661–665.
37. Konishi H, Matsuzaki H, Tanaka M, Ono Y, Tokunaga C, Kuroda S, Kikkawa U. Activation of RAC-protein kinase by heat shock and hyperosmolarity stress through a pathway independent of phosphatidylinositol 3-kinase. Proc Natl Acad Sci USA 1996; 93:7639–7643.
38. Shaw M, Cohen P, Alessi DR. The activation of protein kinase B by H_2O_2 or heat shock is mediated by phosphoinositide 3-kinase and not by mitogen-activated protein kinase-activated protein kinase-2. Biochem J 1998; 336:241–246.
39. Meier R, Thelen M, Hemmings BA. Inactivation and dephosphorylation of protein kinase Balpha (PKBalpha) promoted by hyperosmotic stress. EMBO J 1998; 17:7294–7303.
40. Meier R, Alessi DR, Cron P, Andjeldovik M, Hemmings BA. Mitogenic activation, phosphorylation, and nuclear translocation of protein kinase B beta. J Biol Chem 1997; 272:31515–31524.
41. Cross DA, Alessi DR, Cohen P, Andjelkovich M, Hemmings BA. Inhibition of glycogen synthase kinase-3 by insulin mediated by protein kinase B. Nature 1995; 378:785–789.
42. Zhong H, Chiles K, Feldser D, Laughner E, Hanrahan C, Georgescu MM, Simons JW, Semenza G. Modulation of hypoxia-inducible factor 1alpha expression by the epidermal growth factor/phosphatidylinositol 3-kinase/PTEN/AKT/FRAP pathway in human prostate cancer cells: implications for tumor angiogenesis and therapeutics. Cancer Res 2000; 60:1541–1545.

43. Semenza GL, Wang GL. A nuclear factor induced by hypoxia via de novo protein synthesis binds to the human erythropoietin gene enhancer at a site required for transcriptional activation. Mol Cell Biol 1992; 12:5447–5454.
44. Mazure NM, Chen EY, Laderoute KR, Giaccia AJ. Induction of vascular endothelial growth factor by hypoxia is modulated by a phosphatidylinositol 3-kinase/Akt signaling pathway in Ha-ras-transformed cells through a hypoxia inducible factor-1 transcriptional element. Blood 1997; 90:3322–3331.
45. Hajduch E, Alessi DR, Hemmings BA, Hundal HS. Constitutive activation of protein kinase B alpha by membrane targeting promotes glucose and system A amino acid transport, protein synthesis, and inactivation of glycogen synthase kinase 3 in L6 muscle cells. Diabetes 1998; 47:1006–1013.
46. Pap M, Cooper GM. Role of glycogen synthase kinase-3 in the phosphatidylinositol 3-Kinase/Akt cell survival pathway. J Biol Chem 1998; 273:19929–19932.
47. Frank DA, Greenberg ME. CREB: a mediator of long-term memory from mollusks to mammals. Cell 1994; 79:5–8.
48. Ginty DD, Bonni A, Greenberg ME. Nerve growth factor activates a Ras-dependent protein kinase that stimulates c-fos transcription via phosphorylation of CREB. Cell 1994; 79:713–725.
49. Deisseroth K, Bito H, Tsien RW. Signaling from synapse to nucleus: postsynaptic CREB phosphorylation during multiple forms of hippocampal synaptic plasticity. Neuron 1996; 16:89–101.
50. Pende M, Fisher TL, Simpson PB, Russel JT, Blenis J, Gallo V. Neurotransmitter- and growth factor-induced cAMP response element binding protein phosphorylation in glial cell progenitors: role of calcium ions, protein kinase C, and mitogen-activated protein kinase/ribosomal S6 kinase pathway. J Neurosci 1997; 17:1291–1301.
51. Tan Y, Rouse J, Zhang A, Cariati S, Cohen P, Comb MJ. FGF and stress regulate CREB and ATF-1 via a pathway involving p38 MAP kinase and MAPKAP kinase-2. EMBO J 1996; 15:4629–4642.
52. Iordanov M, Bender K, Ade T, Schmid W, Sachsenmaier C, Engel K, Gaestel M, Rahmsdorf HJ, Herrlich P. CREB is activated by UVC through a p38/HOG-1-dependent protein kinase. EMBO J 1997; 16:1009–1022.
53. Gonzalez GA, Montminy MR. Cyclic AMP stimulates somatostatin gene transcription by phosphorylation of CREB at serine 133. Cell 1989; 59:675–680.
54. Yamamoto KK, Gonzales GA, Biggs WH, Montminy MR. Characterization of a bipartite activator domain in transcription factor CREB. Nature 1988; 334:494–498.
55. Xie H, Rothstein TL. Protein kinase C mediates activation of nuclear cAMP response element-binding protein (CREB) in B lymphocytes stimulated through surface Ig. J Immunol 1995; 154:1717–1723.
56. Xing J, Gitny DD, Greenberg ME. Coupling of the RAS-MAPK pathway to gene activation by RSK2, a growth factor-regulated CREB kinase. Science 1996; 273: 959–963.
57. Beitner-Johnson D, Millhorn DE. Hypoxia induces phosphorylation of the cyclic AMP response element-binding protein by a novel signaling mechanism. J Biol Chem 1998; 273:19834–19839.
58. Kim KS, Lee MK, Carroll J, Joh TH. Both the basal and inducible transcription of the tyrosine hydroxylase gene are dependent upon a cAMP response element. J Biol Chem 1993; 268:15689–15695.

59. Lazaroff M, Patankar S, Yoon SO, Chikaraishi DM. The cyclic AMP response element directs tyrosine hydroxylase expression in catecholaminergic central and peripheral nervous system cell lines from transgenic mice. J Biol Chem 1995; 270:21579–21589.
60. Nagamoto-Combs K, Piech KM, Best JA, Sun B, Tank AW. Tyrosine hydroxylase gene promoter activity is regulated by both cyclic AMP-responsive element and AP1 sites following calcium influx: evidence for cyclic amp-responsive element binding protein-independent regulation. J Biol Chem 1997; 272:6051–6058.
61. Lev S, Moreno H, Martinez R, Canoil P, Peles E, Musacchio JM, Plowman GD, Rudy B, Schlessinger J. Protein tyrosine kinase PYK2 involved in Ca(2+)-induced regulation of ion channel and MAP kinase functions. Nature 1995; 376:737–745.
62. Sasaki H, Nagura K, Ishino M, Tobioka H, Totani K, Sasaki T. Cloning and characterization of cell adhesion kinase beta, a novel protein-tyrosine kinase of the focal adhesion kinase subfamily. J Biol Chem 1995; 270:21206–21219.
63. Yu H, Li X, Marchetto GS, Dy R, Hunter D, Calvo B, Dawson TL, Wilm M, Anderegg RJ, Graves LM, Earp HS. Activation of a novel calcium-dependent protein-tyrosine kinase: correlation with c-Jun N-terminal kinase but not mitogen-activated protein kinase activation. J Biol Chem 1996; 271:29993–29998.
64. Li J, Avraham H, Rogers RA, Sandhya R, Avraham S. Characterization of RAFTK, a novel focal adhesion kinase, and its integrin-dependent phosphorylation and activation in megakaryocytes. Blood 1996; 88:417–428.
65. Dikic I, Tokiwa G, Lev S, Courtneidge SA, Schlessinger J. A role for Pyk2 and Src in linking G-protein-coupled receptors with MAP kinase activation. Nature 1996; 383:547–550.
66. Felsch JS, Cachero TG, Peralta EG. Activation of protein tyrosine kinase PYK2 by the m1 muscarinic acetylcholine receptor. Proc Natl Acad Sci USA 1998; 95:5051–5056.
67. Frank GD, Motley ED, Inagami T, Eguchi S. Unique regulation of c-Jun N-terminal kinase by PYK2/CAK-beta in angiotensin II-stimulated vascular smooth muscle cells. Biochem Biophys Res Commun 2000; 270:761–765.
68. Avraham S, London R, Fu Y, Ota S, Hiregowdara D, Li J, Jiang S, Pasztor LM, White RA, Groopman JE, Avraham H. Identification and characterization of a novel related adhesion focal tyrosine kinase (RAFTK) from megakaryocytes and brain. J Biol Chem 1995; 270:27742–27751.
69. Ohba T, Ishino M, Aoto H, Sasaki T. Interaction of two proline-rich sequences of cell adhesion kinase β with SH3 domains of p130Cas-related proteins and a GTPase-activating protein, Graf. Biochem J 1998; 330:1249–1254.
70. Astier A, Avraham H, Manie SN, Groopman J, Canty T, Avraham S, Freedman AS. The related adhesion focal tyrosine kinase is tyrosinephosphorylated after beta1-integrin stimulation in B cells and binds to p130cas. J Biol Chem 1997; 272:228–232.
71. Della Rocca GJ, van Biesen T, Daaka Y, Luttrell DK, Luttrell LM, Lefkowitz, RJ. Ras-dependent mitogen-activated protein kinase activation by G protein-coupled receptors: convergence of Gi- and Gq-mediated pathways on calcium/calmodulin, Pyk2, and Src kinase. J Biol Chem 1997; 272:19125–19132.
72. Sieg DJ, Iiic D, Jones KC, Damsky CH, Hunter T, Schlaepfer DD. Pyk2 and Src-family protein-tyrosine kinases compensate for the loss of FAK in fibronectin-stimulated signaling events but Pyk2 does not fully function to enhance FAK- cell migration. EMBO J 1998; 17:5933–5947.

73. Charles CH, Abler AS, Lau LF. cDNA sequence of a growth factor-inducible immediate early gene and characterization of its encoded protein. Oncogene 1992; 7:187–190.
74. Brondello JM, Brunet A, Pouyssegur J, McKenzie FR. The dual specificity mitogen-activated protein kinase phosphatase-1 and -2 are induced by the p42/p44MAPK cascade. J Biol Chem 1997; 272:1368–1376.
75. Martell KJ, Angelotti T, Ullrich A. The "VH1-like" dual-specificity protein tyrosine phosphatases. Mol Cells 1998; 8:2–11.
76. Camps M, Nichols A, Gillieron C, Antonsson B, Muda M, Chabert C, Boschert U, Arkinstall S. Catalytic activation of the phosphatase MKP-3 by ERK2 mitogen-activated protein kinase. Science 1998; 280:1262–1265.
77. Hutter D, Chen P, Barnes J, Liu Y. Catalytic activation of mitogen-activated protein (MAP) kinase phosphatase-1 by binding to p38 MAP kinase: critical role of the p38 C-terminal domain in its negative regulation. Biochem J 2000; 352:155–163.
78. Nichols A, Camps M, Gillieron C, Chabert C, Brunet A, Wisbacher J, Cobb M, Pouyssegur J, Shaw JP, Arkinstall S. Substrate recognition domains within extracellular signal-regulated kinase mediate binding and catalytic activation of mitogen-activated protein kinase phosphatase-3. J Biol Chem 2000; 275:24613–24621.
79. Zhou B, Wu L, Shen K, Zhang J, Lawrence DS, Zhang Z-Y. Multiple regions of MAP kinase phosphatase 3 are involved in its recognition and activation by ERK2. J Biol Chem 2001; 276:6506–6515.
80. Cook SJ, Beltman J, Cadwallader KA, McMahon M, McCormick F. Regulation of mitogen-activated protein kinase phosphatase-1 expression by extracellular signal-related kinase-dependent and Ca^{2+}-dependent signal pathways in Rat-1 cells. J Biol Chem 1997; 272:13309–13319.
81. Seta KA, Kim R, Kim H-W, Millhorn DE, Beitner-Johnson D. Hypoxia-induced regulation of MKP-1: identification by subtractive suppression hybridization and cDNA microarray analysis. J Biol Chem 2001; 276:44405–44412.
82. Scimeca JC, Servant MJ, Dyer JO, Meloche S. Essential role of calcium in the regulation of MAP kinase phosphatase-1 expression. Oncogene 1997; 15:717–726.
83. Durham PL, Russo AF. Differential regulation of mitogen-activated protein kinase-responsive genes by the duration of a calcium signal. Mol Endocrinol 2000; 14:1570–1582.
84. Bokemeyer D, Sorokin A, Yan M, Ahn NG, Templeton DJ, Dunn MJ. Induction of mitogen-activated protein kinase phosphatase 1 by the stress-activated protein kinase signaling pathway but not by extracellular signal-regulated kinase in fibroblasts. J Biol Chem 1996; 271:639–642.
85. Li C, Hu Y, Mayr M, Xu Q. Cyclic strain stress-induced mitogen-activated protein kinase (MAPK) phosphatase 1 expression in vascular smooth muscle cells is regulated by Ras/Rac-MAPK pathways. J Biol Chem 1999; 274:25273–25280.
86. Sun H, Charles CH, Lau LF, Tonks NK. MKP-1 (3CH134), an immediate early gene product, is a dual specificity phosphatase that dephosphorylates MAP kinase in vivo. Cell 1993; 75:487–493.
87. Keyse SM. Protein phosphatases and the regulation of mitogen-activated protein kinase signaling. Curr Opin Cell Biol 2000; 12:186–192.
88. Wu J, Lau LF, Sturgill TW. Rapid deactivation of MAP kinase in PC12 cells occurs independently of induction of phosphatase MKP-1. FEBS Lett 1994; 484:189–193.

89. Burgun C, Esteve L, Humbolt N, Aunis D, Zwiller J. Cyclic AMP-elevating agents induce the expression of MAP kinase phosphatase-1 in PC12 cells. FEBS Lett 2000; 484:189–193.
90. Wang GL, Jiang BH, Rue EA, Semenza GL. Hypoxia-inducible factor 1 is a basic-helix-loop-helix-PAS heterodimer regulated by cellular O_2 tension. Proc Natl Acad Sci USA 1995; 92:5510–5514.
91. Norris ML, Millhorn DE. Hypoxia-induced protein binding to O_2-responsive sequences on the tyrosine hydroxylase gene. J Biol Chem 1995; 270:23774–23779.
92. Mishra RR, Adhikary G, Simonson MS, Cherniack NS, Prabhakar NR. Role of c-fos in hypoxia-induced AP-1 cis-element activity and tyrosine hydroxylase gene expression. Brain Res Mol Brain Res 1998; 15:74–83.
93. Semenza GL. HIF-1 and human disease: one highly involved factor. Genes Dev 2000; 14:1983–1991.
94. Semenza GL. HIF-1, O_2, and the 3 PHDs: how animal cells signal hypoxia to the nucleus. Cell 2001; 107:1–3.
95. Tian H, McKnight SL, Russell DW. Endothelial PAS domain protein 1 (EPAS1), a transcription factor selectively expressed in endothelial cells. Genes Dev 1997; 11:72–82.
96. Hipskind RA, Buscher D, Nordheim A, Baccarini M. Ras/MAP kinase-dependent and -independent signaling pathways target distinct ternary complex factors. Genes Dev 1994; 8:1803–1816.
97. Bernstein LR, Ferris DK, Colburn NH, Sobel ME. A family of mitogen-activated protein kinase-related proteins interacts in vivo with activator protein-1 transcription factor. J Biol Chem 1994; 269:9401–9404.
98. Egea J, Espinet C, Comella JX. Calmodulin modulates mitogen-activated protein kinase activation in response to membrane depolarization in PC12 cells. J Neurochem 1998; 70:2554–2564.
99. Egea J, Espinet C, Comella JX. Calcium influx activates extracellular-regulated kinase/mitogen-activated protein kinase pathway through a calmodulin-sensitive mechanism in PC12 cells. J Biol Chem 1999; 274:74–85.
100. Salceda S, Caro J. Hypoxia-inducible factor 1alpha (HIF-1alpha) protein is rapidly degraded by the ubiquitin-proteasome system under normoxic conditions: its stabilization by hypoxia depends on redox-induced changes. J Biol Chem 1997; 272:22642–22647.
101. Ema M, Hirota K, Mimura J, Abe H, Yodoi J, Sogawa K, Poellinger L, Fujii-Kuriyama Y. Molecular mechanisms of transcription activation by HLF and HIF1α in response to hypoxia: their stabilization and redox signal-induced interaction with CBP/p300. EMBO J 1999; 18:1905–1914.
102. Huang LE, Gu L, Schau M, Bunn HF. Regulation of hypoxia-inducible factor 1alpha is mediated by an O_2-dependent degradation domain via the ubiquitin-proteasome pathway. Proc Natl Acad Sci USA 1998; 95:7987–7992.
103. O'Rourke JF, Tian Y, Ratcliffe PJ, Pugh CW. Oxygen-regulated and transactivating domains in endothelial PAS protein 1: comparison with hypoxia-inducible factor-1alpha. J Biol Chem 1999; 274:2060–2071.
104. Janknecht R, Nordheim A. MAP kinase-dependent transcriptional coactivation by Elk-1 and its cofactor CBP. Biochem Biophys Res Commun 1996; 228:831–837.
105. Wiesner MS, Turley H, Allen WE, Willam C, Eckardt KU, Talks KL, Wood SM, Gatter KC, Harris AL, Pugh CW, Ratcliffe PJ, Maxwell PH. Induction of endothelial

PAS domain protein-1 by hypoxia: characterization and comparison with hypoxia-inducible factor-1alpha. Blood 1998; 92:2260–2268.
106. Yuan Y, Beitner-Johnson D, Millhorn DE. Hypoxia-Inducible Factor 2a binds to cobalt in vitro. Biochem Biophys Res Commun 2001; 288:849–854.
107. Maxwell PW, Wiesener MS, Chang G, Clifford SC, Vaux EC, Cockman ME, Wykoff CC, Pugh CW, Maher ER, Ratcliffe PJ. The tumour suppressor protein VHL targets hypoxia-inducible factors for oxygen-dependent proteolysis. Nature 1999; 399:271–275.
108. Epstein, ACR, Gleadle JM, McNeill LA, Hewitson KS, O'Rourke J, Mole DR, Mukherji M, Metzen E, Wilson MI, Dhanda A, Tian Y-M, Masson N, Hamilton DL, Jaakkola P, Barstead R, Hodgkin P, Maxwell PH, Pugh CW, Schofield CJ, Ratcliffe PJ. C elegans EGL-9 and mammalian homologs define a family of dioxygenases that regulate HIF by prolyl hydroxylation. Cell 2001; 107:43–54.
109. Ivan M, Kondo K, Yang H, Kim W, Valiando J, Ohh M, Salic A, Asara JM, Lane WS, Kaelin WG Jr. HIF1α targeted for VHL-mediated destruction by proline hydroxylation: implications for O_2 sensing. Science 2001; 292:464–468.
110. Jaakkola P, Mole DR, Tian Y-M, Wilson MI, Gielbert J, Gaskell SJ, von Kriegsheim A, Hebestreit HF, Mukherji M, Schofield CJ, Maxwell PH, Pugh CW, Ratcliffe PJ. Targeting of HIF-α to the von Hippel-Lindau ubiquitylation complex by O_2-regulated prolyl hydroxylation. Science 2001; 292:468–472.
111. Masson N, William C, Maxwell PH, Pugh CW, Ratcliffe PJ. Independent function of two destruction domains in hypoxia-inducible factor-α chains activated by prolyl hydroxylation. EMBO J 2001; 20:5197–5206.
112. Seta K, Kim H-W, Yuan Y, Lu G, Spicer Z, Kim R, Ferguson T, Pathrose P, Millhorn DE. Gene expression microarrays: a platform for discovery in biological systems. In: Kordal R, ed. Advanced Micro Fabricated Sensors. American Chemical Society 2002; pp 3–21.
113. Diatchenko L, Lukyanov S, Lau YF, Siebert PD. Suppression subtractive hybridization: a versatile method for identifying differentially expressed genes. Methods Enzymol 1999; 303:349–380.

9

Regulation of Tyrosine Hydroxylase Gene Expression by Hypoxia in Neuroendocrine Cells

MARIA F. CZYZYK-KRZESKA, PHILLIP O. SCHNELL, AMY L. BAUER, JUSTIN B. STRIET, JAMES A. NASH, ANNA V. KUZNETSOVA and ANNA S. HUI

University of Cincinnati College of Medicine
Cincinnati, Ohio, U.S.A.

I. Introduction

Tyrosine hydroxylase (TH), the rate-limiting enzyme in the synthesis of catecholamines, is regulated by hypoxia in dopaminergic type I cells of carotid body and in pheochromocytoma (PC12) cells, which are used as an experimental model system for type I cells. Hypoxic induction of TH gene expression occurs at the level of transcription and stability of the TH mRNA. Activation of TH gene transcription during hypoxia involves interaction of the c-fos and JunB transcription factors with the AP1 site on the TH promoter. Three motifs that closely resemble the hypoxia-response element (HRE) are present within the proximal region of the TH promoter. However, the specific functional role of the hypoxia-inducible transcription factors (HIFs) and the p300/CBP coactivators is not fully understood. Regulation of TH gene expression at the level of mRNA stability involves interaction of a polycytidine element located within the 3′ untranslated region of TH mRNA with two poly(C) RNA-binding proteins. Another important regulator of TH gene expression in PC12 cells is the von Hippel–Lindau protein (pVhl). pVhl is part of a multiprotein complex (elongins BC, cullin, Rbx-1) that has E3 ubiquitin ligase activity and thereby ubiquitinates target protein factors. Polyubiquitinated proteins, in turn, are subjected to proteasomal degradation. At present, the major known target proteins for the E3 ubiquitin ligase are alpha subunits of HIFs. Constitutive and hypoxia-induced expression of the TH gene is inversely correlated with the levels of pVHL. pVHL regulates TH gene expression at the level of transcriptional elongation and promoter activity. The role of pVHL in regulation of TH transcription elongation during normoxia and hypoxia is currently being investigated. The effects of pVLH on TH gene expression may result from the actions of pVHL-associated E3-ubiquitin ligase on specific elongation regulatory factors.

II. Regulation of TH Gene Expression by Hypoxia in Catecholaminergic Cells

Catecholamines are essential neurotransmitters that are involved in the adaptive regulatory process during hypoxia. A variety of catecholaminergic neurons are involved in the arterial chemoreceptor pathway, including:

- O_2-sensitive type I cells in the carotid body that detect reduced O_2 tension in the blood (1);
- sensory neurons of petrosal ganglion that convey signals from the carotid body to the nucleus tractus solitari (NTS) in the brainstem (2);
- interneurons within the commissural and medial subnuclei of the caudal NTS and interneurons in the ventrolateral medulla (3–7); and
- postganglionic sympathetic neurons and adrenal medulla cells that directly regulate various cardiovascular parameters (1,8,9).

The O_2 chemoreceptor reflex is critical for survival during hypoxia and catecholaminergic neurotransmitters play a central role in this pathway. Thus, the O_2-dependent regulation of TH, the rate-limiting enzyme in catecholamine synthesis, has been extensively studied. One of the best-characterized examples is the hypoxic regulation of TH in type I, O_2-sensitive cells of the carotid body. Stimulation of the carotid body by hypoxia (for 1–48 hr) causes the release of dopamine from type I cells, augments dopamine synthesis (10,11), increases the enzymatic activity of TH protein (12,13), and induces a severalfold increase in TH mRNA levels (1). This regulation of TH mRNA during hypoxia is an intrinsic property of type I cells (1). Neural or hormonal inputs are not required, and the effect is specific for hypoxia; for example, it is not observed in response to hypercapnia, which is also a physiological stimulus that regulates carotid body activity (1). Short-term hypoxia (6 hr) does not stimulate TH gene expression in postganglionic sympathetic neurons or in catecholaminergic cells of the adrenal medulla (1), although an increase in TH enzymatic activity has been reported in these tissues (13).

Long-term hypoxia increases TH protein content in the adrenal medulla, implying that regulation takes place at the level of gene expression (8,9); this induction, however, requires neural input to the adrenal medulla cells. There is also evidence that hypoxia may affect TH gene expression, protein synthesis, and catecholamine release in the brainstem. For example, long-term hypoxia increases TH protein content in the ventrolateral medulla, in the caudal region of the dorsomedial medulla, and in the locus ceruleus, and TH mRNA is increased in the C2/A2 neurons in the caudal NTS in response to long-term hypoxia (3–7). In contrast, TH protein content is not affected by hypoxia in dopaminergic neurons of substantia nigra (4).

Because of the heterogeneity of catecholaminergic cells in the brain and the scarcity of O_2-sensitive cells in the carotid body, clonal cell lines are needed to study the regulatory mechanisms of hypoxia-mediated TH gene expression. We have identified a dopaminergic, pheochromocytoma-derived cell line, PC12 (14), as a useful experimental model system for this purpose. During hypoxia, PC12 cells depolarize and release dopamine (15). In addition, hypoxia increases TH enzyme activity and protein levels (16), as well as TH gene expression (14). The sensitivity, magnitude, and time course of these responses in PC12 cells are very similar to those measured in O_2-sensitive type I cells of the carotid body (1,14). The hypoxia-induced increase in TH mRNA is mediated by a dual mechanism, involving an increased rate of TH gene transcription and increased stability of the TH mRNA (i.e., a decreased rate of TH mRNA degradation) (14). TH gene expression is also regulated by transsynaptic neuronal activity and by membrane depolarization, calcium ionophores, nerve and epidermal growth factors, glucocorticoids, cAMP analogs, and active phorbol esters, in addition to hypoxia (for review see Ref. 17).

III. Transcriptional Regulation of the TH Gene

The rat TH gene is a single-copy gene (7.3 kb) located on chromosome 11, and it contains 13 exons and 12 introns (18,19). Most of the regulatory elements for the TH gene have been identified within the proximal promoter region, specifically in its first −300 bp (Fig. 1). The rat TH promoter contains multiple *cis*-acting elements, including AP2 motif at −220 to −214 bp relative to the transcriptional start site, an AP1 motif at −206 to −200 bp, and an overlapping E box/dyad at −212 to −185 bp, a POU/OCT site at −176 to −169 bp, a HEPTA site at −159 to −165 bp, an Sp1 site at −120 to −113 bp, and two cAMP-responsive element sites (CREs), with the main site, CRE1, at −45 to −38 bp, and CRE2 at −97 to −90 bp. The rat TH promoter also contains three putative HRE-like sequences located at positions −221 to −225 bp, immediately upstream from the AP2 site, at −109 to −103 bp and at −99 to −94 bp. This last HRE-like element overlaps with the CRE2 site. The AP1 and CRE motifs appear to be the major elements involved in both the constitutive and regulated expression of TH transcription (20–24). However, appropriate tissue-specific expression of the TH gene in a transgenic mouse model requires a large upstream region of the TH gene (−4.8 to −9 kb) (25,26). Additional sequences in the 3′ flanking region of the human TH gene have been reported to be essential for specific TH expression in catecholaminergic cells of the peripheral nervous system, but not in the central nervous system (27).

The effects of hypoxia on TH promoter activity in PC12 cells have been studied by deletion analysis and transient expression of various TH promoter-

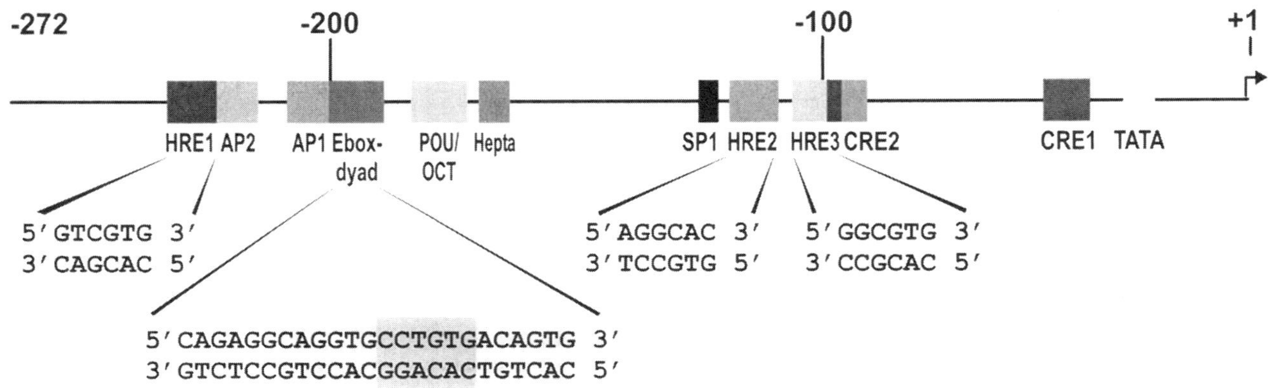

Figure 1 Schematic localization of the regulatory elements within the proximal region of rat TH promoter.

reporter constructs. These studies showed that the −150 to −284 bp fragment of the TH promoter containing the AP1/E box/dyad motifs was critical for activation of TH expression by hypoxia (28,29). Further analysis revealed that hypoxia induces binding of the c-Fos and JunB transcription factors to this element (28,29). It is not clear if the increased activity of the AP1 site during hypoxia results from the depolarization and calcium influx that occurs in PC12 cells during hypoxia or from activation of other hypoxia-specific signal transduction pathways. In that respect, both depolarization and calcium ionophores strongly induce TH gene transcription (30–32). Results from different laboratories show that the primary effect of depolarization and calcium ionophores on TH gene transcription is mediated by the CRE1 motif-binding CREB protein, which is phosphorylated on ser^{133}, and/or possibly other CRE-binding proteins (24,30–32). However, site-directed mutagenesis of the AP1 site or the CRE demonstrated that both elements are necessary for Ca^{2+}-mediated regulation of TH transcription (31). An increase in intracellular calcium leads to increased binding of c-Fos, c-Jun, JunB, and JunD to the AP1 site (31). Analysis of the role of the CRE motif in hypoxia-induced regulation of TH gene expression demonstrated that mutation of the CRE motif attenuated hypoxia-induced activation of a −272-bp TH promoter-reporter construct (33). Hypoxia also induces phosphorylation of CREB at ser^{133}, further suggesting that the CRE may be an important regulatory element in TH gene expression (33). The role of HIF in the hypoxic regulation of the TH gene has been suggested in the literature, but this has not been demonstrated experimentally. Thus, regulation of the TH promoter activity by hypoxia in PC12 cells, and also most likely in the carotid body, is a complex and relatively poorly understood process that appears to involve multiple *trans*-activating factors. These factors can be activated not only by a decrease in pO_2, but also possibly by depolarization and $[Ca^{2+}]$ influx. This suggests that the HRE and AP1 binding factors may interact to confer full hypoxic inducibility. In that respect, a similar relationship between the HRE and AP1 sites has been reported in the case of the promoter for the vascular endothelial growth factor (VEGF) gene in C6 glioma cells (34). Similarly, a synergistic interaction between HIF and the c-Jun or Smad 3 transcription factors has also been reported (35,36).

IV. Hypoxia-Regulated Transcription Factors and Coactivators in Regulation of TH Gene Expression by Hypoxia

A. Activating Protein 1 (AP1) Complexes

AP1 is comprised of a family of basic leucine zipper transcription factors that includes homo- and heterodimers of Jun (v-jun, c-Jun, JunB, JunD) and Fos (v-fos, c-Fos, Fra-1, Fra-2) or activating transcription factors (ATF2, ATF3/LRF1, B-ATF) (37). Jun-Jun and Jun-Fos dimers preferentially bind to the AP-1 consensus site (TGACTCA), whereas ATF or Jun/ATF homo- and heterodimers bind primarily to CRE elements (TGACGTCA). The activity of the AP1 complex is regulated by its composition and by posttranslational modifications. Various extracellular stimuli can cause the AP1 complex to change from Jun-Jun homodimers to Jun-Fos heterodimers (37). Fos proteins are thought to modulate AP1 activity as they cannot homodimerize, but only heterodimerize with members of the Jun-family proteins. The heterodimers have higher DNA-binding affinity and more strongly stimulate transcription (37). The activity of the AP1-binding proteins is regulated by redox potential (38–41). DNA-binding activity of Jun and Fos is inhibited by oxidizing reagents and is partially mimicked by reducing agents (38–41). A single cysteine residue within the DNA-binding domains of Fos and Jun has been identified as playing a role in DNA binding and is subject to redox regulation (38). Ref-1, a dual-function protein that has cysteine-reducing activities as well as DNA repair endonuclease activity, plays a role in this process. In the absence of chemical reducing agents, Ref-1 maintains these cysteine residues in the reduced state (39,40). Interestingly, the activation of Ref-1 and AP1-binding activity have been reported to be early events during hypoxic stimulation of colon cancer cells (41).

Members of the AP1 family of proteins have relatively short half-lives, and at least in the case of c-Fos and c-Jun, new evidence indicates that these proteins are regulated at the level of protein degradation by ubiquitination (42). Ubiquitination has been studied in the cases of c-Jun (42), and also ATF2 and JunB (43). These studies revealed that c-Jun is targeted for ubiquitination by c-Jun kinase (JNK), which is constitutively bound to c-Jun but is inactive as a kinase under basal conditions (43,44). Thus, under basal conditions, the transcription factor is not phosphorylated. In response to various stressors, JNK becomes activated and phosphorylates c-Jun, which protects it from degradation (43,44). The specific ubiquitin-conjugating and ubiquitin ligase enzymes involved in ubiquitination of c-Fos and Juns have not been clearly established (45).

B. Hypoxia-Inducible Factors (HIFs)

The majority of genes regulated by hypoxia contain HREs, which bind HIF transcription factors (for review see Ref. 46). The role of the HIF elements within the TH promoter has not been clearly elucidated. HIFs exist as heterodimers, comprised of the 120-kDa alpha and 90-kDa beta subunits. Both subunits contain a basic helix-loop-helix (bHLH) domain in their N-terminal regions, followed by a PAS (PER-ARNT-SIM) homology domain. Both bHLH and PAS domains are involved in the dimerization of the subunits and in their interaction with DNA. HIF1α and HIF1β are the prototypical HIFs and these proteins were purified based on their ability to bind DNA (47). The HIF1α molecule belongs to class II of the bHLH-PAS proteins. Recently, two new HIFs were identified: HIF2 (48–50) and HIF3 (51). These transcription factors differ in the their alpha subunit, while they contain the same beta subunit (HIF1β, also known as ARNT). HIF2α (also known as endothelial PAS domain protein 1, EPAS1 or HIF1α-like or -related factor, MOP2) has been shown to be highly expressed in endothelial cells (48), in fetal lung, and in catecholamine-producing cells, including the carotid body, suggesting that this transcription factor has specific function in these tissues (49). The expression of

HIF3α is not well known (51). HIF1β, or ARNT, the aryl-hydrocarbon nuclear translocator, belongs to the class I of bHLH-PAS proteins and was first described as binding to the aryl hydrocarbon receptor (AHR), which is also a member of class II of bHLH-PAS proteins (46).

Activation of HIF by hypoxia involves changes in the accumulation and activity of the HIF1α subunit. The primary level of regulation is inhibition of degradation of the HIF1α subunit during hypoxia (52–56). During normoxia, the expression of the HIFα proteins is usually very low, due to ongoing ubiquitination and proteasomal degradation (52–56). The pVHL-associated complex has E3 ligase activity toward the HIF1α (57–63). Ubiquitination and proteasomal degradation of HIFα subunits are inhibited during hypoxia or in response to exposure to iron chelators (deferoxamine) or $CoCl_2$, which mimic the effects of hypoxia (57–63). This causes accumulation of functional HIFα subunits, which dimerize with ARNT, translocate to the nucleus, and transactivate HIF-sensitive genes. The oxygen-dependent degradation domain (ODD) is localized between amino acids 401 and 603 of the HIF1α molecule (53). This region includes PEST (proline-serine-threonine-rich protein stabilization domains) sequences and the hydroxylate proline necessary for interaction with the pVHL (62,63). The EPAS1 (HIF2α) molecule does not seem to have a motif similar to the HIF1α ODD, but putative PEST motifs are located elsewhere on this protein (56). Nuclear translocation, posttranslational modifications [e.g., phosphorylation by MAP kinases (64–65)] and, at least in some tissues, induction of HIFα mRNA (46) represent additional levels of regulation that may regulate the activity of HIF. In addition, the HRE constitutively binds the ATF1 and CREB-1 transcription factors (66). These data indicate that regulation of gene expression by HIFs is a complex process that involves multiprotein complexes with different activities.

C. Coactivators p300/CBP

CBP (CREB-binding protein) and the adenovirus E1A-associated 300-kDa protein are similar proteins that share several important cellular functions (67). They are potent transcriptional coactivators that create a physical bridge between various transcription factors and the basal transcriptional machinery. In addition, CBP/p300 both exhibit histone acetyltransferase activity and have three cysteine-histidine-rich zinc finger domains (CH). The most N-terminal domain (C/H1) binds among other nuclear proteins, including CREB, Jun, MYB, ELK1, SREBP, SAP1, and HIF. The most carboxyl-terminal domain (C/H3) binds E1A, TFIIB and Fos, to name a few. Interestingly, CBP/p300 are present at limiting concentrations and their interaction with one group of specific regulators diminishes their coactivation via another group of regulators. Thus, p300/CBP serves as integrator of multiple signal transduction pathways within the nucleus (67). This competition for limiting amounts of p300/CBP was originally demonstrated between the AP1 and nuclear receptor hormones (68) and between the AP1 and JAK/STAT pathways (69). Importantly, transactivation of specific genes by HIFs involves interaction of the C-terminal transactivation domain with the C/H1 (cysteine-histidine-rich zinc finger motif) region of p300/CBP (70–72). The HIF-p300/CBP interaction can be blocked by a hypoxia-inducible p35srj protein that also interacts with the C/H1 domain of p300. This interaction may further modulate HIF activity (73). It is possible that binding of p300/CBP to the AP1 and HIF proteins on the TH promoter may mediate the gene-specific response to hypoxia of this promoter.

Recently, other transcriptional coactivators harboring histone acetyltransferase activity, namely SRC-1/p160, were demonstrated to enhance hypoxic stimulation (74). This complex also includes the bifunctional Ref-1 protein (74). These protein factors seem to act synergistically with CBP.

V. Posttranscriptional Regulation of TH Gene Expression

Hypoxia-inducible genes are often regulated in a dual fashion, at the level of transcription and at the level of mRNA stability. Such regulation occurs in the case of mRNAs for TH, VEGF, erythropoietin (EPO), and glucose transporter 1 (Glut-1) genes (75). The increase in gene transcription is a fast event that allows for rapid generation of new mRNA. However, prolonged exposure to hypoxia, at least in cell culture models, causes a general decrease in cellular transcription. In that case, an increase in the half-life of mRNA helps to maintain high levels of essential transcripts. In the case of TH mRNA, whose half-life is on the order of several hours, alterations in mRNA stability may contribute substantially to the level of mRNA accumulation in response to hypoxia after approximately 24 hr of exposure. Regulation of mRNA stability involves interactions of specific regulatory factors with *cis*-acting elements within the mRNA. These elements are often found in the 3′ or 5′ untranslated regions, but may also be located in the coding region of the mRNA. These protein–nucleic acid interactions regulate secondary structure of mRNA and the interactions of specific RNA with degrading nucleases. The TH mRNA has a 27-base-long pyrimidine-rich sequence within the 3′ untranslated region (3′UTR) that binds specific proteins in a hypoxia-inducible manner; this region is termed the hypoxia-inducible protein binding sequence (HIPBS) (76–78) (Fig. 2). One such site is represented by the motif (U/C)(C/U)CCCU located within this pyrimidine-rich sequence; it is conserved in TH mRNAs across different species (77). Hypoxia induces protein binding to this sequence in cytoplasmic extracts from PC12 cells, in catecholaminergic cells of superior cervical ganglia, and in dopaminergic cells of the carotid body, but not in extracts from the adrenal medulla cells (78). Thus, the formation of the protein complex associated with the HIPBS site in the 3′UTR of TH mRNA occurs in some, but not all, populations of catecholaminergic cells.

HIPBS regulates the stability of the TH mRNA under both hypoxic and normoxic conditions. A four-point mutation that abolishes the protein binding site within the full-length TH mRNA results in a twofold destabilization of the mutated mRNA and a corresponding twofold decrease in mRNA steady-state levels (79). In addition, mutation of HIPBS abolishes the O_2-dependent regulation of TH mRNA stability (79). A short fragment of the TH 3′UTR containing the wild-type HIPBS is

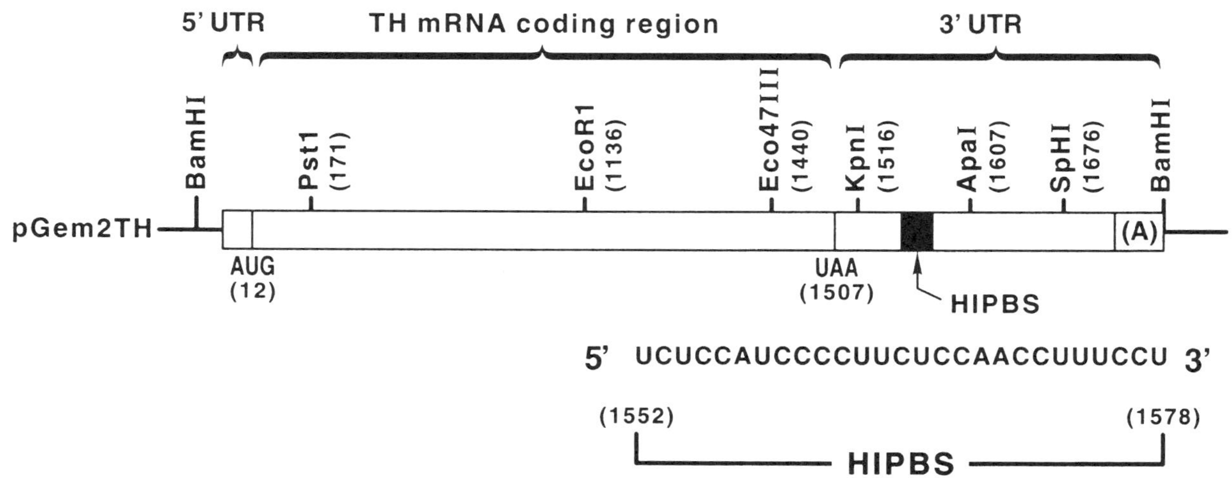

Figure 2 Schematic representation of the rat TH mRNA. Localization of the RNA regulatory sequence hypoxia-inducible protein-binding sequence (HIBS) is shown by the black box within the TH 3′ untranslated region.

sufficient to confer augmented mRNA stability on a heterologous mRNA, which in turn increases steady-state levels of the chimeric mRNA and its derived protein. Conversely, the mutated HIPBS destabilize the chimeric TH mRNA (79).

Analysis of the hypoxia-inducible protein-RNA complexes formed with the HIPBS element revealed the presence of two approximately 40-kDa poly(C)-binding proteins (PCBP1 and PCBP2) (79). The level of PCBP1, but not PCBP2, is increased by approximately twofold in cytoplasmic extracts from hypoxic PC12 cells, as compared to control cells. This correlates with a similar enrichment of PCBP1 in electroeluates from ribonucleoprotein complexes that form with the hypoxic protein extracts and quantitatively closely resembles the hypoxia-induced increase in the complex formation (79). This induction in the PCBP1 levels results from an enrichment in the total level of PCBP1, as similar increases were observed in total cellular lysates, and no intracellular translocation of PCBP1 was observed by immunocytochemistry (MFC-K, unpublished results). The difference in the hypoxic inducibility between the two isoforms suggests that although PCBP1 and PCBP2 share a high degree of sequence homology, they also exhibit some specific functional differences. The PCBPs regulate the stability of other mRNAs, including EPO (80). The 3′UTR of EPO mRNA contains polycytidine elements that bind both PCBP1 and PCBP2. However, in cell lines that express EPO, i.e., Hep3B or HepG2, hypoxia induces neither expression of PCBPs nor formation of the ribonucleoprotein complex associated with EPO mRNA that would involve PCBPs. This indicates that some tissue-specific mechanisms are involved in regulation of mRNA stability by PCBPs (80). The increase in mRNA stability during hypoxia through binding of PCBP to the HIPBS element may be caused by protection of a nuclease cleavage site within the region associated with HIPBS in the TH mRNA, as mutation of the HIPBS region decreases the stability of the chimeric mRNA stability (79). Although HIPBS is necessary, it is not sufficient for the O_2-dependent stabilization of TH mRNA. While it binds protein factors in a hypoxia-inducible manner, it does not intrinsically confer hypoxic regulation to the heterologous mRNA. Additional critical regulatory elements must be located within other regions of the TH mRNA. In that respect, the regulation of VEGF mRNA stability by hypoxia involves elements located within UTRs and within the coding region of VEGF mRNA (81). Thus, studies of both the TH and VEGF genes illustrate that the regulation of mRNA stability is complex and involves multiple factors.

VI. Role of pVhl in Regulation of TH Gene Expression

pVhl was originally identified as a tumor suppressor protein that exhibits loss of function in sporadic and familial renal clear-cell carcinomas (RCCs), hemangioblastomas, and pheochromocytomas (82). The familial form of pVhl-associated cancer syndrome, VHL disease, is an autosomal, dominantly inherited cancer syndrome affecting 1 in 36,000 people with a penetrance of 80% by age of 65 (82). Additional tumors (paragangliomas) can arise from extra-adrenal chromaffin cells, including the carotid body, in VHL disease (83,84).

Pheochromocytomas arise from adrenal medulla chromaffin cells. Pheochromocytomas associated with VHL disease synthesize and secrete increased amounts of catecholamines, predominantly norepinephrine (85). Catecholamines are neurotransmitters that regulate the activity and tone of cardiac and vascular smooth muscle. Thus, pheochromocytomas can cause sustained or episodic arterial hypertension (86–89). There is a strong correlation between catecholamine levels and concentrations of TH mRNA in pheochromocytomas (90–92). TH mRNA levels are two- to sixfold higher in various pheochromocytomas, as compared to the levels measured in normal adrenal medullas (90–92).

The human VHL gene is located on chromosome 3p25–26 and includes three exons (93). The human protein (hpVhl) contains 213 amino acid residues and migrates with an apparent molecular weight of 28–30 kDa (94). The rodent pVhl is significantly shorter (19 kDa), as a pentameric acidic repeat in the N-terminus is absent in rodents (95). The C-terminus of human pVHL is highly conserved among different species, and it is this region that is most frequently mutated in VHL disease–associated neoplasms. pVHL functions in the context of a multiprotein complex (96–100). pVHL binds elongins C and B, the two regulatory subunits of the elongation trimeric factor Elongin, SIII, containing elongins ABC (96–98). The pVHL complex also includes Cullin 2 (Cul2), a member of Cullin family of proteins (99), and Rbx-1, a RING-H2 finger protein, which has E3 ubiquitin ligase activity and strongly stimulates ubiquitination of the E1/E2 ubiquitin activating and conjugating enzymes (100). The role of pVHL in multiprotein complexes is to recognize specific substrates for ubiquitination and subsequent degradation (57). Consistent with this, pVHL-associated protein complex exhibits ubiquitin ligase activity toward the HIF1α subunit (58–63), and loss of VHL gene function leads to high levels of HIF and to up-regulation of various hypoxia-inducible genes during normoxic conditions, such as VEGF and platelet-derived growth factor (PDGF) (58,101). Tumors from patients with VHL disease also express high amounts of VEGF and EPO (102–104).

The role of pVhl in regulation of TH gene expression has been studied in PC12 cells by experimentally manipulating levels of pVhl through either stable overexpression of the wild-type human pVhl (hpVhl) or antisense VHL mRNA, to decrease expression of endogenous pVhl (105,106). PC12 cells expressing endogenous rat pVhl (rpVhl) were used as controls. The hpVhl is expressed at approximately fivefold higher levels than the endogenous rpVhl (Fig. 3). Expression of the VHL antisense mRNA diminished expression of the endogenous rpVhl by approximately five- to ten-fold in several clones (Fig. 3).

The steady-state levels of TH mRNA and TH protein correlated inversely with the levels of pVhl in PC12 cells. Overexpression of hpVhl dramatically decreased steady-state levels of TH protein and TH mRNA in all analyzed clones of transfected cells, as compared to TH expression in clones of regular PC12 cells (Fig. 3). In contrast, down-regulation of endogenous rpVHL induced a two- to threefold stimulation of TH protein and TH mRNA levels (Fig. 3).

pVhl regulates TH gene expression at the level of transcription in a dual mechanism involving efficiency of TH transcript elongation and activation of the TH

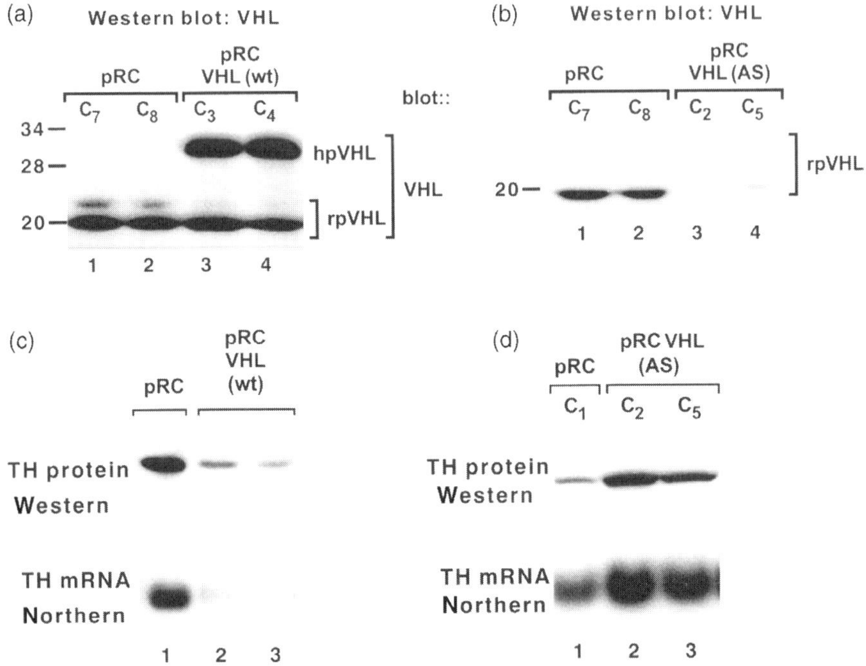

Figure 3 Analysis of TH protein and mRNA steady-state levels in PC12 clones with different amounts of pVhl. (a and b), Western blot analysis of VHL protein in PC12 cells. hpVhl, stably expressed human pVhl; rpVhl endogenous rat pVhl. (c and d), Western and Northern blot analysis of TH protein and mRNA. pRC, regular PC12 cells; pRCVHL(wt), PC12 cells overexpressing human pVHL; pRCVHL(AS), PC12 cells expressing VHL antisense RNA.

promoter. The stability of TH mRNA does not appear to be regulated by pVhl (105). Nuclear run-on assays were used to measure transcription of the TH endogenous gene, and transient transfection assays were used to measure the activity of different deletion constructs of a TH promoter-reporter gene.

In the nuclear run-on assays, transcription is not initiated de novo. Instead, preinitiated RNA transcripts are elongated by RNA polymerase II complexes (RNAPs) for a few hundred bases. The radioactivity incorporated during the run-on experiments represents a "snapshot" of the density of active, elongating RNAPs. Thus, any region of the gene containing pausing or arrest sites for RNAP results in synthesis of fewer radioactive transcripts. The distribution of RNAPs along the TH gene was determined by hybridizing radioactively labeled TH transcripts to fragments of DNA encompassing the full length of the TH gene (Fig. 4). The hybridized signals were quantified by Phosphorimager, and the measurements obtained from run-ons performed on the nuclei from cells with increased and decreased levels of pVHL protein were normalized to the respective signals derived from nuclei of normal PC12 cells. In cells overexpressing pVhl (Fig. 4, left), a

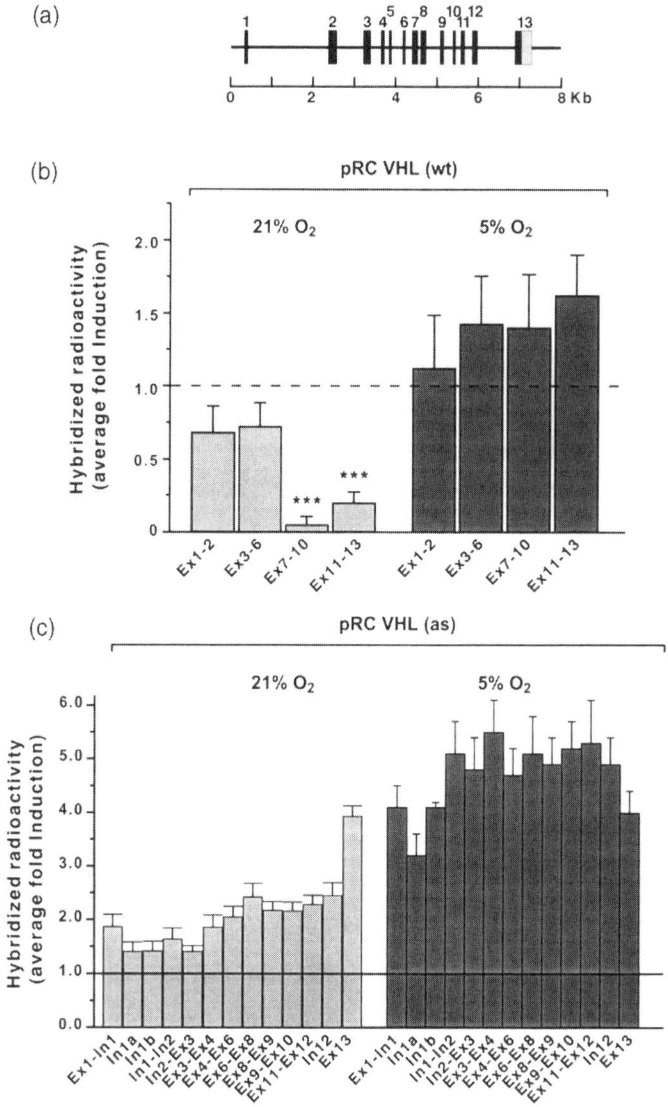

Figure 4 Transcription of TH gene in cells PC12 cells expressing different amounts of pVHL. Schematic representation of the TH gene. Nuclear run-on analysis of TH gene transcription in PC12 clone overexpressing pVHL during normoxia (21% O_2, left) and in hypoxia (5% O_2, right). Hypoxia abolishes transcript elongation block caused by overexpression of pVHL. Nuclear run-on analysis of TH gene transcription in PC12 clone expressing reduced amounts of pVHL. Reduced amount of pVHL increases formation of the full-length transcripts during normoxia (left). Hypoxia additionally stimulates transcription of the full-length transcripts in clones with reduced amounts of pVHL (right). The data represent average data from several experiments and are normalized to the values measured in normoxic control PC12 cells.

substantial decrease in the distribution of RNAPs was measured beyond exon 7, ranging from an 80 to 95% decrease from the value obtained in control cells. In contrast, there was only a slight (25%), nonsignificant decrease in the distribution of RNAPs upstream from exon 7 (Fig. 4, left). These data indicate that the concentration of pVhl constitutively modulates elongation of TH transcript in the 3′ region of the gene, thereby inducing a block in transcription elongation. This suggests that the decrease measured in elongation in the 3′ region of the TH gene does not reflect a general inhibition of TH transcription due to repression of the TH promoter. Rather, this effect appears to be specific for sequences around exons 7–8.

In the case of the cells with decreased amounts of the endogenous pVHL, transcription of the TH gene was measured by nuclear run-on assays where the RNA was hybridized to 13-DNA fragments, each approximately 500 bp long, covering the full length of the TH gene. The decreased concentration of pVhl in cells correlated with an increase in general transcription of the TH gene (Fig. 4, left). This increase in the hybridization signal was only in the range of 50–100% in the 5′ region of the TH gene (exons 1–6), but was 200–400% in the 3′ region of the TH gene (exons 6–13). These data strongly suggest that pVhl plays a role in regulation of transcript elongation in the 3′ region of the TH gene.

Hypoxia affects TH transcript elongation. Exposure of cells overexpressing pVhl to hypoxia stimulated primarily in the 3′ region of the gene, which contains VHL-sensitive pause/arrest sites (Fig. 4, right). Transcription of early transcripts (exons 1–2, probe a) was only modestly induced beyond normoxic (control) levels. When cells with reduced amounts of pVHL were exposed to hypoxia, transcription was induced along the full length of the TH gene, but there was an additional increase in the transcript processivity in the region between exons 3 and 13 (Fig. 4, right). This indicates an increased efficiency of transcript elongation during hypoxia.

Analysis of the activity of various TH promoter-reporter constructs demonstrated that cotransfection of wild-type VHL reduces hypoxic inducibility of the TH promoter, but had little effects on the activity of the TH promoter during normoxia. On the other hand, expression of VHL antisense RNA increased both constitutive and hypoxia-stimulated activity of the TH promoter (Fig. 5). The molecular mechanisms underlying this regulation are at present under investigation. It is logical to predict that at least some aspect of this regulation may be mediated through the effects of pVhl on accumulation of the HIF1α protein.

In conclusion, studies of the effects of pVhl on TH gene trascription revealed a novel mechanism of regulation of expression of this gene by hypoxia, i.e., regulation at the level of efficiency of transcript elongation, or processivity of transcription. pVhl appears to have constitutive inhibitory activity on the elongation of TH transcripts in the downstream region of the gene. This repressive effect is abolished when pVhl levels are significantly reduced. It is also attenuated by hypoxia. Thus, one of the important effects of hypoxia on transcription of the TH gene is increased efficiency of formation of the full-length transcripts, by suppressing pausing of the RNAP complexes. The precise molecular mechanism by which this process is regulated is under continuing investigation. Regulation of transcription at the level of

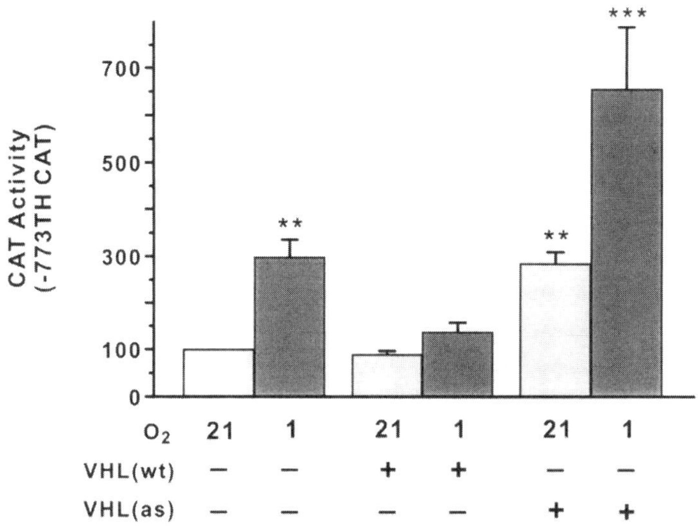

Figure 5 Regulation of the TH promoter activity by hypoxia and pVHL. Cells were transiently transfected with −773 TH −CAT reporter construct and either expression vector containing wild-C type pVHL or VHL cDNA cloned in antisense orientation.

efficiency of transcript elongation is a relatively novel form of regulation that is not well understood. At least in some hypoxia-inducible genes, this mechanism adds to the complexity of hypoxia-induced regulation of gene expression, in addition to the regulation of promoter activity and mRNA stability.

Acknowledgments

This work has been supported by NIH grants HL58687, HL66312, AHA grant-in-aid 9750110N, ACS Research Scholar Grant GMC-101430, and VHL Family Alliance Research Grant. ALB was supported by the training grant T32 HL07571.

References

1. Czyzyk-Krzeska MF, Bayliss DA, Lawson EE, Millhorn DE. Regulation of tyrosine hydroxylase gene expression in the rat carotid body by hypoxia. J Neurochem 1992; 58:1538–1546.
2. Czyzyk-Krzeska MF, Bayliss DA, Lawson EE, Millhorn DE. Expression of messenger RNAs for peptides and tyrosine hydroxylase in primary sensory neurons that innervate arterial baroreceptors and chemoreceptors. Neurosci Lett 1991; 129:98–102.
3. Dahlström A, Fuxe K. Evidence for the existence of monoamine containing neurons in the central nervous system. I. Demonstration of monoamines in the cell bodies of brainstem neurons. Acta Physiol Scand 1964; 62:1–55.

4. Schmitt P, Pequignot J, Garcia C, Pujol JF, Pequignot JM. Regional specificity of the long-term regulation of tyrosine hydroxylase in some catecholaminergic rat brainstem areas. I. Influence of long-term hypoxia. Brain Res 1993; 14:53–60.
5. Soulier V, Dalmaz Y, Cottet-Emard JM, Kitahama K, Pequignot JM. Delayed increase of tyrosine hydroxylase in the rat A2 medullary neurons upon long-term hypoxia. Brain Res 1995; 674:188–195.
6. Pepin JL, Levy P, Garcin A, Feuerstein C, Savasta M. Effects of long-term hypoxia on tyrosine hydroxylase protein content in catecholaminergic rat brainstem areas: a quantitative autoradiographic study. Brain Res 1996; 733:1–8.
7. Roux JC, Pequignot JM, Dumas S, Pascual O, Ghilini G, Pequignot J, Mallet J, Denavit-Saubie M. O_2 sensing after carotid chemodenervation: hypoxic ventilatory responsiveness and upregulation of tyrosine hydroxylase mRNA in brainstem catecholaminergic cells. Eur J Neurosci 2000; 12:3181–3190.
8. Hayashi Y, Miwa S, Lee K, Koshimura K, Hamahata K, Hasegawa H, Fujiwara M, Watanabe Y. Enhancement of in vivo tyrosine hydroxylation in the rat adrenal gland under hypoxic conditions. J Neurochem 1990; 54:1115–1121.
9. Schmitt P, Garcia C, Soulier V, Pujol JF, Pequinot JM. Influence of long-term hypoxia on tyrosine hydroxylase in the rat carotid body and adrenal gland. J Auton Nerv Syst 1992; 40:13–20.
10. Fidone S, Gonzalez C, Yoshizaki K. Effects of low oxygen on the release of dopamine from the rabbit carotid body in vitro. J Physiol (Lond) 1982; 333:93–110.
11. Fishman MC, Greene WL, Platika D. Oxygen chemoreception by carotid body cells in culture. Proc Natl Acad Sci USA 1985; 82:1448–1450.
12. Gonzalez C, Kwok Y, Gibb J, Fidone S. Effects of hypoxia on tyrosine hydroxylase activity in rat carotid body. J Neurochem 1979; 33:713–719.
13. Gonzalez C, Kwok Y, Gibb J, Fidone S. Physiological and pharmacological effects on TH activity in rabbit and cat carotid body. Am J Physiol 1981; 240:R38–R43.
14. Czyzyk-Krzeska MF, Furnari BA, Lawson EE, Millhorn DE. Hypoxia increases rate of transcription and stability of tyrosine hydroxylase mRNA in pheochromocytoma (PC12) cells. J Biol Chem 1994; 269:760–764.
15. Zhu WH, Conforti L, Czyzyk-Krzeska MF, Millhorn DE. Membrane depolarization and dopamine secretion in PC12 cells during hypoxia are regulated by an O_2-sensitive K^+ current. Am J Physiol 1996; 271:C658–C665.
16. Feinsilver SH, Wong R, Rayabin DM. Adaptation of neurotransmitter synthesis to chronic hypoxia in cell culture. Biochem Biophys Acta 1997; 928:56–62.
17. Kumer SC, Vrana KE. Intricate regulation of tyrosine hydroxylase activity and gene expression. J Neurochem 1996; 67:443–462.
18. Brown ER, Coker GT, O'Malley KL. Organization and evolution of the rat tyrosine hydroxylase gene. Biochemistry 1987; 26:5208–5212.
19. Grima B, Lamouroux A, Blanot F, Biguet NF, Mallet J. Complete coding sequence of rat tyrosine hydroxylase mRNA. Proc Natl Acad Sci USA 1985; 82:617–621.
20. Yoon SO, Chikaraishi DM. Tissue specific transcription of the rat tyrosine hydroxylase gene requires synergy between an AP-1 motif and an overlapping E box-containing dyad. Neuron 1992; 9:55–67.
21. Kim K-S, Lee M.K, Carroll J, Joh TH. Both the basal and inducible transcription of the tyrosine hydroxylase gene are dependent upon a cAMP responsive element. J Biol Chem 1993; 268:15689–15695.

22. Tinti C, Yang C, Seo H, Conti B, Kim C, Joh TH, Kim K-S. Structure/function relationship of the cAMP response element in tyrosine hydroxylase gene transcription. J Biol Chem 1997; 272:19158–19164.
23. Tinti C, Conti B, Cubells JF, Kim K-S, Baker H, Joh TH. Inducible cAMP early repressor can modulate tyrosine hydroxylase gene expression after stimulation of cAMP synthesis. J Biol Chem 1996; 271:25375–25381.
24. Ghee M, Baker H, Miller JC, Ziff EB. AP1 CREB and CBP transcription factors differentially regulate the tyrosine hydroxylase gene. Mol Brain Res 1998; 55:101–114.
25. Banerjee SA, Roffler-Tarlov S, Szabo M, Frohman L, Chikaraishi DM. DNA regulatory sequences of the rat tyrosine hydroxylase gene direct correct catecholaminergic cell-type specificity of a human growth hormone reporter in CNS of transgenic mice causing a dwarf phenotype. Mol Brain Res 1994; 24:89–106.
26. Min N, Joh TH, Jkim KS, Peng C, Son JH. 5′ upstream DNA sequence of the rat tyrosine hydroxylase gene directs high-level and tissue specific expression to catecholaminergic neurons in the central nervous system of transgenic mice. Mol Brain Res 1994; 27:281–289.
27. Wong SC, Moffat MA, Coker GT, Merlie JP, O'Malley KL. The 3′ flanking region of the human tyrosine hydroxylase gene directs reporter gene expression in peripheral neuroendocrine tissues. J Neurochem 1995; 65:23–31.
28. Norris ML, Millhorn DE. Hypoxia-induced protein binding to O_2-responsive sequences on the tyrosine hydroxylase gene. J Biol Chem 1995; 270:23774–23779.
29. Mishra RR, Adhikary G, Simonson MS, Cherniack NS, Prabhakar NR. Role of c-fos in hypoxia-induced AP-1 cis-element activity and tyrosine hydroxylase gene expression. Mol Brain Res 1998; 59:74–83.
30. Kilbourne EJ, Nankova BB, Lewis EJ, McMahon AN, Osaka H, Sabban DB, Sabban EL. Regulated expression of the tyrosine hydroxylase gene by membrane depolarization. J Biol Chem 1992; 267:7563–7569.
31. Nagamoto-Combs K, Piech KM, Best JA, Sun B, Tank AW. Tyrosine hydroxylase gene promoter activity is regulated by both cyclic AMP-responsive element and AP1 sites following calcium influx. J Biol Chem 1997; 272:6051–6058.
32. Nankova B, Hiremagalur B, Menezes A, Zeman R, Sabban E. Promoter elements and second messenger pathways involved in transcriptional activation of tyrosine hydroxylase by ionomycin. Mol Brain Res 1996; 35:164–172.
33. Beitner-Johnson D, Millhorn DE. Hypoxia induces phosphorylation of the cyclic AMP response element-binding protein by a novel signaling mechanism. J Biol Chem 1998; 273:19834–19839.
34. Damert A, Ikeda E, Risau W. Activator-protein-1 binding potentiates the hypoxia-inducible factor-1-mediated hypoxia-induced transcriptional activation of vascular-endothelial growth factor expression in C6 glioma cells. Biochem J 1997; 327: 419–423.
35. Sanchez-Elsner T, Botella LM, Velascoe B, Corbi A, Attisano L, Bernabeu C. Synergistic cooperation between hypoxia and transforming growth factor-B pathways on human vascular endothelial growth factor gene expression. J Biol Chem 2001; 276:38527–38535.
36. Alfranca A, Gutierrez MD, Vara A, Aragones J, Vidal F, Landazuri MO. c-Jun and hypoxia inducible factor functionally cooperate in hypoxia induced gene transcription. Mol Cell Biol 2002; 22:12–22.

37. Karin M, Liu ZG, Zandi E. AP-1 function and regulation. Curr Opin Cell Biol 1997; 9:240–246.
38. Abate C, Patel L, Rauscher FJ, Curran T. Redox regulation of fos and Jun DNA binding activity in vitro. Science 1990; 249:1157–1161.
39. Xanthoudakis S, Miao G, Wang YC, Pan E, Curran T. Redox activation of Fos-Jun DNA binding activity is mediated by a DNA repair enzyme. EMBO J 1992; 11:3323–3335.
40. Xanthoudakis S, Curran T. Identification and characterization of Ref-1, a nuclear protein that facilitates AP-1 DNA binding activity. EMBO J 1992; 11:653–665.
41. Yao KS, Xanthoudakis S, Curran T, O'Dwyer PJ. Activation of AP-1 and of a nuclear redox factor, Ref-1, in the response of HT29 colon cancer cells to hypoxia. Mol Cell Biol 1994; 14:5997–6003.
42. Hermida Matsumoto ML, Chock PB, Curran T, Yang DCH. Ubiquitinylation of transcription factors c-Jun and c-fos using reconstituted ubiquitinylating enzymes. J Biol Chem 1996; 271:4930–4936.
43. Fuchs SY, Xie B, Adler V, Fried A, Davis RJ, Ronai Z. NH2 terminal kinases target the ubiquitination of their associated transcription factors. J Biol Chem 1997; 272:32163–32168.
44. Fuchs SY, Fried VA, Ronai Z. Stress-regulated kinases regulate protein stability. Oncogene 1998; 17:1483–1490.
45. Stancovski I, Gonen H, Orian A, Schwartz AL, Ciechanover A. Degradation of the protooncogene product c-fos by ubiquitin proteolytic system in vivo and in vitro: identification and characterization of the conjugating enzymes. Mol Cell Biol 1995; 15:7106–7116.
46. Semenza GL. HIF-1: mediator of physiological and pathophysiological responses to hypoxia. J Appl Physiol 2000; 88:1474–1480.
47. Wang GL, Semenza GL. Purification and characterization of hypoxia inducible factor 1. J Biol Chem 1995; 27:1230–1237.
48. Tian H, McKnight SL, Russell DW. Endothelial PAS domain protein 1 (EPAS1), a transcription factor selectively expressed in endothelial cells. Gen Dev 1997; 11:72–82.
49. Tian H, Hammer RE, Matsumoto AM, Russell DW, McKnight SL. The hypoxia-responsive transcription factor EPAS1 is essential for catecholamine homeostasis and protection against heart failure during embryonic development. Gen Dev 1998; 12:3320–3324.
50. Wiesener MS, Turley H, Allen WE, Wiliam C, Eckardt K-U, Talks KL, Wood SM, Gatter KC, Harris AL, Pugh CW, Ratcliffe PJ, Maxwell PH. Induction of endothelial PAS domain protein 1 by hypoxia: characterization and comparison with hypoxia inducible factor 1á. Blood 1998; 92:2260–2268.
51. Gu YZ, Moran SM, Hogenesch JB, Wartman L, Bradfield CA. Molecular characterization and chromosomal localization of a third-class hypoxia inducible factor subunit, HIF3á. Gene Exp 1998; 7:205–213.
52. Huang LE, Arany Z, Livingston DM, Bunn HF. Activation of hypoxia-inducible transcription factor depends primarily upon redox-sensitive stabilization of its alpha subunit. J Biol Chem 1996; 271:32253–32259.
53. Huang LE, Gu J, Schau M, Bunn HF. Regulation of hypoxia-inducible factor 1á is mediated by an O_2-dependent degradation domain via the ubiquitin-proteasome pathway. Proc Natl Acad Sci USA 1998; 95:7987–7992.

54. Kallio PJ, Wilson WJ, O'Brien S, Makino Y, Poellinger L. Regulation of the hypoxia-inducible transcription factor 1alpha by the ubiquitin-proteasome pathway. J Biol Chem 1999; 274:6519–6525.
55. Sutter CH, Laughner E, Semenza GL. Hypoxia-inducible factor 1alpha protein expression is controlled by oxygen-regulated ubiquitination that is disrupted by deletions and missense mutations. Proc Natl Acad Sci USA 2000; 97:4748–4753.
56. O'Rourke JF, Tian YM, Ratcliffe PJ, Pugh CW. Oxygen-regulated and transactivating domains in endothelial PAS protein 1: comparison with hypoxia inducible factor-1á. J Biol Chem 1999; 274:2060–2071.
57. Iwai K, Yamanaka K, Kamura T, Minato N, Conaway RC, Conaway JW, Klausner DR, Pause A. Identification of the von Hippel–Lindau tumor-suppressor protein as part of an active E3 ubiquitin ligase complex. Proc Natl Acad Sci USA 1999; 96: 12436–12441.
58. Maxwell PH, Wiesener MS, Chang G-W, Clifford SC, Vaux EC, Cockman ME, Wykoff CC, Pugh CW, Maher ER, Ratcliffe PJ. The tumor suppressor protein VHL targets hypoxia-inducible factors for oxygen-dependent proteolysis. Nature 1999; 399:271–275.
59. Cockman ME, Masson N, Mole DR, Jaakkola P, Chang G-W, Clifford SC, Maher ER, Pugh CW, Ratcliffe PJ, Maxwell PH. Hypoxia inducible factor-á binding and ubiquitylation by the von Hippel–Lindau tumor suppressor protein. J Biol Chem 2000; 275:25733–25741.
60. Kamura T, Sato S, Iwai K, Czyzyk-Krzeska MF, Conaway RC, Conaway JW. Activation of HIF1á ubiquitination by a reconstituted von Hippel–Lindau (VHL) tumor suppressor complex. Proc Natl Acad Sci USA 2000; 97:10430–10435.
61. Ohh M, Park CW, Ivan M, Hoffman MA, Kim TY, Huang LE, Pavletich N, Chau V, Kaelin WG. Ubiquitination of hypoxia-inducible factor requires direct binding to the β domain of the von Hippel–Lindau protein. Nat Cell Biol 2000; 2:423–427.
62. Ivan M, Kondo K, Yang H, Kim W, Valiando J, Ohh M, Salic A, Asara JM, Lane WS, Kaelin Jr., WG. HIF alpha targeted for VHL-mediated destruction by proline hydroxylation: implications in oxygen sensing. Science 2001; 292:464–468.
63. Jaakkola P, Mole DR, Tian YM, Wilson MI, Gielbert J, Gaskell SJ, Kriegsheim AV, Hebestreit HF, Mukherji M, Schofield CJ, Maxwell PH, Pugh CW, Ratcliffe PJ. Targeting of HIF-alpha to the von Hippel–Lindau ubiquitylation complex by regulated prolyl hydroxylation. Science 2001; 292:468–472.
64. Richard D, Berra F, Gothie E, Roux D, Poussegur J. p42/p44 mitogen activated protein kinases phosphorylate hypoxia-inducible factor 1 (HIF-1) and enhance the transcriptional activity of HIF-1. J Biol Chem 1999; 274:32631–32637.
65. Conrad PW, Freeman TL, Beitner-Johnson D, Millhorn DL. EPAS1 trans-activation during hypoxia requires p42/p44 MAPK. J Biol Chem 1999; 274:33709–33713.
66. Kvietikova I, Wenger RH, Marti HH, Gassman M. The transcription factors ATF-1 and CREB-1 bind constitutively to the hypoxia-inducible factor-1 (HIF-1) DNA recognition site. Nucl Acids Res 1995; 23:4542–4550.
67. Giles RH, Peters DJM, Breuning MH. Conjunction dysfunction: CBP/p300 in human disease. TIG 1998; 14:5–9.
68. Kamei Y, Xu L, Heinzel T, Torchia J, Kurokawa R, Gloss B, Lin SC, Heyman RA, Rose DW, Glass CK, Rosenfeld MG. A CBP integrator complex mediates transcriptional activation and AP-1 inhibition by nuclear receptors. Cell 1996; 85:403–414.

69. Horvai AE, Xu L, Korzus E, Brard G, Kalafaus D, Muellin T-M, Rose DW, Rosenfeld MG, Glass CK. Nuclear integration of JAK/STAT and Ras/Ap-1 signaling by CBP and p300. Proc Natl Acad Sci USA 1997; 94:1074–1079.
70. Arany Z, Huang LE, Eckner R, Bhattacharya S, Jiang C, Goldberg MA, Bunn HF, Livingstone DM. An essential role for p300/CBP in cellular response to hypoxia. Proc Natl Acad Sci USA 1996; 93:12969–12973.
71. Ebert B, Bunn HF. Regulation of transcription by hypoxia requires a multiprotein complex that induces hypoxia-inducible factor 1, an adjacent transcription factor, and p300/CREB binding protein. Mol Cell Biol 1998; 18:4089–4096.
72. Kallio PJ, Okamoto K, O'Brien S, Carrero P, Makino Y, Tanaka H, Poellinger L. Signal transduction in hypoxic cells: inducible nuclear translocation and recruitment of the CBP/p300 coactivator by the hypoxia inducible factor-1á. EMBO J 1998; 17: 6573–6586.
73. Bhattacharya S, Michels C, Leung MK, Arany Z, Kung AL, Livingstone DM. Functional rote of p35srj, a novel p300/CBP binding protein during transactivation by Hif-1. Gen Dev 1999; 13:64–75.
74. Carrero P, Okamoto K, Coulmailleau P, O'Brien S, Tanaka H, Poellinger L. Redox-regulated recruitment of the transcriptional co-activators CREB-binding protein and SRC-1 to hypoxia-inducible factor 1á. Mol Cell Biol 2000; 20:402–415.
75. Paulding WR, Czyzyk-Krzeska MF. Post-transcriptional regulation of gene expression by hypoxia. Adv Exp Med Biol 2000; 475:111–122.
76. Czyzyk-Krzeska MF, Dominski Z, Kole R, Millhorn DE. Hypoxia stimulates binding of a cytoplasmic protein to a pyrimidine rich sequence in the 3′-untranslated region of rat tyrosine hydroxylase mRNA. J Biol Chem 1994; 269:9940–9945.
77. Czyzyk-Krzeska MF, Beresh JE. Characterization of the hypoxia inducible protein binding site within the pyrimidine rich tract in the 3′ untranslated region of the tyrosine hydroxylase mRNA. J Biol Chem 1996; 271:3293–3299.
78. Czyzyk-Krzeska MF, Paulding WR, Beresh JE, Kroll SL. Posttranscriptional regulation of tyrosine hydroxylase gene expression by oxygen in PC12 cells. Kidney Int 1997; 51:585–590.
79. Paulding WR, Czyzyk-Krzeska MF. Regulation of tyrosine hydroxylase mRNA stability by protein-binding, pyrimidine-rich sequence in the 3′ untranslated region. J Biol Chem 1999; 274:2532–2538.
80. Czyzyk-Krzeska MF, Bendixen AC. Identification of the poly(C) binding protein in the complex associated with the 3′ untranslated region of erythropoietin mRNA. Blood 1999; 93:2111–2120.
81. Dibbens JA, Miller DL, Damert A, Risau W, Vadas MA, Goodall GJ. Hypoxic regulation of vascular endothelial growth factor mRNA stability requires the cooperation of multiple RNA elements. Mol Biol Cell 1999; 10:907–919.
82. Maher ER, Kaelin Jr, WG. Von Hippel–Lindau disease. Rev Mol Med 1997; 76:381–384.
83. Bender BU, Altehofer C, Januszewicz A, Gartner R, Schmidt H, Hoffmann MM, Heidemann PH, Neumann HPH. Functioning thoracic paraganglioma: association with von Hippel–Lindau syndrome. J Clin Endocrinol Metabol 1997; 82:3356–3360.
84. Zanelli M, Van Der Walt JD. Carotid body paraganglioma in von Hippel–Lindau disease. Histopathology 1996; 29:178–181.
85. Eisenhofer G, Walther MM, Huynh T-T, Li S-T, Bornstein SR, Vortmeyer A, Mannelli M, Goldstein DS, Linehan WM, Lenders JWM, Pacak K. Pheochromo-

cytomas in von Hippel–Lindau syndrome and mutiple endocrine neoplasia type 2 display distict biochemical and clinical phenotypes. J Cell Endocrinol Metabol 2001; 86:1999–2008.

86. Kenady DE, McGrath PC, Sloan DA, Schwartz RW. Diagnosis and management of pheochromocytoma. Curr Opin Oncol 1997; 9:61–67.
87. Richard S, Beigelman C, Duclos JM, Fendler JP, Plauchu H, Plouin PR, Resche F, Schlumberger M, Vermesse B, Proye C. Pheochromocytoma as the first manifestation of von Hippel–Lindau disease. Surgery 1994; 116:1076–1081.
88. Crossey PA, Eng C, Ginalska-Malinowska M, Lennard TWJ, Wheeler DD, Ponder BAJ, Maher ER. Molecular genetic diagnosis of von Hippel–Lindau disease in familial pheochromocytoma. J Med Genet 1995; 32:885–886.
89. Atuk NO, Stolle C, Owen Jr, JA, Carpenter JT, Vance ML. Pheochromocytoma in von Hippel–Lindau disease: clinical presentation and mutation analysis in a large multigenerational kindred. J Clin Endocrinol Metabol 1998; 83(1):117–120.
90. Feldman JM, Blalock JA, Zern RT, Wells SA. The relationship between enzyme activity and the catecholamine content and secretion of pheochromocytomas. J Cell Endocrinol Metabol 1979; 49:445–451.
91. Isobe K, Nakai T, Yukimasa N, Nanmoku T, Takekoshi K, Nomura F. Expression of mRNA coding for four catecholamine-synthesizing enzymes in human adrenal pheochromocytoma. Eur J Endocrinol 1998; 138:383–387.
92. Lehnert H. Regulation of catecholamine synthesizing enzyme gene expression in human pheochromocytoma. Eur J Endocrinol 1998; 138:363–367.
93. Latif F, Tory K, Gnarra J, Yao M, Duh FM, Orcutt ML, Stackhouse T, Kuzmin I, Modi W, Geil L, Schmidt L, Zhou F, Li H, Wei MH, Chen F, Glenn G, Choyke P, Walther MM, Weng Y, Duan D-S R, Dean M, Glavac D, Richards FM, Crossey PA, Ferguson-Smith MA, Le Paslier D, Chumakov I, Cohen D, Chinault AC, Maher ER, Linehan WM, Zbar B, Lerman MI. Identification of the von Hippel–Lindau disease tumor suppressor gene. Science 1993; 260:1317–1320.
94. Iliopoulos O, Kibel A, Gray S, Kaelin Jr, WG. Tumour suppression by the human von Hippel–Lindau gene product. Nat Med 1995; 1:822–826.
95. Duan DR, Humphrey JS, Chen DY, Weng Y, Sukegawa J, Lee S, Gnarra JR, Linehan WM, Klausner RD. Characterization of the VHL tumor suppressor gene product: localization, complex formation, and the effect of natural inactivating mutations. Proc Natl Acad Sci USA 1995; 92:6459–6463.
96. Stebbins CE, Kaclin WG, Paveletich NP. Structure of the VHL-elongin C-elongin B complex: implications for VHL tumor suppressor function. Science 1999; 284: 455–461.
97. Kibel A, Iliopoulos O, DeCaprio JA, Kaelin WG. Binding of the von Hippel–Lindau tumor suppressor protein to elongin B and C. Science 1995; 269:1444–1445.
98. Takagi Y, Pause A, Conaway RC, Conaway JW. Identification of elongin C sequences required for interaction with the von Hippel–Lindau tumor suppressor protein. J Biol Chem 1997; 272:27444–27449.
99. Pause A, Lee S, Worrell RA, Chen DYT, Burgess WH, Linehan WM, Klausner RD. The von Hippel–Lindau tumor-suppressor gene product forms a stable complex with human CUL-2, a member of the Cdc53 family of proteins. Proc Natl Acad Sci USA 1997; 94:2156–2161.
100. Kamura T, Koepp DM, Conrad MN, Skowyra D, Moreland RJ, Iliopoulos O, Lane WS, Kaelin WG, Elledge SJ, Conaway RC, Harper JW, Conaway JW. Rbxl, a component

of the VHL tumor suppressor complex and SCF ubiquitin ligase. Science 1999; 284:657–661.

101. Iliopoulos O, Levy AP, Jiang C, Kaelin, Jr, WG, Goldberg MA. Negative regulation of hypoxia-inducible genes by the von Hippel–Lindau protein. Proc Natl Acad Sci USA 1996; 93:10595–10599.
102. Wizigmann-Voos S, Breier G, Risau W, Plate KH. Up-regulation of vascular endothelial growth factor and its receptors in von Hippel–Lindau disease-associated and sporadic hemangioblastomas. Cancer Res 1995; 55:1358–1364.
103. Los M, Aarsman CJ, Terpstra L, Wittebol-Post D, Lips CJ, Blijham GH, Voest EE. Elevated ocular levels of vascular endothelial growth factor in patients with von Hippel–Lindau disease. Ann Oncol 1997; 8:1015–1022.
104. Krieg M, Marti HH, Plate KH. Coexpression of erythropoietin and vascular endothelial growth factor in nervous system tumors associated with von Hippel–Lindau tumor suppressor gene loss of function. Blood 1998; 92:3388–3393.
105. Kroll SL, Paulding WR, Schnell PO, Barton MC, Conaway JW, Conaway RC, Czyzyk-Krzeska MF. Von Hippel–Lindau protein induces hypoxia-regulated arrest of tyrosine hydroxylase transcript elongation in pheochromocytoma cells. J Biol Chem 1999; 274:30109–30114.
106. Bauer AL, Paulding WR, Strict JB, Schnell PO, Czyzyk-Krzeska MF. Endogenous von Hippel–Lindau tumor suppressor protein regulates catecholaminergic phenotype in PC12 cells. Cancer Res 2002; 62:1682–1687.

10

Genome-wide Computational Screen for Candidate HIF Target Genes in *Drosophila melanogaster* and *Caenorhabditis elegans*

THOMAS A. GORR, JOSHUA D. CAHN, and H. FRANKLIN BUNN

Harvard Medical School and
Brigham & Women's Hospital
Boston, Massachusetts, U.S.A.

PAVEL HRADECKY

Harvard Medical School
Boston, Massachusetts, U.S.A.

I. Overview

Using a computational approach, the genomes of *Drosophila melanogaster* and *Caenorhabditis elegans* were searched for candidate hypoxia inducible factor (HIF) target genes. The presence of specific HIF-binding sites or HREs within 600 nucleotides upstream from start and downstream from stop codons was used for identification. Based on this criterion, the screen retrieved 684 HRE-associated genes for *Drosophila*, listed as "fly table," and 405 HRE-associated genes for *Caenorhabditis*, listed as "worm table." Both tables have been posted on the internet (http://bunn.bwh.harvard.edu/). All gene products of the fly and worm tables were used, both as databases and as queries, to conduct a two-way BLASTp batch search to find related genes in fly and worm that could share hypoxia-mediated regulation. Examples of such HRE relatives include heterochromatin proteins and glucose-6-phosphate dehydrogenases. Further analysis of the fly and worm tables should provide insight into the biological scope of oxygen-dependent gene regulation.

II. Introduction

Oxygen homeostasis, the central feature of aerobic metabolism, depends on not only sensing but also translating variations in oxygen tension (pO_2) into differential responses of gene expression. Virtually all aerobic pro- and eukaryotes are thought to have evolved various mechanisms for regulating gene expression in an O_2-dependent fashion (1,2). This allows for adaptations to periods of oxygen deprivation (hypoxia) with a controlled down-regulation of metabolism to a much decreased, yet balanced ATP supply = ATP demand level in the cell (3–5). In this regard, a wide range of animals, from mammals to fruit flies to nematodes, appear to share common oxygen sensing and signaling components. These center around hypoxia-inducible (transcription) factors, or HIFs, as master regulators of hypoxic gene expression (6–8) and oxygen-consuming dioxygenases responsible for modulating the stability and, hence, activity of HIF proteins via a direct oxidation of these substrates (9–11). HIF-controlled target genes in mammals cover physiological responses including activated glycolytic sugar breakdown, enhanced O_2 transport via stimulation of angiogenesis, erythropoiesis, and ventilation, as well as proliferation and survival signals for cells and tissues (10,12).

Many of the hypoxia-induced genes are activated from short *cis*-regulatory motifs called hypoxia-response elements, or HREs. The human erythropoietin (Epo) gene, which encodes a hormone that regulates erythrocyte production, is one of the best-understood models for a robustly activated rate of transcription under hypoxia. Its tripartite 3′ enhancer contains an induction-contributing HRE that was identified as $^{5'}$ TACGTGCT $^{3'}$ octanucleotide (13,14). In 1995, Wang and Semenza purified the transcription factor that binds specifically to this HRE as the heterodimeric HIF-1 protein complex (15). The dimerization partners were later identified as 120-kDa HIF-1α- and 91–94-kDa HIF-1β-subunits (16), both belonging to the bHLH/PAS protein family of transcription/transactivation factors (see Refs. 1 and 17 for review). Many subsequent investigations confirmed that in a growing number of mammalian genes, regulatory HREs are specifically recognized by the HIF-1 heterodimer, a step necessary and often sufficient for hypoxia-inducible activation.

In cultured mammalian cells, the steady-state levels of the α- and β-subunit transcripts as well as the β-subunit protein are all unaffected by change in oxygen tensions (18). Consequentially, the activity of HIF-1 is primarily regulated at the level of the HIF-1α protein through specific oxidative modification. The HIF oxygen sensor is a novel prolyl hydroxylase (9–11) that catalyzes the O_2-dependent hydroxylation of proline residues within the oxygen-dependent degradation domain (ODD) of HIF-1α (19). Once oxidized, HIF-1α rapidly binds to the von Hippel–Lindau (VHL) tumor suppressor product, which enables ubiquitination and degradation by the proteasome (20–24), thus rendering HIF-1 unstable even to a brief exposure to oxygen. Conversely, hypoxia stabilizes HIF-1 due to suppression of proline hydroxylation within the ODD of HIF-1α, enabling it to escape proteolytic degradation, whereupon it dimerizes with HIF-1β. The dimer then translocates into the nucleus where it activates target genes containing HRE-binding sites. This scheme for activation of HIF during hypoxia as direct response of the presence of the α-subunit protein extends from mammals to invertebrates (9,11).

Amongst invertebrates, hypoxia-specific HIF-1-like DNA-binding activities have been found in cells cultured from the fruit fly *D. melanogaster* (25). Moreover, homologs of HIF-1 subunits have been cloned from *Drosophila* [HIF-1α homolog (26); HIF-1β homolog (27–29); see Ref. 30 for review] and from the nematode *C. elegans* [HIF-1β homolog (31), HIF-1α homolog (32)]. As expected for HIF proteins, all four subunits are ubiquitously expressed throughout a wide range of tissues (26,28,32). The stability and activity of *Drosophila*'s HIF-1α homolog, the bHLH/PAS protein similar (aka sima), seems to be regulated in an O_2-dependent fashion very much like the one seen in mammalian HIF-1α subunits. This insight was suggested by similarities in amino acid motifs conferring normoxic instability (33) and further supported by *trans*-activation assays, where sima/Gal4 fusion constructs were able to *trans*-activate a luciferase reporter gene during hypoxia (34). The fact that this *trans*-activation was observed in mammalian cells indicates that sima was able to functionally substitute for human HIF-1α, which provides evidence for the close conservation of these signaling pathways between mammals and insects.

The presence of such O_2-dependent gene regulation in invertebrates should not come as a surprise given that many invertebrate species frequently encounter and withstand severe environmental hypoxia. For example, the free-living soil nematode *C. elegans* maintains its aerobic metabolic rate to a pO_2 as low as 3.6 kPa and shows 100% survival after even 24-hr anoxia [N_2 (35)]. The repertoire of survival strategies that render both fruit flies and *C. elegans* tolerant to hours of anoxia with full recovery afterward include

Flies only:

- Instantaneous loss of coordination resulting in persisting immobility (36)
- Hyperpolarized central nervous system neurons (not depolarized as in mammalian brains) (37,38)

Flies and worms:

- Greatly reduced O_2 consumption to ~20% (flies) or <5% (worms) of normoxic values (38,39)
- Anaerobic, glycogen-fueled metabolism with low pO_2 onset (0.4 kPa, *C. elegans*) and lactate only (all insects), or lactate/succinate/propionate/acetate as major end products (worm) (39,40)
- Arrested development throughout O_2 deprivation (e.g., Dauer larva in *C. elegans*), which in anoxic fly embryos is accompanied by a general and reversible condensation of chromatin and cessation of cell cycle (41,42), and in *C. elegans* Dauer larvae is accompanied by a six- to ninefold reduction of RNA polymerase II-driven transcriptions yielding a residual pool of some 2000 Dauer-specific transcripts [~10% of total number of genes (43,44)].

Another important physiological response in fruit flies and other insects is the long-known hypoxic stimulation of outgrowth and proliferation of tracheal termini

(45–47), a process that poses an intriguing parallel to the O_2 dependence in vertebrate angiogenesis.

Recent reviews present a steadily growing list of known hypoxia-induced mammalian genes (12,48) and give additional information regarding the functionality of HIF-binding sites flanking these genes (49). Functional HREs are defined by being capable of binding HIF in vitro and *trans*-activating expression of reporter genes in vivo above control levels (normoxia) when challenged by hypoxia or any one of the known hypoxia-mimicking stimuli [transition metal ions (Co^{2+}, Ni^{2+}, Mn^{2+}); iron chelators such as desferrioxamine]. In contrast to the wealth of information on mammalian HIF/HRE systems, nothing has been published in regard to hypoxia-inducible genes of known function in *C. elegans*, and only a few cases are known for *Drosophila*. These notable exceptions include glycolytic enzymes [phosphoglycerate kinase (25), lactate dehydrogenase (11)], the tracheogenetic key regulator and FGF homolog branchless (50,51), the heat shock protein Hsp70 (52), and the cyclin-dependent kinase 5 activator molecule [Cdk5α (53)]. Clearly, the completion of the *Drosophila* euchromatin genome (54,55) and *C. elegans* genome (56) has, to the best of our knowledge, yet to be exploited to any large-scale survey of the hypoxia-mediated differential gene expression in either animal.

We conducted a genome-wide computational screen for genes that are flanked by HREs and hence could be regarded as candidate HIF target genes in *Drosophila* and *C. elegans*. Importantly for this project, the binding specificities of *Drosophila* HIF activity have been shown to be very similar, if not identical, to mammalian HIF-1. Wild-type, but not mutated, HRE sequences of the regulatory enhancer region of the human or mouse Epo gene have successfully been used for in vitro binding and EMSA (=gel-shift) analysis of *Drosophila* HIF (25,57). We therefore felt justified in using a list of 25 functional mammalian HREs from a recent review by Wenger and Gassmann (49) as input query for this screen. Table 1 presents a slightly altered version of the Wenger/Gassmann table listing the hypoxia-inducible genes of either human, mouse, or rat, as well as the HRE motifs with locations in relation to the gene and functionality assessment ("+": HRE binds HIF and/or *trans*-activates reporter gene; "−": HRE is *trans*-activation-inactive). As can be seen, functional HREs are almost always located within 600 nucleotides of the transcription start and stop sites in the 5′ and 3′ flanking regions rather than in untranslated (UT) DNA or exons and introns. The resulting HRE consensus of $^{5'}$B(A/G)CGTGVBBB$^{3'}$ fully preserves the essential CGTG core in positions 3–6, which, when mutated, is known to abrogate binding of any HIF complex (with B = all bases except A, V = all bases except T) (13). The complete table, including the original references on the characterization of the 25 functional HREs, is accessible at http://bunn.bwh.harvard.edu/.

A schematic flowchart of our computational screen can be found in Figure 1. DNA sequences of full-length *C. elegans* chromosomes (~88 Mb total) or *D. melanogaster* large genomic scaffolds (~120 Mb total) were downloaded from the NCBI Genomic Biology FTP site (ftp://ncbi.nlm.nih.gov/genbank/genomes/). Sequences longer than 5 Mb were divided into smaller fragments. Whole-genome

Table 1 Functional HREs in Mammalian Hypoxia-Inducible Genes

Hypoxia-inducible gene	Species	Location		HIF-1 binding site (HRE) 1	2	3	4	5	6	7	8	9	10	Fct.	Ref.
Erythropoietin	Human	3′ FS	+3065	T	A	C	G	T	G	C	T	G	T	+	T1-1
Erythropoietin	Mouse	3′ FS	+359	T	A	C	G	T	G	C	T	G	C	+	T1-2
Erythropoietin	Human	5′ FS	−163	G	A	C	G	T	G	C	G	T	G	+	T1-3
Phosphofructokinase L	Mouse	IVS1	+346	T	A	C	G	T	G	C	T	G	C	+	T1-3
Aldolase A	Human	5′ FS	−184	G	A	C	G	T	G	A	C	T	C	+	T1-4
Aldolase A	Human	IVS4	+124	C	A	C	G	T	G	C	G	G	G	−	T1-3
Phosphoglycerate kinase 1	Human	5′ FS	−173	G	A	C	G	T	G	C	G	G	C	+	T1-3
Phosphoglycerate kinase 1	Human	5′ FS	−206	G	A	C	G	T	G	A	C	a	a	N.D.	T1-4
Phosphoglycerate kinase 1	Mouse	5′ FS	−290	C	A	C	G	T	c	C	T	G	C	+	T1-2
Phosphoglycerate kinase 1	Human	5′ UT	+30	G	A	C	G	T	G	C	G	C	T	N.D.	T1-3
Enolase 1	Human	5′ FS	−590	C	A	C	G	T	G	C	G	C	C	−	T1-3
Enolase 1	Human	5′ FS	−413	G	A	C	G	T	G	G	G	G	C	−	T1-4
Enolase 1	Human	5′ FS	−386	T	G	C	G	T	G	C	G	G	G	+	T1-4
Enolase 1	Human	5′ FS	−368	T	A	C	G	T	G	A	C	G	G	+	T1-4
Lactate dehydrogenase A	Mouse	5′ FS	−89	C	A	C	G	T	G	C	G	G	G	+	T1-5
Lactate dehydrogenase A	Mouse	5′ FS	−62	G	A	C	G	T	G	C	G	G	G	+	T1-5
Glucose transporter 1	Mouse	5′ FS	+273	G	G	C	G	T	G	C	C	G	T	+	T1-6
Vascular endothelial growth factor	Human	5′ FS	−978	T	A	C	G	T	G	G	G	C	T	+	T1-7
Vascular endothelial growth factor	Rat	5′ FS	+61	T	A	C	G	T	G	G	G	C	T	+	T1-8
Inducible nitric oxide synthase	Human	5′ FS	−229	T	A	C	G	T	G	C	T	G	C	+	T1-9
Retrotransposon VL30	Rat	5′ FS	−101	T	A	C	G	T	G	C	T	a	−	+	T1-10
Heme oxygenase 1	Mouse	5′ FS	+322	G	A	C	G	T	G	C	T	G	G	+	T1-11
Heme oxygenase 1	Mouse	5′ FS	+350	G	A	C	G	T	G	C	C	a	C	+	T1-11
Transferrin	Human	5′ FS	−201	T	A	C	G	T	G	C	G	C	T	+	T1-12
Transferrin	Human	5′ FS	−185	T	A	C	G	T	G	C	a	G	G	+	T1-12
HIF-1 consensus HRE				B	R	C	G	T	G	V	B	B	B		

Abbreviations used: 3′ FS: 3′ flanking sequence (downstream from poly A signal); 5′ FS: 5′ flanking sequence (upstream from transcription start site); IVS: intervening sequence (intron); UT: untranslated sequence (=utr); HRE location starting with −: element located upstream (5′) from transcription start site (TSS) with HRE-TSS distance given in nucleotides; HRE location starting with +: element located downstream (3′) from polyA site with HRE-polyA site distance given in nucleotides; assessment of functionality of HRE (see text). N.D. = not determined. Within the HRE sequences, mismatching nucleotides relative to the consensus motif are shown in lowercase, while matches with the consensus sequence are capitalized. For references originally cited for these hypoxia-inducible genes, T1-1 to T1-12, see http://bunn.bwh.harvard.edu/. Table reproduced in slightly modified form with permission granted by M. Gassmann.

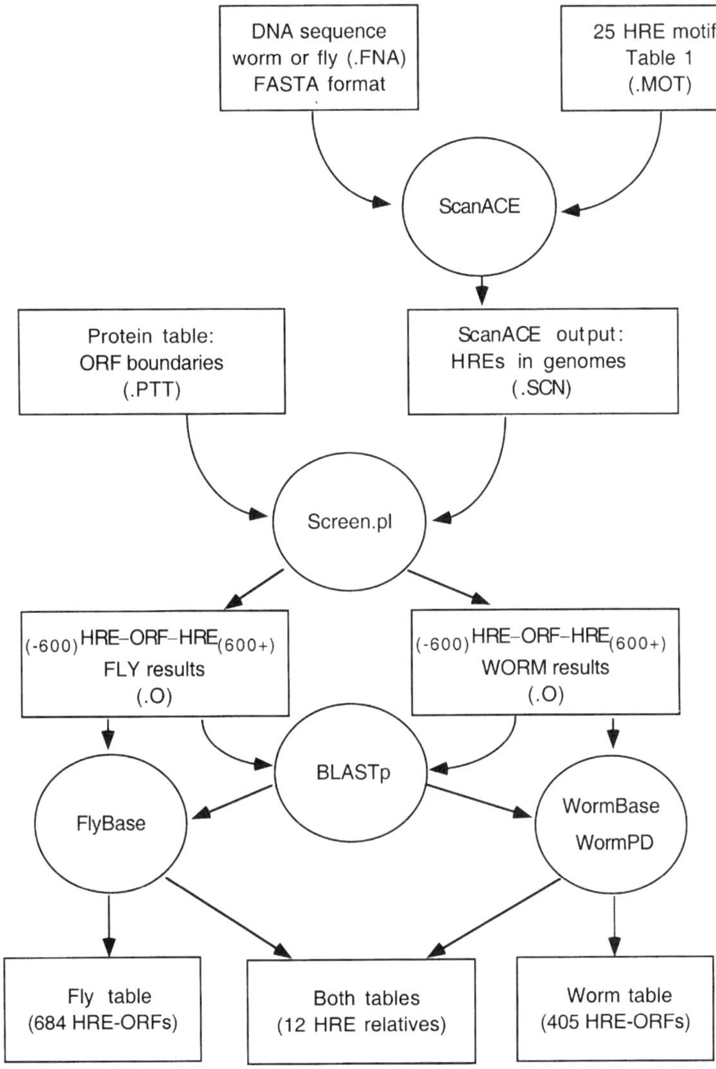

Figure 1 Computational screen of HRE-associated genes in *Drosophilia* (fly) and *C. elegans* (worm) genomes.

scan for the occurrences of HRE motifs was performed using the program ScanACE, a part of the AlignACE package (58). The ScanACE inputs were 5-Mb fragments of genomic DNA from either *C. elegans* or *Drosophila* (.FNA files) and the collection of 25 functional HREs as listed in Table 1 (.MOT file). The output or .SCN files generated by ScanACE were, together with a corresponding protein table (.PTT file, downloaded from NCBI) that describes the boundaries of open reading frames (ORFs), passed into the program Screen.pl written by one of us (PH). To narrow down the number of potential ORFs associated with HREs, Screen.pl positively identified only those ORFs that contained HREs within 600 nt from their start and stop codons. It should be noted that *translational* boundaries, rather than *transcriptional* boundaries as in Table 1, had to be used to characterize HRE-ORF hits, owing to the lack of knowledge concerning the transcription of most of FlyBase or WormBase entries. This 600-nt-long search window in both upstream and downstream regions of ORFs is referred to as $_{(-600)}$HRE—ORF—HRE$_{(600+)}$ in Figure 1. As a result of this restriction, the screen produced 684 and 405 candidate HIF target (=HRE-associated) genes, both confirmed and predicted, for fly and worm, respectively. These two data sets (.O files) obtained by ScanACE/Screen.pl analysis were then subjected to annotation using the resources in FlyBase (http://www.flybase.org), WormBase (http://www.wormbase.org), and WormPD (http://www.proteome.com/databases/index.html) databases generating the fly (684 entries) and worm (405 entries) tables (see Fig. 1).

Furthermore, two files were generated containing the amino acid sequences of 684 fly and 405 worm translation products in FASTA format identified in the above screen. Batch BLASTp analysis of the fly protein set versus the worm protein set and vice versa was performed using locally installed stand-alone BLAST program (59). Genes represented in both data sets, termed HRE relatives, are listed in Table 2 ("Both" table). Selected results for the individual fly and worm data sets can be found in Table 3 ("Fly table") and Table 4 ("Worm table"). The entire fly and worm data sets are available as supplementary information at http://bunn.bwh.harvard.edu/ with entries categorized by function.

III. Results and Discussion

A. Candidate HIF Target Genes Common in *Drosophila* and *Caenorhabditis*

To identify possible fly/worm homologs that share HRE-mediated regulation, a two-way BLASTp batch search was carried out using the protein sequences from the total of 405 worm (w_{405}) and 684 fly (f_{684}) HRE-associated genes (confirmed and predicted) both as databases (db) and as queries (q). The first search, $f_{684}q \rightarrow w_{405}db$, identified HRE-associated genes in worm with similarity to at least one of the 684 fly queries (i.e., worm HRE relative), while the reverse search, $w_{405}q \rightarrow f_{684}db$, identified HRE hits in fly that are similar to at least one worm query (i.e., fly HRE relative). Using an e-value (expectation for chance matches) cutoff of $1e^{-5}$, this two-way screen yielded 12 positive hits, which are listed in Table 2 ("Both table," entries B1–B12). The fly entry is, in columns 2–7, always given above the worm entry. The fly/worm gene pairs presented in B1, B5–B7, and B12 can be considered homologs of candidate hypoxia-responsive genes on the basis that BLASTp searches in GadFly (Genome Annotation Database of *Drosophila*; http://www.fruitfly.org/annot/index.html) were unable to retrieve any more similar sequences than the one shown under HRE relative in Table 2, column 7. Therefore, a parsimonious assumption for these genes would be that their HRE-mediated

Table 2 HRE Relatives Found in Both *Drosophila* and *C. elegans* ("Both" table)

Cat	Fly I.D. Worm I.D.	Motif	HRE/gene	Function and comments	Homologs	Worm HRE relative Fly HRE relative	T2 Ref.
B1	CG15405	TACGTG-CTGC	+423	guanine nucleotide exchange factor for RAS-like GTPases	see HRE relative	R05G6.10 (2e-8)	
	R05G6.10	TACGTG-CTCC	−139	guanine nucleotide exchange factor	see HRE relative	CG15405 (4e-8)	
B2	EG:133E12.2 Hr4 = DHr4	GACGTG-CTGT	+519	steroid hormone receptor; TF with C2C2-type steroid receptor zinc finger DNA-binding domain and ligand-binding domain	m = orphan GCNF (=RTR: 6.8e-40) w = nhr-91 (fragment: 1.9e-63) *Bombyx* homolog, GRF (9.5e-187)	nhr-25 (=F11C1.6) (7e-27)	1, 2
	nhr-25	GACGTG-GGGG	−490	nuclear hormone receptor, FTZ-F1 homolog → embryogenesis	f = FTZ-F1 (1e-45)	EG:133E12.2 (3e-29)	
B3	BG:DS03192.2	GACGTG-CTGC	−234	LRR-rich cell adhesion molecule; putative GPCR function	h = SLIT-3 (=MEGF5: 8.9e-32) w = C56E6.6 (9.1e-59)	C44H4.3 (2e-25)	3
	C44H4.3 sym-1	TACGTG-CCGC	−60	LRR-rich extracellular matrix component; → muscle-to-cuticle attachment	f = CG5195 (2e-29)	BG: DS03192.2 (5e-23)	

B4	chp chaoptic	GACGTG- CGTG	−247	photoreceptor-specific LRR-rich cell adhesion molecule (NCAM); → rhabdomere integrity and microvilli adhesion in photoreceptors	h = SLIT-2 (2.8e-34) w = C56E6.6 (1.6e-44)	C44H3.3 (6e-25)	4, 5
	C44H4.3 sym-1	TACGTG- CCGC	−60	LRR-rich extracellular matrix component; → muscle-to-cuticle attachment	f = CG5195 (2e-29)	BG: DS03192.2 (5e-23)	
B5	CG7041 HP1b	TACGTG- CGAT	−81	chromo-domain (chromatin organization modifier) factor; with chromo-hinge-chromoshadow pattern of domains	m = HP1γ (=CBX3 = M32: 7.1e-43) w = K01G5.2A (4.9e-23)	K08H2.6 (7e-22)	6–10
	K08H2.6 murine-modifier 2 protein-like	CACGTG- CTGG	+310	contains putative chromo (chromatin organization modifier) domain	see HRE relative	CG7041 (3e-19)	
B6	Ets21C Ets at 21C	CACGTG- CCGC	−436	TF, with pointed domain and ETS domain [winged helix structure]	h = ERG 2 (1.4e-56) h = Fli-1 oncogene (6.2e-56) w = see HRE relative	C42D8.4 (3e-45)	11, 12
	C42D8.4	TACGTG- CGCT	+359	DNA binding protein with ETS domain	f = Ets65A (1e-46)	Ets21C (2e-45)	
B7	fd59A = fd3 forkhead domain 59A,	GACGTG- CGAT	+283	TF, with forkhead domain [winged helix structure]	h = TF genesis (=FoxD3/HNF3: 4.8e-52) w = see HRE relative	C47G2.2 (8e-47)	13, 14

(Continued)

	C47G2.2	TACGTG-GGGG	−338	similar to forkhead domain protein	see HRE relative	fd59A (1e-46)	
B8	CG9406	TACGTG-CTGC	+334	Ca-binding EF hand protein	h = calmodulin III (3.8e-16) w = C13C12.1 (=cal-1: 1.9e-14)	T21H3.3 (1e-13)	15, 16
	T21H3.3 cmd-1	GACGTG-CTCT	+207	calmodulin	h = CALM1 (2e-80) f = Cam (4e-81)	CG9406 (1e-13)	
B9	Mlc-c myosin light chain-cytoplasmic	CACGTG-CCGC	−241	nonmuscle myosin essential-light chain (nmELC); 4 EF hand domains comment: MLC3 is major nonmuscle myosin light chain in mammals	h = MLC3nm (6.5e-37) w = T12D8.6 (8.9e-40)	T21H3.3 (2e-32)	17–19
	T21H3.3 cmd-1	GACGTG-CTCT	+207	calmodulin	h = CALM1 (2e-80) f = Cam (4e-81)	Mlc-c (3e-32)	
B10	CG4016	TACGTG-CCCT	+267	serine C-palmitoyltransferase [SPT]; amino transferases class-II; committed rxn in sphingolipid synthesis; → ceramide synthesis	h = SPT1 (1.3e-120) w = C23H3.4 (3.0e-110)	F43H9.2 (2e-36)	20, 21
	F43H9.2	TACGTG-CTCC	−146	lipid and sterol metabolism—member SPT family	h = SPT1 (1e-146)	CG4016 (5e-36)	
B11	CG9796	TACGTG-CTGG	−82	function: ? in analogy to GILT (see homolog), thioredoxin-related reduction of disulfide bonds at lysosomal low pH → organellar protein degradation	h = γ-interferon inducible lysosomal thiol reductase (=GILT = IP-30: 1e-11) w = C02D5.2 (1.0e-17)	ZK669.2 (1e-11)	22, 23

	ZK669.2	TACGTG-CTCT	−134	function: ? in analogy to GILT (see homolog), thioredoxin-related reduction of disulfide bonds at lysosomal low pH → organellar protein degradation	h = IP-30 [=GILT, see above] (1e-5) f = CG9427 (3e-11)	CG9796 (5e-10)	
B12	Zw (alt 2) Zwischenferment	TACGTG-CGGA	+53	glucose-6-phosphate 1-dehydrogenase [G6PD], committed rxn in pentose-phosphate pathway, generates NADPH for reductive processes, e.g., → fatty acid synthesis, → cholesterol synthesis	h = G6PD (3.7e-176) w = see HRE relative	B0035.5 (1e-170)	24–29
	B0035.5	GACGTG-CGAC	+320	putative glucose-6-phosphate dehydrogenase	f = see HRE relative	Zw (1e-162)	

Column 1: Identifies fly-worm gene pairs as B(oth)1–B12.

Columns 2–7: Fly entry given on top, worm entry underneath.

Column 2: Identification of genes by FlyBase or WormBase code, name (underlined), and synonym.

Column 3: HRE motif in $5' \rightarrow 3'$ direction.

Column 4: HRE position in relation to translational boundaries of ORF, i.e., +423 indicates HRE begins 423 nt downstream of stop codon ($3'$ region), −139 indicates HRE begins 139 nt upstream of start codon ($5'$ region).

Column 5: Gives primary and web-based structure/function annotation.

Column 6: Homologs listed for h = human, m = mouse, f = fly, w = worm (nematode). "See HRE relative": no higher scoring BLASTp match than the HRE relative in column 7 was found, suggesting that HRE relatives retrieved from f684q → w405db and w405q → f684db searches can tentatively be considered homologous to each other (HRE homologs).

Column 7: Worm HRE relative to fly entry from f684q → w405db BLASTp searches (above) and fly HRE relative to worm entry from w405q → f684db BLASTp searches (below).

Column 8: For Table 2 references listed (#1–29), see http://bunn.bwh.harvard.edu/.

Abbreviations: TF = transcription factor; → = has role in; LRR = leucine-rich-repeat; GPCR = G-protein-coupled receptor; NCAM = neural adhesion molecule; rxn = reaction.

regulation was already in place in the common ancestor of nematodes and arthropods. Table 2, together with entry-specific references, is accessible at http://bunn.bwh.harvard.edu/.

The fly/worm homologs listed in B5 and B12, with their suggested oxygen-dependent regulation, are very intriguing. Heterochromatin protein 1 (HP1)-like chromo-domain proteins (B5) have been studied in flies and mammals, where they act as dose- and position-dependent silencers of transcription, which in turn was found to be associated with the formation of heterochromatin (60,61). The founding protein of this family—*Drosophila*'s HP1, for example—plays an important role in gene silencing at the borders of eu- and heterochromatin, a phenomenon described as position effect variegation (PEV) (62,63). Mutations in the HP1 gene can cause a suppression of the PEV phenotype. While HP1 specifically associates with heterochromatin, the HP1b protein listed in B5 is a eu- and heterochromatin-binding chromo-hinge-chromoshadow factor (64). It is currently not known whether HP factors cause the formation of heterochromatin or simply maintain it. Their presence, however, correlates with a widespread suppression of transcription through changes in the chromatin structure. The suggested hypoxia-induced expression of HP1b (and related proteins) could be part of the aforementioned global condensation of chromatin in anoxic fly embryos (41). Furthermore, the compacted chromosomes maintain attachment with the nuclear membrane, and HP1 is known to interact with the lamin B receptor that binds nuclear scaffold B-type lamins and double-stranded DNA to the inner nuclear membrane (63).

The homologous glucose-6-phosphate dehydrogenases (G6PD) in fly (=Zwischenferment, Zw) and worm (=B0035.5) could both be oxygen-dependent in their expression (B12 entry). *Drosophila*'s Zw gene has been cloned and the transcription start and stop sites determined for the mRNA yielding the 523-amino-acid polypeptide (65,66). Contrary to the FlyBase annotation ([alt1], [alt2]), no evidence was found for alternatively spliced transcript variants in wild-type or mutant flies (65,66). Using MacVector 6.0.1 gene analysis software (Oxford Molecular Group, 1996) and extended genomic fly and worm G6PD sequences retrieved from Fly- and WormBase, respectively, multiple HREs in both 5′ and 3′ flanks, outside of our ±600-nt search window and outside of utr regions (distinction possible for fly only) were found in both cases (data not shown). G6PD catalyzes the rate-limiting step of the pentose phosphate pathway (=PPP), the irreversible oxidation of glucose-6-phosphate to the intermediate 6-phosphogluconolactone (6PGL) with $NADP^+$ as specific cofactor, thus generating NADPH. 6PGL is subsequently hydrolyzed to 6-phosphogluconate (6PG), which is known to be a potent allosteric effector of several glycolytic enzymes including the rate-limiting phosphofructokinase (67,68), and establishes cross-talk between the PPP and glycolysis. In the subsequent PPP reaction, 6PG is oxidatively decarboxylated by 6-phosphogluconate dehydrogenase (6PGD) under reduction of $NADP^+$. This reaction initiates the production of ribose 5-phosphate, an essential building block in nucleotide syntheses. Taken together, the G6PD and the 6PGD steps are both producing, with NADPH, the reducing power necessary for many biosynthetic reactions, such as the syntheses of fatty acids or cholesterol, therefore linking high-sugar diet with lipogenesis in both vertebrates and invertebrates (69,70) (see below). In support of the prolipogenetic role of an active PPP, mutant *Drosophila* with even slight PPP deficiencies (40% reduction in NADPH output) also reveal a reduced capacity for triglyceride synthesis (73% of wild type) in response to a high-sugar diet (71).

Additionally, NADPH is needed to regenerate reduced and active glutathione or thioredoxin from their respective oxidized forms or is the essential cofactor in many enzymes, notably nitric oxide synthases (72) present in flies but not worms (73,55). O_2/NADPH-dependent production of nitric oxide ($^{\bullet}$NO) could participate in hypoxia-induced $^{\bullet}$NO/cGMP signaling pathways leading to cell cycle arrest at specific checkpoints with enhanced viability in hypoxic fly embryos (74,75) as well as in $^{\bullet}$NO-mediated inhibition of mitochondrial respiration, a process known to be aggravated by low oxygen tensions (76,77). Finally, and most intriguing, G6PD activities across species are stimulated by high-sugar diets (69,70) with the resulting activation of the PPP and fat synthesis crucially depending on aerobic conditions (fat synthesis and breakdown are ATP- and oxygen-dependent pathways). In fact, overall substrate fluxes through the PPP (e.g., 6PG accumulation, NADPH supply), as well as lipogenesis, in *mammalian* tissues are known to be markedly reduced by exposures to both acute and chronic hypoxia (78–82). To reconcile the apparent contradiction between an aerobic PPP and a suggested HIF/hypoxia-mediated regulation of G6PD genes in fly and worm, one needs to realize that nonhypoxic HIF-1-signaling pathways also exist in various mammalian cell lines. Here, insulin, IGF, thrombin, TGF-β1, and other growth factors induce typical hypoxia-responsive genes (GLUT-1, GLUT-3, PGK, VEGF, etc.) under *normoxic* conditions through activated HIF-1 binding to HREs flanking these genes and with reactive oxygen species functioning as second messengers (83–86). The discoveries of such aerobic HIF-1 actions could mean that the HREs flanking fly and worm G6PD genes are in fact serving not (only) as oxygen- but (also) as nutrient/growth factor-signal-dependent docking sites even in invertebrates.

B. Candidate HIF Target Genes in Fly

The screen, limited to 600 nt upstream from the *translation* start and 600 nt downstream from the *translation* stop codons, retrieved 684 HRE-associated FlyBase entries, including 410 annotated ones. The complete table of all 684 hits, categorized according to function and hyperlinked to FlyBase, can be visited at http://bunn.bwh.harvard.edu/. Here, we would like to focus on a few entries specific to fly. An arbitrary selection of 12 mostly well-characterized genes is presented in Table 3, also available at http://bunn.bwh.harvard.edu/, together with the complete list of references cited in this “fly-table” (F-suffix as identifier, column 1). O_2-dependent regulation of these genes is suggested by the presence of flanking HRE motifs (column 3). The entries are homologous to mammalian (human, mouse) and worm sequences (column 6) as indicated by highly negative BLASTp e-scores. Many of these genes are flanked by multiple HREs, often as clusters (see “more” in motif column 3). Such multiple HREs were found in

Table 3 HRE-Associated Fly Genes (Fly Table)

Cat	Fly I.D. (Flybase I.D. or name)	Motif	HRE/gene	Annotation	Homologs	T3 Ref.
F1	ik2 = DmIKKε IκB kinase subunit	TACGTG-CGCG	−367	IκB kinase kinase; Toll/cactus signaling	h = TBK1/NAK (1.6e-118) m = IKKε/IKKi (3.5e-116) w = unc 43 (4.5e-20)	1–4
F2	cact cactus	GACGTG-CGTC CACGTG-CCGG more	−554 +208	IκB homolog; cytoplasmic sequestering of REL (NFκB) proteins (dorsal/dif), Toll signaling; → antifungal response; → dorsoventral patterning	h = IκBα (1.7e-20) m = BCL3 (1.7e-19) w = Y71A12B.4 (5e-16)	5–7
F3	slmb = slimb supernumerary limbs	TACGTG-GTGC more	−95	F-box/WD40 repeat ubiquitin-protein ligase (E3) → ubiquitination/degradation of phospho-IκBα (E3RS-pIκB); → Toll + dpp + Wnt signaling	h = β-TrCP (1.7e-217) w = LIN-23 SCF-E3 (1.6e-173)	8–10
F4	Mt2 = dDnmt2 methyltransferase 2	CACGTG-CCGC more	−179	DNA (5-cytosine)-methyltransferase: generation of DNA/5-methylcytosine; → CpG island methylation; → gene silencing	h = putative DNA cytosine methyltransferase Dnmt2 (6.2e-64) w = no match found	11–15
F5	gcm glial cells missing	TACGTG-CTCT more	+188	hematopoietic TF; GCM motif + PEST motif; → determination of glial cell fate	h = gcm homolog B (1.9e-53) w = no match found	16, 17
F6	CG2781	TACGTG-CGAT	+212	1,3-β-glucan synthase; formation of 1,3β-glucan (callose) → clotting of hemolymph	m = SSC1 (=ELOVL1: 8.1e-42) w = F41H10.8 (1.5e-7)	18, 19

F7	PhKγ[alt1] phosphorylase kinase γ	GACGTG-CGAC	−310	protein serine/threonine kinase; glycogenolytic regulatory enzyme-catalytic subunit	h = PhKγ subunit 2 (3.1e-108) w = Y50D7A.3 (2.4e-64)	20, 21
F8	CG10924 PEPCK (GTP)	TACGTG-CCGC more	−571	phosphoenolpyruvate (PEP) carboxykinase (PEPCK); rate limiting rxn in gluconeogenesis	h = mitochon. PEPCK (1.7e-212) w = R11A5.4 (3e-181)	22–26
F9	CG1907	CACGTG-CTGC more	−336	mitochondrial oxoglutarate/malate antiporter (mtOA/M-AP)	h = mtOA/M-AP (=SLC20A4: 7.3e-93) w = B0432.4 (9e-95)	27
F10	Hmgcr hmg-CoA-reductase	GACGTG-CCGC more	−14	hydroxymethylglutaryl-CoA-reductase (hmgCoAR); commited rxn in isoprenoid + cholesterol syn; requires NADPH	h = hmg-CoA-R (9.9e-189) w = F08F8.2 (hmg-CoA-R-like: 4.7e-89)	28–33
F11	CG6492 Ucp4A	GACGTG-CTCC	−181	mitochondrial uncoupling protein (UCP) 4A	h = UCP 4 (9e-95) w = K07B1.3 (2e-72)	34–37
F12	CG7727 APPl	TACGTG-CCGT more	−579	homolog of β amyloid precursor protein (APP); enriched in growing axons and developing synapses, role in memory formation	h = APP (8.0e-52) w = APL-1 (=C42D8.8: 1.2e-68)	38–42

Column 1: Identifies fly gene entries as F(ly)1–F12.

Column 2: Identification of *Drosophila* candidate HIF target genes by FlyBase code, name (underlined), and synonym.

Columns 3–5: As in Table 2.

Column 6: Homologs listed for h = human, m = mouse, w = worm (nematode) together with BLASTp e-scores.

Column 7: For Table 3 references listed (#1–42), see http://bunn.bwh.harvard.edu/.

Abbreviations: → = has role in; E3RS-pIκB = E3 ubiquitin protein ligase recognition subunit of phosphorylated IκB; TF = transcription factor; rxn = reaction; syn = synthesis.

MacVector analyses of retrieved genomic sequences of these individual FlyBase entries with 5′ and 3′ flanks extended to 2 kb to increase the likelihood of finding HREs in association with *transcription* start and stop sites (outside utr regions). The presence of HRE clusters strengthens the presumption that these elements are functional, possibly even cooperative, and hence, maintained through evolution by selective forces (see Table 1 and Ref. 87 for examples).

Without any hemoglobin-carrying red blood cells, fruit flies also lack erythropoietin homologs. However, entry F5 is suggestive of insect hematopoiesis being controlled by pO_2 variations in a way that could be reminiscent of the hypoxic induction of mammalian red blood cell production. *Drosophila*'s blood cells consist of two major classes: plasmatocytes/macrophages and crystal cells. The production of plasmatocytes crucially depends upon genes activated by the transcription factor called "glial cells missing" (gcm), which in turn is under control of the GATA-like factor "serpent" (88). Gcm levels being modulated by hypoxia, as suggested by multiple HREs in the gcm gene flanking regions (F5), could result in an oxygen-dependent production of plasmatocytes/macrophages, and hence, in the stimulation of cellular immunity in addition to humoral immunity (see below). Moreover, the immunity pathway of Toll/cactus/Rel-protein signaling is also required for blood cell differentiation and density. Mutant fly larvae that lack the Rel-protein/NFκB inhibitor cactus have an overabundance of hemocytes and die prematurely (89). Since cactus itself could be the target of a hypoxia-controlled degradation through its suggested activated phosphorylation (entry F1:ik2) and ubiquitination (entry F3:slimb), maybe low oxygen serves to integrate stimulated humoral immunity (cactus↓, Toll/Rel signaling↑, antifungal defenses↑) with increased hematopoiesis including the production of phagocytotic plasmatocytes (gcm↑ plus cactus↓). Entry F6 (CG2781) lists another interesting example of possible oxygen-dependent changes in the vascular physiology of the fruit fly. 1,3β-glucan synthase generates 1,3β-glucan, aka callose, which in the insect hemolymph aids in the induction of clotting (90). Similarly, in mammals blood clotting is augmented under low oxygen through the hypoxic induction of tissue factor (91).

Hypoxia-mediated metabolic changes in support of activated substrate phosphorylation along with reduced oxidative phosphorylation are indicated by entries F7–F9 and F11. Phosphorylase kinase (PhK) activates glycogen breakdown and at the same time inhibits glycogen synthesis by phosphorylating glycogen phosphorylase (=activation) and glycogen synthase (=inhibition), respectively. Clearly, PhK is a crucial glycogenolytic regulatory enzyme. Since the breakdown of glycogen proceeds without any requirement of oxygen, glycogen is the major energy source during anaerobiosis of many invertebrate species (see Ref. 40 for review) including *C. elegans* (39). Entry F7 adds the recently cloned (92) catalytic γ-subunit of PhK to the list of putative hypoxia-responsive genes. Together with entry B8, the calmodulin (CaM) in worm (T21H3.3), or the CaM-like calcium-binding protein CG9406 in fly utilized as δ-subunit of PhK, two (out of four) subunits of this master regulator of glycogen metabolism have been found to be produced from HRE-associated genes, which seems to correlate well with the anaerobic stimulation of glycogen breakdown.

C. Candidate HIF Target Genes in Worm

The nematode *C. elegans* can successfully adapt to low levels of environmental oxygen and is able to survive anaerobically for extended periods (39). The adaptation to extreme conditions has important consequences for the regulation of intermediary metabolism. In free-living and plant-parasitic nematodes, glycogen is the principal means of energy storage and the primary source for energy production under adverse conditions (93). Under anaerobiosis, glycogen is rapidly metabolized. Glycogen is resynthesized from neutral lipids upon recovery from anaerobiosis (94). Given the importance of carbohydrate catabolism in energy production under anaerobic conditions, it is not surprising to find genes coding for two well-characterized glycolytic enzymes, lactate dehydrogenase and fructose bisphosphate aldolase, among the results of this computational screen as presented in Table 4 (entries W1 and W2). These two enzymes have been shown to be hypoxia-inducible at the transcript level and both have functional HREs in their regulatory sequences in vertebrates (Table 1; Refs. 87,95).

Another group of results raises an intriguing possibility of coordinated regulation of fatty acid metabolism through hypoxia-mediated signaling events. Four genes were identified by the screen code for enzymes involved in beta-oxidation, the principal catabolic pathway of fatty acids (see Ref. 96 for review): F59F4.1 (entry W3), C29F3.1 (W4), C17G10.8 (W5), and F52E4.1 (W6) (see Table 4). Given the recent finding of hypoxia up-regulation of OLE-1, a Δ-9 fatty acid desaturase in yeast (97), we feel the presence of these genes among the screen results deserves attention. Moreover, hypoxia might be a factor in the down-regulation of lipid catabolism by inhibiting both the mitochondrial and peroxisomal beta-oxidation pathways. While the former lipolytic cascade ceases to function without sufficient supply of oxygen acting as terminal electron acceptor in the electron transport chain, the peroxisomal beta-oxidation is directly suppressed by hypoxia due to a HIF-1-mediated silenced expression of its key transcription factor PPARα (98). In addition, hypoxia might inactivate PPARα protein through the feedback inhibition executed by an overexpressed or induced acyl CoA oxidase (99), as suggested by entry W3. This scenario for hypoxia-mediated inhibition of lipolysis agrees with the fact that lipids are not metabolized by free-living and plant parasitic nematodes under anaerobic conditions (100).

IV. Conclusions and Outlook

The genome-wide computational screen presented here yielded, for the first time, a large-scale glimpse of unique as well as shared candidate HIF/hypoxia-responsive genes in *Drosophila* and *C. elegans*. These putative HIF target genes were characterized by the presence of HIF-specific binding elements (HRE motifs) in 5′ and/or 3′ flanking regions. The oxygen-dependent regulation, although confirmed for mammalian homologs of several of the genes discussed, is purely hypothetical at this point. However, the list of genes obtained will allow us a much

Table 4 HRE-Associated Worm Genes (Worm Table)

Cat	Worm I.D. (WormBase I.D.)	Motif	HRE/gene	Annotation	Homologs	T4 Ref.
W1	F13D12.1 ldh-1	TACGTG-CGCC	−409	lactate dehydrogenase → carbohydrate metabolism (glycolysis)	h = LDHB (1e-102) m = Ldh2 (1e-103) f = ImpL3 (1e-115)	1
W2	T05D4.1	TACGTG-CGCC	+56	fructose-bisphosphate aldolase class I → carbohydrate metabolism (glycolysis)	h = ALDOA (1e-124) m = Aldo3 (1e-126) f = Ald (1e-119)	2
W3	F59F4.1	TACGTG-CGAT	+235	putative acyl-CoA oxidase (ACOX); oxidoreductase; → fatty acid metabolism; initial and committed rxn in peroxisomal beta oxidation; catalyzes oxygen-consuming desaturation	h = peroxisomal ACOX1 (1e-147), m = Acox1 (1e-147) f = BcDNA:GH07485 (1e-132) y = Pox1 (7e-56)	3
W4	C29F3.1	TACGTG-CTGT	−567	long-chain enoyl-CoA hydratase (lcECH); lyase → fatty acid metabolism; → mitochondrial beta oxidation in *C. elegans*	h = mitochondrial ECH (E = 0.0, 51% identity, 70% similarity) h = peroxisomal ECH (1e-68) f = BcDNA:GH12558 (E = 0.0, 52% identity, 70% similarity) y = YDR036C (1e-09)	3

W5	C17G10.8 fat-3	CACGTG-CCGT	+424	member of alcohol/ribitol dehydrogenase-like protein family; oxidoreductase; → fatty acid metabolism; → mitochondrial and peroxisomal beta oxidation in *C. elegans*	h = SDR1 (1e-10) m = Cbr2 (1e-11) f = CG5590 (1e-111) y = FOX2p (2e-13)	3
W6	F52E4.1	TACGTG-CTGG	−299	propionyl-CoA carboxylase beta (PCCB); ligase → fatty acid metabolism	h = PCCB (E = 0.0, 76% identity, 85% similarity) f = CG8723 (1e-10)	

Column 1: Identifies *C. elegans* gene entries as W(orm)1–W6.

Column 2: Identification of *C. elegans* candidate HIF target genes by WormBase code and synonym.

Columns 3–5: As in Table 2.

Column 6: Homologs listed for h = human, m = mouse, f = fly, and y = yeast together with BLASTp e-scores.

Column 7: For Table 4 references listed (#1–3), see http://bunn.bwh.harvard.edu/.

Abbreviations: → = has role in; rxn = reaction.

more targeted experimental evaluation of the suggested hypoxia-mediated control of expression.

Acknowledgments

We thank Sophia Kossida, Jason Comander, and Yonatan Grad for their helpful suggestions and discussion of the computational screen design and ScanACE and BLAST analysis. The databases used in this work are Flybase (http://www.flybase.org), WormBase (http://www.wormbase.org), and WormPD (http://www.proteome.com/databases/index.html).

References

1. Bunn HF, Poyton RO. Oxygen sensing and molecular adaptation to hypoxia. Physiol Rev 1996; 76:839–885.
2. Taylor BL, Zhulin IB. PAS domains: internal sensors of oxygen, redox potential, and light. Microbiol Molec Biol Rev 1999; 63:479–506.
3. Schmidt H, Kamp G. The Pasteur effect in facultative anaerobic metazoa. Experientia 1996; 52:440–448.
4. Wegener G. Hypoxia and posthypoxic recovery in insects: physiological and metabolic aspects. In: Hochachka PW, Lutz PL, Sick T, Rosenthal M, van den Thillart G, eds. Surviving Hypoxia: Mechanisms of Control and Adaptation. Boca Raton, FL: CRC Press, 1993:417–434.
5. Hochachka PW, Buck LT, Doll CJ, Land SC. Unifying theory of hypoxia tolerance: molecular/metabolic defense and rescue mechanisms for surviving oxygen lack. Proc Natl Acad Sci USA 1996; 93:9493–9498.
6. Maxwell PH, Pugh CW, Ratcliffe PJ. Inducible operation of the erythropoietin 3′ enhancer in multiple cell lines: evidence for a widespread oxygen-sensing mechanism. Proc Natl Acad Sci USA 1993; 90:2423–2427.
7. Wang GL, Semenza GL. General involvement of hypoxia-inducible factor 1 in transcriptional response to hypoxia. Proc Natl Acad Sci USA 1993; 90:4304–4308.
8. Guillemin K, Krasnow MA. The hypoxic response: huffing and HIFing. Cell 1997; 89:9–12.
9. Epstein ACR, Gleadle JM, McNeill LA, Hewitson KS, O'Rourke J, Mole DR, Mukherji M, Metzen E, Wilson MI, Dhanda A, Tian Y, Masson N, Hamilton DL, Jaakola P, Barstead R, Hodgkin J, Maxwell PH, Pugh CW, Schofield CJ, Ratcliffe PJ. *C. elegans* EGL-9 and mammalian homologs define a family of dioxygenases that regulate HIF by prolyl hydroxylation. Cell 2001; 107:43–54.
10. Semenza GL. HIF-1, O_2, and the 3 PHDs: how animal cells signal hypoxia to the nucleus. Cell 2001; 107:1–3.
11. Bruick RK, McKnight SL. A conserved family of prolyl-4-hydroxylases that modify HIF. Science 2001; 294:1337–1340.
12. Wenger RH. Mammalian oxygen sensing, signalling and gene regulation. J Exp Biol 2000; 203:1253–1263.
13. Semenza GL, Wang GL. A nuclear factor induced by hypoxia via de novo protein synthesis binds to the human erythropoietin gene enhancer at a site required for transcriptional activation. Mol Cell Biol 1992; 12:5447–5454.

14. Blanchard KL, Acquaviva AM, Galson DL, Bunn HF. Hypoxic induction of the human erythropoietin gene: cooperation between the promoter and enhancer, each of which contains steroid receptor response elements. Mol Cell Biol 1992; 12:5373–5385.
15. Wang GL, Semenza GL. Purification and characterization of hypoxia-inducible factor 1. J Biol Chem 1995; 270:1230–1237.
16. Wang GL, Jiang BH, Rue EA, Semenza GL. Hypoxia-inducible factor 1 is a basic-helix-loop-helix-PAS heterodimer regulated by cellular O_2 tension. Proc Natl Acad Sci USA 1995; 92:5510–5514.
17. Semenza GL. Regulation of mammalian O_2 homeostasis by hypoxia-inducible factor 1. Annu Rev Cell Dev Biol 1999; 15:551–578.
18. Huang LE, Arany Z, Livingston DM, Bunn HF. Activation of hypoxia-inducible transcription factor depends primarily upon redox-sensitive stabilization of its α subunit. J Biol Chem 1996; 271:32253–32259.
19. Huang LE, Gu J, Schau M, Bunn HF. Regulation of hypoxia-inducible factor 1α is mediated by an O_2-dependent degradation domain via the ubiquitin-proteasome pathway. Proc Natl Acad Sci USA 1998; 95:7987–7992.
20. Salceda S, Caro J. Hypoxia-inducible factor 1α (HIF-1α) protein is rapidly degraded by the ubiquitin-proteasome system under normoxic conditions. J Biol Chem 1997; 272:22642–22647.
21. Maxwell PH, Wiesener MS, Chang GW, Clifford SC, Vaux EC, Cockman ME, Wykoff CC, Pugh CW, Maher ER, Ratcliffe PJ. The tumour suppressor protein VHL targets hypoxia-inducible factors for oxygen-dependent proteolysis. Nature 1999; 399:271–275.
22. Cockman ME, Masson N, Mole DR, Jaakola P, Chang GW, Clifford SC, Maher ER, Pugh CW, Ratcliffe PJ, Maxwell PH. Hypoxia inducible factor-α binding and ubiquitylation by the von Hippel–Lindau tumor suppressor protein. J Biol Chem 2000; 275:25733–25741.
23. Tanimoto K, Makino, Pereira T, Poellinger L. Mechanism of regulation of the hypoxia-inducible factor-1α by the von Hippel–Lindau tumor suppressor protein. EMBO J 2000; 19:4298–4309.
24. Yu F, White SB, Lee FS. HIF-1α binding to VHL is regulated by stimulus-sensitive proline hydroxylation. Proc Natl Acad Sci USA 2001; 98:9630–9635.
25. Nagao M, Ebert BL, Ratcliffe PJ, Pugh CW. *Drosophila melanogaster* SL2 cells contain a hypoxically inducible DNA binding complex which recognizes mammalian HIF-1 binding sites. FEBS Lett 1996; 387:161–166.
26. Nambu JR, Chen W, Hu S, Crews ST. The *Drosophila melanogaster* similar bHLH-PAS gene encodes a protein related to human hypoxia-inducible factor 1α and *Drosophila* single-minded. Gene 1996; 172:249–254.
27. Ohshiro T, Saigo K. Transcriptional regulation of breathless FGF receptor gene by binding of trachealess/darnt heterodimers to three central midline elements in *Drosophila* developing trachea. Development 1997; 124:3975–3986.
28. Sonnenfeld M, Ward M, Nystrom G, Mosher J, Stahl S, Crews S. The *Drosophila* tango gene encodes a bHLH-PAS protein that is orthologous to mammalian Arnt and controls CNS midline and tracheal development. Development 1997; 124:4571–4582.
29. Ma E, Haddad GG. Isolation and characterization of the hypoxia-inducible factor 1β in *Drosophila melanogaster*. Mol Brain Res 1999; 73:11–16.
30. Crews S. Control of cell lineage-specific development and transcription by bHLH-PAS proteins. Genes Dev 1998; 12:607–620.

31. Powell-Coffman JA, Bradfield CA, Wood WB. *Caenorhabditis elegans* orthologs of the aryl hydrocarbon receptor and its heterodimerization partner the aryl hydrocarbon receptor nuclear translocator. Proc Natl Acad Sci USA 1998; 95:2844–2849.
32. Jiang H, Guo R, Powell-Coffman JA. The *Caenorhabditis elegans* hif-1 gene encodes a bHLH-PAS protein that is required for adaptation to hypoxia. Proc Natl Acad Sci USA 2001; 14:7916–7921.
33. Srinivas V, Zhang LP, Zhu XH, Caro J. Characterization of an oxygen/redox-dependent degradation domain of hypoxia-inducible factor α (HIF-α) proteins. Biochem Biophys Res Commun 1999; 260:557–561.
34. Bacon NCM, Wappner P, O'Rourke JF, Bartlett SM, Shilo B, Pugh CW, Ratcliffe PJ. Regulation of the *Drosophila* bHLH-PAS protein sima by hypoxia: functional evidence for homology with mammalian HIF-1α. Biochem Biophys Res Commun 1998; 249:811–816.
35. Van Voorhies W, Ward S. Broad oxygen tolerance in the nematode *Caenorhabditis elegans*. J Exp Biol 2000; 203:2467–2478.
36. Haddad GG. Mechanisms of anoxia tolerance: a novel approach using a *Drosophila* model system. In: Hudetz, Bruley, eds. Oxygen Transport to Tissue XX. New York: Plenum Press, 1998:273–280.
37. Gu XQ, Haddad GG. *Drosophila* neurons respond differently to hypoxia and cyanide than rat neurons. Brain Res 1999; 845:6–13.
38. Haddad GG. Enhancing our understanding of the molecular responses to hypoxia in mammals using *Drosophila melanogaster*. J Appl Physiol 2000; 88:1481–1487.
39. Föll RL, Pleyers A, Lewandovski GL, Wermter C, Hegemann, Paul RJ. Anaerobiosis in the nematode *Caenorhabditis elegans*. Comp Biochem Physiol Part B 1999; 124:269–280.
40. Grieshaber MK, Hardewig I, Kreutzer U, Pörtner HO. Physiological and metabolic responses to hypoxia in invertebrates. Rev Physiol Biochem Pharmacol 1994; 125:43–147.
41. Foe VE, Alberts BM. Reversible chromosome condensation induced in *Drosophila* embryos by anoxia: visualization of interphase nuclear organization. J Cell Biol 1985; 100:1623–1636.
42. Douglas RM, Xu T, Haddad GG. Cell cycle progression and cell division are sensitive to hypoxia in *Drosophila melanogaster* embryos. Am J Physiol–Reg Int Comp Physiol 2001; 280:R1555–R1563.
43. Dalley BK, Golomb M. Gene expression in the *Caenorhabditis elegans* Dauer Larva: developmental regulation of Hsp90 and other genes. Dev Biol 1992; 151:80–90.
44. Jones SJM, Riddle DL, Pouzyrev AT, Velculescu VE, Hillier L, Eddy SR, Stricklin SL, Baillie DL, Waterson R, Marra MA. Changes in gene expression associated with developmental arrest and longevity in *Caenorhabditis elegans*. Genome Res 2001; 11:1346–1352.
45. Wigglesworth VB. Growth and regeneration in the tracheal system of an insect, *Rhodnius prolixus* (Hemiptera). Q J Microscopical Science 1954; 95:115–137.
46. Locke M. The co-ordination of growth in the tracheal system of insects. Q J Microscopical Science 1958; 99:373–391.
47. Metzger RJ, Krasnow MA. Genetic control of branching morphogenesis. Science 1999; 284:1635–1639.

48. Ebert BL, Bunn HF. Regulation of the erythropoietin gene. Blood 1999; 94:1864–1877.
49. Wenger RH, Gassmann M. Oxygen(es) and the hypoxia-inducible factor-1. Biol Chem 1997; 378:1–8.
50. Sutherland S, Samakovlis, Krasnow MA. Branchless encodes a *Drosophila* FGF homolog that controls tracheal cell migration and the pattern of branching. Cell 1996; 87:1091–1101.
51. Jarecki J, Johnson E, Krasnow MA. Oxygen regulation of airway branching in *Drosophila* is mediated by branchless FGF. Cell 1999; 99:211–220.
52. Ma E, Haddad GG. Anoxia regulates gene expression in the central nervous system of *Drosophila melanogaster*. Mol Brain Res 1997; 46:325–328.
53. Ma E, Haddad G. A *Drosophila* CDK5α-like molecule and its possible role in response to O_2 deprivation. Biochem Biophys Res Commun 1999; 261:459–463.
54. Adams MD. plus 195 authors. The genome sequence of *Drosophila melanogaster*. Science 2000; 287 2185–2195.
55. Rubin GM. plus 53 authors. Comparative genomics of the eukaryotes. Science 2000; 287:2204–2215.
56. The *C. elegans* Sequencing Consortium. Genome sequencing of the nematode *C. elegans*: a platform for investigating biology. Science 1998; 282:2012–2018.
57. Gorr TA. Unpublished results.
58. Roth FP, Hughes JD, Estep PW, Church GM. Finding DNA regulatory motifs within unaligned nocoding sequences clustered by whole-genome mRNA quantitation. Nat Biotechnol 1998; 16:907–908.
59. Altschul SF, Madden TL, Schäffer AA, Zhang J, Zhang Z, Miller W, Lipman DJ. Gapped BLAST and PSI-BLAST: a new generation of protein database search programs. Nucleic Acids Res 1997; 25:3389–3402.
60. Festenstein R, Sharghi-Namini S, Fox M, Roderick K, Tolaini M, Norton T, Saveliev A, Kioussis D, Singh P. Heterochromatin protein 1 modifies mammalian PEV in a dose- and chromosomal-context-dependent manner. Nature Genet 1999; 23:457–461.
61. Nielsen AL, Ortiz JA, You J, Oulad-Abdelghani M, Khechumian R, Gansmuller A, Chambon P, Losson R. Interaction with members of the heterochromatin protein 1 (HP1) family and histone deacetylation are differentially involved in transcriptional silencing by members of the TIF1 family. EMBO J 1999; 18:6385–6395.
62. Elgin SCR. Heterochromatin and gene regulation in *Drosophila*. Curr Opin Gen Dev 1996; 6:193–202.
63. Wallrath LL. Unfolding the mysteries of heterochromatin. Curr Opin Gen Dev 1998; 8:147–153.
64. Smothers JF, Henikoff S. The hinge and chromo shadow domain impart distinct targeting of HP1-like proteins. Mol Cell Biol 2001; 21:2555–2569.
65. Fouts D, Ganguly R, Gutierrez AG, Lucceshi JC, Manning JE. Nucleotide sequence of the *Drosophila* glucose-6-phosphate dehydrogenase gene and comparison with homologous human gene. Gene 1988; 63:261–275.
66. Ito H, Yoshida K, Hori SH. Positive Regulation of the *Drosophila melanogaster* G6PD Gene by an Insertion Sequence. Biochem Gen 1989; 27:379–393.
67. Zera AJ. Inhibition of phosphoglucose isomerase allozymes from the wing polymorphic waterstrider, *Limnoporus canaliculatus*, by pentose shunt metabolites. Biochem Gen 1987; 25:205–223.

68. Sommercorn J, Steward T, Freedland RA. Activation of phosphofructokinase from rat tissues by 6-phosphogluconate and fructose 2,6-bisphosphate. Arch Biochem Biophys 1984; 232:579–584.
69. Wang L, Clark AG. Physiological genetics of the response to a high-sucrose diet by *Drosophila melanogaster*. Biochem Gen 1995; 33:149–165.
70. Yamini S, Staples RC, Hansen CT, Szepesi B. Effect of dietary carbohydrate on liver and kidney enzyme activities and plasma amino acids in the LA/N-cp rat. Int J Obesity 1991; 15:189–203.
71. Geer BW, Lindel DL, Lindel DM. Relationship of the Oxidative Pentose Shunt Pathway to Lipid Synthesis in *Drosophila melanogaster*. Biochem Gen 1979; 17:881–895.
72. Mayer B, Hemmens B. Biosynthesis and action of nitric oxide in mammalian cells. Trends Biol Sci 1997; 22:477–481.
73. Regulski M, Tully T. Molecular and biochemical characterization of dNOS: a *Drosophila* Ca^{2+}/calmodulin-dependent nitric oxide synthase. Proc Natl Acad Sci USA 1995; 92:9072–9076.
74. Wingrove JA, O'Farrel PH. Nitric oxide contributes to behavioral, cellular, and developmental responses to low oxygen in *Drosophila*. Cell 1999; 98:105–114.
75. DiGregorio PJ, Ubersax JA, O'Farrell PH. Hypoxia and nitric oxide induce a rapid, reversible cell cycle arrest of the *Drosophila* syncytial divisions. J Biol Chem 2001; 276:1930–1937.
76. Schweizer M, Richter C. Nitric oxide potently and reversibly deenergizes mitochondria at low oxygen tension. Biochem Biophys Res Commun 1994; 204:169–175.
77. Brown GC. Regulation of mitochondrial respiration by nitric oxide inhibition of cytochrome c oxidase. Biochim Biophys Acta 2001; 1504:46–57.
78. Domanska-Janik K. Hexose monophosphate pathway activity in normal and hypoxic rat brain. Resuscitation 1988; 16: 79–90.
79. Shan X, Yee Aw T, Smith ER, Ingelman-Sundberg M, Mannervik B, Iyanagi T, Jones DP. Effect of chronic hypoxia on detoxication enzymes in rat liver. Biochem Pharmacol 1992; 43: 2421–2426.
80. Tribble DL, Jones DP. Oxygen dependence of oxidative stress. Rate of NADPH supply for maintaining the GSH pool during hypoxia. Biochem Pharmacol 1990; 39:729–736.
81. Strosznajder J, Domanska-Janik K. Effect of anoxia and hypoxia on brain lipid metabolism. Neurochem Res 1980; 5:583–589.
82. Domanska-Janik K, Wideman J. Regulation of thiols in the brain. 2. Effect of hypoxia on the activities of cytoplasmic NADPH-producing enzymes in different parts of the rat brain. Resuscitation 1974; 3:37–41.
83. Zelzer E, Levy Y, Kahana C, Shilo BZ, Rubinstein M, Cohen B. Insulin induces transcription of target genes through the hypoxia-inducible factor HIF-1α/ARNT. EMBO J 1998; 17:5085–5094.
84. Feldser D, Agani F, Iyer NV, Pak B, Ferrier G, Semenza GL. Reciprocal positive regulation of hypoxia-inducible factor 1α and insulin-like growth factor 2. Cancer Res 1999; 59: 3915–3918.
85. Richard DE, Berra E, Pouysségur J. Nonhypoxic pathway mediates the induction of hypoxia-inducible factor 1α in vascular smooth muscle cells. J Biol Chem 2000; 275: 26765–26771.

86. Görlach A, Diebold I, Schini-Kerth VB, Berchner-Pfannschmidt U, Roth U, Brandes RP, Kietzmann T, Busse R. Thrombin activates the hypoxia-inducible factor-1 signaling pathway in vascular smooth muscle cells. Role of the p22^{phox}-containing NADPH oxidase. Circ Res 2001; 89:47–54.
87. Semenza GL, Jiang BH, Leung SW, Passantino R, Concordet JP, Maire P, Giallongo A. Hypoxia response elements in the aldolase A, enolase 1, and lactate dehydrogenase A gene promoters contain essential binding sites for hypoxia-inducible factor 1. J Biol Chem 1996; 271:32529–32537.
88. Lebestky T, Chang T, Hartenstein V, Banerjee U. Specification of *Drosophila* hematopoietic lineage by conserved transcription factors. Science 2000; 288:146–149.
89. Qui P, Pan PC, Govind S. A role for the *Drosophila* Toll/Cactus pathway in larval hematopoiesis. Development 1998; 125:1909–1920.
90. Söderhäll K, Cerenius L. Role of the prophenoloxidase-activating system in invertebrate immunity. Curr Opin Immunol 1998; 10:23–28.
91. O'Rourke JF, Pugh CW, Bartlett SM, Ratcliffe PJ. Idenfication of hypoxically inducible mRNAs in HeLa cells using differential-display PCR. Eur J Biochem 1996; 241:403–410.
92. Bahri SM, Chia W. DPhK-gamma, a putative *Drosophila* kinase with homology to vertebrate phosphorylase kinase gamma subunits: molecular characterization of the gene and phenotypic analysis of loss of function mutants. Mol Gen Genet 1994; 245:588–597.
93. Bolla R. Nematode energy metabolism. In: Zuckerman BM ed. Nematodes as Biological Models. Vol 2. New York: Academic Press, 1980:165–189.
94. Cooper AF, Van Gundy SD. Ethanol production and utilization by *Aphelenchus avenae and Caenorhabditis* sp. J Nematol 1971; 3:205–214.
95. Firth JD, Ebert BL, Pugh CW, Ratcliffe PJ. Oxygen-regulated control elements in the phosphoglycerate kinase I and lactate dehydrogenase A genes: similarities with the erythropoietin 3′ enhancer. Proc Natl Acad Sci USA 1994; 91:6496–6500.
96. Kunau WH, Dommes V, Schulz H. Beta oxidation of fatty acids in mitochondria, peroxisomes and bacteria: a century of continued progress. Prog Lipid Res 1995; 34:267–342.
97. Vasconcelles MJ, Jiang Y, McDaid K, Gilooly L, Wretzel S, Porter DL, Martin CE, Goldberg MA. Identification and characterization of a low oxygen response element involved in the hypoxic induction of a family of *Saccharomyces cerevisiae* genes: implications for the conservation of oxygen sensing in eukaryotes. J Biol Chem 2001; 276:14374–14384.
98. Narrvula S, Colgan SP. Hypoxia-inducible factor-1 mediated inhibition of peroxisome proliferators-activated receptor alpha expression during hypoxia. J Immunol 2001; 166:7543–7548.
99. Fan CY, Pan J, Usuda N, Yeldandi AV, Rao MS, Reddy JK. Steatohepatitis, spontaneous peroxisome proliferation and liver tumors in mice lacking peroxisomal fatty acyl-CoA oxidase. J Biol Chem 1998; 273:15639–15645.
100. Van Gundy SD, Bird AF, Wallace HR. Aging and Starvation of *Maloidogyne javanica* and *Tylenchulus semipenetrans*. Phytopathology 1967; 57:559–571.

Part Two

OXYGEN SENSING IN THE CAROTID BODY, AND OTHER CELLS, ORGANS, AND ORGANELLES

SUKHAMAY LAHIRI and NANDURI R. PRABHAKAR

Hypoxia (a decrease in oxygen availability) is a fundamental physiological stimulus to mammalian cells. The compensatory cardiorespiratory adjustments that occur in response to acute hypoxia critically depend on the oxygen-sensing ability of the peripheral chemoreceptors, especially the carotid bodies. Carotid bodies sense even modest decrease from normoxia in arterial oxygen within a second, by increasing their sensory discharge. Although other mammalian cells cannot respond with such rapidity to lack of oxygen, they do, however, react to sustained hypoxia. The response of the mammalian cells to sustained hypoxia is often reflected in altered gene expression and the resulting encoding proteins that are critical for adaptive behavior. In his overview of Part One, Gregg Semenza summarizes regulation of gene expression and the associated signaling mechanisms during sustained hypoxia in a variety of cellular models including yeast, *Caenorhabditis elegans*, and mammalian cells.

The discovery that carotid bodies are sensory organs for monitoring arterial oxygen attracted not only respiratory physiologists but also neurophysiologists. Even after nearly eight decades of its discovery, the carotid body remains a fascinating organ for unraveling the mechanisms by which changes in oxygen levels are translated to instantaneous neural signals and how carotid bodies contribute to regulation of breathing during hypoxia (1). While the neurophysiologists focus on the mechanisms of sensory transduction and transmission, respiratory physiologists are actively pursuing how carotid bodies regulate breathing during hypoxia, whether it is acute or chronic. Despite the intensive investigations on the carotid body physiology, several fundamental questions still remain unanswered: site(s) of oxygen sensing in the carotid body; the identity of the oxygen sensor; the mechanisms of sensory transduction; identity of neurotransmitter(s) contributing to sensory transmission; developmental and maturation of carotid body function and carotid body adaptation to chronic sustained or intermittent hypoxia (2,3). Eighteen chapters in this volume address various aspects of carotid body's responses and adaptations.

I. Site(s) of Sensory Transduction

The carotid body is primarily composed of two cell types: type I and type II cells. One type of oxygen-sensitive cells depolarize when they are exposed to hypoxia as opposed to those that do not. Carotid body type I cells (also called glomus cells) depolarize leading to calcium influx. The other types are hepatic and renal in origin and they secrete erythropoietin. The latter takes a longer exposure to hypoxia. Type I cells (also called glomus cells) are neural crest in origin; whereas the type II cells (also called sustentacular cells) resemble glial cells of the nervous system. The sensory nerves originating from petrosal ganglion innervate glomus cells. One of the long-standing questions is which of these elements, glomus cells and sensory nerve endings innumerating the glomus cells, serve as the primary site(s) of sensory transduction. Two chapters address this issue. Chapter 19, by Zapata, provides a historical perspective as well as recent data suggesting that type I cells are the primary site of sensory transduction in the carotid body. There have been suggestions in the literature implicating that sensory neurons (more realistically the dendrites) innervating the carotid body are the site of sensory transduction. Chapter 36, by Alcayaga et al., discusses the effects of hypoxia on petrosal neurons by measuring the output of the petrosal neurons. Their data demonstrate that hypoxia has virtually no effect, whereas ATP stimulates the activity of the petrosal neurons.

II. Mechanisms of Sensory Transduction

Although there are several hypotheses regarding the sensory transduction at the carotid body, for the sake of simplicity they can be categorized into two hypotheses. One hypothesis assumes that a heme- and/or a redox-sensitive enzyme is the oxygen sensor and that a biochemical event associated with the heme-containing protein triggers the transduction cascade possibly involving reactive oxygen species (ROS). The other hypothesis suggests that proteins associated with K^+ channels are the primary oxygen sensor and that inhibition of this channel is the seminal event in transduction.

Oxygen binds to heme with high affinity, and several biologically important enzymes, especially those involved in mitochondrial respiration, such as cytochromes, and several other oxidases (e.g., NADPH oxidases) contain heme.

Chapter 15, by Lahiri et al., highlights the importance of heme-containing protein(s) and intracellular free iron in oxygen sensing by the carotid body. In doing so, molecules that activated HIF-1 also affected the outwork K^+ currents of glomus cell membrane. Lahiri et al. wonder how these two events are connected. In Chapter 28, Acker et al. suggest that there is a cross-talk between mitochondrial cytochrome c oxidase (which has low affinity for O_2) and NADPH oxidases in the carotid body. They also assume that ROS levels (especially of H_2O_2) decrease during hypoxia and emphasize the importance of Fenton reaction that utilizes iron in nonenzymatic degradation of H_2O_2. They hypothesize that decrease in H_2O_2 levels modulate the ion channel activity in glomus cells leading to sensory excitation of the carotid body during hypoxia. They do not understand why ROS should increase under hypoxia. ROS is not generated by mitochondria at potential 100 m but it begins to generate when mitochondrial potential is at 140 mV in hyperoxia (4). They also describe optical techniques that can be potentially used in identifying the certain set of oxygen sensor(s) in the carotid body as well as in other nonexcitable cells.

Several other chapters deal with the role of oxidases and ROS in oxygen sensing in cells other than the carotid body. In Chapter 30, Wolin et al. discuss the potential role of NADPH oxidase in oxygen sensing by vascular smooth muscle cells. They point out, however, that there could be multiple oxygen sensors in vascular smooth muscle cells that trigger both ROS-dependent and ROS-independent signaling cascades in the vascular system. In Chapter 29, Archer et al. also discuss the role of NADPH oxidases and ROS in oxygen sensing by pulmonary vascular smooth muscle. Their redox hypothesis is similar to that proposed by Acker in the carotid body. Archer et al. suggest that ROS can diffuse through the plasma membrane, oxidizing and opening K^+ channels in normoxia, resulting in vasodilatation during normoxia. Hypoxia decreases ROS levels in proportion to Po_2, resulting in an increase in reducing equivalents in the cytoplasm, promoting a reduced state that causes K^+ channel inhibition, membrane depolarization, and vasoconstriction. Archer et al. describe participation of NADPH oxidase and complex I and II of the mitochondria, etc., in this oxygen-sensing system. In Chapter 31, Cutz and his group provide evidence for direct participation of NADPH oxidase in oxygen sensing in neuroepithelial body cells (NEB) but not in the pulmonary vasculature and carotid body. However, although NEB may serve as a model for oxygen sensing, their role in the control of breathing in adults remains unexplored. Chromaffin cells of adrenal medulla share several features with carotid body glomus cells, including neural crest origin, catecholaminergic phenotype, etc. Chromaffin cells in adult animals receive sympathetic innervation, whereas in neonates they are not innervated. Neonatal but not adult chromaffin cells release catecholamines by nonneurogenic mechanisms.

Stimulus-secretion coupling in neonatal chromaffin cells by nonneurogenic mechanisms is the focus of the Chapter 32, by Nurse et al. They conclude that in neonatal rat chromaffin cells, hypoxia-induced catecholamine secretion is coupled to inhibition of Ca^{2+}-dependent K^+ currents and a decrease in ROS. The hypothesis that K^+-channel protein might serve as oxygen sensor is attractive and provoked a good deal of interest. However, there is a lack of correlation between K^+-channel inhibition and/or catecholamine secretion (as measured by chronoamperometry) with sensory excitation of the carotid by hypoxia. In view of the conflicting reports on the effects of hypoxia on ROS levels, Gonzalez et al., in Chapter 27, propose a set of criteria for considering ROS as potential messengers mediating the physiological responses to hypoxia. These include: (1) defining the relationship between ROS levels and severity of hypoxia; (2) identifying molecular targets for ROS; (3) defining the cellular responses to various ROS species; (4) defining the mechanisms for generating and terminating the actions of ROS; and (5) characterizing the cellular responses with manipulation of endogenous ROS levels.

On the basis of an original idea by De Castro (5) followed by Heymans (6), it has been proposed that depolarization of glomus cells by hypoxia is central to the transduction process at the carotid body. Until a decade ago very little was known about the cellular basis for the initial depolarization of glomus cells induced by hypoxia. With the advent of patch-clamp technology, considerable advances have been made in the identification of various ionic conductances in glomus cells in general, and K^+ channels in particular (7). In Chapter 17, López-Barneo and Pardal, using carotid body slice preparations, explore the role of K^+-channel inhibition in hypoxia-induced catecholamine secretion from glomus cells. They show that glomus cells also monitor glucose concentrations and conclude that carotid body is a polymodal sensory organ, as originally suggested by Eyzaguirre et al. (8). In Chapter 16, Peers et al. also emphasize the importance of K^+ channels in oxygen sensing by glomus cells. They suggest that TASK-like K^+ channels control the resting membrane potential of the glomus cells and point out the potential use of heterologous expression of K^+-channel proteins to further identify the mechanisms by which hypoxia alters K^+-channel activity.

III. Sensory Transmission: Role of Neurotransmitters

It is amazing that the carotid body, despite being a small tissue, expresses as many neurotransmitters as brain tissue (2). There is little doubt that neurotransmitters are critical for sensory transmission at the carotid body. The question is which one of them is essential for sensory excitation by hypoxia and what is the role(s) of relatively abundant inhibitory neurotransmitters that coexist with the excitatory ones? In Chapter 24, Ganesh Kumar et al. describe the role of excitatory and inhibitory neurotransmitters not only in sensory excitation during acute hypoxia but also their potential contribution to the adaptation of the carotid body to chronic (sustained as well as intermittent) hypoxia. In view of their finding that multiple neurotransmitters are being released simultaneously during hypoxia, they propose that final expression of the sensory response to hypoxia critically depends on the cross-talk between excitatory and inhibitory transmitters involving a "push-pull" type of mechanism(s) seen elsewhere in the biological system.

Chapter 20, by Shirahata et al., and Chapter 21, by Fitzgerald et al., focus on acetyl choline (ACh) as an excitatory neurotransmitter in the carotid body. Shirahata et al. present evidence for coupling of cholinergic receptors to voltage-gated K^+

channels of glomus cells as one of the potential mechanisms. Fitzgerald et al. discuss the potential interactions of ACh with other neurotransmitter systems in the carotid body. Nitric oxide (NO) and carbon monoxide (CO) are gas molecules that are synthesized within the glomus tissue and serve as inhibitory chemical messengers (9). In Chapter 23, analyzing the NOS activity, Iturriaga and Alcayaga emphasize the role of NO as a tonic inhibitor of carotid body chemoreception. The effects of NO are mediated through its action on vascular tone, oxygen delivery, and excitation of glomus cells and petrosal neurons. In Chapter 22, Buerk and Lahiri elaborate on how NO can alter the oxygen sensing in the carotid body via decreasing oxygen uptake, regulating vasculature, modulating neurosecretion, reducing H_2O_2, and influencing gene expression during chronic hypoxia. They present a mathematical model related to the actions of NO and conclude that NO modulates oxygen sensing in the carotid body by altering cytochrome oxidase kinetics.

Many cells, including neurons in the nervous system, are electrically coupled via gap junction proteins. The importance of gap junctions in the carotid body in oxygen sensing is the focus of Chapter 18, by Eyzaguirre and his colleagues.

IV. Developmental Maturity of Carotid Body Oxygen Sensing

It is known at birth that the carotid body is relatively insensitive to hypoxia, and the oxygen-sensing ability develops with age. The Hanson group described the state of chemoreflexes that are underdeveloped in the fetal state, followed by the work of Lagercrantz's group, which beautifully elaborated perinatal transition of oxygen sensing in the peripheral chemoreceptors. Chapter 13, by Carroll, and Chapter 14, by Kumar, deal with functional development of carotid body chemoreception. Carroll describes the changes in intracellular Ca^{2+} in glomus cells and the sensory response to hypoxia in neonatal carotid bodies and how these responses increase with age. An important finding is that chronic hypoxia during development blunts the oxygen-sensing ability of the carotid body leading to blunted ventilatory response to hypoxia. Kumar, on the other hand, emphasizes that CO_2-O_2 interaction does not develop early in life. These interactions develop with age.

V. Adaptation of the Carotid Body to Chronic Hypoxia

It is being increasingly appreciated that carotid bodies are critical for ventilatory adaptations to chronic hypoxia. During chronic hypoxia, the carotid itself undergoes remodeling in addition to the functional change it exerts on the whole body. In Chapter 25, Dinger et al. address the mechanisms associated with functional and morphological alteration in the carotid body by chronic sustained hypoxia. Chapter 26, by Joseph and Pequignot, discusses the alterations in the catecholaminergic neurons in the central nervous system (CNS) related to processing of carotid body inputs during chronic hypoxia. Chapter 26, by Joseph and Pequignot, and Chapter 7, by Agani et al., suggest the possible involvement of HIF-1 transcription factor in the phenotypic alterations in CNS during chronic hypoxia.

VI. Comparative Aspects of Oxygen Sensing by Chemoreceptors

Chapter 37, by Burleson and Milsom, presents a fascinating account of oxygen chemoreception and environmental adaptation in animals other than mammals. These authors began with the premise that comparative studies are the only way to examine the evolution of oxygen chemoreception but resorted to mammals for comparison.

Before the discovery of peripheral chemoreceptors, the search for oxygen chemoreceptors was restricted to the CNS (1,10). More recently, oxygen sensing in the CNS has been revived, it is now known that hypoxia shows both direct depressant and excitatory effects on the CNS, apart from the effects associated with the stimulation of peripheral chemoreceptors.

Chapter 33, by Haddad, presents evidence for multiple oxygen sensors in the central neurons, which participate in acute as well as chronic responses to hypoxia in a variety of species including *Drosophila* and mammals. He states that acute responses to oxygen deprivation involve K^+ efflux and Na^+ influx, whereas chronic responses involve protein alteration in plasma membrane, cytosol, and mitochondria. The effects of hypoxia on central neurons described by Haddad are reminiscent of that described in the carotid body. Haddad concludes that there are multiple oxygen sensors in the central nervous system.

Two chapters describe the effects of hypoxia on systemic responses elicited by putative central oxygen-sensing chemoreceptors. Chapter 34, by Neubauer et al., focuses on the role of heme oxygenase as an oxygen sensor in the rostroventrolateral medulla similar to that described in the carotid body glomus cells (11). Chapter 35, by Solomon and Edelman, emphasizes the role of the pre-Bötzinger complex in eliciting gasping during severe hypoxia. Thus it appears that oxygen-sensing mechanisms at the central neurons share several features similar to that described in the carotid body (especially the kinetics of the response). However, it remains to be established whether the stimulus–response relationships of the central neurons to hypoxia resemble or differ from that described in the carotid body.

References

1. Lahiri S. Historical perspectives of cellular oxygen sensing and responses to hypoxia. J Appl Physiol 2000; 88:1467–1473.
2. Prabhakar NR. Oxygen sensing by the carotid body chemoreceptors. J Appl Physiol 2000; 88:2287–2295.
3. Lahiri S, DiGiulio C, Roy A. Lessons from chronic intermittant and sustained hypoxia at high altitudes. Respir Physiol 2002; 130:223–233.

4. Lee I, Bender E, Arnold S, Kadenbach D. New control of mitochondrial membrane potential and ROS formation—a hypothesis. Biol Chem 2001; 382:1629–1636.
5. De Castro F. Sur la structure et l'innervation de la glande intercarotidienne (Glomus caroticum) de l'homme et des mammifères, et sur un noveau système d'innervation autonome du nerf glossophayngien. Trav Lab Rech Biol Univ Madrid 1926; 24:365–432.
6. Heymans C. Le sinus carotidien, zone réflexogéne régulatrice due tonus vagal cardiaque du tonus neurovasculaire et de l'adrénalosécretion. Arch Int Pharmacodyn Ther 1929; 35:269–306.
7. Lopez-Barneo J, Lopez-Lopez JR, Urena J, Gonzalez C. Chemotransduction in the carotid body: K^+ current modulated by Po_2 in type I chemoreceptor cells. Science 1988; 241:580–582.
8. Eyzaguirre C, Fitzgeraald RS, Lahiri S, Zapata P. Arterial chemoreceptors. In: Shephard JT, Abboud FM, eds. The Cardiovascular System, Handbook of Physiology. Section 2, Vol. III. Bethesda: American Physiol Society, 1983:557–621.
9. Prabhakar NR. NO and CO as second messengers in oxygen sensing in the carotid body. Respir Physiol 1999; 115:161–168.
10. Lahiri S, Forster RE. CO_2 sensing: peripheral and central chemoreception. IJBCB. In press.
11. Prabhakar NR, Dinerman JL, Agani FH, Snyder SH. Carbon monoxide: a role in carotid body chemoreception. Proc Natl Acad Sci USA 1995; 92:1994–1997.

11

Fetal Adaptations to Hypoxia

JAMES P. NEWMAN, MARK A. HANSON, and LUCY R. GREEN

University of Southampton
and Princess Anne Hospital
Southampton, England

I. Introduction

The fetus is both completely dependent upon its mother for oxygen delivery and constantly has a low arterial PO_2 (P_aO_2). Although the fetus is not normally considered to be hypoxic, high fetal levels of HIF-1 (a surrogate maker for oxygenation) compared to the adult indicate that this assumption may be erroneous (1–3). In the absence of significant oxygen stores, both the normal increases in oxygen requirements during gestation and any perturbations of oxygen delivery may require the fetus to undertake adaptive changes to safeguard oxygen delivery to those tissues that have the least tolerance for hypoxia-ischemia. The existence of such adaptive changes requires the presence of a competent system of oxygen sensors locally and/or centrally that are able to detect changes in oxygenation and enact measures to prioritize oxygen supply hierarchically according to need. The nature of the oxygen sensor is a matter of some debate and may lie in the cytosol, in the plasma membrane, or in the mitochondria (4).

The molecular mechanisms underlying oxygen sensing in mammalian cells have been investigated in adult erythropoietin-producing cells (5), in fetal adrenal chromaffin cells (6), in the type I cells of the adult carotid body (7,7a), in adult hepatocytes (8), embryonic cardiomyocytes (9), fetal ductus arteriosus endothelial cells (10), human trophoblast cells (11), fetal pulmonary vessels (12), and a hepatoma cell line in vitro (13). All of these studies provide evidence that certain heme proteins function as oxygen sensors in vivo. The studies in the adult animal or cell lines looking at erythropoietin production provide strong circumstantial evidence that such a sensor is a heme protein (5). Patch clamp studies of adult carotid body plasma membrane indicate that K^+ channel function is PO_2-dependent, suggesting perhaps a plasma-membrane-bound O_2 sensor (7). In the fetus, chromaffin cells have been shown to be O_2-responsive before birth (6), a function that seems to reside in the SK channel population (14). Other studies, however, have focused on the role of the mitochondrion (9) and, smaller still, on cytochrome oxidase (8) as sites for oxygen sensing. Isolated embryonic mitochondria demonstrated a hypoxia-dependent inhibition when incubated at low PO_2 for 3 hr. This effect was quickly reversible upon reoxygenation, suggesting perhaps that metabolic suppression in this model is due to a change of conformation in one or more of the metabolic proteins. Isolated adult rat hepatocytes were used in a similar manner to promote the heme protein cytochrome oxidase as the oxygen sensor (8). Indeed, there is no reason that multiple oxygen-sensing pathways cannot be present within the same cells (6), activated, perhaps, within different intracellular PO_2 microdomains.

Although the exact methods of fetal oxygen sensing and signal transduction have yet to be definitively elucidated, we do know that the fetus is remarkably responsive to acute and chronic alterations in oxygenation well above the level at which oxygen becomes limiting for respiration. This chapter will focus on the fetal integrative physiological adaptations to hypoxia that act to preserve functionality of vital fetal systems and prevent morbidity and mortality.

A range of cardiovascular and metabolic adaptations have been demonstrated to exist in the fetus that can be elicited, depending upon the intensity, duration, and mechanism of oxygen deprivation. In the following sections, these cardiovascular and metabolic adaptations are described in greater detail. Unless otherwise stated, the observations were made in sheep fetuses, the most widely used model for such research. The adaptations have been broadly confirmed in other species, including the human.

II. Cardiovascular Responses to Hypoxia

During hypoxia the fetal combined ventricular output (CVO) is redistributed in favor of the placenta, head, brain, and adrenal glands at the expense of other tissues (15). This finding is similar to human Doppler studies, where chronically hypoxic, intrauterine-growth-restricted fetuses have an abnormal hemodynamic pattern consistent with a redistribution of CVO to vital organs (16). However, the precise pattern of this redistribution depends upon the type of hypoxic challenge used (17), and it is of note that, unlike the adrenal and myocardial circulations, blood flow to the brain and placental circulation is not always increased (Fig. 1).

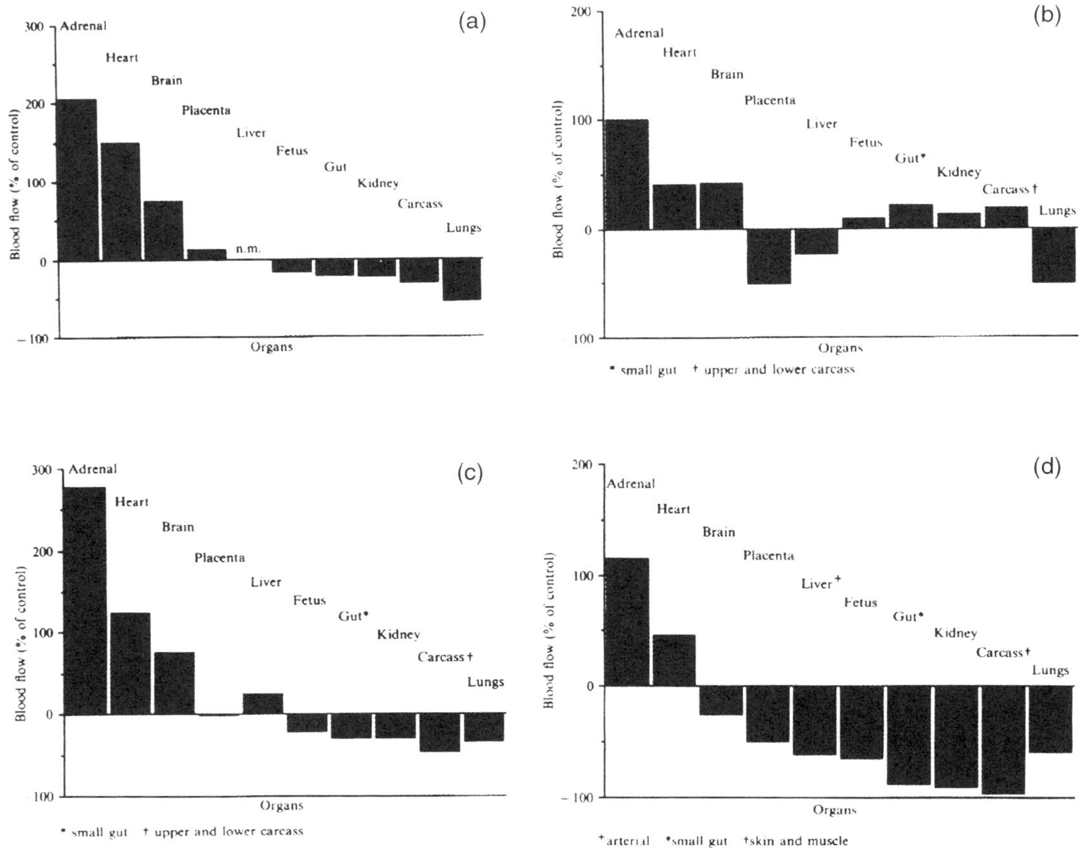

Figure 1 Examples of alterations in distribution of fetal combined ventricular output during experimental hypoxic challenges: (a) maternal hypoxemia, (b) complete umbilical cord occlusion, (c) reduction of uterine artery flow by 50%, and (d) complete arrest of uterine artery flow. (From Ref. 17.)

During hypoxia, the redistribution of CVO is also accompanied by altered distribution of umbilical venous flow. During normoxemia, 55% of umbilical venous flow bypasses the liver through the ductus venosus (18) and enters the inferior vena cava, from where it is preferentially streamed across the foramen ovale (19) and sent to the head and upper limbs (20). In this way the brain and heart always receive the most highly oxygenated blood. During hypoxemia, these processes are augmented such that the percentage of oxygenated blood bypassing the liver increases from 55 to 65% and a greater proportion of umbilical venous blood passes through the foramen ovale to supply the brain and heart (16,20).

Indeed, the function of both brain and heart under all situations has been thought of as paramount in protecting the development of the fetus. However, there are good reasons to believe that the metabolism of the brain is much more plastic than once thought. In the adult at least, the metabolism of the brain is closely linked to the blood flow to it. Maintenance of an adequate oxygen supply to the fetal brain requires matching of blood flow to metabolic demand. While in the adult this can be achieved through a decrease in cerebrovascular resistance and an increase in cardiac output (CO), the fetus is unable to increase CO and thus the ability to alter the distribution of CVO is of particular importance during fetal life (18,19). Alteration of CVO is achieved through control of regional microvascular tone at the local level while plasma concentrations of substrates such as O_2 and glucose (and lactate in the fetus) are kept within tight limits.

A. Reflex Responses

In late gestation, the fetal cardiovascular response to hypoxia is typified by a rapid, transient bradycardia and increased heart rate variability (21,22). The bradycardia reaches a nadir approximately 1–2 min after the onset of the hypoxic insult (15,23) and, presumably, economizes on myocardial oxygen consumption. Fetal hypoxic bradycardia occurs at the onset of hypoxic hypoxia (5), umbilical cord occlusion (24,25), and uterine artery occlusion (26)—the three most common experimental hypoxic challenges.

The rapid onset of the bradycardia in response to hypoxia is suggestive of a neural reflex (Fig. 2). The sensors for this reflex have been identified as the carotid chemoreceptors (27,28), which send afferent fibers in the carotid sinus nerve to a central integrating area in the brainstem (29) and whose efferent arm consists of parasympathetic cholinergic fibers to the heart carried in the vagus nerve (cranial nerve X) and α-adrenergic sympathetic fibers to resistance vessels in peripheral vascular beds (30). The absence of this bradycardic response earlier in gestation (31) is indicative of immaturity of neural reflex mechanisms.

In late gestation, fetal mean arterial pressure responses to hypoxia can be variable. In 1974, Cohn and colleagues reported that there was no significant change in fetal mean arterial pressure with isocapnic hypoxemia, but that mean pressure increased when the fetus became acidemic during the insult (15). Subsequently, other authors have described a significant increase in mean arterial pressure with maternal inhalation hypoxia (27), umbilical cord occlusion (24,25,32,33), and

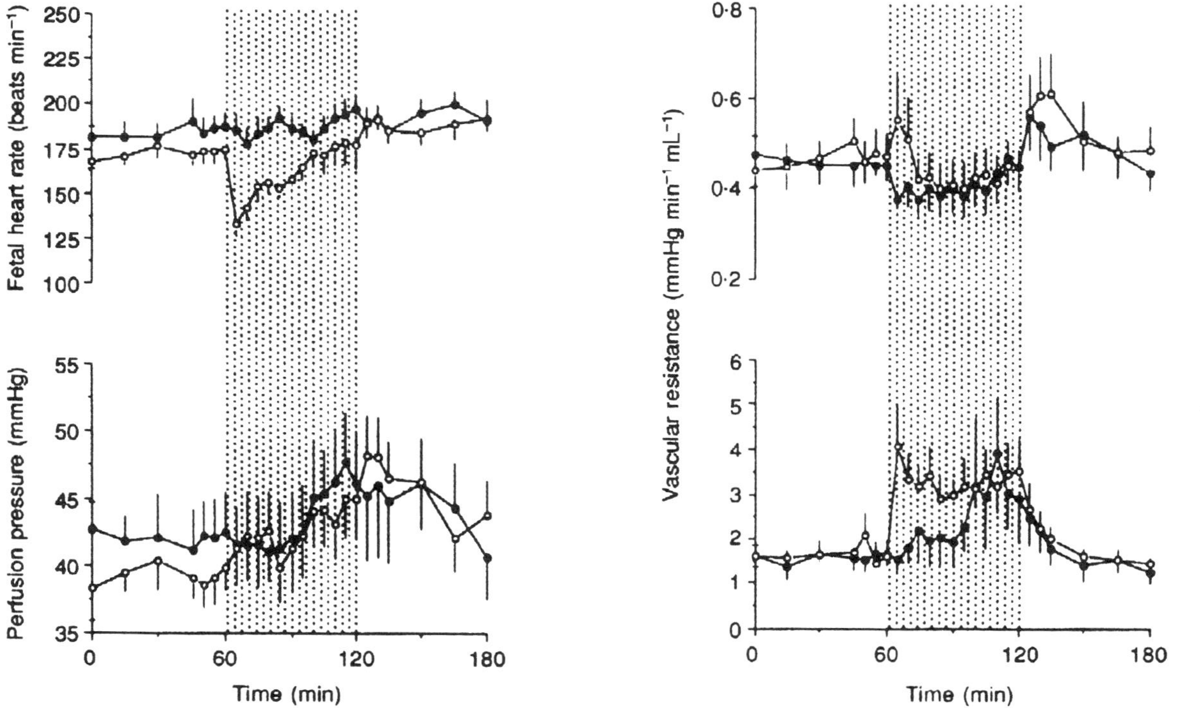

Figure 2 Acute fetal responses to hypoxia in the late-gestation sheep fetus. Right-hand graph shows carotid (upper) and femoral (lower) vascular resistance; left-hand graph shows fetal heart rate (upper) and perfusion pressure (lower). With intact carotid sinus nerves (○), the normal late-gestation fetal responses to hypoxia are present. When the carotid sinus nerves are cut (●), the initial bradycardia is not seen and there is no early increase in femoral vascular resistance. (From Ref. 30.)

interruption of uterine artery flow (34). A hypertensive response to hypoxia is likely to depend in part on the intensity of the insult.

B. Endocrine Responses

Whilst the initial bradycardia at the onset of an hypoxic episode can be explained in terms of neural reflex pathways, if hypoxia is maintained beyond approximately 4 or 5 min, this bradycardia begins to be slowly abolished indicating that slower-acting mechanisms may also be active (Fig. 2 and Ref. 27). Moreover, carotid denervation may abolish the initial bradycardia and increase in peripheral resistance seen at the start of hypoxia (30), but there is still a late-onset rise in peripheral resistance (27).

A number of endocrine hormones, including catecholamines, cortisol, angiotensin II (AngII), and arginine vasopressin (AVP), with known vasoactive properties in the adult are strong candidates for mediating slower cardiovascular responses to hypoxia. Experimental studies of the endocrine responses to hypoxemia show that hormone levels, including those of epinephrine, norepinephrine, AVP, catecholamines, and cortisol, peak at least 2 hr after the onset of sustained hypoxia (35,36) and may either remain high (PGE_2) or return to preinsult levels if the hypoxia is sustained (AVP, epinephrine). Other vasoconstrictor systems stimulated by hypoxia-ischemia include neuropeptide Y (37) and prostaglandins (38). The combined action of these systems maintains peripheral vasoconstriction to vascular beds including those to the lung, kidney, gut, liver, and spleen.

Catecholamines

Circulating catecholamines in the fetus originate from the adrenal medulla (~30%) and the sympathetic nervous system (39,40). It is increased activity of the sympathetic nervous system rather than the adrenal medulla that is thought to account for developmental increases in blood pressure during late gestation (39). During late gestation, the release of catecholamines from adrenal medulla (see Fig. 3) is primarily mediated by splanchnic cholinergic nerves (40). However, before the development of functional innervation the adrenal gland is still capable of secreting catecholamines in response to hypoxia and this suggests the presence of oxygen-sensitive cells in medullary tissue (6). From studies with α-adrenoreceptor blockade, hypertension during hypoxia is believed to originate from α-adrenergic stimulation (41,42). Hypoxic stimulation acting via autonomic nerves causes the circulating levels of epinephrine and norepinephrine to rise 50–100-fold during hypoxia (43), an effect that is abolished by adrenalectomy or chemical sympathectomy (40). The slow abolition of the rapid reflex bradycardia during hypoxia is attributed to increased sympathetic efferent activity and to a direct effect of catecholamines on the heart. The variable response of different vascular beds to adrenergic stimulation is likely to be accounted for by differences in α- and β-receptor populations (44). In the llama α-adrenergic mechanisms are indispensable in mediating the peripheral vasoconstriction during hypoxia (45), whereas in the sheep fetus the contribution of other reflexively released vasoconstrictors is indicated (30).

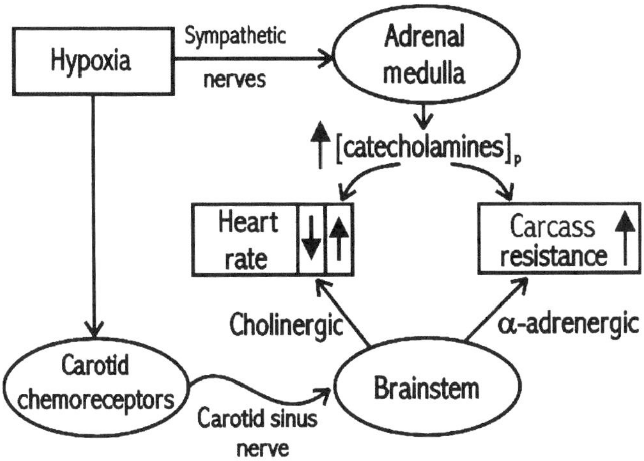

Figure 3 Diagrammatic representation of components of the initial chemoreflex-driven responses to acute isocapnic hypoxaemia in the late-gestation fetal sheep in utero (30). Hypoxia stimulates both carotid chemoreflex and hormonal responses. Catecholamines are released from the adrenal medulla in response to fetal hypoxia and act to increase carcass (e.g., femoral) vascular resistance and to increase fetal heart rate.

Hypothalamo-Pituitary-Adrenal Axis

Cortisol levels are lower in the fetus than in the mother (46), but increase over gestation with a surge in cortisol near to term (47). Exogenously administered cortisol elevates arterial pressure and decreases heart rate in late-gestation fetuses and to a greater extent in younger fetuses perhaps reflecting the immaturity of alternative cardiorespiratory control mechanisms (48–50). Several studies have reported an increase in plasma cortisol during hypoxemia (30,51,52) while others have not (46), possibly reflecting the severity of the insult used. Cortisol progressively inhibits the CRH-elicited ACTH rise during hypoxia by acting at the level of the pituitary (53). An ACTH-independent reflex, with afferents in the carotid sinus nerve and efferents in the splanchnic nerve, appears to exist and contribute to the cortisol response to hypoxia to a greater extent in the llama than in the ovine fetus (30,54). At present, the relative contribution of cortisol to cardiovascular control during hypoxia is unclear, but species differences between the sheep and llama, perhaps as a result of adaptations made to high altitude, suggest that cortisol is an important factor. Aside from its role as a vasoconstrictor, cortisol may modulate the responsiveness of the cardiovascular system to other vasoactive hormones (50,55) and alter gene expression to elicit more generalized adaptations to hypoxic stress (56).

Renin-Angiotensin System

Over late gestation renal content and basal secretion of rennin in vitro increase (57,58) as do plasma levels of angiotensin-converting enzyme (ACE) (59). Overall, AngII levels in the late-gestation fetus are in the same range as those in the maternal circulation, although plasma renin activity in the fetus is much higher (60), indicative of poor conversion of angiotensin I to II by ACE. The predominant AngII receptor subtype during fetal life is AT2, with a developmental switch to the AT1 subtype postnatally. Although present in fetal rat vessels (61,62), AT1 receptors are absent in ovine fetal systemic arteries (63). Thus the hypotensive effect of AT1-receptor antagonism during ovine fetal life may reflect effects on the placental vasculature where AT1 receptors are expressed (63). Under baseline conditions AngII does not appear to contribute substantially to cardiovascular control (64,65). Hypoxia induces rises in both plasma renin activity and AngII levels in the fetal circulation (60,65,66) and administration of exogenous AngII to the fetus caused an elevation in blood pressure and a redistribution of blood flow reminiscent of the cardiovascular response to stress (67). However, AngII does not appear to contribute to the control of renal (68), pulmonary (69), or peripheral vasculature (64) during hypoxia in the late-gestation fetus. In fact, it is not until the fetus is compromised, either by removal of carotid reflex mechanisms or by placental restriction, that a role for AngII in cardiovascular control becomes apparent (64,65)—observations that lend weight to the idea of multiple layers of cardiovascular control.

AVP

AVP is synthesized in the fetal hypothalamus. In addition to its role in altering gene expression (e.g., POMC), plasma AVP levels also increase in response to fetal hypoxia-ischemia (30,70,71). The release of AVP is not mediated by peripheral chemoreceptor mechanisms and therefore is unlikely to contribute to the rapid cardiovascular changes at the onset of hypoxia (see above) (30). Exogenously administered AVP produces hypertension, peripheral vasoconstriction, and bradycardia (71–73), although the contribution of hypoxia-stimulated increases in AVP to the redistribution of fetal CVO is uncertain (74,75). Endothelial V1 receptors mediate an AVP vasodilatory effect, which means that the net effect of AVP on the cardiovascular system is likely to reflect an integrated picture of vasoconstriction from a number of vascular beds, modulated by endothelium-dependent vasodilatory mechanisms (76).

C. Paracrine and Autocrine Responses

As described above, redistribution of CVO achieved through alterations in resistance vessel tone exerted by endocrine and neural factors and modified by local mechanisms, leads to a net increase in vascular resistance in peripheral vascular beds (including gut and femoral beds) and a decrease in vascular resistance in the cerebral, adrenal, and myocardial circulations during hypoxia (31,77–79).

The tissue-specific local conditions, reactivity to vasoactive hormones, and autonomic innervation will therefore determine how each vascular bed reacts to hypoxia, with flow increasing to some while being maintained or falling in others. For example, while there is evidence for a rapid rise in vascular resistance at the onset of hypoxia in femoral (27) and superior mesenteric (24) arteries, blood flow in the carotid arteries shows a variable response and is either increased (30), maintained (26), or decreased (78) depending upon the severity of the insult and its rapidity of onset.

Nitric Oxide

Nitrovasodilators such as nitroglycerine and nitroprusside are among the oldest cardiovascular drugs available for clinical use (80), although it was not until the early 1980s that the mechanism by which they act to relax vascular smooth muscle became apparent. It was first suggested by Furchgott and Zawadzki (81) that these nitrovasodilators mimicked an endogenous paracrine or autocrine agent, whose action was dependent upon the presence of the vascular endothelium. This so-called endothelium-derived relaxing factor (EDRF) was believed to be nitric oxide (NO) or one of its congeners (i.e., nitrogen oxides of formula NO_x), which were known to be produced in many cells types and which subserved a number of physiological functions (82). EDRF was later confirmed to be nitric oxide (83,84), and a number of NO synthase enzymes were identified whose actions are dependent upon NADPH, O_2, and often on calmodulin and, therefore, on Ca^{2+} levels (84).

To date, NO has been reported to be intimately involved in a number of physiological processes including neurotransmission, neuromodulation, intracellular Ca^{2+} homeostasis, and gene expression (82). Owing to its short half-life, NO is well suited to the continuous modulation of vascular tone in a variety of vascular beds and CVO during the latter half of gestation. Nitric oxide contributes to the fall in pulmonary vascular resistance during oxygenation at birth (85) and mediates the increase in carotid and cerebral blood flow during hypoxemia (86,87).

Endothelin-I

Endothelin is produced by both endothelial and smooth muscle cells and exerts vasodilator and vasoconstrictor effects via ETA and ETB receptor subtypes, respectively (88). The net effect of endothelin infused to the fetus is dependent upon the vascular bed in question and the proportion of ETA and ETB receptors: a sustained pulmonary hypertension and modulation of carotid artery responses to hypoxia via action at ETA receptors (64,89,90) and a renal vasodilatation by ETB-receptor mediated NO release (91). As a result, it is likely that the observed lack of effect of specific antagonism of ETA receptors during late gestation on peripheral vascular resistance or arterial pressure reflects the net effect in a number of vascular beds.

Metabolic Factors

A number of products of tissue metabolism are involved in adjusting blood flow at the local tissue level, including NO (see above), carbon monoxide, hydrogen and potassium ions, adenosine, and prostaglandins (see above).

Adenosine has potent cardiovascular actions effected through three ubiquitous classes of adenosine receptor: Al, A2, and A3. These receptors are in turn coupled to either stimulatory or inhibitory cGTP-binding components of adenylate cyclase or to similar proteins modulating Ca^{2+} or K^{+} channels. Adenosine increases flow to myocardial, cerebral, and skeletal muscle vascular beds. It causes vasoconstriction in the renal vascular bed (92–97). Fetal adenosine levels increase in response to systemic hypoxia (98) and adenosine is strongly implicated in mediating the rapid bradycardia, hypertension, and increase in cerebral blood volume during hypoxia (99). Fetal adenosine levels are also believed to be an important determinant of local vascular tone in the fetal brain, where adenosine concentrations are much higher than they are in the adult (100).

As in the adult, the fetal cerebral vasculature is regulated by P_aO_2, by P_aCO_2, by metabolic demand, and by neural factors (101–105). Of these factors, arterial O_2 content seems to be more important than PCO_2 (101) and has been shown to be sensitive to reduction in mean arterial pressure, especially in midgestation when fetal arterial pressure is near the lower end of the autoregulatory range (105). Sympathetic innervation of some areas of the brain is particularly dense and may also be implicated in mediating decreases in cerebrovascular resistance during hypoxia in a regionally selective fashion (103).

Other Hormone Factors

There is growing evidence that several hormones that have endocrine roles also operate at the local tissue level. For example, endothelial and smooth muscle cells contain all the components of the renin-angiotensin system and HPA axis and the potential to synthesize AngII and cortisol (106–108).

III. Metabolic Responses to Hypoxia

Unlike that of the adult, fetal metabolism is geared toward tissue accretion and maturation in preparation for birth (109); however, like the adult, there is an absolute requirement for oxidative metabolism to achieve these ends. The fetus, especially in early to midgestation, has a much lower metabolic rate than its adult counterpart (Fig. 4) and yet, unlike the adult, the fetus is capable of withstanding prolonged periods of profound hypoxia and anoxia.

We have already described how the fetal responses to hypoxic challenges result in a redistribution of CVO in favor of tissues whose metabolism requires the most urgent protection—the heart, adrenal glands, and brain. In addition, the fetus can make behavioral and cellular responses to hypoxia in a way that reduces tissue oxygen demand. This response makes teleological sense since the duration of the hypoxic episode cannot be predicted and the fetus does not have oxygen stores per se. Thus, if the fetal oxygen demand stays high and the hypoxia is extended, tissue PO_2 will fall below the level at which metabolic rate can be sustained and cellular energy stores will become depleted; this is a circumstance that can only

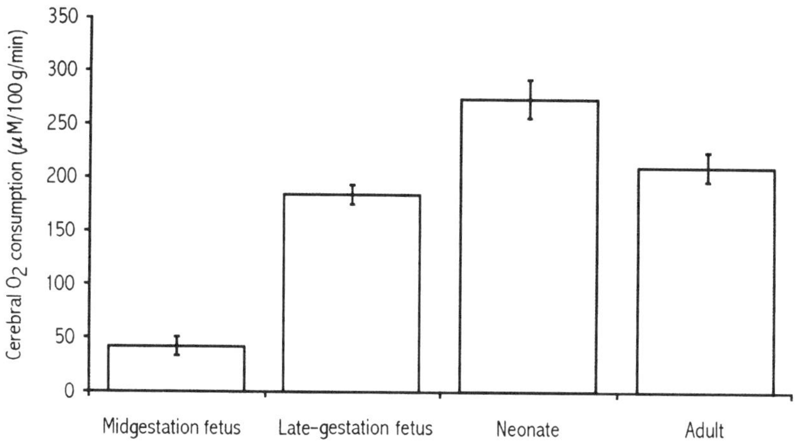

Figure 4 Development of ovine cerebral oxygen consumption. Brain metabolism increases with gestation, reaching a peak after birth before falling to adult levels. (Adapted from Refs. 110 and 111.)

result in death of the most susceptible cells and has disastrous consequences for the fetus.

A. Behavioral Adaptations

In late gestation, the ovine fetus experiences periods of active and quiet "sleep" associated with low- and high-voltage electrocortical states, respectively (112). Fetal breathing and body movements develop with gestation and, when electrocortical states become differentiated at circa 120 days of gestation, these movements are highest in the low-voltage electrocortical state (113,114). During normoxia, fetal breathing and body movements account for about 17% of fetal basal metabolic rate (115–118) and thus can result in a substantial energetic saving to the fetus when they cease in the high-voltage state. In fetuses, there is a rapid decrease in fetal breathing and body movements at the onset of hypoxia, which results in a fall in oxygen consumption (119), in decreased episodes of high-fetal-heart-rate variation and in the prominence of the low-voltage electrocortical state during periods of UCO, which could impact on fetal growth and development (32,120).

B. Cellular Adaptations

As described above, the redistribution of fetal CVO during hypoxia serves to protect the function of essential fetal organs (the heart, brain, and adrenal glands). However, extra protection appears to be provided by the ability of some fetal tissues, at least in vitro, to actively reduce their metabolic rate in response to reduced oxygen delivery (8,9,121–123).

Myocardial Metabolic Adaptations to Hypoxia

Data indicate that fetal sheep myocardium metabolizes both glucose and lactate, with some small contribution from ketone bodies and free fatty acids (124). These substrates subserve a myocardial metabolic rate that is not significantly different from that of the adult but that operates at a much lower P_aO_2 and arterial O_2 content. This necessitates either a higher myocardial blood flow (125) or an increase in myocardial efficiency.

In the low-oxygen fetal environment, mitochondria reduce contractile function via mitochondrially generated reactive oxygen species (9). This metabolic suppression can be augmented during hypoxia such that further reductions in energy requirements, ATP demand, and contractile function can be achieved without any observable energy depletion or ischemia and the increases in lactate production that would ensue. These adaptations are possible only when reductions in myocardial oxygenation are moderate or short-term. Severe reductions in myocardial flow for longer than 20 min are reported to be associated with loss of contractile function leading to irrevocable loss of function (126) and myocardial necrosis. There is no histological evidence for irrevocable myocardial infarction following experimental hypoxic challenges.

Cerebral Metabolic Adaptations to Hypoxia

In fetal sheep, measurement of electrocortical activity indicates that the fetal brain undergoes a progressive reduction in neuronal activity during mild to severe hypoxia. During moderate hypoxia, the electrocorticogram (ECoG) of the late-gestation fetal lamb switches from low-voltage, high-frequency activity to high-voltage, low-frequency activity. High-voltage activity is associated with non-REM sleep in the fetal lamb and is indicative of a low oxygen consumption state (127). Further increases in the severity of hypoxia cause the ECoG to become isoelectric (24,31,128–132), and this state can persist for some time after the cessation of the original insult (14). Abnormalities in the electrocorticogram may persist into neonatal life, providing post hoc markers for brain injury (133,134).

Metabolic data from fetal brain studies indicate that *moderate* hypoxia or asphyxia can be adapted to by increases in cerebral blood flow and/or fractional oxygen extraction (15,111,135–138). Acutely, reductions in cerebral oxygen consumption are small compared to the reduction in arterial oxygen content (34); however, during severe or prolonged asphyxia, reductions in cerebral oxygen consumption become apparent.

Severe, acute reduction in oxygenation, for example by complete arrest of uterine flow or complete umbilical cord occlusion, is not associated with a decrease in cerebrovascular resistance, as is seen in more mild insults. This suggests that the ability of the fetus to adapt to hypoxia has been overwhelmed. The rate at which the fetal arterial O_2 content falls may be a determinant of the response elicited whether it be a rapid onset of ECoG isoelectricity (32,120), reduced regional cerebral blood flows (17), a failure of cardiovascular adaptive responses (139), an increase in cerebral lactate production (140), or a 50% reduction in cerebral metabolic rate

(101). These responses are coupled to a failure to maintain cerebral ATP levels as monitored in chronically instrumented fetal sheep (134,135,142) and by nuclear magnetic resonance spectroscopy in human neonates (141,143). It is the oxidative stress during reoxygenation and subsequent secondary energy failure (144) that is believed to lead to neuronal injury in susceptible cell populations.

In contrast, the late-gestation fetus can accommodate for a *slowly developing* hypoxia for several hours by a combination of increased oxygen extraction, increased oxygen-carrying capacity, behavioral-state alterations and redistribution of CVO (145). A reduction in metabolic rate under these conditions is likely to be influenced by an increased rate of nitric oxide production. In vitro studies of cardiomyocytes, vascular smooth muscle cells, and hepatocytes have shown that there are reversible reductions in metabolic rate in cells held at a PO_2 of 40 mmHg or less for 2 h or more (8,123,146).

This reduction in O_2 consumption may be the result of a functional change in cytochrome oxidase with prolonged hypoxia similar to that seen by Chandel et al. (8), a functional change that is quickly reversed upon reoxygenation. With *severe* prolonged reductions in O_2 availability, increased NO production may begin an irreversible nitrosylation of thiol groups on mitochondrial complex I that could result in *permanent* decreases in mitochondrial function, and thus in O_2 consumption (123). Chandel et al. suggest that exposure of hepatocytes to fetal oxygenation levels for 2 h reversibly depresses respiration and decreases CcO turnover rate at any given degree of electron flux. This may be effected through a depolarization of the mitochondrial membrane potential secondary to inhibition of matrix dehydrogenases or of the transport chain itself. This is in accordance with our own observations of oxidation of cerebral CcO during hypoxia in the sheep fetus in utero as determined by near-infrared spectroscopy (26) (Fig. 5).

IV. Relationship Between Cardiovascular Control and Metabolism

The cardiovascular and metabolic arms of fetal adaptations to hypoxia are not mutually exclusive, and instead it is likely that these processes are integrated and mediated by common mechanisms (Fig. 6). Indeed a number of vasoactive substances, including adenosine and nitric oxide, are also capable of effecting alterations in metabolism and are candidates for the metabolic depression seen in moderate to severe hypoxia-asphyxia.

A. Nitric Oxide

Nitric oxide is a freely diffusible free radical gas that can be generated as needed from L-arginine by three different NO synthase (NOS) isoforms: nNOS, eNOS, or iNOS. As discussed above, NO is a potent modulator of vascular tone. However, the interaction of NO with the mitochondrial electron transfer chain complexes suggests that NO is also able to modulate metabolic rate.

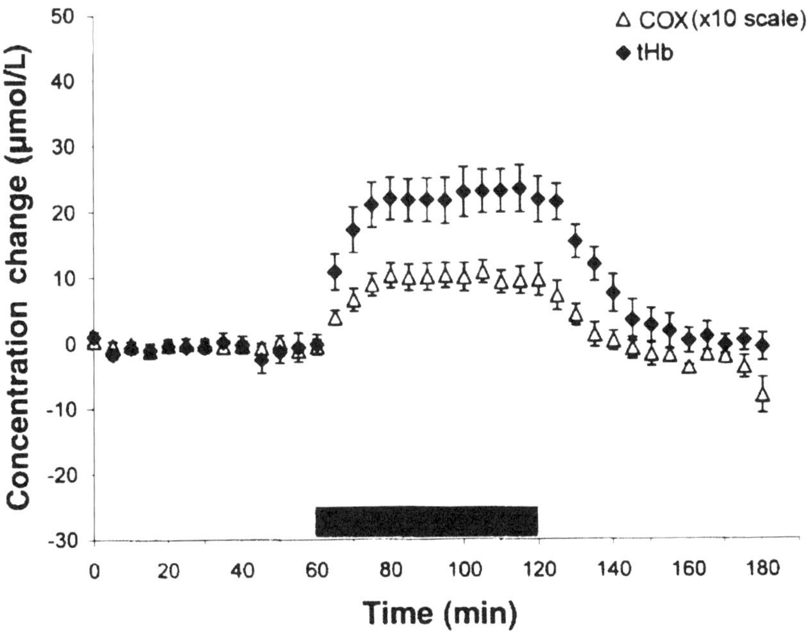

Figure 5 Near-infrared spectroscopy data showing rise in cerebral blood volume (tHb) and cerebral CcO (COX) oxidation state during hypoxia (bar) in the late-gestation sheep fetus in utero. (From Ref. 26.)

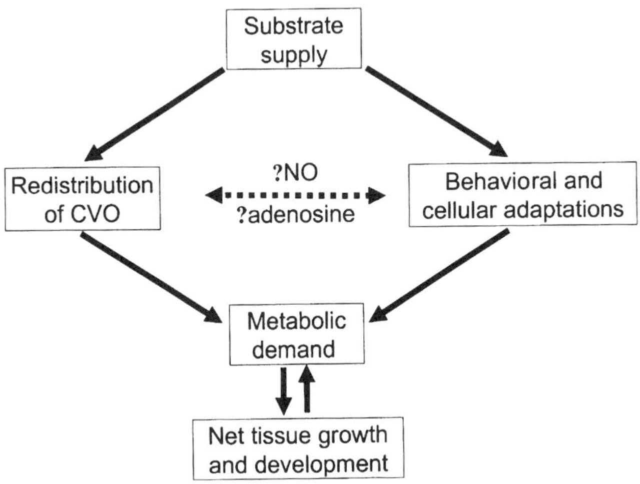

Figure 6 Hypothesized mechanisms integrating the cardiovascular and metabolic adaptations to altered substrate supply.

Several recent publications (147–152) report NO's ability to interact with mitochondrial complexes I, III, and IV (cytochrome oxidase, CcO). During acute increases in NO levels the most rapid interaction appears to be that of NO with the O_2-binding site on mitochondrial complex IV, competitively inhibiting CcO and reducing the rate of transfer of electrons from the electron transfer chain to O_2. In effect, electrons are held at respiratory chain metal centers for an increased period, and as the transfer chain becomes more reduced, mitochondrial O_2 consumption is reduced in a concentration-dependent manner (146). The mechanics and kinetics of the interaction of NO with CcO have been described previously (147,148,151).

From studies in the brain of fetal and adult animals and humans (152,153), total NOS activity is low at 0.48 of gestation, but increases rapidly by 0.75 of gestation, when total NOS activity is not different from adult levels in the frontal lobe and hippocampus. By 0.92 of gestation, all tested areas of the sheep brain have total NOS activity that is not different from that of the adult. There appears to be a biphasic expression of nNOS in the brain areas described in these papers: increasing toward midgestation and then falling to adult levels over time. The different developmental characteristics of total NOS and nNOS suggest that more than one NOS isoform is present during development in utero.

Nitric oxide has been localized within mitochondria in vitro (154) therefore suggesting an active role for NO in modulating cellular metabolic rate. It is thought that by creating microdomains of elevated NO, the NO can exert specific metabolic effects within the cell.

B. Adenosine

Adenosine, a breakdown product of adenine ribonucleotides, is found in the extracellular space and plasma at low concentrations. Aside from potent cardiovascular effects, the adenine nucleotides and adenosine are also intimately involved with energy metabolism. ATP regulates its own formation by negative-feedback inhibition and is an activator of enzymes of purine degradation (AMP deaminase and 5′ nucleotidase), while ADP and AMP are regulators of mitochondrial and substrate-level phosphorylation. The presence of plasma membrane transporters for adenosine indicates that, like cAMP, it may be involved in intracellular communication—perhaps as a signal of cellular energy status. Since adenosine is rapidly broken down in the extracellular space, it is likely that such signaling is autocrine and, perhaps, paracrine.

Fetal plasma adenosine levels are higher than those of the mother (155). Adenosine levels in the fetal cistea magna increase three- to fourfold during hypoxia (155,156). There is an inverse relationship between P_aO_2 and plasma adenosine concentrations during adult (157) and fetal (155) life (Fig. 7). The relatively high plasma concentration of adenosine during fetal life could cause a baseline tonic vasodilatation in many vascular beds and a tonic (although mild) inhibition of energy metabolism, effects that could be enhanced during hypoxic stimulation of adenosine levels.

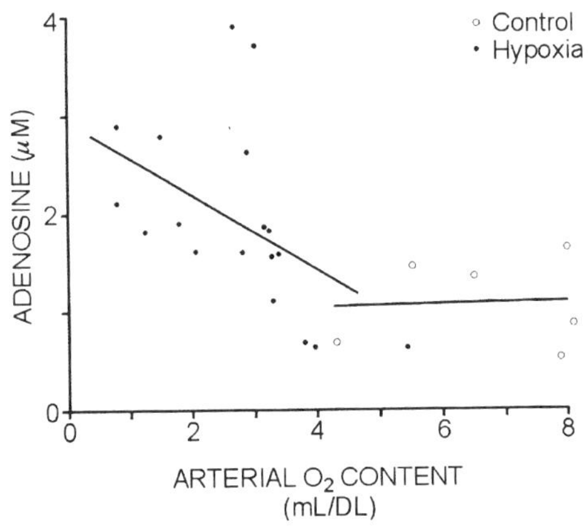

Figure 7 Relationship between arterial O_2 content and plasma adenosine concentration in the near-term fetal sheep in utero. Under normal fetal oxygenation, adenosine remains constant at approximately 1 μM; adenosine rises with falling oxygenation, reaching 3–4 μM. (From Ref. 155.)

By virtue of its dual roles in cardiovascular control and metabolism, adenosine is a strong candidate for integrating these processes during fetal life. Adenosine exerts a hyperpolarizing effect on fetal neuronal plasma membranes (158), which reduces the likelihood of reaching the threshold necessary to initiate an action potential. In addition, adenosine reduces excitatory neurotransmitter release (159,160), shifting the balance of neurotransmitter release in favor of the inhibitory transmitters. These effects of adenosine, alone or in conjunction with direct effects on neurons, contribute to the reduction in fetal breathing and body movements seen with hypoxia and during exogenous infusion of adenosine or adenosine analogs (114,128,155,161–164), which promotes reduction in oxygen consumption (Fig. 8) and thus may contribute to brain temperature regulation (165). A shift away from excitatory to inhibitory neurotransmitters may also be responsible for the decreased incidence of low-voltage electrocortical activity seen during exogenous adenosine infusion (155). All of these effects of hypoxia and adenosine supplementation would act to reduce fetal energy demand.

V. Conclusion and Perspectives

Despite the absolute requirement of the fetus for oxygen to continue growth and differentiation, it has evolved many adaptive processes that serve to minimize the impact of both acute and chronic hypoxia on its development. As in other areas of integrative physiology, hypoxic sensitivity exists at a number of discrete, but interacting, levels—whole-body level, within systems or organs, and locally at the cellular level. Thus hierarchical oxygen supply can be effected by neuronal and hormonal mechanisms according to the needs of individual tissues during fetal normoxia and hypoxia.

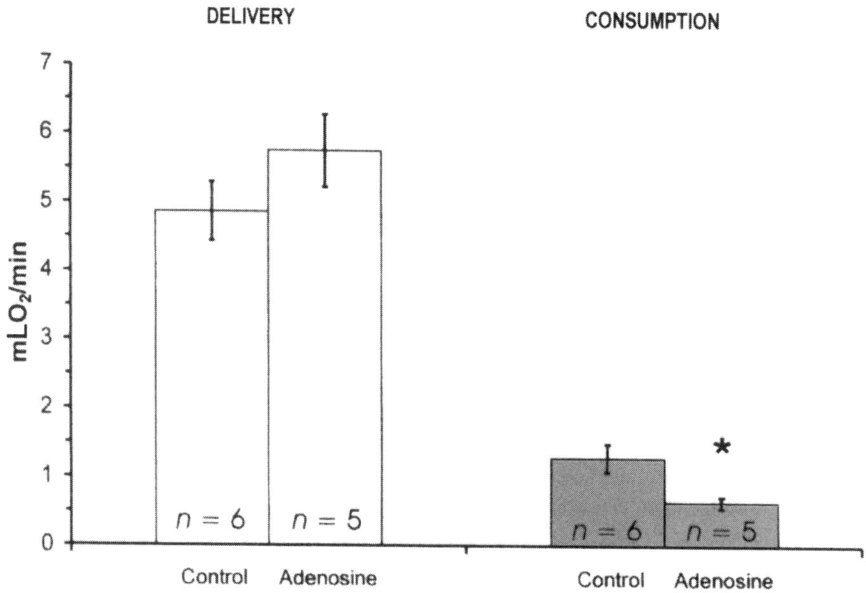

Figure 8 Effects of 20-min adenosine infusion (3 μmol/kg) into an unanesthetized, normoxic midgestation sheep fetus in utero on cerebral oxygen delivery and consumption. While oxygen delivery did not show any significant change, cerebral oxygen consumption fell. $*p < 0.05$, paired *t*-test versus preinfusion level. (From Ref. 161.)

The existence of multiple layers of control is consistent with the observed responses to hypoxia, in which initial fetal adaptations are altered over time in an organ-specific manner and dependent upon the severity and initial cause of the hypoxia. It is believed that these secondary changes are effected by paracrine and autocrine mechanisms and occur as the needs of the local tissues overcome centrally controlled drives to protect the most vital fetal systems. The local mechanisms are needed to prevent ischemic damage to vulnerable tissues, such as the gut.

While currently novel, the idea that cerebral as well as peripheral metabolism can be actively altered during hypoxia-ischemia appears to be well founded. The mechanisms by which reductions in metabolism are achieved are presently unknown but at least two vasoactive compounds important in the hemodynamic response to hypoxia (namely nitric oxide and adenosine) are also able to elicit metabolic alterations and are thus prime candidates for mediation of metabolic adaptations.

Another candidate mediator of hypoxic metabolic depression is cytochrome oxidase, the final oxygen acceptor in the mitochondrial electron transfer chain, whose own oxygen-sensing capabilities are the subject of much speculation. Work must continue to investigate the effects of the chronic hypoxia of the fetus on the function of cytochrome oxidase and its implications for cerebral and peripheral metabolism during pregnancy.

References

1. Madan A, Varma S, Cohen H. Developmental stage-specific expression of the alpha and beta subunits of the HIF-1 protein in the mouse and human fetus. Mol Genet Metab 2002; 75(3):244–249.
2. Wiener C, Booth G, Semenza G. In vivo expression of mRNAs encoding hypoxia-inducible factor 1. Biochem Biophys Res Commun 1996; 225(2):485–488.
3. Yu A, Frid M, Shimoda L, Wiener C, Stenmark K, Semenza G. Temporal, spatial, and oxygen-regulated expression of hypoxia-inducible factor-1 in the lung. Am J Physiol 1998; 275(4 Pt 1):L818–L826.
4. Bunn HF, Poyton RO. Oxygen sensing and molecular adaptation to hypoxia. Physiol Rev 1996; 76(3):839–885.
5. Daghman NA, Elder GE, Savage GA, Winter PC, Maxwell AP, Lappin TR. Erythropoietin production: evidence for multiple oxygen sensing pathways. Ann Hematol 1999; 78(6):275–278.
6. Rychkov G, Adams M, McMillen I, Roberts M. Oxygen-sensing mechanisms are present in the chromaffin cells of the sheep adrenal medulla before birth. J Physiol (Lond) 1998; 509(Pt 3):887–893.
7. Acker H, Xue D. Mechanisms of oxygen sensing in the carotid body in comparison to other oxygen sensing cells. News Physiol Sci 1995; 10:211–216.

7a. Lahiri S, Rozanov C, Roy A, Storey B, Buerk DG. Regulation of oxygen sensing in peripheral arterial chemoreceptors. Int J Biochem Cell Biol 2001; 33:755–774.

8. Chandel N, Budinger G, Choe S, Schumacker P. Cellular respiration during hypoxia: role of cytochrome oxidase as the oxygen sensor in hepatocytes. J Biol Chem 1997; 272(30):18808–18816.
9. Budinger G, Duranteau J, Chandel N, Schumacker P. Hibernation during hypoxia in cardiomyocytes: role of mitochondria as the O_2 sensor. J Biol Chem 1998; 273(6):3320–3326.
10. Reeve HL, Tolarova S, Nelson DP, Archer S, Weir EK. Redox control of oxygen sensing in the rabbit ductus arteriosus. J Physiol 2001; 533(Pt 1):253–261.
11. De Marco CS, Caniggia I. Mechanisms of oxygen sensing in human trophoblast cells. Placenta 2002; 23(suppl A):S58–S68.
12. Porter VA, Rhodes MT, Reeve HL, Cornfield DN. Oxygen-induced fetal pulmonary vasodilation is mediated by intracellular calcium activation of K(Ca) channels. Am J Physiol Lung Cell Mol Physiol 2001; 281(6):L1379–L1385.
13. Daghman NA, Elder GE, Savage GA, Winter PC, Maxwell AP, Lappin TR. Erythropoietin production: evidence for multiple oxygen sensing pathways. Ann Hematol 1999; 78(6):275–278.
14. Keating DJ, Rychkov GY, Roberts ML. Oxygen sensitivity in the sheep adrenal medulla: role of SK channels. Am J Physiol Cell Physiol 2001; 281(5):C1434–C1441.
15. Cohn H, Sacks E, Heymann M, Rudolph A. Cardiovascular responses to hypoxemia and acidemia in fetal lambs. Am J Obstet Gynecol 1974; 120(6):817–824.
16. Kiserud T, Jauniaux E, West D, Ozturk O, Hanson M. Circulatory responses to maternal hyperoxaemia and hypoxaemia assessed non-invasively in fetal sheep at 0.3–0.5 gestation in acute experiments. Br J Obstet Gynaecol 2001; 108(4):359–364.
17. Jensen A, Berger R. Fetal circulatory responses to oxygen lack. J Dev Physiol 1991; 16(4):181–207.

18. Rudolph A, Heymann M. Circulatory changes during growth in the fetal lamb. Circ Res 1970; 26(3):289–299.
19. Barclay A, Franklin K, Prichard M. The foetal circulation and cardiovascular system, and the changes that they undergo at birth. 1st ed. Oxford: Blackwell Scientific Publications, 1944.
20. Reuss M, Rudolph A. Distribution and recirculation of umbilical and systemic venous blood flow in fetal lambs during hypoxia. J Dev Physiol 1980; 2(1–2):71–84.
21. Dalton K, Dawes G, Patrick J. Diurnal, respiratory, and other rhythms of fetal heart rate in lambs. Am J Obstet Gynecol 1977; 127(4):414–424.
22. Parer J, Krueger T, Harris J. Fetal oxygen consumption and mechanisms of heart rate response during artificially produced late decelerations of fetal heart rate in sheep. Am J Obstet Gynecol 1980; 136(4):478–482.
23. Westgate J, Bennet L, Gunn A. Fetal heart rate variability changes during brief repeated umbilical cord occlusion in near term fetal sheep. Br J Obstet Gynaecol 1999; 106(7):664–671.
24. Bennet L, Rossenrode S, Gunning M, Gluckman P, Gunn A. The cardiovascular and cerebrovascular responses of the immature fetal sheep to acute umbilical cord occlusion. J Physiol (Lond) 1999; 517(Pt 1):247–257.
25. Green L, Kawagoe Y, Homan J, White S, Richardson B. Adaptation of cardiovascular responses to repetitive umbilical cord occlusion in the late gestation ovine fetus. J Physiol 2001; 535(Pt 3):879–888.
26. Newman J, Peebles D, Harding S, Springett R, Hanson M. Hemodynamic and metabolic responses to moderate asphyxia in brain and skeletal muscle of late-gestation fetal sheep. J Appl Physiol 2000; 88:82–90.
27. Giussani D, Spencer J, Moore P, Bennet L, Hanson M. Afferent and efferent components of the cardiovascular reflex responses to acute hypoxia in term fetal sheep. J Physiol (Lond) 1993; 461:431–449.
28. Bartelds B, Van Bel F, Teitel D, Rudolph A. Carotid, not aortic, chemoreceptors mediate the fetal cardiovascular response to acute hypoxaemia in lambs. Pediatr Res 1993; 34(1):51–55.
29. Mifflin S. Arterial chemoreceptor input to nucleus tractus solitarius. Am J Physiol 1992; 263(2 Pt 2):R368–R375.
30. Giussani D, McGarrigle H, Spencer J, Moore P, Bennet L, Hanson M. Effect of carotid denervation on plasma vasopressin levels during acute hypoxia in the late-gestation sheep fetus. J Physiol (Lond) 1994; 477(Pt 1):81–87.
31. Iwamoto H, Kaufman T, Keil L, Rudolph A. Responses to acute hypoxemia in fetal sheep at 0.6–0.7 gestation. Am J Physiol 1989; 256(3 Pt 2):H613–H620.
32. Bennet L, Peebles D, Edwards A, Rios A, Hanson M. The cerebral hemodynamic response to asphyxia and hypoxia in the near-term fetal sheep as measured by near infrared spectroscopy. Pediatr Res 1998; 44(6):951–957.
33. Green L, Homan J, White S, Richardson B. Cardiovascular and metabolic responses to intermittent umbilical cord occlusion in the preterm ovine fetus. J Soc Gynecol Invest 1999; 6(2):56–63.
34. Field D, Parer J, Auslender R, Cheek D, Baker W, Johnson J. Cerebral oxygen consumption during asphyxia in fetal sheep. J Dev Physiol 1990; 14(3):131–137.
35. Hooper S, Coulter C, Deayton J, Harding R, Thorburn G. Fetal endocrine responses to prolonged hypoxemia in sheep. Am J Physiol 1990; 259(4 Pt 2):R703–R708.

36. Challis J, Richardson B, Rurak D, Wlodek M, Patrick J. Plasma adrenocorticotropic hormone and cortisol and adrenal blood flow during sustained hypoxemia in fetal sheep. Am J Obstet Gynecol 1986; 155(6):1332–1336.
37. Fletcher A, Goodfellow M, Forhead A, Gardner D, McGarrigle H, Fowden A, Giussani DA. Low doses of dexamethasone suppress pituitary–adrenal function but augment the glycemic response to acute hypoxemia in fetal sheep during late gestation. Pediatr Res 2000; 47(5):684–691.
38. Challis J, Hart I, Louis T, Mitchell M, Jenkin G, Robinson J, Thorburn GD. Prostaglandins in the sheep fetus: implications for fetal function. Adv Prostaglandin Thromboxane Res 1978; 4:115–132.
39. Schuijers J, Walker D, Browne C, Thorburn G. Effect of hypoxemia on plasma catecholamines in intact and immunosympathectomized fetal lambs. Am J Physiol 1986; 251(5 Pt 2):R893–R900.
40. Jones C, Roebuck M, Walker D, Johnston B. The role of the adrenal medulla and peripheral sympathetic nerves in the physiological responses of the fetal sheep to hypoxia. J Dev Physiol 1988; 10(1):17–36.
41. Lewis A, Donovan M, Platzker A. Cardiovascular responses to autonomic blockade in hypoxemic fetal lambs. Biol Neonate 1980; 37(5–6):233–242.
42. Reuss M, Parer J, Harris J, Krueger T. Hemodynamic effects of alpha-adrenergic blockade during hypoxia in fetal sheep. Am J Obstet Gynecol 1982; 142(4):410–415.
43. Jones C, Robinson R. Plasma catecholamines in foetal and adult sheep. J Physiol (Lond) 1975; 248(1):15–33.
44. Shaul P, Muntz K, DeBeltz D, Buja L. Effects of prolonged hypoxia on adenylate cyclase activity and beta-adrenergic receptors in pulmonary and systemic arteries of the rat. Circ Res 1990; 66(6):1526–1534.
45. Giussani D, Riquelme R, Sanhueza E, Hanson M, Blanco C, Llanos A. Adrenergic and vasopressinergic contributions to the cardiovascular response to acute hypoxaemia in the llama fetus. J Physiol 1999; 515(1):233–241.
46. Shepherd R, Stanczyk F, Bethea C, Novy M. Fetal and maternal endocrine responses to reduced uteroplacental blood flow. J Clin Endocrinol Metab 1992; 75(1):301–307.
47. Norman L, Lye S, Wlodek M, Challis J. Changes in pituitary responses to synthetic ovine corticotrophin releasing factor in fetal sheep. Can J Physiol Pharmacol 1985; 63(11):1398–1403.
48. Wood C, Cheung C, Brace R. Fetal heart rate, arterial pressure, and blood volume responses to cortisol infusion. Am J Physiol 1987; 253(6 Pt 2):R904–R909.
49. Dodic M, Wintour E. Effects of prolonged (48 h) infusion of cortisol on blood pressure, renal function and fetal fluids in the immature ovine foetus. Clin Exp Pharmacol Physiol 1994; 21(12):971–980.
50. Tangalakis K, Lumbers E, Moritz K, Towstoless M, Wintour E. Effect of cortisol on blood pressure and vascular reactivity in the ovine fetus. Exp Physiol 1992; 77(5):709–717.
51. Akagi K, Challis J. Hormonal and biophysical responses to acute hypoxemia in fetal sheep at 0.7–0.8 gestation. Can J Physiol Pharmacol 1990; 68(12):1527–1532.
52. Sug Tang A, Bocking A, Brooks A, Hooper S, White S, Jacobs R, Fraher LJ, Challis JR. Effects of restricting uteroplacental blood flow on concentrations of corticotrophin-releasing hormone, adrenocorticotrophin, cortisol, and prostaglandin E2 in the sheep fetus during late pregnancy. Can J Physiol Pharmacol 1992; 70(10):1396–1402.

53. Akagi K, Berdusco E, Challis J. Cortisol inhibits ACTH but not the AVP response to hypoxaemia in fetal lambs at days 123–128 of gestation. J Dev Physiol 1990; 14(6):319–324.
54. Riquelme R, Llanos J, McGarrigle H, Sanhueza E, Hanson M, Giussani D. Chemoreflex contribution to adrenocortical function during acute hypoxemia in the llama fetus at 0.6 to 0.7 of gestation. Endocrinology 1998; 139(5):2564–2570.
55. Forhead A, Broughton-Pipkin F, Fowden A. Effect of cortisol on blood pressure and the renin-angiotensin system in fetal sheep during late gestation. J Physiol (Lond) 2000; 526(Pt 1):167–176.
56. Fraser M, Braems G, Challis J. Developmental regulation of corticotrophin receptor gene expression in the adrenal gland of the ovine fetus and newborn lamb: effects of hypoxia during late pregnancy. J Endocrinol 2001; 169(1):1–10.
57. Siegel S, Fisher D. Ontogeny of the renin–angiotensin–aldosterone system in the fetal and newborn lamb. Pediatr Res 1980; 14(2):99–102.
58. Rawashdeh N, Rose J, Kerr D. Renin secretion by fetal lamb kidneys in vitro. Am J Physiol 1990; 258(2 Pt 2):R388–R392.
59. Forhead A, Melvin R, Balouzet V, Fowden A. Developmental changes in plasma angiotensin-converting enzyme concentration in fetal and neonatal lambs. Reprod Fertil Dev 1998; 10(5):393–398.
60. Broughton-Pipkin F, Lumbers E, Mott J. Factors influencing plasma renin and angiotensin II in the conscious pregnant ewe and its foetus. J Physiol (Lond) 1974; 243(3):619–636.
61. Tufro-McReddie A, Gomez R. Ontogeny of the renin–angiotensin system. Semin Nephrol 1993; 13(6):519–530.
62. Shanmugam S, Monnot C, Corvol P, Gasc J. Distribution of type 1 angiotensin II receptor subtype messenger RNAs in the rat fetus. Hypertension 1994; 23(1):137–141.
63. Cox B, Rosenfeld C. Ontogeny of vascular angiotensin II receptor subtype expression in ovine development. Pediatr Res 1999; 45(3):414–424.
64. Green L, McGarrigle H, Bennet L, Hanson M. Angiotensin II and cardiovascular chemoreflex responses to acute hypoxia in late gestation fetal sheep. J Physiol (Lond) 1998; 507(Pt 3):857–867.
65. Edwards L, Simonetta G, Owens J, Robinson J, McMillen I. Restriction of placental and fetal growth in sheep alters fetal blood pressure responses to angiotensin II and captopril. J Physiol (Lond) 1999; 515(Pt 3):897–904.
66. Robillard J, Nakamura K, DiBona G. Effects of renal denervation on renal responses to hypoxemia in fetal lambs. Am J Physiol 1986; 250(2 Pt 2):F294–F301.
67. Iwamoto H, Rudolph A. Effects of angiotensin II on the blood flow and its distribution in fetal lambs. Circ Res 1981; 48(2):183–189.
68. Robillard J, Schutte B, Page W, Fedderson J, Porter C, Segar J. Ontogenic changes and regulation of renal angiotensin II type 1 receptor gene expression during fetal and newborn life. Pediatr Res 1994; 36(6):755–762.
69. Hyman A, Kadowitz P. Effects of alveolar and perfusion hypoxia and hypercapnia on pulmonary vascular resistance in the lamb. Am J Physiol 1975; 228(2):397–403.
70. Raff H, Kane C, Wood C. Arginine vasopressin responses to hypoxia and hypercapnia in late-gestation fetal sheep. Am J Physiol 1991; 260(6 Pt 2):R1077–R1081.
71. Rurak D. Plasma vasopressin levels during hypoxaemia and the cardiovascular effects of exogenous vasopressin in foetal and adult sheep. J Physiol (Lond) 1978; 277:341–357.

72. Iwamoto H, Rudolph A, Keil L, Heymann M. Hemodynamic responses of the sheep fetus to vasopressin infusion. Circ Res 1979; 44(3):430–436.
73. Nuyt A, Segar J, Holley A, O'Mara M, Chapleau M, Robillard J. Arginine vasopressin modulation of arterial baroreflex responses in fetal and newborn sheep. Am J Physiol 1996; 271(6 Pt 2):R1643–R1653.
74. Perez R, Espinoza M, Riquelme R, Parer J, Llanos A. Arginine vasopressin mediates cardiovascular responses to hypoxemia in fetal sheep. Am J Physiol 1989; 256 (5 Pt 2):R1011–R1018.
75. Piacquadio K, Brace R, Cheung C. Role of vasopressin in mediation of fetal cardiovascular responses to acute hypoxia. Am J Obstet Gynecol 1990; 163(4 Pt 1):1294–1300.
76. Song J, Ram J. Ionic mechanisms of peptide-induced responses in vascular endothelial cells. In: Sowers J, ed. Endocrinology of the Vasculature. Totowa, NJ: Humana Press, 1996:21–36.
77. Ball R, Parer J, Caldwell L, Johnson J. Regional blood flow and metabolism in ovine fetuses during severe cord occlusion. Am J Obstet Gynecol 1994; 171(6):1549–1555.
78. Jensen A, Hohmann M, Kunzel W. Dynamic changes in organ blood flow and oxygen consumption during acute asphyxia in fetal sheep. J Dev Physiol 1987; 9(6):543–559.
79. Yaffe H, Parer J, Block B, Llanos A. Cardiorespiratory responses to graded reductions of uterine blood flow in the sheep fetus. J Dev Physiol 1987; 9(4):325–336.
80. Kelly R, Smith T. Nitric oxide and nitrovasodilators: similarities, differences, and interactions. Am J Cardiol 1996; 77(13):2C–7C.
81. Furchgott R, Zawadzki J. The obligatory role of endothelial cells in the relaxation of arterial smooth muscle by acetylcholine. Nature 1980; 288(5789):373–376.
82. Nathan C, Xie Q. Nitric oxide synthases: roles, tolls, and controls. Cell 1994; 78(6):915–918.
83. Palmer R, Ferrige A, Moncada S. Nitric oxide release accounts for the biological activity of endothelium-derived relaxing factor. Nature 1987; 327(6122):524–526.
84. Palmer RM, Ashton DS, Moncada S. Vascular endothelial cells synthesize nitric oxide from L-arginine. Nature 1988; 333(6174):664–666.
85. Fineman J, Wong J, Morin F, Wild L, Soifer S. Chronic nitric oxide inhibition in utero produces persistent pulmonary hypertension in newborn lambs. J Clin Invest 1994; 93(6):2675–2683.
86. Van Bel F, Sola A, Roman C, Rudolph A. Role of nitric oxide in the regulation of the cerebral circulation in the lamb fetus during normoxemia and hypoxemia. Biol Neonate 1995; 68(3):200–210.
87. Green LR, Bennet L, Hanson MA. The role of nitric oxide synthesis in cardiovascular responses to acute hypoxia in the late gestation sheep fetus. J Physiol (Lond) 1996; 497(Pt 1):271–277.
88. Sakurai T, Yanagisawa M, Masaki T. Molecular characterization of endothelin receptors. Trends Pharmacol Sci 1992; 13(3):103–108.
89. Chatfield B, McMurtry I, Hall S, Abman S. Hemodynamic effects of endothelin-1 on ovine fetal pulmonary circulation. Am J Physiol 1991; 261(1 Pt 2):R182–R187.
90. Jones O, Abman S. Systemic and pulmonary hemodynamic effects of big endothelin-1 and phosphoramidon in the ovine fetus. Am J Physiol 1994; 266(3 Pt 2):R929–R955.
91. Bogaert G, Kogan B, Mevorach R, Wong J, Gluckman G, Fineman J, Heymann MA. Exogenous endothelin-1 causes renal vasodilation in the fetal lamb. J Urol 1996; 156(2 Pt 2):847–853.

92. Schrader J, Haddy F, Gerlach E. Release of (14C) adenosine from the guinea pig heart during autoregulation. Recent Adv Stud Cardiac Struct Metab 1975; 7:171–175.
93. Rubio R, Berne R. Regulation of coronary blood flow. Prog Cardiovasc Dis 1975; 18(2):105–122.
94. Bockman E, Berne R, Rubio R. Release of adenosine and lack of release of ATP from contracting skeletal muscle. Pflügers Arch 1975; 355(3):229–241.
95. Forrester T. Extracellular nucleotides in exercise: possible effect on brain metabolism. J Physiol (Paris) 1978; 74(5):477–483.
96. Rubio R, Berne R, Winn H. Production, metabolism and possible functions of adenosine in brain tissue in situ. Ciba Found Symp 1978; (56):355–378.
97. Gaal K, Forgacs I. Effect of cyclic adenosine monophosphate on renal function and renin secretion. Acta Physiol Acad Sci Hung 1975; 46(1):9–18.
98. Kubonoya K, Power GG. Plasma adenosine responses during repeated episodes of umbilical cord occlusion. Am J Obstet Gynecol 1997; 177:395–401.
99. Koos B, Chau A, Ogunyemi D. Adenosine mediates metabolic and cardiovascular responses to hypoxia in fetal sheep. J Physiol (Lond) 1995; 488(3):761–766.
100. Sawa R, Asakura H, Power GG. Changes in plasma adenosine during simulated birth of fetal sheep. J Appl Physiol 1991; 70(4):1524–1528.
101. Jones M, Sheldon R, Peeters L, Makowski E, Meschia G. Regulation of cerebral blood flow in the ovine fetus. Am J Physiol 1978; 235(2):H162–H166.
102. Bissonnette J, Hohimer A, Richardson B, Machida C. Regulation of cerebral blood flow in the fetus. J Dev Physiol 1984; 6(3):275–280.
103. Busija D, Heistad D. Factors involved in the physiological regulation of the cerebral circulation. Rev Physiol Biochem Pharmacol 1984; 101:161–211.
104. Rudolph A. Distribution and regulation of blood flow in the fetal and neonatal lamb. Circ Res 1985; 57(6):811–821.
105. Helou S, Koehler R, Gleason C, Jones M, Traystman R. Cerebrovascular autoregulation during fetal development in sheep. Am J Physiol 1994; 266(3 Pt 2):H1069–H1074.
106. Brem A, Bina R, King T, Morris D. Localization of 2 11beta-OH steroid dehydrogenase isoforms in aortic endothelial cells. Hypertension 1998; 31(1 Pt 2):459–462.
107. Hatakeyama H, Inaba S, Takeda R, Miyamori I. 11Beta-hydroxysteroid dehydrogenase in human vascular cells. Kidney Int 2000; 57(4):1352–1357.
108. Kifor I, Dzau V. Endothelial renin–angiotensin pathway: evidence for intracellular synthesis and secretion of angiotensins. Circ Res 1987; 60(3):422–428.
109. Bell A, Kennaugh J, Battaglia F, Makowski E, Meschia G. Metabolic and circulatory studies of fetal lamb at midgestation. Am J Physiol 1986; 250(5 Pt 1):E538–E544.
110. Jones MD, Rosenberg AA, Simmons MA, Molteni RA, Koehler RC, Traystman RJ. Oxygen delivery to the brain before and after birth. Science 1982; 216(4543):324–325.
111. Gleason C, Hamm C, Jones M. Cerebral blood flow, oxygenation, and carbohydrate metabolism in immature fetal sheep in utero. Am J Physiol 1989; 256(6 Pt 2): R1264–R1268.
112. Prechtl H. The behavioural states of the newborn infant. Brain Res 1974; 76(2): 185–212.
113. Dawes G. Revolutions and cyclical rhythms in prenatal life: fetal respiratory movements rediscovered. Pediatrics 1973; 51(6):965–971.
114. Walker A, Fleming J, Smolich J, Stunden R, Horne R, Maloney J. Fetal oxygen consumption, umbilical circulation and electrocortical activity transitions in fetal lambs. J Dev Physiol 1984; 6(3):267–274.

115. Rurak D, Gruber N. The effect of neuromuscular blockade on oxygen consumption and blood gases in the fetal lamb. J Dev Physiol 1983; 145:258–262.
116. Rurak D, Gruber N. Increased oxygen consumption associated with breathing activity in fetal lambs. J Appl Physiol 1983; 54(3):701–707.
117. Wilkening R, Boyle D. Effect of neuromuscular blockade on fetal oxidative metabolism. Biol Neonate 1990; 58(1):57–60.
118. Berger P, Soust M, Smolich J, Walker A. Respiratory muscle blood flow in the fetal lamb during apnoea and breathing. Respir Physiol 1994; 97(1):111–121.
119. Bissonnette J, Hohimer A, Willeke G. Effect of asphyxia on respiratory activity in fetal sheep. J Dev Physiol 1989; 12(3):157–161.
120. Kawagoe Y, Green L, White S, Richardson B. Intermittent umbilical cord occlusion in the ovine fetus near term: effects on behavioral state activity. Am J Obstet Gynecol 1999; 181(6):1520–1529.
121. Braems G, Jensen A. Hypoxia reduces oxygen consumption of fetal skeletal muscle cells in monolayer culture. J Dev Physiol 1991; 16(4):209–215.
122. Arai A, Pantely G, Anselone C, Bristow J, Bristow J. Active downregulation of myocardial energy requirements during prolonged moderate ischemia in swine. Circ Res 1991; 69(6):1458–1469.
123. Clementi E, Brown C, Feelisch M, Moncada S. Persistent inhibition of cell respiration by nitric oxide: crucial role of S-nitrosylation of mitochondrial complex I and protective action of glutathione. Proc Natl Acad Sci USA 1998; 95:7631–7636.
124. Bartelds B, Knoester H, Beaufort Krol G, Smid G, Takens J, Zijlstra W, Heymans HS, Kuipers JR. Myocardial lactate metabolism in fetal and newborn lambs. Circulation 1999; 99(14):1892–1897.
125. Fisher D, Heymann M, Rudolph A. Myocardial oxygen and carbohydrate consumption in fetal lambs in utero and in adult sheep. Am J Physiol 1980; 238(3):H399–H405.
126. Heusch G, Schulz R. Myocardial hibernation: adaptation to ischaemia. Eur Heart J 1996; 17(6):824–828.
127. Richardson B, Patrick J, Abduljabbar H. Cerebral oxidative metabolism in the fetal lamb: relationship to electrocortical state. Am J Obstet Gynecol 1985; 153(4):426–431.
128. Boddy K, Dawes G, Fisher R, Pinter S, Robinson J. Foetal respiratory movements, electrocortical and cardiovascular responses to hypoxaemia and hypercapnia in sheep. J Physiol (Lond) 1974; 243:599–618.
129. Ball R, Espinoza M, Parer J, Alon E, Vertommen J, Johnson J. Regional blood flow in asphyxiated fetuses with seizures. Am J Obstet Gynecol 1994; 170(1 Pt 1):156–161.
130. Mallard E, Gunn A, Williams C, Johnston B, Gluckman P. Transient umbilical cord occlusion causes hippocampal damage in the fetal sheep. Am J Obstet Gynecol 1992; 167(5):1423–1430.
131. Mallard E, Williams C, Johnston B, Gluckman P. Increased vulnerability to neuronal damage after umbilical cord occlusion in fetal sheep with advancing gestation. Am J Obstet Gynecol 1994; 170(1 Pt 1):206–214.
132. Bennet L, Quaedackers J, Gunn A, Rossenrode S, Heineman E. The effect of asphyxia on superior mesenteric artery blood flow in the premature sheep fetus. J Pediatr Surg 2000; 35(1):34–40.
133. Watanabe K, Hayakawa F, Okumura A. Neonatal EEG: a powerful tool in the assessment of brain damage in preterm infants. Brain Dev 1999; 21(6):361–372.

134. Connell J, Oozeer R, Regev R, de Vries L, Dubowitz L, Dubowitz V. Continuous four-channel EEG monitoring in the evaluation of echodense ultrasound lesions and cystic leucomalacia. Arch Dis Child 1987; 62(10):1019–1024.
135. Purves M, James I. Observations on the control of cerebral blood flow in the sheep fetus and newborn lamb. Circ Res 1969; 25(6):651–667.
136. Jones M, Sheldon RE, Peeters LL, Meschia G, Battaglia FC, Makowski EL. Fetal cerebral oxygen consumption at different levels of oxygenation. J Appl Physiol 1977; 43(6):1080–1084.
137. Peeters L, Sheldon R, Jones M, Makowski E, Meschia G. Blood flow to fetal organs as a function of arterial oxygen content. Am J Obstet Gynecol 1979; 135(5):637–646.
138. Johnson G, Palahniuk R, Tweed W, Jones M, Wade J. Regional cerebral blood flow changes during severe fetal asphyxia produced by slow partial umbilical cord compression. Am J Obstet Gynecol 1979; 135(1):48–52.
139. Block B, Schlafer D, Wentworth R, Kreitzer L, Nathanielsz P. Intrauterine asphyxia and the breakdown of physiologic circulatory compensation in fetal sheep. Am J Obstet Gynecol 1990; 162(5):1325–1331.
140. Ikeda T, Murata Y, Quilligan E, Choi B, Parer J, Doi S, Park SD. Physiologic and histologic changes in near-term fetal lambs exposed to asphyxia by partial umbilical cord occlusion. Am J Obstet Gynecol 1998; 178(1 Pt 1):24–32.
141. Sibony O, Stempfle N, Luton D, Oury J, Blot P. In utero fetal cerebral intraparenchymal ischemia diagnosed by nuclear magnetic resonance. Dev Med Child Neurol 1998; 40(2):122–123.
142. van Cappellen-van Walsum A, Heerschap A, Nijhuis J, Oeseburg B, Jongsma H. Proton magnetic resonance spectroscopy of fetal lamb brain during hypoxia. Am J Obstet Gynecol 1998; 179(3 Pt 1):756–757.
143. Rutherford M, Pennock J, Schwieso J, Cowan F, Dubowitz L. Hypoxic-ischaemic encephalopathy: early and late magnetic resonance imaging findings in relation to outcome. Arch Dis Child Fetal Neonatal Ed 1996; 75(3):F145–F151.
144. Lorek A, Takei Y, Cady E, Wyatt J, Penrice J, Edwards A, Peebles D, Wylezinska M, Owen-Reese H, Kirkbride V. Delayed ("secondary") cerebral energy failure after acute hypoxia-ischemia in the newborn piglet: continuous 48-hour studies by phosphorus magnetic resonance spectroscopy. Pediatr Res 1994; 36(6):699–706.
145. Richardson B, Rurak D, Patrick J, Homan J, Carmichael L. Cerebral oxidative metabolism during sustained hypoxaemia in fetal sheep. J Dev Physiol 1989; 11:37–43.
146. Clementi E, Brown C, Foxwell N, Moncada S. On the mechanism by which vascular endothelial cells regulate their oxygen consumption. Proc Natl Acad Sci USA 1999; 96:1559–1562.
147. Torres J, Cooper C, Sharpe M, Wilson M. Reactivity of nitric oxide with cytochrome c oxidase: interactions with the binuclear centre and mechanism of inhibition. J Bioenerg Biomembr 1998; 30(1):63–69.
148. Torres J, Cooper C, Wilson M. A common mechanism for the interaction of nitric oxide with the oxidized binuclear centre and oxygen intermediates of cytochrome c oxidase. J Biol Chem 1998; 273(15):8756–8766.
149. Lizasoain I, Moro MA, Knowles RG, Darley UV, Moncada S. Nitric oxide and peroxynitrite exert distinct effects on mitochondrial respiration which are differentially blocked by glutathione or glucose. Biochem J 1996; 314(Pt 3):877–880.
150. Brown G. Nitric oxide regulates mitochondrial respiration and cell functions by inhibiting cytochrome oxidase. FEBS Lett 1995; 369(2–3):136–139.

151. Brown G. Nitric oxide and mitochondrial respiration. Biochim Biophys Acta 1999; 1411(2–3):351–369.
152. Ohyu J, Takashima S. Developmental characteristics of neuronal nitric oxide synthase (nNOS) immunoreactive neurons in fetal to adolescent human brains. Brain Res Dev Brain Res 1998; 110(2):193–202.
153. Northington F, Tobin J, Harris A, Traystman R, Koehler R. Developmental and regional differences in nitric oxide synthase activity and blood flow in the sheep brain. J Cereb Blood Flow Metab 1997; 17(1):109–115.
154. López-Figueroa M, Caamano C, Morano M, Ronn L, Akil H, Watson S. Direct evidence of nitric oxide presence within mitochondria. Biochem Biophys Res Commun 2000; 272(1):129–133.
155. Koos B, Doany W. Role of plasma adenosine in breathing responses to hypoxia in fetal sheep. J Dev Physiol 1991; 16:81–85.
156. Winn H, Rubio R, Beme R. Brain adenosine concentration during hypoxia in rats. Am J Physiol 1981; 241:H235–H242.
157. Mentzer R, Rubio R, Beme R. Release of adenosine by hypoxic canine lung tissue and its possible role in pulmonary circulation. Am J Physiol 1975; 229(6):1625–1631.
158. Dunwiddie T, Fredholm B. Adenosine A1 receptors inhibit adenylate cyclase activity and neurotransmitter release and hyperpolarize pyramidal neurons in rat hippocampus. J Pharmacol Exp Ther 1989; 249(1):31–37.
159. Dolphin A, Archer E. An adenosine agonist inhibits and a cyclic AMP analogue enhances the release of glutamate but not GABA from slices of rat dentate gyrus. Neurosci Lett 1983; 43:49–54.
160. de Mendonca A, Ribeiro J. Adenosine inhibits the NMDA receptor-mediated excitatory postsynaptic potential in the hippocampus. Brain Res 1993; 606(2):351–356.
161. Newman J, Peebles D, Hanson M. Adenosine produces changes in cerebral hemodynamics and metabolism as assessed by near-infrared spectroscopy in late-gestation fetal sheep in utero. Pediatr Res 2001; 50(2):217–221.
162. Runold M, Lagercrantz H, Fredholm B. Ventilatory effect of an adenosine analogue in unanaesthetized rabbits during development. J Appl Physiol 1986; 61(1):255–259.
163. Runold M, Lagercrantz H, Prabhakar N, Fredholm B. Role of adenosine in hypoxic ventilatory depression. J Appl Physiol 1989; 67(2):541–546.
164. Yamamoto M, Nishimura M, Kobayashi S, Akiyama Y, Miyamoto K, Kawakami Y. Role of endogenous adenosine in hypoxic ventilatory response in humans: a study with dipyridamole. J Appl Physiol 1994; 76(1):196–203.
165. Anderson R, Sheehan M, Strong P. Characterization of the adenosine receptors mediating hypothermia in the conscious mouse. Br J Pharmacol 1994; 113:1386–1390.

12

Perinatal Transition of Oxygen Sensing in the Peripheral Chemoreceptors

JEAN-CHRISTOPHE ROUX, JULIE PEYRONNET, and HUGO LAGERCRANTZ

Karolinska Institute
Astrid Lindgren Children's Hospital
Stockholm, Sweden

I. Introduction

The fetus is living at low PO_2 "Mount Everest in utero" and is at birth suddenly exposed to high PO_2. This transition requires a change of setting of oxygen sensitivity. The carotid and aortic bodies are the main peripheral O_2 sensors. While the carotid bodies seem to be the main peripheral chemoreceptors involved in respiratory control, the aortic bodies are more involved in cardiovascular homeostasis in the fetus (1). They have a low hypoxic sensitivity and their involvement in the hypoxic ventilatory response is controversial (2).

Hypoxemia is currently observed in physiology (high altitude, exercise) or in several pathologies (apnea, hypoventilation syndrome) and this is one of the major risk factors in neonatology. The perinatal transition is a critical period in mammals. The carotid chemoafferent pathway is composed of peripheral and central structures representing the first step in the defense against hypoxemia. The carotid body (CB) represents the main oxygen-sensing organ initiating the hypoxic ventilatory response (HVR) (3,4). Afferent cell bodies are located in the petrosal ganglion (PG) and project into discrete areas of the medulla oblongata (4–6).

Hypoxia in the CB may (1) inhibit K^+ current producing depolarization of the chemosensitive cells (glomus cells), (2) lead to the opening of voltage-gated Ca^{2+}

and increase the intracellular Ca^{2+}, and (3) induce secretion of neurotransmitters located in dense-cored granules. The carotid bodies are active in fetal life but at a lower PO_2 level, and perinatal transition is critical for the resetting of oxygen sensing to the atmospheric environment (7). Several factors are involved in the perinatal maturation of these mechanisms, but the oxygen level before, during, and after birth is probably the most important factor. Perinatal stress may modify hypoxic sensing and thus also hypoxic defense.

II. Carotid Chemoreceptor Responses in the Fetus

The carotid bodies originate from neural crest cells in the third brachial arch and appear to be morphologically mature at 17 days and 6 weeks of gestation, respectively, in rat and human fetus (8–11). During gestation, the carotid bodies do not appear to be essential for fetal breathing movements (12), although fetal breathing movements are reduced after carotid sinus denervation in the fetal sheep (13). The threshold of the fetal type I cells is significantly lower ($\sim$30 mmHg) and is adapted to the lower PO_2 in the fetus (7). The activity of arterial chemoreceptors in the fetus has been studied and clearly shows an oxygen-sensing capacity. In the anesthetized, exteriorized fetal lamb, chemoreceptor activity increases suddenly with hypoxemia; however, the magnitude of this activation is comparatively less important than in the adult (14). However, carotid bodies are quiescent in the lamb on the day of birth when PaO_2 has risen well above the threshold. The hypoxic sensitivity of the chemoreceptors is reset from the fetal to the adult range over the next few days (14).

In vitro CB preparations from fetal sheep showed an active increase to hypoxia of the chemoreceptor afferent discharge recorded in the sinus nerve (12). Peripheral chemodenervation decreases the baseline continuous breathing pattern and abolishes the hypoxic response in the lamb (15). Moreover, Breen et al. (16) found in the fetus expression of immediate early genes in the central areas receiving the CB inputs following hypoxic stimulation of the mother. These results suggest that the fetal CB is able to detect a hypoxic level superior to the normal one and stimulate the central autonomic areas. However, the contribution of carotid chemoreceptor to physiological reflexes in the fetus is somewhat controversial. Indeed, although CB are functional during gestation, several central areas are known to induce a strong inhibition on the peripheral chemosensory drive (17) to limit energy utilization. This inhibition might originate from the red nucleus (18), A5 and A7 noradrenergic neurons (19,20) or other pontine structures (16,21).

III. The Stress of Being Born

During parturition, the fetus is squeezed through the birth canal and delivered to a cold extrauterine environment, which is associated with an increase in the oxygen level. Many studies have documented the severity and the duration of asphyxia associated with delivery (22,23). However, newborn mammals are more resistant than the adult to hypoxia/asphyxia (24,25). This transition is associated with major changes of the expression and release of neuroactive agents. There is a surge of catecholamines and other neurotransmitters and hormones (26). The plasma levels of catecholamines in the human newborn are markedly high compared with the normal adult in various stressed conditions (26). The stress at birth, as indicated by the catecholamine surge, may affect the perinatal resetting of the CB. Furthermore, the catecholamines in the fetus are strongly involved in the maturation of the brain and may act as "neurotrophic factors." In the early stage of brain development, norepinephrine is involved in cellular division (27), neuronal maturation (28), and synaptogenesis (29). Therefore, it is possible to speculate that the surge of catecholamine at birth may participate in the postnatal maturation of the CB.

IV. The Functional Role of the Peripheral Chemoreceptors at Birth

The peripheral chemoreceptors were earlier assumed to play an important role in the initiation of the first breath of air at birth (30,31). However, carotid-denervated fetal sheep did not start to breathe later than sham-operated controls (30). The peripheral chemoreceptors are actually silent at birth when the newborn is exposed to a higher PO_2 than in utero since the sensitivity is set at a lower PO_2. After a few days the peripheral chemoreceptors are set at a higher PO_2 level and they fire also during normoxia (32). The situation seems to be the same in the human baby. This has been tested by the hyperoxic test, according to Pierre Dejours (33). When newborn infants were exposed to 100% O_2 they did not decrease their breathing at all, while after 2 days there was a 10% reduction of breathing (34).

If the fetus is asphyxiated at birth, the peripheral chemoreceptors are activated and an increased number of spikes can be recorded from the sinus nerve (35). However, this hypoxic drive is overridden by a central hypoxic depression, although some gasps can be induced.

Birth represents a pivotal transition in CB maturation. Although CB are not involved in the initiation of breathing at birth, they provide a tonic excitatory input to medullary respiratory neurons after resetting. Furthermore, they stabilize rhythmic respiration perhaps more than in the adult (36,37). CB denervation in newborn sheep or rat leads to respiratory failure and, thus, a strong increase in the mortality of these animals (38–42). The respiration of piglet during sleep is also highly affected by carotid chemodenervation (43). Thus, CB seem to be important for survival before the central control has matured. This window of vulnerability linked to the degree of maturity of the animal becomes discernible. The time schedule of this critical window corresponds to the CB resetting occurring just after birth.

In utero the sensitivity of the CB is well adapted to the comparatively hypoxic environment of the fetus and then resets to the higher PaO_2 of the newborn (14,35,44,45) (see Fig. 1). This environmental change in oxygen tension leads to an increase of the carotid body sensitivity and several cellular mechanisms might be involved in this maturational process.

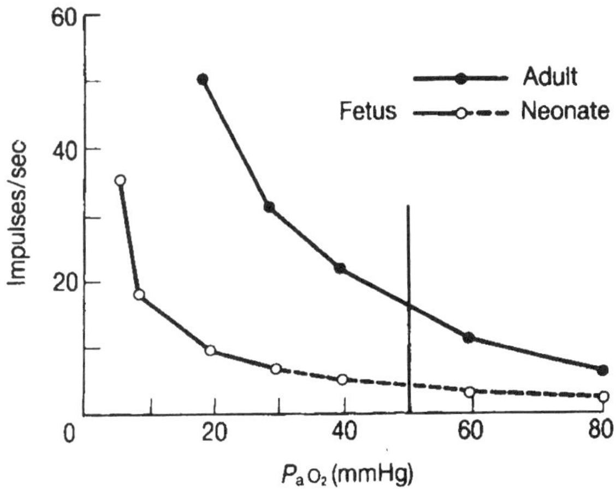

Figure 1 Carotid body response to hypoxia in fetal and adult sheep. Chemoreceptor activity is significantly greater in the older group at all PaO_2 levels. (From Ref. 45a.)

A. Hypoxic Ventilatory Response

In the fetus hypoxia causes a complete inhibition of breathing movement overriding the inputs from the peripheral chemoreceptor. This inhibition seems to originate from a suprapontine level since after midcollicular transection, stimulation of the peripheral chemoreceptors induces vigorous respiratory movements (46).

Newborn animals exhibit a typical biphasic response to acute hypoxia with an increase of the basal ventilation at the first minute of exposure, immediately followed by a decrease around the basal level (14,47–52). The secondary phase with a decrease of ventilation is rather due to a central inhibition than to a decrease of peripheral chemoreceptor activity (48,52–58). Similarly, 1-day-old pups first increase and then decrease their ventilation when exposed to 12% O_2 (47,59). Moreover, the hypoxic ventilatory response is irregular in the newborn. As the pups grow older, their response to hypoxia develops to a more elaborate response to a second hypoxic challenge. In the adult, the ventilatory response to hypoxia is still biphasic, illustrating the persistence of central inhibition (60).

B. Morphology

Around birth the CB increase in size. In the cat, between the late fetal and the early postnatal period, the CB volume increases by over 50% in vascular and nonvascular compartments (61). In the rat, glomus cells occasionally have been reported to undergo mitosis during the late fetal and early postnatal period (62,63). The number of nerve terminals increases about fourfold during the neonatal period (64). These morphological changes could participate in the functional and neurochemical maturation of this tissue.

C. Neurochemical Changes

With a few hours after birth, mRNA for c-Fos, SP, and several enzymes involved in catecholamine synthesis are up-regulated in the brainstem and a number of organs in newborn mice, rats, and rabbits (65–68). Also in the CB neurotransmitter expression and turnover change dramatically during the perinatal period.

Substance P (SP) has been suggested to mediate the hypoxic drive at both the peripheral and central level. In the CB of newborn kitten, SP content is very low and a few weeks later increases significantly (69).

The catecholamine metabolism is also strongly affected by the perinatal transition and is probably involved in resetting of the CB. Thus, the mRNA levels coding for tyrosine hydroxylase (TH), the rate-limiting enzyme in dopamine synthesis, D2 dopamine receptors, and dopamine content occur in parallel (70). The levels of mRNA encoding TH and D2 dopamine receptors in the type I cell of rat are elevated in the fetus and then suddenly decrease at birth (71). In addition, the dopamine content and turnover drop at birth (45,72) and are stabilized first after 3 days. However, Gauda et al. (70) found that the mRNA level for TH peaked 4 hr after birth and subsequently declined, while D_2-dopamine receptor mRNA levels increased during early postnatal maturation (Fig. 2).

In the adult rat, expression of TH mRNA in carotid bodies is elevated under hypoxic conditions (73). It is thus possible that the low PaO_2 maintains the elevated levels of TH mRNA observed in the fetus in utero. At birth, PaO_2 increases, a change that might be responsible for the accompanying reduction in TH mRNA

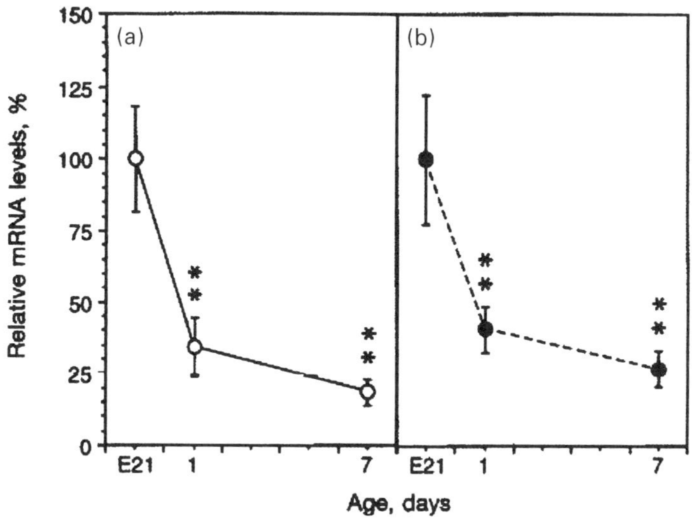

Figure 2 Level of TH (a) and D2R (b) mRNA in the carotid bodies of rat pups on E21, P1, and P7. The levels of TH and D2R mRNA in pups at E21 were set to 100% and used for a relative comparison. The levels of both mRNA were high on E21 but decreased dramatically by P1 and were still low on P7. (From Ref. 74.)

expression (74). These variations in the level of TH mRNA might reflect modification of the rate of the gene transcription and/or alteration of the stability of the mRNA (73). Several studies have reported that dopamine levels in the carotid bodies of newborn are inversely correlated to oxygen chemosensitivity (75). Tomares et al. (76) demonstrated that a treatment with domperidone, an antagonist of the D2-dopamine receptor, almost doubled the hypoxic chemosensitivity of the carotid sinus nerve in neonatal kittens (4–7 days of age). Interestingly, treatment of newborn rabbit with a D2 and/or D1-dopamine antagonist is able to increase the hypoxic ventilatory response (77). However, the D1-dopamine antagonist can act both at a peripheral and central level, but we do not know how the CB are affected per se. Nevertheless, we can assume that the inhibitory role of endogenous dopamine in the CB is adjusted during postnatal development. Furthermore, in vitro preparation of rat CB shows a postnatal increase in the release of catecholamine and the carotid sinus nerve activity due to hypoxic stimulation (78,79) (Fig. 3).

D. Trophic Factors

The transition between prenatal and postnatal life represents a critical step in the trophic maturation of the CB. Neurotrophic factors seem to be crucial for the maturation of the PG ganglion neurons, which represent the main ascending visceral sensory afferents of the CB (80). They are involved in a developmental sequence for the establishment of visceral neurons (81). More particularly, BDNF (brain-derived neurotrophic factor) plays a key role in the innervation of the carotid bodies (82), especially during the perinatal transition. CB has a “trophic” influence in the survival of the catecholaminergic cell located in the petrosal ganglion (PG) since CB denervation leads to a decrease of the neuronal number in the PG (82). BDNF-deficient mice exhibit ventilatory failure associated with an atrophy of PG that may be prevented by the application of BDNF (82,83). Subadult transgenic mice, which lack the BDNF gene, exhibit an altered chemosensitivity that can be compared to that of newborn rats (83). Moreover, in the brain the BDNF protein level changes after focal ischemia and could play a role for survival and plasticity of central neurons (84). Then, changes in oxygen tension around birth could modulate the BDNF expression.

Synaptogenesis is another crucial step in CB maturation. Several factors are implicated in this developmental process. Recently, it has been found that rnx, a

---

Figure 3 Cholinergic and dopaminergic chemotransduction in the carotid body. The hypothesis postulates that the glomus cells release transmitters such as acethylcholine and dopamine. The cholinergic (nicotinic or muscarinic) and dopaminergic postsynaptic receptors are located on the sensory afferent fibers (dendrite N IX) and their activation creates neural noise and some action potentials, which travel via the carotid sinus nerve to the nucleus tractus solitarius. Acethylcholine and dopamine also bind to ‘presynaptic’ autoreceptors on the glomus cells, modulating the further release of these and other transmitters. During hypoxia more acethylcholine and dopamine are released and neural activity is modulated, precipitating the reflex responses. Lower panel shows the perinatal transition. Before and immediately after

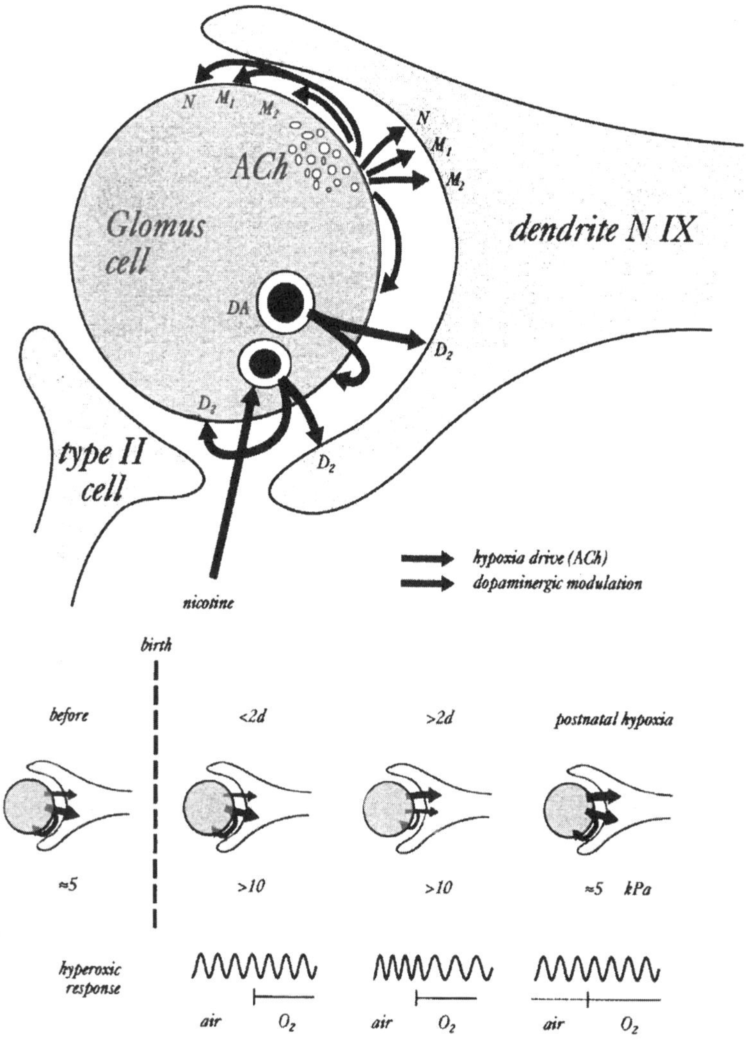

birth the carotid bodies have a low sensitivity, tentatively due to high dopaminergic tonic activity. Therefore, the newborn infant has no ventilatory response to hyperoxia (Dejour test). After 2 days the carotid bodies reset and respond to a milder degree of hypoxia. This is reflected in a positive Dejour test, i.e., decreased ventilatory response to hyperoxia. If the infant is sustaining chronic postnatal hypoxia, the setting of the sensitivity of the peripheral chemoreceptors is retained at a lower level. Finally, a nicotonic environment during the perinatal period seems to influence the cholinergic and dopaminergic activities in the glomus cells. Ach, acethylcholine; DA, dopamine; D_2, DA D_2 receptor; N, nicotinic receptor; M_1, muscarinic 1 receptor; M_2, muscarinic 2 receptor.

member of an orphan homeobox gene family, is required for the formation of PG afferents of the carotid body (85). Rnx-deficient mice die a few hours after birth from respiratory failure and exhibit an aberrant central innervation of primary visceral sensory afferents particularly marked in the early fetal stage preceding delivery (85,86). GAP-43 is involved in synaptogenesis and participates in axonal growth in the CB (87). This regulating factor is strongly expressed in the central nervous system, especially during perinatal maturation. During the first postnatal week in rat, Gap-43 is strongly expressed in the medullary structures involved in cardiorespiratory regulation (88). The in vivo activity of the Gap-43 protein is dependent on the oxygen tension (89) and therefore, Gap-43 could participate after the PaO_2 rise of birth to the postnatal maturation of the CB.

Hypoxia-inducible factor 1 (HIF-1) is a transcriptional factor involved in the cellular and developmental control of O_2 homeostasis (90). HIF-1 exhibits multiple hypoxic responses including the adaptation to anaerobic metabolism, erythropoiesis, angiogenesis, and vasodilatation. Recent studies have demonstrated that HIF-1 is required for both the establishment of the circulatory system during embryonic development and ventilatory responses in postnatal life (91,92). Thus partial HIF-1 deficiency in adult mice has a dramatic effect on CB neural activity and ventilatory adaptation to chronic hypoxia. The increase in oxygen tension just after birth, which is able to regulate the HIF-1 expression, may modify in turn the regulation of the target genes.

E. Chronic Perinatal Hypoxia

Fetal rats that are exposed to prenatal hypoxia during the entire gestation exhibit a significant increase in resting minute ventilation after birth (93–95). However, the hypoxic ventilatory response was not affected by prenatal hypoxia in 1-week-old rats although the subsequent fall leading to the classic biphasic response was abolished (95). The dopamine content (95) and the catecholamine synthesis activity (JC Roux, unpublished data) in the CB of prenatal hypoxic pups were reduced compared to the nonexposed control animals.

Chronic exposure to hypoxia after birth affects the development of the ventilatory response to hypoxia. Resetting of the chemosensitivity of the CB is delayed by hypoxia (96). Five-to-ten-week-old animals maintained under hypoxic conditions after birth show the same response to hypoxia as normoxic controls, but this response is not sustained and exhibits a biphasic, neonatal-like pattern. Rats or cats born in an atmosphere containing 12% O_2 remain unable to respond to hypoxia (72,96). In fact, prolonged hypoxic exposure from birth increases the basal activity of the CB (72,97) and delays the onset of the chemoreflex response to hypoxia (47,72). In neonatal rats, exposure to chronic moderate hypoxia (FiO_2 40.1) for 5 days leads to a subsequent desensitization of the hypoxic response that is still present 50 days after the exposure (94,98,99).

One-day-old pups exposed to postnatal hypoxia (10% O_2) for 6 days demonstrated increased dopamine and norepinephrine content in their CB until they had reached 8 weeks of age (97). Other findings suggest that this increase may reflect

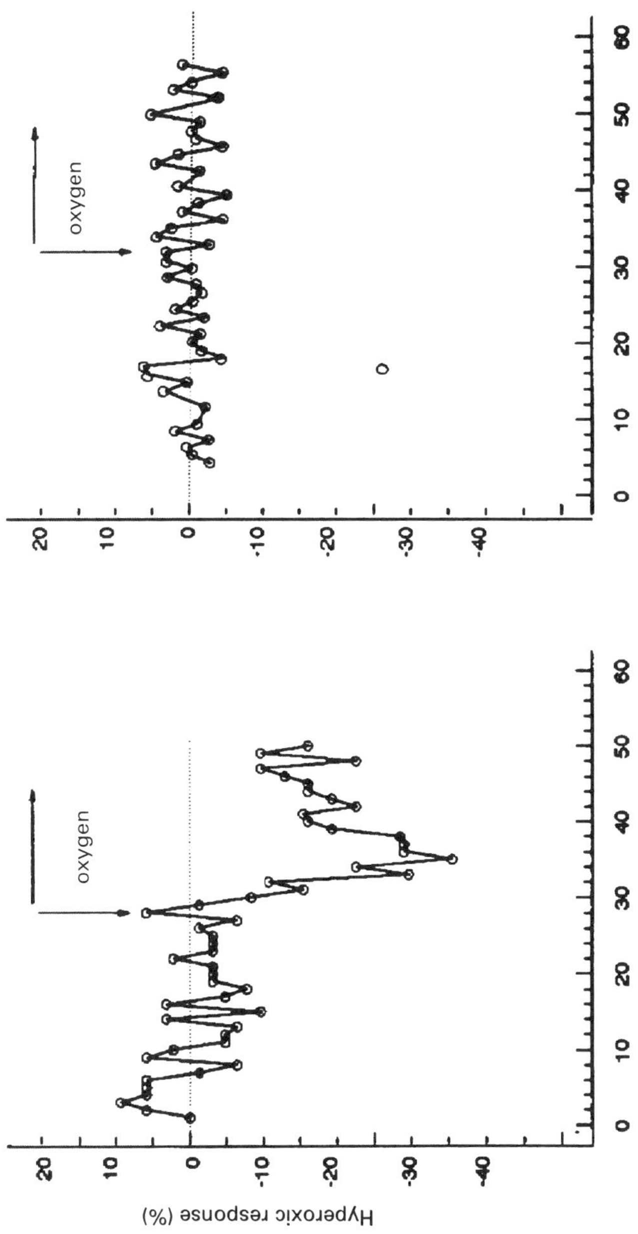

Figure 4 Percentage of deviation from the mean resting ventilation during normoxic breathing and while breathing 100% oxygen for 30 sec in a respiratory distress syndrome with intact response (left) and in a "nonresponding" bronchopulmonary dysplasia infant. (From Ref. 106.)

hypertrophy of and/or a higher rate of metabolism by type I cells, although it was not possible in those studies to determine the rate of dopamine utilization. However, increased storage of dopamine is generally associated with an increase in the rate of its utilization, as well as hypertrophy of the type I cells and an increase in their metabolism. Other reports have also described the impact of hypoxia after birth. Type I cells in the CB of newborn rat pups exposed to acute hypoxia (10% O_2 for 12 hr) exhibited a twofold increase in their level of mRNA coding for TH. When these same animals were then returned to normoxic conditions for 12 hr the level of TH mRNA in their carotid bodies returned to the control value (71). In contrast, the expression of D2 receptor mRNA in the carotid bodies was not affected by acute postnatal exposure to hypoxia. Chronic hypoxia abolishes also the Ca^{2+} response to acute hypoxia in the CB (100). This last result indicates that chronic hypoxia severely impairs the postnatal development of the CB sensitivity to hypoxia.

F. Postnatal Hyperoxia

Hyperoxia leads to a sudden decrease of ventilation due to decline of the CB input (33). The ventilatory reduction is smaller in newborn compared to adult in lamb (38,101), rat (45), and infant (34). However, long-term hyperoxia modification is able to modify the normal setting of the CB. Kittens and rat pups born and maintained in a hyperoxic environment exhibit, respectively, very little increase in the carotid sinus nerve activity or the ventilatory response to hypoxic challenge (72,96,102). The functional impairment seems to result from a persistent deficit in the CB rather than in the central integration of the CB input (103). These physiological results are correlated to a loss of CB volume and cell number and a decrease in the number of axons innervating this tissue (104). This model of "chemical" denervation has long-term effects and spontaneous recovery occurs between the 6th and 15th postnatal weeks.

V. Conclusion and Clinical Implication

Fetal hypoxia might result from several pathophysiological situations, including maternal anemia, reduced uteroplacental blood flow secondary to maternal hypertension, smoking or ethanol consumption, reduced placenta size, or reduced oxygen inhalation by the mother at high altitude. Infants who have died of sudden infant death syndrome have sometimes been found to have a very high level of dopamine in their CB, according to postmortem analyses (105). Preterm infants who had sustained chronic lung disease were often not found to react at all to hyperoxic testing, until several weeks of oxygen treatment. Thus their resetting seemed to be delayed compared with preterm infants without chronic lung disease, who often responded with about a 20% decrease of minute ventilation (106) (Fig. 4). Infants born at high altitude in the Andes were found to have a blunted hyperoxic ventilatory response compared to adults (107,108).

References

1. Dawes GS. Foetal blood-gas homeostasis during development. Proc R Soc Med 1968; 61:1227–1231.
2. Mortola JP. Respiratory Physiology of Newborn Mammals: A Comparative Perspective. Vol. 1, Baltimore and London: Johns Hopkins University Press, 2001:344.
3. Biscoe TJ, Duchen MR. Responses of type I cells dissociated from the rabbit carotid body to hypoxia. J Physiol 1990; 428:39–59.
4. Gonzalez C, Almaraz L, Obeso A, Rigual R. Carotid body chemoreceptors: from natural stimuli to sensory discharges. Physiol Rev 1994; 74:829–898.
5. Blessing WW, Yu Y, Nalivaiko E. Medullary projections of rabbit carotid sinus nerve. Brain Res 1999; 816:405–410.
6. Finley JC, Katz DM. The central organization of carotid body afferent projections to the brainstem of the rat. Brain Res 1992; 572:108–116.
7. Blanco CE, Hanson MA, McCooke HB. Effects on carotid chemoreceptor resetting of pulmonary ventilation in the fetal lamb in utero. J Dev Physiol 1988; 10:167–174.
8. McDonald DM, Mitchell RA. The neural pathway involved in "efferent inhibition" of chemoreceptors in the cat carotid body. J Comp Neurol 1981; 201:457–476.
9. McDonald DM, Blewett RW. Location and size of carotid body-like organs (paraganglia) revealed in rats by the permeability of blood vessels to Evans blue dye. J Neurocytol 1981; 10:607–643.
10. Pearse AG, Polak JM, Rost FW, Fontaine J, Le Lievre C, Le Douarin N. Demonstration of the neural crest origin of type I (APUD) cells in the avian carotid body, using a cytochemical marker system. Histochemie 1973; 34:191–203.
11. Kondo H. A light and electron microscopic study on the embryonic development of the rat carotid body. Am J Anat 1975; 144:275–293.
12. Jansen AH, Purves MJ, Tan ED. The role of sympathetic nerves in the activation of the carotid body chemoreceptors at birth in the sheep. J Dev Physiol 1980; 2:305–321.
13. Murai DT, Lee CC, Wallen LD, Kitterman JA. Denervation of peripheral chemoreceptors decreases breathing movements in fetal sheep. J Appl Physiol 1985; 59:575–579.
14. Blanco CE, Hanson MA, Johnson P, Rigatto H. Breathing pattern of kittens during hypoxia. J Appl Physiol 1984; 56:12–17.
15. Johnston BM, Gluckman PD. Peripheral chemoreceptors respond to hypoxia in pontine-lesioned fetal lambs in utero. J Appl Physiol 1993; 75:1027–1034.
16. Breen S, Rees S, Walker D. Identification of brainstem neurons responding to hypoxia in fetal and newborn sheep. Brain Res 1997; 748:107–121.
17. Bissonnette JM, Knopp SJ. Developmental changes in the hypoxic ventilatory response in C57BL/6 mice. Respir Physiol 2001; 128:179–186.
18. Ackland GL, Moore PJ, Hanson MA. Is the ventilatory decline seen in newborns during hypoxaemia centrally mediated. Adv Exp Med Biol 1994; 360:345–348.
19. Coles SK, Dick TE. Neurones in the ventrolateral pons are required for posthypoxic frequency decline in rats. J Physiol 1996; 497:79–94.
20. Okada Y, Kawai A, Muckenhoff K, Scheid P. Role of the pons in hypoxic respiratory depression in the neonatal rat. Respir Physiol 1998; 111:55–63.
21. Gluckman PD, Johnston BM. Lesions in the upper lateral pons abolish the hypoxic depression of breathing in unanaesthetized fetal lambs in utero. J Physiol 1987; 382:373–383.

22. Koch G, Wendel H. Adjustment of arterial blood gases and acid base balance in the normal newborn infant during the first week of life. Biol Neonat 1968; 12:136–161.
23. Dawes GS, Mott JC. The increase in oxygen consumption of the lamb after birth. J Physiol 1959; 146:295–315.
24. Adolph EF. Regulations during survival without oxygen in infant mammals. Respir Physiol 1969; 7:356–368.
25. Saiki C, Mortola JP. Ventilatory control in infant rats after daily episodes of anoxia. Pediatr Res 1994; 35:490–493.
26. Lagercrantz H, Slotkin TA. The "stress" of being born. Sci Am 1986; 254:100–107.
27. Slotkin TA, Seidler FJ. Adrenomedullary catecholamine release in the fetus and newborn: secretory mechanisms and their role in stress and survival. J Dev Physiol 1988; 10:1–16.
28. Lau C, Cameron A, Antolick L, Slotkin TA. Trophic control of the ornithine decarboxylase/polyamine system in neonatal rat brain regions: lesions caused by 6-hydroxydopamine produce effects selective for cerebellum. Brain Res Dev Brain Res 1990; 52:167–173.
29. Parnavelas JG, Blue ME. The role of the noradrenergic system on the formation of synapses in the visual cortex of the rat. Brain Res 1982; 255:140–144.
30. Jansen AH, Ioffe S, Russell BJ, Chernick V. Effect of carotid chemoreceptor denervation on breathing in utero and after birth. J Appl Physiol 1981; 51:630–633.
31. Purves MJ. The effects of hypoxia in the new-born lamb before and after denervation of the carotid chemoreceptors. J Physiol 1966; 185:60–77.
32. Blanco CE, Dawes GS, Hanson MA, McCooke HB. The response to hypoxia of arterial chemoreceptors in fetal sheep and new-born lambs. J Physiol 1984; 351:25–37.
33. Dejours P. Chemoreflexes in breathing. Physiol Rev 1962; 42:335–358.
34. Hertzberg T, Lagercrantz H. Postnatal sensitivity of the peripheral chemoreceptors in newborn infants. Arch Dis Child 1987; 62:1238–1241.
35. Hanson MA, Eden GJ, Nijhuis JG, Moore PJ. Peripheral chemoreceptors and other O_2 sensors in the fetus and newborn. In: Lahiri S, Forster RE, Davies RO, Pack AI, eds. Chemoreceptors and Reflexes in Breathing. New York: Oxford University Press, 1989:113–120.
36. Hofer MA. Role of carotid sinus and aortic nerves in respiratory control of infant rats. Am J Physiol 1986; 251:R811–817.
37. Forster HV, Pan LG, Lowry TF, Serra A, Wenninger J, Martino P. Important role of carotid chemoreceptor afferents in control of breathing of adult and neonatal mammals. Respir Physiol 2000; 119:199–208.
38. Bureau MA, Begin R. Postnatal maturation of the respiratory response to O_2 in awake newborn lambs. J Appl Physiol 1982; 52:428–433.
39. Bureau MA, Lamarche J, Foulon P, Dalle D. Postnatal maturation of respiration in intact and carotid body-chemodenervated lambs. J Appl Physiol 1985; 59:869–874.
40. Bureau MA, Lamarche J, Foulon P, Dalle D. The ventilatory response to hypoxia in the newborn lamb after carotid body denervation. Respir Physiol 1985; 60:109–119.
41. Hofer MA. Sleep-wake state organization in infant rats with episodic respiratory disturbance following sinoaortic denervation. Sleep 1985; 8:40–48.
42. Serra A, Brozoski D, Hedin N, Franciosi R, Forster HV. Mortality after carotid body denervation in rats. J Appl Physiol 2001; 91:1298–1306.
43. Donnelly DF, Haddad GG. Prolonged apnea and impaired survival in piglets after sinus and aortic nerve section. J Appl Physiol 1990; 68:1048–1052.

44. Mulligan E, Alsberge M, Bhide S. Carotid chemoreceptor recording in the newborn piglet. In: Eyzaguirre C, Fidone SJ, Fitzgerald RS, Lahiri S, McDonald D, eds. Arterial Chemoreception. New York: Springer. 1990:285–289.
45. Hertzberg T, Hellstrom S, Lagercrantz H, Pequignot JM. Development of the arterial chemoreflex and turnover of carotid body catecholamines in the newborn rat. J Physiol 1990; 425:211–225.
45a. Hanson MA. Neonatal, perinatal and developmental aspects of respiratory control. In: Pediatrics and Perinatology, 2nd ed, Gluckman PD and Heyman MA, eds. New York: Oxford University Press 1996; 836–844.
46. Dawes GS, Gardner WN, Johnston BM, Walker DW. Breathing in fetal lambs: the effect of brain stem section. J Physiol 1983; 335:535–553.
47. Eden GJ, Hanson MA. Effects of chronic hypoxia from birth on the ventilatory response to acute hypoxia in the newborn rat. J Physiol 1987; 392:11–19.
48. Fung ML, Wang W, Darnall RA, St John WM. Characterization of ventilatory responses to hypoxia in neonatal rats. Respir Physiol 1996; 103:57–66.
49. Bureau MA, Zinman R, Foulon P, Begin R. Diphasic ventilatory response to hypoxia in newborn lambs. J Appl Physiol 1984; 56:84–90.
50. Mortola JP. Breathing pattern in newborns. J Appl Physiol 1984; 56:1533–1540.
51. Darnall RA, Bruce RD. Effects of adenosine and xanthine derivatives on breathing during acute hypoxia in the anesthetized newborn piglet. Pediatr Pulmonol 1987; 3:110–116.
52. Rigatto H, Wiebe C, Rigatto C, Lee DS, Cates D. Ventilatory response to hypoxia in unanesthetized newborn kittens. J Appl Physiol 1988; 64:2544–2551.
53. Marchal F, Bairam A, Haouzi P, Crance JP, Di Giulio C, Vert P, Lahiri S. Carotid chemoreceptor response to natural stimuli in the newborn kitten. Respir Physiol 1992; 87:183–193.
54. Carroll JL, Bamford OS, Fitzgerald RS. Postnatal maturation of carotid chemoreceptor responses to O_2 and CO_2 in the cat. J Appl Physiol 1993; 75:2383–2391.
55. Hanson G, Jones L, Fidone S. Physiological chemoreceptor stimulation decreases enkephalin and substance P in the carotid body. Peptides 1986; 7:767–769.
56. Waites BA, Ackland GL, Noble R, Hanson MA. Red nucleus lesions abolish the biphasic respiratory response to isocapnic hypoxia in decerebrate young rabbits. J Physiol 1996; 495:217–225.
57. St John WM, Bartlett D, Jr, Knuth KV, Hwang JC. Brain stem genesis of automatic ventilatory patterns independent of spinal mechanisms. J Appl Physiol 1981; 51:204–210.
58. Neubauer JA, Melton JE, Edelman NH. Modulation of respiration during brain hypoxia. J Appl Physiol 1990; 68:441–451.
59. Eden GJ, Hanson MA. Maturation of the respiratory response to acute hypoxia in the newborn rat. J Physiol 1987; 392:1–9.
60. Bissonnette JM. Mechanisms regulating hypoxic respiratory depression during fetal and postnatal life. Am J Physiol Regul Integr Comp Physiol 2000; 278:R1391–1400.
61. Clarke JA, de Burgh Daly M, Ead HW. Comparison of the size of the vascular compartment of the carotid body of the fetal, neonatal and adult cat. Acta Anat 1990; 138:166–174.
62. Fishman MC, Schaffner AE. Carotid body cell culture and selective growth of glomus cells. Am J Physiol 1984; 246:C106–C113.

63. von Dalnok GK, Menssen HD. A quantitative electron microscopic study of the effect of glucocorticoids in vivo on the early postnatal differentiation of paraneuronal cells in the carotid body and the adrenal medulla of the rat. Anat Embryol 1986; 174:307–319.
64. Kondo H. Innervation of the carotid body of the adult rat: a serial ultrathin section analysis. Cell Tissue Res 1976; 173:1–15.
65. Kasik JW, Wan YJ, Ozato K. A burst of c-fos gene expression in the mouse occurs at birth. Mol Cell Biol 1987; 7:3349–3352.
66. Holgert H, Schalling M, Hertzberg T, Lagercrantz H, Hokfelt T. Changes in levels of mRNA coding for catecholamine synthesizing enzymes and neuropeptide Y in the adrenal medulla of the newborn rat. J Dev Physiol 1991; 16:19–26.
67. Ringstedt T, Tang LQ, Persson H, Lendahl U, Lagercrantz H. Expression of c-fos, tyrosine hydroxylase, and neuropeptide mRNA in the rat brain around birth: effects of hypoxia and hypothermia. Pediatr Res 1995; 37:15–20.
68. Srinivasan M, Yamamoto Y, Persson H, Lagercrantz H. Birth-related activation of preprotachykinin-A mRNA in the respiratory neural structures of the rabbit. Pediatr Res 1991; 29:369–371.
69. Scheibner T, Read DJ, Sullivan CE. Distribution of substance P-immunoreactive structures in the developing cat carotid body. Brain Res 1988; 453:72–78.
70. Gauda EB, Shirahata M, Fitzgerald RS. D2-dopamine receptor mRNA in the carotid body and petrosal ganglia in the developing cat. Adv Exp Med Biol 1994; 360:317–319.
71. Holgert H, Hokfelt T, Hertzberg T, Lagercrantz H. Functional and developmental studies of the peripheral arterial chemoreceptors in rat: effects of nicotine and possible relation to sudden infant death syndrome. Proc Natl Acad Sci USA 1995; 92:7575–7579.
72. Hertzberg T, Hellstrom S, Holgert H, Lagercrantz H, Pequignot JM. Ventilatory response to hyperoxia in newborn rats born in hypoxia—possible relationship to carotid body dopamine. J Physiol 1992; 456:645–654.
73. Czyzyk-Krzeska, MF, Bayliss DA, Lawson EE, Millhorn DE. Regulation of tyrosine hydroxylase gene expression in the rat carotid body by hypoxia. J Neurochem 1992; 58:1538–1546.
74. Holgert H, Pequignot JM, Lagercrantz H, Hokfelt T. Birth-related up-regulation of mRNA encoding tyrosine hydroxylase, dopamine beta-hydroxylase, neuropeptide tyrosine, and prepro-enkephalin in rat adrenal medulla is dependent on postnatal oxygenation. Pediatr Res 1995; 37:701–706.
75. Gauda EB, Lawson EE. Developmental influences on carotid body responses to hypoxia. Respir Physiol 2000; 121:199–208.
76. Tomares SM, Bamford OS, Sterni LM, Fitzgerald RS, Carroll JL. Effects of domperidone on neonatal and adult carotid chemoreceptors in the cat. J Appl Physiol 1994; 77:1274–1280.
77. Srinivasan M, Lagercrantz H, Yamamoto Y. A possible dopaminergic pathway mediating hypoxic depression in neonatal rabbits. J Appl Physiol 1989; 67:1271–1276.
78. Kholwadwala D, Donnelly DF. Maturation of carotid chemoreceptor sensitivity to hypoxia: in vitro studies in the newborn rat. J Physiol 1992; 453:461–473.
79. Donnelly DF, Doyle TP. Developmental changes in hypoxia-induced catecholamine release from rat carotid body, in vitro. J Physiol 1994; 475:267–275.
80. Finley JC, Polak J, Katz DM. Transmitter diversity in carotid body afferent neurons: dopaminergic and peptidergic phenotypes. Neuroscience 1992; 51:973–987.

80a. Fitzgerald RS. Oxygen and carotid body chemotransduction: the cholinergic hypothesis—a brief history and new evaluation. Respir Physiol 2000; 120:89–104.
81. ElShamy WM, Ernfors P. Brain-derived neurotrophic factor, neurotrophin-3, and neurotrophin-4 complement and cooperate with each other sequentially during visceral neuron development. J Neurosci 1997; 17:8667–8675.
82. Hertzberg T, Fan G, Finley JC, Erickson JT, Katz DM. BDNF supports mammalian chemoafferent neurons in vitro and following peripheral target removal in vivo. Dev Biol 1994; 166:801–811.
83. Erickson JT, Conover JC, Borday V, Champagnat J, Barbacid M, Yancopoulos G, Katz DM. Mice lacking brain-derived neurotrophic factor exhibit visceral sensory neuron losses distinct from mice lacking NT4 and display a severe developmental deficit in control of breathing. J Neurosci 1996; 16:5361–5371.
84. Kokaia Z, Andsberg G, Yan Q, Lindvall O. Rapid alterations of BDNF protein levels in the rat brain after focal ischemia: evidence for increased synthesis and anterograde axonal transport. Exp Neurol 1998; 154:289–301.
85. Qian Y, Fritzsch B, Shirasawa S, Chen CL, Choi Y, Ma Q. Formation of brainstem (nor)adrenergic centers and first-order relay visceral sensory neurons is dependent on homeodomain protein Rnx/Tlx3. Genes Dev 2001; 15:2533–2545.
86. Shirasawa S, Arata A, Onimaru H, Roth KA, Brown GA, Horning S, Arata S, Okumura K, Sasazuki T, Korsmeyer SJ. Rnx deficiency results in congenital central hypoventilation. Nat Genet 2000; 24:287–290.
87. Shughrue PJ, Dorsa DM. The ontogeny of GAP-43 (neuromodulin) mRNA in postnatal rat brain: evidence for a sex dimorphism. J Comp Neurol 1994; 340:174–184.
88. Dani JW, Armstrong DM, Benowitz LI. Mapping the development of the rat brain by GAP-43 immunocytochemistry. Neuroscience 1991; 40:277–287.
89. Huang CC. Perinatal hypoxic/ischemic encephalopathy: clinical challenge and experimental implications. Zhonghua Min Guo Xiao Er Ke Yi Xue Hui Za Zhi 1998; 39:157–165.
90. Guillemin K, Krasnow MA. The hypoxic response: huffing and HIFing. Cell 1997; 89:9–12.
91. Jain S, Maltepe E, Lu MM, Simon C, Bradfield CA. Expression of ARNT, ARNT2, HIF1alpha, HIF2alpha and Ah receptor mRNAs in the developing mouse. Mech Dev 1998; 73:117–123.
92. Kline DD, Peng YJ, Manalo DJ, Semenza GL, Prabhakar NR. Defective carotid body function and impaired ventilatory responses to chronic hypoxia in mice partially deficient for hypoxia-inducible factor 1alpha. Proc Natl Acad Sci USA 2002; 99:821–826.
93. Gleed RD, Mortola JP. Ventilation in newborn rats after gestation at simulated high altitude. J Appl Physiol 1991; 70:1146–1151.
94. Okubo S, Mortola JP. Control of ventilation in adult rats hypoxic in the neonatal period. Am J Physiol 1990; 259:R836–R841.
95. Peyronnet J, Roux JC, Geloen A, Tang LQ, Pequignot JM, Lagercrantz H, Dalmaz Y. Prenatal hypoxia impairs the postnatal development of neural and functional chemoafferent pathway in rat. J Physiol 2000; 524 Pt 2:525–537.
96. Hanson MA, Kumar P, Williams BA. The effect of chronic hypoxia upon the development of respiratory chemoreflexes in the newborn kitten. J Physiol 1989; 411:563–574.

97. Soulier V, Dalmaz Y, Cottet-Emard JM, Lagercrantz H, Pequignot JM. Long-term influence of neonatal hypoxia on catecholamine activity in carotid bodies and brainstem cell groups of the rat. J Physiol 1997; 498:523–530.
98. Okubo S. [Long-term respiratory effects of neonatal hypoxia in rats]. Nihon Kyobu Shikkan Gakkai Zasshi 1991; 29:16–23.
99. Okubo S, Mortola JP. Long-term respiratory effects of neonatal hypoxia in the rat. J Appl Physiol 1988; 64:952–958.
100. Sterni LM, Bamford OS, Wasicko MJ, Carroll JL. Chronic hypoxia abolished the postnatal increase in carotid body type I cell sensitivity to hypoxia. Am J Physiol 1999; 277:L645–L652.
101. Purves MJ. Respiratory and circulatory effects of breathing 100 per cent oxygen in the new-born lamb before and after denervation of the carotid chemoreceptors. J Physiol 1966; 185:42–59.
102. Ling L, Olson EB, Jr, Vidruk EH, Mitchell GS. Attenuation of the hypoxic ventilatory response in adult rats following one month of perinatal hyperoxia. J Physiol 1996; 495:561–571.
103. Ling L, Olson EB, Jr, Vidruk EH, Mitchell GS. Integrated phrenic responses to carotid afferent stimulation in adult rats following perinatal hyperoxia. J Physiol 1997; 500:787–796.
104. Erickson JT, Mayer C, Jawa A, Ling L, Olson EB, Jr, Vidruk EH, Mitchell GS, Katz DM. Chemoafferent degeneration and carotid body hypoplasia following chronic hyperoxia in newborn rats. J Physiol 1998; 509:519–526.
105. Perrin DG, Cutz E, Becker LE, Bryan AC. Ultrastructure of carotid bodies in sudden infant death syndrome. Pediatrics 1984; 73:646–651.
106. Katz-Salamon M, Jonsson B, Lagercrantz H. Blunted peripheral chemoreceptor response to hyperoxia in a group of infants with bronchopulmonary dysplasia. Pediatr Pulmonol 1995; 20:101–106.
107. Sorensen SC, Severinghaus JW. Irreversible respiratory insensitivity to acute hypoxia in man born at high altitude. J Appl Physiol 1968; 25:217–220.
108. Lahiri S, Delaney RG, Brody JS, Simpser M, Velasques T, Motoyama EK, Polgar G. Relative role of environmental and genetic factors in respiratory adaptation to high altitude. Nature 1976; 261:133–135.

13

Postnatal Maturation of the Carotid Chemoreceptor O_2 Sensitivity at the Cellular Level

JOHN L. CARROLL

University of Arkansas for Medical Sciences
and Arkansas Children's Hospital
Little Rock, Arkansas, U.S.A.

I. Overview: Carotid Chemoreceptors and Postnatal Development

In mammals, the main sensors of arterial oxygen level are the carotid chemoreceptors, which provide the main drive underlying the ventilatory response to hypoxia and mediate, at least in part, other defenses during hypoxic stress (1–7). Carotid denervation in neonates, but not adults, leads to profound abnormalities of respiratory control and high mortality rates (1,8,9), suggesting a vulnerable period during mammalian postnatal maturation during which functioning carotid chemoreceptors are important for normal maturation of breathing control and, in some species, survival.

Perhaps surprisingly, given their importance in the developing infant, the carotid chemoreceptors have minimal sensitivity to hypoxia at birth and become more sensitive over the first few days or weeks of life (10–16). Increasing hypoxia sensitivity of the arterial chemoreceptors after birth is termed "resetting," and it occurs in both carotid and aortic chemoreceptors (17). Carotid chemoreceptor resetting appears to be modulated by the ~4-fold rise in arterial O_2 tension that occurs at birth (18–22), raising the possibility that peri- and postnatal hypoxia may impair carotid chemoreceptor development. In addition, perinatal hyperoxia causes lifelong blunting of the ventilatory response to acute hypoxia in rats, in part by adversely affecting carotid chemoreceptor maturation (23–25).

Developing infants are particularly vulnerable to hypoxic stress. Peri- and postnatal hypoxia may lead to impaired cognitive development and abnormalities in cardiovascular function, breathing control maturation, and lung function (26–28). Hypoxia and abnormal O_2 sensing have been implicated in several disorders with high morbidity and/or mortality rates in infants such as bronchopulmonary dysplasia (BPD), severe bradycardia, apparent life-threatening events (ALTE), and the sudden infant death syndrome (SIDS) (29–33). In addition, infants with lung disease of prematurity often experience intermittent hypoxia or hyperoxia, both of which may lead to delayed or abnormal maturation of O_2 chemosensitivity (21,31,34).

Postnatal carotid chemoreceptor maturation and mechanisms of development were recently reviewed (35) and are discussed elsewhere in this volume. This chapter will therefore focus on postnatal developmental changes occurring at the level of the type I or glomus cells, believed to be the O_2-sensing element in the carotid body.

II. O_2 Sensing Mechanisms and Development

A. Type I Cell

To understand maturation, it is necessary to understand the function of the mature carotid body. However, fundamental questions about mechanisms of O_2 sensing by the carotid chemoreceptors remain unanswered. It is widely speculated that hypoxia is transduced by type I cells, which somehow signal apposed carotid sinus nerve (CSN) terminals. The generally accepted O_2 chemotransduction scheme is as follows (numbers refer to steps in Fig. 1): An O_2 sensor, probably by inhibiting O_2-sensitive K^+ currents (step 1a), causes depolarization (step 2) of the type I cell, leading to Ca^{2+} influx through voltage-gated calcium channels (step 3). The rise in intracellular calcium ($[Ca^{2+}]_i$) (step 4) results in release of neurotransmitter(s) and/or neuromodulator(s) (step 5), which are believed to cause firing of action potentials in the adjacent carotid sinus nerve terminal (steps 6a and 7).

Although there is strong evidence that type I cells are necessary for O_2 chemoreception (36), the above transduction pathway is not fully established. Key elements, such as the identity of the O_2 sensor, the mechanism(s) by which hypoxia leads to type I cell depolarization, the identification of an excitatory neurotransmitter, and the mechanisms of afferent nerve excitation, remain unknown. The following discussion will focus on developmental changes at the cellular level, assuming the conventional model shown in Figure 1.

B. Role of Type I Cell Depolarization and Ca^{2+} Influx

Type I cell depolarization appears to be a critical step in O_2 transduction and therefore is a major potential site for maturational changes. Type I cells in vitro (37) and in situ depolarize in response to hypoxia and, in the absence of a type I cell depolarization, hypoxia does not induce catecholamine secretion from the carotid

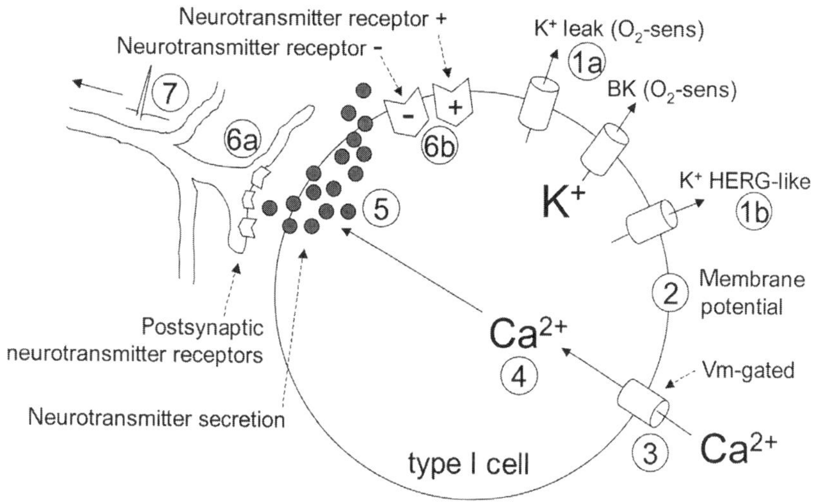

Figure 1 Elements of carotid chemoreceptor O_2 transduction pathway. See text for explanation.

body (38). In addition, the type I cell $[Ca^{2+}]_i$ response to hypoxia is markedly reduced by clamping membrane potential (Vm) at resting membrane potential (37).

Ca^{2+} influx is also a critically important step in type I cell O_2 chemotransduction, based on the following evidence: (1) Ca^{2+}-free solution with EGTA eliminates the carotid body neural response to hypoxia (39); (2) Cd^{2+} blocks the neural response to hypoxia (40–42); (3) Ca^{2+}-free solutions reduce the catecholamine secretion response to hypoxia by 95% (43); (4) voltage-gated Ca^{2+} channel blockers reduce the carotid body neural response to hypoxia (39,43); (5) voltage-gated calcium channel enhancers (BayK 8644) increase neural response to hypoxia (39); and (6) voltage-gated Ca^{2+} channel blockers inhibit catecholamine secretion in response to hypoxia and elevated K^+ (40–43). Therefore, the case for a critical role for Ca^{2+} influx in O_2 chemotransduction is strong. In addition, the type I cell $[Ca^{2+}]_i$ response (e.g., to hypoxia, temperature, acid, elevated extracellular K^+, etc.) has been shown to mimic the neural response of the carotid body (37,44). Therefore, $[Ca^{2+}]_i$ can be used as a marker to study chemoreceptor response at the type I cell level.

C. Potential Sites of Developmental Changes

In every species studied to date, the carotid chemoreceptor response to hypoxia increases with postnatal age. In in vitro carotid body preparations, the peak nerve discharge in response to a potent hypoxia stimulus increases about fourfold in neonatal rats during the first month, with most of that change occurring by ~2 weeks (10,12,45) (Fig. 2). Similar patterns of postnatal maturation of carotid sinus nerve

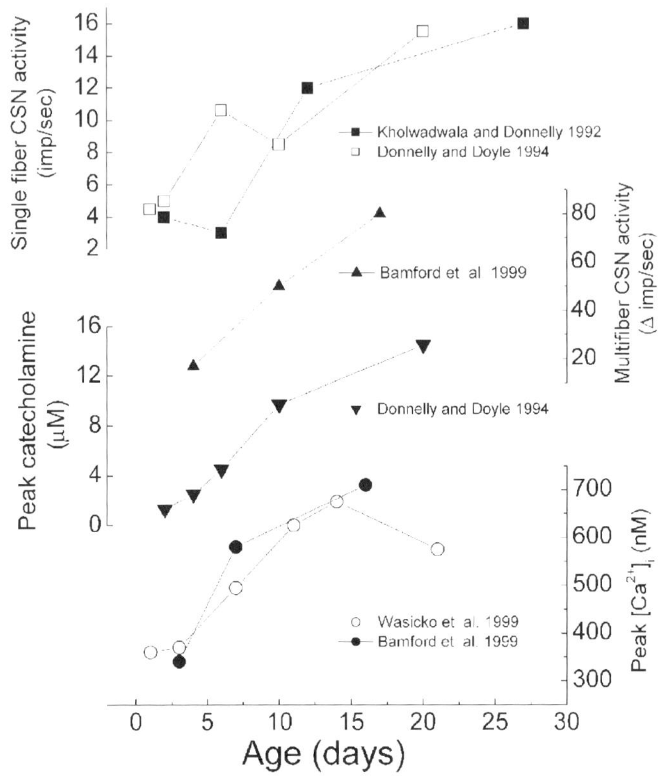

Figure 2 Summary of rat carotid chemoreceptor maturation. All plots depict responses to a strong hypoxia stimulus. Hypoxia-induced CSN activity, catecholamine secretion, and intracellular calcium responses all increase over approximately the same time course. (Data adapted from Refs. 10,12,45,68.)

activity have been measured in other species, using in vivo recording techniques (11,14).

Numerous potential sites for developmental changes in carotid chemotransduction, recently reviewed (35,46,47), are listed in Table 1. Humoral and circulatory factors are unlikely to play a significant role because a large developmental increase in the CSN response to hypoxia is observed using in vitro carotid chemoreceptor preparations (10,12). Some of the maturational increase in CSN activity could be due to an age-related ~fourfold increase in the ratio of afferent nerve terminals to parent axon (48). However, the developmental increase in carotid chemoreceptor O_2 sensitivity cannot be fully explained by anatomical maturation because hypoxia-induced carotid body catecholamine secretion also increases during postnatal maturation in rats (10) and rabbits (49). The type I cell neurosecretory response is presynaptic to CSN nerve terminals suggesting that carotid chemoreceptor O_2 response maturation occurs, at least in part, at the type I cell level.

Table 1 Potential Sites of Postnatal Carotid Chemoreceptor Developmental Changes

Carotid body anatomy	Blood flow to the carotid body Flow patterns within carotid body Blood vessel control
Humoral factors	Alterations in serum K^+, Na^+, osmolarity Endorphins Inhibitory humoral modulators
Morphological changes in synapses and nerve terminals	Number of synapses between type I cells and nerve terminals
O_2 sensor	O_2 sensitivity of sensor(s)
pH/CO_2	pH sensitivity CO_2 sensitivity O_2–CO_2 interaction
Type I cell membrane depolarization	Mechanism(s) linking O_2 sensor to mediator of type I cell depolarization
K^+ currents regulating type I cell membrane potential	Large-conductance K^+ currents Background ("leak") K^+ currents HERG-like K^+ currents Other K^+ currents O_2 sensitivity of K^+ currents regulating type I cell membrane potential
Other ionic conductances regulating type I cell membrane potential	Cl^- currents Na^+ currents
$[Ca^{2+}]_i$ responses of type I cells	Voltage-gated Ca^{2+} currents O_2 sensitivity of Ca^{2+} channels Variation in types of Ca^{2+} channels Ca^{2+} release from intracellular stores
Neurosecretory responses	Tonic release of inhibitor neuromodulator (e.g., dopamine) Magnitude of secretory response to hypoxia Glomus cell/nerve terminal stimulus-secretion coupling Identity/mix of excitatory neurotransmitter(s) released in response to hypoxia Identity/mix of inhibitory neuromodulator(s) released in response to hypoxia
Autoreceptors	Type I cell inhibitory D2 autoreceptors Other type I cell neurotransmitter receptors
Postsynaptic receptors	Excitatory postsynaptic neurotransmitter receptors Inhibitory postsynaptic neurotransmitter receptors
Nerve terminals	Membrane properties of chemoreceptor afferent neurons Excitability of CSN nerve terminals

D. A Caveat for Developmental Studies

That elements like dissociated type I cells may function differently in vivo versus in vitro is obvious; dissociated cells are removed from their in vivo ionic microenvironment and neurotransmitter milieu, which may vary during development. In vitro whole carotid body preparations lack blood flow, lack efferent CSN activity, and have artificial O_2 gradients that vary with carotid body size. Even with in vivo nerve recording, the CSN is usually severed, potentially altering efferent modulation mechanisms that may vary during development. In addition, when carotid bodies or cells are harvested into room air (relatively hyperoxic $PO_2 \sim 150\,mmHg$), hypoxia-modulated gene regulation and other cellular functions may change rapidly. Therefore, the time frame of studies postharvesting cells/tissue may be critically important. Such factors should be taken into account when designing or interpreting developmental chemoreceptor studies.

E. Mechanisms of Type I Cell O_2 Response Maturation

Maturational changes in type I cell O_2 transduction could occur at any point in the cascade shown in Figure 1. In the following discussion, developmental changes will be considered in reverse, starting with nerve activity and working backward to the O_2 sensor.

CSN Activity and Nerve Terminal Properties

Although beyond the scope of this chapter, it should be noted that developmental changes may occur in the properties of the carotid sinus nerve endings. A recent study revealed little maturational change in the properties of chemoreceptor afferent neurons in the petrosal ganglion, making this possibility unlikely (50). However, very little is known about postsynaptic receptors on CSN nerve terminals generally and virtually nothing is known about developmental changes in the identity, density, or function of these receptors.

Type I Cell Neurosecretion and Neuromodulators

The excitatory neurotransmitter(s) shown in step 5 of Figure 1 have not been identified and the role of catecholamines in the carotid body is controversial. Catecholamines are released from the carotid body in response to hypoxia in a graded fashion (51). However, dopamine (DA) antagonists generally lead to an increase in carotid sinus nerve activity and administration of exogenous DA inhibits afferent nerve activity in most studies (52). In addition, type I cells possess DA D2 autoreceptors, which inhibit catecholamine secretion. This has led to the general hypothesis that DA is inhibitory in the carotid body and the specific hypothesis that maturational changes in O_2 chemosensitivity may reflect developmental changes in DA metabolism, release, D2 autoreceptor function, and/or autoreceptor density.

Catecholamine Content and Release

In rat carotid bodies DA turnover rate is high at birth and then declines (20,53). However, for changes in DA neurosecretion to explain carotid body maturational changes, DA levels should be high in newborns and decline with age on a time course consistent with the known developmental time course of the hypoxia response (~2 weeks in rats). Closer examination of the time course showed that DA turnover was high within 6 hr after birth, fell ~75% by 12 hr, and changed little thereafter up to 7 days (20,53). Donnelly and Doyle used carbon fiber electrodes in rat carotid bodies, in vitro, to study free tissue catecholamine levels in rats aged 1, 2, 6, 10, and 20–30 days (10). The results indicated that baseline levels of catecholamine were very low from 1 to 6 days, approximately doubled by 10 days, and approximately doubled again by 20–30 days.

Tyrosine hydroxylase (TH), a key enzyme in DA synthesis, also changes during development. An early study of rat carotid body development showed that the level of TH mRNA expression was greatest in the term fetus, decreased ~60% by 10 hr postnatal age, and declined another ~20% by 4–7 days (54). Similarly, a recent study reported that the level of rat carotid body TH mRNA expression was greatest at birth, significantly decreased by 48 hr postnatal age, and remained decreased at 14 and 21 postnatal days (55). Thus, consistent with the above studies on DA levels, TH mRNA drops rapidly within hours of birth and changes little during the time frame of carotid sinus nerve activity maturation.

In an in vitro rabbit carotid body preparation, Bairam et al. found that hypoxia-induced DA release was minimal in 1–15-day-old rabbits and only became significant after 25 days of age (49). Similarly, using carbon fiber electrodes in rat carotid bodies, peak free tissue catecholamine levels in response to anoxia were low in newborns and increased >10 fold over the first month of life (10). Thus, studies to date indicate low baseline DA levels and increasing DA release during the first weeks of life in rabbits and rats, which does not support the hypothesis that declining carotid body catecholamine levels account for the maturational increase in nerve activity or O_2 sensitivity. However, catecholamine release by nerve terminals (56) may be altered in an in vitro cut-CSN preparation and, therefore, the question has not been fully resolved.

Catecholamine Autoreceptors and Development

DA D1 and D2 receptors are present in the carotid bodies of rabbits and rats (57–59) (Fig. 1, step 6b). However, major controversy still surrounds the issues of where they are located and what their function is. In rat type I cells, D2-DA receptor and TH messenger RNAs are colocalized (55) likely in type I cells. D2-DA receptor messenger RNA levels significantly increase with postnatal age in both rat and rabbit carotid body (55,60). However, the location of D1 receptors and their role is less clear (57).

Several studies indicate that DA release in rabbit carotid bodies in vitro is modulated by D2 autoreceptors in an age-dependent fashion. Specifically, under normoxic and hypoxic conditions, the D2 receptor antagonist domperidone increases DA release with lower concentrations in carotid bodies from adults versus newborns, suggesting that this mechanism is less developed in neonates (58,61). An in vivo study, using carotid sinus nerve recording in newborn versus mature cats, reported that domperidone increased CSN activity at all oxygen tensions in mature cats, but only in severe hypoxia in newborns (62). Thus, based on in vivo and in vitro studies, DA D2 autoreceptor-mediated inhibition of carotid chemoreceptor activity appears to be better developed in mature subjects of several species.

It is clear that dopaminergic aspects of carotid chemoreceptor function change with age. However, the role of catecholamines in *modulating* chemoreceptor development is less than straightforward. Studies to date indicate that DA functions as an inhibitory neuromodulator in the carotid body, baseline and stimulated catecholamine release increase with age, D2 autoreceptor mRNA increases with age, and inhibitory D2 autoreceptor function seems to be more effective in the mature carotid body. These findings, taken together, make it difficult to envisage how dopaminergic mechanisms could be responsible for maturational changes in carotid chemoreceptor function.

Other Neurotransmitters/Modulators

Type I cells express a variety of other neurotransmitter receptors for adenosine, acetylcholine (63), substance P, and other substances that may enhance or inhibit excitability (Fig. 1, step 6b). Exogenous adenosine stimulates carotid body nerve activity (64,65). Although the mechanism remains unclear, it appears that adenosine inhibits 4-AP-sensitive outward (K^+) current at Vm positive to -40 mV, which would likely enhance type I cell excitability (65). In contrast, adenosine acting on A_{2A} receptors inhibits the rat type I cell $[Ca^{2+}]_i$ response to hypoxia and has been shown to decrease voltage-dependent Ca^{2+} currents (66). The latter result raises the interesting possibility that adenosine could act as an inhibitory modulator of chemoreceptor maturation. In support of this hypothesis, Gauda and colleagues showed that adenosine A2 receptors are present in the rat carotid body at birth, change little up to 6 days, and then decline ~25% by 14 days (66). However, until the role of adenosine in chemotransduction becomes clearer, it is difficult to assess whether it has a role in carotid body maturation. Similarly, although substance P appears to modulate carotid chemoreceptor O_2 sensitivity (67), its role in maturation remains unclear. The potential role of nicotinic and muscarinic acetylcholine and other receptors in type I cell maturation remains unknown.

Intracellular Calcium Response

Hypoxia-induced neurosecretion from glomus cells is mediated by a rise in intracellular calcium, which resembles the CSN activity response to hypoxia in its hyperbolic shape (37,44,68), PO_2 range, and relationship to neurosecretion (69,70). Because neurosecretion and (presumably) afferent nerve activity depend on the type I cell $[Ca^{2+}]_i$ response to hypoxia, age-related changes in this crucial transduction step could account for maturation of O_2 sensitivity. The original hypothesis of the first studies in this area was simply that a major site of chemoreceptor resetting and development of oxygen sensitivity, within the carotid body, lies in the type I cell.

$[Ca^{2+}]_i$ was used as a marker for the type I cell response to hypoxia during development.

Magnitude and Shape of $[Ca^{2+}]_i$ Response to Hypoxia

Type I cells were enzymatically dissociated from the carotid chemoreceptors of newborn and adult rabbits and studied using the fluorescent dye fura-2 (71). $[Ca^{2+}]_i$ responses to hypoxia were three- to fivefold larger in type I cells from adult versus newborn rabbits, while $[Ca^{2+}]_i$ responses to ionomycin (a Ca^{2+} ionophore) did not differ. Thus, type 1 cells from immature rabbits demonstrated three- to fivefold smaller $[Ca^{2+}]_i$ responses than cells from mature rabbits, even when removed from the usual in vivo milieu of neurotransmitters and modulators, influences of blood flow, humoral factors, and modulation by carotid body efferent innervation (71). These results indicated that carotid chemoreceptor development was not a matter of simple age-related changes in type I cell modulation (e.g., by dopamine).

A more detailed study of $[Ca^{2+}]_i$ response maturation was undertaken in clusters of type 1 cells isolated from near-term fetal rats and rats 1, 3, 7, 11, 14, and 21 days old (68). The $[Ca^{2+}]_i$ response to maximal hypoxic stimulation (anoxia) was low in term-fetal cells and increased with age (Fig. 3), consistent with previous CSN recordings both in vivo (11,14) and in vitro (10,12,45). In addition, the submaximal portion of the type I cell O_2 response curve for $[Ca^{2+}]_i$ increased approximately fourfold during postnatal development (Fig. 4). Thus, as with afferent nerve activity, graded maximal and submaximal type I cell $[Ca^{2+}]_i$ responses depend on the level of postnatal maturity.

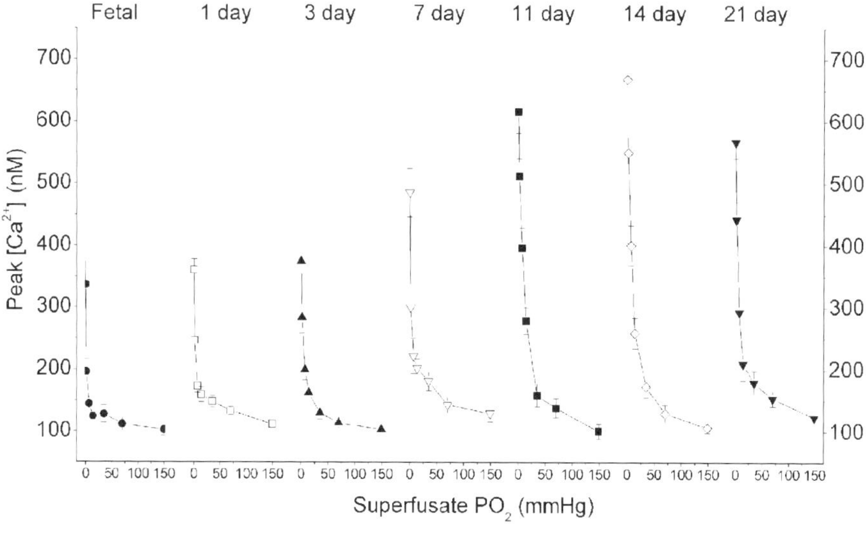

Figure 3 Rat type I cell response to graded hypoxia during development. Each point represents the mean $\pm$SEM of peak $[Ca^{2+}]_i$ values for a given superfusate PO_2. (Adapted from Ref. 68.)

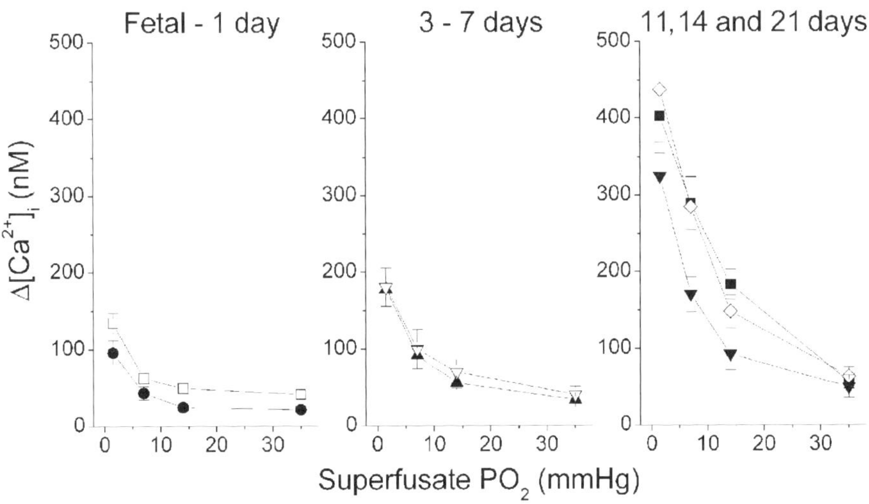

Figure 4 Rat type I cell Δ $[Ca^{2+}]_i$ for PO_2 levels between ~2 and 35 mmHg versus age. ●, fetal; □, 1 day; ▲, 3 days; ▽, 7 days; ■, 11 days; ◇, 14 days; ▼, 21 days. O_2 sensitivity increases mostly between 7 and 11–14 days. (Adapted from Ref. 68.)

The PO_2 at half-maximal response shifted from ~1–2 mmHg in the young age groups to ~6 mmHg in the 11–21-day age group (arrows, Fig. 5) suggesting that O_2 sensitivity matures at the level of the type I cell. In striking contrast to the maturation of $[Ca^{2+}]_i$ responses to hypoxia, there were *no age-related changes* in the $[Ca^{2+}]_i$ response to 10, 20, or 40 mM extracellular K^+ ($[K^+]_o$) (68). This finding strongly suggested that immature cells were not depolarizing as much as mature cells, for a given level of hypoxia, consistent with the hypothesis that resetting involves maturation of the process causing hypoxia-induced depolarization, rather than developmental changes in Ca^{2+} influx for a given degree of depolarization.

Time Course of $[Ca^{2+}]_i$ Response Maturation

If development of type I cell O_2 sensitivity plays a role in maturation of the CSN response to hypoxia, then the developmental time course of the two should match for a given species. Several studies show that $[Ca^{2+}]_i$ responses of type I cells from term-fetal and 1-day-old rats are small and increase with age between ~3 and 11–14 days of age (45,68), which matches the time course of rat postnatal carotid body nerve activity from three separate studies (Fig. 2) (10,12,45). Such close temporal matching between the time course of type I cell O_2 sensitivity and CSN neural response maturation supports the hypothesis that maturation at the type I cell level underlies carotid chemoreceptor development.

Neurotransmitter Modulation of Type I Cell $[Ca^{2+}]_i$ Response

As noted above, several studies suggest that dopaminergic modulation of carotid chemoreceptor O_2 sensitivity changes with age. We therefore studied the effects of the D2 receptor agonist quinpirole (QP) on type I cell $[Ca^{2+}]_i$ responses to hypoxia

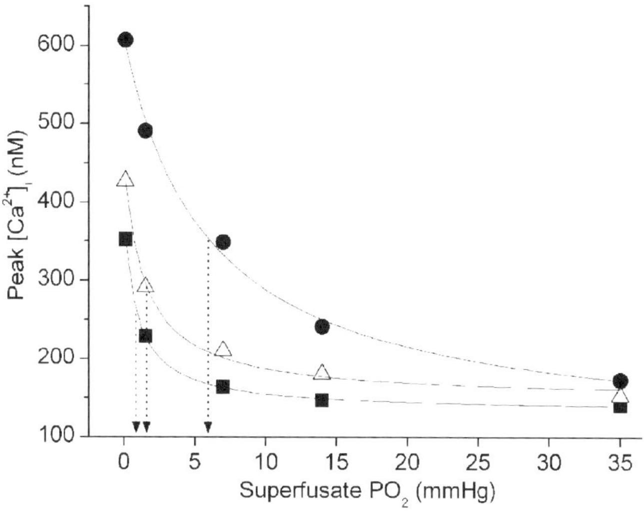

Figure 5 Rightward shift in the submaximal portion of rat type I cell O_2 response profile with age. Mean peak $[Ca^{2+}]_i$ responses of type I cells from three age groups. ■, fetal–1 day ($n = 28$); △, 3–7 days ($n = 15$); ●, 11–21 days ($n = 34$). Data in each age group fitted with a hyperbolic function. Arrows indicate PO_2 at $\frac{1}{2}$ maximal $[Ca^{2+}]_i$ response. (From Ref. 68.)

during postnatal development. Using $[Ca^{2+}]_i$ measurements, preliminary data show that 10 μM quinpirole inhibits the $\Delta[Ca^{2+}]_i$ response to hypoxia in cells from 12–14-day-old rats but not 1-day-old rats and inhibition is blocked by the D2 receptor antagonist sulpiride. These findings, if confirmed, suggest that the inhibitory effect of extracellular dopamine on type I cell Ca^{2+} influx increases with postnatal maturation.

O_2-Sensitive $[Ca^{2+}]_i$ Currents

Previous studies show that carotid chemoreceptor type I cell Ca^{2+} currents are either insensitive to acute hypoxia (72) or exhibit mild inhibition by hypoxia (70). However, hypoxia caused a reversible ~50% inhibition of peak Ca^{2+} currents in type I cells from 1–3-day-old rats but no inhibition in type I cells from 11–14-day-old rats (73). Hypoxic inhibition of Ca^{2+} influx in immature type I cells that disappears with age may contribute to the weak $[Ca^{2+}]_i$ response to hypoxia observed in cells from newborn rats. The developmental time course and mechanism of this finding are unknown.

Maturation of $[Ca^{2+}]_i$ Responses to CO_2 and O_2–CO_2 Interaction

Several studies (in vivo) have shown that the carotid chemoreceptor neural response to CO_2 is also weak in the newborn and increases with age (11,14). Therefore, it was proposed that maturation of carotid chemoreceptor hypoxic sensitivity may be related to development of O_2–CO_2 interaction (74). In an early study of rats 1–14 days of age, type I cell $[Ca^{2+}]_i$ responses to CO_2 were found to increase with age between 3 and 11 days, although multiplicative O_2–CO_2 interaction was not evident. Significant O_2–CO_2 interaction in CSN neural responses was not detected until after 14 days of age (45). A more recent study of type I cell $[Ca^{2+}]_i$ responses in 11–16-day-old rats did find weak multiplicative O_2–CO_2 interaction, particularly in individual cells (75). The authors suggested that the O_2 sensor and pH/CO_2 sensor are probably separate and converge at the level of type I cell depolarization and the resulting Ca^{2+} influx. Taken together, these results suggest that O_2–CO_2 interaction would not be a likely explanation for carotid body O_2 response maturation between birth and 14 days in rats, although it could play a role after 14 days.

Type I Cell Membrane Depolarization

It is now widely accepted that the rise in type I cell $[Ca^{2+}]_i$ induced by hypoxia is due to cell membrane depolarization and the resulting Ca^{2+} influx via voltage-gated Ca^{2+} channels. Therefore, a likely hypothesis for maturational changes in the $[Ca^{2+}]_i$ response to hypoxia is that hypoxia-induced Vm depolarization is small in cells of newborns and increases with age.

Hypoxia-Induced Cell Membrane Depolarization

The resting membrane potential of enzymatically dissociated rat type I cells from 1–3 versus 11–14-day-old rats is ~ -50 to -55 mV and does not differ with age (73). However, in response to a strong hypoxia stimulus voltage-clamped type I cells from 1–3-day-old rats depolarized only ~8 mV compared with ~18–19 mV depolarization in cells from 11–14-day-old rats (73). Just as the rat type I cell $[Ca^{2+}]_i$ response to elevated extracellular K^+ was large and not significantly different from 1 to 21 days (68), the type I cell Δ depolarization in response to high $[K^+]_o$ was large (~30 mV) and did not vary with age (73). Thus, rat type I cells depolarize to the same degree in response to a nonspecific depolarizing stimulus such as elevated $[K^+]_o$. The finding that hypoxia-induced depolarization of cells from newborn rats is much smaller than the mature response suggests a developmental change specifically in the mechanism causing depolarization in response to hypoxia. These findings raise the question of what mediates type I cell hypoxia-induced Vm depolarization, which is not fully understood even for mature carotid body, and whether this mechanism changes with age.

Resting Membrane Potential

Type I cells exhibit resting K^+ and Cl^- permeability (76), both of which likely play a role in resting membrane potential and excitability. Numerous studies show that elevated extracellular K^+ depolarizes isolated type I cells (37,43,68), implicating resting K^+ permeability. Carotid chemoreceptor type I cells express a variety of K^+ channel types that are inhibited by hypoxia (77–84). Large-conductance O_2-sensitive K^+ channels are not active at type I cell resting membrane potential in enzymatically dissociated cells, and blockers of these channels, such as tetraethylammonium (TEA), 4-aminopyridine (4-AP), and charybdotoxin, do not stimulate carotid sinus nerve activity or cause type I cell depolarization (84–89). Although TEA was reported in a carotid body slice preparation to stimulate catecholamine secretion (90), most evidence to date indicates that large-conductance K^+ currents are minimally active at resting membrane potentials. Other "background" or "leak" K^+ currents *are* present at resting Vm in type I cells and therefore more likely candidates for mediating hypoxia-induced depolarization.

Role of Large-Conductance K^+ Currents in Maturation

Although they are not likely to initiate hypoxia-induced type I cell depolarization, Ca^{2+}-sensitive K^+ (K_{Ca}) currents do show maturational changes. K^+ current density in type I cells from 4-day-old rats was only ~40% that of cells from adult rats (91). In addition, hypoxia-induced inhibition of K_{Ca} currents was significantly less in cells from 4-day versus adult rats (91). At least in enzymatically dispersed type I cells, these large-conductance K^+ currents are not active at membrane potentials negative to −40 mV. In addition, their current-voltage relationship suggests that they would buffer Vm depolarization and likely assist in repolarization. Therefore, it is difficult to see how the reported developmental increase in these K^+ currents, or their O_2 sensitivity, could explain maturational changes in the type I cell depolarization or $[Ca^{2+}]_i$ response to hypoxia.

"Background" K^+ Currents and Development

Type I cells possess several "background" or "resting" K^+ currents, active in the range of −45 to −60 mV, the typical resting membrane potential of type I cells. In 1997 Buckler reported the existence of a novel background K^+ current in neonatal rat type I cells (89), which appears to be a member of the newly described family of tandem-pore, 4-transmembrane domain (2P/4TM) potassium channels. This O_2-sensitive current has been shown to be voltage insensitive, not blocked by the K^+ channel blockers TEA and 4-AP, blocked by mM concentrations of Ba^{2+}, active at negative potentials in the range of resting potential, but inhibited in a hyperbolic graded fashion in the same PO_2 range as the type I cell and intact carotid body (89). Recently, Williams and Buckler (92) reported that this conductance is inhibited by 1 mM quinidine and activated two- to threefold by anesthetics such as halothane, which also abolished the $[Ca^{2+}]_i$ response of type I cells. In cell-attached studies, anoxia inhibited channel activity by about 50% but had no effect on channel activity in excised patches. Single-channel recordings of rat type I cell background K^+ conductance revealed a flickering type of channel with ~14 pS single-channel conductance in cell-attached mode and ~12 pS single-channel conductance in excised patches. Single-channel recordings also show that this O_2-sensitive background K^+ channel is insensitive to TEA and 4-AP and blocked by Ba^{2+} (92). All of these features suggest that type I cells possess a TASK-like O_2-sensitive K^+ channel, possibly TASK-1 or TASK-3.

Whether a TASK-like O_2-sensitive background K^+ current is responsible for hypoxia-induced type I cell depolarization is unknown, as there are no specific pharmacological blockers of this channel. In any case, if it mediates hypoxia-induced depolarization, it would be an excellent candidate site for developmental changes in type I cell O_2 sensitivity. Data from type I cells of 1–3 versus

11–14-day-old rats indicate the presence of a background K^+ conductance as described by Buckler and colleagues (89,92), which was a similar magnitude background K^+ current in both age groups. However, this background or "leak" K^+ current was significantly inhibited by hypoxia in type I cells from the 11–14-day-old but not the 1–3-day-old rats (73). If confirmed, a developmental increase in hypoxic inhibition of a background K^+ current, combined with hypoxic-inhibition of voltage-gated Ca^{2+} currents in newborns that disappears with age, suggests two potentially important mechanisms for maturation changes in hypoxia-induced depolarization.

Another background K^+ current described in rabbit type I cells is a HERG-like K^+ current (93) (Fig. 1, step 1b). HERG K^+ channels (*H*uman *E*ther-a-go-go *R*elated *G*ene product) are potassium channels believed to play a major role in maintaining resting membrane potential and controlling repolarization in several excitable cell types, including neurosecretory cells (94–96). HERG-like K^+ currents in rabbit carotid body type I cells were blocked by dofetilide (a HERG K^+ channel blocker) and mM concentrations of Ba^{2+}. Dofetilide depolarized current-clamped type I cells, caused a rise in $[Ca^{2+}]_i$, and, in an in vitro CB preparation, increased CSN activity and attenuated the hypoxia response (93). While it seems clear that a HERG-like K^+ current plays a role in regulating membrane potential in carotid body type I cells, it remains unclear whether this HERG-like K^+ current is O_2-sensitive. If so, then a HERG-like K^+ current would be another potential candidate underlying maturation changes in the depolarization and $[Ca^{2+}]_i$ response to hypoxia.

O_2 Sensor

An obvious possibility to explain carotid chemoreceptor maturation is that the sensitivity of the O_2 sensor itself increases with age. This question cannot be addressed at this time because the identity of the type I cell O_2 sensor remains unknown, even for the mature carotid body. As discussed above, available evidence suggests that type I cell membrane potential is regulated, in large part, by several K^+ currents, some of which are inhibited by hypoxia (89,92,97,98) (Fig. 1). Whether the K^+ channels themselves are O_2 sensitive or linked to a separate O_2 sensor is unknown. Maturational changes in O_2 sensitivity of hypoxia-sensitive K^+ currents could occur by shifts in K^+ channel subunits, developmental shifts in expression of K^+ channel subtypes with different O_2 sensitivity, and a variety of other mechanisms. However, these possibilities have not been explored.

III. Chronic Hypoxia During Postnatal Development

Chronic hypoxia (CH) from birth blunts the ventilatory response to hypoxia in rats (21) and other species by unknown mechanisms. Although CH alters the VE response to hypoxia at the central nervous system level (99), CH from birth also affects maturation of carotid body type I cell O_2 sensitivity. Type I cells dissociated from 11-day-old rats reared from birth in FiO_2 0.12 demonstrated marked blunting of $[Ca^{2+}]_i$ responses to acute hypoxia compared with controls reared in normoxia

(Fig. 6) (22). In type I cells of rats reared from birth to 11 days in hypoxia and then in room air from day 12–18, $[Ca^{2+}]_i$ responses to acute hypoxia showed partial recovery to normal levels (Fig. 6, panel c). Although the mechanism remains unknown, the absence of type I cell $[Ca^{2+}]_i$ response to hypoxia in rats reared in CH is probably not related to impaired Ca^{2+} influx because the $[Ca^{2+}]_i$ response to elevated extracellular K^+ was largely preserved (22) and, in another study, CH from birth was found to increase voltage-gated Ca^{2+} current density (100). The finding that CH appears to delay onset of "resetting" suggests that CH from birth perpetuates the fetal state of type I cell function. This is consistent with previous studies showing that the rise in PaO_2 after birth modulates carotid chemoreceptor resetting (18) and suggests that "resetting" occurs, at least in part, at the type I cell level.

IV. Fetal Arterial Oxygen Tension: Implications for Carotid Chemoreceptor Maturation

Fetal sheep carotid sinus nerve (13) and fetal rat type I cell $[Ca^{2+}]_i$ (68) responses to hypoxia are weak compared to mature postnatal responses. The consensus view is that carotid chemoreceptor O_2 sensitivity is suppressed or adapted to the normally low PaO_2 of the mammalian fetal environment (~23–27 mmHg). Nature's design makes sense, from a teleological perspective: to disable carotid chemoreceptor

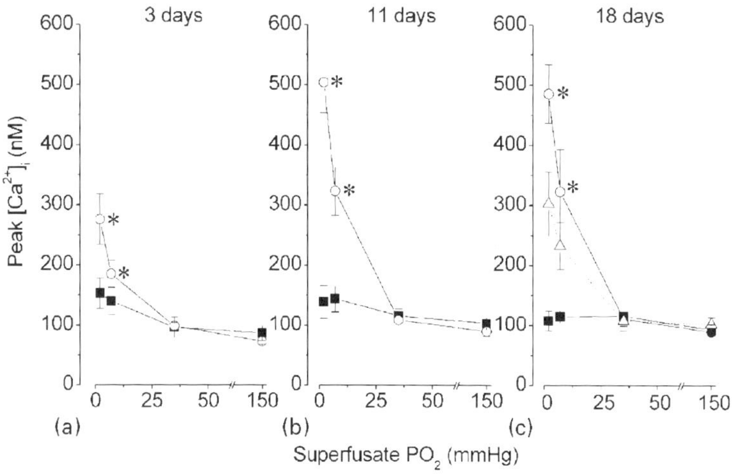

Figure 6 Effects of chronic hypoxia (CH) on type I cell $[Ca^{2+}]_i$ responses to acute hypoxia challenge at 3, 11, and 18 days from CH (■) and control (○) rats. In panel c, △ represents data from a "recovery" group reared in CH from birth to 11 days and then in room air from 11 to 18 days. CH from birth abolished postnatal development of $[Ca^{2+}]_i$ responses to hypoxia. After 11 days in hypoxia, returning to room air for 1 week resulted in partial recovery of $[Ca^{2+}]_i$ response to hypoxia (c). * indicates significant difference from CH ($p < 0.05$). (Adapted from Ref. 22.)

function in the low PO_2 fetal environment, while allowing anatomical and biochemical maturation, and provide a mechanism to "reset" or disinhibit O_2 sensitivity after birth. Birth per se does not initiate resetting; rather, the rise in PaO_2 at birth modulates carotid chemoreceptor "resetting" and chronic hypoxia from birth can delay "resetting." Consistent with this view, in humans the O_2 chemoreflex is active in premature infants born as early as 28 weeks gestational age (101), suggesting that the peripheral chemoreceptors reach anatomical and physiological maturity quite early in utero, but their function is minimal until they are reset by the change in PO_2 at birth.

Thus, the low O_2 sensitivity of the fetal and newborn carotid body is likely modulated by the low PaO_2 environment of the fetus, raising possibilities such as regulation of "resetting" by hypoxia-inducible factor 1 (HIF-1) and other oxygen-responsive modulators (e.g., AP-1). HIF-1 is a transcriptional activator that functions as a global regulator of O_2 homeostasis, modulating embryonic development as well as a variety of postnatal physiological adaptations to hypoxia (102,103). For example, hypoxia up-regulates Shaker-type K^+ channels and TH genes but down--regulates others, such as cytochrome c oxidases and the dopamine D1 receptor (104). Indeed, mice partially deficient in HIF-1α have impaired carotid body responses to acute hypoxia and an abnormal adaptation to chronic hypoxia (105). Nitric oxide and carbon monoxide also modulate the carotid chemoreceptor O_2 response (106) and interact with HIF-1 (107).

Although it is tempting to speculate that hypoxia-inducible transcription regulators may play a role in chemoreceptor resetting, it is important to note that the effects of chronic hypoxia are age-dependent. CH in mature mammals *increases* the VE response to hypoxia while, during early postnatal maturation, CH *blunts* VE response and carotid chemoreceptor resetting (22,108). Therefore, the effects of low arterial PO_2 in the fetus on chemoreceptor resetting are likely related to complex, age-dependent interactions of several hypoxia-inducible transcriptional and physiological regulators.

V. Summary

Postnatal maturation of the carotid body and resetting of O_2 sensitivity appear to involve numerous aspects of the O_2 transduction cascade at the cellular level. Maturational changes have been described in type I cell neurotransmitter secretion, neuromodulator function, neurotransmitter receptor expression, type I cell depolarization in response to hypoxia, $[Ca^{2+}]_i$ responses to hypoxia, and K^+ and Ca^{2+} channel O_2 sensitivity. However, very little is known about the mechanisms regulating maturation of these numerous aspects of chemotransduction. Progress is likely to come from studies of genetically altered mice and hypoxia-inducible gene regulators and their interactions during fetal and postnatal development.

References

1. Bureau MA, Lamarche J, Foulon P, Dalle D. Postnatal maturation of respiration in intact and carotid body–chemodenervated lambs. J Appl Physiol 1985; 59(3):869–874.
2. Bureau MA, Lamarche J, Foulon P, Dalle D. The ventilatory response to hypoxia in the newborn lamb after carotid body denervation. Respir Physiol 1985; 60(1):109–119.
3. Hofer MA. Role of carotid sinus and aortic nerves in respiratory control of infant rats. Am J Physiol 1986; 251(4 Pt 2):R811–R817.
4. Hofer MA. Lethal respiratory disturbance in neonatal rats after arterial chemoreceptor denervation. Life Sci 1984; 34(5):489–496.
5. Fewell JE, Taylor BJ, Kondo CS, Dascalu V, Filyk SC. Influence of carotid denervation on the arousal and cardiopulmonary responses to upper airway obstruction in lambs. Pediatr Res 1990; 28(4):374–378.
6. Fewell JE, Kondo CS, Dascalu V, Filyk SC. Influence of carotid-denervation on the arousal and cardiopulmonary responses to alveolar hypercapnia in lambs. J Dev Physiol 1989; 12(4):193–199.
7. Fewell JE, Kondo CS, Dascalu V, Filyk SC. Influence of carotid denervation on the arousal and cardiopulmonary response to rapidly developing hypoxemia in lambs. Pediatr Res 1989; 25(5):473–477.
8. Cote A, Porras H, Meehan B. Age-dependent vulnerability to carotid chemodenervation in piglets. J Appl Physiol 1996; 80(1):323–331.
9. Donnelly DF, Haddad GG. Prolonged apnea and impaired survival in piglets after sinus and aortic nerve section. J Appl Physiol 1990; 68(3):1048–1052.
10. Donnelly DF, Doyle TP. Developmental changes in hypoxia-induced catecholamine release from rat carotid body, in vitro. J Physiol 1994; 475(2):267–275.
11. Carroll JL, Bamford OS, Fitzgerald RS. Postnatal maturation of carotid chemoreceptor responses to O_2 and CO_2 in the cat. J Appl Physiol 1993; 75(6):2383–2391.
12. Kholwadwala D, Donnelly DF. Maturation of carotid chemoreceptor sensitivity to hypoxia: in vitro studies in the newborn rat. J Physiol 1992; 453:461–473.
13. Blanco CE, Dawes GS, Hanson MA, McCooke HB. The response to hypoxia of arterial chemoreceptors in fetal sheep and new-born lambs. J Physiol 1984; 351:25–37.
14. Marchal F, Bairam A, Haouzi P, Crance JP, Di Giulio C, Vert P, Marchal F, Bairam A, Haouzi P, Crance JP, Di Giulio C, Vert P, Lahiri S. Carotid chemoreceptor response to natural stimuli in the newborn kitten. Respir Physiol 1992; 87(2):183–193.
15. Biscoe TJ, Purves MJ. Carotid body chemoreceptor activity in the new-born lamb. J Physiol 1967; 190(3):443–454.
16. Mulligan E, Lahiri S. Cat carotid body chemoreceptor responses before and after nicotine receptor blockade with alpha-bungarotoxin. J Auton Nerv Syst 1987; 18(1):25–31.
17. Kumar P, Hanson MA. Re-setting of the hypoxic sensitivity of aortic chemoreceptors in the new-born lamb. J Dev Physiol 1989; 11(4):199–206.
18. Blanco CE, Hanson MA, McCooke HB. Effects on carotid chemoreceptor resetting of pulmonary ventilation in the fetal lamb in utero. J Dev Physiol 1988; 10(2):167–174.
19. Hanson MA, Kumar P, Williams BA. The effect of chronic hypoxia upon the development of respiratory chemoreflexes in the newborn kitten. J Physiol 1989; 411:563–574.

20. Hertzberg T, Hellstrom S, Holgert H, Lagercrantz H, Pequignot JM. Ventilatory response to hyperoxia in newborn rats born in hypoxia—possible relationship to carotid body dopamine. J Physiol 1992; 456:645–654.
21. Eden GJ, Hanson MA. Effects of chronic hypoxia from birth on the ventilatory response to acute hypoxia in the newborn rat. J Physiol 1987; 392:11–19.
22. Sterni LM, Bamford OS, Wasicko MJ, Carroll JL. Chronic hypoxia abolished the postnatal increase in carotid body type I cell sensitivity to hypoxia. Am J Physiol 1999; 277(3 Pt 1):L645–L652.
23. Ling L, Olson EB, Vidruk EH, Mitchell GS. Developmental plasticity of the hypoxic ventilatory response. Respir Physiol 1997; 110(2–3):261–268.
24. Fuller D, Bavis RW, Vidruk EH, Wang ZY, Olson EB, Jr., Bisgard GE, Fuller DD, Bavis RW, Vidruk EH, Wang ZY, Olson EB Jr, Bisgard GE, Mitchell GS. Life-long impairment of hypoxic phrenic responses in rats following 1 month of developmental hyperoxia. J Physiol 2002; 538(Pt 3):947–955.
25. Erickson JT, Mayer C, Jawa A, Ling L, Olson EB, Vidruk EH, Erickson JT, Mayer C, Jawa A, Ling L, Olson EB Jr, Vidruk EH, Mitchell GS, Katz DM. Chemoafferent degeneration and carotid body hypoplasia following chronic hyperoxia in newborn rats. J Physiol 1998; 509(Pt 2):519–526.
26. Nyakas C, Buwalda B, Luiten PG. Hypoxia and brain development. Prog Neurobiol 1996; 49(1):1–51.
27. Hudlicka O, Brown MD. Postnatal growth of the heart and its blood vessels. J Vasc Res 1996; 33(4):266–287.
28. Okubo S, Mortola JP. Long-term respiratory effects of neonatal hypoxia in the rat. J Appl Physiol 1988; 64(3):952–958.
29. Calder NA, Williams BA, Smyth J, Boon AW, Kumar P, Hanson MA. Absence of ventilatory responses to alternating breaths of mild hypoxia and air in infants who have had bronchopulmonary dysplasia: implications for the risk of sudden infant death. Pediatr Res 1994; 35(6):677–681.
30. Garg M, Kurzner SI, Bautista D, Keens TG. Hypoxic arousal responses in infants with bronchopulmonary dysplasia. Pediatrics 1988; 82(1):59–63.
31. Katz-Salamon M, Jonsson B, Lagercrantz H. Blunted peripheral chemoreceptor response to hyperoxia in a group of infants with bronchopulmonary dysplasia. Pediatr Pulmonol 1995; 20(2):101–106.
32. Samuels MP, Poets CF, Southall DP. Abnormal hypoxemia after life-threatening events in infants born before term. J Pediatr 1994; 125(3):441–446.
33. Marshall JM. Chemoreceptors and cardiovascular control in acute and chronic systemic hypoxia. Braz J Med Biol Res 1998; 31(7):863–888.
34. Katz-Salamon M, Lagercrantz H. Hypoxic ventilatory defence in very preterm infants: attenuation after long term oxygen treatment. Arch Dis Child Fetal Neonatal Ed 1994; 70(2):F90–F95.
35. Carroll JL, Donnelly DF. Postnatal development of chemoreceptor function. In: Loughlin GM, Carroll JL, Marcus CL, eds. Sleep and Breathing in Children: A Developmental Approach. New York: Marcel Dekker, 2000:231–260.
36. Verna A, Roumy M, Leitner LM. Loss of chemoreceptive properties of the rabbit carotid body after destruction of the glomus cells. Brain Res 1975; 100(1):13–23.
37. Buckler KJ, Vaughan-Jones RD. Effects of hypoxia on membrane potential and intracellular calcium in rat neonatal carotid body type I cells. J Physiol 1994; 476(3):423–428.

38. Weiss N, Donnelly DF. Depolarization is a critical event in hypoxia-induced glomus cell secretion. Adv Exp Med Biol 1996; 410:181–187.
39. Shirahata M, Fitzgerald RS. Dependency of hypoxic chemotransduction in cat carotid body on voltage-gated calcium channels. J Appl Physiol 1991; 71(3):1062–1069.
40. Rozanov C, Roy A, Mokashi A, Wilson DF, Lahiri S, Acker H. Chemosensory response to high pCO is blocked by cadmium, a voltage-sensitive calcium channel blocker. Brain Res 1999; 833(1):101–107.
41. Urena J, Fernandez-Chacon R, Benot AR, Alvarez de Toledo GA, Lopez-Barneo J. Hypoxia induces voltage-dependent Ca^{2+} entry and quantal dopamine secretion in carotid body glomus cells. Proc Natl Acad Sci USA 1994; 91(21):10208–10211.
42. Hatton CJ, Peers C. Electrochemical detection of K(+)-evoked quantal secretory events from isolated rat type I carotid body cells. Exp Physiol 1997; 82(2):415–418.
43. Obeso A, Rocher A, Fidone S, Gonzalez C. The role of dihydropyridine-sensitive Ca^{2+} channels in stimulus-evoked catecholamine release from chemoreceptor cells of the carotid body. Neuroscience 1992; 47(2):463–472.
44. Biscoe TJ, Duchen MR. Responses of type I cells dissociated from the rabbit carotid body to hypoxia. J Physiol 1990; 428:39–59.
45. Bamford OS, Sterni LM, Wasicko MJ, Montrose MH, Carroll JL. Postnatal maturation of carotid body and type I cell chemoreception in the rat. Am J Physiol 1999; 276(5 Pt 1):L875–L884.
46. Gauda EB, Lawson EE. Developmental influences on carotid body responses to hypoxia. Respir Physiol 2000; 121(2–3):199–208.
47. Donnelly DF. Developmental aspects of oxygen sensing by the carotid body. J Appl Physiol 2000; 88(6):2296–2301.
48. Kondo H. An electron microscopic study on the development of synapses in the rat carotid body. Neurosci Lett 1976; 3:197–200.
49. Bairam A, Basson H, Marchal F, Cottet-Emard JM, Pequignot JM, Hascoet JM, Bairam A, Basson H, Marchal F, Cottet-Emard JM, Pequignot JM, Hascoet JM, Lahiri S. Effects of hypoxia on carotid body dopamine content and release in developing rabbits. J Appl Physiol 1996; 80(1):20–24.
50. Donnelly DF. Developmental changes in membrane properties of chemoreceptor afferent neurons of the rat petrosal ganglia. J Neurophysiol 1999; 82(1):209–215.
51. Gonzalez C, Vicario I, Almaraz L, Rigual R. Oxygen sensing in the carotid body. Biol Signals 1995; 4(5):245–256.
52. Zapata P. Effects of dopamine on carotid chemo- and baroreceptors in vitro. J Physiol 1975; 244(1):235–251.
53. Hertzberg T, Hellstrom S, Lagercrantz H, Pequignot JM. Development of the arterial chemoreflex and turnover of carotid body catecholamines in the newborn rat. J Physiol 1990; 425:211–225.
54. Holgert H, Hertzberg T, Dagerlind A, Hokfelt T, Lagercrantz H. Neurochemical and molecular biological aspects on the resetting of the arterial chemoreceptors in the newborn rat. Adv Exp Med Biol 1993; 337:165–170.
55. Gauda EB, Bamford O, Gerfen CR. Developmental expression of tyrosine hydroxylase, D2-dopamine receptor and substance P genes in the carotid body of the rat. Neuroscience 1996; 75(3):969–977.
56. Almaraz L, Wang ZZ, Stensaas LJ, Fidone SJ. Release of dopamine from carotid sinus nerve fibers innervating type I cells in the cat carotid body. Biol Signals 1993; 2(1):16–26.

57. Almaraz L, Perez-Garcia MT, Gonzalez C. Presence of D1 receptors in the rabbit carotid body. Neurosci Lett 1991; 132(2):259–262.
58. Bairam A, Kinkead R, Marchal F. Age-dependent effect of domperidone on dopamine release by the hypoxic carotid body in the rabbit. Biol Neonate 2001; 80(3):235–238.
59. Bairam A, Frenette J, Dauphin C, Carroll JL, Khandjian EW. Expression of dopamine D1-receptor mRNA in the carotid body of adult rabbits, cats and rats. Neurosci Res 1998; 31(2):147–154.
60. Bairam A, Dauphin C, Rousseau F, Khandjian EW. Expression of dopamine D2-receptor mRNA isoforms at the peripheral chemoreflex afferent pathway in developing rabbits. Am J Respir Cell Mol Biol 1996; 15(3):374–381.
61. Bairam A, Neji H, De Grandpre P, Carroll JL. Autoreceptor mechanism regulating carotid body dopamine release from adult and 10-day-old rabbits. Respir Physiol 2000; 120(1):27–34.
62. Tomares SM, Bamford OS, Sterni LM, Fitzgerald RS, Carroll JL. Effects of domperidone on neonatal and adult carotid chemoreceptors in the cat. J Appl Physiol 1994; 77(3):1274–1280.
63. Dasso LL, Buckler KJ, Vaughan-Jones RD. Muscarinic and nicotinic receptors raise intracellular Ca^{2+} levels in rat carotid body type I cells. J Physiol 1997; 498 (Pt 2):327–338.
64. McQueen DS, Ribeiro JA. Pharmacological characterization of the receptor involved in chemoexcitation induced by adenosine. Br J Pharmacol 1986; 88(3):615–620.
65. Vandier C, Conway AF, Landauer RC, Kumar P. Presynaptic action of adenosine on a 4-aminopyridine-sensitive current in the rat carotid body. J Physiol 1999; 515 (Pt 2):419–429.
66. Gauda EB, Northington FJ, Linden J, Rosin DL. Differential expression of a(2a), A(1)-adenosine and D(2)-dopamine receptor genes in rat peripheral arterial chemoreceptors during postnatal development. Brain Res 2000; 872(1–2):1–10.
67. Kumar GK, Kou YR, Overholt JL, Prabhakar NR. Involvement of substance P in neutral endopeptidase modulation of carotid body sensory responses to hypoxia. J Appl Physiol 2000; 88(1):195–202.
68. Wasicko MJ, Sterni LM, Bamford OS, Montrose MH, Carroll JL. Resetting and postnatal maturation of oxygen chemosensitivity in rat carotid chemoreceptor cells. J Physiol 1999; 514(Pt 2):493–503.
69. Rumsey WL, Iturriaga R, Spergel D, Lahiri S, Wilson DF. Optical measurements of the dependence of chemoreception on oxygen pressure in the cat carotid body. Am J Physiol 1991; 261(4 Pt 1):C614–C622.
70. Montoro RJ, Urena J, Fernandez-Chacon R, Alvarez dT, Lopez-Barneo J. Oxygen sensing by ion channels and chemotransduction in single glomus cells. J Gen Physiol 1996; 107(1):133–143.
71. Sterni LM, Bamford OS, Tomares SM, Montrose MH, Carroll JL. Developmental changes in intracellular Ca^{2+} response of carotid chemoreceptor cells to hypoxia. Am J Physiol 1995; 268(5 Pt 1):L801–L808.
72. Carpenter E, Bee D, Peers C. Ionic currents in carotid body type I cells isolated from normoxic and chronically hypoxic adult rats. Brain Res 1998; 811(1–2):79–87.
73. Wasicko MJ, Carroll JL. Role of background K^+ conductance in the development of carotid body type I cell O_2 chemosensitivity (abstract). Oxygen Sensing: Molecule to Man—Meeting of the Internation Society of Arterial Chemoreception, Philadelphia, 1999.

74. Pepper DR, Landauer RC, Kumar P. Postnatal development of CO_2-O_2 interaction in the rat carotid body in vitro. J Physiol 1995; 485(Pt 2):531–541.
75. Dasso LL, Buckler KJ, Vaughan-Jones RD. Interactions between hypoxia and hypercapnic acidosis on calcium signaling in carotid body type I cells. Am J Physiol Lung Cell Mol Physiol 2000; 279(1):L36–L42.
76. Oyama Y, Walker JL, Eyzaguirre C. The intracellular chloride activity of glomus cells in the isolated rabbit carotid body. Brain Res 1986; 368(1):167–169.
77. Lopez-Lopez JR, Gonzalez C, Perez-Garcia MT. Properties of ionic currents from isolated adult rat carotid body chemoreceptor cells: effect of hypoxia. J Physiol 1997; 499(Pt 2):429–441.
78. Lopez-Barneo J, Lopez-Lopez JR, Urena J, Gonzalez C. Chemotransduction in the carotid body: K^+ current modulated by PO_2 in type I chemoreceptor cells. Science 1988; 241(4865):580–582.
79. Lopez-Lopez J, Gonzalez C, Urena J, Lopez-Barneo J. Low pO_2 selectively inhibits K channel activity in chemoreceptor cells of the mammalian carotid body. J Gen Physiol 1989; 93(5):1001–1015.
80. Peers C. Hypoxic suppression of K^+ currents in type I carotid body cells: selective effect on the $Ca^{2(+)}$-activated K^+ current. Neurosci Lett 1990; 119(2):253–256.
81. Peers C. Effects of D600 on hypoxic suppression of K^+ currents in isolated type I carotid body cells of the neonatal rat. FEBS Lett 1990; 271(1–2):37–40.
82. Delpiano MA, Hescheler J. Evidence for a PO_2-sensitive K^+ channel in the type-I cell of the rabbit carotid body. FEBS Lett 1989; 249(2):195–198.
83. Stea A, Nurse CA. Whole-cell and perforated-patch recordings from O_2-sensitive rat carotid body cells grown in short- and long-term culture. Pflügers Arch 1991; 418(1–2):93–101.
84. Donnelly DF. Are oxygen dependent K^+ channels essential for carotid body chemotransduction? Respir Physiol 1997; 110(2–3):211–218.
85. Doyle TP, Donnelly DF. Effect of Na^+ and K^+ channel blockade on baseline and anoxia-induced catecholamine release from rat carotid body. J Appl Physiol 1994; 77(6):2606–2611.
86. Cheng PM, Donnelly DF. Relationship between changes of glomus cell current and neural response of rat carotid body. J Neurophysiol 1995; 74(5):2077–2086.
87. Osanai S, Buerk DG, Mokashi A, Chugh DK, Lahiri S. Cat carotid body chemosensory discharge (in vitro) is insensitive to charybdotoxin. Brain Res 1997; 747(2):324–327.
88. Lahiri S, Roy A, Rozanov C, Mokashi A. K^+-current modulated by PO_2 in type I cells in rat carotid body is not a chemosensor. Brain Res 1998; 794(1):162–165.
89. Buckler KJ. A novel oxygen-sensitive potassium current in rat carotid body type I cells. J Physiol 1997; 498(Pt 3):649–662.
90. Pardal R, Ludewig U, Garcia-Hirschfeld J, Lopez-Barneo J. Secretory responses of intact glomus cells in thin slices of rat carotid body to hypoxia and tetraethylammonium. Proc Natl Acad Sci USA 2000; 97(5):2361–2366.
91. Hatton CJ, Carpenter E, Pepper DR, Kumar P, Peers C. Developmental changes in isolated rat type I carotid body cell K^+ currents and their modulation by hypoxia. J Physiol 1997; 501(Pt 1):49–58.
92. Williams BA, Buckler KJ. Identification of an oxygen-sensitive potassium channel in neonatal rat carotid body type I cells. Adv Exp Med Biol 2000; 475:419–424.

93. Overholt JL, Ficker E, Yang T, Shams H, Bright GR, Prabhakar NR. HERG-Like potassium current regulates the resting membrane potential in glomus cells of the rabbit carotid body. J Neurophysiol 2000; 83(3):1150–1157.
94. Pancrazio JJ, Ma W, Grant GM, Shaffer KM, Kao WY, Liu QY, Pancrazio JJ, Ma W, Grant GM, Shaffer KM, Kao WY, Liu QY, Manos P, Barker JL, Stenger DA. A role for inwardly rectifying K^+ channels in differentiation of NG108–15 neuroblastoma x glioma cells. J Neurobiol 1999; 38(4):466–474.
95. Schafer R, Wulfsen I, Behrens S, Weinsberg F, Bauer CK, Schwarz JR. The erg-like potassium current in rat lactotrophs. J Physiol 1999; 518(Pt 2):401–416.
96. Bauer CK, Engeland B, Wulfsen I, Ludwig J, Pongs O, Schwarz JR. RERG is a molecular correlate of the inward-rectifying K current in clonal rat pituitary cells. Receptors Channels 1998; 6(1):19–29.
97. Wyatt CN, Wright C, Bee D, Peers C. O_2-sensitive K^+ currents in carotid body chemoreceptor cells from normoxic and chronically hypoxic rats and their roles in hypoxic chemotransduction. Proc Natl Acad Sci USA 1995; 92(1):295–299.
98. Buckler KJ. Background leak K^+-currents and oxygen sensing in carotid body type 1 cells. Respir Physiol 1999; 115(2):179–187.
99. Tatsumi K, Pickett CK, Weil JV. Decreased carotid body hypoxic sensitivity in chronic hypoxia: role of dopamine. Respir Physiol 1995; 101(1):47–57.
100. Hempleman SC. Increased calcium current in carotid body glomus cells following in vivo acclimatization to chronic hypoxia. J Neurophysiol 1996; 76(3):1880–1886.
101. Rigatto H, Brady JP, Rigatto H, Brady JP, de la Torre Verduzco R. Chemoreceptor reflexes in preterm infants. II. The effect of gestational and postnatal age on the ventilatory response to inhaled carbon dioxide. Pediatrics 1975; 55(5):614–620.
102. Semenza GL. Hypoxia-inducible factor 1: oxygen homeostasis and disease pathophysiology. Trends Mol Med 2001; 7(8):345–350.
103. Semenza GL, Agani F, Iyer N, Kotch L, Laughner E, Leung S, Yu A. Regulation of cardiovascular development and physiology by hypoxia-inducible factor 1. Ann NY Acad Sci 1999; 874:262–268.
104. Prabhakar NR, Fields RD, Baker T, Fletcher EC. Intermittent hypoxia: cell to system. Am J Physiol Lung Cell Mol Physiol 2001; 281(3):L524–L528.
105. Kline DD, Peng YJ, Manalo DJ, Semenza GL, Prabhakar NR. Defective carotid body function and impaired ventilatory responses to chronic hypoxia in mice partially deficient for hypoxia-inducible factor 1alpha. Proc Natl Acad Sci USA 2002; 99(2):821–826.
106. Prabhakar NR, Kumar GK, Chang CH, Agani FH, Haxhiu MA. Nitric oxide in the sensory function of the carotid body. Brain Res 1993; 625(1):16–22.
107. Semenza GL. HIF-1 and mechanisms of hypoxia sensing. Curr Opin Cell Biol 2001; 13(2):167–171.
108. Hanson MA. Role of chemoreceptors in effects of chronic hypoxia. Comp Biochem Physiol A Mol Integr Physiol 1998; 119(3):695–703.

14

Maturation of Chemoreceptor O_2 and CO_2 Sensitivity

PREM KUMAR

The Medical School
University of Birmingham
Birmingham, England

I. Introduction

The maintenance of an adequate tissue oxygen supply requires coordination of a number of local and systemic responses to hypoxia. Hypoxemia can originate naturally with ascent to high altitude. It may also arise through a series of clinical disorders, which include inadequate respiration, as seen, for example, in sleep apnea or chronic obstructive diseases of the lung, but also through an inadequate O_2 supply as might occur in heart failure or stroke or even through inadequate O_2 extraction as in a number of metabolic myopathies or sepsis. The precise value for any tissue bed will depend upon the inspired O_2 tension, the O_2 carriage capacity of the blood, the vascular structure and blood flow of the tissue, as well as upon a specific set of circumstances including the local diffusion conditions and the O_2 consumption of the tissue. While local responses can act to increase O_2 delivery to a point, ultimate survival to systemic hypoxia usually depends primarily upon cardiorespiratory reflexes initiated by arterial hypoxemia. These have as their ultimate aim the establishment of a high, arterial driving pressure of O_2 and have their origins at specialized chemical transducers. While oxygen sensing is increasingly beginning to appear to be a ubiquitous feature of cell physiology (1), with all cells responding in some way to falls in Po_2 once severe enough to compromise cell activity, the systemic chemoreceptors are distinguished from other cell types by their relatively low threshold to arterial hypoxia, often beginning to respond at partial pressures of O_2 just a few mmHg below normal arterial levels, as well as by the fact that the effect of their stimulation is observable at a systemic level. Classically, peripheral chemoreceptors, located in specialized cells within the carotid bodies, the aortic bodies, and the abdominal viscera, as well as pulmonary smooth muscle cells and erythropoeitin-secreting cells of the kidney, have fulfilled these criteria and to these should now perhaps be added the neuroepithelial bodies of the respiratory airways. In recent years, it is becoming clear that chemoreceptor cells can also exhibit a high degree of plasticity, adapting their stimulus response characteristics to match prevailing conditions, whether these occur naturally, pathologically or are artificially applied. This brief review will focus on the former and will discuss the postnatal maturational changes in O_2 sensing observed in the peripheral chemoreceptors of the carotid body and speculate briefly on possible sites for the developmental process. Ultimately, any proposed mechanism of chemotransduction should be able to account for the findings at the level of chemoafferent discharge and for the maturational changes that occur postnatally.

II. Chemical Control of Fetal Breathing

That the fetus makes breathing movements (FBM) in utero has long been known. These are rapid, shallow, irregular movements of the thoracic musculature, including the diaphragm, and originate from central, brainstem neuronal activity (2). While these clearly do not participate in the control of in utero blood gas tensions, they do appear essential for fetal lung growth and development (3). In late gestation, as electrocortical activity (ECoG) becomes differentiated into high- and low-voltage states, FBM become entrained to the same cycle frequency as the ECoG, occurring now only during the low-voltage state, concomitant with rapid eye movements and often a decreased postural tone (4) and being absent, possibly through an active inhibition of respiratory related neural activity (5,6), during the high-voltage state. At this stage of development, hypocapnia abolishes (7) and hypercapnia augments (6) FBM during low-voltage ECoG suggesting possible cortical influences upon central sensitivity to CO_2. In contrast to these essentially adult-like respiratory reflex responses to changes in arterial CO_2 (ECoG state dependency, notwithstanding), in utero hypoxemia in late gestation leads to the rapid and sustained inhibition of FBM (8). This inhibition has its origins in a central neural site, distinct from that mediating the apneas of high-voltage ECoG (9), located through the application of discrete bilateral lesions in and around the vicinity of the rostrolateral pons (10). In these lesioned sheep fetuses, hypoxia now increased FBM in an adult-like fashion and in a carotid chemoreceptor-dependent manner (11), demonstrating that activation of carotid chemoreceptors by hypoxia was occurring in utero but that its effects upon respiratory reflexes were being masked by the greater influence of a central inhibition. Indeed, recording from carotid chemoreceptor afferents in exteriorized fetal sheep had previously demonstrated a slight, but tonic, discharge at the prevalent arterial Po_2 of 25 mmHg that was increased by decreases in oxygen tension and/or by increases in CO_2 tension (12) that led primarily to cardiovascular responses (13).

III. Postnatal Respiratory and Chemoreceptor Responses to Hypoxia

After cord occlusion at birth, breathing in the neonate, appropriately, becomes essentially continuous and independent of cyclic changes in ECoG. The ventilatory response to an acute hypoxic challenge now becomes characteristically biphasic (14) with a rapid-onset (seconds) but short-lasting (1–2 min) increase giving way to a sustained decrease toward or even below control, euoxic levels. As might be expected, the initial, excitatory phase is accepted as being mediated through peripheral chemoreceptor stimulation as confirmed by carotid sinus denervation (15) and by direct electrophysiological recording of chemoreceptor afferents (16). A number of mechanisms have been postulated for the secondary decline in ventilation during hypoxia, ranging from fatigue of respiratory muscles to depression of central neural activity. A simple measurement of respiration in conscious neonatal rats exposed to a mild and then to a more severe hypoxia illustrates that these "failure of the system" mechanisms are not likely to be the main reasons for the secondary decline as the animal is able to sustain a further increase in ventilation despite the greater intensity of the hypoxic stimulus. The most compelling evidence thus seems to point to the persistence of fetal suprapontine inhibitory site(s) (17–19). This decline does not appear to be due to adaptation or failure of peripheral chemoreceptor discharge (20), and indeed there is a suggestion that the maintained input from peripheral chemoreceptors is an essential requirement for this secondary fall in ventilation during hypoxia (21). In addition, there may also be a contribution to the total response from hypoxic hypometabolism (22). With increasing postnatal age, the adult response of a relatively sustained increase in ventilation during hypoxia is acquired through the gradual increase in the initial phase and a gradual diminution in the magnitude of the secondary phase, a maturational process observable in a number of mammalian species studied including rat (23) and human (24), with the rate of maturation appearing somewhat proportional to the degree of maturity at birth.

The maturation of the initial respiratory responses to hypoxia, and, to a lesser extent, to hypercapnia, can be explained by the postnatal development of peripheral chemoreceptor afferent nerve sensitivity. This has been observed most commonly for chemoafferents of the carotid chemoreceptors (25) but is also demonstrable in the aortic chemoafferents (26). This "resetting," as it has been termed, is observable at the level of single-fiber chemoafferents as well as in multifiber preparations and also in in vitro preparations (27) and thus must represent inherent changes in the sensitivity of the chemotransduction process rather than through alterations in the delivery of O_2 to chemosensitive tissue or through less direct mechanisms such as fiber recruitment, with increasing stimulus intensity, although it is clear that this latter effect does occur. Decreases in Po_2, in vivo, at all stages of development give rise to increases in afferent activity that follow characteristic curves best described by single exponentials with offset or, in the case of the aortic chemoreceptors, by hyperbolic functions. The effect of maturation is to shift these response curves rightward and upward, an effect that not only increases baseline chemoreceptor activity but also increases hypoxic sensitivity at all levels of Po_2. Indeed, if Po_2 is decreased isocapnically in lambs older than 21 days, chemoafferent discharge can begin to fail at levels of arterial Po_2 around 30–40 mmHg, a level at which discharge in 3–5-day-old lambs is essentially just beginning to increase (Fig. 1). Given the almost fourfold postnatal increase in arterial Po_2 from approximately 25 mmHg in utero toward 100 mmHg in the adult that occurs over a number of days from the moment of air breathing at birth (28), the effect of resetting will be to reposition maximal O_2 sensitivity toward the prevalent arterial levels.

IV. CO_2/O_2 Interaction

In contrast to the relatively complex postnatal maturation of the ventilatory responses to hypoxia and given the similarity of the fetal response to hypercapnia to that of the adult, namely a sustained increase in breathing movements (6), as well as the linearity of the CO_2 stimulus response curve, the postnatal maturation of responses to hypercapnia might appear, initially, a more straightforward process. It is

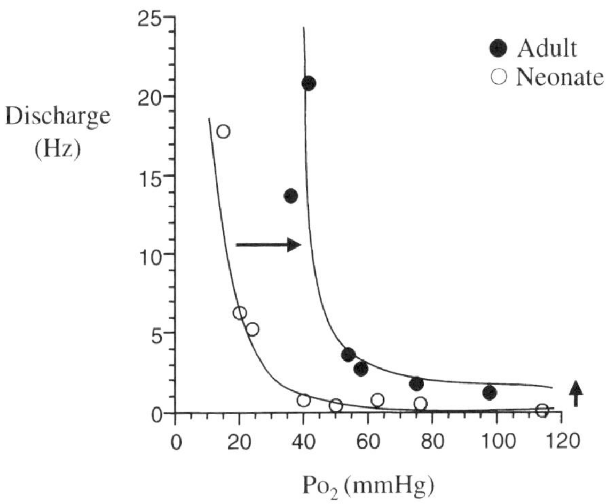

Figure 1 Single-fiber chemoreceptor afferent discharge recorded from in vivo carotid body preparations of a neonatal (3-day-old) and adult (4-week-old) lamb during steady-state decreases in isocapnic arterial levels of Po_2. Data points are fitted with single exponentials with offset. With increasing postnatal age, O_2 sensitivity increases as seen by a rightward and upward shift in the position of the response curves (arrows). Note the discharge at the lowest Po_2 in the adult animal is beginning to decrease at a hypoxia level at which, in the younger animal, discharge is essentially beginning to increase from baseline levels.

obvious, though, that any blood gas changes in Pco_2 must occur in a background of Po_2 and as, in the adult, these two stimuli show a greater-than-additive interaction at the level of single-fiber afferents (29) and ventilation (30), than the complexities of the Po_2 maturational process must impact upon the maturation to CO_2. Given the additional complicating effects of central and peripheral responses with different delays and gains, it is perhaps not surprising that reflex studies have proved unequivocal. Thus, a developmental increase in peripheral CO_2 sensitivity has been reported (31,32) but was not confirmed in newborn pigs (33), and when considering the effect of hypoxia upon CO_2 sensitivity in the newborn, the interaction has been shown to be either positive (33,34), as in the adult, or even negative, with hypoxia decreasing CO_2 sensitivity (35,36). More conclusively, direct recordings in an in vitro rat carotid body preparation demonstrated the clear impact of hypoxia upon CO_2 sensitivity at the peripheral chemoreceptor (37,37a). Thus, as expected, in adult animals, increasing hypoxia increased peripheral CO_2 chemoafferent sensitivity, while in neonatal animals, the linear, steady-state chemoafferent stimulus response curve to CO_2, although increasingly offset upward by increasing hypoxia, showed no measurable change in the slope, or sensitivity, of the curve. In other words, in the absence of the multiplying stimulus of hypoxia, the sensitivity to CO_2 remains at its hyperoxic level. The impact of resetting of hypoxia sensitivity thus converts an additive steady-state CO_2–O_2 interaction into a multiplicative one (Fig. 2).

Unlike the simply proportional response to hypoxia, however, the adult carotid chemoreceptor response to a rapid change in CO_2 also exhibits a dynamic sensitivity with the initial response giving way after a few seconds to a lower, adapted steady state (38). In the adult, this dynamic sensitivity was shown to be independent of Po_2 (39), a finding confirmed in newborn lambs by direct recordings of chemoafferent discharge responses to alternating hypercapnic stimuli in vivo (40). Thus, with the steady-state and dynamic sensitivities to CO_2 being Po_2 dependent and independent, respectively, it follows that the degree of adaptation must be Po_2 dependent (Fig. 3). An early hypothesis to account for this Po_2-dependent adaptive behavior (39,41) has not been verified experimentally in full, but this area may prove an interesting avenue for future studies.

Interestingly, the process of resetting, which acts to increase Po_2 sensitivity of peripheral chemoreceptors, is itself an O_2-dependent process. Maintaining newborn animals in an hypoxic environment from birth acts to greatly reduce the normal increase in arterial Po_2 levels and thus maintains it, postnatally, at a more fetal level. In these animals it has been shown that the fetal phenotype of a relatively reduced O_2 sensitivity is maintained at the level of both chemoafferent discharge (42) and ventilation (43,44). As expected, this retention of the fetal response to hypoxia also prevented the normal increase in CO_2–O_2 interaction (42), an effect we later demonstrated (45) to be both reversible upon the return to normal inspiratory Po_2 levels and age-dependent, with animals from 3 weeks of age being immune to the blunting effects of chronic hypoxemia (Fig. 2). Certainly, the effect of chronic hypoxia must be viewed in terms not only of its severity and duration but also in terms of the maturational age at which it is applied with a weighting biased to the early neonatal period (46). Given the increased neonatal morbidity associated with

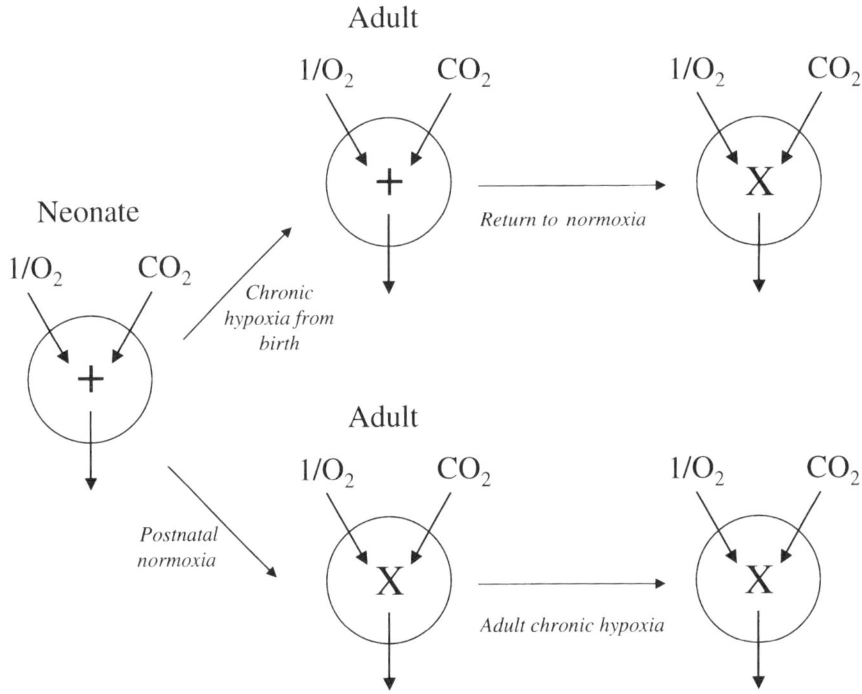

Figure 2 Schematic representation of the maturation of CO_2 and hypoxia ($1/O_2$) sensitivities in the carotid body. The additive (+) interaction of the newborn becomes multiplicative (×) in the adult when it is raised in normoxia from birth but is blunted and remains additive when chronic hypoxia from birth (FiO_2 0.12) is imposed. This blunting is reversible upon a return to normoxia, and adult animals with a developed multiplicative interaction between CO_2 and hypoxia do not revert to the neonatal phenotype when chronic hypoxia is imposed in adulthood.

sinus nerve denervation (47–49), it is tempting to speculate that peripheral chemoreceptor afferent information is of greater importance in determining ventilation in early postnatal life than in adulthood and a failure to reset might therefore play some role in the etiology of sudden infant death where postmortem studies have revealed some changes in carotid body morphology and chemistry (50,51). Certainly, infants exhibiting chronic hypoxemia as a consequence of bronchopulmonary dysplasia demonstrated a much reduced sensitivity to an applied hypoxia challenge than age-matched controls (52). More recently, plasticity in chemosensitivity has been examined with the use of intermittent hypoxia as a model of more pathological situations such as recurrent apnea (53), and these studies illustrate that intermittent hypoxia, like sustained hypoxia, can also augment hypoxia sensitivity, although this may be via a separate mechanism involving reactive O_2 species, as in intermittent hypoxia, unlike chronic hypoxia, there will always be a regular increase in Po_2 levels.

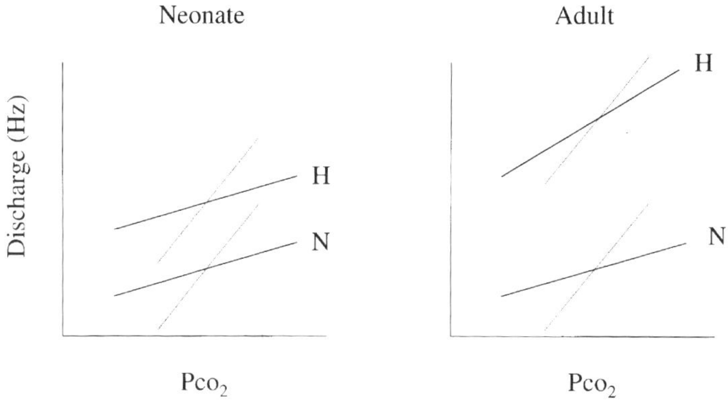

Figure 3 Schematic representation of the effect of hypoxia upon CO_2 chemosensitivity in the carotid body. In the neonate, CO_2–O_2 interaction is additive and the steady-state CO_2 sensitivity (solid lines) is thus independent of O_2 and is shown as parallel lines in both normoxia (N) and hypoxia (H). Dynamic sensitivity (dashed lines) is steeper than the steady sensitivity and is also independent of O_2 at this age and therefore the degree of adaptation (the vertical distance between the dynamic and steady-state responses) is also independent of O_2. In the adult, the development of a multiplicative interaction between CO_2 and O_2, with the maintained O_2 independence of the dynamic sensitivity, gives rise at this age to an O_2-dependent adaptation, with the degree of adaptation decreasing proportionally with increasing hypoxia.

V. Mechanisms

Clearly, any mechanism invoked to account for the sensitivity changes to blood gas stimuli during the postnatal period must be inherently tied to the chemotransduction process. The number of studies examining the transduction process in adult tissue (or, more recently, in model cell lines) greatly exceeds the number of studies designed specifically to examine developmental processes, yet, as mentioned earlier, the latter could certainly inform the former. Even fewer studies have been designed to examine the maturation of CO_2–O_2 interaction. Despite some early controversies (54), there now appears to be a consensus developing that places the inhibition of K channels located in the type I cell of the carotid body as central to the hypoxia and hypercapnia transduction processes. These channels must be involved in setting the resting membrane potential but may, or, perhaps more likely, may not be, the actual sensors of the hypoxia/hypercapnic stimuli and thus must be linked through a direct mechanism to the actual sensor or via an intracellular signaling pathway. A number of heme proteins have been suggested as sensors of chemoreceptor cell hypoxia (55) with signaling to the channel occurring perhaps through a redox-sensitive, intracellular mechanism that may not be NADPH oxidase (56) but may be some other associated heme protein (57). Of course, cytochromes associated with mitochondrial respiration remain an attractive alternative hypothesis for O_2 sensing. Inhibitors of mitochondrial oxidative phosphorylation, including cyanide, have long been known to act as powerful carotid body chemostimulants (58,59), but less intense stimulation with graded hypoxia also reveals a unique high sensitivity to Po_2 in carotid body mitochondria (60,61). With regard to CO_2–O_2 interaction, the potential for allosteric modification of such a heme protein by changes in intracellular pH is an attractive proposition that has its roots in ideas expressed many years previously (62), but no tenable hypothesis has been proposed to account for acid sensing by mitochondria.

K-channel inhibition should be of significant magnitude to induce membrane depolarization and of sufficient amplitude to initiate voltage-gated Ca^{2+} entry and, subsequently, neurosecretion of transmitter substance(s) to cause afferent nerve terminal depolarization and spike generation in the sinus nerve. This view, although popular, is not yet fully established and alternative hypotheses involving perhaps a transducer role for the afferent nerve terminal itself or perhaps even electrical synaptic connections cannot yet be ruled out. Clearly, according to the more conventional model of transduction, resetting could occur at one or more sites, and maturational processes linked in time with changes in chemosensitivity have been described at the K-channel, intracellular-Ca^{2+}, and neurosecretion level.

There appears to be a clear species difference in the nature of the voltage-gated, K channel inhibited by hypoxia (63). In the rat, a large-conductance, Ca^{2+}-dependent K channel (BK) was shown to exhibit a specific maturation in its density in isolated type I cells over a period of postnatal time at which resetting occurs, and this was concomitant with an increase in the hypoxia sensitivity of the whole-cell current (64). No change in the density of the whole-cell Ca^{2+} current was detected over this period. In addition, cells harvested from animals raised in chronic hypoxia showed a decrease in the density of these BK channels (65). While this correlation with resetting and the effect of chronic hypoxia is interesting, it is certainly not definitive, and there also remains the concern that this particular current may not be sufficiently activated at the resting membrane potential in normoxia, at least in isolated cells. This seemed to be corroborated by the finding that block of this channel by the venom charybdotoxin fails to increase carotid body chemoafferent discharge in normoxia when applied to the whole organ and yet increased discharge in hypoxia (66,66a). In adult type I cells, while a clear inhibition of the BK channel by either hypoxia or acidosis was evident, there was no evidence for a greater-than-additive response to these two stimuli when given together (67). The presence of a background, or leak, voltage-insensitive K current belonging to the TASK family of tandem pore domain K channels (68,69) that is activated at the resting membrane potential might prove a more interesting avenue for study, but no full published studies have yet demonstrated a similar increase in the density of this channel type with increasing postnatal age. A new complexity arises, however, with the finding that thin-cut slices of the carotid body do exhibit voltage-sensitive K channels that can be blocked pharmacologically in normoxia (70) and may thus cast some doubt upon the use of results obtained from isolated type I cells as representative of the whole organ response. No developmental studies have yet been performed using this novel preparation.

Relatively few studies have looked at the maturational responses to hypoxia or hypercapnia in terms of the Ca^{2+} response. The most definitive of these (71,72) show a clear increase in the maximal amplitude of the intracellular Ca^{2+} signal with age as well as an increased sensitivity to hypoxia such that the hypoxia stimulus-response curve was shifted rightward in an exact analogous way to that described for afferent nerve resetting. As the only study to examine the combined effect of hypoxia and acidosis upon intracellular Ca^{2+} had previously demonstrated, at one relatively mature postnatal age, a potentiation of the amplitude of the response to hypoxia by acidosis (73), it may be that the maturational effect in some way reflected an increasing interaction between the two stimuli. Given, however, that no change in Ca^{2+}-channel density had been found over this age period (64) and that Ca^{2+} entry subsequent to nonspecific cell depolarization by an increased extracellular K^+ concentration showed no maturational increase (71), it follows that the maturational process of resetting may be located at a prior point to the intracellular Ca^{2+} signal in the chemotransduction cascade, perhaps at the site of O_2 sensing.

Interesting, on this latter point, are the data recently obtained using the mitochondrial inhibitor carbon monoxide (CO). At high partial pressures relative to oxygen, CO decreases carotid body O_2 consumption and acts like hypoxia to increase adult carotid body chemoafferent neural discharge through a Ca^{2+}-dependent and photolabile action upon mitochondrial cytochromes (74–80). CO can also act like hypoxia to increase carotid chemoafferent CO_2 sensitivity (81,81a) in the adult carotid body. Using perforated patch-clamp techniques to make whole-cell recordings of isolated rat type I cells, we have shown that high tensions of CO could induce a significant depolarization and a decrease in the amplitude of an outward current around and above the resting membrane potential. These effects were not present in conventional whole-cell preparations and were, surprisingly, not photolabile. As they were present also when a voltage-dependent K "blocking solution" containing TEA, 4-AP, Ni^{2+}, and zero extracellular Ca^{2+} was used, this strongly suggested an effect upon a background, leak conductance that requires an intracellular mediator acting between mitochondrion and cell membrane in a similar fashion to that previously described for other mitochondrial uncouplers and inhibitors (82,83). These effects may well account for the nonphotolabile effect of CO upon the release of dopamine from type I cells (84). Interesting, with respect to maturational processes, was the finding that high tensions of CO did not increase baseline chemodischarge in neonatal (<4 days old) rat carotid bodies (85), an age when hypoxia, but not hypercapnic, sensitivity is also low (Fig. 4). Given that a similar CO stimulus intensity in adult animals was sufficient to increase baseline discharge by more than 10-fold, it may well be, therefore, that in immature neonatal rats either the O_2 sensor is not yet fully mature and/or the signaling pathway between mitochondrion and plasma membrane is not yet established.

The potential for maturational changes in the sensitivity of a putative O_2 sensor does not, of course, exclude the possibility of other maturational changes occurring at sites more distal to the sensor. Indeed, a significant body of evidence has been gathered to demonstrate a postnatal elevation in catecholamine, specifically dopamine, mRNA levels in carotid body and petrosal ganglion tissue (86,87). Additionally, the amount of catecholamine released with hypoxia (88,89) but not with elevated extracellular K^+ concentration (89) increases with postnatal age,

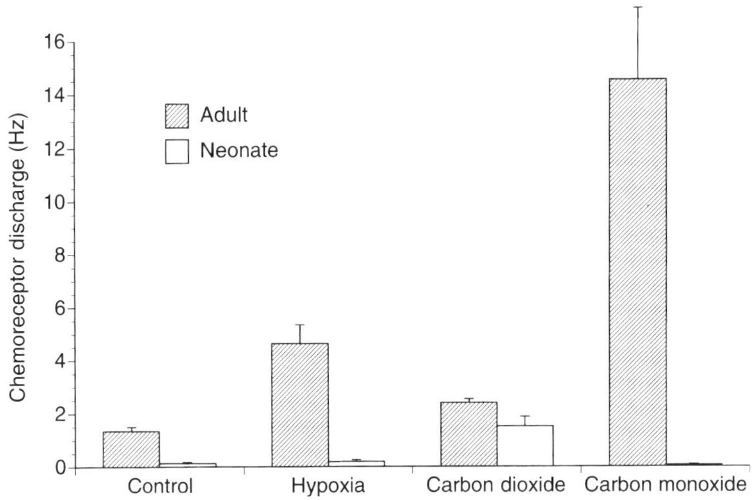

Figure 4 Carbon monoxide at high tensions does not stimulate carotid chemoafferents recorded from neonatal (<4 days) rat carotid body preparations in vitro. Data shown are steady mean ± SEM chemodischarge recorded from eight adult (dashed bars) and five neonatal (open bars) single-fiber chemoafferents in response to control ($Po_2 > 400$ mmHg, Pco_2 40 mmHg), hypoxia ($Po_2 < 100$ mmHg, Pco_2 40 mmHg), hypercapnia ($Po_2 >$ 400 mmHg, Pco_2 80 mmHg), and carbon monoxide (Po_2 160 mmHg, Pco_2 40 mmHg, Pco 320 mmHg) superfusate. Discharge is significantly increased by all three stimuli in adult but only by CO_2 in neonatal preparations.

which, again, tends to localize the key maturational process prior to membrane depolarization.

The data regarding catecholamines and resetting are however, complicated by the observation that dopamine levels in the carotid body might actually be inversely related to the hypoxia sensitivity (90,91). The role of neurotransmitters, other than dopamine, upon resetting warrants further investigation. Of these possible other transmitters, the purinoceptor agonist ATP seems worthy of further examination, given the demonstrated ability of ATP-receptor antagonists to block hypoxia sensitivity in the carotid body (92,93). We have recently shown that the large, photolabile component of CO-mediated chemoexcitation in the adult appears to be generated through a direct action of CO at the afferent nerve terminal. This excitation could be blocked by the P_2-receptor antagonist pyridoxalphosphate-6-azophenyl-2′,4′-disulfonic acid (PPADS), and it would seem, therefore, that the action of CO at the afferent nerve terminal may be mediated through the release of ATP, which may then act upon receptors on the afferent nerve as well as upon purinoceptors on the type I cell to perhaps amplify its effect. We suggest that the source of this ATP is the nerve terminal itself and that release is subsequent to photolabile block of mitochondrial electron transport. The relatively high density of mitochondria located in these nerve terminals and their proximity to the plasma membrane (94), as well as the presence there of transmitter-like substances (95), adds further circumstantial evidence to this hypothesis. To what extent these findings might implicate a dual site of action for hypoxia itself remains to be resolved, as does the precise nature of the transmitter(s) and their mechanism of action, but these data do provide additional evidence for a role for the afferent nerve in chemotransduction and therefore possibly in the maturation of sensitivity to hypoxia and/or hypercapnia. Adding further interest to this idea is our observation that ATP can augment in vitro CO_2 afferent nerve chemosensitivity (93), an effect observed also at the level of respiration (96,97). If these data are looked at in a different way, it could be stated that elevations in CO_2 (i.e., decreased pH) can augment ATP sensitivity, and when this is measured we find that a fall in external pH from 7.38 to around 7.2 can increase afferent nerve sensitivity to 100 μM ATP by three- to fourfold, an effect previously described using recombinant purinoceptors in an oocyte expression system (98). Could modulation of ATP receptor sensitivity by pH underlie the maturation of CO_2–O_2 interaction? To date, the ATP sensitivity of neonatal carotid bodies has not been measured, but these data should prove of great interest to many.

The phenomenon of resetting is well established and is a clear demonstration of the plasticity in function inherent in this organ. Yet, clearly, there remains much unknown regarding the precise mechanisms that might underlie the postnatal maturation in hypoxic and hypercapnic sensitivities of the carotid body and their interaction. Many features of the chemotransduction process undergo postnatal changes, and the task is to ascertain which of these are primary and which secondary to the development of an increased sensitivity. Our understanding of this natural process must progress with our understanding of the chemotransduction process itself. This understanding will have clinical impact as well as being of fundamental importance. While the type I cell has reached prominence as the site of chemotransduction in recent years, it is apparent that this might be an oversimplification and new approaches may yet need to be designed before the story is complete.

References

1. Bunn HF, Poyton RO. Oxygen sensing and molecular adaptation to hypoxia. Physiol Rev 1996; 76:839–885.
2. Bahoric A, Chernick V. Electrical activity of phrenic nerve and diaphragm in utero. J Appl Physiol 1975; 39:513–518.
3. Wigglesworth JS, Desai R. Effect on lung growth of cervical cord section in the rabbit fetus. Early Hum Dev 1979; 3:51–65.
4. Dawes GS, Fox GE, Leduc MB, Liggins GC, Richards RT. Respiratory movements and REM sleep in the fetal lamb. J Physiol 1972; 220:119–143.
5. Dawes GS, Gardner WN, Johnston BM, Walker DW. Breathing in fetal lambs: the effect of brain stem section. J Physiol 1983; 335:535–553.
6. Johnston BM, Gluckman PD. Lateral pontine lesions affect central chemosensitivity in unanaesthetised fetal lambs. J Appl Physiol 1989; 67:1113–1118.
7. Kuipers I, Maertzdorf W, De Jong D, Hanson M, Blanco C. Effects of mild hypocapnia on fetal breathing and behaviour in unanaesthetised, normoxic fetal lambs. J Appl Physiol 1994; 76:1476–1480.

8. Boddy K, Dawes GS, Fisher R, Pinter S, Robinson JS. Foetal respiratory movements: electrocortical and cardiovascular responses to hypoxaemia and hypercapnia in sheep. J Physiol 1974; 243:599–618.
9. Dawes GS. The central control of fetal breathing and skeletal muscle movements. J Physiol 1984; 346:1–18.
10. Gluckman PD, Johnson BM. Lesions in the upper lateral pons abolish the hypoxic depression of breathing in unanaesthetised fetal lambs in utero. J Physiol 1987; 382:373–383.
11. Johnston BM, Gluckman PD. Peripheral chemoreceptors respond to hypoxia in pontine-lesioned fetal lambs in-utero. J Appl Physiol 1993; 75:1027–1034.
12. Blanco CE, Dawes GS, Hanson MA, McCooke HB. The response to hypoxia of arterial chemoreceptors in fetal sheep and newborn lambs. J Physiol 1984; 351:25–37.
13. Guissani D, Spencer J, Moore P, Bennet L, Hanson M. Afferent and efferent components of the cardiovascular reflex responses to acute hypoxia in term fetal sheep. J Physiol 1993; 461:431–449.
14. Cross KW, Oppe TE. The effect of inhalation of high and low concentrations of oxygen on the respiration of the newborn infant. J Physiol 1952; 117:38–55.
15. Schwieler GH. Respiratory regulation during postnatal development in cats and rabbits and some of its morphological substrates. Acta Physiol Scand Suppl 1968; 304.
16. Rigatto H. Control of ventilation in the newborn. Annu Rev Physiol 1984; 46:661–667.
17. Ackland GL, Noble R, Hanson MA. Red nucleus inhibits breathing during hypoxia in neonates. Respir Physiol 1997; 110:251–260.
18. Moore PJ, Ackland GL, Hanson MA. Unilateral cooling in the region of locus coeruleus blocks the fall in respiratory output during hypoxia in anaesthetised neonatal sheep. Exp Physiol 1996; 81:983–994.
19. Martin-Body RL. Brain transections demonstrate the central origin of hypoxic ventilatory depression in carotid body-denervated rats. J Physiol Lond 1988; 407:41–52.
20. Kumar P, Thomas T, Elnazir B. The reflex ventilatory responses of conscious, newborn rats to alternations of inspiratory oxygen concentration. J Dev Physiol 1992; 17:109–118.
21. Schramm CM, Grunstein MM. Respiratory influence of peripheral chemoreceptor stimulation in maturing rabbits. J Appl Physiol 1987; 63:1671–1680.
22. Frappell P, Saiki C, Mortola JP. Metabolism during normoxia, hypoxia and recovery in the newborn kitten. Respir Physiol 1991; 86:115–124.
23. Eden GJ, Hanson MA. Maturation of the respiratory response to acute hypoxia in the newborn rat. J Physiol 1987; 392:1–9.
24. Williams BA, Smyth J, Boon AW, Hanson MA, Kumar P, Blanco CE. Development of respiratory chemoreflexes in response to alternations of fractional inspired oxygen in the newborn-infant. J Physiol Lond 1991; 442:81–90.
25. Blanco CE, Dawes GS, Hanson MA, McCooke HB. The response to hypoxia of arterial chemoreceptors in fetal sheep and newborn lambs. J Physiol 1984; 351:25–37.
26. Kumar P, Hanson MA. Re-setting of the hypoxic sensitivity of aortic chemoreceptors in the newborn lamb. J Dev Physiol 1989; 11:199–206.
27. Kholwadwala D, Donnelly DF. Maturation of carotid chemoreceptor sensitivity to hypoxia—in vitro studies in the newborn rat. J Physiol 1992; 453:461–473.
28. Koch G, Wendel H. Adjustment of arterial blood gases and acid balance in the normal newborn infant during the first week of life. Biol Neonate 1968; 12:136–161.

29. Lahiri S, DeLaney RG. Stimulus interaction in the responses of carotid body chemoreceptor single afferent fibers. Respir Physiol 1975; 24:249–266.
30. Nielsen M, Smith H. Studies on the regulation of respiration in acute hypoxia. Acta Physiol Scand 1952; 24:293–313.
31. Guthrie RD, Standaert TA, Hodson WA, Woodrum DE. Sleep and maturation of eucapnic ventilation and CO_2 sensitivity in the premature primate. J Appl Physiol 1980; 48:347–354.
32. Rigatto H, Brady JP, Verduzco RT. Chemoreceptor reflexes in preterm infants: I. The effect of gestational and postnatal age on the ventilatory response to inhalation of 100% and 15% oxygen. Pediatrics 1975; 55:604–613.
33. Wolsink JG, Berkenbosch A, Degoede J, Olievier CN. The effects of hypoxia on the ventilatory response to sudden changes in CO_2 in newborn piglets. J Physiol 1992; 456:39–48.
34. Albershelm S, Boychuk R, Seshia MMK, Cates D, Rigatto H. Effects of CO_2 on immediate ventilatory response to O_2 in preterm infants. J Appl Physiol 1976; 41:609–611.
35. Rigatto H, De La Torre Verduzco R, Cates DB. Effects of O_2 on the ventilatory response to CO_2 in preterm infants. J Appl Physiol 1975; 39:896–899.
36. Guthrie R, LaFramboise WA, Standaert TA, VanBelle G, Woodrum DE. Ventilatory interaction between oxygen and carbon dioxide in the preterm primate. Pediatr Res 1985; 19:528–533.
37. Pepper DR, Landauer RC, Kumar P. Postnatal development of CO_2-O_2 interaction in the rat carotid body in vitro. J Physiol 1995; 485:531–541.
37a. Roy A, Rozanov C, Mokashi A, Lahiri S. Po_2-Pco_2 stimulus interaction in $[Ca^{2+}]_I$ and CSN activity in the adult rat carotid body. Respir Physiol 2000; 122:15–26.
38. Black AMS, McCloskey DI, Torrance RW. The responses of carotid body chemoreceptors in the cat to sudden changes of hypercapnic and hypoxic stimuli. Respir Physiol 1971; 13:36–49.
39. Torrance RW, Bartels EM, McLaren A. Update of the bicarbonate hypothesis. In: Data PG, Acker H, Lahiri S, eds. Neurobiology and Cell Biology of Chemoreception: Advances in Experimental Medicine and Biology. Vol. 337. New York: Plenum Press, 1993:241–250.
40. Calder NA, Kumar P, Hanson MA. Development of carotid chemoreceptor dynamic and steady-state sensitivity to CO_2 in the newborn lamb. J Physiol 1997; 503:187–194.
41. Hanson MA, Nye PCG, Torrance RW. The exodus of an extracellular bicarbonate theory of chemoreception and the genesis of an intracellular one. In: Belmonte C, Pallot DJ, Acker H, Fidone S, eds. Arterial Chemoreceptors. Leciester: Leicester University Press, 1981:403–416.
42. Landauer RC, Pepper DR, Kumar P. Effect of chronic hypoxaemia from birth upon chemosensitivity in the adult rat carotid body in vitro. J Physiol 1995; 485:543–550.
43. Eden GJ, Hanson MA. Effects of chronic hypoxia from birth in the ventilatory response to acute hypoxia in the newborn rat. J Physiol 1987; 392:11–19.
44. Hanson MA, Kumar P, Williams BA. The effect of chronic hypoxia upon the development of respiratory chemoreflexes in the newborn kitten. J Physiol 1989; 411:563–574.
45. Landauer RC, Conway AF, Pepper DR, Kumar P. Age-dependent and reversible effects of chronic hypoxia upon carotid body CO_2–O_2 interaction. J Physiol 1996; 479.P:25–26P.

46. Okuba S, Mortola JP. Long-term respiratory effects of neonatal hypoxia in the rat. J Appl Physiol 1986; 64:952–958.
47. Hofer MA. Lethal respiratory disturbance in neonatal rats after arterial chemoreceptor denervation. Life Sci 1984; 34:489–496.
48. Donnelly DF, Haddad GG. Prolonged apnea and impaired survival in piglets after sinus and aortic nerve section. J Appl Physiol 1990; 68:1048–1052.
49. Bureau MA, Lamarche J, Foulon P, Dalle D. The ventilatory response to hypoxia in the newborn lamb after carotid-body denervation. Respir Physiol 1985; 60:109–119.
50. Naeye RL, Fisher R, Ryser M, Whalen P. Carotid body in sudden infant death syndrome. Science 1976; 191:567–569.
51. Perrin DG, Becker LE, Madapallimatum A, Cutz E, Bryan AC, Sole MJ. Sudden infant death syndrome—increased carotid-body dopamine and noradrenaline content. Lancet 1984; 2:535–537.
52. Calder NA, Williams BA, Smyth J, Boon AW, Kumar P, Hanson MA. Absence of ventilatory responses to alternating breaths of mild hypoxia and air in infants who have had bronchopulmonary dysplasia—implications for the risk of sudden infant death. Pediatr Res 1994; 35:677–681.
53. Prabhakar NR. Physiological and genomic consequences of intermittent hypoxia—Invited review: Oxygen sensing during intermittent hypoxia: cellular and molecular mechanisms. J Appl Physiol 2001; 90:1986–1994.
54. Donnelly DF. Are oxygen dependent K^+ channels essential for carotid body chemotransduction? Respir Physiol 1997; 110:211–218.
55. Prabhakar NR. Oxygen sensing by the carotid body chemoreceptors. J Appl Physiol 2000; 88:2287–2295.
56. He L, Chen J, Dinger B, Sanders K, Sundar K, Hoidal J, Fidone S. Characteristics of carotid body chemosensitivity in NADPH oxidase-deficient mice. Am J Physiol 2002; 282:C27–C33.
57. Riesco-Fagundo AM, Perez-Garcia MT, Gonzalez C, Lopez-Lopez JR. O_2 modulates large-conductance Ca^{2+}-dependent K^+ channels of rat chemoreceptor cells by a membrane-restricted and CO-sensitive mechanism. Circ Res 2001; 89:430–436.
58. Mulligan E, Lahiri S, Storey BT. Carotid-body O_2 chemoreception and mitochondrial oxidative-phosphorylation. J Appl Physiol 1981; 51:438–446.
59. Anichkov SV, Belen'kii ML. Pharmacology of the Carotid Body Chemoreceptors. New York: Macmillan, 1963.
60. Duchen MR, Biscoe TJ. Mitochondrial-function in type-i cells isolated from rabbit arterial chemoreceptors. J Physiol Lond 1992; 450:13–31.
61. Duchen MR, Biscoe TJ. Relative mitochondrial-membrane potential and $[Ca^{2+}]_i$ in type-i cells isolated from the rabbit carotid-body. J Physiol Lond 1992; 450:33–61.
62. Lloyd BB. Discussion. In: Cunningham DJC, Lloyd BB, eds. The Regulation of Human Respiration. Oxford: Blackwell, 1963:360.
63. Shirahata M, Sham J. Roles of ion channels in carotid body chemotransmission of acute hypoxia. Jpn J Physiol 1999; 49:213–228.
64. Hatton CJ, Carpenter E, Pepper DR, Kumar P, Peers C. Developmental changes in isolated rat type I carotid body cell K^+ currents and their modulation by hypoxia. J Physiol 1997; 501:49–58.
65. Wyatt CN, Wright C, Bee D, Peers C. O_2-sensitive k^+ currents in carotid-body chemoreceptor cells from normoxic and chronically hypoxic rats and their roles in hypoxic chemotransduction. Proc Natl Acad Sci USA 1995; 92:295–299.

66. Pepper DR, Landauer RC, Kumar P. Effect of charybdotoxin on hypoxic chemosensitivity in the adult rat carotid-body, in vitro. J Physiol 1995; 487P:177P–178P.
66a. Osanai S, Buerk DG, Mokashi A, Chugh DK, Lahiri S. Cat carotid body chemosensory discharge (in vitro) is insensitive to charybdotoxin. Brain Res 1997; 747:324–327.
67. Pepper DR, Kumar P. Inhibition of adult rat carotid body type I cell K^+ currents by combined hypoxic and acidotic stimuli. J Physiol 1997:P202–P203.
68. Buckler KJ. A novel oxygen-sensitive potassium current in rat carotid body type I cells. J Physiol 1997; 498:649–662.
69. Buckler KJ, Williams BA, Honore E. An oxygen-, acid- and anaesthetic-sensitive TASK-like background potassium channel in rat arterial chemoreceptor cells. J Physiol 2000; 525:135–142.
70. Pardal R, Ludewig U, Garcia-Hirschfeld J, Lopez-Barneo J. Secretory responses of intact glomus cells in thin slices of rat carotid body to hypoxia and tetraethylammonium. Proc Natl Acad Sci USA 2000; 97:2361–2366.
71. Wasicko MJ, Sterni LM, Bamford OS, Montrose MH, Carroll JL. Resetting and postnatal maturation of oxygen chemosensitivity in rat carotid chemoreceptor cells. J Physiol Lond 1999; 514:493–503.
72. Sterni LM, Bamford OS, Tomares SM, Montrose MH, Carroll JL. Developmental changes in intracellular Ca^{2+} response of carotid chemoreceptor cells to hypoxia. Am J Physiol Lung Cell Mole Physiol 1995; 268:L801–L808.
73. Dasso LLT, Buckler KJ, Vaughn-Jones RD. Interactions between hypoxia and hypercapnic acidosis on calcium signaling in carotid body type I cells. Am J Physiol Lung Cell Mole Physiol 2000; 279:L36–L42.
74. Joels N, Neil E. The idea of a sensory transmitter. In: Torrance RW, ed. Arterial Chemoreceptors. Oxford: Blackwell Scientific Publications, 1968:153–178.
75. Joels N, Neil E. The action of high tensions of carbon monoxide on the carotid chemoreceptors. Arch Int Pharmacodyn Ther 1962; 139:528–534.
76. Lahiri S, Buerk DG, Chugh D, Osanai S, Mokashi A. Reciprocal photolabile O_2 consumption and chemoreceptor excitation by carbon-monoxide in the cat carotid-body—evidence for cytochrome a(3) as the primary O_2 sensor. Brain Res 1995; 684:194–200.
77. Lahiri S, Buerk DG, Mokashi A, Chugh DK, Osanai S. Respiratory cytochrome-oxidase vs nonrespiratory heme protein as O_2 sensor in the cat carotid-body. FASEB J 1995:A431–A431.
78. Wilson DF, Mokashi A, Chugh D, Vinogradov S, Osanai S, Lahiri S. The primary oxygen sensor of the cat carotid-body is cytochrome alpha(3) of the mitochondrial respiratory-chain. FEBS Lett 1994; 351:370–374.
79. Mills E, Jobsis FF. Mitochondrial respiratory chain of carotid body and chemoreceptor response to changes in oxygen tension. J Neurophysiol 1972; 35:405–428.
80. Rozanov C, Roy A, Mokashi A, Wilson DF, Lahiri S, Acker H. Chemosensory response to high pCO is blocked by cadmium, a voltage-sensitive calcium channel blocker. Brain Res 1999; 833:101–107.
81. Conway AF, Pepper DR, Kumar P. Dose-dependent modulation of carotid body CO_2–O_2 interaction by carbon monoxide. J Physiol Lond 1997; P504:P201–P202.
81a. Osanai S, Chugh DK, Mokashi A, Lahiri S. Stimulus interaction between CO and CO_2 in the cat carotid body chemoreception. Brain Res 1996; 711:56–63.

82. Buckler KJ, VaughanJones RD. Effects of mitochondrial uncouplers on intracellular calcium, pH and membrane potential in rat carotid body type I cells. J Physiol Lond 1998; 513:819–833.
83. Wyatt CN, Buckler KJ. The effect of mitochondrial inhibitors on membrane currents in isolated neonatal rat carotid body type I cells. J Physiol Lond 2000; 527:121P–122P.
84. Buerk DG, Chugh DK, Osanai S, Mokashi A, Lahiri S. Dopamine increases in cat carotid body during excitation by carbon monoxide: implications for a chromophore theory of chemoreception. J Auton Nerv Syst 1997; 67:130–136.
85. Al-Hashem F, Conway AF, Kumar P. Calcium- and age-dependent effects of carbon monoxide upon chemodischarge in the rat carotid body in vitro. J Physiol Lond 2000; 523P:296P.
86. Bairam A, Dauphin C, Rousseau F, Khandjian EW. Expression of dopamine d2-receptor messenger-rna isoforms at the peripheral chemoreflex afferent pathway in developing rabbits. Am J Respir Cell Mol Biol 1996; 15:374–381.
87. Gauda EB, Bamford O, Gerfen CR. Developmental expression of tyrosine-hydroxylase, d-2-dopamine receptor and substance-p genes in the carotid-body of the rat. Neuroscience 1996; 75:969–977.
88. Donnelly DF, Doyle TP. Developmental-changes in hypoxia-induced catecholamine release from rat carotid-body, in-vitro. J Physiol Lond 1994; 475:267–275.
89. Rigual R, Almaraz L, Gonzalez C, Donnelly DF. Developmental changes in chemoreceptor nerve activity and catecholamine secretion in rabbit carotid body: possible role of Na^+ and Ca^{2+} currents. Pflügers Arch Eur J Physiol 2000; 439:463–470.
90. Tomares SM, Bamford OS, Sterni LM, Fitzgerald RS, Carroll JL. Effects of domperidone on neonatal and adult carotid chemoreceptors in the cat. J Appl Physiol 1994; 77:1274–1280.
91. Hertzberg T, Hellstrom S, Lagercrantz H, Pequignot JM. Development of the arterial chemoreflex and turnover of carotid-body catecholamines in the newborn rat. J Physiol Lond 1990; 425:211–225.
92. Zhang M, Zhong HJ, Vollmer C, Nurse CA. Co-release of ATP and ACh mediates hypoxic signalling at rat carotid body chemoreceptors. J Physiol Lond 2000; 525:143–158.
93. Al-Hashem F, Barbe C, Kumar P. Interaction of ATP and CO_2 upon carotid body chemosensitivity in vitro. J Physiol 2001; 536.P:142P.
94. McDonald DM. Peripheral chemoreceptors. Structure–function relationships of the carotid body. In: Hornbein TF, ed. Regulation of Breathing. Vol. 17. Lung Biology in Health and Disease. New York: Dekker, 1981:105–319.
95. Osborne MP, Butler PJ. New theory for receptor mechanisms of carotid body chemoreceptors. Nature 1975; 254:701–703.
96. Thomas T, Spyer KM. ATP as a mediator of mammalian central CO_2 chemoreception. J Physiol–Lond 2000; 523:441–447.
97. Al-Hashem F, Kumar P. Purinergic modulation of peripheral CO_2 ventilatory chemosensitivity. J Physiol 2001; 531.P:96P.
98. King BF, Ziganshina LE, Pintor J, Burnstock G. Full sensitivity of P2X purinoceptor to ATP revealed by changing extracellular pH. Br J Pharmacol 1996; 117:1371–1373.

15

Further Evidence That Oxygen Sensing in the Carotid Body Involves Iron and Heme Proteins

SUKHAMAY LAHIRI, ARIJIT ROY, ANIL MOKASHI, PETER A. DAUDU, JINQUING LI, SANTHOSH M. BABY, and DONALD G. BUERK

University of Pennsylvania School of Medicine
Philadelphia, Pennsylvania, U.S.A.

I. Introduction

Although O_2 is sensed by almost all mammalian cells, it is in the carotid body, strategically placed, that O_2 sensing takes place for the global benefit of the rest of the body. Here, in the carotid body, a tiny organ, situated at the bifurcation of the carotid artery close to the heart, it captures P_{O_2} in the flow of blood on the way to the brain. With each inspiration (few seconds), alveolar P_{O_2} is renewed. The new P_{O_2} from the lung circulation is pumped by the heart into the carotid artery and carotid body. This is sensed instantaneously and carried by the chemoreceptor afferents to the brain and then by the efferents to the respiratory and cardiovascular reflex systems. By the next inspiration, the event is repeated from breath to breath, and life is sustained, by the feedback control system.

This semireview chapter focuses on O_2 sensing in the carotid body chemoreceptor complex (glomus cell plus afferent endings). By definition, this chemosensing must occur within a few seconds, involving the cellular membrane and cytoplasm, and may not have enough time to get involved directly in the hypoxia-inducible factor 1 (HIF-1) and gene expression. This chapter will deal with the reactions of ferrous iron and preformed heme proteins in oxygen sensing. The evidence will include: afferent discharge pattern resembling O_2-Hb equilibrium curves; CO-induced excitation and inhibition; action spectrum of respiratory chain of the carotid body; iron chelator effects on sensory discharge and on glomus cell K^+ currents; transition metal effects; NO and NOS effects.

II. Chemosensory Discharge Pattern Resembling O_2-Hb Equilibrium Reactions

The carotid chemosensory discharge pattern resembled O_2-Hb equilibrium curves mimicking the "receptor" discharge with a Bohr effect (Fig. 1). We assumed that the receptor is located in the cell membrane and named it a chromophore because it appeared to contain a heme group (1).

III. High P_{co} Effects in Normoxia: Excitation

High P_{co} ($P_{co} = 550$ torr) in normoxia stimulated the chemosensory discharge (Fig. 2a). This CO effect is due to heme protein interaction (2), which diminished

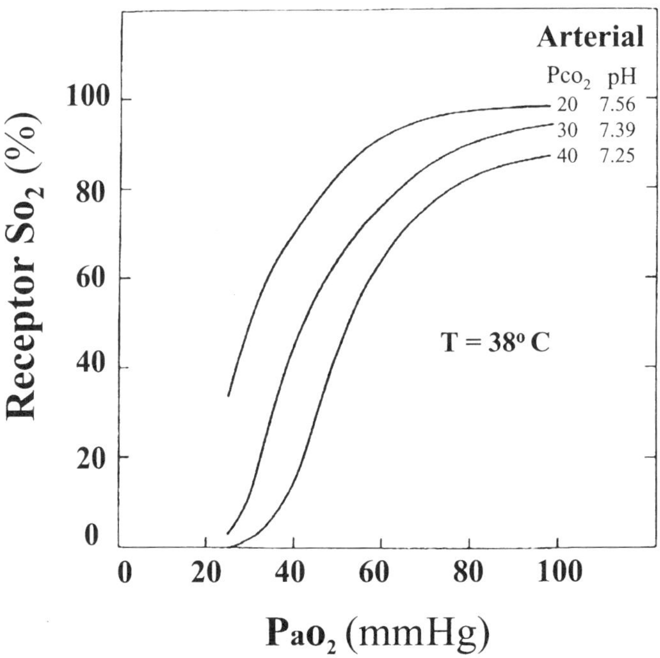

Figure 1 Carotid chemoreception activities plotted against P_{O_2} resemble Hb-O_2 dissociation curves. Oxygen equilibrium curves (SO_2 %) corresponding to activity of carotid chemoreceptors at three measurements of arterial P_{CO_2} and pH plotted against arterial P_{O_2}. The maximal activity at pHa 7.25 and PaO_2 of 25 torr was taken as 100% desaturation of a hypothetical heme-containing chromophore group. The curves resemble O_2 equilibrium curves of a polymeric chromophore group with a Bohr effect (1).

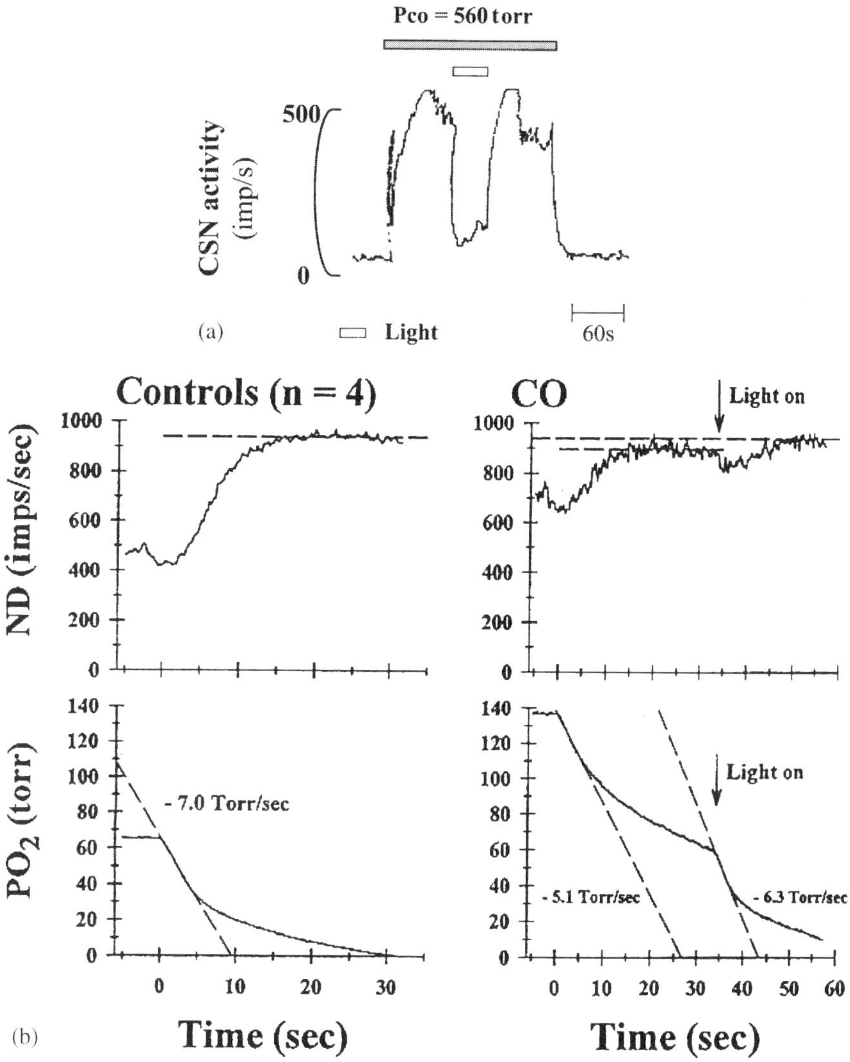

Figure 2 (a) Effect of high P_{co} on carotid chemosensory activity (CSN) plotted against time. Light reversed the CSN activity. (b) Effect of high P_{co} on oxygen consumption rate of carotid body perfused in vitro. The top panel shows carotid chemoreception nerve discharge (ND), and the bottom panel displays carotid body tissue P_{O_2} against time. Flow interruption was followed by a disappearance of P_{O_2}. The slope indicated the oxygen consumption of the carotid body along with the increased activity of chemosensory activity (control). The right-hand panel shows the effect of high P_{co}, which increased the chemoreceptor activity along with an increased tissue P_{O_2}. Flow interruption was followed by a further increase in the activity along with a fall in P_{O_2} in both panels. With CO administration, ND increased along with P_{O_2} (right-hand panel). At some point, bright light during flow interruption was accompanied by a sudden increase of P_{O_2} fall, i.e., increased oxygen consumption. This rate of P_{O_2} fall was similar to the control rate.

oxygen consumption of carotid body as shown by a diminished slope of P_{O_2} fall upon interruption of perfusion of the carotid body (Fig. 2b). Light reversed (increased slope), increasing the fall of P_{O_2} (right-hand panels). Note that the chemosensory discharge did not shut down with light unlike that illustrated in Figure 2a. This shows that the chemosensory discharge rate was not dependent on the oxygen consumption rate. When the carotid body was prefused with calcium-free solution, the chemosensory response to CO was abolished without any effect on oxygen consumption, consistent with the foregoing conclusion (3). The implication of the relationship is not clear at this time.

IV. High P_{co} Effects in Hypoxia: Inhibition

Using high P_{co} (550 torr), excitation of chemosensory discharge is seen in normoxia but the same P_{co} diminished the response to hypoxia ($P_{O_2} = 50$ torr) (3) (Fig. 3). Lopez-Lopez and Gonzalez (4) showed, using whole-cell (glomus) recording, that suppression of K^+ currents by hypoxia ($P_{O_2} = 40$ torr) could be fully reversed by P_{co} of 70 torr. This reversal of K^+ currents by CO is equivalent to injecting "hyperoxia." That is, the chemosensory discharge by hypoxia would be inhibited by high P_{co}. These results are consistent with those of CO effects offsetting hypoxic effects on the expression VEGF (5), platelet-derived growth factor and endothelin-1 (6), phosphoenolpyruvate carboxykinase (7), and hypoxic activation of HIF-1 (8), suggesting the involvement of heme protein.

V. Activation of Chemosensory Discharge by Transition Metal: Co^{2+}

Co^{2+} induced excitation of chemosensory discharge (9) and caused induction of Epo by transition metals (10).

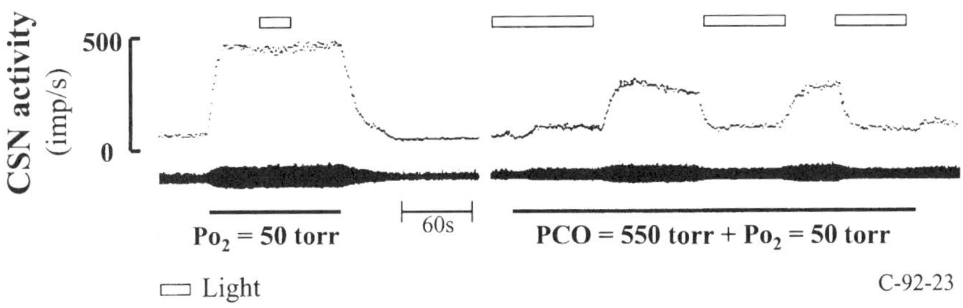

Figure 3 Hypoxia diminished the high CO-induced chemoreceptor activity. From left to right: hypoxic ($P_{O_2} = 50$ torr) increased the CSN activity with no light effect; high P_{co} plus P_{O_2} of 50 torr with light showed very slight increase in the activity; without light the CSN activity increased but remained lower than that of the CO effect; light exposure diminished the activity and then without light the activity increased, which was again diminished with light.

VI. Stimulation of Growth of Glomus Cell by Chronic Co^{2+} Administration

Chronic Co^{2+} administration led to an increase of glomus cell size. This was associated with the increase of red blood cell concentration (11). Similar responses are seen in chronic hypoxia (12). Chronic Co^{2+} should have increased catecholamine synthesis, increasing tyrosine hydroxylase activity. These are reminiscent of HIF induction (13), providing further evidence that heme protein is involved.

VII. Stimulation of Chemosensory Discharge by Iron Chelation

Chelation of Fe^{2+} is expected to decrease OH radical (ROS) by inhibiting the Fenton reaction (14,15). This decrease in ROS production is expected to stimulate the chemosensory discharge (16). Also, chelation of iron is expected to stabilize the cellular HIF-1α, thereby mimicking hypoxia (17). By both counts, chelation of Fe^{2+} is to mimic hypoxia.

We used ciclopirox (CPX), a permeating Fe^{2+} chelator (18), in the perfused carotid body and measured the chemosensory discharge. CPX stimulated chemosensory dis-charge in a dose-dependent manner in normoxia (Fig. 4), mimicking hypoxia. Also, CPX effects never exceeded the hypoxia effects, which suggests that CPX might interfere with the putative heme oxygen sensor. Also, the effect was blocked after addition of ferrous iron (not shown), adding weight to the foregoing conclusion (19).

CPX also suppressed the outward whole-cell K^+ currents of rat glomus cells during normoxia (Fig. 5), mimicking the hypoxic effects (not shown). A similar result has been found in the PC-12 cells (20). In the PC-12 cells, the suppresssed K^+

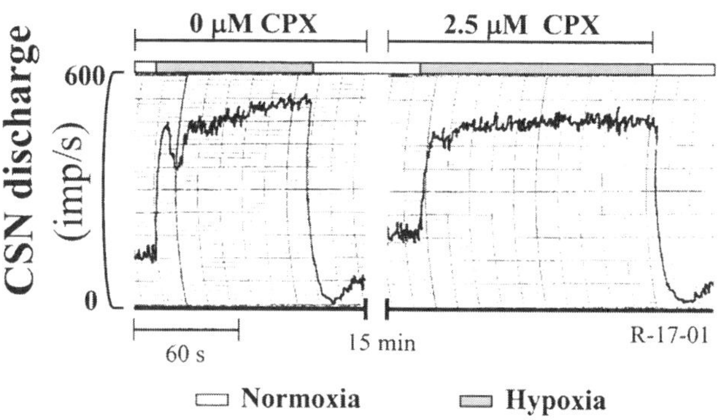

Figure 4 Iron chelator CPX effects on chemosensory discharges. The left-hand panel shows the effect of hypoxia ($P_{O_2} = 22$ torr) plotted against time. The right-hand panel shows the effects of CPX plus the effects of hypoxia. CPX without hypoxia stimulated the chemosensing discharge. Superimposition of the same hypoxia produced the same effect without adding to the effects of CPX, as if the hypoxia effect was replaced by CPX effect.

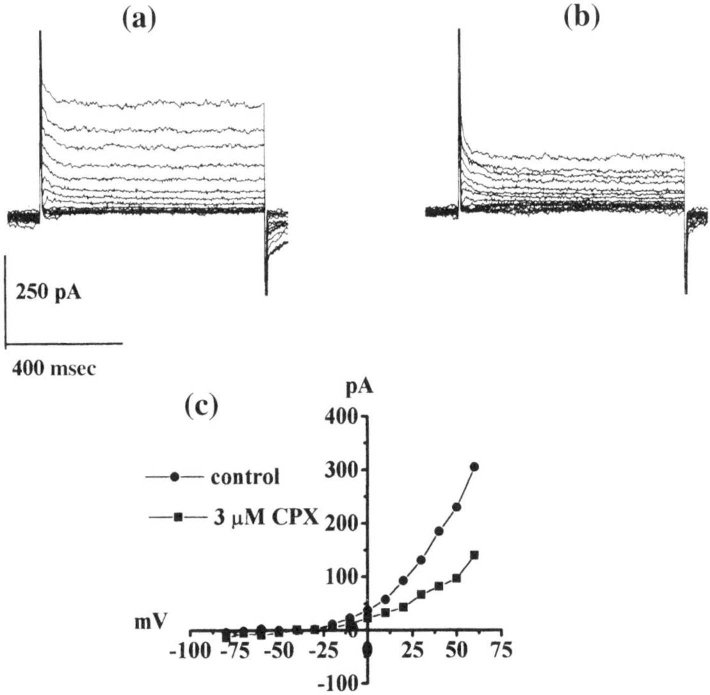

Figure 5 The outward whole-cell K^+ current was suppressed by 3 μM CPX in glomus cell. Bath solution contained (in mM): NaCl 140, KCl 5.0, $MgCl_2$ 1.0, $CaCl_2$ 1.8, HEPES 10, glucose 10, pH 7.4. The pipette solution was composed of (in mM): K aspartate 100, KCl 20, $MgCl_2$ 2.0, $CaCl_2$ 1.0, HEPES 10, EGTA 10, glucose 10, ATP 0.05, pH 7.2. Membrane potential was held at −85 mV. Depolarizing voltage steps from −80 to +60 mV with 10-mV increments. (a) the recording trace showing the outward K^+ current; (b) the recording trace in the presence of 3 μM CPX from the same experiment. The outward whole-cell K^+ current was significantly suppressed by 3 μM CPX; (c) the current voltage relationship in the absence (solid circles) and the presence (solid squares) of CPX.

currents by CPX were restored by adding $FeSO_4$ (Li J, Ray A, Lahiri S, unpublished observations). These preliminary results, if confirmed, show the importance of intracellular free iron in oxygen sensing in the glomus cells.

VIII. Nitric Oxide–Related Effects

Nitric oxide decreases oxygen consumption, and the effect is incremental with hypoxia (21,22). This inhibition appears to result from S-nitrosylation of the critical thiol group in complex IV of cytochrome oxidase and is immediately reversed by exposure to high-intensity light (23). This is analogous to CO effects on the respiratory chain (2,3,24), and this presumably is the basis of O_2 sensing by NO.

NOS is also a heme protein, and NO activates gunylate cyclase, also a heme protein. Both are affected by NO. NOS is localized in nerve fiber and vascular endothelium in the carotid body, and hypoxia inhibits the NOS activity (25,26). Thus, during hypoxia, a decrease in NO synthesis occurs and this may contribute to augmentation of sensory activity (25,27).

NO also causes inhibition of HIF-1 (8,23). This inhibition results from blocking of activation of HIF-1 by hypoxia. Thus, NO like CO presumably acts as heme ligand binding to O_2.

IX. Hypoxia-Inducible Factor

Schematically, hypoxia and H^+ bind to membrane heme and interact with K^+ channels (not the pore) followed by membrane depolarization. This membrane phenomenon will take place in a split second. Down the cascade, HIF-α is stabilized, which can form a heterodimer with constitutively expressed HIF-β in the cytosol, activating HIF-1. This then is translocated to the nucleus and binds to hypoxia response elements in inducible genes. In normoxia, HIF-α is formed continually and is oxidatively modified and then is degraded by the proteasomal pathway. Ferrous iron is needed at two points, in the generation and degradation of HIF-α, although HIF does not contain Fe^{2+}. Without Fe^{2+}, these reactions will not occur (see Refs. 28,29).

Two domains of HIF-1 are known to respond to hypoxia: one is the oxygen-dependent degradation N-terminal domain which is subject to prolyl hydroxylation and the other is COOH-terminal activation domain which is dependent on asparagine hydroxiation. This has been described more recently (30). A full induction of HIF-1 relies on the abrogation on both proline and asparagine hydroxylation in the presence of oxygen, ferrous iron, 2-oxogluterate, and ascorbic acid. Iron chelation and inhibition of 2-oxogluterate are both expected to inhibit the degradation of HIF-1, and will provide a pharmacological tool for their investigation.

X. Simple Heme Protein Models

That heme proteins play a critical role in oxygen sensing and signal transduction has been found in nitrogen-fixing bacteria *Rhizobium*, whose oxygen sensor is a simple oxygen-binding heme protein (31,32).

XI. Summary

The information provided adds to the evidence that it is not only heme proteins that are involved in oxygen sensing in various subcellular and cellular compartments but also intracellular free iron featured prominently in the process in the carotid body chemoreception. In doing so, molecules that activated HIF-α also activated (suppressed) the outward K^+ currents of the glomus cell membrane. The connection of the two processes is intriguing.

Acknowledgments

This work was supported in part by grants R37-HL43413-12, RO1-HL50180-7, T32-HL07027-27*, and US Navy N00014-01-1-0948. The secretarial assistance of Mary Pili is gratefully acknowledged.

References

1. Lahiri S, DeLaney RG. Stimulus interaction in the response of carotid chemoreceptor single afferent fibers. Respir Physiol 1975; 24:249–266.
2. Lahiri A, Iturriaga R, Mokashi A, Ray DK, Chugh D. CO reveals dual mechanisms of O_2 chemoreception in the cat carotid body. Respir Physiol 1993; 94:227–240.
3. Lahiri S, Buerk DG, Osanai S, Mokashi A, Chugh DK. Effect of CO on VO_2 of carotid body and chemoreception with and without Ca^{2+}. J Autonom Nerv Syst 1997; 66:1–6.
4. Lopez-Lopez JR, Gonzalez C. Time-course of K^+ current initiation by low oxygen in the chemoreceptor cells of adult rabbit carotid body. FEBS Lett 1992; 299:251–254.
5. Goldberg MA, Schneider TJ. Similarities between the oxygen-sensing mechanisms regulating the expression of vascular endothelial growth factor and erythropoietin. J Biol Chem 1994; 269:4355–4359.
6. Morita T, Kourembanas S. Endothelial cell expression of vasoconstrictors and growth factors is regulated by smooth muscle cell-derived carbon monoxide. J Clin Invest 1995; 96:2676–2682.
7. Kietzmann T, Schmidt H, Unthan-Fechner K, Probst I, Jungermann K. A ferro-heme protein senses oxygen levels, which modulate the glucagon dependent activation of the phosphoenolpyruvate carboxykinase in rat hepatocyte cultures. Biochem Biophys Commun 1993; 195:792–798.
8. Huang LE, Willmore W, Gu J, Goldberg MA, Bunn HF. Inhibition of HIF-1 activation by carbon monoxide and nitric oxide; implications for oxygen sensing and signaling. J Biol Chem 1999; 274:9038–9044.
9. Di Giulio C, Huang W-X, Lahiri S, Mokashi A, Buerk DG. Cobalt stimulates carotid body chemoreceptors. J Appl Physiol 1990; 68:1844–1849.
10. Goldberg MA, Dunning SP, Bunn HF. Regulation of the erythropoietin gene: evidence that the oxygen sensor is a heme protein. Science 1988; 242:1412–1415.
11. Di Giulio C, Datta PG, Lahiri S. Chronic cobalt causes hypertrophy of glomus cells in the rat carotid body. Am J Physiol 1991; 261:C102-C105.
12. McGregor KH, Gil J, Lahiri S. A morphometric study of the carotid body in chronically hypoxic rats. J Appl Physiol 1984; 57:1430–1438.
13. Wang CL, Semenza GL. General involvement of hypoxia-inducible factor 1 in transcriptional response to hypoxia. Proc Natl Acad Sci USA 1993; 90:4304–4308.
14. Ho VT, Bunn HF. Effects of transition metals on the expression of the erythropoietin gene: further evidence that the oxygen sensor is a heme protein. Biochem Biophys Res Commun 1996; 223:175–180.

15. Wang GL, Semenza GL. Desferrioxamine induces erythropoietin gene expression and hypoxia-inducible factor 1 DNA-binding activity; implications for models of hypoxia signal transduction. Blood 1993; 82:3610–3615.
16. Acker H. Mechanisms and meaning of cellular oxygen sensing in the organism. Respir Physiol 1994; 95:1–10.
17. Maxwell PH, Pugh CW, Ratcliffe PJ. Inducible operation of the erythropoietin 3′ enhancer in multiple cell lines: Evidence for a widespread oxygen-sensing mechanism. Proc Natl Acad Sci USA 1993; 90:2423–2427.
18. Wanner RM, Spielman P, Stroka DM, Camenisch G, Camenisch I, Scheid A, Houck DR, Bauer C, Gassman M, Wenger RW. Epolones induce erythropoietin expression via hypoxia-inducible factor 1α activation. Red Cells 2000; 96:1558–1565.
19. Daudu PA, Roy A, Rozanov C, Mokashi A, Lahiri S. Extra and intracellular free iron and carotid body responses. Respir Physiol. 2002; 130:21–31.
20. Li J, Roy A, Daudu P, Lahiri S. Ciclopirox olamine induced inhibition of K^+ current in PC-12 cells. Exp Biol 2002. In press.
21. Beltrán B, Mathur A, Duchen MR, Eresalinsky JD, Moncada S. The effect of nitric oxide on cell respiration: a key to understanding its role in cell survival and death. Proc Natl Acad Sci USA 2000; 97:14602–14607.
22. Buerk DG, Lahiri S. Evidence that nitric oxide plus a role in O_2 sensing from tissue NO and P_{O_2} measurements in cat carotid body. In: Lahiri S, Prabhakar NR, Forster RE II, eds. Oxygen Sensing: Molecule to Man. Adv Exp Med Biol 2000; 475:337–348.
23. Sogawa K, Numayama-Tsunita K, Ema M, Abe M, Abe H, Fusü-Kuriyama Y. Inhibition of hypxoia-inducible factor 1 by nitric oxide donors in hypoxia. Proc Natl Acad Sci USA 1998; 95:7368–7373.
24. Wilson DF, Mokashi A, Chugh D, Vinogradov, Osanai S, Lahiri S. The primary oxygen sensor of the ct carotid body is cytochrome a_3 of the mitochondrial respiratory chain. FEBS Lett 1994; 351:370–374.
25. Prabhakar NR. NO and CO as second messengers in oxygen sensing in the carotid body. Respir Physiol 1999; 115:161–165.
26. Grimes PA, Mokashi A, Stone RA, Lahiri S. Nitric oxide synthesis in autonomic innovation of the cat carotid body. J Autonom Nerv Syst 1995; 54:80–86.
27. Chugh DK, Katayama M, Mokashi A, Bebout DE, Ray DK, Lahiri S. Nitric oxide related inhibition of carotid chemosensory nerve activity in the cat. Respir Physiol 1994; 97:147–156.
28. Lahiri S, DiGiulio C, Roy A. Lessons from chronic intermittent and sustained hypoxia at high altitude. Respir Physiol 2002; 130:223–233.
29. Lahiri S. Historical perspectives of cellular oxygen sensing and responses to hypoxia. J Appl Physiol 2000; 88:1467–1473.
30. Lando D, Peet DJ, Whelan DA, Gorman JJ, Whitelaw ML. Asparagine hydroxylation of the HIF transactivation domain: a hypoxic switch. Science 2002; 295:858–861.
31. Gilles-Gonzalez MA, Ditta GS, Helinski DR. A hemeprotein with kinase activity encoded by the oxygen sensor of *Rhizobium meliloti*. Nature 1991; 350:170–172.
32. Lukat-Rodgers GS, Rogers KR. Characterization of ferrous FixL-nitric oxide adducts by resonance Raman spectroscopy. Biochemistry 1997; 36:4178–4187.

16

O_2-Sensitive K^+ Channels Controlling Cell Excitability

CHRIS PEERS, ANTHONY LEWIS, LEIGH D. PLANT, HUGH A. PEARSON, and PAUL J. KEMP

University of Leeds
Leeds, England

I. O_2-Sensitive K^+ Channels

K^+ channels are the largest and most widely distributed family of ion channels. In excitable cells, their main role is to regulate membrane potential and so control cell excitability. This can be achieved by altering action potential waveform and duration (during the repolarizing phase), by altering the rate at which threshold for action potential generation is achieved (via their contribution to pacemaker potentials), or more simply by providing a tonic hyperpolarizing influence on resting membrane potential. Any external stimulus that can modulate K^+ channel activity will, therefore, have a profound effect on cell function. This is particularly true for cellular functions that are dependent on Ca^{2+} influx through voltage-gated Ca^{2+} channels; by alteration of membrane potential (via modulation of K^+ channel activity), the degree of Ca^{2+} channel activity is finely controlled, hence Ca^{2+}-dependent functions such as contraction or exocytosis can be indirectly regulated via modulation of K^+ channel activity.

In 1988, O_2 was added to the list of agents capable of modulating K^+ channel activity (1): when O_2 was removed from the environment of glomus cells isolated from the rabbit carotid body, K^+ channel activity was rapidly and reversibly suppressed. This immediately provided a simplistic model for the long-standing mystery of how hypoxia excited the carotid body—through hypoxic inhibition of K^+ channels, these cells could depolarize, permitting voltage-gated Ca^{2+} influx and hence triggering exocytotic transmitter release. Although alternative scenarios have been suggested, this proposed mechanism (often referred to as the "membrane hypothesis" of chemotransduction) has gained support from numerous independent research groups. In the intervening years, the presence of O_2-sensitive K^+ channels has been reported in a variety of other tissues, such as pulmonary (but not systemic) vascular smooth muscle (2), airway chemoreceptors [neuroepithelial bodies (NEBs (3)] and their immortalized counterparts, the small cell lung carcinoma cell line, H146 (4)), chromaffin cells (5) and their immortalized counterparts, the pheochromocytoma cell line, PC12 (6), and central neurons (7). What is clear from these studies is that numerous different K^+ channel types are susceptible to regulation by local O_2 levels (see Ref. 8 for recent review). What remains unclear, however, is how O_2 (or, more accurately, a lack of O_2) is sensed. In this chapter, we attempt to review specific aspects of O_2 sensing by K^+ channels. Due to the fact that this field is rapidly expanding, and that a recent article has reviewed a number of O_2-sensitive, voltage-gated K^+ channels (9), this chapter focuses on two important O_2-sensitive K^+ channels, their influence in three distinct cell types, and their properties when studied in a heterologous expression system.

II. O_2-Sensitive K^+ Channels in the Carotid Body

The original description of O_2-sensitive K^+ currents in rabbit carotid body type I cells was soon supported by detailed single-channel recordings. Using excised membrane patches, Ganfornina and Lopez-Barneo reported a voltage-dependent, inactivating, Ca^{2+}-insensitive K^+ channel with a conductance of 40 pS that was reversibly inhibited by hypoxia (10); this channel has been termed the K_{O_2} channel. The observation that hypoxia could inhibit this channel in excised patches immediately led to the conclusion that diffusible, cytosolic factors were not required. Interestingly, these authors also demonstrated that channel activity could be reversibly inhibited by exposure of patches to reduced glutathione (GSH) (11). Reduced glutathione is the major diffusible cytosolic pool of thiol-containing agents. It exists in equilibrium with the oxidized (GSSG) form and, as such, is an essential component of the cellular defense mechanism that protects against oxidative damage. Under hypoxic conditions, one might anticipate an increase in GSH levels that would inhibit K_{O_2} channels, suggesting that redox modulation of the channels underlies hypoxic inhibition. However, GSH is not necessary for hypoxic inhibition of the K_{O_2} channels (10). This implies that inhibition by GSH may represent a secondary mechanism by which the activity of these channels can be suppressed during hypoxia. However, a clear rise in GSH during hypoxia remains to be demonstrated and available evidence, in the carotid body at least, suggests that GSH/GSSG ratio remains relatively stable even during prolonged periods of hypoxia (12).

In an elegant study, Lopez-Lopez and Gonzalez provided the first, and to this date most convincing, evidence to shed light on the mechanism of O_2 sensing in the carotid body. Using whole-cell recordings, they demonstrated that the suppression of K^+ currents by hypoxia could be fully reversed by carbon monoxide (CO), and indeed that CO could prevent hypoxic inhibition of K^+ currents (13). This suggested strongly that O_2 sensing involved a heme protein and, coupled to the single-channel observations described earlier, this heme protein must be closely associated with the plasma membrane, since hypoxic inhibition was apparent in excised membrane patches. Furthermore, this observation is in keeping with the fact that CO suppresses hypoxic excitation of the intact carotid body (14). In the same study, Lopez-Lopez and Gonzalez also demonstrated that hypoxic inhibition of K^+ currents was extremely rapid (faster than blockade of voltage-gated Na^+ channels by tetrodotoxin), an observation in keeping with the rapid response of the intact organ to hypoxia (13). Although there are no further details concerning the mechanism of O_2 sensing in this preparation, the physiological consequences are clear: available data indicate that these cells fire action potentials spontaneously during normoxia, and that during hypoxia these action potentials increase in frequency (14), since the activity of the K_{O_2} channels—which normally serve to delay the cell resting potential from reaching threshold for firing another action potential—is suppressed. This in turn would lead to accumulation of Ca^{2+} within the cell, permitting increased neurotransmitter release (15,16).

The studies described above were all performed using rabbit carotid body type I cells. Contemporary studies from other groups indicated important species- and age-related differences in the nature of the O_2-sensitive K^+ channels present in type I cells. In cultured embryonic rabbit type I cells, a K^+ channel with strikingly different properties (inward rectification, conductance of 137 pS) has been described (17), although its physiological role—with regard to embryonic O_2 sensing—has yet to be established. In the rat type I cell, two distinct K^+ channel types have been shown to be O_2 sensitive: a high-conductance, Ca^{2+}-sensitive K^+ channel (maxiK channel) (18) and an acid-sensitive, voltage-independent K^+ channel that is a member of the tandem-P domain K^+ channel family, known as TASK1 (19).

Rat type I cells are electrically silent, but depolarize and can generate Ca^{2+}-dependent spikes under hypoxic conditions (20,21). Thus, the hypoxia-sensitive K^+ channel appears to contribute to resting membrane potential (which is around -50 mV) under normoxic conditions. TASK1 seems an ideally suited channel for this role, since it is does not require depolarization for activation, and it is additionally acid-sensitive (as is the intact carotid body) (19,20). This latter point, however, requires further examination since acid sensing in the carotid body appears more dependent on reduction in intracellular pH (23), while TASK1 is sensitive to changes in extracellular pH (24). The maxiK channel, by contrast, does not appear ideally suited to the function of determining resting membrane potential, especially when one considers the fact that ultimately hypoxia will raise $[Ca^{2+}]_i$ to trigger exocytosis. This rise in $[Ca^{2+}]_i$ would be expected to reactivate maxiK channels and so terminate secretory responses through hyperpolarization. According to the membrane hypothesis, this would result in a transient excitation of the intact carotid body; in reality, hypoxic excitation is sustained. However, the experimental evidence that maxiK channels do indeed serve this role is strong: selective inhibitors of this channel (charybdotoxin and iberiotoxin) cause depolarization of these cells (20) and, in both a novel carotid body slice preparation (25) and a cultured type I cell cluster preparation (26), evoke catecholamine secretion. In addition, the activity of these channels is tightly regulated by intracellular pH (27), and postnatal, developmental expression of maxiK is concurrent with postnatal maturation of the O_2 sensitivity of the intact organ (28). Finally, the expression of maxiK channels is suppressed in animals born and reared in chronic hypoxia—a condition that prevents postnatal developmental increase of sensitivity to acute hypoxia (20). In all probability, O_2 sensing by the rat carotid body relies on these two channels acting in concert to produce the physiological response to hypoxia, i.e., membrane depolarization, as recently suggested (29).

While little is known of the mechanisms underlying hypoxic inhibition of K^+ channels in the rat carotid body, some unexpected observations have arisen. First, attempts to mimic hypoxic inhibition of maxiK channels at the single-channel level using excised membrane patches were unsuccessful (30), suggesting the involvement of cytosolic factors. This lack of effect in excised patches was also reported in recent studies of TASK channels in rat type I cells; again, an intact intracellular milieu was required (19). Furthermore, inhibition by hypoxia in these cells was mimicked by mitochondrial uncouplers (31), raising the possibility that mitochondria may be central to O_2 sensing in these cells. However, unlike in cells of young rats, a most recent report has demonstrated convincingly that maxiK channels in adult rat type I cells are O_2 sensitive in excised membrane patches, and this effect could be reversed by carbon monoxide, suggesting the involvement of a membrane-associated heme protein (32). The only difference between this latter study and those reported earlier was the age of the animals used. The earlier studies had employed cells from young rats (ca. 10–14 days old), while the latter used adult rats. These discrepancies might indicate further postnatal development in the coupling of sensor to channel. Interestingly, these studies support the idea that the sensor could be the mitochondrion itself.

III. O_2-Sensitive K^+ Channels in Airway Chemoreceptor Cells

Some 5 years after the first report of O_2-sensitive K^+ channels in the carotid body, similar findings were reported for cells of the neuroepithelial bodies (NEBs) of lung (3). These cells appear to act as airway O_2 sensors: they are found in discreet clusters throughout the airway (33) and possess transmitter-filled vesicles (34), much like type I carotid body cells. Their excitability, and the hypoxic enhancement of excitability through inhibition of a K^+ current, suggested that they share many similarities, at least superficially, with carotid body type I cells. While the specific identity of the O_2-sensitive K^+ currents requires more detailed studies, it is clear that there are likely more than one type coexisting in these cells. Thus, using a neonatal rabbit lung slice preparation, both Ca^{2+}-dependent and Ca^{2+}-independent components of the whole-cell K^+ current have been identified, both of which display O_2 sensitivity (3,35,36).

Investigating the process of O_2 sensing by NEBs has proved difficult, owing to the low numbers of NEBs in the lung and their relative inaccessibility. However, a number of human cell lines are believed to be derived from the same precursor pool as NEBs (37). We have exploited one of these small cell lung carcinoma lines, H146, to investigate airway O_2 sensing. Initial approaches indicated that an O_2-sensitive K^+ channel in H146 cells was resistant to conventional, broad-spectrum K^+ channel blockers and, importantly, contributed to resting membrane potential (4). Further pharmacological analysis suggested the possible presence of tandem-P domain K^+ channels (38), and RT-PCR showed that mRNA for several members of this family of channels were present in these cells (39). To distinguish which, if any, of these channels was selectively inhibited by acute hypoxia, antisense oligodeoxynucleotides directed against TASK1 and TASK3 were applied to H146 cells and found to block completely the hypoxic inhibition of K^+ currents (39). Subsequent demonstration that Zn^{2+} blockade did not prevent hypoxic inhibition indicated clearly that TASK3 was the O_2-sensitive K^+ current in H146 cells (39). This finding raises the question of how good a model for NEB cells is the H146 line. At present, there is no published information concerning the presence or otherwise of TASK in NEBs, but electrophysiological studies to date have employed whole-cell patch-clamp recordings—conditions in which the intracellular milieu is dialyzed—and other studies have shown that TASK currents run down rapidly under such conditions (40). Therefore, the presence of an O_2-sensitive TASK-like channel in NEBs has yet to be discounted.

Despite the fact that NEB cells have been far less studied than the carotid body, progress toward understanding the molecular nature of O_2 sensing by these cells (and indeed H146 cells) has been more rapid. NADPH oxidase had long been proposed to act as an O_2 sensor in the carotid body (41), utilizing O_2 to generate superoxide that is rapidly dismutated to H_2O_2. This H_2O_2 may then regulate K^+ channels via a redox mechanism, to promote channel opening (Fig. 1a). Thus, under hypoxia, H_2O_2 production would be limited and the channel would tend toward a reduced (inactive) state (Fig. 1b). The identification in both NEB cells (42) and H146 cells (43) of mRNAs encoding some of the subunits of NADPH oxidase, and also mRNA for a voltage-gated K^+ channel (Kv3.3a) (42), the activity of which is enhanced by H_2O_2, strongly supported the idea that this enzyme/channel complex functioned as O_2 sensor and effector in these chemoreceptive cells. In H146 cells, we exploited the fact that NADPH oxidase can be activated by protein kinase C (PKC) and demonstrated that PKC activation suppressed hypoxic inhibition of K^+ currents, most likely by changing the affinity of the oxidase for O_2 and in so doing maintaining cellular H_2O_2 levels as available O_2 declines (Fig. 1c) (43). Together, these data provide convincing evidence for NADPH oxidase to act as the airway chemoreceptor O_2 sensor, and this was subsequently firmly established by the report of complete lack of O_2 sensitivity of K^+ currents recorded from NEB cells of knockout mice lacking a major subunit of NADPH oxidase (44). Interestingly, carotid body (45) and

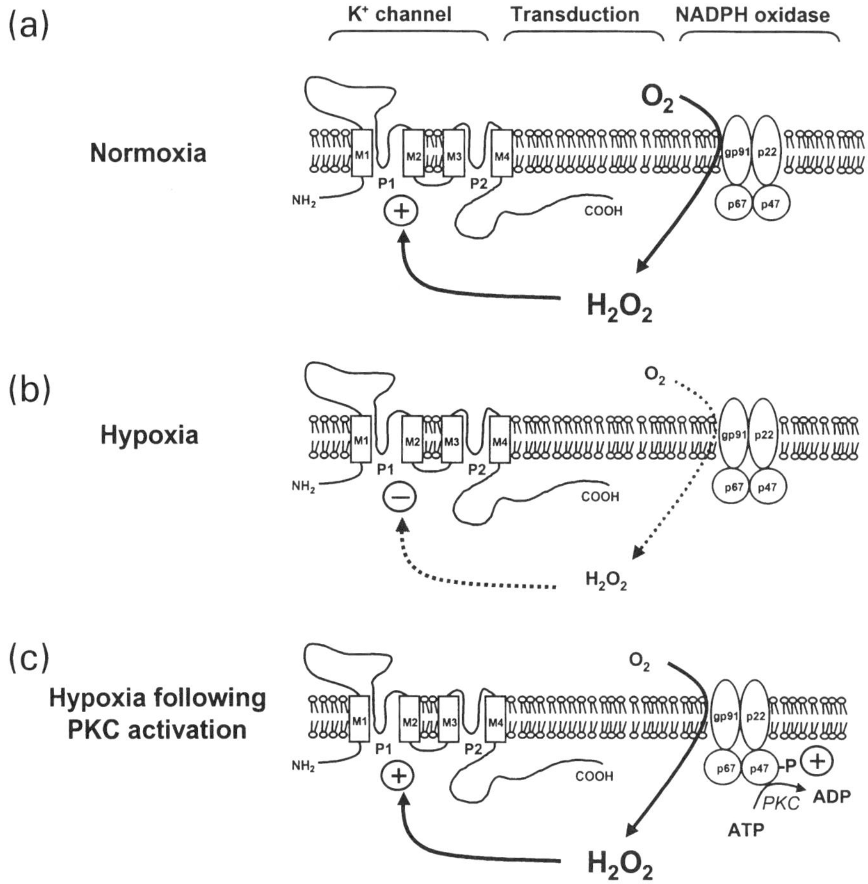

Figure 1 Schematic representation of regulation by NADPH oxidase of K^+ channel activity. (a) In normoxia, NADPH oxidase tonically generates hydrogen peroxide (H_2O_2) from atmospheric O_2. This H_2O_2 promotes channel activity. (b) In hypoxia, O_2 levels are limited and H_2O_2 production is reduced, resulting in channel closure and cell depolarization. (c) Upon activation of NADPH oxidase by protein kinase C (PKC)-mediated phosphorylation, substrate (molecular O_2) affinity is increased with the result that cellular H_2O_2 production is sustained during hypoxia and K^+ channel activity is maintained.

pulmonary smooth muscle (46) O_2 sensitivity was unaffected in these mice, suggesting that oxygen sensing by NADPH oxidase is unique to NEB cells.

IV. An O_2-Sensitive K^+ Channel in Central Neurons

As noted earlier, O_2-sensitive K^+ channels have also been reported in central neurons. In particular, Haddad and colleagues have described a K^+ channel that is regulated by intracellular Ca^{2+}, ATP, and pH and is also reversibly inhibited by hypoxia—clearly this channel represents a more general sensor of cellular metabolic status (47–49). Specific sensing of O_2 occurs via a membrane-bound protein that probably contains a metal-binding (although not heme) region. Unusually, hypoxic modulation of these channels is biphasic; an initial increase in amplitude may be due to relief of tonic block as ATP levels decline. This increase in current amplitude then reverses to become an inhibitory response since hypoxia causes a slower reduction in channel open state probability. However, the relationship of these responses (recorded necessarily under voltage clamp) to changes in cell membrane potential and excitability remains to be clarified.

More recently, tandem-P domain K^+ channels have been reported in central neurons (40,50), and their lack of voltage sensitivity would immediately suggest that they can exert profound effects on cell membrane potential and excitability. Specifically, TASK channels have been identified in cerebellar granule neurons (40,50); these channels underlie the standing outward K^+ current, which can be isolated by holding cell membrane potential at a relatively depolarized level to inactivate other, voltage-gated, K^+ channels. Previous studies in cerebellar and other neurons have shown this current to be inhibited by extracellular acidosis and, importantly, regulated by a number of neurotransmitters, including serotonin, norepinephrine, substance P, thyrotropin-releasing hormone, and acetylcholine acting via muscarinic receptors (40,51). The convergence of such transmitters on this channel effector suggests that it is a primary site of regulation of neuronal excitability.

Our recent studies, using cerebellar granule neurons, indicate that this channel is TASK1, based on its biophysical properties and the fact that it can be inhibited by the endocannibanoid anadamide (52), thereby distinguishing it from TASK3. Figure 2a indicates that it can be rapidly and reversibly inhibited by acute hypoxia. Furthermore, inhibition of the TASK current by low extracellular pH prevents this hypoxic response (Fig. 2b), indicating that both hypoxia and acidity inhibit the same channel. Owing to its lack of obvious voltage dependence, this channel exerts an important influence on resting membrane potential, and so hypoxia causes marked depolarization and increased excitability (Fig. 2c). Thus, while the underlying mechanisms remain undetermined (as they do in the carotid body), the inhibition of TASK1 in central neurons is likely to have a profound influence on neuronal responses to hypoxic or ischemic episodes.

V. Recombinant O_2-Sensitive K^+ Channels

Understanding the mechanisms underlying regulation of K^+ channel activity by acute reduction in environmental O_2 has been hampered by the fact that in O_2-sensing tissues these channels often represent only one of a whole gamut of native, expressed K^+ channel proteins, as well as other ion channel types. Furthermore, pharmacological and/or biophysical dissection of specific K^+ channel types is not always possible. Clearly, to study in detail the precise mode of hypoxic K^+ channel inhibition, it would be desirable to record activity from a pure

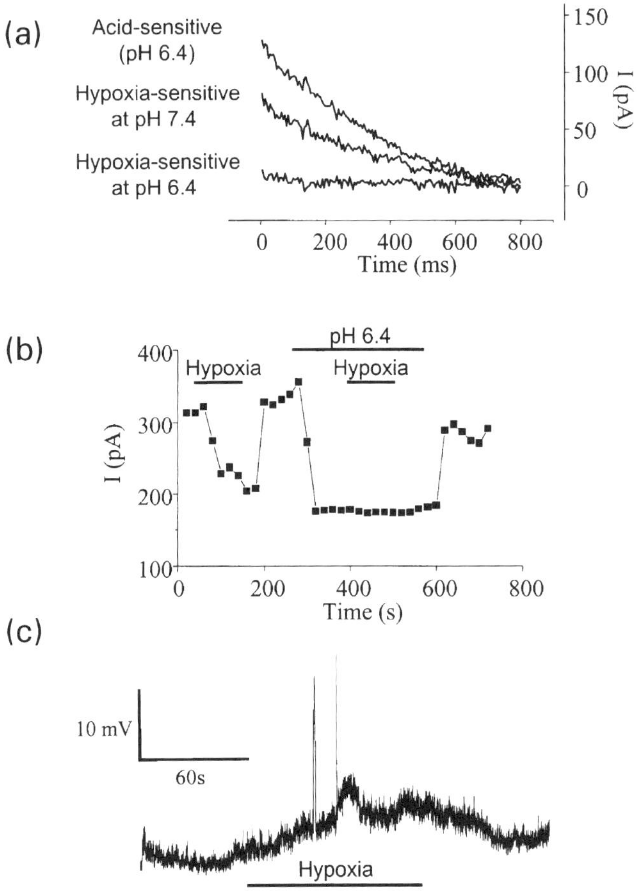

Figure 2 Characterization of TASK1 in rat cerebellar granule neurons. (a) Subtracted K^+ currents evoked by ramp hyperpolarizations from a resting potential of −20 mV to −100 mV on the time scale on the X-axis showing the acid- and hypoxia-sensitive components as indicated. Also shown is the negligible component that is hypoxia-sensitive under acidic conditions. (b) Exemplar time course of acid and hypoxia sensitivity of K^+ currents measured at −20 mV. (c) Current clamp recording of the effect of hypoxia on membrane potential. All recordings were made employing the perforated patch technique.

population of channels. To this end, a number of potentially important O_2-sensitive K^+ channels have been investigated recently using various heterologous expression systems, including *Xenopus laevis* oocytes as well as various mammalian cell lines, chosen because they express relatively low levels of native K^+ channels. These studies are still in their infancy and have yet to provide information concerning the molecular basis for O_2 sensing. Patel and Honore (9) have provided a recent review highlighting the diversity of O_2-sensitive, voltage-gated K^+ channels and their O_2

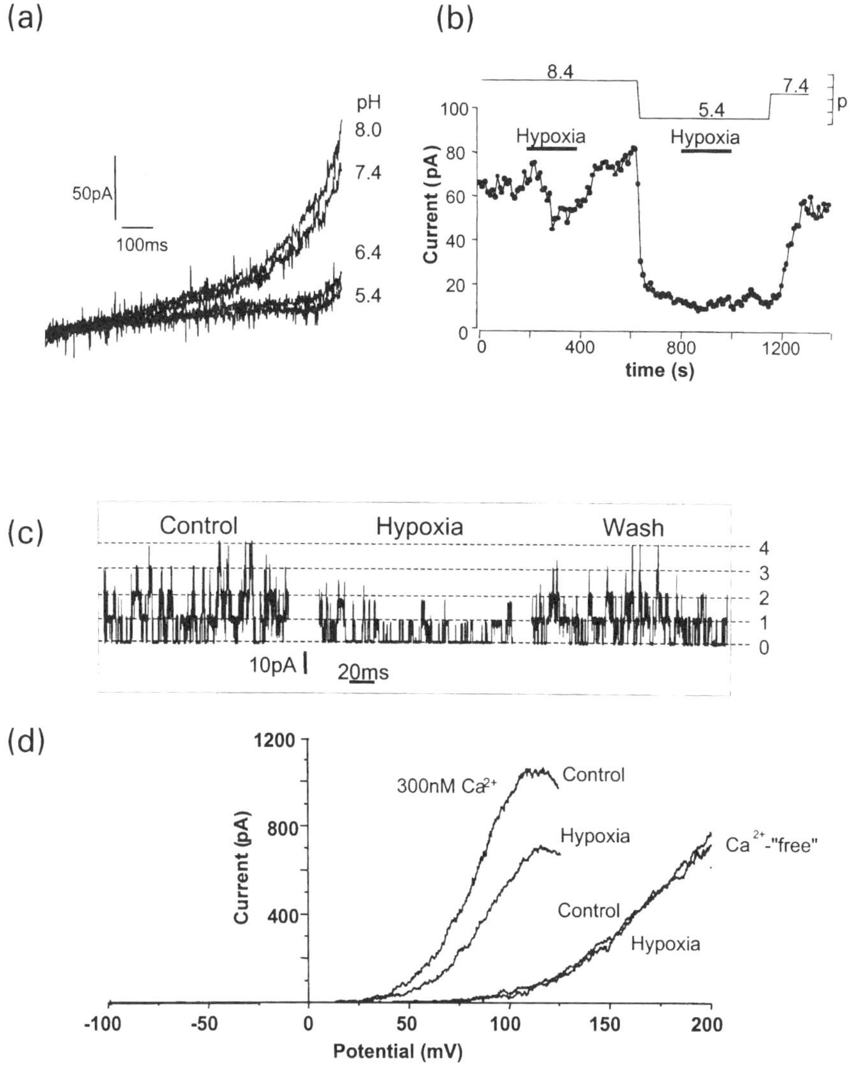

Figure 3 Recombinant TASK1 and maxiK channel regulation by hypoxia. (a) Whole-cell K^+ currents evoked by ramp depolarizations from −110 mV to +10 mV in a HEK293 cell stably expressing hTASK1 in the presence of different extracellular pH levels as indicated. (b) Time series plot of hypoxic inhibition of TASK1 K^+ currents under alkaline and acid conditions. K^+ current amplitudes were measured at 0 mV. (c) Exemplar recordings of recombinant maxiK channels during normoxia (left), hypoxia (middle), and upon return to normoxia (right). The patch was held at +50 mV (−Vp) and channel openings are upward deflections. Cytosolic Ca^{2+} buffered to 300 nM. (d) Currents evoked by ramp depolarizations applied to an excised, inside-out patch containing numerous recombinant maxiK channels under normoxic and hypoxic conditions where cytosolic Ca^{2+} buffered to either 300 nM or <4 nM (Ca^{2+}-"free").

sensitivity in heterologous expression systems. Clearly, this is a rapidly expanding field, and our own work has added to the awareness of O_2-sensitive, recombinant K^+ channels. We have shown directly that TASK1 is O_2-sensitive when expressed in human embryonic kidney (HEK293) cells, while still retaining its pH sensitivity (53). Indeed, acidification of the extracellular medium suppressed or prevented hypoxic inhibition, indicating that the same channel was both O_2 and pH sensitive (Fig. 3a,b).

Bearing in mind that Ca^{2+}-activated K^+ channels have been implicated directly in carotid body chemotransduction, we have also examined the effects of hypoxia on recombinant maxiK channels in HEK293 cells. In this study, cells were stably cotransfected with the α- and β-subunits of a human brain maxiK channel, and hypoxia caused inhibition of these channels even when studied in excised membrane patches (Fig. 3c,d). MaxiK channels are both Ca^{2+}- and voltage-dependent: in the virtual absence of Ca^{2+} bathing the intracellular face of the channel, strong depolarizations can still elicit activity (Fig. 3d). When Ca^{2+} is present, far less depolarization is required for their activation (Fig. 3d). Interestingly, channels opened by voltage alone were resistant to hypoxia, while in the presence of Ca^{2+}, hypoxic inhibition was apparent (Fig. 3c,d). This suggests strongly that hypoxic inhibition of maxiK channels occurs by inhibition of Ca^{2+} activation, a possibility that has important physiological consequences. In the carotid body, the idea that maxiK channels are central to chemoreception has been criticized on the basis that following depolarization and Ca^{2+} influx, reactivation of the channels would occur and transmitter release would be transient. However, if maxiK channels in the carotid body behave as recombinant human brain maxiK channels (Fig. 3d), then decreased Ca^{2+} sensitivity would prevent reactivation and sustain the response of the type I cell to hypoxia.

Although information obtained from such recombinant studies is likely to increase, the full potential for learning about the molecular determinants underlying regulation of K^+ channels by O_2 has yet to be realized. However, an idea of the potential for such studies can be appreciated from recent investigations into the O_2 sensitivity of Ca^{2+} channels. In 1995, Franco-Obregon and colleagues described the O_2 sensitivity of L-type Ca^{2+} channels in myocytes of the systemic vasculature (54). Their observation that hypoxia caused a rapid and voltage-dependent inhibition of these channels provided a clear and simple mechanism to account for hypoxic vasodilation. Working with a recombinant human L-type Ca^{2+} channel α-subunit, α_{1C} [originally cloned from human heart, but also found in vascular tissue (55)], we found that when expressed in HEK293 cells, the responses of these channels to acute hypoxia were identical to those of native vascular smooth muscle (56). This immediately indicated that auxiliary subunits (β, γ, α_2-δ) were not required for O_2 sensing. Three splice variants of the α_{1C}-subunit used in these studies occur naturally, and they vary only in a specific region of the C-terminus (Fig. 4a). The longest variant, containing a 71-amino-acid insert (Fig. 4b, hHT), was the one used originally in these studies. The two other variants either lacked this insert completely (Fig. 4c, rHT) or possessed a smaller, structurally unrelated sequence (Fig. 4d, fHT); neither of these variants displayed any sensitivity to acute hypoxia (Fig. 4c,d),

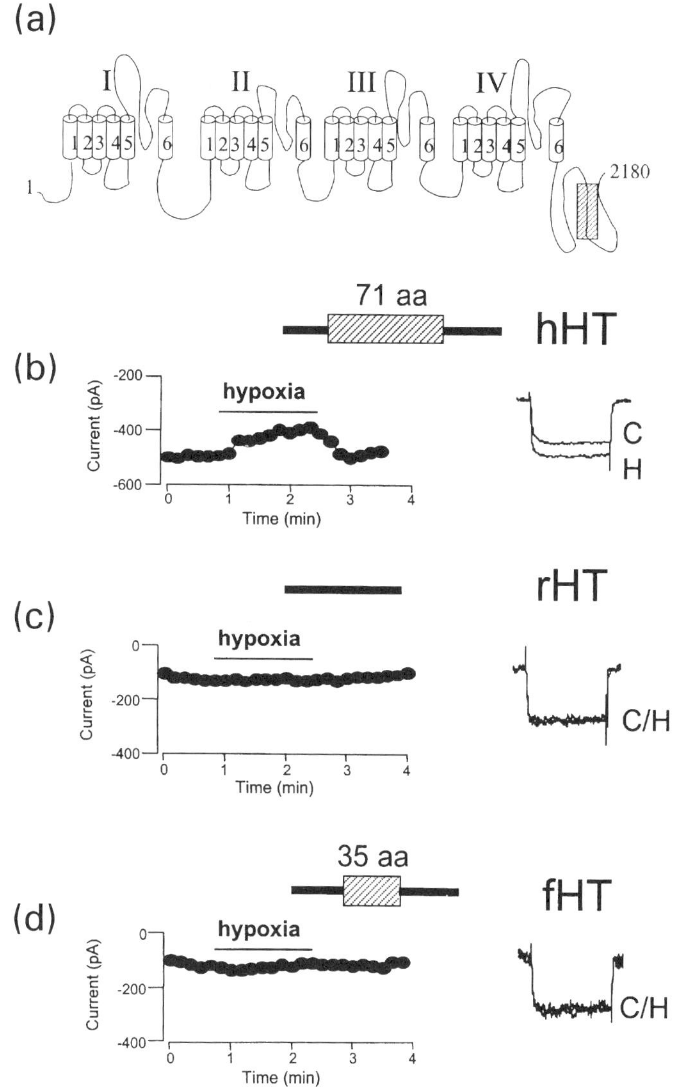

Figure 4 Differential O_2 sensitivity of L-type Ca^{2+} channel α_{1C}-subunit splice variants. (a) Schematic representation of the full-length α_{1C}-subunit of the L-type Ca^{2+} channel, showing the four repeating domains, each containing six transmembrane alpha helices. The hatched box at the intracellular C-terminal indicates the only site of splice variation. (b–d) Whole-cell Ca^{2+} currents (right panels) evoked in HEK293 cells transiently transfected with α_{1C}-subunit splice variants hHT (b), rHT (c), and fHT (d) when stepping from a holding potential of -80 mV to 0 mV for 100 msec (C = control normoxia, H = hypoxia). In the left-hand panels are shown representative time-series plots for each splice variant before, during, and after perfusion with hypoxic solution; horizontal bar indicates duration of exposure to hypoxia. Also shown pictorially in each panel is the length of the splice insert.

indicating that this insert site was an absolute requirement for O_2 sensing by L-type channels (57). This was supported by loss- and gain-of-function of mutants, and the insert region was restricted further in mutational studies to a 39-amino-acid region (57). These studies were the first to reveal a part of an ion channel protein required for O_2 sensing and serve to illustrate the potential value of recombinant ion channels in understanding the molecular basis for O_2 sensing. However, it should be noted that modulation of L-type Ca^{2+} channels in rabbit glomus cells may differ: while one report indicates hypoxic inhibition in these cells similar to the above-described effects (15), others indicate that hypoxia augments L-type Ca^{2+} channel activity via a mechanism involving activation of protein kinase C (58).

VI. Concluding Remarks

Since their discovery in 1988, there has been a remarkable increase in our awareness of O_2-sensitive K^+ channels, not only in chemoreceptor cells but in a diverse array of tissues. What has also become apparent over this time is that O_2 sensing is not a property of any specific K^+ channel type, and that structurally diverse channel subfamilies can be regulated by the O_2 content of the local environment. While it remains to be demonstrated definitively, available evidence indicates that O_2 sensing is not an intrinsic property of the K^+ channel protein and that other sensory proteins mediate O_2 sensing. Exactly what these proteins are will require further study, but again they may vary in different tissues: this is clear from the fact that NADPH oxidase serves the role of O_2 sensor in airway chemoreceptors, but is not involved in O_2 sensing by the carotid body or the pulmonary vasculature. Finally, while molecular details of structural requirements for O_2 sensing are lacking at present, the discovery that recombinant K^+ channels retain their O_2 sensitivity in heterologous expression systems holds great promise for future studies into this important physiological process.

Acknowledgments

The authors' own contributions to the studies described in this article are supported by the Wellcome Trust, the British Heart Foundation, and the Medical Research Council.

References

1. Lopez-Barneo J, Lopez-Lopez JR, Urena J, Gonzalez C. Chemotransduction in the carotid body: K^+ current modulated by PO_2 in type I chemoreceptor cells. Science 1988; 241:580–582.
2. Weir EK, Archer SL. The mechanism of acute hypoxic pulmonary vasoconstriction; the tale of two channels. FASEB J 1995; 9:183–189.
3. Youngson C, Nurse CA, Yeger H, Cutz E. Oxygen sensing in airway chemoreceptors. Nature 1993; 365:153–155.

4. O'Kelly I, Peers C, Kemp PJ. Oxygen-sensitive K^+ channels in neuroepithelial body-derived small cell carcinoma cells of the human lung. Am J Physiol 1998; 275:L709–L716.
5. Thompson RJ, Nurse CA. Anoxia differentially modulates multiple K^+ currents and depolarizes neonatal rat adrenal chromaffin cells. J Physiol 1998; 512:421–434.
6. Zhu WH, Conforti L, Czyzyk-Krzeska MF, Millhorn DE. Membrane depolarization in PC12 cells during hypoxia is regulated by an O_2-sensitive K^+ current. Am J Physiol 1996; 271:C658–C665.
7. Jiang C, Haddad GG. Oxygen deprivation inhibits potassium channel independently of cytosolic factors in rat central neurons. J Physiol 1994; 481:15–26.
8. Lopez-Barneo J, Pardal R, Ortega-Saenz P. Cellular mechanism of oxygen sensing. Annu Rev Physiol 2001; 63:259–287.
9. Patel AJ, Honore E. Molecular physiology of oxygen-sensitive channels. Eur Respir J 2001; 18:221–227.
10. Ganfornina A, Lopez-Barneo J. Potassium channel types in arterial chemoreceptor cells and their selective modulation by oxygen. J Gen Physiol 1992; 100:401–426.
11. Lopez-Barneo J, Pardal R, Montoro RJ, Smani T, Garcia-Hirschfeld J, Urena J. K^+ and Ca^{2+} channel activity and cytosolic $[Ca^{2+}]$ in oxygen-sensing tissues. Respir Physiol 1999; 115:215–227.
12. Sanz-Alfayate G, Obeso A, Agapito MT, Gonzalez C. Reduced to oxidised glutathione ratio and oxygen sensing in carotid body chemoreceptor cells. J Physiol 2001; 537:209–220.
13. Lopez-Lopez JR, Gonzalez C. Time course of K^+ current inhibition by low oxygen in chemoreceptor cells of adult rabbit carotid body: effects of carbon monoxide. FEBS Lett 1992; 299:251–254.
14. Lahiri S, Iturriaga R, Mokashi A, Ray DK, Chugh D. CO reveals dual mechanisms of O_2 chemoreception in the cat carotid body. Respir Physiol 1993; 94: 227–240.
15. Montoro RJ, Urena J, Fernandez-Chacon R, Alvarez de Toledo G, Lopez-Barneo J. Oxygen sensing by ion channels and chemotransduction in single glomus cells. J Gen Physiol 1996; 107:133–143.
16. Urena J, Fernandez-Chacon R, Benot AR, Alvarez de Toledo G, Lopez-Barneo J. Hypoxia induces voltage-dependent Ca^{2+} entry and quantal dopamine secretion in carotid body glomus cells. Proc Natl Acad Sci USA 1994; 91:10208–10211.
17. Delpiano MA, Hescheler J. Evidence for a P_{O_2}-sensitive K^+ channel in the type-I cell of the rabbit carotid body. FEBS Lett 1989; 249:195–198.
18. Peers C. Hypoxic suppression of K^+ currents in type-I carotid-body cells—selective effect on the Ca^{2+}-activated K^+ current. Neurosci Lett 1990; 119:253–256.
19. Buckler KJ, Williams BA, Honore E. An oxygen-, acid- and anaesthetic-sensitive TASK-like background potassium channel in rat arterial chemoreceptor cells. J Physiol 2000; 525:135–142.
20. Wyatt CN, Wright C, Bee D, Peers C. O_2-sensitive K^+ currents in carotid-body chemoreceptor cells from normoxic and chronically hypoxic rats and their roles in hypoxic chemotransduction. Proc Natl Acad Sci USA 1995; 92:295–299.
21. Buckler KJ, Vaughan-Jones RD. Effects of hypoxia on membrane potential and intracellular calcium in rat neonatal carotid body type I cells. J Physiol 1994; 476:423–428.
22. Lesage F, Lazdunski M. Molecular and functional properties of two-pore-domain potassium channels. Am J Physiol 2000; 793:F793–F801.

23. Buckler KJ, Vaughan-Jones RD, Peers C, Nye PCG. Intracellular pH and its regulation in isolated type-I carotid body cells of the neonatal rat. J Physiol 1991; 436:107–129.
24. Duprat F, Lesage F, Fink M, Reyes R, Heurteaux C, Lazdunski M. TASK, a human background K^+ channel to sense external pH variations near physiological pH. EMBO J 1997; 16:5464–5471.
25. Pardal R, Ludewig U, Garcia-Hirschfeld J, Lopez-Barneo J. Secretory responses of intact glomus cells in thin slices of rat carotid body to hypoxia and tetraethylammonium. Proc Natl Acad Sci USA 2000; 97:2361–2366.
26. Jackson A, Nurse CA. Role of acetylcholine receptors and dopamine transporter in regulation of extracellular dopamine in rat carotid body cultures grown in chronic hypoxia or nicotine. J Neurochem 1998; 70:653–662.
27. Green FK, Peers C. Inhibition of Ca^{2+}-activated K^+ currents by intracellular acidosis in isolated type I cells of the neonatal rat carotid body. J Physiol 1991; 437:589–602.
28. Hatton CJ, Carpenter E, Pepper DR, Kumar P, Peers C. Developmental changes in isolated rat type I carotid body cell K^+ currents and their modulation by hypoxia. J Physiol 1997; 501:49–58.
29. Peers C, Kemp PJ. Acute oxygen sensing: diverse but convergent mechanisms in airway and arterial chemoreceptors. Respir Res 2001; 2:145–149.
30. Wyatt CN, Peers C. Ca^{2+}-activated K^+ channels in isolated type-I cells of the neonatal rat carotid-body. J Physiol 1995; 483:559–565.
31. Buckler KJ, Vaughan-Jones RD. Effects of mitochondrial uncouplers on intracellular calcium, pH and membrane potential in rat carotid body type I cells. J Physiol 1998; 513:819–833.
32. Riesco-Fagundo AM, Perez-Garcia MT, Gonzalez C, Lopez-Lopez JR. O_2 modulates large-conductance Ca^{2+} dependent K^+ channels of rat chemoreceptor cells by a membrane-restricted and CO-sensitive mechanism. Circ Res 2001; 89:430–436.
33. Cutz E, Jackson A. Neuroepithelial bodies as airway oxygen sensors. Respir Physiol 1999; 115:201–214.
34. Lauweryns JM, Cokeleare M. Hypoxia sensitive neuroepithelial bodies intrapulmonary secretory neuroreceptors, modulated by CNS. Z Zellforsch Mikrosk Anat 1973; 145:521–540.
35. Fu XW, Nurse C, Wang YT, Cutz E. Selective modulation of membrane currents by hypoxia in intact airway chemoreceptors from neonatal rabbit. J Physiol 1999; 514:139–150.
36. Youngson C, Nurse, Yeger H, Cutz E. Characterization of membrane currents in pulmonary neuroepithelial bodies: hypoxia-sensitive airway chemoreceptors. In O'Regan RG, Nolan P, McQueen DS, Paterson DJ, eds. Arterial Chemoreceptors. Cell to System. New York and London: Plenum Press, 1994:179–182.
37. Gazdar AF, Helman LJ, Israel MA, Russell EK, Linnoila RI, Mulshine JL, Schuller HM, Park GJ. Expression of neuro-endocrine cell markers L-DOPA decarboxylase, chromogranin-A, and dense core granules in human-tumors of endocrine and nonendocrine origin. Cancer Res 1988; 48:4078–4082.
38. O'Kelly I, Stephens RH, Peers C, Kemp PJ. Potential identification of the O_2-sensitive K^+ current in a human neuroepithelial body-derived cell line. Am J Physiol 1999; 276:L96–L104.
39. Hartness ME, Lewis A, Searle GJ, O'Kelly I, Peers C, Kemp PJ. Combined antisense and pharmacological approaches implicate hTASK as an airway O_2 sensing K^+ channel. J Biol Chem 2001; 276:26499–26508.

40. Millar JA, Barratt L, Southan AP, Page KM, Fyffe RE, Robertson B, Mathie A. A functional role for the two-pore domain potassium channel TASK-1 in cerebellar granule neurons. Proc Natl Acad Sci USA 2000; 97:3614–3618.
41. Cross AR, Herderson L, Jones OTG, Delpiano MA, Hentschel J, Acker H. Involvement of an NAD(P)H oxidase as a pO_2 sensor protein in the rat carotid body. Biochem J 1990; 272:743–747.
42. Wang D, Youngson C, Wong V, Yeger H, Dinauer MC, Vega-Saenz de Miera E, Rudy B, Cutz E. NADPH-oxidase and hydrogen peroxide sensitive K^+ channel may function as an oxygen sensor complex in airway chemoreceptors and small cell lung carcinoma cell lines. Proc Natl Acad Sci USA 1996; 93:13182–13187.
43. O'Kelly I, Lewis A, Peers C, Kemp PJ. O_2 sensing by airway chemoreceptor-derived cells: protein kinase C activation reveals functional evidence for involvement of NADPH oxidase. J Biol Chem 2000; 275:7684–7692.
44. Fu XW, Wang D, Nurse CA, Dinauer MC, Cutz E. NADPH oxidase is an O_2 sensor in airway chemoreceptors: evidence from K^+ current modulation in wild type and oxidase-deficient mice. Proc Natl Acad Sci USA 2000; 97:4374–4379.
45. Roy A, Rozanov C, Mokashi A, Daudu P, Al-Mehdi AB, Shams H, Lahiri S. Mice lacking in gp91 phox subunit of NAD(P)H oxidase showed glomus cell $[Ca^{2+}]_i$ and respiratory responses to hypoxia. Brain Res 2000; 872:188–193.
46. Archer SL, Reeve HL, Michelakis E, Puttagunta L, Waite R, Nelson DP, Dinauer MC, Weir EK. O_2 sensing is preserved in mice lacking the gp91 phox subunit of NADPH oxidase. Proc Natl Acad Sci USA 1999; 96:7944–7949.
47. Haddad GG, Jiang C. O_2-sensing mechanisms in excitable cells: role of plasma membrane K^+ channels. Annu Rev Physiol 1997; 59:23–42.
48. Jiang C, Haddad GG. Oxygen deprivation inhibits a K^+ channel independently of cytosolic factors in rat central neurons. J Physiol 1994; 481:15–26.
49. Jiang C, Haddad GG. A direct mechanism for sensing low oxygen levels by central neurons. Proc Natl Acad Sci USA 1994; 91:7198–7201.
50. Brickley SG, Revilla V, Cull-Candy SG, Wisden W, Farrant M. Adaptive regulation of neuronal excitability by a voltage-independent potassium conductance. Nature 2001; 409:88–92.
51. Talley EM, Lei Q, Sirois JE, Bayliss DA. TASK-1, a two-pore domain K^+ channel, is modulated by multiple neurotransmitters in motoneurons. Neuron 2000; 25:399–410.
52. Maingret F, Patel AJ, Lazdunski M, Honore E. The endocannabinoid anandamide is a direct and selective blocker of the background K^+ channel TASK-1. EMBO J 2001; 20:47–54.
53. Lewis A, Hartness ME, Chapman CG, Fearon IM, Meadows HJ, Peers C, Kemp PJ. Recombinant hTASK1 is an O_2-sensitive K^+ channel. Biochem Biophys Res Commun 2001; 285:1290–1294.
54. Franco-Obregon A, Urena J, Lopez-Barneo J. Oxygen-sensitive calcium channels in vascular smooth muscle and their possible role in hypoxic arterial relaxation. Proc Natl Acad Sci USA 1995; 92:4715–4719.
55. Klockner U, Mikala G, Eisfeld J, Iles DE, Strobeck D, Mershon JL, Schwartz A, Varadi G. Properties of three COOH-terminal splice variants of a human cardiac L-type Ca^{2+} channel α-subunit. Am J Physiol 1997; 272:H1372–H1381.
56. Fearon IM, Palmer ACV, Balmforth AJ, Ball SG, Mikala G, Schwartz A, Peers C. Hypoxia inhibits the recombinant α_{1C} subunit of the human cardiac l-type Ca^{2+} channel. J Physiol 1997; 500:551–556.

57. Fearon IM, Varadi G, Koch S, Isaacsohn I, Ball SG, Peers C. Splice variants reveal the region involved in oxygen sensing by recombinant human L-type Ca^{2+} channels. Circ Res 2000; 87:537–539.
58. Summers BA, Overholt JL, Prabhakar NR. Augmentation of L-type calcium current by hypoxia in rabbit carotid body glomus cells: evidence for a PKC-sensitive pathway. J Neurophysiol 2000; 84:1636–1644.

17

Carotid Body Thin Slices

New Answers for Old Questions

JOSÉ LÓPEZ-BARNEO and RICARDO PARDAL

University of Seville
Seville, Spain

I. Introduction

Oxygen is sensed by almost all living organisms (1,2). In mammals, survival in acute hypoxia requires fast respiratory and cardiocirculatory adjustments to ensure sufficient O_2 supply to the most critical organs such as the brain or the heart (3). The main O_2 sensor mediating the acute responses to hypoxia is the carotid body, a minute bilateral organ located in the bifurcation of the carotid artery, which contains afferent nerve fibers that activate the brainstem respiratory centers to produce hyperventilation. A major advance in the understanding of carotid body physiology was the discovery that the neuroectodermal-derived glomus cells, the O_2-sensitive elements in the carotid body, are electrically excitable (4,5) and have O_2-regulated K^+ channels in their plasma membrane (5–10). It is broadly accepted that hypoxia signaling in glomus cells involves inhibition of K^+ channels of the plasma membrane leading to cell depolarization, external Ca^{2+} influx, and activation of neurotransmitter release, which, in turn, stimulates the afferent sensory fibers. Low O_2 tension (P_{O_2}) has also been reported to augment Ca^{2+} current amplitude in rabbit glomus cells (11). This "membrane model" of chemotransduction, suggested by the electrophysiological experiments, has been confirmed by monitoring cytosolic $[Ca^{2+}]$ and quantal catecholamine secretion in single cells (12–17).

The mechanism of acute O_2 sensing based on the regulation of membrane K^+ channels seems to be a rather general phenomenon that has been shown to operate in other neurosecretory systems, such as the cells in the neuroepithelial bodies of the lung (18), chromaffin cells of the adrenal medulla (19), or PC 12 cells

(20), as well as in the pulmonary arterial smooth muscle (21–23). However, some investigators believe that the O_2-sensitive membrane electrical events in glomus cells are not directly involved in sensory transduction. The main argument supporting this belief is that inhibitors of the O_2-sensitive K^+ current, such as tetraethylammonium (TEA), 4-aminopyridine (4-AP), or charybdotoxin (CTX), do not increase the action potential firing frequency of the afferent sensory fibers or the secretory activity in the whole carotid body preparations used in these studies (24–26). Although it was shown that CTX can depolarize dialyzed rat glomus cells (27), it has also been reported that neither TEA nor 4-AP influences the membrane potential of the same cells (10). In contrast with these observations, a more recent study has demonstrated that 4-AP reversibly regulates the resting membrane potential of perforated-patched rabbit glomus cells (28).

Given the discrepancies among observations reported by different laboratories using dispersed glomus cells and the contradictions between findings in isolated cells and those in whole carotid bodies, we have developed a slice preparation of these organs to study the O_2 sensitivity of glomus cells in the best possible physiological conditions (17,29). We also attempted to obtain a preparation of glomus cells with consistent properties since O_2 sensitivity is a labile characteristic easily disrupted by the enzymatic treatment and/or mechanical dispersion of the cells (30). The carotid body thin slice has been found to be an excellent preparation to study the cellular bases for the sensitivity of glomus cells to hypoxia and has served to demonstrate that, as suggested in the old literature (31,32), the carotid body is a polimodal sensory organ activated by changes of multiple chemical and physical variables. In this chapter, we summarize our recent studies on rat carotid body cells in slices using patch-clamp recordings and single-cell amperometry to monitor secretion (17,29,33). Our work gives full support to the membrane model of chemotransduction in the carotid body and further stresses the importance of voltage-gated K^+ channels as effector molecules in the acute response of glomus cells to hypoxia. We have also found that glomus cells are highly sensitive, low-glucose detectors. This novel observation strongly suggests that the carotid bodies have a major role in blood glucose homeostasis.

II. Carotid Body Thin Slices

A. Methodology

The procedures followed to make carotid body slices are described in detail elsewhere (17). Briefly, carotid bodies were isolated from Wistar rats of ages between postnatal day 15 and 30 and then included in 3% (weight/volume) low-melting-point agarose. We used animals older than 15 days to allow for complete postnatal maturation of carotid body sensitivity to hypoxia (34,35). After quick cooling on ice, the agarose blocks were glued to the chamber of a Lancer vibratome and the carotid bodies sectioned in slices $\approx 150\,\mu m$ thick and $\approx 400\,\mu m$ in diameter. The resulting three or four slices per carotid body were washed with sterile PBS and then cultured with DMEM (10% FBS and 1% penicillin-streptomycin) at 37°C in a 5% CO_2

incubator. In healthy slices, clusters of glomus cells were clearly distinguished from the surrounding tissue (Fig. 1A). These clusters, of similar appearance as the glomeruli described in histological preparations of the carotid body (36), contained numerous ovoid cells $\approx 12\,\mu m$ in diameter (Fig. 1A). The similarity between fresh and fixed carotid body preparations can be appreciated by observing slices immunostained with antibodies against tyrosine hydroxylase (TH), where glomus cells appear in clusters with intensely stained thin cytoplasmic layers and big clear nuclei (Fig. 1B). The usual incubation time of slices in our experiments was between 48 and 72 h. Although the longevity of the slices was not studied in detail, these were kept for up to 6 days in culture without signs of deterioration. For the experiments a

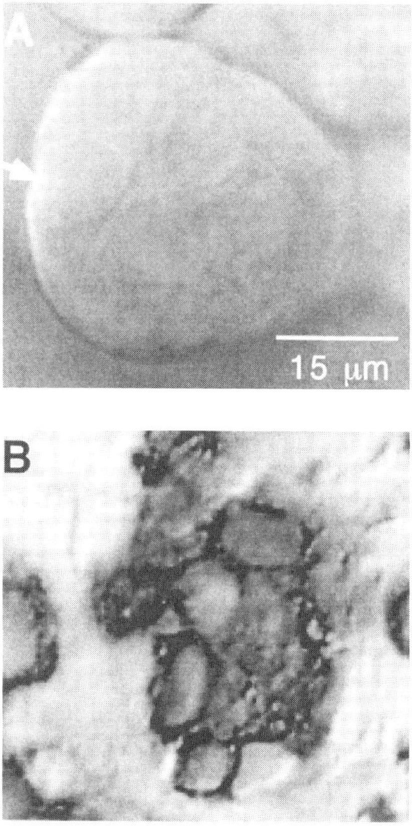

Figure 1 Morphological appearance of a glomerulus within a cultured rat carotid body thin slice. (A) Typical glomerulus in a slice maintained for 72 hr in culture. A well-defined single cell is indicated by the arrow. (B) Carotid body slice immunostained with antibodies against tyrosine hydroxilase. The carotid body was fixed, then sliced and stained. Note the typical appearance of glomus cells with large nuclei and a thin layer of stained cytoplasm. The organization of type I cells in glomeruli is similar to that seen in fresh slices. (Modified from Ref. 17.)

slice was transferred to the recording chamber mounted on the stage of an upright microscope, where it was continuously perfused by gravity (flow 1–2 mL/min) with a solution containing (in mM) 117 NaCl, 4.5 KCl, 23 $NaHCO_3$, 1 $MgCl_2$, 2.5 $CaCl_2$, and 10 glucose. The recording electrodes (either patch pipette or carbon fiber) were placed adjacent to a well-identified cell within a glomerulus, like the one indicated by the arrow in Figure 1A. The entire recording setup was maintained at a temperature between 34 and 37°C.

B. Electrophysiology of Cells in the Slices

Stable recordings of membrane currents can easily be obtained from glomus cells in the slices using either the whole-cell or the excised patch configuration of the patch-clamp technique as adapted in our laboratory (37,38) (Fig. 2). We normally used low-resistance pipettes (1–3 MΩ) filled with a solution containing (in mM) 125 KCl, 4 $MgCl_2$, 4 MgATP, 10 Hepes, and 10 EGTA (pH adjusted to 7.2 with KOH). Ionic currents were recorded with series resistance compensation (between 40 and 50%) and studied after subtraction of linear leakage and capacitive currents. In the slices cell input resistance (between 0.2 and 1 GΩ) was smaller than in isolated rat and rabbit carotid body cells (10,37). These differences could simply reflect the higher temperature used in the present experiments but they could also be due to the existence of some electrical coupling between neighboring cells in the slices. Thin slices are a good preparation for future studies on the existence of electrical coupling between glomus cells and on how this influences the physiology of the carotid body.

The majority of the cells recorded in the slices (77% from a total number of 110) had inward and outward voltage-dependent currents qualitatively similar to those described in enzymatically dispersed rat glomus, or type I, cells (27,39). The remaining 23% of cells ($n = 26$) had no measurable currents or presented a small transient outward current, thus suggesting that they were type II or sustentacular cells, present in the carotid body (4,37). Glomus cells had typically small inward currents followed by outward currents (Fig. 2b) that in some cases had a characteristic transient current component (Fig. 2c). The outward currents were highly sensitive to the application of external TEA (5 mM caused an inhibition of $67.5 \pm 6\%$ at +20 mV, mean ± standard deviation, $n = 4$ cells) or the selective calcium-dependent K^+ channel blocker iberiotoxin (IbTX, 200 nM, caused an inhibition of 65% and 45% at +20 mV in two cells tested). Therefore, as described in isolated cells (27,39), a large proportion of the outward current was primarily due to the activity of maxi-K^+ voltage- and calcium-dependent channels. A small percentage (9%) of the glomus cells studied in the slices had a considerable amount of inward current (Fig. 2d). Although we have not yet studied these currents in detail, they are very similar to the Na^+ currents reported in a small population of dispersed rat glomus cells (39). We have also recorded several classes of single-channel K^+ currents from outside-out excised membrane patches. An example of maxi-K^+ single-channel currents recorded from a patch containing at least four channels is illustrated in Figure 2f and g.

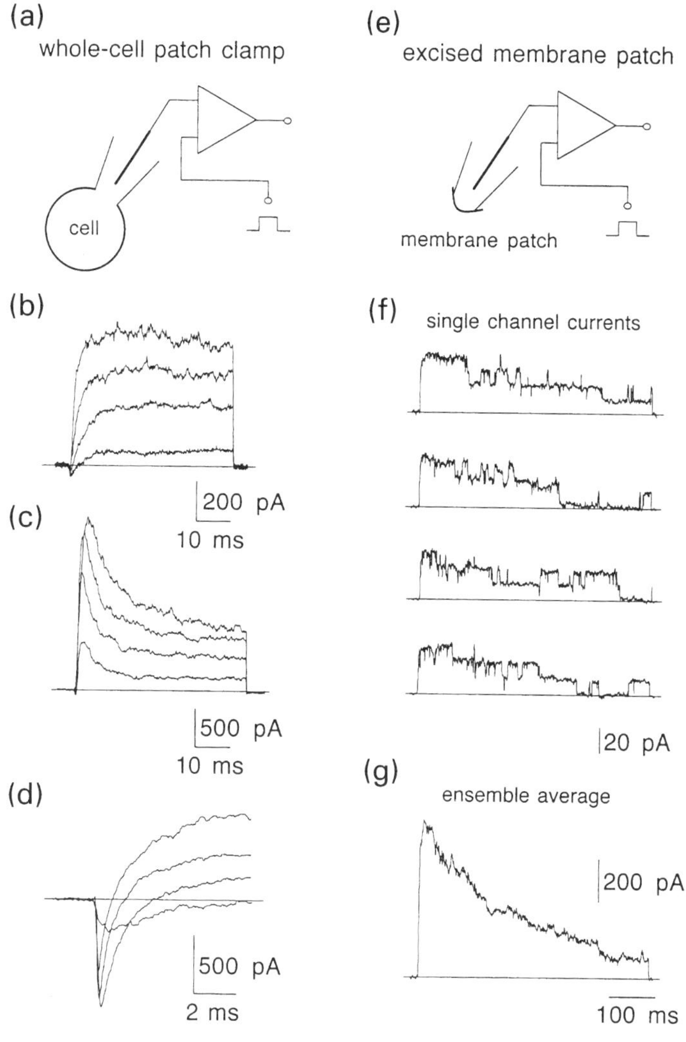

Figure 2 Patch-clamp recordings from single cells in the slices. (a) Diagram of the whole-cell configuration of the patch-clamp technique used to record macroscopic voltage-dependent currents. (b–d) Recordings obtained from single cells in the slices and elicited by depolarizing pulses from the holding potential of −80 mV to −20, 0, +20, and +40 mV (in each case the four pulses are superimposed). (e) Diagram of the outside-out excised patch configuration of the patch-clamp technique used to record single-channel currents. (f) Traces of single-channel currents obtained by depolarizing steps to +40 mV from −80 mV. (g) Average of 20 traces similar to those shown in (f). (Modified from Ref. 29.)

C. Monitoring Secretion from Single Cells

Glomus cells have numerous secretory vesicles containing the transmitters (mainly ATP and acetylcholine) that, once released, activate the afferent sensory fibers of the sinus nerve (40). Dopamine is also highly concentrated in the secretory granules but its precise role is under discussion. It is possible that the major action of dopamine released in the carotid body is autocrine since this agent inhibits selectively the Ca^{2+} current in glomus cells (41). We have found the slice preparation to be particularly convenient to study responsiveness of intact glomus cells to hypoxia and other stimuli using the amperometric detection of catecholamines (13,16,17). Quantal transmitter release can be monitored with a polarized 8–12-μm carbon-fiber electrode positioned near the surface of a glomus cell (Fig. 3a). The fiber is connected to the input of a high-gain current-to-voltage converter and polarized to +750 mV, a value more positive than the redox potential of dopamine, the most abundant catecholamine in glomus cells. A representative secretory response of a glomus cell to depolarization by a high external K^+ solution is shown in Figure 3b (top). Spike-like signals, with the typical shape observed at an expanded time scale in Figure 3c, resulted from the fusion of single vesicles. When the secretory activity was high, individual exocytotic events fused in a concentration envelope. The area under the individual spikes (quantal charge) yields an estimate of the number of catecholamine molecules released assuming that two electron charges are transferred to the fiber during the oxidation of a catecholamine molecule. Average quantal charge estimated from high K^+-induced events was 40.3 ± 17.6 fC ($n = 293$ spikes in seven cells), corresponding to $\approx 125{,}000 \pm 50{,}000$ molecules per vesicle. These values, obtained from cells in slices, are comparable to those previously described in dispersed rat and rabbit glomus cells (13,16,17). The magnitude of the secretory responses of the cells was estimated either from the number of spikes in the minute after 90 sec of exposure to the stimulus (secretion rate in events/min) or from the sum of all quantal charges measured in the same time period and expressed as femtocoulombs per minute (fC/min; cumulative secretion in Fig. 3b, bottom). This last variable represents the total amount of catecholamine molecules released per minute at the peak of the response.

The majority of glomus cells in the slices showed no secretory activity at rest, or less than one event per minute. However, some cells were spontaneously more active, with a secretion rate of about 10 or more spikes per minute. Although we have not studied the causes of this secretion in detail, it could be that these are cells damaged by mechanical contact with the carbon fiber that have acquired an unusually high intracellular $[Ca^{2+}]$. Another possibility is that these are electrically active cells, generating action potentials spontaneously, and therefore with high cytoplasmic $[Ca^{2+}]$.

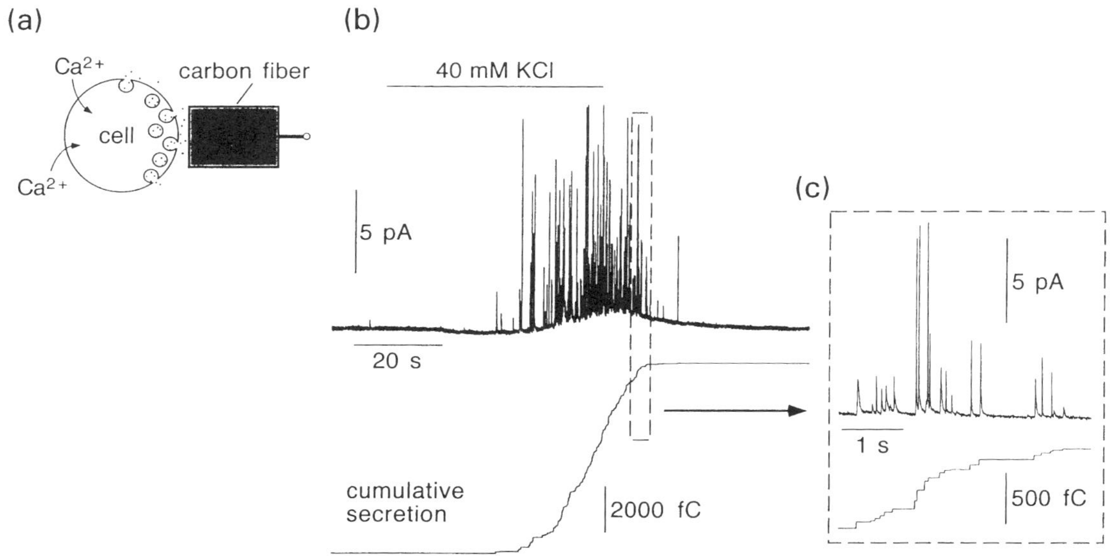

Figure 3 Monitoring of secretion from single cells by amperometry. (a) Diagram of the electrochemical detection of catecholamine released by exocytosis from a glomus cell with a polarized carbon-fiber electrode. (b) Top: Secretory response of a glomus cell to high extracellular potassium. Each spike in the recording corresponds to the oxidation of catecholamines released from a single vesicle. (b) Bottom: Cumulative secretion signal (in femtocoulombs) resulting from the time integral of the amperometric recording. (c) View of the traces in the box in (b) at an expanded time scale. Note the typical shape of the secretory events with fast-rising phase and slower decay. (Modified from Ref. 29.)

III. Responses of Glomus Cells to Hypoxia and Potassium Channel Blockers

In experiments designed to test the effect of hypoxia, the control (normoxic) solution was bubbled with a gas mixture of 5% CO_2, 20% O_2, and 75% N_2 ($P_{O_2} \approx 150$ mmHg), and the hypoxic solution with 5% CO_2, and 95% N_2 (P_{O_2} in the chamber ≈ 20 mmHg). After switching from normoxia to hypoxia, complete equilibration of the new solution in the chamber required between 1 and 2 min. In slices with well-defined glomeruli, low P_{O_2} induced consistently a progressive increase in the frequency of secretory events (Fig. 4a), reaching a value of 48.6 ± 19 spikes/min ($n = 24$ cells) and a cumulative secretion rate of 1710 ± 270 fC/min ($n = 17$ cells). At the peak of the response to hypoxia the secretory events fused into a broad concentration envelope that quickly declined after switching to the control, normoxic, solution. All the glomus cells that responded to hypoxia were also activated by solutions with high external K^+ (Fig. 4a), as expected from electrically excitable cells. Interestingly, we also observed glomus cells that were unresponsive to hypoxia but were activated by depolarization with high external K^+. Cells insensitive to hypoxia were more frequently observed in slices that appeared not too healthy, possibly owing to damage during the experimental protocol. One of the possible explanations is that these are cells with the O_2 sensor uncoupled from the membrane ion channels. In all cells tested ($n = 10$), the neurosecretory response to hypoxia was completely abolished by the addition of the voltage-dependent calcium channel blocker cadmium (Fig. 4b) or the removal of extracellular calcium with EGTA (17). This observation has confirmed previous data on dispersed rabbit carotid body cells (13,15).

As described in dispersed rabbit (5,6,8,28) and rat (7,9,27,39) glomus cells, the amplitude of macroscopic voltage-dependent currents was reduced when exposed to low P_{O_2}. Typical recordings of O_2-sensitive currents are shown in Figure 4c. Although our work on slices has focused so far on the study of secretion rather than on the recording of ionic currents, reversible reduction of K^+ current amplitude by hypoxia is a response seen less consistently than the increase of secretory activity monitored by amperometry. These observations could mean that in patch-clamped glomus cells the O_2-sensing mechanism is altered and the sensitivity to low P_{O_2} decreased possibly owing to intracellular dialysis. It is also possible that, besides voltage-gated K^+ channels, other conductances, mediated by voltage-gated Ca^{2+} (11) or K^+-selective leaky (10) channels, not studied so far in slices, also contribute to mediate the low P_{O_2}-induced secretory response.

The major contributors to the O_2-sensitive macroscopic K^+ currents in rat glomus cells are voltage- and Ca^{2+}-dependent maxi-K^+ channels (7,27,39). Because these channels are blocked by TEA or iberiotoxin (IbTX), we have studied whether, like hypoxia, addition of these agents to the external solution induces Ca^{2+} entry and secretion from glomus cells. In most cells studied (33 of 34), application of 5 mM TEA to the bath solution elicited an increase in the secretory activity similar to that triggered by hypoxia (Fig. 5a). The response to the blocker reached a secretion rate of 42 ± 17 spikes/mm ($n = 6$ cells), with a cumulative secretion of

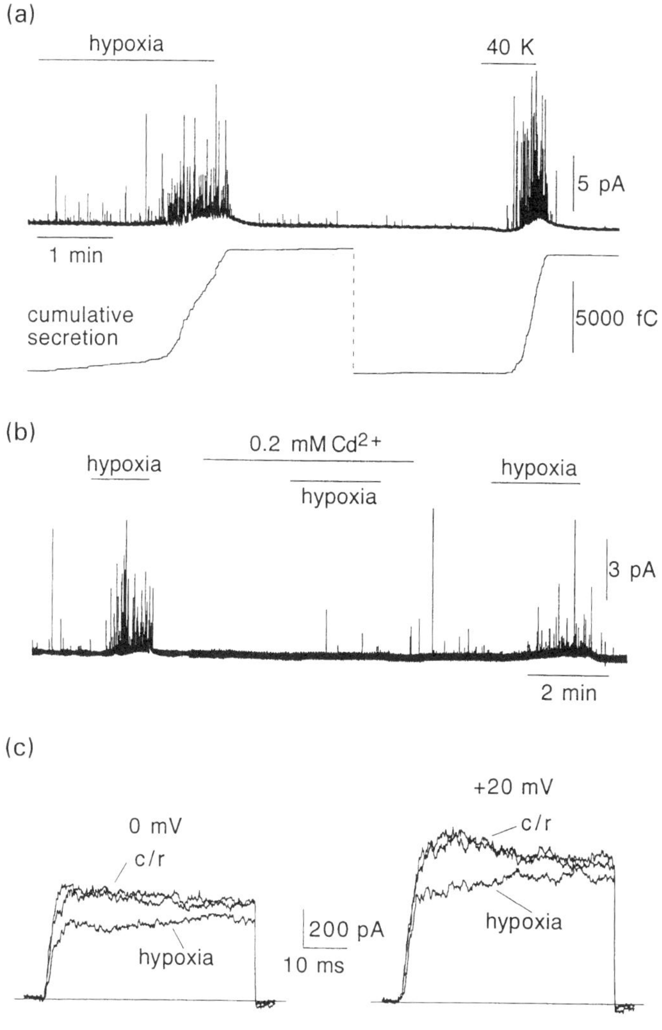

Figure 4 Secretory responses to hypoxia of glomus cells in the slices. (a) Amperometric recording from an O_2-sensitive glomus cell illustrating the increase of secretory activity elicited by hypoxia ($P_{O_2} \approx 20$ mmHg). Note the typical response of the cell to high extracellular potassium. The bottom panel shows the cumulative secretion signal. Resetting of the integrator used to calculate cumulative secretion is indicated by dotted lines. (b) Secretory activity recorded from an O_2-sensitive glomus cell to illustrate the reversible abolishment of the response to hypoxia during the blockade of Ca^{2+} channels by addition of 0.2 mM cadmium to the extracellular solution. (Modified from Ref 17.) (c). Superimposed macroscopic K^+ currents from a glomus cell elicited by depolarizing pulses from -80 mV to the indicated voltage in the three experimental conditions (control, low P_{O_2}, and recovery). Note the reversible reduction of the current by hypoxia ($P_{O_2} \approx 20$ mmHg). (Modified from Ref. 29.)

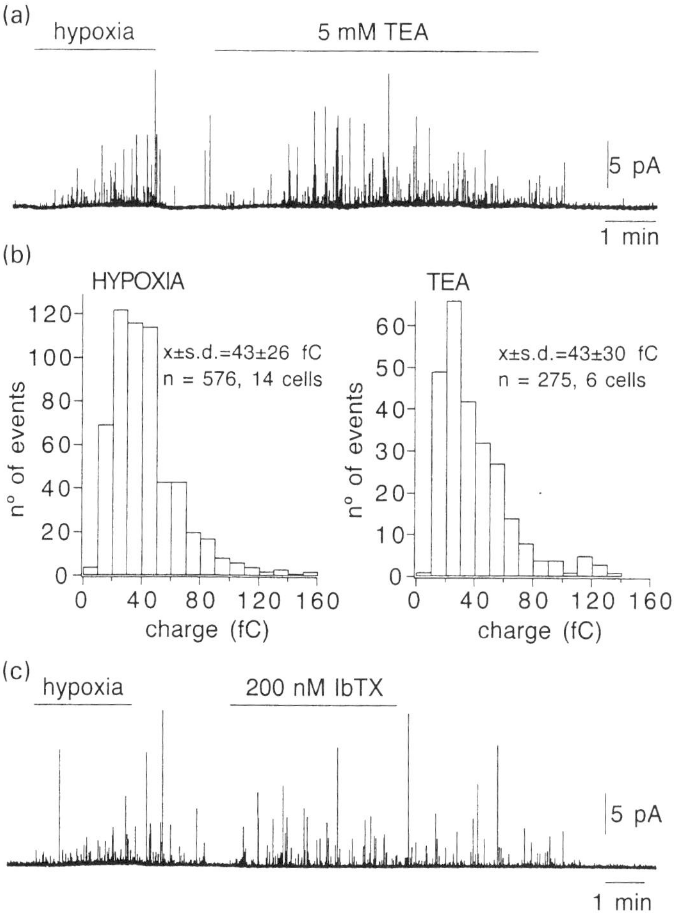

Figure 5 Secretory responses of intact glomus cells exposed to K^+ channel blockers. (a) Amperometric recording from a glomus cell illustrating the similar effects elicited by low P_{O_2} and the application of 5 mM TEA. (b) Frequency histograms of the quantal charge of events elicited by hypoxia and TEA. Note that the parameters of the distributions are the same in the two experimental conditions. (c) Secretory activity induced in a glomus cell by hypoxia and 200 nM IbTX. (Modified from Ref. 17.)

1878 ± 470 fC/min ($n = 6$). These values are not significantly different from the respective ones obtained in low P_{O_2} (Student's t-test, $\alpha = 0.05$). The average quantal charge of events induced by TEA was 43 ± 30 fC ($n = 275$ spikes in six cells). This value and the distribution of quantal events (Fig. 5b) are also similar to those estimated with events elicited by hypoxia (43 ± 26 fC; $n = 576$ spikes in 14 cells), suggesting that both stimuli can trigger the release of vesicles from the same cellular pool. Although we have recorded one cell that responded to hypoxia but was insensitive to TEA alone, when TEA and low P_{O_2} were applied together their effects were normally additive (17). This result was expected since, besides maxi-K^+, TEA is known to block other voltage-gated K^+ channels. The effect of TEA was observed even in quiescent cells, without any measurable spontaneous quantal release, as well as in O_2-insensitive glomus cells. We have also tested the effect of IbTX, a selective blocker of Ca^{2+}- and voltage-activated maxi K^+ channels (42). Figure 5c illustrates the increase of secretory activity in a glomus cell exposed to 200 nM IbTX. The response is similar to those obtained with TEA or hypoxia, although the recovery phase seems to be somewhat longer possibly owing to slower washout of IbTX. All these observations indicate that direct blockade of the O_2-sensitive K^+ channels with TEA or IbTX can elicit secretion from rat glomus cells in the slices. Whether the leak (10) and the TEA-sensitive (7) channels are present in the same rat glomus cell type and whether their functions overlap the same range of P_{O_2} values are unsolved questions. Our data also make it difficult to understand why K^+ channel blockers do not activate whole carotid body preparations, in which the glomus cell–afferent fiber synapses are maintained intact (25,26). As suggested before (17), a possible explanation is that the blockers do not diffuse at the appropriate concentration into the extracellular space of the carotid bodies either superfused by the bath solution or perfused through the carotid artery.

IV. Glomus Cells Are Combined Glucose and Oxygen Sensors

Long ago the carotid body was proposed to participate in glucose homeostasis (31,43), and more recently it has been shown that resection of the carotid bodies and surrounding tissues results in impairment of the insulin-induced counterregulatory responses to mild hypoglycemia (44). However, there is no evidence that any of the elements in the carotid body can respond directly to changes of extracellular glucose concentration. We have recently studied the effect of external glucose removal on the secretory activity of glomus cells in carotid body slices (33). In healthy preparations, exposure to 0 glucose consistently produced a marked and reversible increase of cell secretory activity (Fig. 6a). The average rate of secretion during the last minute of exposure to low glucose (1870 ± 386 fC/min, $n = 14$ cells) was over 20 times that of the control condition (88 ± 45 fC/min, $n = 14$). The size distribution and mean area of quantal events triggered by low glucose (Fig. 6b) were similar to those previously observed in glomus cells activated by high K^+, hypoxia, or TEA (see above), further suggesting that all these stimuli induce the release of a common vesicle pool. The effect of low glucose on glomus cells was concentration-dependent and additive with the effects of hypoxia. At normal air O_2 tension ($P_{O_2} \approx 150$ mmHg), secretion was evoked only when glucose decreased below 2 mM; however, at a P_{O_2} of 90 mmHg (a value close to the normal O_2 tension in arterial blood) glomus cell secretory activity was significantly modulated by glucose in the concentration range (2–5 mM) that includes the values observed in common hypoglycemic situations (33). Catecholamine secretion induced by glucose-free solutions was totally suppressed by the addition of 0.2 mM Cd^{2+} to the extracellular

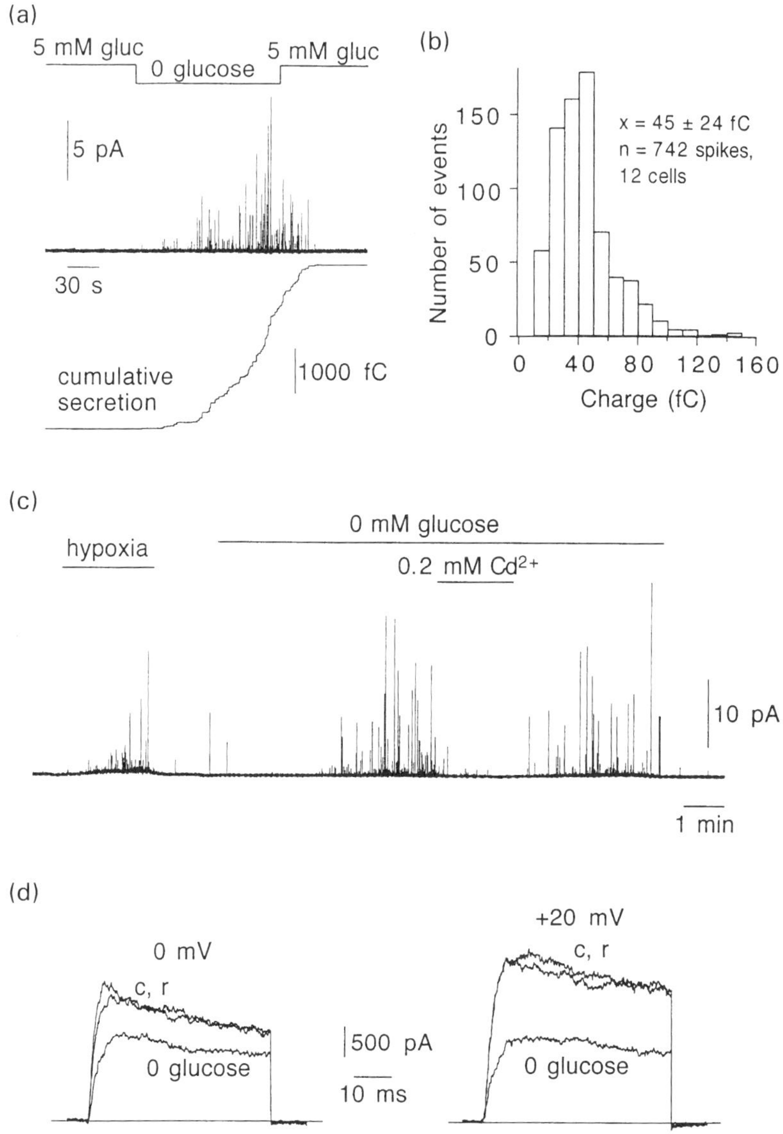

Figure 6 Response of glomus cells to low glucose. (a) Top: Amperometric signal illustrating the increase of secretory activity in a glomus cell exposed to glucose-free solution. Bottom: Cumulative secretion signal (in femtocoulombs) resulting from the time integral of the amperometric recording. (b) Histogram representing the distribution of the area of exocytotic events in low glucose. (c) Reversible suppression of low glucose-evoked secretory activity by application of 0.2 mM Cd^{2+} to the extracellular solution. (d) Recordings of outward K^+ currents from a patch-clamped glomus cell depolarized to 0 and +20 mV and exposed to 0 mM glucose. The control (c) and recovery (r) external solutions contained 5 mM glucose. (Modified from Ref. 33.)

solution in all cells tested ($n=3$; Fig. 6c). Because 0.2 mM extracellular Cd^{2+} completely blocks voltage-gated Ca^{2+} channels in glomus cells (13), this indicates that low glucose induces transmitter secretion when membrane electrical events result in depolarization and Ca^{2+} influx through voltage-dependent channels. This idea was confirmed by experiments on patch-clamped cells dialyzed with an internal solution containing 4 mM Mg-ATP. Glucose deficiency produced a reversible reduction of peak outward K^+ current amplitude (Fig. 6d) that at +20 mV had an average value of 38 ± 12% ($n=7$ cells). Low glucose appeared to act selectively on voltage-dependent K^+ channels since it had no effect on the small inward current characteristic of most rat glomus cells (data not shown). In addition, the input resistance measured in patch-clamped cells held at −80 mV (286 ± 11 MΩ, mean ± SD, $n=9$), although relatively low possibly owing to the high temperature in the experiments or to electrical coupling, was unaltered by glucose removal (267 ± 11 MΩ, $n=9$). Thus, these data suggested that leakage or K_{ATP} channels were not appreciably influenced by the changes of glucose concentration.

The electrophysiological and amperometric data obtained from glomus cells in carotid body slices strongly suggest that these cells are physiological low-glucose detectors capable of transducing glucose levels into variable rates of transmitter release. The low-glucose signaling pathway in glomus cells appear to be initiated by inhibition of voltage-gated K^+ channel activity, which leads to membrane depolarization, Ca^{2+} influx through voltage-gated Ca^{2+} channels, and transmitter release. Therefore, in glomus cells low glucose and hypoxia converge to raise cytosolic $[Ca^{2+}]$ and to release transmitters, which stimulate afferent sensory fibers and evoke sympathoadrenal activation. These observations help to explain previous reports on anesthetized animals describing rapid increase in the output of hepatic glucose after activation of the carotid body with sodium cyanide (31,43), alterations of carbohydrate metabolism in acute hypoxia (45), or impairment of insulin-induced counterregulatory response to mild hypoglycemia in carotid body–resected dogs (44). Although the existence of peripheral glucosensors presumably located in the liver or portal vein has been proposed (46,47), the strategically located carotid bodies may be of special importance for brain homeostasis as neurons are particularly vulnerable to the simultaneous lack of glucose and oxygen (48). The function of glomus cells as combined O_2 and glucose sensors, in which the two stimuli potentiate each other, is surely advantageous to facilitate activation of the counterregulatory measures in response to small reductions of any of the regulated variables. Impairment of low-glucose sensing by carotid body glomus cells might contribute to the susceptibility of insulin-dependent diabetic patients to hypoglycemia.

V. Conclusion

In this chapter, we describe the properties of cells in carotid body thin slices. This preparation has allowed us to study the responses of glomus cells to hypoxia and K^+ channel blockers in almost optimal physiological conditions. Exposure of cells in the slices to K^+ channel blockers, like TEA or IbTX, induced a secretory activity that resembled the effect of low P_{O_2}. These results suggest that the direct blockade of the O_2-sensitive voltage-dependent K^+ currents, which in these cells are those inhibited by TEA and IbTX, is sufficient to induce secretion. We have also demonstrated that glomus cells are integrated metabolic sensors activated to release transmitter in response to lowering external glucose. Our data indicate that glomus cells are highly sensitive, low-glucose detectors possibly required for the proper sympathoadrenal activation in conditions of hypoglycemia. The thin slice preparation may help to prove that, as suggested in the old literature, the carotid body is a polimodal sensory organ that responds not only to changes in the gases of the blood but to alterations of multiple chemical or physical variables.

Acknowledgments

This research was supported by grants from the Spanish Ministry of Science and Technology (1FD97-1614) and Fundaciones La Caixa and Ramón Areces. José López-Barneo received the "Ayuda a la investigación 2000" of the Juan March Foundation.

References

1. Bunn HF, Poyton RO. Oxygen sensing and molecular adaptation to hypoxia. Physiol Rev 1996; 76:839–885.
2. Lahiri S, Prabhakar NR, Forster RE. Oxygen Sensing: Molecule to Man. New York: Kluwer Academic/Plenum Publishers, 2000.
3. López-Barneo J, Pardal R, Ortega-Sáenz P. Cellular mechanisms of oxygen sensing. Annu Rev Physiol 2001; 63:259–287.
4. Duchen MR, Caddy KWT, Kirby GC, Patterson DL, Ponte J, Biscoe TJ. Biophysical studies of the cellular elements of the rabbit carotid body. Neuroscience 1988; 26: 291–311.
5. López-Barneo J, López-López JR, Ureña J, González C. Chemotransduction in the carotid body: K^+ current modulated by P_{O_2} in type I chemoreceptor cells. Science 1988; 241:580–582.
6. Delpiano MA, Hescheler J. Evidence for a P_{O_2}-sensitive K^+ channel in the type-I cell of the rabbit carotid body. FEBS Lett 1989; 249:195–198.
7. Peers C. Hypoxic suppression of K^+ currents in type I carotid body cells: selective effect on the Ca^{2+}-activated K^+ current. Neurosci Lett 1990; 119:253–256.
8. Ganfornina MD, López-Barneo J. Single K^+ channels in membrane patches of arterial chemoreceptor cells are modulated by O_2 tension. Proc Natl Acad Sci USA 1991; 88:2927–2930.
9. Stea A, Nurse CA. Whole-cell and perforated-patch recordings from O_2-sensitive rat carotid body cells grown in short- and long-term culture. Pflügers Arch 1991; 418:93–101.
10. Buckler KJ. A novel oxygen-sensitive potassium current in rat carotid body type I cells. J Physiol 1997; 498:649–662.

11. Summers BA, Overholt JL, Prabhakar NR. Augmentation of L-type calcium current by hypoxia in rabbit carotid body glomus cells: evidence for a PKC-sensitive pathway. J Neurophysiol 2000; 84:1636–1644.
12. López-Barneo J, Benot AR, Ureña J. Oxygen sensing and the electrophysiology of arterial chemoreceptor cells. News Physiol Sci 1993; 8:191–195.
13. Ureña J, Fernández-Chacón R, Benot AR, Álvarez de Toledo G, López-Barneo J. Hypoxia induces voltage-dependent Ca^{2+} entry and quantal dopamine secretion in carotid body glomus cells. Proc Natl Acad Sci USA 1994; 91:10208–10211.
14. Buckler KJ, Vaughan-Jones RD. Effects of hypoxia on membrane potential and intracellular calcium in rat neonatal carotid body type I cells. J Physiol 1994; 476:423–428.
15. Montoro RJ, Ureña J, Fernández-Chacón R, Álvarez de Toledo G, López-Barneo J. Oxygen sensing by ion channels and chemotransduction in single glomus cells. J Gen Physiol 1996; 107:133–143.
16. Carpenter E, Hatton CJ, Peers C. Effects of hypoxia and dithionite on catecholamine release from isolated type I cells of the rat carotid body. J Physiol 2000; 523:719–729.
17. Pardal R, Ludewig U, García-Hirschfeld I, López-Barneo J. Secretory responses of intact glomus cells in thin slices of rat carotid body to hypoxia and tetraethylammonium. Proc Natl Acad Sci USA 2000; 97:2361–2366.
18. Youngson C, Nurse C, Yeger H, Cutz E. Oxygen sensing in airway chemoreceptors. Nature 1993; 365:153–155.
19. Thompson RJ, Nurse CA. Anoxia differentially modulates multiple K^+ currents and depolarizes neonatal rat adrenal chromaffin cells. J Physiol 1998; 512:421–434.
20. Zhu WH, Conforti L, Czyzyk-Krzeska MF, Millhorn DE. Membrane depolarization in PC12 cells during hypoxia is regulated by an O_2-sensitive K^+ current. Am J Physiol 1996; 271:C658–C665.
21. Post JM, Hume JR, Archer SL, Weir EK. Direct role for potassium channel inhibition in hypoxic pulmonary vasoconstriction. Am J Physiol 1992; 262:C882–C890.
22. Yuan XJ, Goldman WF, Tod ML, Rubin LJ, Blaustein MP. Hypoxia reduces potassium currents in cultured rat pulmonary but not mesenteric arterial myocytes. Am J Physiol 1992; 264:L116–L123.
23. Osipenko ON, Evans AM, Gurney AM. Regulation of the resting potential of rabbit pulmonary artery myocytes by a low threshold, O_2-sensing potassium current. Br J Pharmacol 1997; 120:1461–1470.
24. Doyle TP, Donnelly DF. Effect of Na^+ and K^+ channel blockade on baseline and anoxia induced catecholamine release from rat carotid body. J Appl Physiol 1994; 77:2606–2611.
25. Osanai S, Buerk DG, Mokashi A, Chugh DK, Lahiri S. Cat carotid body chemosensory discharge (in vitro) is insensitive to charybdotoxin. Brain Res 1997; 747:324–327.
26. Lahiri S, Roy A, Rozanov C, Mokashi A. K^+-current modulated by P_{O_2} in type I cells in rat carotid body is not a chemosensor. Brain Res 1998; 794:162–165.
27. Wyatt CN, Peers C. Ca^{2+}-activated K^+ channels in isolated type I cells of the neonatal rat carotid body. J Physiol 1995; 483:559–565.
28. Pérez-García T, López-López JR, Riesco AM, Hoppe UC, Marbán E, González C, Johns DC. Viral gene transfer of dominant negative Kv4 construct suppresses an O_2 sensitive K^+ current in chemoreceptor cells. J Neurosci 2000; 20:5689–5695.
29. Pardal R, López-Barneo J. Carotid body thin slices: responses of glomus cells to hypoxia and K^+ channel blockers. Respir Physiol Neurobiol 2002; 132:69–79.

30. López-Barneo J, Pardal R, Montoro RJ, Smani T, García-Hirschfeld J, Ureña J. K^+ and Ca^{2+} channel activity and cytosolic [Ca^{2+}] in oxygen-sensing tissues. Respir Physiol 1999; 115:215–227.
31. Alvarez-Buylla R, Alvarez-Buylla ER. Carotid sinus receptors participate in glucose homeostasis. Respir Physiol 1988; 72:347–360.
32. Eyzaguirre C, Zapata P. Perspectives on carotid body research. J Appl Physiol 1984; 57:931–957.
33. Pardal R, López-Barneo J. Identification of low-glucose sensing cells in the carotid body. Nat Neurosci 2002; 5:197–198.
34. Donnelly DF, Doyle TP. Developmental changes in hypoxia-induced catecholamine release from rat carotid body in vitro. J Physiol 1994; 475:267–275.
35. Hatton CJ, Carpenter E, Pepper DR, Kumar P, Peers C. Developmental changes in isolated rat type I carotid body cell K^+ currents and their modulation by hypoxia. J Physiol 1997; 501:49–58.
36. Fidone S, González C. Initiation and control of chemoreceptor activity in the carotid body. In: Cherniack NS, Widdicombe FG, eds. Handbook of Physiology, Section 3: The Respiratory System, Vol. II: Control of Breathing, part 2. Bethesda: American Physiological Society, 1986:247–312.
37. Ureña J, López-López JR, González C, López-Barneo J. Ionic currents in dispersed chemoreceptor cells of the mammalian carotid body. J Gen Physiol 1989; 93:979–999.
38. Ganfornina MD, López-Barneo J. Potassium channel types in arterial chemoreceptor cells and their selective modulation by oxygen. J Gen Physiol 1992; 100:401–426.
39. López-López JR, González C, Pérez-García MT. Properties of ionic currents from isolated adult rat carotid body chemoreceptor cells: effect of hypoxia. J Physiol 1997; 499:429–441.
40. Zhang M, Zhong H, Vollmer C, Nurse CA. Co-release of ATP and ACh mediates hypoxic signalling at rat carotid body chemoreceptors. J Physiol 2000; 525:143–158.
41. Benot A, López-Barneo J. Feedback inhibition of Ca^{2+} currents by dopamine in glomus cells of the carotid body. Eur J Neurosci 1990; 2:809–812.
42. Gálvez A, Gimenez-Gallego G, Reuben JP, Roy-Contancin L, Feigenbaum P, Kaczorowski GJ, García ML. Purification and characterization of a unique, potent, peptidyl probe for the high conductance calcium-activated potassium channel from venom of the scorpion *Buthus tamulus.* J Biol Chem 1990; 265:11083–11090.
43. Alvarez-Buylla R, Alvarez-Buylla ER. Changes in blood glucose concentration in the carotid body sinus modify brain glucose retention. Brain Res 1994; 654:167–170.
44. Koyama Y, Coker RH, Stone EE, Lacy DB, Jabbour K, Williams PE, Wasserman DH. Evidence that carotid bodies play an important role in glucoregulation in vivo. Diabetes 2000; 49:1434–1442.
45. Zinker BA, Namdaran K, Wilson R, Lacy DB, Wasserman DH. Acute adaptation of carbohydrate metabolism to decreased arterial P_{O_2}. Am J Physiol 1994; 266:E921–E929.
46. Donovan CM, Hamilton-Wessler M, Alter JB, Bergman RN. Primacy of liver glucosensors in the sympathetic response to progressive hypoglycemia. Proc Natl Acad Sci USA 1994; 91:2863–2867.
47. Hevener AL, Bergman RN, Donovan CM. Novel glucosensor for hypoglycemia detection localized to the portal vein. Diabetes 1997; 46:1521–1525.
48. Martin RL, Lloyd HG, Cowan AI. The early events of oxygen and glucose deprivation: setting the scene for neuronal death? Trends Neurosci 1994; 17:251–257.

18

Electric and Dye Coupling Between Rat Carotid Body Cells and Between These Cells and Carotid Nerve Endings

CARLOS EYZAGUIRRE

University of Utah School of Medicine
Salt Lake City, Utah, U.S.A.

RUGANG JIANG

Tianjin Medical University
Tianjin, People's Republic of China

VERÓNICA ABUDARA

University of Montevideo Medical School
Montevideo, Uruguay

I. Introduction

Intercellular coupling, detected by injected electric currents or intracellular dye injections, is widespread in different organisms. It occurs in plants and animals alike. The intercellular communications that allow transfers of currents and dyes are possible because of the presence of "gap junctions." These junctions, which would be better named cribriform or perforated, are formed by small pores surrounded by a hexamer of six proteins (connexins) on each side of the intercellular membranes. The total complex on both cellular sides becomes a dodecameric protein complex (connexon) surrounding the intercellular pore. This configuration allows passage of ions (transferring currents) and relatively small molecules (transferring dyes) across cell membranes. The configuration of the connexons allows the pores to be open or shut (completely or partially). When the pores are open, there is free flow of ions and molecules between adjoining cells and the cells are well coupled, leading to a very small intercellular resistance or high conductance. When the pores are partially occluded, there is low intercellular conductance or high resistance between the cells and the flow of ions and molecules is more difficult. If the pores are totally closed, there is no coupling and no intercellular communication. These junctions open and close like the shutter of a photographic camera (1).

In many organs (exocrine and endocrine glands, epithelia, neurons in the central nervous system and retina, smooth muscle, etc.,) there are different degrees of coupling between the cells in the organs. Adequate or physiological stimuli can reduce intercellular coupling (uncoupling or decoupling) or increase it, depending on the type of tissue or its physiological condition. Exogenous applications of transmitters or other chemicals can produce similar effects (2). We became interested in exploring the possibility of intercellular coupling in the carotid body because it is a secreting organ, synaptically connected to the terminals of the carotid nerve. The glomus cells contain and release transmitters toward the carotid nerve terminals to modulate the sensory discharges controlling ventilation in the nervous system (Fig. 1). This interest was spurred when we found that intracellular injections of the cell-impermeant dye Procion Navy Blue spread to adjoining glomus and, perhaps, to sustentacular cells (3). This finding was supported by McDonald's discovery of gap junctions between glomus cells (4).

Later, it became clear that glomus cells in the intact carotid body, and when clustered in cultures, behaved differently from isolated cultured cells. For instance, in whole carotid bodies and in cultured but clustered glomus cells, the membrane potential (E_m) was not dependent on $[K^+]_o/[K^+]_i$ whereas the E_m of isolated

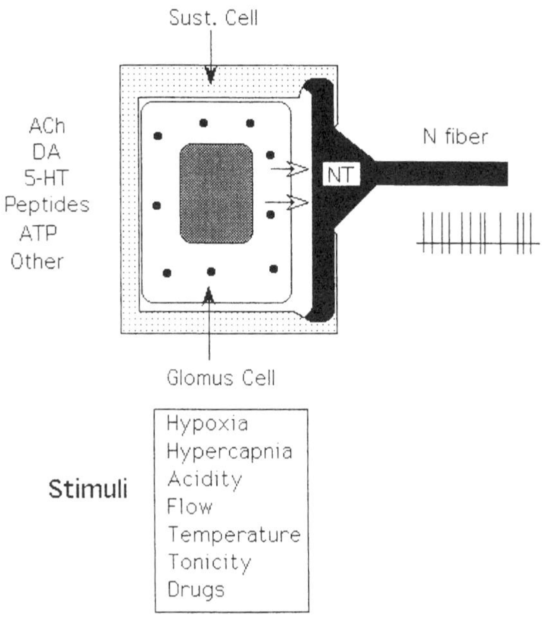

Figure 1 Schematic diagram illustrating the basic features of chemotransduction from glomus cells to carotid nerve terminals (NT). The glomus cell is packed with dark-core vesicles and contains a plethora of chemical agents (list on the left). Natural stimulation (list at the bottom) releases these agents toward the endings of the carotid nerve (NT). The chemicals cross the synaptic cleft between glomus cells and nerve endings (horizontal open arrows).

cultured cells was, behaving like many other cells (5,6). Furthermore, the intracellular pH of cultured and clustered cells ($pH_{i\,clst} = 6.32$) was significantly lower than that of isolated cells ($pH_{i\,sgl} = 6.8$) (7). However, intracellular sodium, $[Na^+]_i$, had statistically similar values (14.1 vs. 12.6 mM) in clustered and isolated glomus cells (8). During hypoxia induced by sodium dithionite (Na-DTN, $Na_2S_2O_4$) there were differences in the behavior of these cells. More than 80% of clustered cells depolarized and the input resistance (R_o) decreased whereas only 40% of isolated cells depolarized with decreased R_o (9). Intracellular potassium followed this pattern since $[K^+]_i$ decreased in 85% of clustered cells but in only 38% of single ones (6). Na-DTN also produced different results concerning pH_i. The reducing agent increased and decreased pH_i of clustered cells almost equally whereas the pH_i of isolated cells decreased substantially in most instances (7).

The different behavior of grouped and isolated glomus cells has been attributed to the presence of sustentacular cells within and surrounding the clusters (7,9). Sustentacular cells are glia-like (10), and glial cells possess a sophisticated machinery that can modify the behavior of enveloped neurons (11) through interconnecting gap junctions. Likewise, the different behavior of grouped and isolated glomus cells could have been due to a coupling mechanism mediated by the gap junctions between them (12,13). Thus, we have started exploring possible coupling between glomus and sustentacular cells.

Another important point to consider is that carotid nerve terminals are not in direct contact with glomus cells. They do so across 'chemical synapses" where the presynaptic elements (glomus cells) are separated from the nerve endings (postsynaptic elements) by a synaptic cleft (4). It is generally agreed that the "natural" stimuli (low pO_2, high pCO_2 and acidity) excite the carotid nerve terminals by transmitters released from the glomus cells. Many substances are contained in and released from these cells, such as ACh, catecholamines [especially dopamine (DA)], serotonin (5-HT), enkephalins, prostaglandins, ATP, substance P, and peptides (cholecystokinins, atrial natriuretic peptide-ANP), as shown by many authors (for references see Refs. 14,15). In addition, second messengers such as cAMP and cGMP seem to play an important role in chemotransduction since their levels change during hypoxia (16–18).

The rich variety of possible transmitter substances has complicated the originally designated "transmitter hypothesis of chemoreception," which was proposed in simpler times when only ACh and DA were known to occur in and be released from glomus cells. If multiple transmitters are released during stimulation, we may begin to understand why specific blockers for ACh and DA do not block but, at best, depress the effects of natural stimuli, while being very effective against exogenous applications of these transmitters. Another complication arose when Kondo and Iwasa (19) found gap junctions within the chemical synapses linking glomus cells and carotid nerve terminals. If these electric junctions are functional, one cannot expect synaptic blockers to be effective against transmitters released from the presynaptic glomus cells. Their effectiveness against exogenous applications of the same transmitters could be explained by a blocking action on the membrane of the nerve endings where the transmitters should act.

Our aim has been to study the role of gap junctions on chemotransduction, focusing first on intercellular communications between glomus cells. We have concluded from these studies that at rest, most glomus cells are coupled because numerous intercellular channels are permeant. During stimulation, most glomus cells uncouple and release their contents toward the terminals. However, other cells undergo tighter coupling because they are "recharging" and accumulate transmitters for further use. During prolonged stimulation, the recharged cells will uncouple and release their contents. The previously uncoupled cells will begin to tighten intercellular coupling and recharge. During intense and prolonged stimulation this push-pull system allows the carotid body to maintain sustained nerve activity, without exhausting its reserves. In fact, during chronic or prolonged acute hypoxia, the carotid nerve is able to sustain a high-frequency discharge for long periods. This would be impossible if a strong stimulus activates all glomus cells at once and deplete their reserves (20).

More recently, we have tackled the problem of possible electrical connections between glomus cells and carotid nerve terminals and have started to study communications between glomus and sustentacular cells. Thus, results presented below are a synopsis of these studies in hopes of attaining a better understanding of the complexities of transduction in these receptors.

II. Methods

A. Electrical Recordings

Carotid bodies were excised from rats anesthetized with intraperitoneal injections of sodium pentobarbital (50 mg/kg). The organ, or clusters of cultured glomus cells, were placed in a lucite chamber through which flowed physiological solutions equilibrated with different gas mixtures. The saline composition varied, depending on whether we used the whole organ (21) or cultured cells (22,23). The pO_2 of the control medium was about 300 torr at a pH of 7.43. Acidity was obtained with lactic acid at pH 6.3. Hypoxia was provoked by superfusing with 100% N_2, which lowered pO_2 to about 30 torr. When sodium dithionite (Na-DTN or $Na_2S_2O_4$) was used, pO_2 fell to 10 torr. Different concentrations of chemicals or drugs were superfused with the solutions. The preparations were mounted on the stage of an inverted microscope and viewed with phase-contrast or Hoffmann optics. For stimulation and recording, two microelectrodes filled with 3M KCl (tip resistance 20–40 MΩ), with a tip separation of 10–15 μm, were mounted on the same micromanipulator and lowered into the tissue for intracellular penetration of adjoining glomus cells. The electrodes were connected to independent electronic devices that allowed voltage or current delivery to the cells, and also the recording of voltages or currents produced in the cells. The behavior of intercellular junctions is independent of the membrane potentials (E_m) of the cells within physiological parameters. Under current clamping (or voltage recordings), a pulse delivered to Cell 1 elicited a voltage drop (V_1) in this cell. In coupled cells, voltages were also detected in coupled Cell 2 because a voltage drop occurred across the intercellular junction (ΔV_j). Consequently, the electrode lodged in Cell 2 registered V_2, which was smaller than V_1. The degree of coupling was established as the coupling coefficient (K_c) equal to V_2/V_1. To record intercellular currents, voltage clamping was employed. Under these conditions, the cells were voltage-clamped at levels intermediate between their resting potentials. Pulses delivered to Cell 1 elicited a current in this cell (I_1) that was transferred to Cell 2 across the intercellular junction and recorded in this cell as I_2 or I_j, the junctional current. The polarity (positive or negative) of I_2 or I_j was the opposite of that recorded as I_1. To establish the total conductance of the junction (macroconductance), we converted intercellular current to conductance as I_j/V_j.

To record the activity of intercellular channels, pulses, and more frequently DC voltages, were delivered to Cell 1 after voltage clamping of both cells as described above. These procedures produced "flickering" (multiple current activities) in the intercellular channels of both cells (i_1 and i_2), which were of similar amplitudes and opposite polarities. Inward channel currents in one coupled cell were taken as the i_j currents. The intercellular junctional conductance of channels (microconductance or g_j) was calculated as i_j/V_j.

B. Morphological Identification of Recorded Structures

Preparation

Wistar rats (both male and female) weighing 60–200 g were used. Segments of both carotid arteries with attached carotid bodies were removed from rats anesthetized with sodium pentobarbital. For dissection, the organs were placed in an ice-cold physiological solution (98 mM NaCl; 47 mM Na glutamate; 4.6 mM KCl; 3.0 mM $CaCl_2$; 1.1 mM $MgCl_2$ 7.0 mM glucose; 5.0 mM HEPES. NaOH was added to reach a pH 7.43. The carotid bodies were cut into 100–150-μm slices and kept on ice before being used. The slices were then immersed in a lucite chamber filled with the same solution, bubbled with 100% O_2, and kept at 30–32°C. The chamber was mounted on the stage of an inverted microscope (Olympus IMT2-RFL).

Dye Ejection

Electrodes were made from two-barrel borosilicate glass tubing with filaments (WPI). One barrel was filled with 3 M KCl (2–10 MΩ) for stimulation and recording. For microejections, the other barrel was filled with 5% Lucifer Yellow (Lucifer Yellow CH, lithium salt, Molecular Probes) in 150 mM LiCl or with 10 mM Alexa fluor 488 (hydrazide, sodium salt, Molecular Probes) in 200 mM KCl. When the double-barrel microelectrode was inserted into a cell, dye ejection was accomplished with 100-msec negative current pulses (1–50 nA) delivered through the dye-filled barrel and applied for 1–5 min at 100-msec intervals. After dye ejection the tissue was fixed.

Immunofluorescence Staining for TH

Tissues were fixed in 4% paraformaldehyde for 15–18 hr at 4°C or 1 hr at room temperature. After being washed with TBST (50 mM Tris-HCl, 150 mM NaCl, 0.3%

Triton X-100, pH 7.6), the tissue was treated with 2% goat serum in TBST for 1 hr at room temperature. Then, the tissues were incubated in the primary antibody, which was rabbit anti–tyrosine hydroxylase (CHEMOCON), for 15–18 hr at 4°C. This antibody was diluted 1:500 in TBST plus 2% goat serum. The secondary antibody was Texas Red goat anti rabbit IgG (Vector Laboratories), diluted 1:100 in 50 mM TBS (Tris-HCl and 150 mM NaCl, pH 7.6). Tissues were incubated in the secondary antibody for 30 min at room temperature and washed with TBS before being covered with Vectashield placed on a cover glass slip.

Confocal Imaging

With a confocal microscope (Zeiss LSM 510), images were acquired in both the green and red channels by Argon and HeNe lasers. Many images were viewed as series of optical sections > 0.8-μm optical sections. Other images were viewed as optical sections > 2 or 4 μm. All images were processed with LSM 5 Examiner and Adobe PhotoDeluxe. The left view in Figure 2 is a confocal image of carotid body glomera stained for tyrosine hydroxylase (TH) and viewed through the red channel. The brilliantly stained rings (red in the original) of cytoplasm surround unstained and dark nuclei. The main glomerulus seems to ring around a blood vessel. The right view was taken from a cell injected with Lucifer Yellow (LY) and viewed through the green channel of the confocal microscope. Because of its shape and size it appears to be a carotid nerve ending.

III. Results and Discussion

Electrophysiological experiments have amply confirmed the idea that gap junctions occur between glomus cells using current and voltage clamping. Voltages or currents

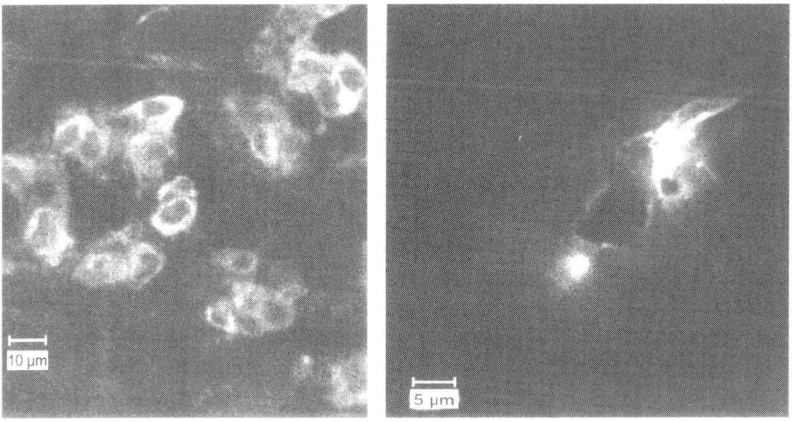

Figure 2 Left, carotid body glomera stained for TH and viewed through a confocal microscope. The cytoplasm of glomus cells appears white (originally it was red) and surrounds unstained and large nuclei. Cal. 10 μm. Right, cell (probably a nerve terminal) stained with Lucifer Yellow ejected from a microelectrode. Cal. 5 μm.

elicited in one cell spread to adjoining and coupled glomus cells. Voltage or current transfers across the junctions varied in different pairs of cells depending on the degree of intercellular coupling. At this point it is important to emphasize that when we talk about coupling between pairs of cells it is because this has been established only by using cell pairs, because of limitations in the recording methods. In fact, coupling is likely to be much more extensive involving many cells, as will be shown in morphological studies.

A. Intercellular Junctions at Rest and During Stimulation

In other tissues, many junction channels are present between coupled cells. Numbers vary from very few to hundreds or thousands. Furthermore, the nature of the connexins surrounding the intercellular pores is variable. In many instances there are hybrid channels (different connexins in one channel) or channels of different nature within one junction (24). This is important to understand the effects of different stimuli on intercellular coupling since a given stimulus does not necessarily act in identical form on different connexins. Furthermore, intercellular channels do not behave in an all-or-none manner since some may be totally or partially open as well as totally or partially closed.

Concerning the carotid body, only connexin43 (Cx43) has been identified as the protein surrounding the intercellular pores between carotid body glomus cells, between sustentacular cells, and between glomus and sustentacular cells (12,13). Other connexins have not been identified yet, although they may be present in these junctions. Nevertheless, the wide distribution of Cx43 presents some interesting problems. For instance, dye and current transfers are likely to spread between several cellular elements in the carotid body, which may behave more like a syncytium rather than as independent entities. We know that glomus cells in the whole organ and when clustered in cultures behave differently than isolated glomus cells, probably because of the presence or absence of the sustentacular cell envelope. This envelope is present in the whole organ and in clustered cells in culture. It is absent in isolated and cultured glomus cells. We know almost nothing about the behavior of sustentacular cells. However, it is something that should be carefully explored because of the presence of gap junctions between glomus and sustentacular cells. Thus, if currents or dyes are injected into one glomus cell, they may be transferred directly to other coupled glomus cells across the gap junctions. However, in addition to this pathway we must also consider the possibility of an indirect transfer via a more tortuous route involving the enveloping sustentacular cell processes. This very real possibility may explain the low coupling indices between glomus cells obtained when one cell is activated by voltages or currents and recordings are made from another adjoining and coupled glomus cell. In fact, coupling may not occur directly but through a cascade of events involving intermediate structures.

Another aspect of the carotid body cellular syncytium is that there is a distinct possibility of direct connections between the carotid body cells (glomus and sustentacular) and the nerve endings of the carotid nerve. This study, still in its infancy, has been prompted by the discovery of Kondo and Iwasa (19) who found gap junctions between glomus cells and carotid nerve terminals. Connexins between these two elements, essential in chemotransduction, have not been identified yet. However, if there are direct cell-nerve connections, dyes and currents should spread between them.

We have used the carotid body of the rat, which is thin and flat, and have impaled two adjoining and coupled cellular elements to study the effects of voltage injections into one cell and its effect on the other. In this type of study it is important to identify the stimulated and recorded cells. Identification of an impaled carotid nerve chemosensory ending is simple because this structure shows multiple small depolarizing potentials (s.d.p.s) and action potentials (spikes) originating from the larger s.d.p.s. (25,26). Usually, nerve terminals thus identified have shown large input resistances, which is not unusual because of their small size. Glomus cells, on the other hand, have usually shown low input resistances when impaled with microelectrodes. This probably happens because they are coupled to other glomus and sustentacular cells, resulting in a large surface area.

B. Coupling Between Glomus Cells

This aspect of carotid body functions has been extensively studied in the whole organ and in clusters of cultured glomus cells. Experiments with current and voltage clamping have revealed that intercellular coupling between these structures is fairly common. This indicates that the number of tight junctions between glomus cells is much higher than that expected from McDonald's work (4) showing about 30% of these junctions in his EM studies. The junctions are basically of the resistive type, with little or no rectification. An interesting aspect, similar to what occurs in other systems, was that intercellular coupling was independent of the resting potential of the cells within a range of ± 60 mV. In other words, within physiological or reasonable limits the cell's E_m does not affect the junctional potential or resistance values. This is important because at times the E_m of cells impaled with microelectrodes is relatively low.

Figure 3 illustrates physiological and morphological examples of coupling between glomus cells. Figure 3a and b present results obtained during voltage clamping of two adjoined and coupled cells. In (a), the holding potential (E_H) was -20 mV, and negative voltage pulses applied to Cell 1 (V, in upper trace) elicited inward currents in this cell (I_1) leading to a junction current (I_j) detected in Cell 2. Note that the polarity of I_j is opposite to that of I_1. Figure 3b shows the activity of junction channels when two coupled glomus cells were voltage-clamped at different levels (E_{H1} was -30 mV and E_{H2} was -180 mV) creating a transjunctional voltage (ΔV_j of 150 mV). This situation tends to close previously opened, thus silent, intercellular channels. However, channel closure is not all-or-none since rapid openings and closures occur before complete shutdown. This flickering reflects total channel activity at this moment and is used to estimate junction channel currents (i_js) that permits calculating total channel conductances (g_js) as $i_j \Delta V_j$. The figure depicts channel currents, which are of similar magnitude and of opposite polarities in Cells 1 and 2. The bottom of Figure 3 shows two confocal microscope images of a cell

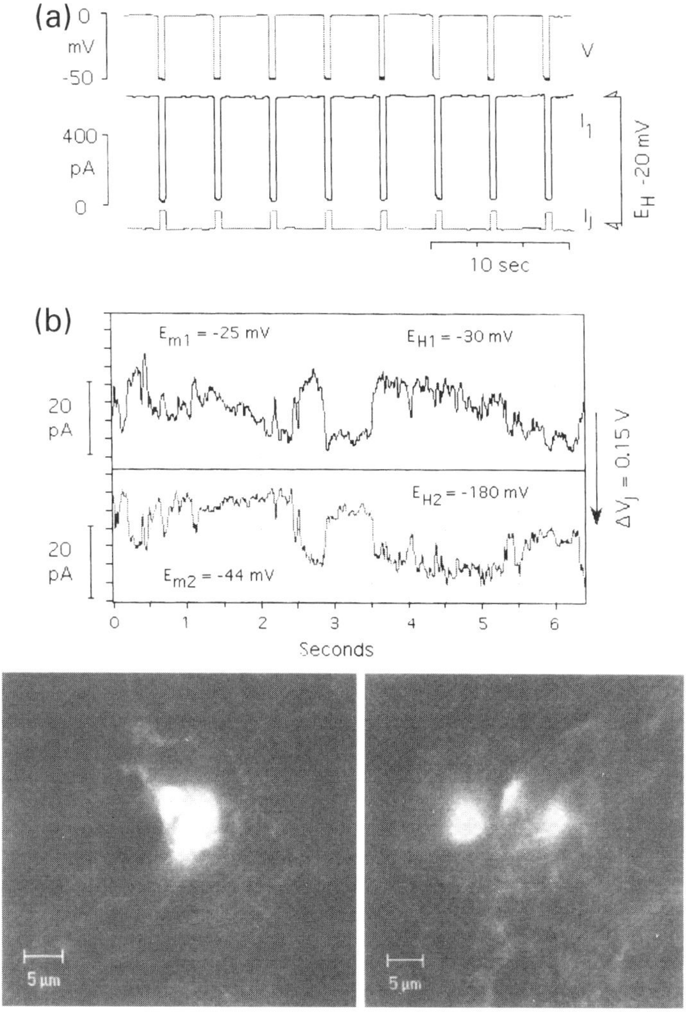

Figure 3 Two coupled glomus cells dually voltage-clamped (E_H) at -20 mV. (a) 50-mV negative pulses were delivered to Cell 1 (upper trace). The middle trace shows inward currents (I_1) elicited in this cell, which are a mixture of junctional and nonjunctional currents. The lower trace shows the currents leaving coupled Cell 2 mostly through the gap junctions (I_j). (b) Intercellular channel activity (currents) elicited by voltage-clamping Cells 1 and 2 at different levels ($E_{H_1} = -30$ mV and $E_{H_2} = -180$ mV) creating a transjunctional voltage difference (ΔV_j) of 150 mV. Note that intercellular channel activity is shown by current deflections of equal amplitude and opposite polarity (mirror image). Currents leave one cell to enter its coupled partner. Bottom: Superimposed confocal images from the red and green channels of preparation stained for TH after one glomus cell was injected with Lucifer Yellow. Left: Fluorescence detected in two cells in one optical plane. Right: A third structure shows fluorescence at different optical plane. Faint background shows TH staining of glomus cell cytoplasm. Calibrations, 5 µm.

injected with Lucifer Yellow. On the left, dye injection into one glomus cell spreads to an adjoining and coupled cell through the gap junctions. The right view shows an image obtained from a different focal plane. One can see that the dye injection spread to at least two cells, suggesting the possibility of widespread dye diffusion because of multiple coupling.

"Natural" stimuli (low pO_2, high pCO_2, and acidity) uncoupled most (70–80%) of the cells whereas the rest showed tighter coupling. However, there were some quantitative differences. Hypoxia induced by superfusion with 100% N_2 that lowered the bath pO_2 to about 30 torr was more effective in both uncoupling and tightening coupling than 1 mM of sodium dithionite (Na-DTN or $Na_2S_2O_4$) that lowered pO_2 to about 10 torr. We do not know why this was the case since pH_o of both solutions was about the same (7.43). Acidity produced by lactic acid or 5% CO_2 was quite effective in uncoupling glomus cells although some pairs showed increased coupling (21,22).

Similar results were obtained with exogenous applications of the naturally occurring transmitters such as acetylcholine (ACh) and dopamine (DA). DA (a good uncoupler in most systems) had variable and marginally significant effects on coupling whereas ACh also had variable effects. However, other cholinergic agents such as nicotine and bethanechol proved to be effective uncouplers. Figure 4 illustrates two examples of the uncoupling effects of bethanechol (a) and dopamine (b) under current-clamp conditions. The inset shows the experimental procedure. Positive voltage pulses delivered to Cell 1 (V_1) elicited a smaller and positive voltage drop in Cell 2 (E_2) whereas negative pulses delivered to Cell 2 (V_2) produced a smaller and negative voltage in Cell 1 (E_1). The ratio between the resulting voltage in the coupled cell and the voltage applied to the command cell is the coupling coefficient (K_c). The K_c resulting from the positive pulses (1 to 2) is shown by the open circles. K_c obtained from the negative pulses (2 to 1) is represented by the open squares. In (a), the control K_c fluctuated between 0.5 and 0.7 (50–70%) and was rapidly reduced to about 10% by bethanechol in both instances. In (b), dopamine decreased K_cs from 30–50% in the controls to 10–20% although this effect was rather protracted.

These experiments illustrate that positive and negative pulses were equally effective in the electric coupling between glomus cells showing lack of rectification; hence intercellular glomus junctions are resistive in nature.

C. Activity of Junction Channels

The activity of intercellular junction channels was studied by voltage clamping the cells in a coupled pair using a steady voltage that created a ΔV_j of about 100 mV. This procedure produced intermittent activity (flickering) of equal or similar amplitude and opposite polarity in both cells (23; also, Fig. 3).

During normoxia (pO_2 about 300 torr) and normal pH (7.43), dual voltage clamping elicited multiple junction channel activity with a mean microconductance (g_j) of about 100 pS. Single-channel conductances, calculated as variance/mean g_j, gave a mean value of around 17 pS. Manual measurements of single-channel activity

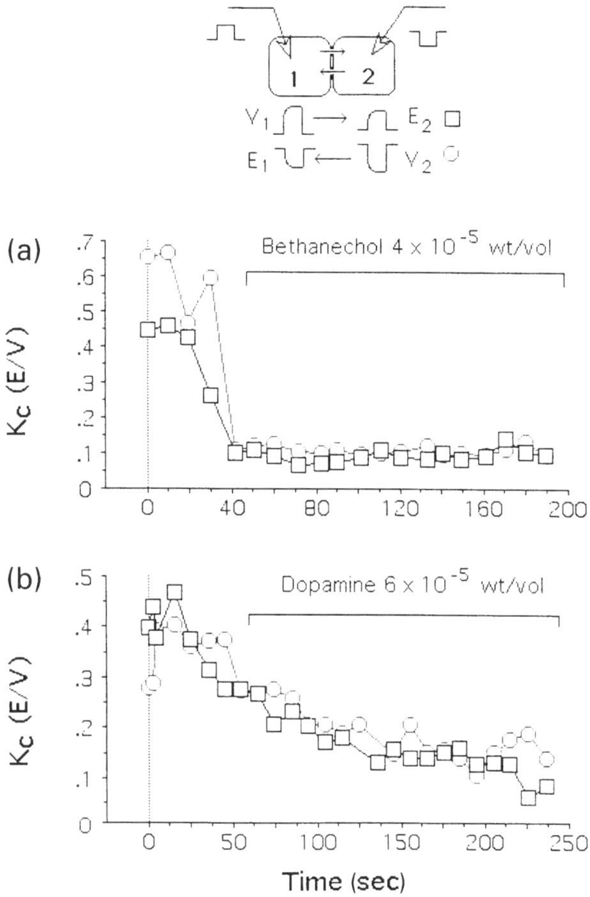

Figure 4 Effects of bethanechol (a) and dopamine (b) on intercellular coupling of two current-clamped glomus cells. The inset shows the experimental setup where Cell 1 was stimulated with positive pulses (V_1) and Cell 2 with negative pulses (V_2). V_1 produces a smaller deflection (E_2) in coupled Cell 2. V_2 elicits a smaller deflection (E_1) in coupled Cell 1. Ordinates, coupling coefficients (K_c). Open squares, coupling from 1 to 2. Open circles, coupling from 2 to 1. Abscissa, time in seconds.

gave larger values, showing a mean g_j of 22–34 pS. Computer analysis of the noise spectral density distribution gave a channel mean open time of about 12 msec. The number of junction channels, estimated in each experiment from G_j/single channel g_j, showed a range of 1–2471 channels.

Acidification of the medium with lactic acid 1 mM (pH 6.3) induced a significant decrease in mean g_j (to about 80 pS) and in single-channel conductance ($g_j = 12.8$ pS in computer analyses and 17.2 pS when measured by hand). Acidity induced variable changes in open times. Thus, the mean open time did not change (12.8 msec). The number of junction channels did not change during acidity. Thus, saline acidification to pH 6.3 depresses the conductance of junction channels. Hypoxia was induced by Na-DTN or 100% N_2. $Na_2S_2O_4$ reduced the saline pO_2 to about 10 torr whereas 100% N_2 reduced ambient oxygen to about 60 torr. Na-DTN did not change significantly the total intercellular microconductance, the conductance of single channels, the number of channels, or their mean open time. This was due to the fact that this reducing agent produced almost equal decreases and increases in these values. However, hypoxic hypoxia (100% N_2) significantly depressed junctional conductance, the number of intercellular channels, and their mean open time. Thus, 100% N_2 significantly uncoupled glomus cells by reducing intercellular channel conductances.

Figure 5 shows an example of the depressant effects of 1 mM Na-dithionite on intercellular channel currents in a voltage-clamp experiment. The experimental conditions are presented at the bottom of the figure, showing an imposed ΔV_j of 40 mV. (a) and (b) show channel activity during a 12-sec period in the controls (a) and during superfusion with 1 mM Na-dithionite (b). In each recording the upper sweep is the activity of Cell 1 and the lower trace shows the activity of Cell 2. When the traces separate, the channels opened. Sodium dithionite significantly depressed the intercellular channel currents, a phenomenon accompanying intercellular de-

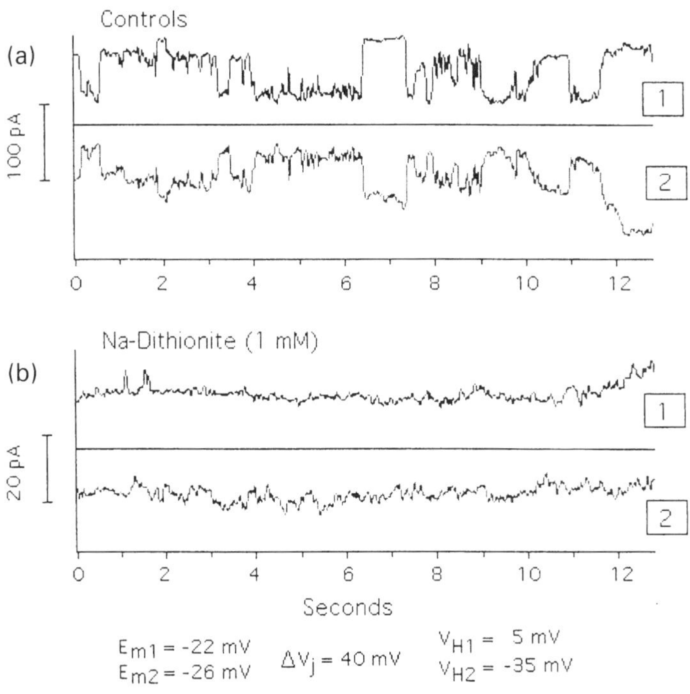

Figure 5 Effect of 1 mM Na-dithionite on intercellular channel currents. (a) control activity and (b) during superfusion with dithionite. Note deflections of equal amplitude and opposite polarity. In squares, cell numbers. Numbers at bottom indicate experimental parameters.

coupling. Our measurements of single channel conductances (around 17 and 22 pS when measured by computer and by hand) resembled those reported in other cells (27–30). However, the junctional conductance between glomus cells may be conditioned by several connexins in addition to connexin43, which has been identified in the rat carotid body (12,13).

The mechanisms responsible for coupling changes during acidity and hypoxia may be multiple. Both stimuli increase intracellular acidity and $[Ca^{2+}]_i$, (7,31,32), which contribute to cell uncoupling in many tissues (33–36) including the carotid body (21,22). Therefore, both high $[Ca^{2+}]_i$ and low pH_i may act in synchrony. However, 100% N_2 is more effective than $Na_2S_2O_4$ in uncoupling glomus cells while inducing weaker pH_i changes and similar increases in $[Ca^{2+}]_i$. Therefore, coupling changes by these agents may also be influenced by other factors.

The glomus cells contain and release chemicals during stimulation [ACh, catecholamines (mostly dopamine (DA)], serotonin (5-HT), enkephalins, prostaglandins, ATP, substance P, and peptides (cholecystokinins, atrial natriuretic peptide-ANP) (14,15). Also, hypoxia increases intracellular cAMP while decreasing cGMP (16–18), substances that affect coupling elsewhere and possibly in the carotid body. ACh and DA are released from glomus cells during acidity and hypoxia, and when given exogenously, uncouple most glomus cells (21). When released, these substances may be reincorporated to disturb intercellular coupling as happens when they are administered to other cells (37–39). Exogenous db-cAMP increases coupling between glomus cells (22), as in other tissues (40–42). An increase in intracellular cAMP during hypoxia would tend to tighten coupling between glomus cells, and since this is accompanied by a decrease in cGMP, an uncoupler in other tissues (43), coupling would tend to be even tighter. Thus, hypoxia releases two opposing forces on glomus cell coupling, one trying to decouple the cells (ACh and DA) and another having the opposite effect (increased cAMP and decreased cGMP). Therefore, the variable effects of hypoxia on glomus cell coupling may depend on which one predominates at a given time. Other cytoplasmatic substances may also play a role in glomus cell coupling:

1. ATP is contained in and released by exocytosis from the dense-core granules of glomus cells (also containing DA) during hypoxia. The released ATP is partly converted into adenosine by the ectonucleotidases. ATP and adenosine may act on nearby cells because they have P_1 purinergic receptors for ATP and P_2 for adenosine (for references see Refs. 14,15). ATP release increases junction permeability in neighboring cells of other tissue, by extracellular action (44–48), and a similar mechanism may occur in glomus cell junctions, leading to increased coupling during hypoxia.

2. Cyclo-oxygenases and prostaglandin E_2 (PGE_2) may also play a role in glomus cell coupling and its changes during hypoxia although this has not been studied. Hypoxia increases the synthesis of endogenous PGE_2 and its exogenous application inhibits catecholamine release from glomus cells during hypoxia, inhibiting inward calcium currents. In other tissues, PGE_2 is essential for intercellular communication across gap junctions when mechanically activated.

3. Applications of cholecystokinin octapeptide (CCK-8), a secretagogue, increases the chemosensory discharge in the carotid nerve after a period of depression. In pancreatic acini CCK-8 induces electrical uncoupling (49).

4. Delivery of atrial natriuretic peptide (ANP) to the cat and rabbit carotid bodies depresses or inhibits the increased discharge obtained during hypoxia. This effect may be produced by increases in cGMP since administrations of the cell-permeant form of cGMP had a similar effect. Interestingly, de Mello (50) found that delivery of ANP (10^{-8} M) to myocytes of cardiomyopathic hamsters decreased g_j as did applications of db-cGMP (10^{-4} M). Similar experiments with coupling between glomus cells have not been done.

The information just presented shows how complex the intimate mechanisms behind glomus cell coupling or uncoupling may be. Much more work is required to understand these processes.

D. Coupling Between Glomus Cells and Nerve Terminals

Kondo and Iwasa (19), with morphological techniques, described the presence of gap junctions between glomus cells and carotid nerve terminals. These junctions appeared within the more commonly observed "chemical" synapses between these structures.

Encouraged by these findings we have begun exploring the possibility of electric and dye coupling between glomus cells and carotid nerve terminals. In this type of exploration, as said before, it is crucial to have proper identification of the recorded elements. In a first series of experiments we probed the feasibility of frequently impaling the nerve terminals since in the past we found it extremely difficult to do so (25). Figure 6 shows the result of one of these efforts using double-barrel microelectrodes; one was filled with 3 M KCl for recording and the other with Lucifer Yellow. The upper part of the figure shows three traces obtained in one of these impalements where a series of small deflections can be seen. These potentials undoubtedly were small depolarizing potentials (s.d.p.s) previously described by Hayashida et al. (25) as originating in carotid nerve terminals. After the recordings, Lucifer Yellow was ejected from the second pipette and the tissue prepared for histology and TH staining. The illustrations at the bottom of Figure 6 were taken by superimposing the green and red channels in the confocal microscope. This technique allowed seeing the bright injected site against a fainter background showing TH fluorescence in the cytoplasm of glomus cells. The left illustration shows two bright spots corresponding to the injected nerve terminal. Only one spot corresponds to the injected site. However, we know since De Castro's work (51) that a carotid nerve fiber branches (sometimes profusely) close to the innervation sites on the glomus cells. Therefore, the dye in an injected terminal bouton may easily spread to nearby boutons through the interconnecting nerve filaments. The illustration on the right was taken from the same specimen but at another optical plane. It shows dye diffusion toward a glomus cell. As shown immediately below, it may be due to coupling between the nerve ending and the glomus cell.

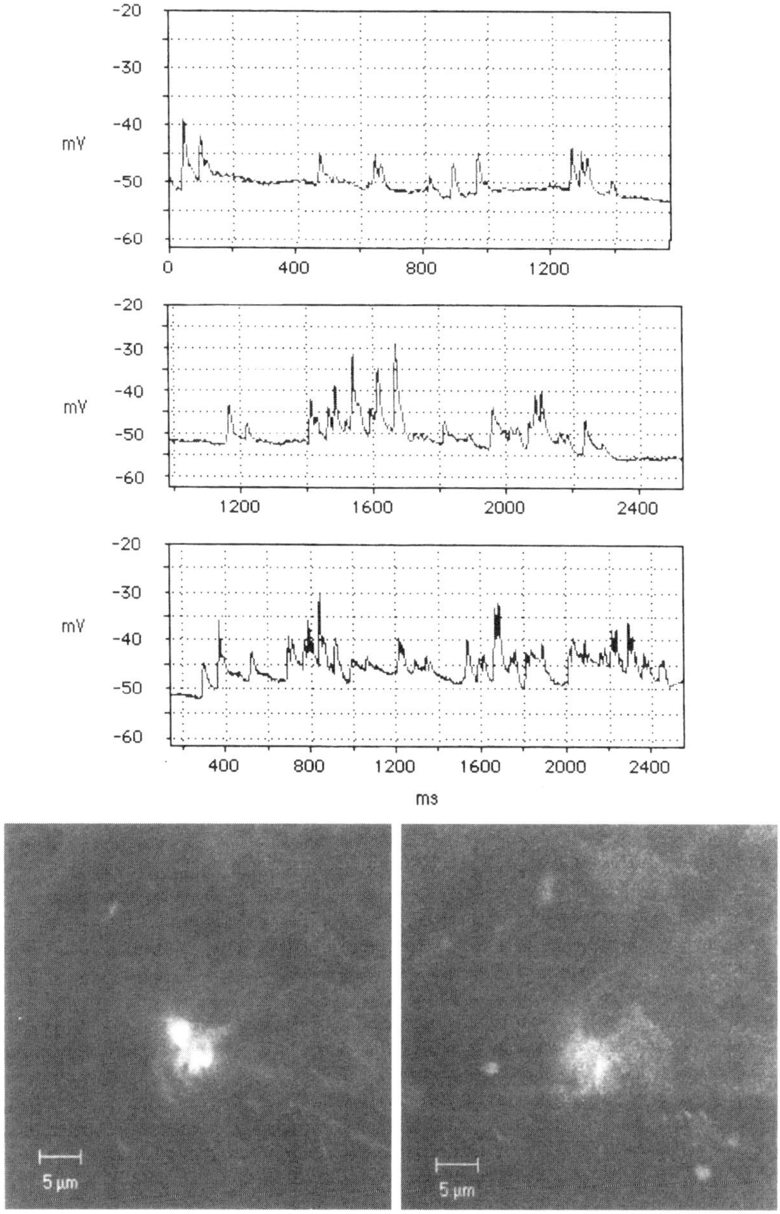

Figure 6 Upper part: Intracellular records obtained from impalement of chemoreceptor nerve terminal. Note occurrence of multiple small depolarizing potentials (s.d.p.s). Lower part: Superimposed confocal images showing TH staining of glomus cells (faint background) and fluorescence from Lucifer Yellow injection into the recorded nerve ending. Bottom left: Two bright fluorescing spots separated by fainter strip. Bottom right: Same observation at a different optical plane. Note spread of dye into glomus cells. Calibrations, 5 µm

Figure 7, taken from another experiment, shows electric and dye coupling between a nerve terminal and adjacent glomus cells. The left part of the figure shows the effects of dual stimulation and recordings from penetrations with electrodes filled with 3 M KCl. A 100-msec positive-voltage pulse was applied to one cell close to the beginning of the traces and another positive pulse was delivered to the other cell near the end of the traces. The larger responses evoked in each cell correspond to the direct effects of stimulation. The upper trace in (a) is the response of a putative glomus cell with low input resistance (48 MΩ). The lower trace shows the recording

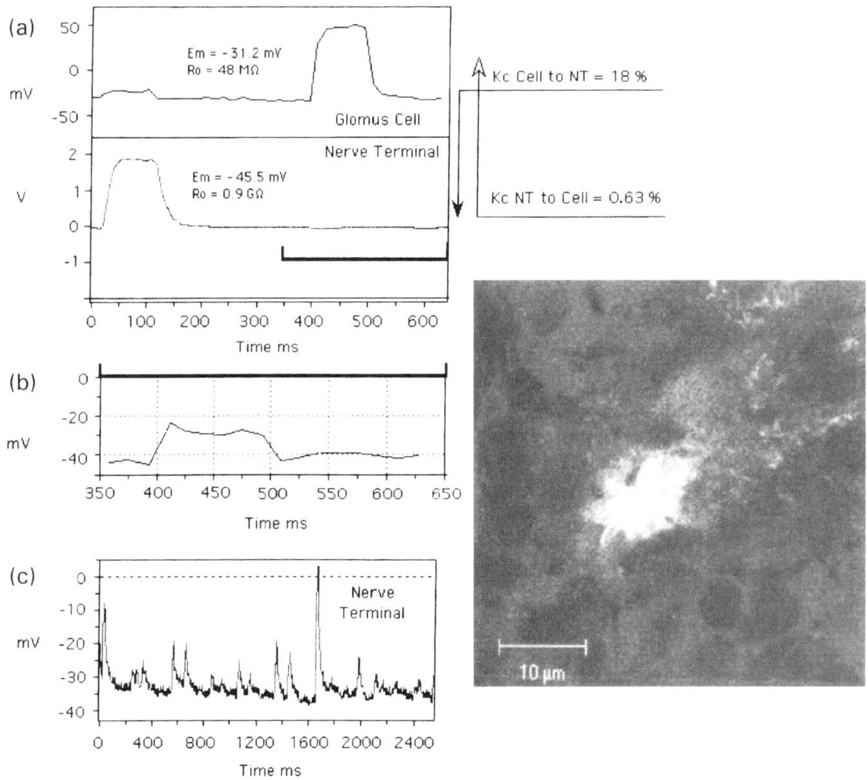

Figure 7 Left: Chart showing intracellular records from impalement of a chemoreceptor nerve terminal coupled to a glomus cell. (a) The upper trace is recorded from the glomus cell whereas the lower trace is from the nerve ending. Note the low resistance ($R_o = 48$ MΩ) of the glomus cell and high input resistance (0.9 GΩ) recorded from the nerve ending. The voltage elicited in the nerve ending is transferred very attenuated ($K_c = 0.63\%$) to the glomus cell. (b) Higher amplification of part of lower trace in (a) (marked by bracket) showing that voltage applied to glomus cell was transferred to the nerve ending with a K_c of 18%. (c) Spontaneous activity of nerve ending represented by multiple s.d.p.s. Illustration on right: Superimposed confocal images showing TH staining of glomus cells (background) and bright fluorescence from Lucifer Yellow injection into the nerve ending. Note that fluorescence does not stay within the ending but is transferred to the glomus cells.

from a carotid nerve terminal with a high input resistance (0.9 GΩ). Stimulation of the nerve ending gave rise to a large deflection [lower trace in (a)] that spread to the putative glomus cell giving rise to a small voltage deflection [upper trace in (a)]. This maneuver indicated that there was electric coupling from the nerve ending toward the glomus cell. When the glomus cell was stimulated [large deflection in the upper trace of (a)] there was voltage spread to the nerve ending, which could not be seen in the lower trace of (a) because of low amplifier gain. However, at higher gain (b), coupling from glomus cell to nerve ending could be seen. In fact, the coupling coefficient (K_c) was much higher (18%) from cell to nerve than in the other direction (0.63%). (c) shows the spontaneous electrical activity of the nerve ending, consisting of s.d.p.s and one spike. The confocal microscope illustration on the right shows that a Lucifer Yellow injection into a nerve terminal (identified by the presence of s.d.p.s), diffuses quite readily toward (or into) surrounding glomus cells.

These experiments show that electric and dye coupling occur between glomus cells and nerve endings of the carotid nerve. However, other experimental details should be discussed now. Stimulation of a nerve terminal with voltage pulses gave rise to a large voltage drop across the membrane because of the high resistance of the terminal. Very often, however this large depolarization did not show any indications of action potentials, suggesting that the stimulated element was "inexcitable," following Grundfest's terminology (52). He used this term to mean that membrane patches responsible for sensory reception do not generate action potentials, as is the case in most sensory receptors. Concerning the carotid body, the glomus cell–nerve ending junction is essentially sensory or centripetal; thus the nerve ending is the postsynaptic element of this synapse. It appears to be "inexcitable." In some cases, the nerve terminal depolarization gave rise to small spikes that appeared to be generated at some distance from the point of stimulation. It would be no different from the nerve endings of other receptors where the spikes are generated away from the unmyelinated terminal and in the first node of Ranvier.

Depolarization or hyperpolarization of the carotid nerve terminal was transferred to the glomus cell. In some cases (one just described), coupling was resistive since the cell followed the applied pulse, as did the stimulated element, but with considerable attenuation. Often, however, this coupling was mainly capacitative because an initial deflection at the on and off declined toward the baseline. Stimulation of putative glomus cells induced similar effects on the nerve ending; sometime coupling was resistive, but often it was capacitative. Therefore, it appears that gap junctions between these structures have noticeable capacity between them. This may not be the case in the chemical part of these unions.

These findings have an interesting corollary. Slow potential changes on either side of the junction may have difficulty in being transferred from the cells to nerve and vice versa. However, it would be easier for rapid voltage changes, like action potentials, to be transferred (although attenuated) from one side to the other of the junctions. Therefore, action potentials generated in glomus cells could affect the nerve ending resting polarization. Similarly, when nerve spikes can reach the endings they could end up affecting the glomus cell membrane potential. However,

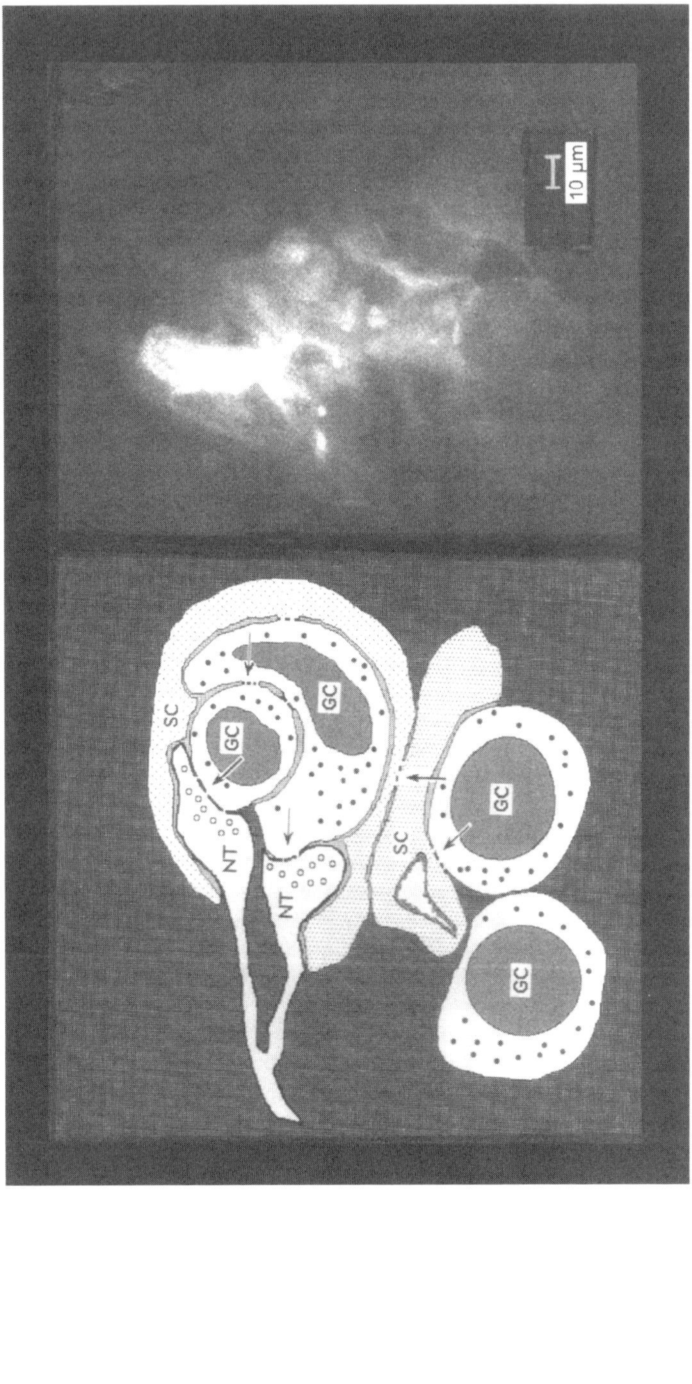

Figure 8 Left: Diagram summarizing intercellular coupling between different structures in the carotid body. GC, glomus cells; SC, sustentacular cells; NT, carotid nerve terminal. Arrows show points (gap junctions) where there is transfer of information from one cell to another. Note that all structures are interconnected. Right: A Lucifer Yellow microinjection into a cell with a complex shape stains this element and dye diffuses to many other structures in the organ. Clearly, some are glomus cells. The injected structure may be a sustentacular cell.

judging from the experiment presented in Figure 7, resistive coupling toward the glomus cells is not very marked.

Our observations and those of Kondo and Iwasa have important physiological and pharmacological implications. One of the difficulties in fully accepting the "transmitter hypothesis of chemoreception" is that specific synaptic blockers, while blocking a given transmitter exogenously applied, usually depress but do not block the effects of natural stimuli such as hypoxia or acidity. Two possible explanations come to mind to interpret this problem. One is that glomus cells release multiple transmitters or other substances, as shown by many authors (for references see Ref. 15). Furthermore, recent work by Zhang et al. (53) clearly show that ATP and ACh act as cotransmitters in cocultures of glomus cells and petrosal ganglion neurons. Further work on this important problem may reveal that cotransmission is a widespread phenomenon. If this is the case, the transmitter hypothesis would be on solid ground. The other explanation concerning poor chemical block is that there are direct electric connections between glomus cells and nerve terminals across the gap junctions described by Kondo and Iwasa (19). We have shown that this electric mechanism for intercellular communication is present. We do not know, however, which one of these two options is more important, or whether both occur together or alternatively during stimulation.

E. The Carotid Body May Operate as a Syncytium

Studies by McDonald (4), by Kondo and Iwasa (19), and by us have revealed that all cellular elements involved in carotid body chemotransduction are interconnected by gap junctions. However, all these studies have been done in the rat and we do not know if these findings can be extended to other animal species. There is only one observation showing that an intracellular dye injected into a cat glomus cell spread to its neighbors (3).

The schematic drawing in Figure 8 attempts to summarize our present knowledge about coupling in the carotid body. There are intercellular communications between glomus cells, between glomus and sustentacular cells, between sustentacular cells, and between glomus cells and nerve endings of the carotid nerve. These widespread connections may have contributed to the results shown on the right side of Figure 8. A microinjection of Lucifer Yellow resulted in intense staining of an irregularly shaped cell and spread of dye to a number of neighboring structures, some of which are clearly glomus cells. It is possible that the widespread dye diffusion resulted from injection into a sustentacular cell that is amply connected in the carotid body.

References

1. Hille B. Ionic Channels of Excitable Membranes. Sunderland, MA: Sinauer, 1992.
2. Bennett MVL, Spray DC. Gap Junctions. Cold Spring Harbor, NY: Cold Spring Harbor Laboratory, 1985.
3. Baron M, Eyzaguirre C. Effects of temperature on some membrane characteristics of carotid body cells. Am J Physiol Cell Physiol 1977; 2:C35–C46.

4. McDonald DM. Peripheral chemoreceptors: structure-function relationships of the carotid body. In: Hornbein TF, ed. Regulation of Breathing. Lung Biology in Health and Disease, Lenfant C, ed. New York: Marcel Dekker, 1981:105–320.
5. Oyama Y, Walker JL, Eyzaguirre C. Intracellular potassium activity, potassium equilibrium potential and membrane potential of carotid body glomus cells. Brain Res 1986; 381:405–408.
6. Zhang XQ, Pang L, Eyzaguirre C. Effects of hypoxia on the intracellular K^+ of clustered and isolated glomus cells of mice and rats. Brain Res 1995; 676:413–420.
7. Pang L, Eyzaguirre C. Hypoxia affects differently the intracellular pH of clustered and isolated glomus cells of the rat carotid body. Brain Res 1993; 623:349–355.
8. Jiang RG, Zhang XQ, Eyzaguirre C. Hypoxia induced by $Na_2S_2O_4$ increases $[Na^+]_i$ in mouse glomus cells, an effect depressed by cobalt: experiments with Na^+-selective microelectrodes and voltage-clamping. Brain Res 1998; 797:197–208.
9. Pang L, Eyzaguirre C. Different effects of hypoxia on the membrane potential and input resistance of isolated and clustered carotid body glomus cells. Brain Res 1992; 575:167–173.
10. Kondo H, Iwanaga T, Nakajima T. Immunocytochemical study on the localization of neuron-specific enolase and S-100 protein. Cell Tiss Res 1982; 227:291–295.
11. Kuffler SW, Nicholls JG. The physiology of neuroglial cells. Ergebn Physiol 1966; 57:1–90.
12. Abudara V, Garcés G, Sáez JC. Cells of the carotid body express connexin43 which is up-regulated by cAMP. Brain Res 1999; 849:25–33.
13. Abudara V, Eyzaguirre C, Sáez, JC. Short- and long-term regulation of rat carotid body gap junctions by cAMP: identification and regulation of connexin43, a gap junction subunit. Adv Exp Med Biol 2000; 475:359–370.
14. González C, Dinger B, Fidone SJ. Functional significance of chemoreceptor cell neurotransmittes. In: González C, ed. The Carotid Body Chemoreceptors. New York: Springer; Austin: Landes, 1997:47–64.
15. Zapata P. Chemosensory activity in the carotid nerve: effects of pharmacological agents. In: González C, ed. The Carotid Body Chemoreceptors. New York: Springer; Austin: Landes, 1997:119–146.
16. Wang WJ, Cheng GF, Dinger BG, Fidone SJ. Effects of hypoxia on cyclic nucleotide formation in rabbit carotid body in vitro. Neurosci Lett 1989; 105:164–168.
17. Perez-Garcia MT, Almaraz L, González C. Effects of different types of stimulation on cyclic AMP content in the rabbit carotid body: functional significance. J Neurochem 1990; 55:1287–1293.
18. Delpiano MA, Acker H. Hypoxia increases the cycle AMP content of the cat carotid body in vitro. J Neurochem 1991; 57:291–297.
19. Kondo H, Iwasa H. Re-examination of the carotid body ultrastructure with special attention to intercellular membrane appositions. Adv Exp Med Biol 1996; 410:45–50.
20. Eyzaguirre C, Abudara V. Carotid body glomus cells: chemical secretion and transmission (modulation?) across cell-nerve ending junctions. Respir Physiol 1999; 115:135–149.
21. Monti-Bloch L, Abudara V, Eyzaguirre C. Electrical communication between glomus cells of the rat carotid body. Brain Res 1993; 622:119–131.
22. Abudara V, Eyzaguirre C. Modulation of junctional conductance between rat carotid body glomus cells by hypoxia, cAMP and acidity. Brain Res 1998; 792:114–125.

23. Abudara V, Jiang RG, Eyzaguirre C. Acidic regulation of junction channels between glomus cells in the rat carotid body: possible role of $[Ca^{2+}]_i$. Brain Res 2001; 916:50–60.
24. Srinivas M, Rozental R, Kojima T, Dermietzel R, Mehler M, Condorelli DF, Kessler JA, Spray DC. Functional properties of channels formed by the neuronal gap junction protein connexin36. J Neurosci 1999; 15:9848–9855.
25. Hayashida Y, Koyano H, Eyzaguirre C. An intracellular study of chemosensory fibers and endings. J. Neurophysiol 1980; 44:1077–1088.
26. Zhong H, Zhang M, Nurse CA. Synapse formation and hypoxic signaling in co-cultures of rat petrosal neurones and carotid body type I cells. J Physiol 1997; 503:599–612.
27. Elenes S, Rubart M, Moreno AP. Junctional communication between isolated pairs of canine atrial cells is mediated by homogeneous and heterogeneous gap junction channels. J Cardiovasc Electrophysiol 1999; 10:990–1004.
28. Rook MB, van Ginneken AC, de Jonge B, el Aoumari A, Gross D, Jongsma HJ. Differences in gap junction channels between cardiac myocites, fibroblasts, and heterogeneous pairs. Am J Physiol Cell Physiol 1992; 263:C959–C977.
29. Somogyi R, Kolb HA. Cell-to-cell channel conductance during loss of gap junctional coupling in pairs of pancreatic acinar cells and Chinese hamster ovary cells. Pflügers Arch 1988; 412:54–65.
30. Spray DC, Chanson M, Moreno AP, Dermietzel R, Meda P. Distinctive gap junction channel types connect WB cells, a clonal cell line derived from rat liver. Am J Physiol Cell Physiol 1991; 260:C513–C527.
31. He SF, Wei JY, Eyzaguirre C. Intracellular pH and some membrane characteristics of cultured carotid body glomus cells. Brain Res 1991; 547:258–266.
32. He SF, Wei JY, Eyzaguirre C. Effects of relative hypoxia and hypercapnia on intracellular pH and membrane potential of cultured carotid body cells. Brain Res 1991; 556:333–338.
33. Francis D, Stergiopoulos K, Ek-Vitorin JF, Cao FL, Taffet SM, Delmar M. Connexin diversity and gap junction regulation by pH_i. Dev Genet 1999; 24:123–136.
34. Lazrak A, Peracchia C. Gap junction gating sensitivity to physiological internal calcium regardless of pH in Novikoff hepatoma cells. Biophys J 1993; 65:2002–2012.
35. Obaid AL, Soccolar SJ, Rose B. Cell to cell channels with two independently regulated gates in series: analysis of junctional channel modulation by membrane potential, calcium and pH. J Membr Biol 1983; 73:69–89.
36. White RL, Doeller JE, Verselis VK, Wittenberg BA. Gap junctional conductance between pairs of ventricular myocites is modulated synergistically by H^+ and Ca^{++}. J Gen Physiol 1990; 95:1061–1075.
37. He S, Weiler R, Vaney DI. Endogenous dopaminergic regulation of horizontal cell coupling in the mammalian retina. J Comp Neurol 2000; 418:33–40.
38. Piccolino M, Neyton J, Gerschenfeld HM. Decrease of gap junction permeability induced by dopamine and cyclic adenosine 3′:5′-monophosphate in horizontal cells of turtle retina. J Neurosci 1984; 10:2477–2488.
39. Randriamampita C, Giaume C, Neyton J, Trautmann A. Acetylcholine-induced closure of gap junction channels in rat lacrimal glands is probably mediated by protein kinase C. Pflügers Arch 1988; 412:462–468.
40. van Rijen HV, van Veen TA, Hermans MM, Jongsma HJ. Human connexin40 gap junction channels are modulated by cAMP. Cardiovasc Res 2000; 45:941–951.

41. Chanson M, White NM, Garber SS. cAMP promotes gap junctional coupling in T84 cells. Am J Physiol Cell Physiol 1996; 271:C53–C59.
42. Romanello M, Moro L, Pirulli D, Crovella S, D'Andrea P. Effects of cAMP on intercellular coupling and osteoblast differentiation. Biochem Biophys Res Commun 2001; 282:1138–1144.
43. Kwak BR, Saez JC, Wilders R, Chanson M, Fishman GI, Hertzberg EL, Spray DC, Jongsma HJ. Effects of cGMP-dependent phosphorylation on rat and human connexin43 gap junction channels. Pflügers Arch 1995; 430:770–778.
44. Isakson BE, Evans WH, Boitano S. Intercellular Ca^{2+} signaling in alveolar epithelial cells through gap junctions and by extracellular ATP. Am J Physiol Lung Cell Mol Physiol 2001; 28:L221–L228.
45. Cotrina ML, Lin JH, López-Garcia JC, Naus CC, Nedergaard M. ATP-mediated glia signaling. J Neurosci 2000; 20:2835–2844.
46. Sauer H, Hescheler J, Wartenberg M. Mechanical strain-induced Ca^{2+} waves are propagated via ATP release and purinergic receptor activation. Am J Physiol Cell Physiol 2000; 279:C295–C307.
47. Guthrie PB, Knappenberger J, Segal M, Bennett MV, Charles AC, Kater SB. ATP released from astrocytes mediates glial calcium waves. J Neurosci 1999; 19:520–528.
48. Homolya L, Steinberg TH, Boucher RC. Cell to cell communication in response to mechanical stress via bilateral release of ATP and UTP in polarized epithelia. J Cell Biol 2000; 150:1349–1360.
49. Ngezahayao A, Kolb HA. Gap junctional conductance tunes phase difference of cholecystokinin evoked calcium oscillations in pairs of pancreatic acinar cells. Pflügers Arch 1993; 422:413–415.
50. de Mello WC. Atrial natriuretic factor reduces cell coupling in the failing heart, an effect mediated by cyclic GMP. J Cardiovasc Pharmacol 1998; 32:75–79.
51. De Castro, F. Sur la structure de la synapse dans les chemorecepteurs: leur mécanisme d'excitation et rôle dans la circulation sanguine locale. Acta Physiol Scand 1951; 22:14–43.
52. Grundfest H. Evolution of electrophysiological varieties among sensory receptor systems. In: Essays on Physiological Evolutions. New York: Pergamon, 1964:107–138.
53. Zhang M, Zhong H, Vollmer C, Nurse CA. Co-release of ATP and ACh mediate hypoxic signalling at rat carotid body chemoreceptors. J Physiol 2000; 525:143–158.

19

From Oxygen Sensing to Chemosensory Activity
The Mediator Role of Glomus Cells

PATRICIO ZAPATA

Catholic University of Chile
Santiago, Chile

I. Introduction

The very nature of the carotid body has been a matter of much discussion. The names initially given to this organ pointed to conjectures that it was either a ganglion or a gland. Only the exhaustive microscopic studies of De Castro (1,2) led to his proposal that the carotid body was a receptor organ. This presumption was based on the structural relationships found between capillaries and glomus cells (then described as "epithelioid cells") on one side, and between glomus cells and nerve endings (whose sensory nature was revealed in those papers) on the other side.

The above seminal studies gave a preponderant role to glomus cells as intermediary elements between what is sensed (chemicals in the blood) and how neural information is generated. But, the need to count on glomus cells for the transduction of chemical stimuli into nerve impulses has been subjected to much debate and it is still an open problem. In a book devoted to oxygen sensing, this subject deserves a closer examination.

To put the problem in the simplest way is to consider the carotid bodies (as well as the aortic bodies) either as "simple receptors," in which sensory nerve terminals are directly excited by the chemical stimuli carried by the blood, or as "composite receptors," in which chemical stimuli only reach and act upon specialized receptor cells (glomus cells in our case) and these cells must transmit information about the stimuli to the nerve endings of primary sensory neurons. In the first situation, nerve terminals could properly qualify as "chemosensory" per se; in the second situation, they would be postsynaptic elements to receptor (glomus) cells.

To solve the problem presented above, two convergent strategies have been followed: (1) to demonstrate whether chemosensory nerve endings can sense physiological chemical stimuli when separated from glomus cells, and (2) to find out if chemoreception fails after destruction of glomus cells.

II. Carotid Bodies Resection

Holton and Wood (3) reported the permanent disappearance of the reflex hyperventilation evoked by hypoxia in two asthmatic patients after removal of both carotid bodies (glomectomy). Baroreflexes were also abolished in these subjects, but recovered within a few weeks. Thus, reinnervation of the carotid sinuses by regenerating barosensory fibers resulted in recovery of baroreceptor function, while a similar timing and degree of regeneration of chemosensory fibers failed to reinstate ventilatory chemoreflexes. This means that carotid body cells—presumably glomus cells—are required for chemoreceptor function.

Later reports (4–9) have confirmed that carotid chemoreceptor function is permanently lost after resection of the carotid bodies in humans. Therefore, these reports provide a first indication that the carotid body parenchyma is required for oxygen sensing. In other words, carotid nerve chemosensory fibers by themselves are unable to detect changes in O_2 level.

III. Reestablishment of Chemoreceptor Function by Regenerating Chemosensory Fibers

After section of the carotid (sinus) nerve, chemosensory discharges are recorded from the peripheral stump for not longer than 18 hr in the cat (10); this corresponds to the time of early degenerative changes in the chemosensory nerve endings within the carotid body (10,11). However, regeneration of carotid nerve fibers proceeds rapidly and chemosensory discharges are recordable again from the central stump of this nerve by the end of a week (12).

It must be noted that the distance between the site of nerve section and the neural pole of the carotid body is important for chemoreceptor function recovery. When the regenerating central stump of the nerve has to travel a longer path to reach carotid body tissue, the reappearance of chemoreceptor discharges is delayed (12). Moving forward from this observation, we studied whether the time allowed for regeneration of chemosensory fibers by itself is determinant or whether the access of regenerating fibers to glomus tissue is the important factor for recovery of chemoreceptor function. This will be discussed later in this chapter.

Mitchell et al. (13) and Kienecker et al. (14) reported that the neuromata resulting from carotid nerve section attained recovery of chemosensory responses, pointing out that carotid nerve fibers—if allowed enough time for regeneration—were capable of detecting hypoxic challenges in the absence of their connections with glomus tissue. However, these reports were criticized on the basis that carotid bodies had not been excised at the time when carotid neurotomies were performed and, therefore, the possibility that regenerating fibers would have achieved contact with glomus tissue had not been discarded.

In fact, later studies by Smith and Mills (15) showed that regenerated carotid nerves in cats failed to recover chemoreceptor function when the carotid bodies had been removed simultaneously with the carotid neurotomies, leading the authors to conclude that chemoreceptivity was not a property of carotid nerve chemosensory axons. Furthermore, restoration of reflex ventilatory response to hypoxia occurred after a month of removal of the carotid bodies in the cat, a response unaffected by recutting the carotid nerves but completely abolished by section of the vagus nerves, including the aortic nerves (16). Thus, central reorganization of the gain of chemoreflex pathways was concluded and the recovery of chemosensitivity by regenerating carotid nerves was discarded.

In a similar vein, Belmonte et al. (17) observed that carotid nerves regenerated into superior cervical ganglia, organs extremely well vascularized, also failed to regain chemoreceptor activity. Therefore, all evidence available suggests that regenerating chemosensory nerve fibers recover chemoreception only when reestablishing contact with carotid body tissue.

We (18) studied the timing for recovery of chemosensory nerve activity and for regeneration of nerve fibers into carotid body parenchyma after crushing the carotid nerve in cats. We observed a clear correlation between reappearance of chemosensory nerve responses and reestablishment of close contacts between regenerating nerve endings and glomus cells. This was evidenced by a series of experiments in which both carotid nerves were simultaneously crushed, one of them at the neural pole of the carotid body, and the contralateral one at its emergence from the glossopharyngeal nerve. The carotid nerve that had to regenerate a shorter path to reach carotid body tissue recovered its chemoreceptor function more promptly than the contralateral one, and this was correlated with ultrastructural observations on the earlier appearance of extensive appositions between regenerating axon sprouts and glomus cells, contacts that were later wrapped by sustentacular cells.

IV. Are Glomus Cells Capable of Reestablishing Chemosensory Function When Reinnervated by Foreign Nerves?

In an attempt to find out if glomus cells would reestablish chemosensitivity when constructing artificial synapses with other nerve, De Castro (19) sutured the central stump of the vagus nerve sectioned below the nodose ganglion with the distal stump of the glossopharyngeal nerve. The idea was that regenerating vagal sensory fibers would follow the path of the glossopharyngeal and carotid nerves, to reach the carotid body. To avoid contamination with vagal motor fibers, De Castro had severed the vagal roots within the cranium. Not disposing of electrophysiological equipment to record nerve discharges from the central processes of sensory fibers, he anastomosed the vagus nerve stump cephalad to the nodose ganglion to the preganglionic trunk of the superior cervical ganglion, to replace an oscilloscope by the own pupil of the cat. After allowing enough time for fiber regeneration and synaptic reorganization, De Castro observed that perfusion of the reinnervated carotid bifurcation with hypercapnic or acidic solutions resulted in sympathetically mediated ipsilateral pupillary dilatation. Unfortunately, he did not test hypoxic solutions. Nevertheless, he concluded that vagal fibers regenerating into the carotid body and making contact with its parenchymal cells had been transformed into fibers carrying chemosensory information, thus suggesting that glomus ("epithelioid") cells were the chemoreceptor elements.

The above observations were later criticized on two grounds: the nodose ganglion normally contains the perikarya of the chemosensory fibers of the aortic nerve (20), as well as of bronchopulmonary C-fibers responsive to CO_2 (21). Thus, responses described by De Castro (19) in his double anastomosed preparations could result from reinnervation of the carotid body by originally chemosensitive vagal fibers.

A different approach consisted in reinnervating the carotid body with the superior laryngeal nerve (22). This nerve normally contains a large population of mechanosensory fibers (23) and a small population of chemosensory fibers responding to changes in CO_2, but not in O_2 (24). By electrical recording from the superior laryngeal nerve reinnervating the carotid bifurcation, we observed increased discharges in response to hypoxia and asphyxia in preparations that, when later studied ultrastructurally, showed intimate connections between regenerated nerve endings and glomus cells (22). Thus, vagal nerve fibers had been turned responsive to hypoxia after contacting carotid body tissue.

Kondo (25) transplanted rat carotid bodies into the anterior chamber of the eye, observing 3-mo survival of glomus and sustentacular cells, with cytological characteristics similar to those of intact carotid bodies, but with axon-like processes growing from glomus cells. These transplants were reinnervated by iridial nerve fibers, some of them of sympathetic nature and others probably of sensory nature. However, synapses were rarely seen between transplanted glomus cells and nerve fibers, both elements usually being separated by sustentacular cells.

Monti-Bloch et al. (26) transplanted the carotid body into the tenuissimus muscle of cats. After several months, recording of impulse activity from the tenuissimus muscle nerve revealed afferent discharges whose frequency was increased by hypoxia and hypercapnia. Some single afferents retained their mechanosensitivity (probably through a branch innervating muscle spindles) at the same time that they revealed chemosensitivity (attributed to other branches reinnervating the transplanted carotid body). Ultrastructural studies of these preparations revealed some glomus cells making contacts with axons.

Carotid bodies have also been transplanted into the previously damaged striatum of monkeys (27). The idea was to replace central dopaminergic neurons by glomus cells, capable of reestablishing dopaminergic function. Survival of tyrosine hydroxylase immunoreactive glomus cells was indeed observed in the striatum. These autografts allowed a recovery of motility and reduction of tremor in the contralateral extremities of these parkinsonian monkeys.

Carotid bodies and nodose ganglia from rats have also been cotransplanted to the chorioallantois of chicken-fertilized eggs (28). In this preparation in ovo, the carotid body is highly vascularized by chorioallantoic vessels and is reinnervated by regenerating axons from nodose neurons. Carotid body tissues may also be explanted and grown in tissue culture. Interestingly, nodose ganglion cells co-cultured with glomus tissue from carotid bodies reveal different properties from those present when cultured alone (29,30). This points to a reconstitution in culture of chemoreceptor complexes between glomus cells and ganglion sensory neurons. The nodose ganglion does not normally provide innervation to the carotid body in mammals; otherwise, explanted carotid bodies contained not only glomus cells, but also sustentacular and mesenchymal cells.

More recently, successful reinnervation of carotid body glomus cells, changes in the electrical properties, pharmacological sensitivity, and reappearance of hypoxic chemosensitivity in petrosal ganglion cells have been reported when explants of both organs are cultured together (31–35). Petrosal ganglion neurons acutely disconnected from the carotid body and superfused in vitro are not excited by hypoxic hypoxia (36).

Most of the above observations indicate that foreign nerves may acquire hypoxic chemosensitivity when put into contact with carotid body tissue. One interpretation—the most commonly presented—is that glomus cells release an excitatory transmitter, which initiates impulse activity from adjacent sensory nerve endings. But there is also another interpretation less commonly discussed: that peripheral nerve endings are turned chemosensitive when put into close contact with glomus cells, i.e., that nerve ending chemosensitivity is induced by a trophic factor released from glomus cells. One could ask: why not released from sustentacular cells? But this is unlikely because sustentacular cells are considered glial elements of the same lineage as Schwann cells wrapping the sensory nerve fibers, and therefore adjacent to the growing tips of regenerating axons all along the peripheral trajectory of the nerves.

On the other hand, glomus cells retain their cytological characteristics (abundant dense-cored vesicles, formaldehyde-induced fluorescence), as well as their neurotransmitter content (at least ACh and DA), after sensory denervation (37,38). Only the presynaptic specializations of glomus cells (electron-dense material and short projections in cell membrane apposed to nerve endings, with clustering of dense-cored and small, clear vesicles nearby) rapidly disappear after carotid nerve section (i.e., postsynaptic deprivation), but they also reappear shortly after reestablishment of apposition with regenerating nerve sprouts (39). With regard to sustentacular cells, upon sensory denervation, they rapidly grow profuse interdigitations to occupy the space left vacant by the retraction of degenerating nerve endings (10,12). Upon reinnervation, it is not clear whether sustentacular cells represent an initial barrier for reestablishment of synapses between glomus cells and regenerating nerves or provide guidance for nerve sprouts toward glomus cells.

V. Does Chemoreception Fail After Destruction of Glomus Cells?

A first attempt to destroy glomus cells was reported by Verna et al. (40). It consisted of cryocoagulation of the carotid body in rabbits. After allowing 3–5 mo for regeneration of carotid nerve tips into the scar tissue, they failed to record chemosensory impulse activity from the carotid nerve, except when some glomus cells had been spared by congelation, had probably multiplied by mitosis (41), and the resulting islands had established synapses with regenerating nerves. This was confirmed by a later study allowing up to 1 year for regeneration of sensory nerve fibers and where chemosensitivity was controlled by measuring ventilatory reflexes elicited by hypoxic and cytotoxic hypoxia (42). These authors concluded that regenerated afferents "did not show any chemosensory property in absence of [glomus] type I cells" and that "these [glomus] cells appear consequently necessary for a normal chemosensitivity".

A second attempt to destroy glomus cells made use of 6-hydroxy-dopamine (6-OH-DA), a drug extensively used to destroy central dopaminergic neurons (43). This agent was taken up by glomus cells and incorporated into dense-cored vesicles (44), followed by a decreased number of these electron-dense granules and a marked reduction of formaldehyde-induced fluorescence, but no degenerative changes in glomus cells (45). However, neonatal rats receiving 6-OH-DA reveal disappearance of over 90% of noradrenergic autonomic ganglion cells, without changes in fluorescence of dopaminergic ganglion interneurons (46) and carotid body glomus cells (47). Dopamine and norepinephrine levels in the carotid body of rats were unaffected by treatment with 6-OH-DA (38). When recording chemosensory discharges from the carotid nerves in cats, neither basal levels nor responses to asphyxia or cytotoxic hypoxia were modifed for up to 48 hr after intracarotid injections of 6-OH-DA (48). Therefore, 6-OH-DA did not impair carotid body chemoreceptor function.

The third attempt to destroy glomus cells consisted of subjecting the carotid body to ischemia. The cat carotid body receives its vascular supply from the occipital and ascending pharyngeal arteries, which in turn originate from the bifurcation of the common carotid into the external and internal carotids (49). The ischemia provoked by occlusion of the occipitopharyngeal trunk produced within 1 hr depletion of dense-cored vesicles and swollen mitochondria in glomus cells and marked degenerative changes of these cells (vacuolation, decreased cytoplasmic electron density, and nuclear shrinkage) within 3 hr of ischemia, while the ultrastructure of nerve endings was essentially preserved (50). Electrical recordings of the carotid nerves of these preparations revealed that—after a 15-min increase of basal discharges—chemosensory responses to cytotoxic hypoxia and asphyxia were markedly depressed within 1 hr of ischemia, and such responses were minimal or absent after 3 hr of such conditions. Restoring the circulation after 1–3 hr of ischemia resulted in limited recovery of chemosensory responses. In a later study, Monti-Bloch et al. (51) reported that 2 hr or longer ischemia of the carotid body—with or without subsequent reestablishment of circulation by removal of the ligature—produced irreversible damage followed by disappearance of glomus cells as well as of identifiable sustentacular cells, but regenerating unmyelinated axons were observed after 1–2 wk. Electrical recording of the carotid nerves of these preparations showed that they were unresponsive to hypoxic and cytotoxic hypoxia, in contrast with responses elicited by mechanical stimulation of the carotid bifurcation. Thus, permanent damage of glomus cells appears to eliminate hypoxic chemoreception.

VI. Final Considerations on the Chemoreceptor Complex

The evidence presented above indicates that the presence of glomus cells is a requisite for hypoxic chemoreception. However, some limitations in the interpretation of the above considerations must be presented here:

1. That glomus cells exhibit membrane and cytoplasmic changes in response to hypoxia is very well documented (see Refs. 52–54), but the only way to demonstrate that hypoxia-induced glomus cell excitation initiates the full chemoreceptor process is by recording sensory discharges from the adjacent nerve fibers.
2. Local destruction of glomus cells is commonly accompanied by elimination—or at least dedifferentiation—of sustentacular cells, whose actions have seldom been searched for and whose importance is poorly understood.
3. Elimination of glomus cells suppresses not only their transduction mechanisms for natural chemical stimulation but also their possible trophic effects upon adjacent sensory nerve endings.
4. Destruction of the parenchymal cells of the carotid body (both glomus and sustentacular cells)—by either cryocoagulation or by ligature-induced ischemia—is accompanied by a profound distortion of the rich vascularization of the carotid body, whose consequences may contribute to the abolition of the hypoxic chemoreception observed.
5. It is puzzling to observe that glomus cells are rapidly destroyed by circulatory arrest of the carotid body, while these cells may survive after excision of the carotid body and its transplant into another organ.

It has recently been taken for granted that the availability of a given cellular mechanism—e.g., an ion channel, a transport system, a metabolic pathway, or a secretory mechanism—is enough proof of the physiological role played by the cell disposing of such a tool. Furthermore, changes in the level of a given substance at a certain locus, even when in strict correlation with the functioning of such locus, may be necessary causal steps for such function, but they may also be consequences of that. Therefore, at a time of very recent, abundant information on molecular, subcellular, and cellular mechanisms observed in arterial chemoreceptor organs, the intention of this chapter was to critically review some older literature on the complex cellular interactions operative at the level of carotid body tissue, information that may be useful as a frame for a balanced understanding of the physiology of this organ.

Acknowledgments

Thanks are due to Mrs. Carolina Larrain for her helpful assistance and valuable comments during the preparation of this document. The work of the author of this chapter is presently supported by grant 1010951 from the National [Chilean] Fund for Scientific and Technological Development (FONDECYT).

References

1. De Castro F. Sur la structure et l'innervation de la glande intercarotidienne (glomus caroticum) de l'homme et des mammifères, et sur un nouveau système d'innervation autonome du nerf glossopharyngien. Études anatomiques et expérimentales. Trab Lab Invest Biol Univ Madrid 1926; 24:365–432.
2. De Castro F. Sur la structure et l'innervation du sinus carotidien de l'homme et des mammifères. Nouveaux faits sur l'innervation et la fonction du glomus carotidien. Études anatomiques et physiologiques. Trab Lab Invest Biol Univ Madrid 1928; 25:331–380.
3. Holton P, Wood JB. The effects of bilateral removal of the carotid bodies and denervation of the carotid sinuses in two human subjects. J Physiol (Lond) 1965; 181:365–378.
4. Lugliani R, Whipp BJ, Seard C, Wasserman K. Effect of bilateral carotid body resection on ventilatory control at rest and during exercise in man. N Engl J Med 1971; 285:1105–1111.
5. Swanson GD, Whipp BJ, Kaufman RD, Aqleh KA, Winter B, Bellville JW. Effect of hypercapnia on hypoxic ventilatory drive in carotid body-resected man. J Appl Physiol 1978; 45:971–977.
6. Bellville JW, Whipp BJ, Kaufman RD, Swanson KA, Aqleh KA, Wiberg DM. Central and peripheral chemoreflex loop gain in normal and carotid body-resected subjects. J Appl Physiol 1979; 46:843–853.
7. Honda Y, Watanabe S, Hashizume I, Satomura Y, Hata N, Sakakibara Y, Severinghaus JW. Hypoxic chemosensitivity in asthmatic patients two decades after carotid body resection. J Appl Physiol 1979; 46:632–638.
8. Honda Y, Myojo S, Hasegawa S, Hasegawa T, Severinghaus JW. Decreased exercise hyperpnea in patients with bilateral carotid chemoreceptor resection. J Appl Physiol 1979; 46: 908–912.
9. Vermeire P, de Backer W, van Maele R, Bal J, van Kerckhoven W. Carotid body resection in patients with severe chronic airflow limitation. Bull Eur Physiopathol Respir Clin Respir Physiol 1987; 23:165S–169S.
10. Hess A, Zapata P. Innervation of the cat carotid body: normal and experimental studies. Fed Proc 1972; 31:1365–1382.
11. Abbott CP, Daly M de B, Howe A. Early ultrastructural changes in the carotid body after degenerative section of the carotid sinus nerve in the cat. Acta Anat 1972; 83:161–183.
12. Zapata P, Stensaas LJ, Eyzaguirre C. Axon regeneration following a lesion of the carotid nerve: electrophysiological and ultrastructural observations. Brain Res 1976; 113:235–253.
13. Mitchell RA, Sinha AK, McDonald DM. Chemoreceptive properties of regenerated endings of the carotid sinus nerve. Brain Res 1972; 43:681–685.

14. Kienecker EW, Knoche H, Bingman D. Functional properties of regenerating sinus nerve fibres in the rabbit. Neuroscience 1978; 3:977–988.
15. Smith PG, Mills E. Physiological and ultrastructural observations on regenerated carotid sinus nerves after removal of the carotid bodies in cats. Neuroscience 1979; 4:2009–2020.
16. Smith PG, Mills E. Restoration of reflex ventilatory response to hypoxia after removal of carotid bodies in the cat. Neuroscience 1980; 5:573–580.
17. Belmonte C, Rigual R, Gallego R. Responses of carotid nerve fibres regenerating into the superior cervical ganglion. In: Belmonte C, Pallot D, Acker H, Fidone S, eds. Arterial Chemoreceptors. Leicester: Leicester Univ Press, 1981:125–132.
18. Zapata P, Stensaas LJ, Eyzaguirre C. Recovery of chemosensory function of regenerating carotid nerve fibers. In: Acker H, Fidone S, Pallot D, Eyzaguirre C, Lübbers DW, Torrance RW, eds. Chemoreception in the Carotid Body. Berlin: Springer-Verlag, 1977:44–50.
19. De Castro F. Sur la structure de la synapse dans les chemocepteurs: leur mécanisme d'excitation et rôle dans la circulation sanguine locale. Acta Physiol Scand 1951; 22:14–43.
20. Coleridge HM, Coleridge JC, Howe A. A search for pulmonary arterial chemoreceptors in the cat, with a comparison of the blood supply of the aortic bodies in the newborn and adult animal. J Physiol (Lond) 1967; 191:353–374.
21. Delpierre S, Grimaud CH, Jammes Y, Mei N. Changes in activity of vagal bronchopulmonary C fibres by chemical and physical stimuli in the cat. J Physiol (Lond) 1981; 316:61–74.
22. Zapata P, Hess A, Eyzaguirre C. Reinnervation of carotid body and sinus with superior laryngeal nerve fibers. J Neurophysiol 1969; 32:215-228.
23. Sampson S, Eyzaguirre C. Some functional characteristics of mechanoreceptors in the larynx of the cat. J Neurophysiol 1964; 27:464–480.
24. Boushey HA, Richardson PS, Widdicombe JG, Wise JCM. The response of laryngeal afferent fibres to mechanical and chemical stimuli. J Physiol (Lond) 1974; 240:153–175.
25. Kondo H. Fine structure of the rat carotid body transplanted into the anterior chamber of the eye. J Neurocytol 1978; 7:505–516.
26. Monti-Bloch L, Stensaas, LJ, Eyzaguirre C. Carotid body grafts induce chemosensitivity in muscle nerve fibers of the cat. Brain Res 1983; 270:77–92.
27. Luquin MR, Montoro RJ, Guillen J, Saldise L, Insausti R, Del Rio J, López-Barneo J. Recovery of chronic parkinsonian monkeys by autotransplants of carotid body cell aggregates into putamen. Neuron 1999; 22:743–750.
28. Gual A, Eugenín J, Alcayaga J, Stensaas LJ, Eyzaguirre C. The chick chorioallantoic membrane promotes survival of co-transplanted rat carotid bodies and nodose ganglia. Brain Rcs 1991; 556:139–144.
29. Goldman WF, Sato M, Stensaas LJ, Eyzaguirre C. Acetylcholine- and dopamine-induced excitation of cultured newborn rabbit nodose ganglion neurons: effects of co-culture with carotid body fragments. In: Ribeiro JA, Pallot DJ, eds. Chemoreceptors in Respiratory Control. London: Croom Helm, 1987:284–295.
30. Alcayaga J, Eyzaguirre C. Electrophysiological evidence for the reconstitution of chemosensory units in co-cultures of carotid body and nodose ganglion neurons. Brain Res 1990; 534:324–328.
31. Zhong H, Nurse C. Co-cultures of rat petrosal neurons and carotid body type I cells: a model for studying chemosensory mechanisms. Adv Exp Med Biol 1996; 410:189–193.

32. Alcayaga J, Arroyo J. Responses of cat petrosal ganglion neurons are modified by the presence of carotid body cells in tissue cultures. Adv Exp Med Biol 1996; 410:195–201.
33. Zhong H, Zhang M, Nurse C. Synapse formation and hypoxic signaling in co-cultures of rat petrosal neurones and carotid body type 1 cells. J Physiol (Lond) 1997; 503: 599–612.
34. Nurse CA, Zhang M. Acetylcholine contributes to hypoxic chemotransmission in co-cultures of rat type 1 cells and petrosal neurons. Respir Physiol 1999; 115:189–199.
35. Zhang M, Zhong H, Vollmer C, Nurse CA. Co-release of ATP and ACh mediates hypoxic signalling at rat carotid body chemoreceptors. J Physiol (Lond), 2000; 525: 143–158.
36. Alcayaga J, Varas R, Arroyo J, Iturriaga R, Zapata P. Responses to hypoxia of petrosal ganglia in vitro. Brain Res 1999; 845:28–34.
37. Hess A. Chronically denervated rat carotid bodies. Acta Anat 1977; 97:307–316.
38. Hellström S. Putative neurotransmitters in the carotid body. mass fragmentographic studies. Adv Biochem Psychopharmacol 1977; 16:257–263.
39. Kondo H, Pappas GD. The effect of postsynaptic deprivation on the presynaptic chief cell in the rat carotid body. Brain Res 1981; 215:125–133.
40. Verna A, Roumy AM, Leitner L-M. Loss of chemoreceptive properties of the rabbit carotid body after destruction of the glomus cells. Brain Res 1975; 100:13–23.
41. Verna A. Dense-cored vesicles and cell types in the rabbit carotid body. In: Acker H, Fidone S, Pallot D, Eyzaguirre C, Lübbers DW, Torrance RW, eds. Chemoreception in the Carotid Body. Berlin: Springer-Verlag, 1977:216–220.
42. Verna A, Roumy M, Leitner L-M. Role of the carotid body cells: long-term consequences of their cryodestruction. Neurosci Lett 1980; 16:281–285.
43. Hökfelt T, Ungerstedt U. Specificity of 6-hydroxydopamine induced degeneration of central monoamine neurons: an electron and fluorescence microscopic study with special reference to intracerebral injection on the nigro-striatal dopamine system. Brain Res 1973; 60:269–297.
44. Hess A. The effects of 6-hydroxydopamine on the appearance of granulated vesicles in glomus cells of the rat carotid body. Tissue Cell 1976; 8:381–397.
45. Lassmann H, Böck P. Die Wirkung von 6-Hydroxydopamine auf den Katecholamingehalt des Glomus Caroticum der Ratte. Z Zellforsch Mikrosk Anat 1972; 127: 220–229.
46. Eränkö L, Eränkö O. Effect of 6-hydroxy-dopamine on the ganglion cells and the small intensely fluorescent cells in the superior cervical ganglion of the rat. Acta Physiol Scand 1972; 84:115–124.
47. Aloe L, Levi-Montalcini R. Comparative studies on the effects elicited by pre- and postnatal injections of anti-NGF, guanethidine, and 6-hydroxydopamine in chromaffin and ganglion cells of the adrenal medulla and carotid body in infant rats. Adv Biochem Psychopharmacol 1980; 25:221–226.
48. Zuazo A, Zapata P. Effects of 6-hydroxy-dopamine on carotid body chemosensory activity. Neurosci Lett 1978; 9:323–328.
49. Seidl E. On the variability of form and vascularization of the cat carotid body. Anat Embryol 1976; 149:79–86.
50. Nishi K, Okajima Y, Ito H, Sugahara K. Alteration of chemoreceptor responses and ultrastructural features of ischemic carotid body of the cat. Jpn J Physiol 1981; 31:677–694.
51. Monti-Bloch L, Stensaas LJ, Eyzaguirre C. Effects of ischemia on the function and structure of the cat carotid body. Brain Res 1983; 270:63–76.

52. López-Barneo J, Pardal R, Ortega-Saenz P. Cellular mechanisms of oxygen sensing. Annu Rev Physiol 2001; 63:259–287.
53. Prabhakar NR, Overholt JL. Cellular mechanisms of oxygen sensing at the carotid body: heme proteins and ion channels. Respir Physiol 2001; 122:209–221.
54. Shirahata M, Sham JSK. Roles of ion channels in carotid body chemotransmission of acute hypoxia. Jpn J Physiol 1999; 49:213–228.

20

Excitation of Glomus Cells

Interaction Between Voltage-Gated K^+ Channels and Cholinergic Receptors

MACHIKO SHIRAHATA, TOMOKO HIGASHI, SERABI HIRASAWA, SHIGEKI YAMAGUCHI, ROBERT S. FITZGERALD, and BORIS LANDE

Johns Hopkins University
Baltimore, Maryland, U.S.A.

I. Introduction

Systemic hypoxia is a potentially lethal situation for the animal. To protect major organs from irreversible damage, the carotid body, a primary sensory organ for arterial hypoxia, sends a message to the central nervous system and induces various responses in cardiovascular, respiratory, renal, and endocrine systems (1). This is a unique feature of the carotid body: i.e., the consequences of oxygen sensing are not confined within the organ, but are used to serve other organs. Decades of investigations have tried to clarify characteristics of this unique organ, such as oxygen-sensing mechanisms, transduction pathways, and transmission of the information between the glomus cells of the carotid body and the sensory afferent neurons. At present, most investigators propose a model in which glomus cells, the chemosensory cells, release neurotransmitters as a result of an increase in intracellular Ca^{2+} ($[Ca^{2+}]i$) in response to stimuli such as hypoxia. The resulting interaction of neurotransmitters and their receptors generates action potentials in the chemosensory afferent fiber. In this schema, neurotransmitters, their receptors, and ion channels are critical components. To date, much information has been gathered regarding the effects of hypoxia on several types of ion channels in glomus cells.

However, it is still difficult to obtain a unified view of the mechanisms of chemoreception and chemotransduction in the carotid body. Perhaps a deeper understanding of the mechanisms involved in the hypoxic excitation of the carotid body can be gained by considering the fact that all channels and receptors are operating together, and their cross-talk, not the action of single component, determines the excitability of glomus cells. In this chapter we present an example of such interactions. In brief, we have found that cholinergic receptors modify the activity of voltage-gated K^+ channels (Kv channels).

II. Mechanisms of Ca^{2+} Influx to Glomus Cells of the Carotid Body

The release of neurotransmitters from the glomus cell is assumed to be regulated by $[Ca^{2+}]i$, and much experimental data support the concept. A close correlation between $[Ca^{2+}]i$ level and catecholamine release has been shown in cultured adult rabbit glomus cells (2,3). Further, the influx of Ca^{2+} from an extracellular milieu appears essential for the release of neurotransmitters during hypoxia, because the removal of extracellular Ca^{2+} inhibits the release of catecholamines (4–8) and substance P (9). Several reports indicate that the influx of Ca^{2+} via L-type voltage-gated Ca^{2+} channels is responsible for catecholamine release (5,7,10). However, glomus cells express several types of Ca^{2+} channels (11,12), and N-type calcium channels, in addition to L-type Ca^{2+} channels, appear responsible for the release of substance P (9). Contrariwise, agents that mobilize Ca^{2+} from intracellular stores do not affect catecholamine release (13).

Since voltage-gated Ca^{2+} channels are activated by depolarization of the plasma membrane, mechanisms involved in depolarizing the glomus cell have been the focus of investigation. López-Barneo et al. first reported that Kv channels of adult rabbit glomus cells were inhibited by hypoxia (14). A series of their studies and others have revealed characteristics of the Kv channels and their O_2 sensitivity in rabbit, rat, and cat glomus cells (15–29). These results coalesced into a hypothesis that hypoxic inhibition of Kv channels induces depolarization of glomus cells. Significant variability, however, was seen among species. In the rabbit, glomus cells fire spontaneously (22). Hypoxic inhibition of Kv channels, which would be periodically activated, may be the cause of increasing firing frequencies of glomus cells (30). In the rat carotid body, a major O_2-sensitive K^+ channel is the large conductance Ca^{2+}-activated K^+ channel (maxi-K channel). Charybdotoxin, a blocker specific for the maxi-K channel, depolarizes rat glomus cells (28). In the cat, delayed rectified K^+ channels are sensitive to hypoxia (15,16). The blocker for these channels, 4-aminopyridine, increased $[Ca^{2+}]i$ and depolarized adult cat glomus cells (Fig. 1).

However, some investigators have questioned the role of Kv or maxi-K channels in the hypoxic excitation of glomus cells. The activation thresholds of these channels are approximately -30 mV. Therefore, most channels would be closed at the normal resting membrane potential (~ -50 mV) (31). Hypoxic inhibition of

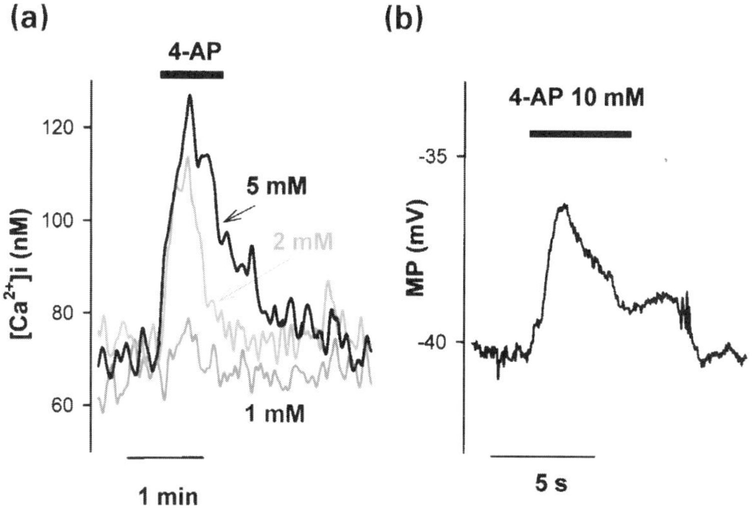

Figure 1 Effects of 4-AP, an effective Kv channel blocker of the cat glomus cell (16), on $[Ca^{2+}]i$ and membrane potential of cat glomus cells. Carotid bodies were harvested from adult cats that were euthanized with an overdose of pentobarbital and decapitated. Cells were dissociated and cultured as described previously (63). During experiments cells were continuously perfused with Krebs. The composition of Krebs was (mM): NaCl 118.3, KCl 4.7, $CaCl_2$ 1.8, $MgSO_4 \cdot 7H_2O$ 1.2, $KH_2PO_4$1.2, EDTA 0.0016, $NaHCO_3$ 25, and glucose 10, pH 7.4 equilibrated with 5% CO_2/air. (a) 4-AP dose-dependently increased $[Ca^{2+}]i$. $[Ca^{2+}]i$ was measured from a cluster of carotid body cells with microfluorometric methods using Indo-1 (43). The Ca^{2+} indicator Indo-1 was added in the culture medium (4 μM) for 1 hr to load the dye. To measure $[Ca^{2+}]i$ Indo-1 in the cell was repeatedly excited at a wavelength of 395 nm for 35 msec at every 2 sec. Two wavelengths of emission light (405 nm and 495 nm) were recorded from a restricted microscope field with a dual-emission fluorometer (Biomedical Instrumentation Group, University of Pennsylvania). A PC-based computer and pCLAMP software (Axon Instruments, Inc.) were used for acquisition of the data. $[Ca^{2+}]i$ was calculated with a standard method (64). The cells were exposed to 4-AP included in the Krebs as indicated. (b) 4-AP depolarized a cat glomus cell. Membrane potential was recorded from a patched glomus cell with a conventional whole-cell configuration under a current clamp mode. The composition of the internal solution was (mM): K gluconate 90, KCl 33, NaCl 10, EGTA 10, MgATP 5, $CaCl_2$, 1, HEPES 10, pH = 7.2. A PC-based computer and pCLAMP software were used for acquisition of the data. 4-AP was dissolved in Tyrode solution and applied with a puff pipette located close to the patched cell. Application of simple Tyrode did not change membrane potential (data not shown). The composition of Tyrode was (mM): NaCl 143.3, KCl 4.7, $CaCl_2$ 1.8, $MgSO_4 \cdot 7H_2O$ 1.2, KH_2PO_4 1.2, and EDTA 0.0016, pH 7.4 with HEPES 10 mM.

these channels that are mostly closed may not significantly influence membrane potential. In addition, experimental results using Kv channel and maxi-K channel blockers are controversial (32–35). Recently, it was reported that hypoxia inhibited TASK-like background K^+ channels (35,36). Also, an involvement of voltage-gated

HERG-like channels was suggested (37). These channels are active at resting membrane potential, and the inhibition of these channels has been proposed to initiate the depolarization of glomus cells. In summary, it is still controversial how glomus cells are depolarized in response to hypoxia.

In a previous study we found that cat glomus cells were gradually depolarized (from −55 mV to −27 mV) in response to hypoxia (PO_2 25 mmHg) (15). However, the speed and the degree of depolarization may not be sufficient to activate voltage-gated Ca^{2+} channels at mild hypoxia. It is well known that afferent neural activity from the carotid body starts increasing at mild hypoxia. This increase in neural output, in a current model, assumes a release of neurotransmitters and a preceding $[Ca^{2+}]i$ increase. Therefore, we asked whether depolarization of glomus cells is a prerequisite for the increase in $[Ca^{2+}]i$ and the following neurotransmitter release at the beginning of hypoxia. Our recent immunocytochemical and molecular biological studies indicate that α3, α4, and β2 subunits of neuronal nicotinic ACh receptors (nAChRs) are localized in cat glomus cells (31,38,39). Neuronal nAChRs are ligand-gated cation channels with high Ca^{2+} permeability (40–42). We previously showed that the activation of nAChRs of cat glomus cells increased $[Ca^{2+}]i$ (43). This increase in $[Ca^{2+}]i$ was partially blocked by an antagonist for α4β2 nAChRs, dihydro-β-erythroidine (DHβE; Fig. 2). These data suggest that at least some Ca^{2+} influx occurs via α4β2 nAChRs. Now it is well known that the activation of α4β2 nAChRs is voltage-dependent, and that the inward current via α4β2 nAChRs is larger at more negative membrane potentials (42). Taken together, an increase in $[Ca^{2+}]i$ via α4β2 nAChRs can occur without a preceding depolarization of glomus cells.

Recently we have observed that ACh is tonically released from cat glomus cells under the normoxic condition (44). Therefore, there is always a basal amount of ACh in the extracellular environment of glomus cells, presumably acting on α4β2 nAChRs and allowing the entry of extracellular Ca^{2+} and Na^+ at resting membrane potential. The influx of these cations starts depolarizing the glomus cell. This depolarization decreases the influx of cations through nAChRs, because the current via nAChRs is voltage-dependent and depolarization significantly reduces the current. Hence, during normoxia, the interaction between ACh and nAChRs might be "self-regulatory," and $[Ca^{2+}]i$ of glomus cells is stable. However, hypoxia would disrupt the balance. At the beginning of hypoxia the influx of Ca^{2+} via nAChRs may play an important role for the further release of neurotransmitters (see below for further discussion). With progress of hypoxia, the influx of Ca^{2+} via nAChRs would decrease, but it would continue via voltage-gated Ca^{2+} channels that would be activated by the hypoxic depolarization of the glomus cell.

III. Cholinergic Modulation of Kv Channels

Autoradiographic studies of the cat (45) and rabbit carotid body (45–47) showed that glomus cells contain not only nAChRs, but also muscarinic ACh receptors (mAChRs). Physiological studies suggested that the M1 and M2 mAChRs can alter

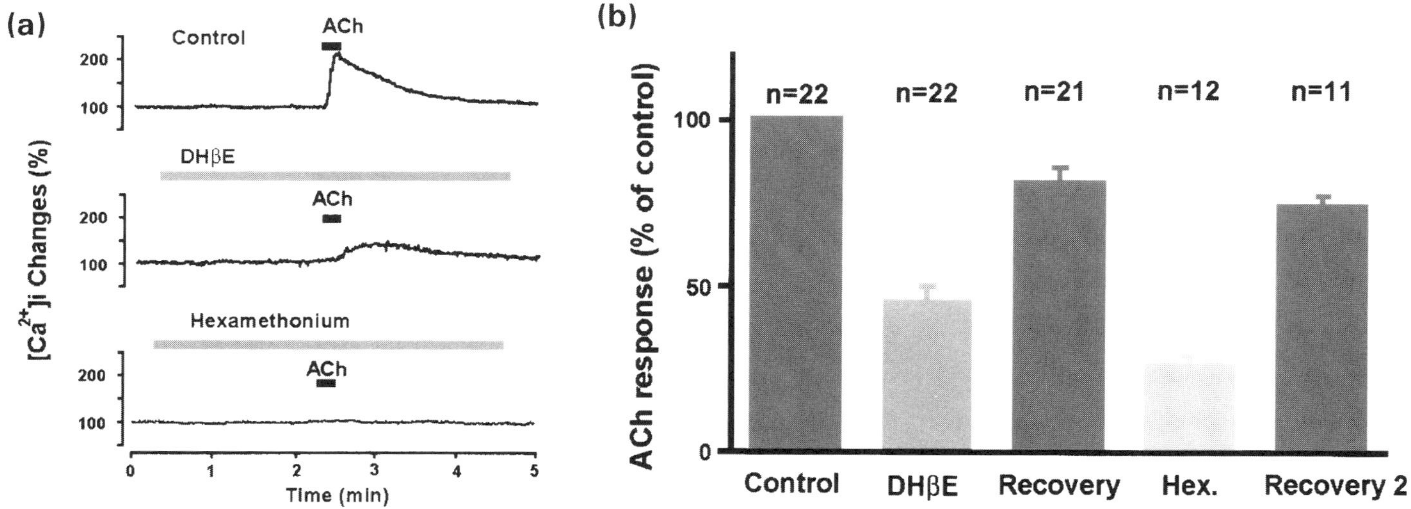

Figure 2 ACh-induced $[Ca^{2+}]i$ changes in cat glomus cells. Cells were cultured as described before (63). (a) An application of ACh increased $[Ca^{2+}]i$. Dihydro-β-erythroidine (DHβE), an antagonist for $\alpha4\beta2$ nAChRs, and hexamethonium, a nonspecific nAChR blocker, attenuated $[Ca^{2+}]i$ response. Cultured cells were continuously perfused with Krebs at 37°C. ACh and blockers, DHβE and hexamethonium, were dissolved in Tyrode and applied via a puff pipette. Bars indicate the duration and the timing of the chemical applications. $[Ca^{2+}]i$ was estimated as a fluorescent ratio with microfluorometric techniques using Indo-1 and was expressed as % of the basal level before chemical applications. (b) Summarized data showing the contribution of nAChRs for ACh-induced increase in $[Ca^{2+}]i$. ACh was applied to the same clusters during Krebs perfusion (Control), DHβE application (DHβE), Krebs perfusion (Recovery), hexamethonium application (Hex.), and Krebs perfusion (Recovery 2). All experiments were performed in this order. DHβE inhibited approximately 50% of ACh-induced increase in $[Ca^{2+}]i$, indicating that $\alpha4\beta2$-type nAChRs are the responsible receptors. Hexamethonium further inhibited the response, indicating some other type(s) of nAChRs may also be involved. These nAChRs may contain $\alpha3$ subunits. Some clusters were washed away during experiments, and therefore, the numbers of experiments (n in the figure) decreased toward the end of the protocol.

the response of the cat carotid body to hypoxia (48,49). Recently, we have found that the mRNA for M1 and for M2 mAChRs are expressed in the cat carotid body (Fitzgerald et al., Chapter 21 in this book). These cholinergic receptors (nicotinic and muscarinic) influence not only the release of ACh but also the release of catecholamines (50–53). A part of these effects may be through modifying some channel activities.

ACh affects Kv currents in many tissues. In sympathetic neurons, muscarinic (M1 and M3) inhibition of K^+ current (M current) is important to increase the excitability of the neuron (54). In other tissues M1 receptor activation inhibits several types of Kv channels (54–58). Nicotinic inhibition of delayed rectified K^+ current in cardiac cells (59) and vascular smooth muscle cells (60) has been reported. Since cat glomus cells express nicotinic and muscarinic AChRs, and Kv channels (Fig. 3), cholinergic modulation of Kv channels may occur in cat glomus cells as well. We examined this possibility and found that low doses of ACh (100 nM–1 μM) enhanced while high doses of ACh (100 μM–1 mM) inhibited Kv current in cat glomus cells (Fig. 4).

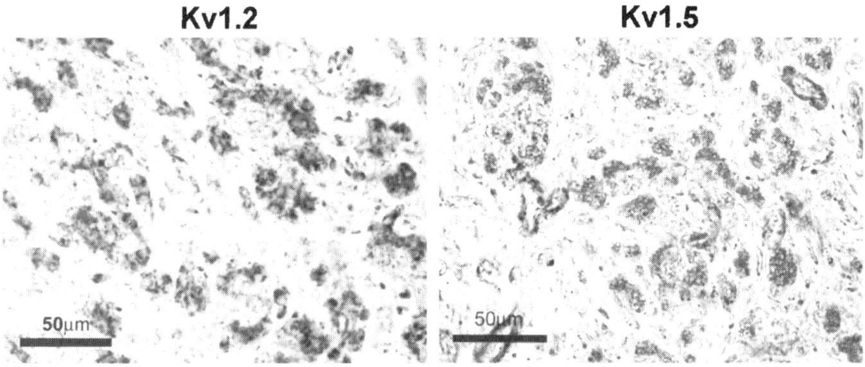

Figure 3 Expression of Kv1.2 (left) and Kv1.5 (right) channel proteins in the cat carotid body. The carotid bodies were harvested from adult cats and fixed with zinc fixative at room temperature for 6 hr. Tissues were embedded in paraffin and sectioned (4–5 μm). After deparaffinization, tissues were boiled with citric acid buffer for antigen retrieval. Subsequently, endogenous peroxidase was quenched with 1% H_2O_2 in phosphate-buffered saline (PBS); nonspecific bindings and endogenous biotin were blocked with normal goat serum (1:10) including casein and with avidin-biotin blocking kit (Vector), respectively. Antibodies for Kv1.2 and Kv1.5 (made in the rabbit; Alomone Lab) were applied overnight at room temperature, and then biotinylated goat anti-rabbit IgG was applied for 1 hr at room temperature. VECTASTAIN *Elite* ABC (Vector) kits were used for peroxidase reaction, and SG was used as the chromogen. Purified normal rabbit IgG was used as negative control. Immunoreactivities for Kv1.2 and Kv1.5 were seen in glomus cells. Negative control did not show any staining (data not shown). It is worth noting here that Kv1.2 in PC12 cells (65) and human Kv1.2 expressed in mouse L-cell line (66) are inhibited by hypoxia. Kv1.5 in pulmonary arterial myocytes contributes to hypoxic pulmonary vasoconstriction and hypoxia-induced $[Ca^{2+}]i$ increase (67).

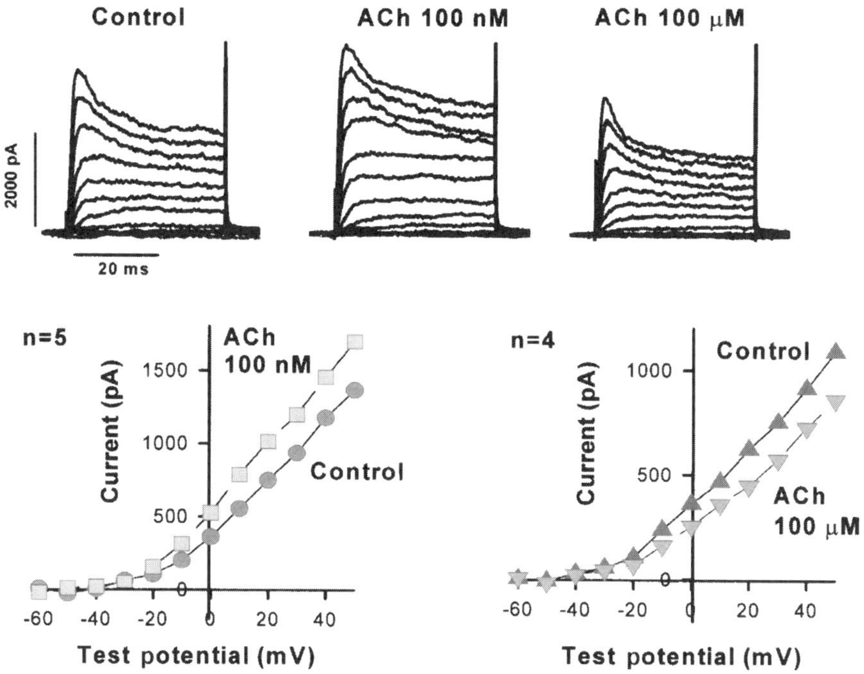

Figure 4 Dual effects of ACh on K^+ current. Glomus cells were cultured as described above. Cells were continuously perfused with Krebs equilibrated with 5% CO_2/air. A conventional whole-cell configuration was achieved. K^+ current was induced by applying step pulses from -60 mV to $+50$ mV with $+10$ mV increments for 40 msec (top). ACh was applied topically via a puff pipette. Tyrode was used as control for ACh application. Low doses of ACh (100 nM–1 μM) increased the Kv current, and high doses of ACh (100 μM–1 mM) decreased the current. The effects of ACh were reversible (data not shown). Current-voltage curves (bottom) were constructed with values at 35 msec from the start of voltage pulses.

This dual effect of ACh on Kv current may be mediated via different cholinergic receptors. For example, nicotine inhibited Kv current, suggesting that neuronal nAChRs are involved in the ACh-induced inhibition of Kv channels (Fig. 5). The mechanisms of this inhibition are not clear from these experiments. One possibility is that the increase in $[Ca^{2+}]i$ via $\alpha4\beta2$ nAChRs, as seen in Figure 2, may be linked to the inhibition of Kv channels in glomus cells, since it has been previously indicated that Ca^{2+} inhibits delayed rectified K^+ channels via activation of tyrosine kinase (57,61). Nicotinic inhibition of Kv channels may have a significant impact on the glomus cell function. Activation of Kv channels normally repolarizes the cell. However, the binding of ACh to nAChRs not only depolarizes the glomus cell by the influx of cations via nAChRs, but also inhibits the repolarization process by inhibiting Kv channels.

On the other hand, lower doses of ACh augmented Kv current in glomus cells. This effect is mediated by mAChRs. Muscarine, a mAChR agonist, enhanced Kv

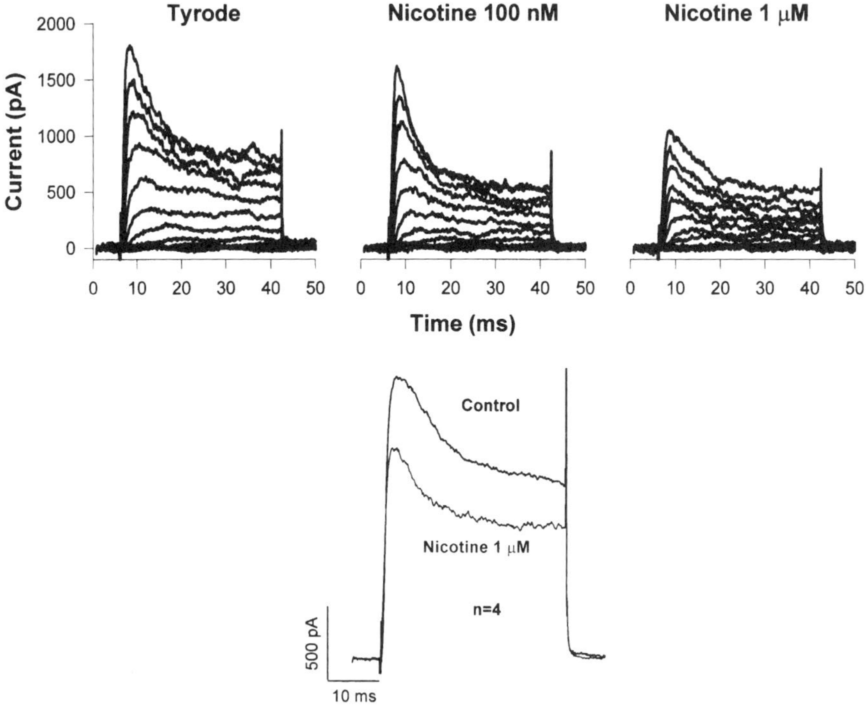

Figure 5 The inhibitory effect of nicotine on Kv current. The experimental conditions were the same as those described in Figure 4. K^+ current was induced by applying step pulses from $-60\,mV$ to $+50\,mV$ with $+10$-mV increments for 40 msec (top). In the summary data (bottom) K^+ current evoked by a step pulse from $-60\,mV$ to $+50\,mV$ was shown. Nicotine was dissolved in Tyrode and applied topically via a puff pipette, and control K^+ current was obtained during application of Tyrode. The effect of nicotine was reversible (data not shown).

current. Gallamine, a M2 mAChR antagonist, totally inhibited the muscarine- or ACh-induced enhancement of Kv current (Fig. 6). Such results suggest that M2 mAChR activation enhances the activity of Kv channels. The effect of M2 receptor activation on Kv channels would stabilize the membrane potential of glomus cells. Several investigators have reported that M2 receptor activation attenuated the tonic (62) or hypoxia-induced increase in the release of catecholamines (53) and of ACh (52). Augmentation of Kv current by M2 mAChR activation may be, at least partly, an underlying mechanism for reducing the release of these neurotransmitters.

It is well known that mAChRs can induce cellular responses without an agonist (constitutive activity). We explored whether the constitutive activity of mAChRs influenced the Kv current. Application of atropine (1 μM), a nonspecific antagonist for mAChRs, significantly enhanced Kv current in cat glomus cells. No agonist was present (Fig. 7). We recorded the Kv current from single isolated glomus cells to avoid any possible influence of endogenously released ACh from neighboring glomus cells. One might argue that ACh released from the patched

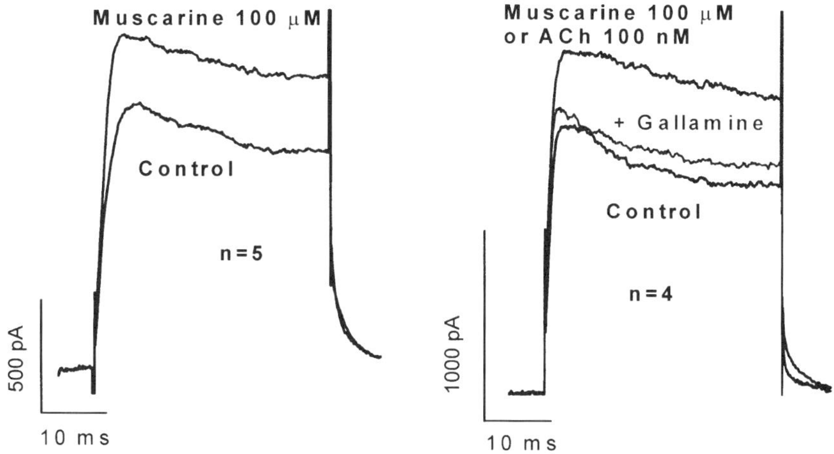

Figure 6 Muscarine-augmented Kv current (left). Gallamine (4 μM) inhibited the augmentation of K^+ current caused by muscarine or ACh (right). The experimental conditions were the same as those described in Figure 4. K^+ current was induced by applying a step pulse from −60 mV to +50 mV. All cholinergic agents were dissolved in Tyrode and applied topically via a puff pipette, and control K^+ current was obtained during application of Tyrode. The mean currents from 5 cells (left) or 4 cells (right) were shown.

cell could affect Kv current. However, this possibility may be remote, because the cells were continuously perfused with Krebs solution with a relatively high speed (2–3 mL/min). Therefore, these results suggest that the constitutive activity of mAChRs attenuates Kv current in cat glomus cells. In other tissues, M1 receptor activation suppresses K current (55–58,61) by activating tyrosine kinase. We

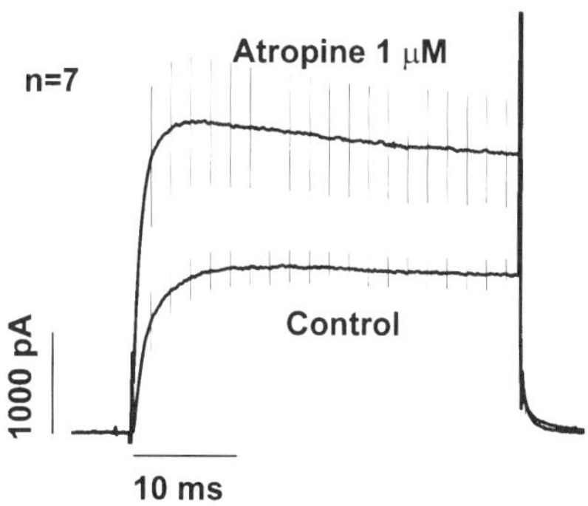

Figure 7 The effect of atropine perfusion (1 μM) on Kv current. Atropine significantly increased Kv current. The experimental conditions were the same as those described in Figure 4. K^+ current was induced by applying a step pulse from −60 mV to +50 mV. Atropine was added in Krebs solution. No cholinergic agonists were used. The mean current from 7 cells was shown. Vertical bars indicate SEM.

recently found the expression of mRNA for M1 and M2 mAChRs in the cat carotid body. In glomus cells, M2 receptor activation enhanced Kv current (Fig. 6). Thus, it is possible in cat glomus cells that constitutive activity of M1 receptors inhibits Kv current, and this action may be blocked by atropine.

IV. Our View Regarding Cholinergic Modulation of K^+ Channels and Excitation of Glomus Cells

Our data indicate that cholinergic receptors modulate Kv channels. Even though the data do not fully provide the underlying mechanisms of this interaction, they have prompted a modification of our view as to how glomus cells are excited. Neurotransmitters appear to contribute to the excitability of glomus cells not only by their immediate and direct action on their receptors, but also by their modulating effects on ion channel activities.

Figure 8 summarizes the data and presents a model that explains the fine-tuning of the excitability of glomus cells by the interaction among cholinergic receptors and Kv channels. Our data support the presence of $\alpha4\beta2$ type nAChRs and possibly another type including $\alpha3$ subunits in cat glomus cells. We found mRNA for M1 and M2 mAChRs in the cat carotid body. Functional M1 and M2 mAChRs seem present in glomus cells (Figs. 6 and 7). Kv channels (Kv1.2 and Kv1.5) are localized on glomus cells (Fig. 3).

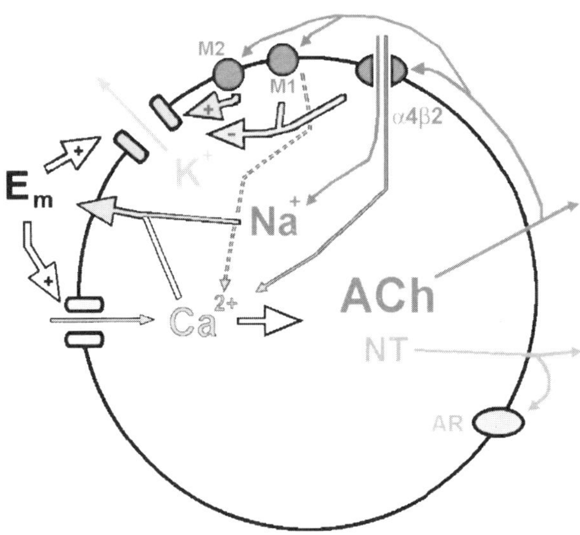

Figure 8 A model illustrating the interaction between cholinergic receptors and Kv channels. Depending on the amount of ACh released the interaction shifts one way or another. For the details see the text. $\alpha4\beta2$, $\alpha4\beta2$-type nAChR; M1, M1 mAChR; M2, M2 mAChR; NT, neurotransmitter; AR, autoreceptor; Em, membrane potential.

We have data showing that during normoxia a modest amount of ACh is released tonically from glomus cells (44). Based on the data showing the enhancing effect of lower doses of ACh on Kv current (Fig. 4) and the augmenting effect of M2 mAChRs on Kv current (Fig. 6), we postulate that a modest amount of ACh released during normoxia predominantly activates M2 mAChRs. ACh binding to nAChRs may be minimal during normoxia. Although the effect may be small, the minimal activation of nAChRs could still allow the influx of Na^+ and Ca^{2+}. This, per se, promotes a small depolarization of the cell and a small increase in $[Ca^{2+}]i$. Such a small depolarization would begin to activate Kv channels, and this activation would repolarize the glomus cell. Further, M2 mAChR activation induced by a small amount of ACh would enhance Kv current. This, too, would promote the process of glomus cell repolarization, or even cause hyperpolarization. At this point, the constitutive activity of the M1 mAChRs would reduce Kv current, inhibiting hyperpolarization. Depolarization by nAChR activation, repolarization by Kv channel activation, enhancement of Kv current mediated by M2 mAChRs, and inhibition of Kv current by constitutive M1 mAChR activity are balanced during normoxia.

During hypoxia, this balance is disrupted. Our data show that during hypoxia an increased amount of ACh is released (44). The influx of a large amount of Ca^{2+} and Na^+ influx via nAChRs causes a large increase in $[Ca^{2+}]i$ and a large depolarization of the glomus cell. The increase in $[Ca^{2+}]i$ triggers a larger release of neurotransmitters. Depolarization also activates Kv channels. During normoxia Kv channel activation repolarizes the glomus cell, but during hypoxia repolarization hardly takes place. Hypoxia inhibits Kv channel activity, and larger amounts of ACh also inhibit the Kv current via nAChRs. M1 mAChR activity (constitutive and ACh-induced) also inhibits Kv current. Thus, depolarization of the glomus cell persists. Depolarization attenuates Ca^{2+} influx via nAChRs, but Ca^{2+} influx continues via voltage-gated Ca^{2+} channels. The release of ACh, together with other neurotransmitters, persists.

This model, of course, does not explain every aspect of glomus cell excitation. For example, activation of nAChRs enhances further release of ACh. How, during normoxia, is this positive-feedback loop regulated? Is the enhancement of Kv current by M2 mAChR activation sufficient? How well are nAChRs "self-regulated"? Although experimental evidence is not yet available, it is possible that other mechanisms, such as background K^+ channels or D2 receptors, are also influencing the stability of cat glomus cells. Another important issue is the hypoxia-induced increase in ACh release. How does this happen? We propose that mild hypoxia enhances the activity of nAChRs by virtue of some action of oxygen molecules on the nAChR itself or on structures/processes that control the activity of nAChRs. Some data support our proposal (31). With recent technological advancements we have gained much information regarding the roles of several ion channels, neurotransmitters, and their receptors in the carotid body. Further investigation of interactions among these elements would provide a better picture of chemoreception and chemotransduction in the carotid body.

Acknowledgments

This work was supported by AHA 9951284U, HL61596, and HL50712.

References

1. Fitzgerald RS, Shirahata M. Systemic responses elicited by stimulating the carotid body: primary and secondary mechamisms. In: Gonzalez C, ed. Carotid Body Chemoreceptors. Barcelona: Springer-Verlag, 1997:171–202.
2. Montoro RJ, Urena J, Fernandez-Chacon R, Alvarez de Toledo G, López-Barneo J. Oxygen sensing by ion channels and chemotransduction in single glomus cells. J Gen Physiol 1996; 107(1):133–143.
3. Urena J, Fernandez-Chacon R, Benot AR, Alvarez de Toledo GA, López-Barneo J. Hypoxia induces voltage-dependent Ca^{2+} entry and quantal dopamine secretion in carotid body glomus cells. Proc Natl Acad Sci USA 1994; 91(21):10208–10211.
4. Donnelly DF. Electrochemical detection of catecholamine release from rat carotid body in vitro. J Appl Physiol 1993; 74(5):2330–2337.
5. Shaw K, Montague W, Pallot DJ. Biochemical studies on the release of catecholamines from the rat carotid body in vitro. Biochim Biophys Acta 1989; 1013(1):42–46.
6. Fishman MC, Greene WL, Platika D. Oxygen chemoreception by carotid body cells in culture. Proc Natl Acad Sci USA 1985; 82(5):1448–1450.
7. Obeso A, Rocher A, Fidone S, Gonzalez C. The role of dihydropyridine-sensitive Ca^{2+} channels in stimulus-evoked catecholamine release from chemoreceptor cells of the carotid body. Neuroscience 1992; 47(2):463–472.
8. Pardal R, Ludewig U, Garcia-Hirschfeld J, López-Barneo J. Secretory responses of intact glomus cells in thin slices of rat carotid body to hypoxia and tetraethylammonium. Proc Natl Acad Sci USA 2000; 97(5):2361–2366.
9. Kim DK, Oh EK, Summers BA, Prabhakar NR, Kumar GK. Release of substance P by low oxygen in the rabbit carotid body: evidence for the involvement of calcium channels. Brain Res 2001; 892(2):359–369.
10. Jackson A, Nurse C. Dopaminergic properties of cultured rat carotid body chemoreceptors grown in normoxic and hypoxic environments. J Neurochem 1997; 69(2):645–654.
11. e Silva MJ, Lewis DL. L- and N-type Ca^{2+} channels in adult rat carotid body chemoreceptor type I cells. J Physiol (Lond) 1995; 489(Pt 3):689–699.
12. Overholt JL, Prabhakar NR. Ca^{2+} current in rabbit carotid body glomus cells is conducted by multiple types of high-voltage-activated Ca^{2+} channels. J Neurophysiol 1997; 78(5):2467–2474.
13. Vicario I, Obeso A, Rocher A, Lopez-Lopez JR, Gonzalez C. Intracellular Ca^{2+} stores in chemoreceptor cells of the rabbit carotid body: significance for chemoreception. Am J Physiol Cell Physiol 2000; 279(1):C51-C61.
14. López-Barneo J, Lopez-Lopez JR, Urena J, Gonzalez C. Chemotransduction in the carotid body: K^+ current modulated by PO_2 in type I chemoreceptor cells. Science 1988; 241 (4865):580–582.
15. Chou C-L, Schofield B, Sham JSK, Shirahata M. Electrophysiological and immunological demonstration of cell-type specific responses to hypoxia in the adult cat carotid body. Brain Res 1998; 789:229–238.

16. Chou CL, Shirahata M. Two types of voltage-gated K channels in carotid body cells of adult cats. Brain Res 1996; 742(1–2):34–42.
17. Delpiano MA, Hescheler J. Evidence for a PO_2-sensitive K^+ channel in the type-I cell of the rabbit carotid body. FEBS Lett 1989; 249(2):195–198.
18. Ganfornina MD, López-Barneo J. Single K^+ channels in membrane patches of arterial chemoreceptor cells are modulated by O_2 tension. Proc Natl Acad Sci USA 1991; 88(7):2927–2930.
19. Ganfornina MD, López-Barneo J. Potassium channel types in arterial chemoreceptor cells and their selective modulation by oxygen. J Gen Physiol 1992; 100(3):401–426.
20. Ganfornina MD, López-Barneo J. Gating of O_2-sensitive K^+ channels of arterial chemoreceptor cells and kinetic modifications induced by low PO_2. J Gen Physiol 1992; 100(3):427–455.
21. Hescheler J, Delpiano MA, Acker H, Pietruschka F. Ionic currents on type-I cells of the rabbit carotid body measured by voltage-clamp experiments and the effect of hypoxia. Brain Res 1989; 486(1):79–88.
22. Lopez-Lopez J, Gonzalez C, Urena J, López-Barneo J. Low PO_2 selectively inhibits K channel activity in chemoreceptor cells of the mammalian carotid body. J Gen Physiol 1989; 93(5):1001–1015.
23. Lopez-Lopez JR, De Luis DA, Gonzalez C. Properties of a transient K^+ current in chemoreceptor cells of rabbit carotid body. J Physiol (Lond) 1993; 460:15–32.
24. Lopez-Lopez JR, Gonzalez C, Perez-Garcia MT. Properties of ionic currents from isolated adult rat carotid body chemoreceptor cells: effect of hypoxia. J Physiol (Lond) 1997; 499(Pt 2):429–441.
25. Peers C. Hypoxic suppression of K^+ currents in type I carotid body cells: selective effect on the Ca^{2+}-activated K^+ current. Neurosci Lett 1990; 119(2):253–256.
26. Peers C. Effects of D600 on hypoxic suppression of K^+ currents in isolated type I carotid body cells of the neonatal rat. FEBS Lett 1990; 271(1–2):37–40.
27. Stea A, Nurse CA. Whole-cell and perforated-patch recordings from O_2-sensitive rat carotid body cells grown in short- and long-term culture. Pflügers Arch 1991; 418(1–2):93–101.
28. Wyatt CN, Peers C. Ca^{2+}-activated K^+-channels in isolated type I cells of the neonatal rat carotid body. J Physiol (Lond) 1995; 483:559–565.
29. Wyatt CN, Wright C, Bee D, Peers C. O_2-sensitive K^+ currents in carotid body chemoreceptor cells from normoxic and chronically hypoxic rats and their roles in hypoxic chemotransduction. Proc Natl Acad Sci USA 1995; 92(1):295–299.
30. Gonzalez C, Lopez-Lopez JR, Obeso A, Perez-Garcia MT, Rocher A. Cellular mechanisms of oxygen chemoreception in the carotid body. Respir Physiol 1995; 102(2–3):137–147.
31. Shirahata M, Sham JS. Roles of ion channels in carotid body chemotransmission of acute hypoxia. Jpn J Physiol 1999; 49(3):213–228.
32. Cheng PM, Donnelly DF. Relationship between changes of glomus cell current and neural response of rat carotid body. J Neurophysiol 1995; 74(5):2077–2086.
33. Donnelly DF. Modulation of glomus cell membrane currents of intact rat carotid body. J Physiol (Lond) (1995); 489(Pt 3):677–688.
34. Roy A, Rozanov C, Buerk DG, Mokashi A, Lahiri S. Suppression of glomus cell K^+ conductance by 4-aminopyridine is not related to $[Ca^{2+}]$i, dopamine release and chemosensory discharge from carotid body. Brain Res 1998; 785(2):228–235.

35. Buckler KJ. A novel oxygen-sensitive potassium current in rat carotid body type I cells. J Physiol (Lond) 1997; 498(Pt 3):649–662.
36. Buckler KJ, Williams BA, Honore E. An oxygen-, acid- and anaesthetic-sensitive TASK-like background potassium channel in rat arterial chemoreceptor cells. J Physiol 2000; 525 (Pt 1):135–142.
37. Overholt JL, Ficker E, Yang T, Shams H, Bright GR, Prabhakar NR. Chemosensing at the carotid body. Involvement of a HERG-like potassium current in glomus cells. Adv Exp Med Biol 2000; 475:241–248.
38. Ishizawa Y, Fitzgerald RS, Shirahata M, Schofield B. Localization of nicotinic acetylcholine receptors in cat carotid body and petrosal ganglion. Adv Exp Med Biol 1996; 410:253–256.
39. Hirasawa S, Chen L-M, Schofield B, Fitzgerald RS, Shirahata M. Alpha4 and beta2 subunits of neuronal nicotinic ACH receptors (nAChRs) in the cat chemosensory unit. Am J Respir Crit Care Med 2001; 163:A308.
40. Albuquerque EX, Pereira EF, Castro NG, Alkondon M, Reinhardt S, Schroder H, Maelicke A. Nicotinic receptor function in the mammalian central nervous system. Ann NY Acad Sci 1995; 757:48–72.
41. Lindstrom J, Anand R, Peng X, Gerzanich V, Wang F, Li Y. Neuronal nicotinic receptor subtypes. Ann NY Acad Sci 1995; 757:100–116.
42. McGehee DS, Role LW. Physiological diversity of nicotinic acetylcholine receptors expressed by vertebrate neurons. Annu Rev Physiol 1995; 57:521–546.
43. Shirahata M, Fitzgerald RS, Sham JSK. Acetylcholine increases intracellular calcium of arterial chemoreceptor cells of adult cats. J Neurophysiol 1997; 78(5):2388–2395.
44. Fitzgerald RS, Shirahata M, Wang HY. Acetylcholine release from cat carotid bodies. Brain Res 1999; 841(1–2):53–61.
45. Hirano T, Dinger B, Yoshizaki K, Gonzalez C, Fidone S. Nicotinic versus muscarinic binding sites in cat and rabbit carotid bodies. Biol Signals 1992; 1(3):143–149.
46. Dinger BG, Hirano T, Fidone SJ. Autoradiographic localization of muscarinic receptors in rabbit carotid body. Brain Res 1986; 367(1–2):328–331.
47. Dinger BG, Almaraz L, Hirano T, Yoshizaki K, Gonzalez C, Gomez-Nino A, Fidone SJ. Muscarinic receptor localization and function in rabbit carotid body. Brain Res 1991; 562(2):190–198.
48. Fitzgerald RS, Shirahata M, Ide T. Further cholinergic aspects of carotid body chemotransduction of hypoxia in cats. J Appl Physiol 1997; 82(3):819–827.
49. Fitzgerald RS. Oxygen and carotid body chemotransduction: the cholinergic hypothesis Respir Physiol 2000; 120(2):89–104.
50. Fidone SJ, Gonzalez C, Obeso A, Gomez-Nino A, Dinger B. Biogenic amines and neuropeptide transmitters in carotid body chemotransmission: experimental findings and perspectives. In: Sutton JR, Coates G, Remmers JE, eds. Hypoxia: The Adaptations. Toronto: BC Deker, Inc., 1990:116–126.
51. Fidone SJ, Gonzalez C, Almaraz L, Dinger B. Cellular mechanisms of peripheral chemoreceptor function. In: Crystal RG, West JB, eds. The Lung. Philadelphia: Lippincott-Raven, 1997:1725–1746.
52. Fitzgerald RS, Shirahata M, Wang HY. Acetylcholine is released from in vitro cat carotid bodies during hypoxic stimulation. Adv Exp Med Biol 2000; 475:485–494.
53. Wang HY, Fitzgerald RS. Muscarinic modulation of hypoxia-induced release of catecholamines from the cat carotid body. Brain Res 2002; 927(2):122–137.

54. Brown DA, Abogadie FC, Allen TG, Buckley NJ, Caulfield MP, Delmas P, Haley JE, Lamas JA. Muscarinic mechanisms in nerve cells. Life Sci 1997; 60(13–14):1137–1144.
55. Brown DA. G-proteins and potassium currents in neurons. Annu Rev Physiol 1990; 52:215–242.
56. Fukuda K, Higashida H, Kubo T, Maeda A, Akiba I, Bujo H, et al. Selective coupling with K^+ currents of muscarinic acetylcholine receptor subtypes in NG108–15 cells. Nature 1988; 335(6188):355–358.
57. Huang XY, Morielli AD, Peralta EG. Tyrosine kinase-dependent suppression of a potassium channel by the G protein-coupled m1 muscarinic acetylcholine receptor. Cell 1993; 75(6):1145–1156.
58. Kitai ST, Surmeier DJ. Cholinergic and dopaminergic modulation of potassium conductances in neostriatal neurons. Adv Neurol 1993; 60:40–52.
59. Wang H, Shi H, Wang Z. Nicotine depresses the functions of multiple cardiac potassium channels. Life Sci 1999; 65(12):L143-L149.
60. Tang G, Hanna ST, Wang R. Effects of nicotine on K^+ channel currents in vascular smooth muscle cells from rat tail arteries. Eur J Pharmacol 1999; 364(2–3):247–254.
61. Lev S, Moreno H, Martinez R, Canoll P, Peles E, Musacchio JM, Plowman GD, Rudy B, Schlessinger J. Protein tyrosine kinase PYK2 involved in Ca^{2+}-induced regulation of ion channel and MAP kinase functions. Nature 1995; 376(6543):737–745.
62. Bairam A, Neji H, Marchal F. Cholinergic dopamine release from the in vitro rabbit carotid body. J Appl Physiol 2000; 88(5):1737–1742.
63. Shirahata M, Schofield B, Chin BY, Guilarte TR. Culture of arterial chemoreceptor cells from adult cats in defined medium. Brain Res 1994; 658(1–2):60–66.
64. Grynkiewicz G, Poenie M, Tsien R. A new generation of Ca^{2+} indicators with greatly improved fluorescence properties. J Biol Chem 1985; 260(6):3440–3450.
65. Conforti L, Millhorn DE. Selective inhibition of a slow-inactivating voltage-dependent K^+ channel in rat PC12 cells by hypoxia. J Physiol (Lond) 1997; 502(Pt 2):293–305.
66. Hulme JT, Coppock EA, Felipe A, Martens JR, Tamkun MM. Oxygen sensitivity of cloned voltage-gated K^+ channels expressed in the pulmonary vasculature. Circ Res 1999; 85(6):489–497.
67. Archer SL, Souil E, Dinh-Xuan AT, Schremmer B, Mercier JC, El Yaagoubi A, Nguyen-Huu L, Reeve HL. Molecular identification of the role of voltage-gated K^+ channels, Kv1.5 and Kv2.1, in hypoxic pulmonary vasoconstriction and control of resting membrane potential in rat pulmonary artery myocytes. J Clin Invest 1998; 101(11):2319–2330.

21

Some Neurotransmitter Relationships in the Carotid Body's Response to Hypoxia

ROBERT S. FITZGERALD, SERABI HIRASAWA, and MACHIKO SHIRAHATA

Johns Hopkins University
Baltimore, Maryland, U.S.A.

HAY-YAN WANG

National Institute on Drug Abuse
National Institutes of Health
Baltimore, Maryland, U.S.A.

I. Introduction and Background

The carotid body is the primary sensor of hypoxemia across many species. When stimulated by a decrease in P_aO_2 or hypoxic perfusates, it promotes many cardiopulmonary responses. Ventilation, static lung volumes, airway resistance, and airway secretions increase. The hypoxic pulmonary vasoconstrictor response and the systemic vascular vasodilator responses are attenuated; bronchial vascular resistance decreases (for review see Ref. 1). A current question is: How does the carotid body transduce a low P_aO_2 stimulus into increased neural output traveling to the nucleus tractus solitarii? Here the increased neural traffic is processed and distributed to various other nuclei responsible for the increased breathing and the autonomically regulated changes in the cardiopulmonary system.

Early attempts to answer this question suggested a key role was played by neurotransmitters contained in the carotid body. Corneille Heymans, who won the 1938 Nobel Prize in Physiology or Medicine for his discovery of the cardiopulmonary consequences of stimulating the carotid and aortic mechanisms, used acetylcholine (ACh) to stimulate the carotid body of dogs (2). The possibility that ACh was the excitatory neurotransmitter was reinforced by the later studies of Schweitzer and Wright (3) and of Liljestrand and his Swedish colleagues (4,5). But an excitatory role for ACh during hypoxia was dismissed by other investigators (6,7) and, ironically enough, eventually by Heymans and Neil (8). In the 1960s and 1970s, however, the studies of Eyzaguirre and his many colleagues reintroduced the possibility that ACh did play an excitatory role in the hypoxia-induced increase in carotid body neural output (9; for review see Ref. 10). Again, however, not all agreed (11,12). Some recent studies have yet once more furnished support for an excitatory role for ACh during a hypoxic stimulus at least in the cat (13,14). And very recently studies in the rat have reported an excitatory role for ACh during hypoxia and, introduced what appears to be an equally important role for adenosine triphosphate (ATP) (15).

The question of the role of the various neurotransmitters in the carotid body has always been controversial, proceeding from contradictory data. And these contradictions quite possibly are due at least in part to the variety of species, preparations, and methodologies used in trying to gain deeper insight into the role of the neurotransmitters during the application of a hypoxic stimulus. Within this variety of preparations and methodologies perhaps a further source of controversy is found in a failure to understand as completely as necessary the degree of involvement of extra- and intrasynaptic receptors. Quite possibly exogenously delivered pharmacological agonists and antagonists would primarily stimulate or inhibit carotid body neural output by way of the extrasynaptic receptors whereas the endogenously released neurotransmitters would operate primarily on the intrasynaptic receptors. Intrasynaptic concentrations of ACh in some sympathetic ganglia are calculated to reach 300–800 μM (16,17). In the central nervous system synapses neurotransmitters have been calculated to reach concentrations in the 1–5 mM range (18). Hence, an antagonist (e.g., when administered intravenously) might arrive at the carotid body in the lower micromolar ranges (2–15 μM) and be able to block the effect of an exogenously delivered agonist in the lower concentration ranges acting primarily on extrasynaptic receptors, but not be at all sufficient to block the endogenously released neurotransmitter released in the high micromolar to millimolar concentrations present intrasynaptically.

II. Model of Carotid Body Chemotransduction

A widely accepted chronology of events occurring in the carotid body's chemotransduction of hypoxia is as follows (Fig. 1): (1) Hypoxia provokes an initial depolarization, perhaps by acting on an oxygen-sensitive membrane protein in the glomus cell or perhaps by somehow "sensitizing" the glomus cells' nicotinic receptors allowing the entry of extracellular Na^+ and Ca^{2+} ions (19); (2) at a certain level of membrane potential the effect of hypoxia on K^+ channels is to reduce the outward movement of K^+ ions, further depolarizing the glomus cell; (3) voltage-gated Ca^{2+} channels open, allowing yet a further increase in intracellular Ca^{2+}; intracellular stores of Ca^{2+} could also contribute to the rise in cytosolic Ca^{2+}; (4) Ca^{2+}-bearing vesicles containing neurotransmitters move to the specialized protein in the glomus cell membrane at the glomus cell–sensory fiber synapse; (5) the neurotransmitter is released into the synaptic cleft and proceeds to the appropriate

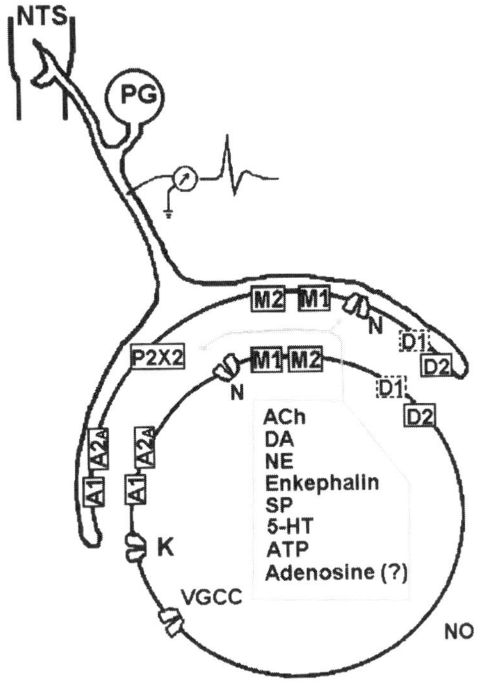

Figure 1 Model of cat carotid body's components thought to be involved in carotid body chemotransduction. NTS: nucleus tractus solitarii in the brainstem; PG: petrosal ganglion; A1, $A2_A$: adenosine receptors; $P2X_2$: purinoceptor; M1, M2: types of muscarinic receptors; N: nicotinic receptors; D1, D2: dopamine receptors; K: potassium channels; VGCC: voltage-gated calcium channels; Ach: acetylcholine; DA: dopamine; NE: norepinephrine; SP: substance P; ATP: adenosine triphosphate; NO: nitric oxide. The glomus cell, embraced by the calyx type sensory afferent fiber, contains several putative neurotransmitters. It is highly unlikely that every glomus cell contains all the listed neurotransmitters. Presumably the neurotransmitter can act wherever the appropriate receptors are located, postsynaptically as well as presynaptically. See text for postulated steps in the release of the neurotransmitters.

postsynaptic and presynaptic receptors; (6) binding to the postsynaptic receptor, the excitatory neurotransmitter initiates the depolarization of the sensory fiber, whereas the inhibitory neurotransmitter initiates a hyperpolarization of the sensory fiber; (7) binding to the presynaptic autoreceptors on the glomus cells, both the excitatory and inhibitory neurotransmitters can enhance or attenuate their own release and probably the release of other neurotransmitters or neuromodulators as well. Finally, this entire process is subject to modulation by neural and endocrine factors outside of the present description. For example, nitric oxide (NO) is known to attenuate the hypoxia-induced increase in carotid body neural output. In the cat this could be by virtue of reducing the release of ACh or increasing the release of dopamine (DA). Or it could be due to an effect of NO on cholinergic and/or dopaminergic receptors, or both.

III. Carotid Body Neurotransmitters

Data support the presence and activity of several neurotransmitters in the carotid body: ACh, DA, ATP, substance P (SP), adenosine, NO, to name those most frequently studied. Our interest up to the present has focused primarily on the first two. In the cat carotid body ACh functions as an excitatory neurotransmitter. An overwhelming amount of evidence supports an inhibitory role for DA in the cat carotid body, although under certain quite limited conditions exogenously administered DA can excite or give a biphasic response. In the rat ATP appears to play as important an excitatory role as ACh (15). Adenosine and NO appear to play important modulatory roles at least. SP can excite the carotid body, but in the cat this effect can be reduced by the use of cholinergic antagonists (14). It is not uncommon for SP to be coreleased with and modulate the effect of ACh in other tissues under cholinergic control (20,21).

Clearly the behavior patterns adopted for survival and procreation, and passed down over the eons of evolution, show great differences among the dog, cat, rat, and rabbit. Oxygen, however, remains for each the most essential substrate to be captured from the environment. And the carotid body in each still functions as the principal sensor of oxygen in the arterial blood. If a process as fundamental as the repair of nuclear DNA in *Escherichia coli*, in *Saccharomyces cervesiae*, and in *Homo sapiens* is essentially the same, simply requiring more players in one than in the other, it would not be at all surprising to find the fundamental process of chemotransduction of hypoxia in different species is the same, with the dominant excitatory neurotransmitter differing in different species as well as the roles of the other neurotransmitters. The complete picture across species of the role these neurotransmitters play in the excitation or inhibition of hypoxia-induced carotid body neural output will require some decades of further investigation.

IV. Cholinergic and Dopaminergic Receptors in the Cat Carotid Body

One approach to a deeper understanding of the roles of and interrelationships between ACh and DA in the excitation and inhibition of carotid body neural output is first to locate the receptors for ACh and DA in the carotid body chemotransductive unit (i.e., glomus cell, sensory afferent fiber, petrosal ganglion). In the pursuit of nicotinic receptors Chen and his colleagues (22,23) located α bungarotoxin (αBgt)-binding sites on carotid body glomus cells in the rat, and Dinger and his colleagues in the cat (24). The binding sites were not observed on nerves in the carotid body. But in a later study (25) the carotid sinus nerve of the cat was ligated in situ for 10 h. This procedure allowed orthograde and retrograde axoplasmic transport to deliver αBgt-binding sites down the carotid sinus nerve from the petrosal ganglion to the ligature, or back toward the ganglion from the carotid body. Positive signal for αBgt-binding sites was found on the carotid sinus nerve on both sides of the ligature. Subsequently we have located with immunocytochemical techniques several

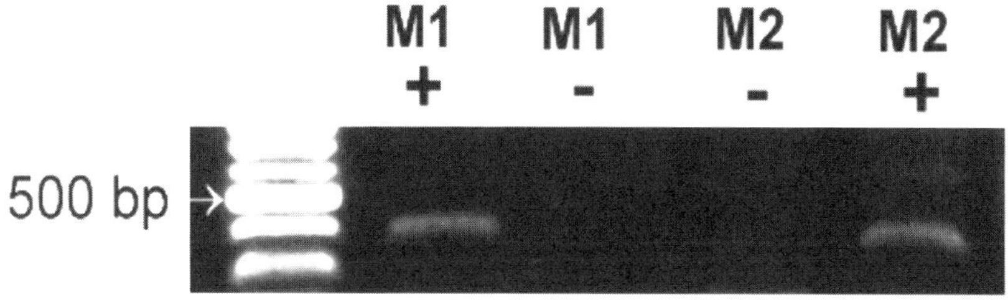

Figure 2 RT-PCR analysis of mRNA signal for M1 and M2 muscarinic receptors in the cat carotid body. The bands of the expected sizes of partial cDNA for M1 and M2 receptors were observed (RT+). As controls (RT−) total RNA was added instead of cDNA. Carotid bodies were harvested from adult cats and immersed in Trizol reagent (Invitrogen, Carlsbad, CA), a RNA extraction reagent. Total RNA was isolated according to the manufacturer's protocol. Samples were treated with RQI DNase I (Promega, Madison, WI) to eliminate a possible genomic DNA contamination. First-strand cDNA was obtained using Ready to Go You-Prime First Strand Beads (Amersham Phamacia Biotech, Inc., Piscataway, NJ) with random primers. Oligonucleotides for PCR reactions were designed to include the conserved regions of DNA sequences among human and rat M1 and M2 mAChRs. PCR amplification was performed using Ready to Go PCR Beads (Amersham Pharmacia Biotech, Inc.) and the PCR products were separated by electrophoresis on 2% agarose gel.

subunits of the neuronal nicotinic receptor—α7 on the sensory afferents (26), α4, α3, and β2 in the carotid body and petrosal ganglia (unpublished observations). There is a correlation between the α7 and α8 subunit-containing neuronal nicotinic receptors and αBgt-binding sites (27).

Carotid body muscarinic receptors were uncovered by Hirano and associates (28) using ^{3}H-quinuclidinyl benzilate (^{3}H-QNB). However, this agent selects for all five of the muscarinic receptors. Recently we have found mRNA signal for the M1 and M2 muscarinic receptors in the carotid body of the cat (Fig. 2). The M2 receptor has also been immunocytochemically located in glomus cells.

Regarding dopamine receptors, Almaraz and her associates (29) restricted the DA D_1 receptors in the rabbit carotid body to the vasculature. Bairam and her colleagues (30) also found DA D_1 mRNA in the carotid body of rats, rabbits, and cats; they found it in the petrosal ganglion and in the superior cervical ganglion as well. DA D_2 receptors have been located on carotid body glomus cells, afferent nerves, and petrosal ganglia in rabbit and rat (31–33) and in the cat (unpublished observations). Recently Wang et al. (34) have reported the presence of a DA D_3 in the carotid body of the goat where both the DA D_2 and DA D_3 receptors inhibit carotid body neural output in response to DA in a dose range of 10–50 μg/kg.

V. ACh Regulates Its Own Release from Glomus Cells

In explaining the potential interrelationships between ACh and DA in the cat carotid body two assumptions have substantial data supporting them. Namely, ACh can stimulate and DA can inhibit cat carotid body neural output. Second, nicotinic, muscarinic, and dopaminergic receptors act in the cat carotid body as they do elsewhere. Thus, exocytotically released ACh and DA should stimulate and inhibit, respectively, postsynaptic neural traffic in the carotid sinus nerve. Presynaptically (i.e., on the glomus cells) the presence of the receptors would suggest that—as in other species—they can regulate by positive/negative feedback their own release from the glomus cells and, perhaps, the release of other neurotransmitters.

Figure 3 shows an experiment (average of two runs) in which a pair of cat carotid bodies were incubated in a hypoxic medium with gallamine—an M2-receptor antagonist. The gallamine-treated carotid bodies release more ACh than the antagonist-free control carotid body. Such a result is consistent with the presence of M2 receptors on glomus cells that are functioning as they do on the presynaptic component of other cholinergic synaptic-like junctions. Stimulated,

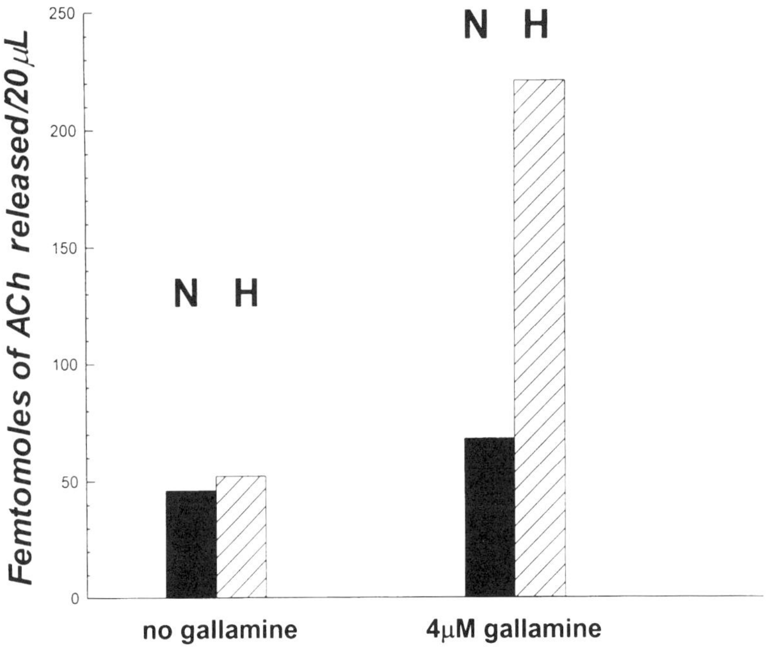

Figure 3 Effect of an M2 receptor antagonist (4 μM gallamine) on the release of acetylcholine from paired cat carotid bodies. Acetylcholine (ACh) released from two carotid bodies (average of two runs) was incubated in Krebs Ringer bicarbonate solution at 37°C for 15 min under normoxic conditions (N; solution bubbled with $21\%O_2/5\%CO_2$) and then for 15 min under hypoxic conditions (H; solution bubbled with $4\%O_2/5\%CO_2$). Gallamine appears to increase the release of ACh during normoxia and particularly during hypoxia. This pair of carotid bodies released only about 25% of the usual amount released. Presumably the ACh released by the carotid bodies acts on glomus cell autoreceptors to attenuate the further release of ACh. But the block by gallamine inhibits the inhibiting receptor; more ACh is released.

these receptors attenuate the further release of ACh. Hence, blocked with gallamine, the inhibiting receptor is inhibited, and more ACh is released. These results are supportive of some previous neural recordings in which the hypoxic perfusate containing AFDX 116, a quite specific M2-receptor antagonist, produced a larger neural response than the nonantagonized control carotid body (Fig. 4). An interpretation of these data could include a block of the glomus cell M2 thus releasing more ACh from the glomus cell and/or a block of the M2 receptor on the postsynaptic sensory fiber. This would attenuate or abolish the slow inhibitory postsynaptic potential, thus making the sensory afferent fibers more excitable.

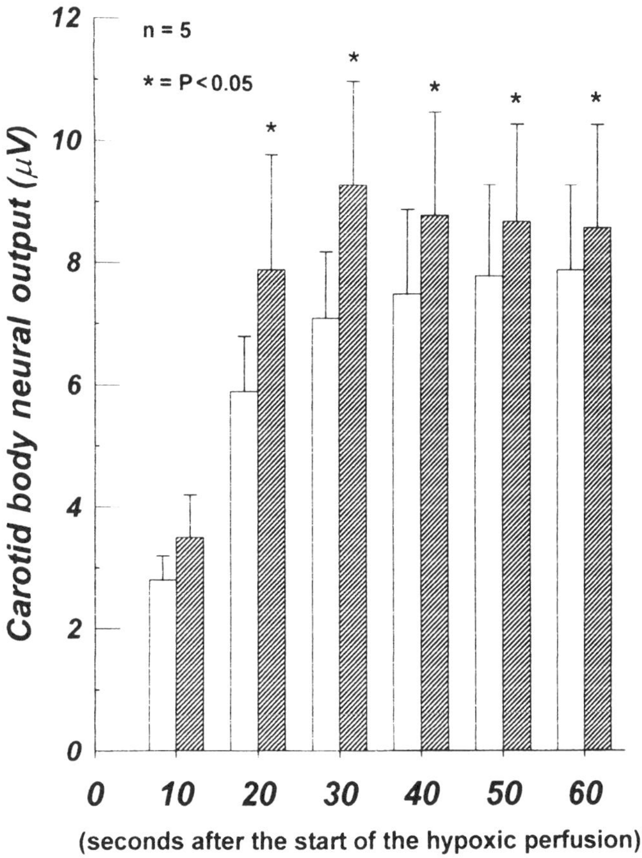

Figure 4 In situ recording of neural traffic in the whole carotid sinus nerve. Cat carotid body is responding to a perfusion of hypoxic Krebs Ringer bicarbonate solution without (open bars) and with (hatched bars) 4 μM AFDX 116, an M2 receptor inhibitor (mean ± SEM). This neural response could be due to an increase in ACh release because glomus cell M2-inhibiting autoreceptors are inhibited, releasing a greater amount of ACh. Or the postsynaptic M2 receptors (responsible for the slow inhibitory postsynaptic potential) are inhibited, making the postsynaptic sensory afferent neuron more excitable, or both processes.

Finally, it remains to be seen if the glomus cells' M1 and nicotinic receptors will influence the release of glomus cell ACh.

VI. Nicotinic Action of ACh Appears to Regulate the Release of Catecholamines in Hypoxia-Challenged Cat Carotid Bodies

Obeso and her colleagues have reported that αBgt reduced the nicotine-induced release of catecholamines by 50% while mecamylamine virtually abolished it in the cat carotid body (35). Other studies as well have shown that the carotid body releases catecholamines in response to nicotinic stimulation (36,37). Since previously the α4β2 subunits of the neuronal nicotinic receptor had been immunocytochemically detected on the cat carotid body (unpublished observations), Wang and his

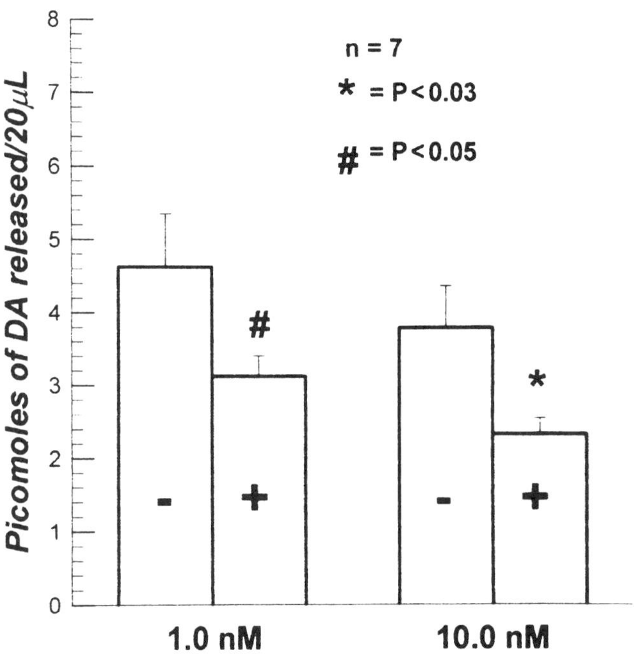

Figure 5 Effect of dihydro-β-erythroidine (DHβE) on hypoxia-induced release of DA (mean ± SEM). One cat carotid body (left bars; negative sign) was incubated in hyperoxic Krebs Ringer bicarbonate solution at 37°C for 15 min and then in a second volume of the same solution bubbled with $4\%O_2/5\%CO_2$ for 15 min. The contralateral mate was treated in the same manner except that the solution was either 1.0 nM or 10.0 nM for DHβE, an antagonist specific for the neuronal nicotinic receptor containing the α4β2 subunit combination (right bars; plus sign). Neuronal nicotinic receptors having the α4β2 subunit combination have been immunocytochemically located on cat carotid body glomus cells. The data show the involvement of glomus cell nicotinic receptors in the hypoxia-induced release of DA.

colleagues have recently studied the effect of dihydro-β-erythroidine (DHβE), a nicotinic antagonist that is quite specific for the $\alpha 4\beta 2$ subunit combination, on the hypoxia-induced release of catecholamines. Figure 5 shows that low doses of this nicotinic receptor blocker also reduce the hypoxia-induced release of DA. From these three nicotinic blockers it appears quite clear that in the cat there is significant control of DA release exercised by nicotinic mechanisms.

VII. Muscarinic Action of ACh Appears to Regulate the Release of Catecholamines in Hypoxia-Challenged Carotid Bodies

Bairam et al. (38) have recently explored in the rabbit the effect of muscarinic agents on DA and catecholamine release from the carotid body under normoxic conditions. In these carotid bodies 1 μM carbachol and the M2 blocker AFDX 116 appeared to increase the basal release. Pirenzepine, the M1-receptor blocker, clearly reduced the basal release of the catecholamines. Shaw et al. (39) have reported that the hypoxia-induced or carbachol-induced (10 μM) release of newly synthesized catecholamines from rat carotid bodies was significantly inhibited by atropine (100 μM). They interpreted this as muscarinic activity. But at high concentrations both carbachol and

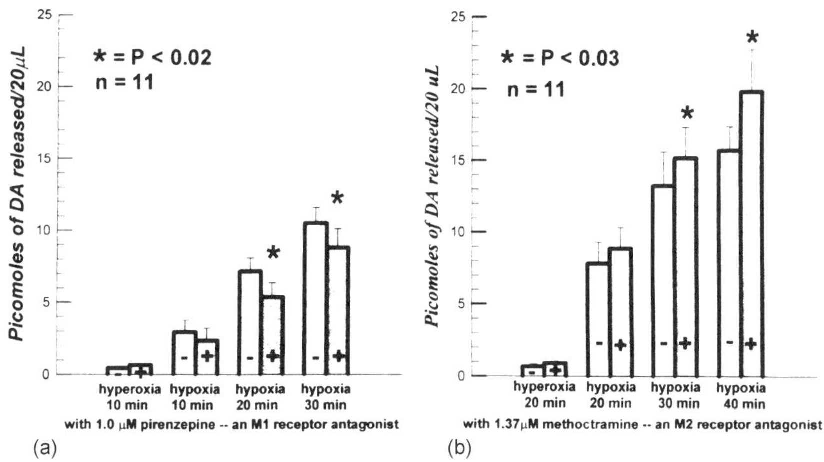

Figure 6 Effect of (a) the M1-receptor blocker (1 μM pirenzepine) and (b) the M2-receptor blocker (1.37 μM methoctramine) on the hypoxia-induced release of DA (mean ± SEM). (a) One of a pair of cat carotid bodies was incubated as above (Fig. 3) without the blocker (open bars; negative sign) while its contralateral mate was incubated with the blocker (shaded bars; plus sign) first in a hyperoxic medium (95%O_2/5%CO_2) for 10 min and then in the hypoxic solution (4%O_2/5%CO_2) for 10, 20, or 30 min. (b) With a second group of 11 cats the carotid bodies were tested as in (a). Here, however, the M2 antagonist (1.37 μM methoctramine) was used, and the incubation periods were somewhat longer. The data suggest ACh, acting on M1 and M2 glomus cell autoreceptors, is able to modify the hypoxia-induced release of DA.

atropine can have significant nicotinic action. It would be difficult to assign all of the inhibition they observed solely to the muscarinic action of these agents. Recently Wang et al., extending studies in which hypoxia evoked an increased release of ACh from cat carotid bodies (40), explored the impact of muscarinic blockers on the hypoxia-induced release of DA (41). Given the facts that (1) hypoxia increased the release of ACh and (2) the above-described nicotinic mechanisms seemed to be involved in the hypoxia-induced release of DA, they hypothesized that muscarinic mechanisms would also be involved in the hypoxia-induced release of DA. Indeed, the effect of pirenzepine was to reduce the hypoxia-induced release of DA (Fig. 6a), whereas the M2 blocker methoctramine enhanced the release of the hypoxia-induced release of DA (Fig. 6b). They concluded that pirenzepine and methoctramine exercised their effect by inhibiting the glomus cell M1 and M2 autoreceptors.

VIII. Other Interrelationships

Though several interrelationships are much clearer, many important interrelationships remain to be explored. For example, adenosine is known to stimulate increased

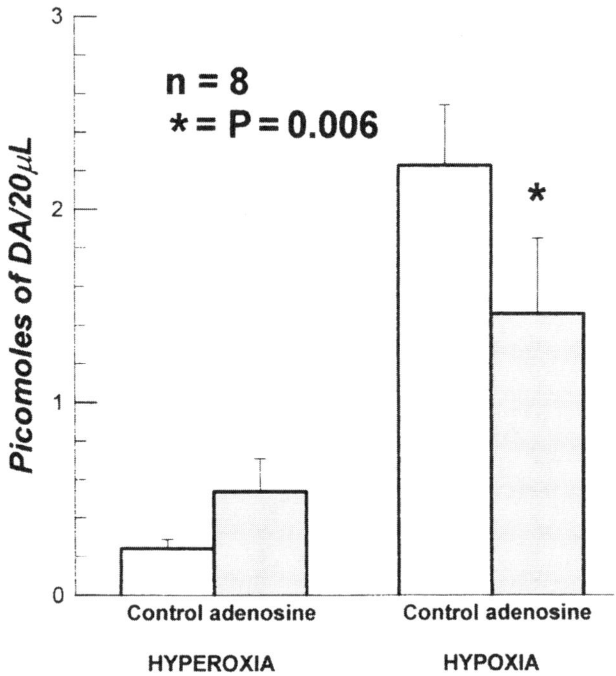

Figure 7 Effect of adenosine on the release of DA. Cat carotid bodies responding to $4\%O_2/5\%CO_2$ (for 15 min) after being bubbled with $95\%O_2/5\%CO_2$ (15 min) in the absence (open bars) or in the presence (shaded bars) of 100 μM adenosine. Significant inhibition ($\sim 36\%$) of the hypoxia-induced release of DA suggests the presence of A1 receptors on the glomus cells.

neural output from the carotid body (42,43). The possible involvement of adenosine with the "classic neurotransmitters"—ACh and DA—is not clear. But recently we have observed that adenosine reduced the hypoxia-induced release of DA (Fig. 7) suggesting the possibility that A1 purinoceptors were present on the cat glomus cells. These receptors exert inhibitory effects.

IX. Conclusion

Our understanding of the components, mechanisms, and their interrelationships in the chemotransductive process remains fragmentary. Nonetheless, substantial progress has been made over the last two decades in at least recognizing what constitutes a significant player in the process, and what are some of the interrelationships. A major triumph will come when we can understand fully the basic mechanisms of chemotransduction. Capturing oxygen from the environment is a requirement essential for the organsim's survival. Knowing how this tiny organ operates to send appropriate signals to the effector mechanisms of the cardiopulmonary control system could allow the manufacture of highly specific drugs able to up- or down-regulate carotid body chemotransductive processes as needed.

References

1. Fitzgerald RS, Shirahata M. Systemic responses elicited by stimulating the carotid body: primary and secondary mechanisms. In: González C, ed. Carotid Body Chemoreceptors. Barcelona: Springer-Verlag, 1997:171–202.
2. Heymans C, Bouckaert JJ, Farber S, Hsu FJ. Influence réflexogène de l'acétylcholine sur les terminaisons nerveuses chimiosensitives du sinus carotidien. Arch Int Pharmacodyn Ther 1936; 54:129–135.
3. Schweitzer A, Wright S. Action of prostigmine and acetylcholine on respiration. Q J Exp Physiol 1938; 28:33–47.
4. Landgren S, Liljestrand G, Zotterman Y. The effect of certain autonomic drugs on the action potentials of the sinus nerve. Acta Physiol Scand 1952; 26:264–290.
5. Liljestrand G. The action of certain drugs on respiration. Br Med J 1951; 2:623–627.
6. Moe GK, Capo LR, Peralta B. Action of tetraethylammonium on chemoreceptor and stretch receptor mechanisms. Am J Physiol 1948; 153:601–605.
7. Douglas WW. The effect of a ganglionic-blocking drug, hexamethonium, on the response of the cat's carotid body to various stimuli. J Physiol (Lond) 1952; 118:373–383.
8. Heymans C, Neil E. Reflexogenic Areas of the Cardiovascular System. Boston: Little, Brown, 1958:190–192.
9. Eyzaguirre C, Koyano H, Taylor JR. Presence of acetylcholine and transmitter release from carotid body chemoreceptors. J Physiol (Lond) 1965; 178:463–476.
10. Eyzaguirre C, Fitzgerald RS, Lahiri S, Zapata P. Arterial chemoreceptors. In: Shepherd JT, Abboud FM, eds. Handbook of Physiology. Section 2: The Cardiovascular System, vol. 3: Peripheral Circulation and Organ Blood Flow, part 2. Bethesda, MD: American Physiological Society, 1983:557–621.

11. Sampson SR. Effects of mecamylamine in responses of carotid body chemoreceptors in vivo to physiological and pharmacological stimuli. J Physiol (Lond) 1971; 212:655–666.
12. McQueen DS. A quantitative study of the effects of cholinergic drugs on carotid chemoreceptors in the cat. J Physiol (Lond) 1977; 273:515–532.
13. Fitzgerald RS, Shirahata M. Acetylcholine and carotid body excitation during hypoxia in the cat. J Appl Physiol 1994; 76:1566–1574.
14. Fitzgerald RS, Shirahata M, Ide T. Further cholinergic aspects of carotid body chemotransduction of hypoxia in cats. J Appl Physiol 1997; 82:819–827.
15. Zhang M, Zhong H, Vollmer C, Nurse C. Co-release of ATP and ACh mediates hypoxic signalling at rat carotid body chemoreceptors. J Physiol (Lond) 2000; 525:143–158.
16. Nishi S, Soeda H, Koketsu K. Release of acetylcholine from sympathetic preganglionic nerve terminals. J Neurophysiol 1967; 30:114–134.
17. Udgaonkar J, Hess G. Acetylcholine receptor kinetics: chemical kinetics. J Membr Biol 1986; 93:93–109.
18. Clements, J. Transmitter time course in the synaptic cleft: its role in central synaptic function. Trends Neurosci 1996; 19:163–172.
19. Shirahata M, Sham JSK. Roles of ion channels in carotid body chemotransmission of acute hypoxia. Jpn J Physiol 1999; 49:213–228.
20. Boska P, Livett B. Substance P protects against densensitization of the nicotinic response in isolated adrenal chromaffin cells. J Neurochem 1984; 42:618–627.
21. Role L. Substance P modulation of acetylcholine-induced currents in embryonic chicken sympathetic and ciliary ganglion neurons. Proc Natl Acad Sci USA 1984; 81:1092–1096.
22. Chen I-L, Mascoro J, Yates R. Autoradiographic localization of α-bungarotoxin binding sites in the carotid body of the rat. Cell Tiss Res 1981; 219:609–618.
23. Chen I-L, Yates R. Two types of glomus cells in the rat carotid body as revealed by α-bungarotoxin binding. J Neurophysiol 1984; 13:281–302.
24. Dinger BG, González C, Yoshizaki K, Fidone SJ. Alpha-bungarotoxin binding in cat carotid body. Brain Res 1981; 205:187–193.
25. Fitzgerald RS, Shirahata M. The role of acetylcholine in chemoreception of hypoxia by the carotid body. In: Eyzaguirre C, Fidone SJ, Fitzgerald RS, Lahiri S, McDonald DM, eds. Arterial Chemoreception. New York: Springer-Verlag, 1990:124–130.
26. Shirahata M, Ishizawa Y, Rudisill M, Schofield B, Fitzgerald RS. Presence of nicotinic acetylcholine receptors in cat carotid body afferent system. Brain Res 1998; 814:213–217.
27. Lindstrom J, Anand R, Peng X, Gerzanich V, Wang F, Li Y. Neuronal nicotinic subtypes. Ann NY Acad Sci 1995; 757:100–116.
28. Hirano T, Dinger BG, Yoshizaki K, González C, Fidone SJ. Nicotinic versus muscarinic binding sites in cat and rabbit carotid bodies. Biol Signals 1992; 1:143–149.
29. Almaraz L, Perez-Garcia MT, González C. Presence of D1 receptors in the rabbit carotid body. Neurosci Lett 1991; 132:259–262.
30. Bairam A, Frenette J, Dauphin C, Carroll J, Khandjian E. Expression of dopamine D1-receptor mRNA in the carotid body of rabbits, cats and rats. Neurosci Res 1998; 31:147–154.
31. Czyzyk-Krzeska MF. Molecular aspects of oxygen sensing in physiological adaptation to hypoxia. Respir Physiol 1997; 110:99–111.

32. Bairam A, Dauphin C, Rousseau F, Khandjian EW. Expression of dopamine D2-receptor mRNA isoforms at the peripheral chemoreflex afferent pathway in developing rabbits. Am J Respir Cell Mol Biol 1996; 15:374–381.
33. Gauda EB, Bamford O, Gerfen CR. Developmental expression of tyrosine hydroxylase, D2-dopamine receptor and substance P genes in the carotid body of the rat. Neuroscience 1996; 52:969–977.
34. Wang Z-Y, Herman JK, O'Halloran KD, Keith IM, Bisgard GE. Pharmacological and immunochemical evidence of the dopamine D3 receptor in the goat carotid body. In: Poon, C-S, Kazemi H., eds. Frontiers in Modeling and Control of Breathing. Adv Exp Med Biol 2001; 499:49–53.
35. Obeso A, Gómez-Niño MA, Almaraz L, Dinger B, Fidone SJ, González C. Evidence for two types of nicotinic receptors in the cat carotid body chemoreceptor cells. Brain Res 1997; 754:298–302.
36. Dinger B, González C, Yoshizaki K, Fidone SJ. Localization and function of cat carotid body nicotinic receptors. Brain Res 1985; 339:295–304.
37. Gómez-Niño A, Dinger B, González C, Fidone SJ. Differential stimulus coupling to dopamine and norepinephrine stores in rabbit carotid body type I cells. Brain Res 1990; 525:160–164.
38. Bairam A, Neji H, Marchal F. Cholinergic dopamine release from the in vitro rabbit carotid body. J Appl Physiol 2000; 88:1737–1742.
39. Shaw K, Montague W, Pallot DJ. Biochemical studies on the release of catecholamines from the rat carotid body in vitro. Biochim Biophys Acta 1989; 1013:42–46.
40. Fitzgerald RS, Shirahata M, Wang H-Y. Acetylcholine release from cat carotid bodies. Brain Res 1999; 841:53–61.
41. Wang H-Y, Fitzgerald RS. Muscarinic modulation of hypoxia-induced release of catecholamines from the cat carotid body. Brain Res 2002; 927:122–137.
42. Monteiro EC, Ribeiro JA. Ventilatory effects of adenosine mediated by carotid body chemoreceptors in the rat. Naunyn-Schmiedebergs Arch Pharmacol 1987; 335:143–148.
43. McQueen DS, Ribeiro JA. Pharmacological characterization of the receptor involved in chemoexcitation induced by adenosine. Br J Pharmacol 1986; 88:615–620.

22

Effects of Nitric Oxide on Carotid Body Oxygen Consumption at Low PO_2

DONALD G. BUERK and SUKHAMAY LAHIRI

University of Pennsylvania School of Medicine
Philadelphia, Pennsylvania, U.S.A.

I. Introduction

We have reviewed mechanisms of O_2 sensing by the carotid body (CB), including possible roles for nitric oxide (NO) (1) and current mathematical modeling approaches to characterize the complex behavior of NO in biological systems (2). The presence of constitutive (Ca^{2+} and calmodulin dependent) NO synthase (NOS) enzymes in CBs of rats (3,4) and cats (5) has been verified from histochemical and immunological studies. The endothelial isoform (eNOS) is found in endothelial cells lining the CB microvasculature. The neuronal isoform (nNOS) is found in the petrosal ganglion, and in nerve plexuses that innervate the CB vascular supply. Over half of the nNOS fibers ($\sim$60%) encircle glomus cells with the rest directly innervating blood vessels (6). There is no evidence that glomus cells contain any NOS enzymes, and the inducible isoform (iNOS) is not normally present in the CB. Both nNOS and eNOS have been implicated in O_2 sensing by the CB (1,7). Ventilatory responses to hypoxia and cyanide were found to be augmented in nNOS knockout mice (8) and blunted in eNOS knockout mice (9) compared to wild-type mice. These observations could be due to combined effects of NO on the CB and lower brainstem respiratory centers.

Endogenous NO appears to play an inhibitory role in CB O_2 sensing. Our laboratory reported that a NOS inhibitor (L-NAME) augmented hypoxic responses in perfused cat CBs (10). Enhanced hypoxic responses were reversed with L-arginine, the required precursor for NO production. Furthermore, increased chemosensory excitation after L-NAME could be eliminated by the NO donor sodium nitroprusside (SNP). Also, other investigators have reported that the CB response to hypoxia is inhibited after adding NO gas to the perfusate (11).

We have investigated effects of NO on CB O_2 consumption using recessed PO_2 microelectrodes to measure the tissue PO_2 disappearance rate ($-dPO_2/dt$) after stopping flow to perfused cat CBs (12). We measured faster O_2 disappearance rates after inhibiting NOS with L-NAME and slower rates when CBs were perfused with SNP (NO donor). In this study, we examined O_2 disappearance measurements made in perfused cat CBs under control conditions and after perfusion with L-NAME and applied an analysis based on known effects of NO on O_2 consumption, using a two-pathway model combining a high-O_2-affinity (low K_m) cytochrome oxidase with a second, low-O_2-affinity (high K_m) oxidase.

II. Theory

NO has a direct, reversible effect on O_2 consumption (see Refs. 2,13 for reviews) by competing with O_2 at the heme site of cytochrome oxidase (14–18). Mathematical models for these effects will be developed in this chapter.

A. Single Oxidase Model with Inhibition by NO

The variation in O_2 consumption rate (RO_2) with PO_2 for a single oxidase with Michaelis-Menten enzyme kinetics is given by

$$RO_2 = [RO_2]_{max} PO_2/(PO_2 + K_m) \tag{1}$$

where $[RO_2]_{max}$ is the maximum possible rate. The apparent K_m for O_2 (PO_2 at half-maximal consumption rate) can be characterized as a linear function of NO (2,16)

$$K_m = K_m(0)[1 + \beta_1 C_{NO}] \tag{2}$$

where $K_m(0)$ is the Michaelis constant at zero NO concentration, and β_1 is the slope. There is also a linear decrease in the maximum O_2 consumption rate with increasing NO, given by

$$[RO_2]_{max} = [RO_2]_{max}(0)[1 - \beta_2 C_{NO}] \tag{3}$$

where $[RO_2]_{max}(0)$ is the maximum O_2 consumption rate in the absence of NO, and β_2 is the slope. Predicted effects of increasing amounts of NO on the O_2 dependence of experimentally measured tissue PO_2 disappearance rates ($-dPO_2/dt$) are shown in Figure 1a. For this figure, we set $K_m = 1$ torr and the maximum disappearance rate $= -6$ torr/sec in the absence of NO (circle). Effects of NO at concentrations of 100 (triangle), 250 (diamond), and 500 nM (square) on the maximum disappearance rate (top panel), apparent K_m (middle panel), and the resulting O_2-dependent tissue

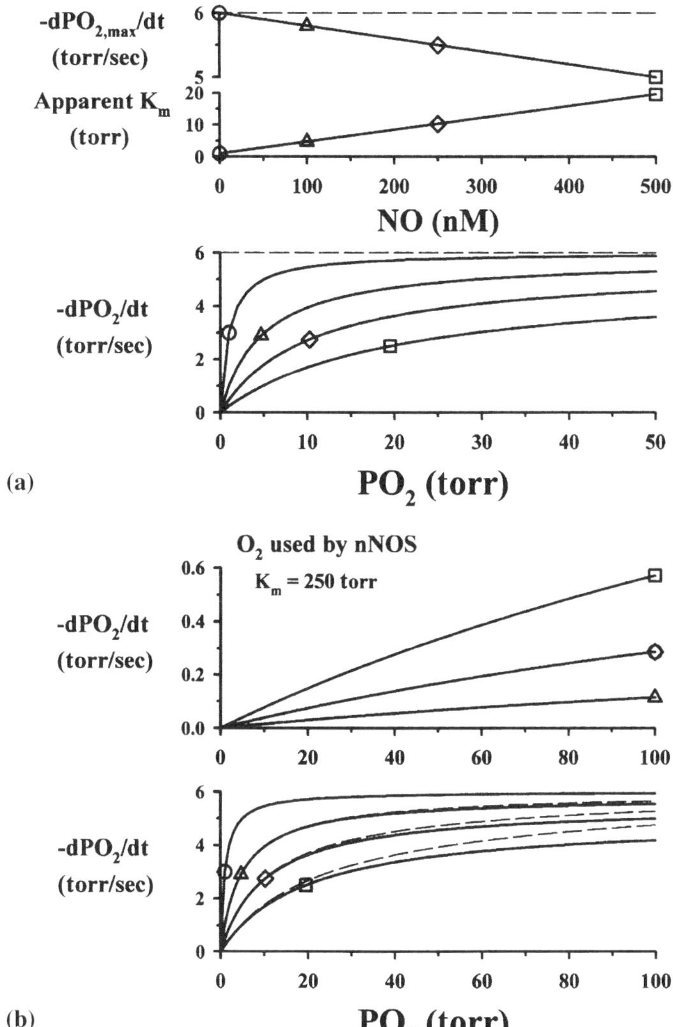

Figure 1 Models for analyzing tissue PO_2 disappearance rates measured with PO_2 microelectrodes after stopping perfusate flow to the carotid body. (a) Inhibitory effects of NO on a single enzyme model for cytochrome oxidase (high-affinity pathway) are shown, based on a decrease in maximum tissue PO_2 disappearance rate (top panel) and increase in K_m (middle panel) with NO. Predicted O_2-dependent PO_2 disappearance rates (bottom panel) for NO concentrations of 0 (circle), 100 (triangle). 250 (diamond), and 500 nM (square) are shown. (b) The single-oxidase model was modified by adding a second, low-affinity (high K_m) enzyme (top panel) for O_2-dependent production of NO by neuronal nitric oxide synthase (nNOS). Predicted O_2-dependent PO_2 disappearance rates (bottom panel) are shown with the additional amount of O_2 consumed by the low-affinity pathway (dashed lines) over that required by the high-affinity pathway (solid lines) for each NO concentration (symbols same as above).

PO_2 disappearance rates (bottom panel) were predicted using slopes $\beta_1 = 0.035\,nM^{-1}$ and $\beta_2 = 0.0002\,nM^{-1}$. The first slope is based on literature reports (14,16), but the second is an arbitrary value for the computer simulation.

B. Two-Oxidase Model Incorporating NOS Enzyme Kinetics and Inhibition by NO

Mills and Jöbsis (19) were the first to hypothesize that there could be a low-affinity (high K_m) metabolic pathway in addition to cytochrome oxidase in the CB. The nonlinear time course of O_2 disappearance curves measured in vivo with PO_2 microelectrodes in cat CBs (20–23) is consistent with this possibility. Buerk et al. (20–23) developed two-oxidase models to interpret in vivo measurements. Similar nonlinear O_2 disappearance curves have been measured in vitro from perfused cat CBs (24,25). The two-oxidase model can be written as

$$RO_2 = [RO_2]_{max}\{f\ PO_2/(PO_2 + K_{m,1}) + (1 - f)PO_2/(PO_2 + K_{m,2})\} \qquad (4)$$

where f is the fraction of the high-O_2-affinity cytochrome oxidase pathway ($K_{m,1}$) and (1-f) is the contribution of the low-O_2-affinity pathway ($K_{m,2}$). We will consider the possibility that NOS could be a second pathway for O_2 utilization, since O_2 is required along with L-arginine to synthesize NO and all three NOS isoforms are known to be O_2-dependent (see Ref. 2 for review). O_2 is thought to influence NOS activity by mediating heme ferrous-to-ferric conversion at the active site. The K_m for O_2 of NO production by nNOS is higher than for eNOS (26) and has been reported to be as high as 260 (27) to 400 μM (28). A K_m value of 250 torr, near the high range in the literature, was used to represent O_2 requirements for nNOS in the top panel of Figure 1b, which have been added to the predicted single-oxidase relationships (shown previously in Fig. 1a) in the bottom panel. This simplified model predicts that greater amounts of nNOS in tissue or higher activity rates will generate more NO, with a corresponding decrease in $[RO_2]_{max}$ and increase in the K_m for cytochrome oxidase. Futhermore, combined O_2 utilization for both oxidases (dashed lines) will be higher than that for cytochrome oxidase alone (solid lines).

III. Methods

Cats were anesthetized with sodium pentobarbital (35 mg/kg, i.p.). The carotid bifurcation was then exposed and isolated as described for previous in vitro studies (12,24,25). The CB was removed and mounted in a temperature-regulated (37°C) perfusion chamber. Tyrode's solution containing (in mM) Na^+ 112; K^+ 4.7; Ca^{2+} 2.2; Mg^{2+} 1.1; sodium glutamate 22; glucose 5; $NaHCO_3$ buffer 21.4; HEPES buffer 5; and dextran 5 g/L; equilibrated with 20% O_2 and 5% CO_2 (pH 7.40, osmotic pressure 320 mOsm) was used as the control perfusate. Neural discharge (ND) was measured from the intact carotid sinus nerve, passing the signal through a 60-Hz notch fiber, amplitude discriminator, and spike counter.

Recessed gold microelectrodes (29) with ~5-μm tips were used to measure tissue PO_2. Disappearance curves were obtained by briefly stopping perfusate flow until zero PO_2 was reached while simultaneously measuring neural discharge (ND). After several control stop flow measurements were made, CBs were perfused with Tyrode's containing L-NAME for 10–15 min, and then disappearance measurements were repeated. Initially, a low dose (100 or 300 μM) of L-NAME was given, followed by a higher dose (600 μM or 1.2 mM). Low PO_2 ranges (<40 torr) of control and L-NAME tissue PO_2 disappearance measurements were compared. Similar measurements were made during perfusion with 1 μM SNP, usually given after CBs had been perfused with L-NAME.

Tissue PO_2 and ND measurements were acquired by computer at either 2 or 5 Hz sampling rates at 12 bit resolution. Data from three or more measurements were averaged together, starting at a time point when the tissue PO_2 was close to 40 torr. After averaging curves together, derivatives ($-dPO_2/dt$) were computed. Model parameters were obtained by fitting the derivatives for each experimental condition from individual CBs (SigmaPlot, Jandel Corp.), and averaged for statistical comparison by paired t-test, with $p < 0.05$ considered to be significant.

IV. Results

Results from an individual perfused cat CR experiment are illustrated in Figure 2. Mean ± SD values were obtained by averaging four separate measurements of ND and PO_2 in each group. Data for control perfusate (a), with 100 μM (b) or 500 μM (C) L-NAME in the perfusate, are shown in the left panels. Derivatives of the experimental tissue PO_2 disappearance curves and the two-oxidase model fit (solid curves) for each data set are shown in the right panels. The contribution of the high-affinity pathway (low K_m) is represented by dashed curves (right panels). Tissue PO_2 disappearance curves were progressively faster with increasing L-NAME concentration. Also, there was a small rightward shift in the relationship between ND and tissue PO_2 (insets in left panels). Tissue PO_2 at 50% of the peak increase in ND increased from 32 torr for controls, 33 torr with 100 μM L-NAME, and 36 torr with 500 μM L-NAME in this CB. There was a progressive increase in maximum $-dPO_2/dt$ for the high-affinity pathway, from −2.49 torr/sec for controls to −2.68 and −3.05 torr/sec, respectively, for the two L-NAME doses. These results suggest that endogenous NO during control perfusion is inhibiting O_2 consumption by at least 19% in this CB. Also, a progressive decrease in K_m was estimated for the low-affinity pathway, from 8.9 torr for controls to 5.5 torr at the highest L-NAME dose. This is consistent with an expected decrease in CB tissue NO levels after inhibiting NOS. However, the second low-affinity pathway was not completely eliminated in this experiment.

Overall averages ± SE from seven perfused cat CBs are illustrated in Figure 3a for control measurements and the highest L-NAME concentration tested. The high-affinity pathway maximum $-dPO_2/dt$ was -5.97 ± 1.73 torr/sec for control CBs. There was a consistent increase with L-NAME in all experiments, averaging -6.67 ± 1.85 torr/sec ($p < 0.05$). Also, there was a consistent decrease in the high-affinity K_m after L-NAME, from 28.5 ± 11.6 torr to 16.5 ± 4.6 torr ($p < 0.05$). Since

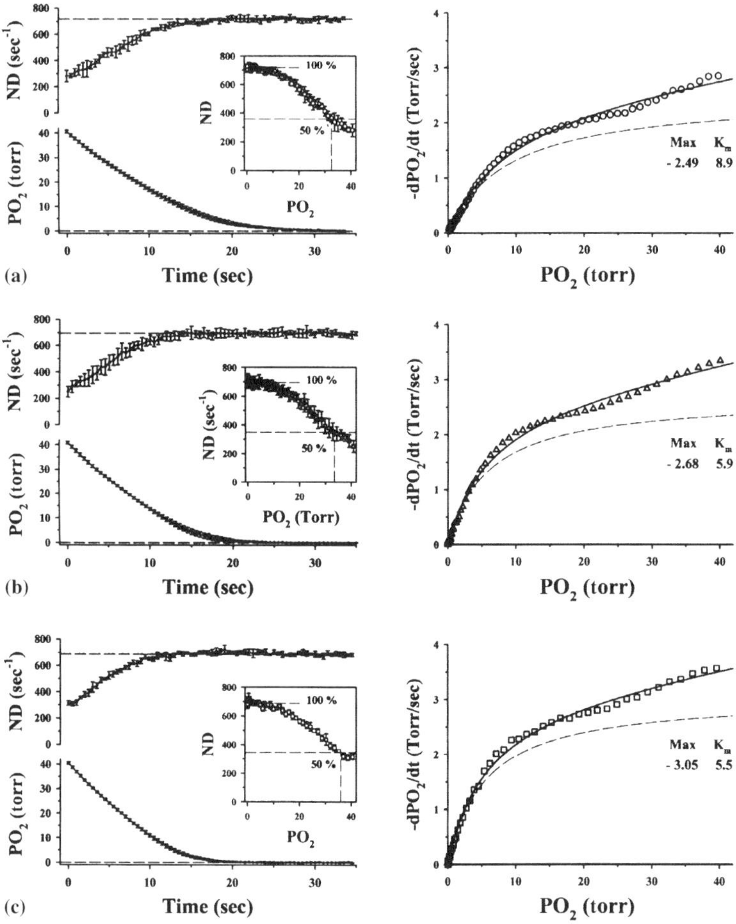

Figure 2 Experimental results from an individual perfused cat carotid body experiment (mean ± SD for four measurements in each group). (a) Average neural discharge (ND, top) and lower range (<40 torr) of the tissue PO_2 disappearance curve (bottom), with relationship between tissue PO_2 and ND (inset) are shown for stop flow measurements with control perfusate (left panel). The derivative of the tissue PO_2 disappearance curve (circles) and two-oxidase-model fit (solid curve), with estimated high-affinity pathway (dashed curve) and values for the maximum rate and K_m, are shown (right panel). (b) Average ND and PO_2 measurements after stopping flow with 100μM L-NAME in the perfusate (left panel) with derivative (triangles), two-oxidase-model fit (solid curve), and low affinity pathway (dashed curve) are shown (right panel). Compared to control, both the PO_2 at 50% of maximum ND and maximum $-dPO_2/dt$ for the high affinity pathway are higher, and the K_m is lower. (c) Average ND and PO_2 measurements after stopping flow with 500 μM L-NAME in the perfusate (left panel) with the derivative (squares), two-oxidase-model fit (solid curve), and low-affinity pathway (dashed curve) are shown (right panel). Compared to control and 100-μM L-NAME measurements, there is an additional increase in the PO_2 at 50% of the maximum ND and maximum $-dPO_2/dt$ for the low-affinity pathway, and decrease in K_m.

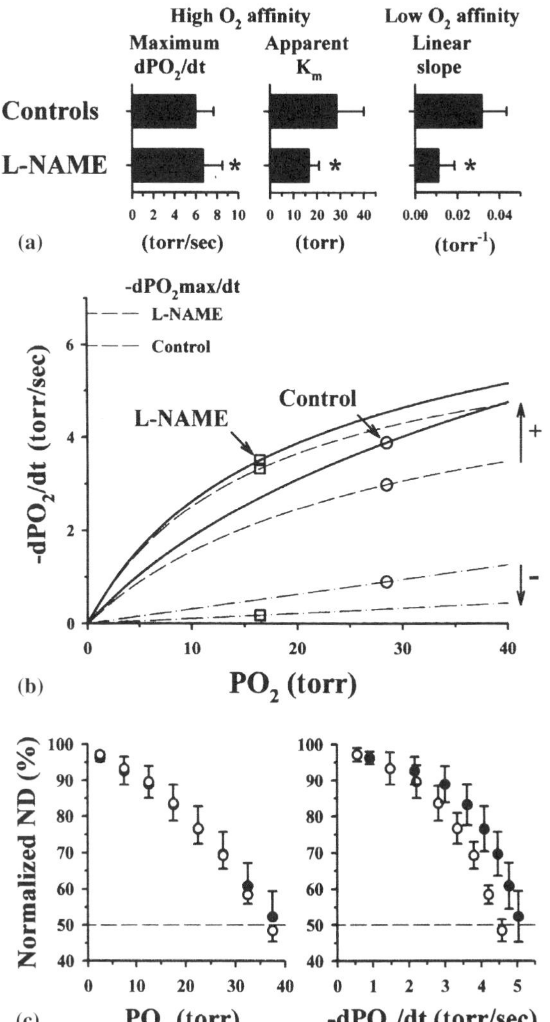

Figure 3 Summary of experimental results and two-oxidase-model parameters. (a) Mean ± SE of parameters (maximum $-dPO_2/dt$ and K_m) for the high-affinity pathway and linear slope of the low-affinity (high K_m) pathway are shown for controls and highest L-NAME doses from seven perfused cat CB experiments. (b) Two-oxidase-model fit for the tissue PO_2 disappearance curves after L-NAME (upper solid curve) is shifted to the left of control in the tissue PO_2 range >40 torr. There is an increase in the contribution of the high-affinity pathway (dashed lines) and decrease in the low-affinity pathway (dash-dot lines). Symbols (control, circles; L-NAME, squares) mark the high-affinity K_m on each component of the two-oxidase-model fit. (c) Normalized neural discharge (ND ± SE) averaged over 5-torr increments in tissue PO_2 (left panel) for control CBs (open circles) and after L-NAME (solid circles) were the same, but the relationship with the disappearance rate (right panel) is shifted to the right with the higher rate of O_2 consumption after inhibiting NO production.

it was difficult to directly estimate low-affinity model parameters, an approximation was made using a linear slope to represent the low-affinity pathway in the low-PO_2 range. This slope decreased after L-NAME from $0.032 \pm 0.012\,torr^{-1}$ to $0.011 \pm 0.008\,torr^{-1}$ ($p < 0.05$). Resulting PO_2 disappearance rates as a function of tissue PO_2 are shown in Figure 3b for the average two-oxidase-model parameter values before and after L-NAME (solid curves). The low-affinity pathway (dash-dot lines) decreases as marked by the arrow (bottom right). The high-affinity pathway (dashed curves) increases, with maximum $-dPO_2/dt$ values indicated by dashed lines (top left), and the increase at 40 torr marked by an arrow (top right). The high-affinity K_m is marked by symbols for control (circles) and L-NAME (squares) on each component of the model. These experimental results are consistent with a reduction in the inhibitory effect of NO on O_2 consumption after L-NAME. The smaller low-affinity component is consistent with less O_2, usage by the NOS enzyme pathway after it is inactivated. The shift in kinetics (lower K_m) is also consistent with other studies in the literature.

Normalized ND curves before (open circles) and after L-NAME (solid circles) for all experiments are shown in Figure 3c as a function of tissue PO_2 (left panel) averaged over 5 torr increments, and as a function of the two-oxidase-model disappearance rate (right panel). There was no significant difference in ND with tissue PO_2, with only a slight rightward shift for $PO_2 > 30$ torr. The rightward shift in ND after L-NAME illustrated in the right panel of Fig. 3b reflects the higher rate of O_2 used by the high-affinity pathway after inhibiting NO production.

Experimental results from an individual CB experiment shown in Figure 4 illustrate the opposite effects of SNP. Control ND and tissue PO_2 averaged from five measurements are shown in Fig. 4a (left panel) with the derivative (open circles), two-oxidase fit (solid curve), and high-affinity pathway (dashed curve) with maximum $-dPO_2/dt$ and K_m indicated. In this experiment, the control period was immediately followed by perfusion with 1 μM SNP. Averages for 3 measurements during SNP perfusion are shown in Fig. 4b (left panel) with the derivative (solid circles), two-oxidase fit (solid curve), and high-affinity pathway (dashed curve) (right panel). Maximum $-dPO_2/dt$ for the high-affinity pathway decreased from -1.73 to -1.59 torr/sec and K_m increased from 10.4 to 15.6 torr as indicated. Measurements during SNP perfusion were also made in 4 CBs after they had been perfused with L-NAME (data not shown). SNP caused a consistent increase in tissue PO_2 during steady-state perfusion, with slower O_2 disappearance curves and higher K_m values compared to L-NAME measurements just prior to SNP.

V. Discussion

It is of interest that the effect of inhibiting NO production in the CB is more pronounced at low PO_2. If the two-oxidase model curves in Figure 3 are extrapolated to higher tissue PO_2 ranges, they would intersect at a tissue PO_2, around 57.8 torr, and then the L-NAME curve would fall below control at higher values. The reduced effect at higher tissue PO_2 is primarily due to the decrease in K_m, but also reflects

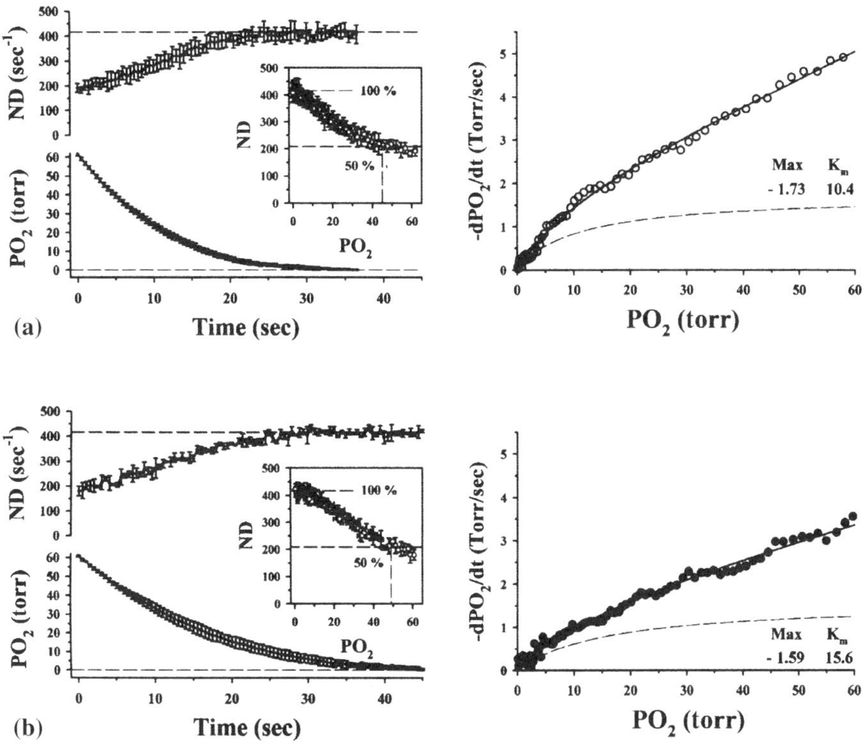

Figure 4 Experimental results from an individual perfused cat carotid body experiment demonstrating NO donor (sodium nitroprusside) effect. (a) Average ND and tissue PO_2 disappearance curve for five measurements with control perfusate and the relationship between tissue PO_2 and ND (inset) are shown (left panel). The derivative of the tissue PO_2 disappearance curve (open circles) and two-oxidase-model fit (solid curve), with estimated low-affinity pathway (dashed curve) and maximum rate and K_m, are shown (right panel). (b) Average ND and tissue PO_2 disappearance curve for three measurements with 1 μM SNP in the perfusate and relationship between tissue PO_2 and ND (inset) are shown (left panel). The derivative of the tissue PO_2 disappearance curve (open circles) and two-oxidase-model fit (solid curve), with estimated low-affinity pathway (dashed curve) and maximum rate and K_m are shown (right panel).

that the model predicts that less O_2 would be required by NOS to synthesize NO. Additional experimental data are needed at higher tissue PO_2 levels to verify this, and the effects of selective NOS isoform inhibitors on tissue PO_2 disappearance curves should be investigated. Since perfusate flow decreased and tissue PO_2 declined after L-NAME, our analysis was limited to the lower PO_2 range. If higher perfusate PO_2 values were used to compensate for the reduction in flow, we might be able to obtain better parameter estimates for the low-affinity pathway before and after inhibiting NOS. Also, we cannot rule out the possibility that there are other low-O_2-affinity pathways (e.g., NADPH oxidase, heme oxygenase) that should be included in CB O_2 consumption kinetics, which may explain why there was still a low-affinity component after L-NAME.

Previously, one of the major arguments against mitochondrial cytochrome a_3 serving as an acute O_2, sensor is that its affinity for O_2, has been considered to be too high. O_2 uptake studies with isolated mitochondria typically yield very high O_2, affinities with $K_m < 1$ torr (30). If cytochrome a_3 is operating as an acute O_2 sensor with conventional high-O_2-affinity Michaelis-Menton kinetics, it is very difficult to explain how physiological blood PO_2 values could excite the CB. One argument to support this mechanism is diffusion-limited O_2 transport with very steep O_2 gradients within glomus cells due to the spatial arrangement of mitochondria (31). Another, perhaps more likely possibility is that mitochondrial oxidative phosphorylation in intact tissues has a much higher dependence on PO_2 compared to isolated mitochondria (32,33). In addition to these factors, our studies demonstrate that the apparent K_m for CB O_2 consumption is much higher than found for isolated mitochondria, and that NO profoundly alters O_2 consumption kinetics.

Besides cytochrome a_3, there are other hypothesized mechanisms for acute O_2 sensing by the CB (see Ref. 1 for review). One possibility is based on signaling associated with reactive O_2 species (ROS) generated by the mitochondrial respiratory chain, cytochrome P-450, or NADPH oxidase. ROS production mediated by NADPH-oxidase is hypothesized to decrease during hypoxia (34). Accordingly, with less superoxide anion production (O_2^-), less H_2O_2 is formed. Since NO also scavenges O_2^-, it could also directly impact this mechanism. Peroxynitrite formed by this reaction could also have an effect. The fall in H_2O_2 is thought to suppress K^+ conductance of glomus cell membranes, leading to a rise in cytosolic Ca^{2+}, increased secretion of neurotransmitters, and greater chemosensory activity. To test this hypothesis, we performed experiments in transgenic mice deficient in gp^{91}phox, a membrane-bound subunit of NADPH oxidase (35). Changes in cytosolic Ca^{2+} in isolated glomus cells and whole-animal ventilatory responses to hypoxia were measured. Since both responses to hypoxia were normal, we concluded that NADPH oxidases were not essential (35).

ROS and reactive nitrogen species (RNS) could influence ion channels in the CB. Eu et al. (36) reported evidence for a skeletal muscle Ca^{2+} release channel/ryanodine receptor (R_yR_1) that appears to be a thiol-based O_2 sensor. At low PO_2 levels, nanomolar levels of NO can activate this channel by S-nitrosylating one of the cysteine residues of the channel protein. Much higher NO concentrations are required to activate the channel at high PO_2. Thus, NO can modulate the O_2 sensitivity of the R_yR_1 channel. Although it is not known whether glomus cells have R_yR_1 channels, there is evidence that glomus cell L-type Ca^{2+} channel currents are inhibited by NO (37). Another factor to consider is the possible effect of circulating S-nitrosothiols (SNOs) in the bloodstream. A direct effect on ventilation has been demonstrated in rodents by injecting SNOs into the nucleus tractus solitarius of the lower brainstem (38). Circulating SNOs might alter CB tissue NO levels or directly modulate CB O_2, sensitivity.

The O_2 dependence of NO production in the brain has been investigated under acute hyperbaric conditions in rats with specific nNOS inhibitors and other pharmacological agents, and in wild-type, eNOS, and nNOS knockout mice (39). High tissue PO_2 at 2.8 atmospheres exposure for 30–45 min caused large increases in NO in the cortex, primarily due to increased nNOS activity, and an increase in heat shock protein 90 associated with nNOS. Low tissue PO_2 would be expected to acutely decrease activities of all NOS isoforms, although increases in intracellular Ca^{2+} with hypoxia could compensate for this or even lead to an increase in NO production by the constitutive NOS isoforms. Chronic hypoxia appears to have variable effects on gene expression of different NOS isoforms in the CB (see Ref. 40 for review). However, there is increasing evidence that NO plays a role in long-term O_2 sensing by modulating transcription of several different genes. Both NO and carbon monoxide (CO) inhibit hypoxia-inducible factor 1 (HIF-1) (41,42) and NO is involved in hypoxic induction of vascular endothelial growth factor (VEGF) (43,44). The role of NO and the importance of eNOS in angiogenesis mediated by VEGF have been investigated in collagen gels implanted over the mouse brain, including eNOS knockout mice (45). Finally, mechanisms for inactivating NO once it is produced may also play an important role in long-term regulation. Scavenging by hemoglobin has been thought to be the major mechanism for removing NO, although there may also be mechanisms that conserve NO for release elsewhere. There is recent evidence for another cellular mechanism that could inactivate NO before levels are high enough to inhibit respiration (46). Although the exact mechanism is not known, it appears to be exhaustible. It is clear that NO acts in a complex manner and can have an impact on a large number of physiological processes that are linked to O_2 sensing and long-term O_2 homeostasis. We are probably only in the early stages of discovering these interactions.

In summary, we found that NO inhibits O_2 consumption in the normal perfused cat CB and can account for the unusual Michaelis-Menten kinetics that have been observed experimentally. Therefore, endogenous NO production can directly modulate O_2 sensitivity as well as influencing many other physiological processes in the CB.

Acknowledgments

This work is supported by R-37-HL-43413 from NIH and N00014-001-1-0948 from ONR.

References

1. Lahiri S, Rozanov C, Roy A, Storey B, Buerk DG. Regulation of oxygen sensing by peripheral arterial chemoreceptors. Int J Biochem Cell Biol 2001; 33:755–774.
2. Buerk DG. Can we model NO biotransport? A survey of mathematical models for a simple diatomic molecule with surprisingly complex biological activities. Annu Rev Biomed Eng 2001; 3:109–143.
3. Wang ZZ, Bredt DS, Fidone SJ, Stensaas LJ. Neurons synthesizing nitric oxide innervating the carotid body. J Comp Neurol 1993; 336:419–432.

4. Hohler B, Mayer B, Kummer W. Nitric oxide synthase in the rat carotid body and carotid sinus. Cell Tissue Res 1994; 276:559–564.
5. Grimes PA, Mokashi A, Stone RA, Lahiri S. Nitric oxide synthase in autonomic innervation of the cat carotid body. J Autonom Nerv Sys 1995; 54:80–86.
6. Kusakabe T, Matsuda H, Harada Y, Hayashida Y, Gono Y, Kawakami T, Takenaka T. Changes in the distribution of nitric oxide synthase immunoreactive nerve fibers in the chronically hypoxic rat carotid body. Brain Res 1998; 795:292–296.
7. Gozal D, Gozal E, Gozal YM, Torres JE. Nitric oxide synthase isoforms and peripheral chemoreceptor stimulation in conscious rats. Neuroreport 1996; 7:1145–1148.
8. Kline DD, Yang T, Huang PL, Prabhakar NR. Altered respiratory responses to hypoxia in mutant mice deficient in neuronal nitric oxide synthase. J Physiol 1998; 511:273–287.
9. Kline DD, Yang T, Prekumar DRD, Thomas AJ, Prabhakar NR. Blunted respiratory responses to hypoxia in mutant mice deficient in nitric oxide synthase-3. J Appl Physiol 2000; 88:1496–1508.
10. Chugh DK, Katayama M, Mokashi A, Bebout DE, Ray DK, Lahiri S. Nitric oxide–related inhibition of carotid chemosensory nerve activity in the cat. Respir Physiol 1994; 97:147–156.
11. Iturriaga R, Mosqueira M, Villanueva S. Effects of nitric oxide gas on cat carotid body chemosensory response to hypoxia. Brain Res 2000; 855:282–286.
12. Buerk DG, Lahiri S. Evidence that nitric oxide plays a role in O_2 sensing from tissue NO and PO_2 measurements in cat carotid body. In: Lahiri S, Prabhakar NR, Forster RE, eds. Oxygen Sensing: Molecule to Man. London: Plenum Publishers 2000:337–347.
13. Brown GC. Nitric oxide and mitochondrial respiration. Biochim Biophys Acta 1999; 1411:351–369.
14. Brown GC, Cooper CE. Nanomolar concentrations of nitric oxide reversibly inhibit synaptosomal respiration by competing with oxygen at cytochrome oxidase. FEBS Lett 1994; 356:295–298.
15. Shen W, Hintze TH, Wolin MS. Nitric oxide. An important signaling mechanism between vascular endothelium and parenchymal cells in the regulation of oxygen consumption. Circulation 1995; 92:3505–3512.
16. Koivisto A, Matthias A, Bronnikov G, Nedergaard J. Kinetics of the inhibition of mitochondrial respiration by NO. FEBS Lett 1997; 417:75–80.
17. Boveris A, Costa LE, Cadenas E, Poderoso JJ. Regulation of mitochondrial respiration by adenosine diphosphate, oxygen, and nitric oxide. Meth Enzymol 1999; 301:188–198.
18. Clementi E, Brown GC, Foxwell N, Moncada S. On the mechanism by which vascular endothelial cells regulate their oxygen consumption. Proc Natl Acad Sci USA 1999; 96:1559–1562.
19. Mills E, Jöbsis FF. Mitochondrial respiratory chain of carotid body and chemoreceptor response to changes in oxygen tension. J Neurophysiol 1972; 35:405–428.
20. Buerk DG, Nair PK, Bridges EW, Hanley TR. Interpretation of O_2 disappearance curves measured in blood perfused tissue. In: Longmuir IA, ed. Oxygen Transport to Tissue. Vol. VIII. New York: Plenum Press 1986:151–161.
21. Nair PK, Buerk DG, Whalen WJ. Cat carotid body oxygen metabolism and chemoreception described by a two cytochrome model. Am J Physiol 1986; 250:H202–H207.
22. Buerk DG, Nair PK, Whalen WJ. Two-cytochrome model for carotid body PO_2 and chemosensitivity changes after hemorrhage. J Appl Physiol 1989; 67:60–66.

23. Buerk DG, Nair PK, Whalen WJ. Evidence for second metabolic pathway for O_2 from $P_{ti}O_2$ measurements in denervated cat carotid body. J Appl Physiol 1989; 67:1578–1584.
24. Buerk DG, Itturiaga R, Lahiri S. Testing the metabolic hypothesis of O_2 chemoreception in the cat carotid body in vitro. J Appl Physiol 1994; 76:1317–1323.
25. Lahiri S, Buerk DG, Chugh DK, Osanai S, Mokashi A. Reciprocal photolabile O_2 consumption and chemoreceptor excitation by carbon monoxide in the cat carotid body: Evidence for cytochrome a_3 as the primary O_2 sensor. Brain Res 1995; 684:194–200.
26. Rengasamy A, Johns RA. Determination of K_m for oxygen of nitric oxide synthase isoforms. J Pharmacol Exp Ther 1996; 276:30–33.
27. Elayan IM, Axley MJ, Prasad PV, Ahlers ST, Auker CR. Effect of hyperbaric oxygen treatment on nitric oxide and oxygen free radicals in rat brain. J Neurophysiol 2000; 83:2022–2029.
28. Abu-Soud HM, Rousseau DL, Stuehr DJ. Nitric oxide binding to the heme of neuronal nitric-oxide synthase links its activity to changes in oxygen tension. J Biol Chem 1996; 271:32515–32518.
29. Whalen WJ, Riley J, Nair P. A microelectrode for measuring intracellular PO_2. J Appl Physiol 1967; 23:798–801.
30. Sugano T, Oshino N, Chance B. Mitochondrial functions under hypoxic conditions. The steady states of cytochrome c reduction and energy metabolism. Biochim Biophys Acta 1974; 347:340–358.
31. Chance B, Quistorff B, Sugano T, Mayevsky A. Study of tissue oxygen gradients by single and multiple indicators. In: Silver IA, Ericinska M, Bicher HI, eds. Oxygen Transport to Tissue. Vol. III. New York: Plenum Press 1978:331–338.
32. Wilson DF, Rumsey WL, Green TJ, Vanderkooi JM. The oxygen dependence of mitochondrial oxidative phosphorylation measured by a new optical method for measuring oxygen concentration. J Biol Chem 1988; 263:2712–2718.
33. Wilson DF, Mokashi A, Chugh D, Vinogradov C, Osanai S, Lahiri S. The primary oxygen sensor of the cat carotid body is cytochrome a_3 of the mitochondrial respiratory chain. FEBS Lett 1994; 351:370–374.
34. Cross A, Henderson L, Jones OTG, Delpiano MA, Hentschel J, Acker H. Involvement of an NAD(P)H oxidase as a PO_2 sensor protein in the rat carotid body. Biochem J 1990; 272:743–747.
35. Roy A, Rozanov C, Mokashi A, Daudu P, Al Medhi AB, Shams H, Lahiri S. Mice lacking in gp^{91} phox subunit of NAD(P)H oxidase showed glomus cell $[Ca^{2+}]_i$ and respiratory responses to hypoxia. Brain Res 2000; 872:188–193.
36. Eu JP, Sun J, Xu L, Stamler JS, Meissner G. The skeletal muscle calcium release channel: coupled O_2 sensor and NO signaling functions. Cell 2000; 102:499–509.
37. Summers BA, Overholt JL, Prabhakar NR. Nitric oxide inhibits L-type CA^{2+} currents in glomus cells of the rabbit carotid body via cGMP-independent mechanisms. J Neurophys 1999; 81:1449–1457.
38. Lipton AJ, Johnson MA, MacDonald T, Lieberman MW, Gozal D, Gaston B. S-nitrosothiols signal the ventilatory response to hypoxia. Nature 2001; 413:171–174.
39. Thom SR, Bhopale V, Fisher D, Manevich Y, Huang PL, Buerk DG. Stimulation of nitric oxide synthase in cerebral cortex due to elevated partial pressures of oxygen: an oxidative stress response. J Neurobiol 2002; 51:85–100.
40. Prabhakar NR. NO and CO as second messengers in oxygen sensing in the carotid body. Respir Physiol 1999; 115:161–168.

41. Huang LE, Willmore WG, Gu J, Goldberg MA, Bunn HF. Inhibition of hypoxia-inducible factor 1 activation by carbon monoxide and nitric oxide: implications for oxygen sensing and signaling. J Biol Chem 1999; 274:9038–9044.
42. Sogawa K, Numayama-Tsuruta K, Eme M, Abe M, Abe H, Fujii-Kuriyama Y. Inhibition of hypoxia-inducible factor 1 activity by nitric oxide donors in hypoxia. Proc Natl Acad Sci USA 1998; 95:7368–7373.
43. Liu Y, Christou H, Morita T, Laughner EB, Semenza GL, Kourembanas S. Carbon monoxide and nitric oxide suppress the hypoxic induction of vascular endothelial growth factor gene via the 5′ enhancer. J Biol Chem 1998; 273:15257–15262.
44. Kimura J, Weisz A, Kurashima Y, Hashimoto K, Ogura T, D'Acquisto F, Addeo R, Makuuchi M, Eshumi H. Hypoxia response element of the human vascular endothelial growth factor gene mediates transcriptional regulation by nitric oxide: control of hypoxia-inducible factor-1 activity by nitric oxide. Blood 2000; 95:189–197.
45. Fukamura D, Gohongi T, Kadambi A, Ang J, Yun C-O, Buerk DG, Huang PL, Jain RK. Predominant role of endothelial nitric oxide synthase in VEGF-induced angiogenesis and vascular permeability. Proc Natl Acad Sci USA. 2001; 98:2604–2609.
46. Griffiths C, Garthwaite J. The shaping of nitric oxide signals by a cellular sink. J Physiol 2001; 536:855–862.

23

Nitric Oxide and Carotid Body Chemoreception

Multiple Target Sites

RODRIGO ITURRIAGA

Catholic University of Chile
Santiago, Chile

JULIO ALCAYAGA

University of Chile
Santiago, Chile

I. Introduction

The carotid body (CB) is the main arterial chemoreceptor organ, which senses the partial pressure of blood gases and pH and contributes to the respiratory and cardiovascular reflex regulation. Hypoxia, hypercapnia, and acidosis increase the frequency of carotid chemosensory discharges in the carotid sinus nerve, while hyperoxia, hypocapnia, and alkalosis decreases it (1,2). The current model of carotid chemoreception proposes that glomus (type I) cells are the primary sites for chemical transduction. In response to natural stimuli, glomus cells are expected to release, in a Ca^{2+}-dependent manner, one—or more—excitatory transmitter(s), which in turn increases the frequency of discharges in nerve terminals of chemosensory petrosal neurons. Among several molecules present in the CB, dopamine, acetylcholine, substance P, and adenosine nucleotides have been proposed as excitatory transmitters in the junctions between glomus cell and nerve terminals (1,2). However, in addition to the excitatory transmitters, others molecules produced within the CB may regulate the chemosensory process. In the last decade, the proposition that the gaseous molecule nitric oxide (NO) is a tonic inhibitory modulator of carotid chemoreceptor activity has received great attention (3–7).

II. Nitric Oxide Synthase Localization in the Carotid Body and Petrosal Ganglion

NO synthase (NOS) immunoreactivity and NADPH diaphorase activities have been found in the CB and in the petrosal ganglion of several species (Table 1). The endothelial isoform of NOS is expressed in blood vessels of the cat CB (8). In addition, NOS immunoreactivity is present in a plexus of nerve fibers innervating the CB blood vessels and encircling glomus cells, but not in type II cells or smooth muscle cells (5,6,9). Wang et al. (5) reported that NOS-immunoreactive nerve terminals associated with blood vessels in the cat CB are unaffected by section of the carotid sinus nerve and removal of the superior cervical ganglion. On the contrary, the section of the carotid sinus nerve selectively eliminated the NOS-immunoreactive nerve terminals encircling glomus cells. Thus, they proposed two neural sources of NO production in the CB (5,6). One potential site for NO production are the parasympathetic neurons controlling the vascular tone of the CB blood vessels, and the second site are the terminals of petrosal neurons. On the contrary, Grimes et al. (9) found that NOS immunoreactivity and NADPH diaphorase activity localize in the cat CB predominantly in nerve fibers associated with blood vessels and occasionally lying close to glomus cells. They reported that NOS-positive fibers are originated from autonomic ganglion cells scattered within and around the CB and in the glossopharyngeal nerve. In the superior cervical ganglion, they found that NOS and diaphorase activities localize to many preganglionic fibers and to a small population of vasoactive intestinal peptide-positive neurons, presumably cholinergic ganglion cells. Accordingly, Grimes et al. (9) proposed that the NOS-positive innervation of the CB is mainly originated from a population of dispersed

Table 1 Distribution of NOS Immmunoreactivity and NADPH Diaphorase Activity in the CB and Petrosal Ganglion

Localization	Species	Refs.
Endothelial cells	Cat	8
Microganglial neurons in the CB Nerve fibers innervating blood vessels Petrosal neuron's terminals associated with glomus cells	Cat	4,5
Nerve fibers innervating blood vessels Nerve fibers innervating glomus cells (occasionally) Neurons in the CB periphery	Cat	9
Nerve fibers innervating blood vessels Nerve plexus associated with glomus cells (occasionally)	Rat	10
Nerve fibers associated with glomus cells Some glomus cells	Guinea pig	11

parasympathetic neurons. However, they proposed that NO released from these parasympathetic fibers could affect glomus cell activity directly or indirectly by regulating the vascular tone. Hohler et al. (10) found a similar pattern of NOS distribution in the rat CB. Most of the NOS-positive varicose nerve fibers innervate the arterial vascular vessels and, to a lesser extent, the glomoids. They found that the NOS-positive fibers persisted after the section of the carotid sinus nerve, and those fibers are probably derived from intrinsic neurons. The glomus cells of the cat and rat CBs apparently do not contain NOS immunoreactivity (3,5,6,9), but some glomus cells are positive for NOS in the guinea pig CB (11).

In the cat petrosal ganglion, Wang et al. (5,6) found the presence of NOS immunoreactivity in a population of neurons innervating the CB. Most of these fibers in the carotid sinus nerve arise from small-diameter neurons (10–15 μm) located in the core of the petrosal ganglion. These fibers constitute a small part (~10%) of the axons of the carotid sinus nerve and correspond to C-fibers. It is noteworthy that the cat petrosal ganglion has a similar NOS activity—measured as [^{3}H]-citruline formation—to the intact CB and about five times larger than that observed in the CB after section of the carotid nerve (5). On the contrary, Grimes et al. (9) did not find NOS activity in the petrosal ganglion and very rare occurrence in the nodose ganglion.

III. Inhibitory Effects of NO on Carotid Body Chemoreception

The administration of the NO precursor L-arginine (5), NO donor's molecules, such as sodium nitroprusside (SNP), nitroglycerine, S-nitroso-*N*-acetylpenicillamine (SNAP), and 6-(2-hydroxy-1-methyl-nitrosohydrazino)-*N*-methyl-1-hexanamine (NOC-9), to the CB partially reduces the chemosensory response to hypoxia. (4,5,12,13). More recently, we used NO in low concentration (25 ppm NO in N_2) to evaluate its effect on cat CB chemoreception to hypoxia (14). At a PO_2 of about 30 torr, the reaction of 25 ppm NO with O_2 dissolved in water is rather slow, with an expected half-time of several hours for NO disappearance. We found that during steady hypoxic chemosensory excitation, bolus injections or perfusion of NO-equilibrated Tyrode reduced the increased frequency of carotid chemosensory discharges. Perfusion with NO-equilibrated Tyrode reduced the rate of rise and the maximal amplitude of the normal chemosensory response to hypoxia. These results provided direct evidence that NO—in low concentration—acts as an inhibitory modulator of CB hypoxic chemoreception. On the other hand, the inhibition of NOS activity with *N*-ω-nitro-L-arginine methyl ester (L-NAME) and L-nitro-ω-arginine (L-NNA) increases basal chemosensory discharges (3–6) and enhances chemosensory responses to hypoxic hypoxia in the cat CB in vitro (4) and in situ (15). Sodium nitroprusside reverses the excitatory effect in the cat CB perfused in vitro (12), but not in situ (15). Taken together the pharmacological and physiological observations suggest that endogenously generated NO acts as an inhibitory modulator in the CB.

Little is known about the effect of NO on chemosensory responses to other stimuli (16). In the cat CB in vitro, we found that superfusion with Tyrode supplemented with the NO donor SNP did not increase basal chemosensory discharge, but reversibly reduced the chemosensory responses elicited by large doses of NaCN and nicotine (17). In anesthetized and paralyzed cats, we studied the effects of L-NAME and SNP on chemosensory responses to intravenous injections of NaCN, dopamine, and to hyperoxic ventilation (15). L-NAME increased basal chemosensory activity and potentiated the responses to NaCN and dopamine. SNP increased basal chemosensory activity, but reduced NaCN-induced increases of chemosensory discharges baseline and the transient inhibitions induced by dopamine, but not those induced by hyperoxia. These results indicate that besides the known inhibitory effect of NO on chemosensory responses to low PO_2, NO also modulates the chemosensory response to other stimuli.

IV. Target Sites and Mechanisms of NO Action in the Carotid Body

NO is a reactive molecule that may modulate the chemoreception process in the CB at different levels and by several mechanisms. Table 2 shows the proposed mechanisms of NO action in the CB.

A. Vascular Smooth Muscle Dilatation

According to the distribution of NOS immunoreactivity in the cat CB, Wang et al. (5,6) proposed two neural mechanisms dependent on cGMP production to mediate the inhibitory actions of NO on hypoxic chemoreception. The first mechanism involves the production and release of NO from autonomic parasympathetic neurons located in the CB, which control the vascular smooth muscle tone. They found that the administration of L-NAME to the cat CB perfused in vitro evoked a larger excitatory effect on basal chemosensory discharge than in the superfused CB preparation, where vascular effects are absent. Thus, a predominant part of the inhibitory effect of NO on hypoxic chemoreception appears to be vascularly mediated, presumably by increasing the cGMP content in smooth muscle cells as in other tissues. In fact, Prabhakar et al. (3) found that cGMP levels were lower in the cat CB treated with NOS inhibitors than in the untreated CBs, suggesting that the

Table 2 Mechanisms of Action of NO in the CB

Mechanism	Refs.
Vascular smooth muscle dilatation	4,5,9,18
Retrograde inhibition of glomus cell's excitability	6
Inhibition of glomus cell's calcium channels	21
Withdrawal of an inhibitory NO tone during hypoxia	3,13,23
Modulation of petrosal ganglion neuron's excitability	26
Inhibition of mitochondrial metabolism	13,33

actions of NO in the CB are coupled to cGMP. More recently, Lahiri and Buerk (18), using an in vitro perfused preparation of the cat CB, found that SNP infusion increases CB tissue PO_2 and reduces basal chemosensory discharges, supporting the proposal that part of the inhibitory effect of NO on CB chemoreception is mediated by vasodilatation.

B. Retrograde Inhibition of Glomus Cell Excitability

In addition to the cGMP-mediated vascular effects, Wang et al. (6) proposed that NO produced and released from petrosal nerve C-fiber terminals produces a retrograde inhibition of the glomus cell activity. They studied the effects of antidromic stimulation of the cat carotid sinus nerve on the production of NO measured by the accumulation of $[^3H]$-citrulline synthesized from $[^3H]$-arginine in the CB. They found that electrical stimulation of carotid sinus nerve C-fibers evoked a Ca^{2+}-dependent increase in $[^3H]$-citrulline accumulation in the CB, which was blocked by the NOS inhibitor L-NAME. To separate the vascular and neuronal effects of NO, Wang et al. (6) used an in vitro preparation of the cat CB that allows perfusion or superfusion of the organ. When the CB was perfused, the electrical stimulation of the carotid sinus nerve at C-fiber intensities increased cGMP levels and inhibited basal chemosensory discharges, an effect that was blocked by L-NAME. However, when the CB was superfused, the carotid sinus nerve C-fiber stimulation failed to modify the chemosensory discharge and the cGMP level. Under these conditions, prolonged nerve stimulation for 5 min attenuated the chemosensory response to hypoxia, an effect that was blocked by L-NAME.

C. Inhibition of Glomus Cell's Calcium Channel

Glomus cells from several species have a voltage-activated K^+ conductance, evoked by an imposed depolarization, which is reversibly reduced by hypoxia (19). According to the membrane hypothesis of chemoreception, the glomus cell response to hypoxia is initiated by a depolarization caused by low PO_2, which opens voltage-gated Ca^{2+} channels and raises the intracellular Ca^{2+} (19). NO increased cGMP in glomus cells, but it is not clear how the increased cGMP could reduce the excitability of the glomus cells, because the PO_2-dependent K^+ current is unaffected by cAMP or cGMP (20). Moreover, a direct effect of NO on the PO_2-dependent K^+ conductance has not been found. In fact, Hatton and Peers (20) and Summers et al. (21) found that application of NO donors SNAP and SNP did not modify the PO_2-dependent K^+ currents in rat and rabbit glomus cells. However, Summers et al. (21) reported that NO donors such as SNP and spermine inhibit L-type Ca^{2+} currents in rabbit glomus cells through a cGMP-independent mechanism mediated by a direct modification of the thiol groups of calcium channel proteins. The inhibition of the Ca^{2+} current by NO appears to be a direct effect on the Ca^{2+} channel protein, because it could be abolished by *N*-ethylmalemide, which prevents the NO-mediated nitrosylation of proteins (21).

D. Withdrawal of an Inhibitory NO Tone During Hypoxia

It has been proposed that CB chemosensory excitation induced by hypoxia may be the result of a decreased availability of an inhibitory messenger such as NO (3,22). Since the inhibition of NOS activity increases basal chemosensory discharges and NOS activity is reduced at low PO_2, related to normoxic controls (3), it is possible that low concentrations of NO may exert a tonic inhibitory effect on chemosensory discharges during normoxia. Thus, it is likely that the increased chemosensory activity induced by hypoxia resulted from a reduced production of NO. However, electrochemical measurements of [NO] changes in the CB with NO-sensitive carbon electrodes did not support this proposal. We measured simultaneously the chemosensory responses and the [NO] changes in the cat CB produced by hypoxia (13). We found that [NO] remained constant during 2–5 min of hypoxic stimulation in most CB, but [NO] increased at the end of the hypoxic challenge in few CBs. More recently, Fung et al. (23) reported that hypoxia increased [NO] by 17 ± 2 nM in the rat CB. The amount of NO released during hypoxia was augmented by L-arginine and it was abolished by pretreatment with L-NAME. These results suggest that endogenous NO production in the CB increases during hypoxia. Fung et al. (23) concluded that acute hypoxia increases NO generation in the rat CB, and the elevated levels of NO may reduce the chemoreceptor activity during hypoxia. Thus, NO play an active inhibitory role in the control of carotid chemoreceptor activity during hypoxia. This observation agrees with the results obtained by Di Guilio et al. (24) that chronic hypoxia induced an increase in NOS activity in the rat CB, suggesting that increased NO release during chronic hypoxia causes an inhibitory effect on carotid chemosensory discharge of the CB.

E. Modulation of Petrosal Ganglion Neurons' Excitability

Petrosal ganglion neurons are another target site for the action of NO. In view of the fact that NO is produced not only in the nerve terminals within the CB, but also in perikarya of petrosal ganglion neurons (6), it is possible that NO may modulate the excitability of the petrosal neurons. Since we found (25) that a population of petrosal neurons projecting through the carotid nerve is selectively activated by acetylcholine, we (26) studied the effects of SNP and L-NAME on the responses evoked in the carotid sinus nerve by acetylcholine applied to the petrosal ganglion superfused in vitro. Acetylcholine increased the frequency of discharges recorded from the carotid sinus nerve in a dose-dependent manner. SNP reduced the sensitivity and amplitude of the response to acetylcholine, although the maximal response appears less affected. L-NAME slightly increased the sensitivity of the acetylcholine-induced responses, an effect that persisted after L-NAME withdrawal. These results show that NO may modulate the activity of a population of sensory neurons of the cat petrosal ganglion activated by acetylcholine. This observation suggests that chemosensory information carried by primary sensory neurons may be locally modulated within the petrosal ganglia. Thus, NO may play a role as a local modulator in this autonomic primary sensory ganglion.

F. Inhibition of Oxidative Mitochondrial Metabolism

When we studied the effects of SNP on carotid chemosensory responses to excitatory and inhibitory stimuli in paralyzed and artificially ventilated cats (15), we found that SNP increased basal chemosensory discharges. This result was completely unexpected, since NO donors reduced or had no effect on basal chemoreceptor activity in the cat CB superfused or perfused in vitro (3–6,16). A major difference between CB preparations in situ and in vitro is the presence of large amounts of endothelium and vascular smooth muscle tissue in situ. Since endothelium and smooth muscle cells are required to activate SNP for releasing NO (27), it is likely that large amounts of NO released from SNP in situ may account for the increased chemosensory activity.

We studied the effects of NO released by spontaneous NO donors on cat CB chemosensory activity during normoxia and hypoxia. We measured the chemosensory responses and the changes in [NO] with NO-selective carbon-fiber microelectrodes inserted into the CB. The injection of the NO donors SNAP and NOC-9 into the CB transiently reduced the hypoxic-augmented chemosensory discharges in a dose-dependent manner. However, during normoxia the injection of the same NO donors increased the chemosensory discharges in a dose-dependent manner, showing a dual effect of NO on carotid chemoreception depending on PO_2 levels (13). Accordingly, we proposed that a high [NO] or its metabolite peroxynitrite released from the NO donors might account for the increased chemosensory discharge, because NO and peroxynitrite inhibit the electron transport chain and oxidative phosphorylation (28,29). The probable target molecules for the actions of NO in the CB are the soluble enzymes guanidyl cyclase and cytochrome oxidase a_3, because they are extremely sensitive to NO (see Ref. 28 for review). The activation of guanidyl cyclase in the CB vascular smooth muscle is expected to produce vasodilatation and a related reduction in chemosensory activity. By contrast, the blockade of cytochrome oxidase a_3 activity and oxidative phosphorylation is expected to increase chemosensory activity (30,31).

It is well known that NO and peroxynitrite reversibly inhibit mitochondrial respiration at different levels, reducing oxygen consumption. Indeed, NO at low and medium concentrations ($<5\,\mu M$) specifically and reversibly inhibits cytochrome oxidase in complex IV in competition with O_2 (28,29). However, higher [NO] concentrations may inhibit other respiratory chain complexes (28). We found that the administration of several large doses of SNAP and NOC-9 increased the basal chemosensory activity up to the maximal discharge in half of the CB studied. Similarly to what we found in situ (15), the increased chemosensory activity returned to baseline when the CBs were perfused with 100% O_2. Thus, our studies indicate that NO has a dual effect on carotid chemoreception depending on the oxygen level. During hypoxia, NO is predominantly an inhibitory modulator of carotid chemoreception, while in normoxia NO increased the activity. The excitatory effect produced by NO is probably mediated by an impairment of mitochondrial electron transport and oxidative phosphorylation. The dual effect of NO supports the main role played by mitochondrial metabolism in CB chemoreception to oxygen.

Furthermore, a new and original possibility is that the mitochondrial NO/O_2 ratio may be crucial to regulate the respiratory rate, playing a physiological role in oxygen sensing. Recently the mitochondria have been found to contain a new NOS isoform that produces significant amounts of NO, enough for regulating their own respiration (32), suggesting that NO may be important for the regulation of energy metabolism. Thus, it is possible that NO inhibition of cytochrome a_3 could be involved in the physiological regulation of carotid chemosensory sensing of PO_2. Recently, Buerk and Lahiri (33) studied the role of NO in the cat CB measuring tissue PO_2 with recessed microelectrodes in cat CB perfused in vitro before and after NOS inhibition. They compared the O_2 disappearance curves during stop flow. They found that L-NAME reduced the tissue PO_2 from 74.5 ± 8.7 to 37.4 ± 6.9 torr and reduced the maximum rate of O_2 consumption by 18%. As expected, SNP increased the CB tissue PO_2, but reduced the rate of O_2 disappearance by 15%, indicating that NO inhibits O_2 consumption. Thus, they proposed that NO may play a role in oxygen sensing in the CB. Buerk and Lahiri (33) found an unusually high [NO] in the CB ($\approx$300 nM), which may explain the high K_m of about 35 torr for oxygen consumption.

V. Contribution of Endothelial and Neuronal NOS Isoforms

Three types of isoforms of NOS have been characterized (34): the neuronal (NOS-1), the inducible (NOS-2), and the endothelial (NOS-3) isoforms. In the CB, the NOS-1 isoform is present in neuronal structures (5), while the NOS-3 is localized to the endothelium of the blood vessels (8). As we mentioned before, the NO produced by the endothelial NOS-3 isoform may modulate CB activity through the regulation of blood flow and subsequent changes in tissue PO_2 within the chemoreceptors (see Fig. 1). Gozal et al. (35,36) studied the effect of acute blockade of the NOS-1 and NOS-3 isoforms on the ventilatory responses to CB chemoreceptor stimulation in freely moving and chronically instrumented rats. The rats were treated with the nonselective NOS inhibitor L-NAME and S-methyl-L-thiocitrulline, a selective neuronal NOS inhibitor. Gozal et al. (35) found that the selective neuronal NOS inhibitor did not modify the NaCN dose-response curve. However, L-NAME significantly enhanced the ventilatory responses to NaCN. Western blots of equivalent amounts of protein from CB tissue homogenates showed higher levels of endothelial NOS-3 than of neuronal NOS-1 isoform. Thus, they concluded that endothelial NOS provides the major source for NO within the CB and exerts a down-regulatory effect upon peripheral chemoreceptor responsiveness. More recently, Kline et al. (37) studied the role of NO generated from the endothelial NOS-3 isoform in the control of respiration during hypoxia and hypercapnia using mutant mice deficient in NOS-3. The wild-type mice responded with greater increases in respiration during hypoxia than mutant mice. Respiratory responses to hyperoxic-hypercapnia were comparable between both groups of mice. Respiratory responses to NaCN and brief hyperoxia were attenuated in mutant compared with wild-type mice, indicating a reduced peripheral chemoreceptor sensitivity. These

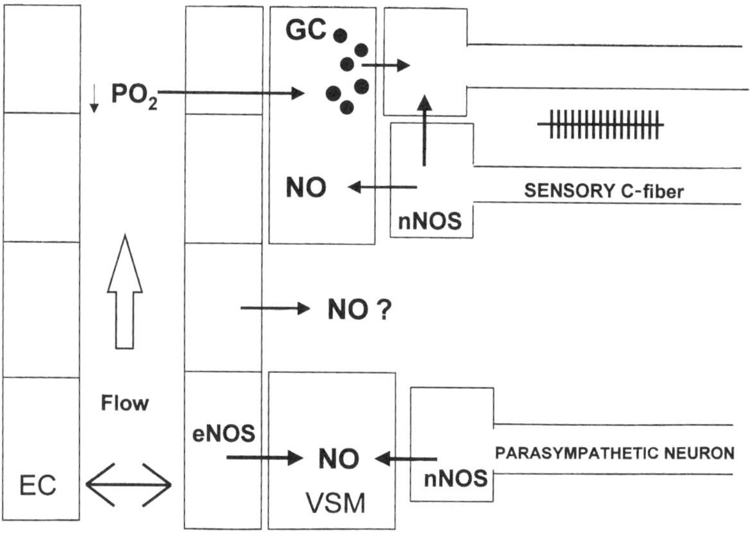

Figure 1 Diagram of possible sources for NO production in the carotid body. EC, endothelial cell; GC, glomus cell; VSM, vascular smooth muscle; nNOS, neuronal nitric oxide synthase; eNOS, endothelial nitric oxide synthase.

observations showed that NOS-3 mutant mice exhibit selective blunting of the respiratory responses to hypoxia, but not to hypercapnia, and supported the idea that NO generated by NOS-3 is an important physiological modulator of respiration during hypoxia.

The results obtained by Kline et al. (37) are in contradiction with previous reports by Gozal et al. (35,36) showing that ventilatory responses to hypoxia and to NaCN were unaffected after systemic administration of a selective NOS inhibitor, whereas a general NOS inhibitor potentiated the stimulatory effects of hypoxia and NaCN. Kline et al. (37) speculated that the acute physiological consequences of blockade of NOS-3 might differ from chronic deficiency of the NOS-3 protein. They proposed that in mutant mice, the gene encoding NOS-3 protein is functionally defective since birth. Consequently, it is expected that the CB will receive less blood flow and will be subjected to persistent tissue hypoxia. It is known that persistent tissue hypoxia, such as that which occurs in the later stages of chronic hypertension, may cause the CB to be insensitive to hypoxia. Consistent with such an idea, NOS-3 mutant mice were found to have higher blood pressures than control mice do. Thus, the blunted peripheral chemosensitivity in NOS-3 mutant mice might be secondarily due to chronic hypertension.

Using the same approach, Kline et al. (38) studied the role of endogenous NO generated by neuronal NOS-1 in the control of respiration during hypoxia and hypercapnia using mutant mice deficient in NOS-1. Unanesthetized mutant mice exhibited greater respiratory responses during hypoxia than wild-type controls. Respiratory responses were associated with significant decreases in oxygen consumption in both groups of mice, and the magnitude of change was greater in mutant than wild-type mice. Similar augmentation of respiratory responses during hypoxia was also observed in anesthetized mutant mice. Respiratory responses to brief hyperoxia (Dejours test) and to NaCN were more pronounced in mutant mice, suggesting augmented peripheral chemoreceptor sensitivity. The magnitude of respiratory responses to hypercapnia was comparable in both groups of mice in the awake and anaesthetized conditions. These results suggest that the hypoxic chemosensory responses were selectively augmented in mutant mice deficient in NOS-1. Peripheral as well as central mechanisms may contribute to the altered responses to hypoxia.

VI. Conclusions

In conclusion, the distribution of NOS immunoreactivity and the pharmacological and physiological evidence analyzed and discussed here support the proposal that CB endogenous NO exerts a tonic inhibitory effect on CB chemoreception. Clearly, NO in low concentration is a broad inhibitory modulator of the chemoreception process in the CB. NO may modulate chemoreception by regulating the CB vascular tone, the oxygen delivery to the chemoreceptor cells, and the excitability of glomus cells and petrosal sensory neurons.

Acknowledgments

This work was in part supported by grants FONDECYT 198-0965 and 199-0030 and by the Direction of Research and Graduate Studies (DIPUC) of the P. Catholic University of Chile. We would like to thank Mrs Carolina Larraín for her assistance in the preparation of this chapter.

References

1. Eyzaguirre C, Zapata P. Perspectives in carotid body research. J Appl Physiol 1984; 57:931–957.
2. González C, Almaraz L, Obeso A, Rigual R. Carotid body chemoreceptors: from natural stimuli to sensory discharges. Physiol Rev 1994; 74:829–898.
3. Prabhakar NR, Kumar CK, Chang CH, Agani FH, Haxhiu MA. Nitric oxide in the sensory function of the carotid body. Brain Res 1993; 625:16–22.
4. Chugh DK, Katayama M, Mokashi A, Debout DE, Ray DK, Lahiri S. Nitric oxide-related inhibition of carotid chemosensory activity in the cat. Respir Physiol 1994; 97:147–152.
5. Wang Z-Z, Stensaas LJ, Bredt DS, Dinger BG, Fidone SD. Localization and actions of nitric oxide in the cat carotid body. Neuroscience 1994; 60:275–286.
6. Wang ZZ, Stensaas LJ, Dinger BG, Fidone SJ. Nitric oxide mediates chemoreceptor inhibition in the cat carotid body. Neuroscience 1995; 65:217–229.
7. Trzebski A, Sato Y, Susuki A, Sato A. Inhibition of nitric oxide synthesis potentiates the responsiveness of carotid chemoreceptors to systemic hypoxia in the rat. Neurosci Lett 1995; 190:29–33.

8. Wang Z-Z, Bredt DS, Fidone SJ, Stensas LJ. Neurons synthesizing nitric oxide innervating the carotid body. J Comp Neurol 1993; 336:419–432.
9. Grimes PA, Lahiri S, Stone R, Mokashi A, Chugh D. Nitric oxide synthase occurs in neurons and nerve fibers of the carotid body. Adv Exp Med Biol 1994; 360:221–224.
10. Hohler B, Mayer B, Kummer W. Nitric oxide synthase in the rat carotid body and carotid sinus. Cell Tissue Res 1994; 6:559–564.
11. Tanaka K, Chiba T. Nitric oxide synthase containing neurons in the carotid body and sinus of the guinea pig. Microsc Res Tech 1994; 29:90–93.
12. Katayama M, Chugh DK, Mokashi A, Ray DK, Bebout DE, Lahiri S. NO mimics O_2 in the carotid body chemoreception. Adv Exp Med Biol 1994; 360:225–227.
13. Iturriaga R, Villanueva S, Mosqueira M. Dual effects of nitric oxide on carotid body chemoreception. J Appl Physiol 2000; 89:1005–1012.
14. Iturriaga R, Mosqueira M, Villanueva S. Effects of nitric oxide gas on carotid body chemosensory response to hypoxia. Brain Res 2000; 855:282–286.
15. Iturriaga R, Alcayaga J, Rey S. Sodium nitroprusside blocks the cat carotid chemosensory inhibition induced by dopamine, but not that by hyperoxia. Brain Res 1998; 799:26–34.
16. Iturriaga R, Villanueva S, Alcayaga J. Nitric oxide modulation of carotid chemoreception. Adv Exp Med Biol 2000; 475:761–768.
17. Alcayaga J, Iturriaga R, Ramirez J, Readi R, Quezada C, Salinas S. Cat carotid body chemosensory response to non-hypoxic stimuli are inhibited by sodium nitroprusside both in situ and in vitro. Brain Res 1997; 767:384–387.
18. Lahiri S, Buerk DG. Vascular and metabolic effects of nitric oxide synthase inhibition evaluated by tissue PO_2 measurements in carotid body. In: Hudetz AG, Bruley DF, eds. Oxygen Transport to Tissue XX. New York: Plenum Press, 1998:455–460.
19. Lopez-Barneo J, Pardal R, Ortega-Saenz P. Cellular mechanism of oxygen sensing. Annu Rev Physiol 2001; 63:259–287.
20. Hatton CJ, Peers C. Hypoxic inhibition of K^+ currents in isolated rat type I carotid body cells: Evidence against the involvement of cyclic nucleotides. Pflügers Arch 1996; 433:129–135.
21. Summers BA, Overholt LJ, Prabhakar NR. Nitric oxide inhibits L-type Ca^{2+} current in glomus cells of the rabbit carotid body via a cGMP-independent mechanism. J Neurophysiol 1999; 81:1449–1457.
22. Prabhakar NR. NO and CO as second messengers in oxygen sensing in the carotid body. Respir Physiol 1999; 115:161–168.
23. Fung ML, Ye JS, Fung PC. Acute hypoxia elevates nitric oxide generation in rat carotid body in vitro. Pflügers Arch 2001; 442:903–909.
24. Di Giulio C, Grilli A, De Lutiis MA, Di Natale F, Sabatino G, Felaco M. Does chronic hypoxia increase rat carotid body nitric oxide? Comp Biochem Physiol Mol Integr Physiol 1998; 120:243–247.
25. Alcayaga J, Iturriaga R, Varas R, Arroyo J, Zapata P. Selective activation of carotid nerve fibers by acetylcholine applied to the cat petrosal ganglion in vitro. Brain Res 1998; 786:47–54.
26. Alcayaga J, Barrios M, Bustos F, Miranda G, Molina V, Iturriaga R. Modulatory effect of nitric oxide on acetylcholine-induced activation of cat petrosal ganglion neurons in vitro. Brain Res 1999; 825:194–198.
27. Kowaluk E, Seth E, Fung H. Metabolic activation of sodium nitroprusside to nitric oxide in vascular smooth muscle. J Pharmacol Exp Ther 1992; 262:916–922.

28. Brown CG. Nitric oxide and mitochondrial respiration. Biochim Biophys Acta 1999; 1411:351–369.
29. Cassina A, Radi R. Differential inhibitory action of nitric oxide and peroxynitrite on mitochondrial electron transport. Arch Biochem Biophys 1996; 328:309–316.
30. Mulligan E, Lahiri S, Storey BT. Carotid body O_2 chemoreception and mitochondrial oxidative phosphorylation. J Appl Physiol 1981; 51:438–466.
31. Mulligan E, Lahiri S. Separation of carotid body chemoreceptor responses to O_2 and CO_2 by oligomycin and by actinomycin A. Am J Physiol Cell Physiol 1982; 242:C200-C206.
32. Ghafourifar P, Richter C. Nitric oxide synthase activity in mitochondria. FEBS Lett 1997; 418:291–296.
33. Buerk DG, Lahiri S. Evidence that nitric oxide plays a role in O_2 sensing from tissue NO and PO_2 measurements in cat carotid body. Adv Exp Med Biol 2000; 475:337–347.
34. Moncada S, Palmer RM, Higgs EA. Nitric oxide: Physiology, pathophysiology, and pharmacology. Pharmacol Rev 1991; 43:109–141.
35. Gozal D, Gozal E, Gozal YM, Torres JE. Nitric oxide synthase isoforms and peripheral chemoreceptor stimulation in conscious rats. Neuroreport 1996; 26:1145–1148.
36. Gozal D, Torres JE, Gozal YM, Littwin SM. Effect of nitric oxide synthase inhibition on cardiorespiratory responses in the conscious rat. J Appl Physiol 1996; 81:2068–2077.
37. Kline DD, Yang T, Premkumar DR, Thomas AJ, Prabhakar NR. Blunted respiratory responses to hypoxia in mutant mice deficient in nitric oxide synthase-3. J Appl Physiol 2000; 88:1496–1508.
38. Kline DD, Yang T, Huang PL, Prabhakar NR. Altered respiratory responses to hypoxia in mutant mice deficient in neuronal nitric oxide synthase. J Physiol (Lond) 1998; 511:273–287.

24

Multiple Roles of Neurotransmitters in the Carotid Body

Involvement in Sensory Transmission and Adaptation to Hypoxia

GANESH K. KUMAR, JEFFERY L. OVERHOLT, and NANDURI R. PRABHAKAR

Case Western Reserve University School of Medicine
Cleveland, Ohio, U.S.A.

I. Introduction

Since their discovery in 1930s, it is being increasingly appreciated that carotid bodies are critical for triggering the ventilatory and circulatory changes during acute hypoxia. There has also been a substantial body of evidence implicating the carotid bodies in ventilatory adaptations to chronic sustained hypoxia (2). Possibly of more clinical importance, investigations on human subjects with recurrent apneas provided compelling evidence for the involvement of carotid bodies in causing circulatory and ventilatory abnormality in response to chronic intermittent hypoxia (69). While the mechanisms of sensory transduction are being intensely debated (10,20,44,68), it is fairly certain that transmitters are critical for sensory transmission at the carotid body. Indeed, the carotid bodies express as many types of neurotransmitters as the mammalian brain. It is likely that some of the transmitters, in addition to and/or independent of their participation in sensory transmission, also play critical roles in the adaptation of the carotid bodies to chronic sustained or intermittent hypoxia. The purpose of this chapter is to highlight recent advances in the role of transmitters in sensory transmission and summarize their potential involvement in inducing functional and structural plasticity of the chemoreceptors by chronic sustained or intermittent hypoxia.

II. Conventional and Unconventional Neurotransmitters in the Carotid Body

The mammalian carotid body expresses a variety of neurochemicals and their receptors (see Tables 1 and 2). The transmitters expressed in the carotid body can be classified into two major categories; conventional and unconventional. The conventional class of neurotransmitters includes those that are stored in vesicles and exert their effect via activation of specific receptors. Examples of this class include catecholamines, acetylcholine, and neuropeptides, whereas gas molecules like nitric oxide (NO) and carbon monoxide (CO) form the class of unconventional neurotransmitters. These molecules are generated spontaneously by way of enzymatic reactions and mediate their biological actions by either activation

Table 1 Neurotransmitter Expression in the Carotid Body

Neurotransmitters[a]	Species	Localization
Conventional neurotransmitters		
Biogenic amines		
1. Acetylcholine (12,53)	Rat, cat, rabbit	Type I cells
2. Dopamine (10,20,91)	Rat, cat, human, rabbit	Type I cells, nerve fibers
3. Norepinephrine (10,18)	Rat, cat, rabbit	Type I cells and nerve fibers
4. Serotonin (56,91,93)	Rat, cat, human	Type I cells and mast cells
Neuropeptides		
1. Enkephalin (10,91)	Cat	Type I cells
2. Substance P (28,66,73)	Cat, rabbit, rat, human	Type I cells and nerve fibers
3. Neurokinin-A (73)	Cat	Type I cells
4. Endothelin (23)	Rat	Type I cells
5. Atrial natriuretic Peptide (90)	Cat	Type I cells
6. Galanin (25)	Cat	Nerve fibers
7. Neuropeptide Y (10,66)	Rabbit and rat	Nerve fibers
8. Calcitonin-gene Related peptide (10,66)	Guinea pig	Type I cells and nerve fibers
Amino acids		
1. Glutamate (86)	Cat	Type I and type II cells
2. GABA (56)	Mouse	Type I cells
Unconventional neurotransmitters[b]		
1. NO (67,76,88)	Rat, cat, rabbit, mouse	Nerve fibers, endothelium
2. CO (67,77)	Rat, mouse	Type I cells

[a]Pertinent references are given in parenthesis.

[b]NO and CO are generated spontaneously from enzymatic reactions involving NO synthase and heme oxygenase, respectively, and enzymes responsible for their synthesis are localized.

Table 2 Neurotransmitter Receptor Expression in the Carotid Body

Receptors	Subtypes[a]	Localization	Species[b]
1. Cholinergic-R	Nicotinic (8,55)	Type I cells	Cat, rabbit, rat (Phar)
	α-7 subunit (84)	Nerve terminals	Cat (IC)
	Muscarinic (9,10)	Type I cells	Cat, rat (Phar)
2. Dopaminergic-R	D1 (20)	N.D.	Rat, cat, rabbit (Phar)
	D2 (20)	Type I cells	Rabbit, cat, rat (Phar)
3. Adrenergic-R	$\alpha 2$ (39,57)	N.D.	Cat (Phar)
	$\beta 1$	N.D.	Cat (Phar)
4. Serotonergic-R	5-HT2, 5HT-3 (33)	Type I cells	Rat (Phar)
	5-HT5a (92)	Type I cells	Rat (IC)
5. Purinergic-R	P2X2-P2X3 (79)	Afferent nerve	Rat (IC)
	ATP-R (50a)	Type I cells	Rat (Phar)
6. Adenosine-R	A2a (16,38)	Type I cells	Rat, cat (ISH)
	A1 (16,38)	Afferent nerve fibers	Rat (ISH)
7. Neurokinin-R	NK-1 (66)	Afferent nerve	Cat (Auto R)
8. Opiate-R	Delta-type (32)	N.D.	Cat (Phar)
9. Endothelin-R	ET-A (5)	N.D.	Cat, rat, rabbit (Phar)
10. VEGF-R	fit-1 (27)	Type I cells	Human (IC)
	flk-1 (3)	Type I cells	Rat (IC)

[a]Pertinent references are given in parentheses.
[b]Key: N.D., not determined; Phar, from pharmacological analysis; IC, from immunocytochemical studies; Auto R, from autoradiographic analysis; ISH, from in situ hybridization studies.

of downstream enzymes coupled to signaling pathways or direct modification of proteins (67,76,77,88).

Morphological and physiological studies indicate that glomus cells (also called type I cells) are the primary site(s) of sensory transduction and satisfy several criteria for presynaptic cells (10,68). Currently it is thought that hypoxia (also perhaps hypercapnia) releases one or more excitatory transmitters from the glomus cells that, by way of depolarizing the nearby afferent nerve terminals, increase the sensory discharge. The following section summarizes the recent progress toward understanding the role(s) of transmitters in the sensory transmission of the carotid body during hypoxia.

A. Conventional Neurotransmitters

Acetylcholine (ACh)

At the time when carotid bodies were discovered as sensory organs, there was an intense debate on whether synaptic transmission is mediated by endogenous chemicals such as ACh. In 1938, Schweitzer and Wright (83) made the observation that ACh stimulates the carotid body similar to hypoxia. These findings initiated a surge of interest to determine whether ACh also functions as a sensory transmitter in the carotid body. Now it is fairly established that glomus cells express cholineacetyltransferase (CAT), the enzyme associated with ACh synthesis, and acetylcholinesterase (AChE), the enzyme that terminates the actions of ACh (53). These findings indicate that carotid bodies have enzymatic machinery for the generation and inactivation of ACh. Several studies attempted to assess the role of ACh in sensory transmission by measuring its release from the carotid body in response to hypoxia. Earlier reports on ACh release from carotid bodies were inconclusive, mainly owing to the nonspecific nature of the methods employed in these studies (see Refs. 10,12 for references). Recently, ACh release from the cat carotid body was reexamined using HPLC-ECD technique (13). Results from this study showed a modest release of ACh in response to hypoxia (1% O_2 + 5% CO_2, pH 7.4), whereas a greater release was noted when hypoxic stimulus (4% O_2) was applied in combination with 2% CO_2 and alkaline pH (7.8; see Figure 3 of Ref. 13). Based on these results, the authors concluded that hypoxia stimulates ACh release from the cat carotid body. In sharp contrast, studies on rat and rabbit carotid bodies employing identical protocols and methodologies used for cat carotid bodies showed that hypoxia *inhibits*, whereas hypercapnia (10% CO_2 + 90% O_2; pH = 7.4) or acidosis (pH = 6.8) *stimulates* ACh release (29,31). These results are similar to several studies on central and peripheral nervous system that showed inhibition of ACh release by hypoxia (6,17). Thus, it remains uncertain whether hypoxia releases ACh from carotid bodies in all mammalian species.

Consistent with the earlier observations of Schweitzer and Wright (83), several studies documented the excitatory actions of ACh in cat and rat carotid bodies (10,12,20,66). However, in rabbit ACh inhibits the sensory activity (10,64,68). The excitatory effects of ACh in the carotid body are mediated by nicotinic acetylcholine receptors (nAChR). Pharmacological studies suggest that glomus cells express at least two types of nAChR receptors, one sensitive to α-bungarotoxin and the other to mecamylamine (8,55). Afferent nerve terminals appear to express nAChR that contains $\alpha 7$ subunits (84). On the other hand, muscarinic cholinergic receptors (mAChRs) seem to be coupled to inhibition of the sensory activity (9,10). The relative abundance of nAChR and mAChR in the carotid body varies among species (24), which may explain the species-dependent inhibition or excitation of the sensory discharge by ACh. Cholinergic antagonists that block the action of exogenous ACh have little influence on chemosensory response to hypoxia (10,20,66,68). However, a cocktail of cholinergic blockers attenuates the sensory response to acute hypoxia (12). Recently, Zhang and co-workers reported that hexamethonium or mecamylamine in combination with suramin, a nonselective purinergic receptor antagonist, prevents the sensory excitation by hypoxia in cocultures of rat glomus cells with petrosal neurons (94). These authors suggested that ACh, in concert with ATP, mediates the sensory excitation by hypoxia. Interestingly, combined application of cholinergic blockers and suramin also attenuates the sensory response of the carotid body to hypercapnia (94). Thus, although ACh has long been considered the primary sensory transmitter in the carotid body for hypoxia, because of the uncertainty of its release by low oxygen and the inability of the cholinergic antagonists to block the sensory response to hypoxia, its role still remains equivocal.

Catecholamines

The morphology of the carotid body resembles chromaffin tissues expressing catecholamines. Now it is fairly established that carotid bodies express dopamine and norepinephrine, whereas there is no convincing evidence for epinephrine. Type I cells from a variety of species express tyrosine hydroxylase (TH) and dopamine β hydroxylase (DBH), the enzymes responsible for the synthesis of dopamine (DA) and norepinephrine (NE), respectively (10,20,91). In addition, nerve fibers (of sensory as well as autonomic origin) and ganglion cells also show TH immunoreactivity (91). The actions of DA and NE are terminated by a reuptake mechanism involving specific transporters. However, evidence for DA and/or NE transporters in the carotid body is lacking.

Several studies have demonstrated that hypoxia releases DA from cat, rabbit, and rat carotid bodies (1,11,18,19,54,59,87). Recent studies using chronoamperometry in carotid body slices have provided evidence that much of the DA released during hypoxia comes from glomus cells (59). DA release by hypoxia is Ca^{2+} dependent and involves at least voltage-gated L-type Ca^{2+} channels (54). Acidosis (pH = 7.2–6.6) as well as high CO_2 (20% CO_2, pH 7.4) also stimulates DA release from the rabbit carotid body in a Ca^{2+}-dependent manner (80). Although type I cells express NE, hypoxia preferentially releases DA more than NE (20,87).

The data derived from a number of species suggest that DA is inhibitory to carotid body activity, because exogenous administration of DA inhibits and blockade of dopaminergic receptors augments the baseline activity and potentiates the response to hypoxia. Although earlier studies reported stimulation of the carotid body by NE (10,50), recent studies suggest that NE, under physiological conditions, functions as an inhibitory messenger, similar to DA (see Ref. 60 for references). The inhibitory actions of NE are coupled to α_2-adrenergic receptors. Blockade of α_2-receptors stimulates the baseline activity and further augments the sensory response to hypoxia (39). The actions of α_2-receptors are coupled to inhibition of Ca^{2+} current via G-protein-coupled voltage-gated Ca^{2+} channels in the glomus cells (57). The sensory excitation seen with pharmacological doses of NE is conceivably due to vasoconstriction of carotid body vessels mediated by β-adrenergic receptors (10,20,66,68).

5-Hydroxytryptamine (5-HT)

5-HT-like immunoreactivity has been reported in type I cells of rat and mouse carotid bodies (56,93). Exogenous application of 5-HT results in brief excitation followed by inhibition of carotid body activity in the cat (52) and dog (10,66). In the cat, the initial excitatory phase is mediated by 5-HT3 receptors, whereas the delayed excitation is coupled to 5-HT2 receptors (33). On the other hand, in rats, in addition to the biphasic response, 5-HT causes a delayed slow increase in the sensory discharge. Recently, it has been reported that 5-HT produces spontaneous rhythmic activation of isolated rat glomus cell clusters, and this effect is prevented by the 5-HT2 receptor antagonist ketansarin, but not by MDL 72222, a purported blocker of 5-HT3 receptors (93). Wang et al. (92) reported 5-HT5a receptor expression in rat glomus cells as well as in the neurons of the petrosal ganglion. However, the significance of 5-HT5a receptors remains to be studied.

Although 5-HT has marked effects on sensory discharge, its significance in sensory transmission during hypoxia is less clear. Neither 5-HT-2 nor 5-HT-3 receptor blockers had any significant influence on the sensory response to hypoxia (33). So far, there are no studies demonstrating the release of 5-HT during hypoxia. However, 5-HT may play a more important role in inducing functional plasticity during chronic intermittent hypoxia (see below).

Neuropeptides

Pearse (60) reported that type I cells of the carotid body resemble the amine precursor uptake decarboxylation (APUD) cell system and suggested that they secrete a polypeptide, provisionally named "glomin," that could be important in chemoreception. Now it is well recognized that carotid bodies express a variety of peptides that serve as transmitters or modulators elsewhere in the nervous system (Table 1).

Substance P (SP)

The following lines of evidence suggest that carotid bodies express SP as well as the enzymatic machinery required for its synthesis and inactivation. First, SP-like immunoreactivity is expressed in glomus cells as well as nerve fibers of sensory and autonomic origin in the carotid body (4,28,43). Second, glomus cells express preprotachykinin (PPT) mRNA that encodes the precursor of SP (66). Third, carboxypeptidase-E (CPE), an enzyme for SP processing from PPT, is seen in the carotid body glomus cells (19,40). Fourth, the enzyme neutral endopeptidase (NEP), responsible for terminating the actions of SP, is found in the interstitial spaces surrounding type I cells and nerve terminals (41).

A recent report by Kim et al. (28) shows that hypoxia releases SP from rabbit carotid bodies in a stimulus-dependent manner. This hypoxia-evoked SP release requires extracellular Ca^{2+} and activation of N and L-type Ca^{2+} channels. On the other hand, hypercapnia has no effect on the release of SP. Exogenous administration of SP or its analogs augments the sensory discharge of the carotid body in a number of species (7,42,46,51,73) with the exception of goat (62) and potentiates the hypoxic sensory response (73). More importantly, SP receptor antagonists, when given in nanomolar concentration, not only prevent the excitatory effects of SP, but also abolish the sensory response to hypoxia while leaving the sensory response to hypercapnia unaffected in cats (70–72,75) and rats (7).

Autoradiographic analysis has also revealed SP-binding sites in the cat carotid body (66). Neurokinin-1 receptor mediates the actions of SP elsewhere in the nervous system. Blockade of NK-1 receptors in the cat carotid body prevented hypoxia-evoked sensory excitation (75), suggesting that SP, acting on NK-1 receptors, participates in the sensory transmission in this species. However, in rats, although SP stimulates the carotid body activity (7), there is no evidence for NK-1 receptor mRNA expression either in glomus cells or on the petrosal neuron (15). Thus, in rats, the actions of SP may be mediated by other neurokinin receptors (for example, NK-2 or NK-3). Alternatively, there are studies suggesting that SP may directly influence mitochondrial metabolism by acting as a protonophore (74). Inhibitors of the mitochondrial respiratory chain (for example, rotenone and antimycin A) prevent the sensory response to SP and hypoxia, but not that caused by lobeline or nicotine (66; Prabhakar and Kumar, unpublished observations). These observations suggest that SP causes sensory excitation perhaps by a dual mechanism; one involves receptor activation and the other via acting on mitochondria. Thus, SP-like peptide(s) may participate in the sensory transmission during hypoxia by mechanisms not strictly adherent to the criteria proposed for conventional synaptic transmission.

Endothelin-1 (ET-1)

Under normoxic conditions ET-1 expression is either low or undetectable in type I cells. However, chronic sustained hypoxia markedly increases ET-1 expression (23). Exogenous administration of ET-1 stimulates carotid body activity by acting on ET-A receptors (5). While ET-1 has no effect on basal $[Ca^{2+}]_i$ in glomus cells, hypoxia-evoked increases in $[Ca^{2+}]_i$ are augmented. Likewise, ET-1 also potentiated hypoxia-induced elevation in cAMP in the carotid body. Based on these studies, it has been suggested that ET, by promoting the phosphorylation of Ca^{2+} channel proteins, leads to potentiation of the sensory response to hypoxia. These studies suggest that ET is a potent modulator of the sensory response to hypoxia. Recent observations also indicate that ET-1 plays an important role in the sensory adaptation of the carotid body during chronic sustained hypoxia (see below).

Enkephalins (ENK)

Immunocytochemical analysis revealed that nearly 98% of glomus cells express ENK-like immunoreactivity (10,91). Hanson et al. (22) reported a decrease in met-ENK content following 30 min of hypoxia in rabbit carotid bodies indicating release of ENK by hypoxia. The following observations are consistent with the notion that ENK exerts an inhibitory influence on carotid body activity. First, exogenous ENK inhibits the carotid body activity. Second, both naloxone and delta-opiate receptor antagonist augment the baseline carotid body activity as well as the sensory response to hypoxia (63). On the other hand, delta-opiate receptor antagonist reduced the sensory response to hypercapnia (32).

Other Peptides

In addition to the peptides mentioned above, type I cells also express atrial natriuretic peptide (ANP) (90). Wang et al. (90) showed that ANP inhibits the carotid body activity and its actions are mediated by cGMP. Galanin, neuropeptide Y (NPY), and calcitonin-gene-related peptide (CGRP)-like immunoreactivities are also reported in the nerve fibers innervating the glomus tissue (10,25,66). However, the significance of these peptides in sensory transmission is not known.

ATP, Adenosine, and Purinergic Receptors

The fact that hypoxia influences mitochondrial function and that ATP is the product of oxidative metabolism prompted the idea that it might participate in the sensory response to hypoxia. In recent years, it has also become increasingly evident that ATP is indeed stored along with other transmitters in secretory vesicles and participates as a cotransmitter by acting on purinergic receptors. The effects of ATP on the sensory activity were initially investigated in anesthetized cats (48) and the reported effects include prompt excitation followed by inhibition of the sensory activity. Studies by McQueen and Ribeiro (48) suggest that the excitatory actions of ATP result from its hydrolysis product adenosine. On the other hand, ATP by itself seems to inhibit the sensory activity. Later studies have shown that adenosine indeed stimulates carotid (49) as well as aortic chemoreceptors (81,82). Two groups of investigators examined the expression of adenosine receptors in the rat carotid body. Type I cells express A2a adenosine receptors (16,38), whereas A1 adenosine receptor expression seems confined to neurons of the petrosal ganglion (16). The following observations are consistent with the idea that A2a receptor might function as an autoreceptor in the glomus cells. Selective A2a receptor antagonist prevents adenosine-evoked inhibition of voltage-gated Ca^{2+} currents and modulatory effects of adenosine on hypoxia-evoked elevation of $[Ca^{2+}]_i$ in glomus cells (38). The functional significance of A1 receptors, however, is unclear.

With regard to the purinergic receptors that mediate the actions of ATP, recent studies by Prasad and co-workers, using both immunocytochemical and molecular biological approaches, provide convincing evidence for P2X2-P2X3 heteromultimeric receptor expression in glomus cells as well as petrosal neurons that make synaptic contact with glomus cells (79). Purinergic receptor blockade did not affect the sensory response to hypoxia in in vivo cat carotid bodies (48,49). However, suramin, a nonselective purinergic blocker given in combination with hexamethonium or mecamylamine (nicotinic receptor antagonist), prevented the sensory response to hypoxia and hypercapnia in cocultures of glomus cell and petrosal neuronal preparations in vitro (79,94). Studies by Lahiri and De Laney (45) have shown that O_2 and CO_2 interact synergistically leading to a greater sensory response than either stimulus alone. Studies by Prasad et al. (79) suggest that purinergic receptors might contribute to O_2-CO_2 interaction at the rat carotid body. Thus, ATP and perhaps its metabolites modulate the sensory transmission during hypoxia and hypercapnia in cooperation with other transmitters rather than acting as transmitters themselves. However, direct release of ATP and/or its metabolites from the carotid body by hypoxia remains to be demonstrated. Thus, both ATP and adenosine seem to function as potent modulators of sensory transmission during hypoxia.

Amino Acid Transmitters

A variety of amino acids function as transmitters in the nervous system. For instance, glutamate functions as an excitatory, whereas glycine and gamma amino butyric acid (GABA) serve as inhibitory transmitters. Carotid bodies express both glutamate and GABA (56,86). Glutamate-like immunoreactivity is associated with type I cells and to a lesser extent with type II cells or afferent axon (86). Many glomus cells express GABA-like immunoreactivity in the mouse carotid body (56). Whether the amino acid transmitters are released in response to hypoxia and, if so, participate in the sensory transmission remain to be investigated.

B. Unconventional Neurotransmitters

It is becoming increasingly recognized that gas molecules such as NO and CO function as chemical messengers in various physiological systems. Reviewed below are the studies that examined the function of NO and CO in sensory transmission at the carotid body.

Nitric Oxide (NO)

Of the three NO synthase (NOS) isoforms, NOS-1 and NOS-3 are constitutively expressed in nerve fibers of sensory and autonomic origin and vascular endothelial cells of the carotid body, respectively. There is no evidence for expression of NOS isoforms in type I cells. NOS contains heme and requires molecular oxygen for NO synthesis (67). When carotid body extracts were exposed to hypoxia, NOS activity was markedly inhibited, suggesting that hypoxia might decrease NO level in the carotid body (67,76). Recently, two groups of investigators examined the effects of hypoxia on NO production in the carotid bodies using amperometric techniques. Iturriaga et al. (26) reported no alterations in NO levels during hypoxia in an ex vivo cat carotid body preparation. On the other hand, using a similar preparation, Fung et al. (14) observed a dramatic increase in NO levels during hypoxia. The increases in NO levels during hypoxia reported by Fung et al. (14) are difficult to reconcile with the biochemical studies showing inhibition of NOS activity by hypoxia. It is likely that NO measurements in this study might have been confounded by variables such as flow, temperature, and pH as well as by catecholamines undergoing oxidation at the potentials used for NO measurements. Therefore, whether hypoxia increases or decreases NO levels in the carotid body remains unclear.

Endogenous NO exerts a tonic inhibitory influence on carotid body activity. Evidence includes: (1) NOS inhibitors augment baseline activity and potentiate the sensory response to hypoxia, and (2) NO donors inhibit the sensory activity (see Ref. 67 for references). Furthermore, targeted deletion of the NOS-1 isoform leads to an enhanced peripheral chemoreceptor sensitivity to hypoxia (34) and this is further confirmed by direct measurements of carotid body activity (35). On the other hand, hypoxic sensitivity of the carotid body seems to be depressed in NOS-3-deficient mice (36). It has been suggested that chronic vasoconstriction in NOS-3-deficient mice may render the carotid body insensitive to hypoxia. Reported cellular mechanisms associated with the actions of NO include: (1) elevation of cGMP, (2) nitrosylation of Ca^{2+} channel proteins leading to altered Ca^{2+} homeostasis, and (3) influence of NO on mitochondrial function (67,85). In addition, NO has been shown to modulate the actions of other neurotransmitters elsewhere in the nervous system (67) and recent studies suggest that NO inhibits hypoxia-evoked SP release from the carotid bodies (30).

Carbon Monoxide (CO)

Heme oxygenase 1 and 2 (HO-1 and HO-2) catalyze the formation of CO and molecular oxygen is required for CO synthesis (67). HO-2 is constitutively expressed in type I cells and inhibition of HO-2 by zinc protoporphyrin-9 augments the sensory activity of the carotid body (67). These observations indicate that type I cells of the carotid bodies are capable of producing CO and that CO is inhibitory to the carotid body activity. Consistent with this notion is the findings of Lahiri and Acker (44), who reported that low doses of CO inhibit chemosensory activity of the isolated rat carotid bodies. Future studies may provide insight as to the role of CO in the carotid body function.

III. Interactions Among the Transmitters; Push-Pull Mechanism

A. Colocalization and Corelease; Evidence for Auto- and/or Paracrine Regulation of Transmitter Release

Neurotransmitters are often colocalized in the carotid body. Examples include colocalization of SP/TH, SP/ENK, SP/NKA, SP/galanin, SP/5-HT, galanin/CGRP, TH/NOS-1, TH/DBH, and TH/HO-2 (77,91). The transmitters are not only colocalized, but they are also coreleased during hypoxia. For example, hypoxia evokes simultaneous release of DA and NE (18) and DA and SP (Kim et al., unpublished observations) from rabbit carotid bodies. Pharmacological studies also suggest that ACh and ATP are coreleased during hypoxia (94). In view of these studies, the long-held notion that sensory transmission during hypoxia involves release of a single neurotransmitter is no longer valid. To make the matter even more complicated, there are studies suggesting that once a transmitter is released by hypoxia it may modulate the release of others by acting in an autocrine or paracrine fashion. For instance, cholinergic receptor agonists potentiate hypoxia-evoked DA release (1,87) whereas NO inhibits hypoxia-evoked SP release (30). It follows that the magnitude of release of a transmitter, for a given hypoxic challenge, is regulated via autocrine and/or paracrine mechanisms. The fact that glomus cells express a variety of transmitter receptors, perhaps even more than the afferent nerve ending, might be consistent with such an idea. In fact, the interaction of the transmitters at the level of the glomus cells may be more critical to the sensory transmission than hitherto appreciated.

How do inhibitory transmitters contribute to sensory excitation by hypoxia? Hypoxia ubiquitously augments the sensory discharge of the carotid bodies irrespective of the species studied. Therefore, much attention has been given to identify the excitatory transmitter mediating the sensory response. However, when one examines the transmitter profile in the carotid body (see Table 1), it can be seen that glomus tissue expresses a large number of inhibitory transmitters along with excitatory ones. It is noteworthy that the inhibitory transmitters are often colocalized and coreleased with excitatory messengers during hypoxia. A notable example is the release of dopamine by hypoxia from the carotid body in every mammalian species examined thus far. What could be the significance of corelease of inhibitory messengers in the sensory transmission? It has been well established that the increase in sensory discharge evoked by hypoxia is maintained more or less during the entire duration of the stimulus. Thus, the sensory response of the carotid body to hypoxia is analogous to the behavior of a slowly adapting sensory receptor. If excitatory transmitter alone participates in the sensory transmission, then one would expect only a brief excitation followed by prompt return to baseline discharge, despite maintaining the stimulus. In other words, sensory excitation will no longer be maintained during the entire period of hypoxia. On the other hand, if the inhibitory messengers are coreleased then they aid in producing sustained excitation by preventing the overexcitation caused by excitatory transmitters. Thus, excitatory and inhibitory messengers act in concert like a "push-pull" mechanism. Indeed, many biological processes are regulated by such a "push-pull" mechanism involving interactions between excitatory and inhibitory messenger molecules. Therefore, it is conceivable that initiation and maintenance of the carotid body sensory response to hypoxia depends on complex interplay between excitatory and inhibitory neurochemicals, as suggested earlier (65).

IV. Adaptation of the Carotid Body During Chronic Hypoxia: Role of Neurochemicals

It is being increasingly appreciated that carotid bodies play an important role in maintaining homeostasis during chronic hypoxia (2,69). Perhaps the reflexes triggered by carotid bodies may be more important for survival during chronic hypoxic situations than they are during acute hypoxia. Two patterns of chronic hypoxia are encountered in a variety of physiological and pathophysiological situations. For example, chronic sustained hypoxia is associated with high altitude sojourn, whereas an intermittent pattern of hypoxia is seen in recurrent apneas associated with obstructive sleep apnea syndrome as well as central hypoventilation. The following section summarizes how various neurochemicals contribute to the altered carotid body function during both types of chronic hypoxia, i.e., sustained and intermittent.

A. Chronic Sustained Hypoxia

Chronic sustained hypoxia leads to hypertrophy and hyperplasia of type I and II cells and angiogenesis in experimental animals as well as in humans (3,10,27,69). These morphological alterations are reversible. Recent studies have shown that the carotid body expresses a variety of growth factors including VEGF and PD-ECGF (3,27). While VEGF expression is seen in type I cells (3,27), PD-ECGF expression is confined to extracellular stroma (27). The expression of growth factors is seen only after chronic sustained hypoxia or in human carotid body tumors. In addition, carotid bodies also express Fit-1 and Flk-1 (3,27), the receptors associated with VEGF actions. It is likely that VEGF, acting on the Flk-1, may regulate hyperplasia of type I cells, whereas it promotes neovascularization acting on Fit-1. On the other hand, PD-ECGF might contribute more to angiogenesis, as it does elsewhere in the body including the lung (27). Peptides like ET-1, in addition to their function as transmitter, also act as potent mitogens (57). Since ET-1 is up-regulated after sustained hypoxia (23), it may also contribute to hyperplasia. The role of ACh, DA, or NO, if any, in the remodeling of the carotid body during chronic sustained hypoxia remains to be explored.

The role(s) of various neurotransmitters contributing to the augmented hypoxic sensitivity during chronic sustained hypoxia has been recently reviewed (2). Much attention has been focused on the idea that chronic hypoxia might alter the balance between excitatory and inhibitory transmitters leading to an augmented sensory response. Consequently, it was thought that sustained hypoxia up-regulates excitatory and down-regulates inhibitory transmitters. However, following chronic sustained hypoxia, inhibitory transmitters such as DA (21) and NO (67,78) are up-regulated, whereas SP-like excitatory chemicals are down-regulated (89), in the carotid bodies. Thus, these studies seem to suggest that chronic hypoxia does not necessarily lead to coordinated regulation of "so-called" excitatory and inhibitory chemical messengers in the carotid body.

It is likely additional transmitter(s), which are normally absent but are recruited during chronic hypoxia, might contribute to the augmented sensory response of the carotid body. Such a possibility seems to be supported by recent studies by Fidone and co-workers (5,23), who reported upregulation of ET-1 and ET-1 receptors in the carotid body during chronic sustained hypoxia. More importantly, these investigators have shown that blockade of ET-1 receptors prevented the enhanced hypoxic sensory response in chronic hypoxic animals without affecting their basal sensory response to acute hypoxia. While these are important observations, further studies are needed to identify other transmitters that may also contribute to the altered sensory response during sustained hypoxia.

Chronic hypoxia up-regulates certain genes in the carotid body, including TH and ET-1 (see Ref. 2 for references). This upregulation of gene expression involves activation of specific transcription factors. For example, hypoxia-inducible factor-1 (HIF-1) mediates activation of more than three dozen downstream genes, including ET-1 (37). Recently, Kline et al. (37) reported markedly impaired carotid body sensory response to hypoxia and absence of ventilatory adaptation to chronic sustained hypoxia in mice partially deficient in HIF-1α, the regulatory subunit of HIF-1 complex. Further studies are needed to determine whether the altered carotid body function in HIF-1 mutant mice is due to lack of ET-1 and/or other downstream genes.

B. Chronic Intermittent Hypoxia

Chronic episodic hypoxia is associated with recurrent apneas and often leads to disturbances in the cardiorespiratory system causing morbidity and mortality (69). Moreover, studies have implicated the carotid bodies in these disturbances. Glomectomized patients with recurrent apneas no longer develop high blood pressure. Also, chronic ablation of sinus nerves in experimental animals prevents elevations in arterial blood pressures and sympathetic nerve activity caused by chronic intermittent hypoxia (see Ref. 69 for references). These studies highlight the importance of the carotid bodies in pathophysiological changes associated with recurrent apneas. Recently, Peng et al. (61) examined the effects of chronic intermittent hypoxia on carotid body activity in rats. These authors found an enhanced sensory response to hypoxia and persistent elevation of baseline activity after acute repetitive hypoxic episodes (long-term facilitation, LTF) in intermittent hypoxia-conditioned rats. However, these alterations in carotid body function caused by episodic hypoxia were not associated with either hyperplasia or hypertrophy of the carotid bodies. The effects of intermittent hypoxia on the carotid body seem to be due to enhanced generation of reactive oxygen species, such as $O_2^{\cdot -}$, because an $O_2^{\cdot -}$ scavenger prevented the enhanced hypoxic sensitivity as well as LTF. Further, there was a marked reduction in carotid body aconitase activity supporting enhanced $O_2^{\cdot -}$ generation. Methysergide and/or ketanserin, blockers of 5-HT receptors, prevented the LTF but not the enhanced sensory response to hypoxia. Thus, although both ROS and 5-HT may play a minor role(s) in the sensory transmission under normal conditions, they seem to play vital roles in reshaping the carotid body function during chronic intermittent hypoxia.

Acknowledgments

This work is supported by grants from the National Institutes of Health, Heart, Lung and Blood Institute HL-25830, HL-46462, and HL-66448.

References

1. Bairam A, Neji H, Kinkead R, Marchal F. Carbachol effect on carotid body dopamine in vitro release in response to hypoxia in adult and pup rabbit. Neurosci Res 2001; 40:183–188.
2. Bisgard GE. Carotid body mechanisms in acclimatization to hypoxia. Res Physiol 2000; 121:237–246.
3. Chen J, Dinger B, Jyung R, Stensaas L, Fidone S. Upregulation of vascular endothelial growth factor (VEGF) and VEGF receptor (Flk-1) in chronically hypoxic rat carotid body, Abstract presented at the national meeting of the Society for Neuroscience, New Orleans, Louisiana, October 25–30, 1997.
4. Chen IL, Yates RD, Hansen JT. Substance P-like immunoreactivity in rat and cat carotid bodies: light and electron microscopic studies. Histol Histopathol 1986; 1:203–212.
5. Chen J, He L, Dinger B, Fidone S. Cellular mechanisms involved in rabbit carotid body excitation elicited by endothelin peptides. Respir Physiol 2001; 121:13–23.

6. Chleide E, Ishikawa K. Hypoxia-induced decrease of brain acetylcholine release detected by microdialysis. Neuroreport 1990; 1:197–199.
7. Cragg PA, Runold M, Kou YR, Prabhakar NR. Tachykinin antagonists in carotid body responses to hypoxia and substance P in the rat. Respir Physiol 1994; 95:295–310.
8. Dinger B, Gonzalez C, Yoshizaki K, Fidone S. Localization and function of cat carotid body nicotinic receptors. Brain Res 1985; 339:295–304.
9. Dinger BG, Hirano T, Fidone SJ. Autoradiographic localization of muscarinic receptors in rabbit carotid body. Brain Res 1986; 367:328–331.
10. Fidone S, Gonzalez C. Initiation and control of chemoreceptor activity in the carotid body. In: Cherniack NS, Widdicombe JG, eds. Handbook of Physiology. Section 3: The Respiratory System. Vol. II. Bethesda: American Physiological Society, 1986: 267–312.
11. Fidone S, Gonzalez C, Yoshizaki K. Effects of low oxygen on the release of dopamine from the rabbit carotid body in vitro. J Physiol (Lond) 1982; 333:93–110.
12. Fitzgerald RS. Oxygen and carotid body chemotransduction: the cholinergic hypothesis—a brief history and new evaluation. Respir Physiol 2000; 120:89–104.
13. Fitzgerald RS, Shirahata M, Wang HY. Acetylcholine release from cat carotid bodies. Brain Res 1999; 841:53–61.
14. Fung ML, Ye JS, Fung PC. Acute hypoxia elevates nitric oxide generation in rat carotid body in vitro. Pflügers Arch 2001; 442:903–909.
15. Gauda EB, Bamford OS, Northington FJ. Lack of induction of substance P gene expression by hypoxia and absence of neurokinin 1-receptor mRNA in the rat carotid body. J Auton Nerv Syst 1998; 74:100–108.
16. Gauda EB, Northington FJ, Linden J, Rosin DL. Differential expression of a(2a), A(1)-adenosine and D(2)-dopamine receptor genes in rat peripheral arterial chemoreceptors during postnatal development. Brain Res 2000; 872:1–10.
17. Gibson GE, Peterson C. Decreases in the release of acetylcholine in vitro with low oxygen. Biochem Pharmacol 1982; 31:111–115.
18. Gomez-Nino A, Dinger B, Gonzalez C, Fidone S. Differential stimulus coupling to dopamine and norepinephrine stores in rabbit carotid body type I cells. Brain Res 1990; 525:160–164.
19. Gonzalez-Guerrero PR, Rigual R, Gonzalez C. Opioid peptides in the rabbit carotid body: identification and evidence for co-utilization and interactions with dopamine. J Neurochem 1993; 60:1762–1768.
20. Gonzalez C, Almaraz L, Obeso A, Rigual R. Carotid body chemoreceptors: from natural stimuli to sensory discharges. Physiol Rev 1994; 74:829–898.
21. Gonzalez-Guerrero PR, Rigual R, Gonzalez C. Effects of chronic hypoxia on opioid peptide and catecholamine levels and on the release of dopamine in the rabbit carotid body. J Neurochem 1993; 60:1769–1776.
22. Hanson G, Jones L, Fidone S. Physiological chemoreceptor stimulation decreases enkephalin and substance P in the carotid body. Peptides 1986; 7:767–769.
23. He L, Chen J, Dinger B, Stensaas L, Fidone S. Endothelin modulates chemoreceptor cell function in mammalian carotid body. Adv Exp Med Biol 1996; 410:305–311.
24. Hirano T, Dinger B, Yoshizaki K, Gonzalez C, Fidone S. Nicotinic versus muscarinic binding sites in cat and rabbit carotid body. Biol Signals 1992; 1:143–149.
25. Ichikawa H, Helke CJ. Distribution, origin and plasticity of galanin immunoreactivity in the rat carotid body. Neuroscience 1993; 52:757–767.
26. Iturriaga R, Villanueva S, Mosqueira M. Dual effects of nitric oxide on cat carotid body chemoreception. J Appl Physiol 2000; 89:1005–1012.

27. Jyung RW, LeClair EE, Bernat RA, Kang TS, Ung F, McKenna MJ, Tuan RS. Expression of angiogenic growth factors in paragangliomas. Laryngoscope 2000; 110:161–167.
28. Kim D, Oh EK, Summers BA, Prabhakar NR, Kumar GK. Release of substance P by low oxygen in the rabbit carotid body: evidence for the involvement of calcium channels. Brain Res 2001; 892:359–369.
29. Kim D, Summers BA, Prabhakar NR, Kumar GK. Neurotransmitter release from the rabbit carotid body: differential effects of hypoxia on substance P and acetylcholine release. Adv Exp Med Biol 2001; 499:39–43.
30. Kim D, Summers BA, Prabhakar NR, Kumar GK. Nitric oxide modulates substance P release in the carotid body. FASEB J 2001; 1:A150.
31. Kim D, Summers BA, Prabhakar NR, Kumar GK. Inhibition of acetylcholine release by hypoxia in the carotid body. FASEB J 2001; 15:A150.
32. Kirby GC, McQueen DS. Characterization of opioid receptors in the cat carotid body involved in chemosensory depression in vivo. Br J Pharmacol 1986; 88:889–898.
33. Kirby GC, McQueen DS. Effects of the antagonist MDL 72222 and ketanserin on responses of cat carotid body chemoreceptors to 5-hydroxytryptamine. Br J Pharmacol 1984; 83:259–269.
34. Kline DD, Yang T, Huang PL, Prabhakar NR. Altered respiratory responses to hypoxia in mutant mice deficient in neuronal nitric oxide synthase. J Physiol 1998; 511:273–287.
35. Kline DD, Prabhakar NR. Peripheral chemosensitivity in mutant mice deficient in nitric oxide synthase. Adv Exp Med Biol 2000; 475:571–579.
36. Kline DD, Yang T, Premkumar DR, Thomas AJ, Prabhakar NR. Blunted respiratory responses to hypoxia in mutant mice deficient in nitric oxide synthase-3. J Appl Physiol 2000; 88:1496–1508.
37. Kline DD, Peng YJ, Manalo DJ, Semenza GL, Prabhakar NR. Defective carotid body function and impaired ventilatory responses to chronic hypoxia in mice partially deficient for hypoxia-inducible factor 1alpha. Proc Natl Acad Sci USA 2002; 99:821–826.
38. Kobayashi S, Conforti L, Millhorn DE. Gene expression and function of adenosine A(2A) receptor in the rat carotid body. Am J Physiol Lung Cell Mol Physiol 2000; 279:L273–282.
39. Kou YR, Ernsberger P, Cragg PA, Cherniack NS, Prabhakar NR. Role of α_2-adrenergic receptors in the carotid body response to isocapnic hypoxia. Respir Physiol 1991; 83:353–364.
40. Kumar GK. Neuropeptide processing enzymes of the carotid body: biochemical and immunological characterization of carboxypeptidase activity. Adv Exp Med Biol 1996; 410:319–323.
41. Kumar GK. Peptidases of the peripheral chemoreceptors: biochemical, immunological, in vitro hydrolytic studies and electron microscopic analysis of neutral endopeptidase-like activity of the carotid body. Brain Res 1997; 748:39–50.
42. Kumar GK, Kou YR, Overholt JL, Prabhakar NR. Involvement of substance P in neutral endopeptidase modulation of carotid body sensory responses to hypoxia. J Appl Physiol 2000; 88:195–202.
43. Kusakabe T, Kawakami T, Tanabe Y, Fujii S, Takenaka T. Distribution of substance P-containing and catecholaminergic nerve fibers in the rabbit carotid body: an immunohistochemical study in combination with catecholamine fluorescent histochemistry. Arch Histol Cytol 1994; 57:193–199.

44. Lahiri S, Acker H. Redox-dependent binding of CO to heme protein controls PO_2-sensitive chemoreceptor discharge of the rat carotid body. Respir Physiol 1999; 115:169–177.
45. Lahiri S, DeLaney RG. Stimulus interaction in the response of carotid body chemoreceptor single afferent fibers. Respir Physiol 1975; 24:249–266.
46. McQueen DS. Effects of substance P on carotid chemoreceptor activity in the cat. J Physiol (Lond) 1980; 302:31–47.
47. McQueen DS, Ribeiro JA. Inhibitory actions of methionine-enkephalin and morphine on the cat carotid chemoreceptors. Br J Pharmacol 1980; 71:297–305.
48. McQueen DS, Ribeiro JA. On the specificity and type of receptor involved in carotid body chemoreceptor activation by adenosine in the cat. Br J Pharmacol 1983; 80:347–354.
49. McQueen DS, Ribeiro JA. Pharmacological characterization of the receptor involved in chemoexcitation induced by adenosine. Br J Pharmacol 1986; 88:615–620.
50. Milsom WK, Sadig T. Interaction between norepinephrine and hypoxia on carotid body chemoreception in rabbits. J Appl Physiol 1983; 55:1893–1898.
50a. Mokashi A, Rozanov C, Roy A, Lahiri S. ATP induces $Ca^{2+}]_I$ rise in rat carotid body cultured glomus cells. FASEB J 2001; 15:A154.
51. Monti-Bloch L, Eyzaguirre C. Effects of methionine and substance P on the chemosensory discharge of the cat carotid body. Brain Res 1985; 338:297–307.
52. Nishi K. The action of 5-hydroxytryptamine on chemoreceptor discharges of the cat's carotid body. Br J Pharmacol 1975; 55:27–40.
53. Nurse CA, Zhang M. Acetylcholine contributes to hypoxic chemotransmission in co-cultures of rat type I cells and petrosal neurons. Respir Physiol 1999; 115:189–199.
54. Obeso A, Rocher A, Fidone S, Gonzalez C. The role of dihydropyridine-sensitive Ca^{2+} channels in stimulus-evoked catecholamine release from chemoreceptor cells of the carotid body. Neuroscience 1992; 47:463–472.
55. Obeso A, Gomez-Nino MA, Almaraz L, Dinger B, Fidone S, Gonzalez C. Evidence for two types of nicotinic receptors in the cat carotid body chemoreceptor cells. Brain Res 1997; 754:298–302.
56. Oomori Y, Nakaya K, Tanaka H, Iuchi H, Ishikawa K, Satoh Y, Ono K. Immunocytochemical and histochemical evidence for the presence of noradrenaline, serotonin and gamma-aminobutyric acid in chief cells of the mouse carotid body. Cell Tissue Res 1994; 278:249–254.
57. Overholt JL, Prabhakar NR. Norepinephrine inhibits a toxin resistant Ca^{2+} current in carotid body glomus cells: evidence for a direct G protein mechanism. J Neurophysiol 1999; 81:225–233.
58. Paciga M, Vollmer C, Nurse C. Role of ET-1 in hypoxia-induced mitosis of cultured rat carotid body chemoreceptors. Neuroreport 1999; 10:3739–3744.
59. Pardal H, Ludewig U, Garcia-Hirschfeld J, Lopez-Barneo J. Secretory responses of intact glomus cells in thin slices of rat carotid body to hypoxia and tetraethylammonium. Proc Natl Acad Sci USA 2000; 97:2361–2366.
60. Pearse AGE. The cytochemistry and ultrastructure of polypeptide hormone-producing cells of the APUD series and the embryologic, physiologic and pathologic implications of the concept. J Histochem Cytochem 1969; 17:303–313.
61. Peng Y, Kline DD, Hamadani A, Prabhakar NR. Induction of long term facilitation in the carotid body activity by intermittent hypoxia. FASEB J 2001; 15:A153.

62. Pizarro J, Ryan ML, Hedrick MS, Xue DH, Keith IM, Bisgard GE. Intracarotid substance P infusion inhibits ventilation in the goat. Respir Physiol 1995; 101:11–22.
63. Pokorski M, Lahiri S. Effects of naloxone on carotid body chemoreception and ventilation in the cat. J Appl Physiol 1981; 51:1533–1538.
64. Ponte J, Sadler CL. Interactions between hypoxia, acetylcholine and dopamine in the carotid body of rabbit and cat. J Physiol (Lond) 1989; 410:595–610.
65. Prabhakar NR. Significance of excitatory and inhibitory neurochemicals in hypoxic chemotransmission of the carotid body. In: Control of Breathing and Its Modeling Perspective. Honda Y, Miyamoto Y, Konno K, Widdicombe JG, eds. New York: Plenum Press, 1992:205–212.
66. Prabhakar NR. Neurotransmitters in the carotid body. Adv Exp Med Biol 1994; 360:57–69.
67. Prabhakar NR. NO and CO as second messengers in oxygen sensing in the carotid body. Respir Physiol 1999; 115:161–168.
68. Prabhakar NR. Oxygen sensing by the carotid body chemoreceptors. J Appl Physiol 2000; 88:2287–2295.
69. Prabhakar NR. Oxygen sensing during intermittent hypoxia: cellular and molecular mechanisms. J Appl Physiol 2001; 90:1986–1994.
70. Prabhakar NR, Runold M, Yamamoto Y, Lagercrantz H, von Euler C. Effect of substance P antagonist on the hypoxia-induced carotid chemoreceptor activity. Acta Physiol Scand 1984; 121:301–303.
71. Prabhakar NR, Mitra J, Cherniack NS. Role of substance P in hypercapnic excitation of carotid chemoreceptors. J Appl Physiol 1987; 63:2418–2425.
72. Prabhakar NR, Runold M, Yamamoto Y, Lagercrantz H, Cherniack NS, von Euler C. Role of the vagal afferents in substance P-induced respiratory responses in anaesthetized rabbits. Acta Physiol Scand 1987; 131:63–71.
73. Prabhakar NR, Landis SC, Kumar GK, Mullikin-Kilpatrick D, Cherniack NS, Leeman S. Substance P and neurokinin A in the cat carotid body: localization, exogenous effects and changes in content in response to arterial PO_2. Brain Res 1989; 481:205–214.
74. Prabhakar NR, Runold M, Kumar GK, Cherniack NS, Scarpa A. Substance P and mitochondrial oxygen consumption: evidence for a direct intracellular role for the peptide. Peptides 1989; 10:1003–1006.
75. Prabhakar NR, Cao H, Lowe JAD, Snider RM. Selective inhibition of the carotid body sensory response to hypoxia by the substance P receptor antagonist CP-96,345. Proc Natl Acad Sci USA 1993; 90:10041–10045.
76. Prabhakar NR, Kumar GK, Chang CH, Agani FH, Haxhiu MA. Nitric oxide in the sensory function of the carotid body. Brain Res 1993; 625:16–22.
77. Prabhakar NR, Dinerman JL, Agani FH, Snyder SH. Carbon monoxide: a role in carotid body chemoreception. Proc Natl Acad Sci USA 1995; 92:1994–1997.
78. Prabhakar NR, Pieramici SF, Premkumar DR, Kumar GK, Kalaria RN. Activation of nitric oxide synthase gene expression by hypoxia in central and peripheral neurons. Brain Res Mol Brain Res 1996; 43:341–346.
79. Prasad M, Fearon IM, Zhang M, Laing M, Vollmer C, Nurse CA. Expression of P2X2 and P2X3 receptor subunits in rat carotid body afferent neurones: role in chemosensory signaling. J Physiol 2000; 537(3):667–677.
80. Rigual R, Lopez-Lopez JR, Gonzalez C. Release of dopamine and chemoreceptor discharge induced by low pH and high PCO_2 stimulation of the cat carotid body. J Physiol 1991; 433:519–531.

81. Runold M, Cherniack NS, Prabhakar NR. Effect of adenosine on chemosensory activity of the cat carotid body. Respir Physiol 1990; 80:299–306.
82. Runold M, Cherniack NS, Prabhakar NR. Effect of adenosine on isolated and superfused cat carotid body activity. Neurosci Lett 1990; 113:111–114.
83. Schweitzer A, Wright S. Action of prostigmin and acetylcholine on respiration. Q J Exp Physiol 1938; 28:33–47.
84. Shirahata M, Ishizawa Y, Rudisill M, Schofield B, Fitzgerald RS. Presence of nicotinic acetylcholine receptors in cat carotid body afferent system. Brain Res 1998; 814:213–217.
85. Summers BA, Overholt JL, Prabhakar NR. Nitric oxide inhibits L-type Ca^{2+} current in glomus cells of the rabbit carotid body via a cGMP-independent mechanism. J Neurophysiol 1999; 81:1449–1457.
86. Torrealba F, Bustos G, Montero VM. Glutamate in the glomus cells of the cat carotid body: immunocytochemistry and in vitro release. Neurochem Int 1996; 28:625–631.
87. Vicario I, Rigual R, Obeso A, Gonzalez C. Characterization of the synthesis and release of catecholamine in the rat carotid body in vitro. Am J Physiol Cell Physiol 2000; 278:C490–C499.
88. Wang ZZ, Bredt DS, Fidone SJ, Stensaas LJ. Neurons synthesizing nitric oxide innervate the mammalian carotid body. J Comp Neurol 1993; 336:419–432.
89. Wang ZZ, Dinger B, Fidone SJ, Stensaas LJ. Changes in tyrosine hydroxylase and substance P immunoreactivity in the cat carotid body following chronic hypoxia and denervation. Neuroscience 1998; 83:1273–1281.
90. Wang ZZ, He L, Stensaas LJ, Dinger BG, Fidone SJ. Localization and in vitro actions of atrial natriuretic peptide in the cat carotid body. J Appl Physiol 1991; 70:942–946.
91. Wang ZZ, Stensaas LJ, Dinger B, Fidone SJ. The co-existence of biogenic amines and neuropeptides in the type I cells of the cat carotid body. Neuroscience 1992; 47:473–480.
92. Wang ZY, Keith IM, Beckman MJ, Brownfield MS, Vidruk EH, Bisgard GE. 5-HT_{5a} receptors in the carotid body chemoreception pathway of rat. Neurosci Lett 2000; 278:9–12.
93. Zhang M, Nurse CA. Does endogenous 5-HT mediate spontaneous rhythmic activity in chemoreceptor clusters of rat carotid body? Brain Res 2000; 872:199–203.
94. Zhang M, Zhong H, Vollmer C, Nurse CA. Co-release of ATP and ACh mediates hypoxic signaling at rat carotid body chemoreceptors. J Physiol 2000; 525(1):143–158.

25

Mechanisms of Morphological and Functional Plasticity in the Chronically Hypoxic Carotid Body

BRUCE DINGER, LIANG HE, JIA CHEN, L. J. STENSAAS, and SALVATORE J. FIDONE

University of Utah School of Medicine
Salt Lake City, Utah, U.S.A.

I. Introduction

Chronic exposure in a low-O_2 environment (i.e., chronic hypoxia, CH) initiates a multitude of physiological adjustments that tend to mitigate the adverse effects of hypoxia. These homeostatic mechanisms involve multiple sensory and hormonal factors acting locally and systemically to increase the capacity and efficiency of O_2 delivery and utilization. Among the most important of the changes is increased ventilation, evident as elevated tidal volume and/or frequency of breathing. During acute hypoxia, breathing increases to a new steady-state level. However, if hypoxia continues for hours or days, a further progressive increase in ventilation occurs, until a second steady-state level is established (1). Such ventilatory acclimatization to hypoxia (VAH), develops in rats and humans over a 4–5-day period, and the elevated breathing persists for at least 14 days (2). Recent data indicate that VAH involves adjustments both in the central processing of sensory input (3,4) and the generation of chemoreceptor activity in the carotid body. Indeed, quantitative studies of carotid sinus nerve (CSN) activity have demonstrated increased carotid body hypoxic chemosensitivity following prolonged exposure to hypoxia (5–8). Importantly, the time course of these changes in carotid body function parallels the development of VAH (9).

The chemosensory apparatus comprising the carotid body consists of three highly integrated functional elements. First, an extensive highly permeable microvascular network of fenestrated sinusoidal capillaries and small veins constitutes a remarkable vascular bed that supports the highest per weight blood flow and O_2 consumption of any organ (see Ref. 10). Second, a sensory transduction system consists of aggregates or lobules of highly specialized type I cells that release multiple neuroactive and paracrine agents in response to hypoxia, hypercapnia, and acidosis and that are enveloped by glial type II cells. Finally, sensory neurons from the PG give rise to axons that form synaptic terminals on the type I cells and convey chemoreceptor impulse traffic into the nucleus tractus solitarius (NTS) of the medulla. In the following sections we consider the involvement of these diverse tissue components in carotid body adaptation to CH. An exhaustive review is beyond the scope of this chapter; however, the presentation is intended to highlight recent findings that elucidate cellular and molecular mechanisms involved in regulating the adaptive process. Other recent reviews of carotid body function and adaptation are available (1,11,11a).

II. Chronic Hypoxia-Induced Tissue Reshaping and Remodeling in the Carotid Body

The carotid body rapidly and markedly enlarges in response to CH, a change that involves hypertrophy and hyperplasia of type I/type II cells and extensive dilation and/or transformation of capillaries and sinusoidal blood vessels (12–14). Morphometric studies by Pequignot and Helistrom in rat carotid body revealed a fourfold increase in organ size and a 2.3-fold increase in microvascular-capillary volume density (% of total organ volume) following a 2-week period of 10% O_2 breathing (15). The total volume occupied by microvascular endothelial cells was estimated to increase by threefold, although this elegant study left unresolved critical issues of whether such vascular changes involve extensive remodeling induced by CH or are the product of genuine angiogenesis and the formation of new blood vessels. In a recent study (16), we investigated the time course of CH-induced vascular changes in rats exposed to hypobaric hypoxia at 380 torr (equivalent to 5500 m). Morphometric data indicated that after 3, 7, and 14 days of CH the vascular volume density was elevated 1.45-, 1.87-, and 2.72-fold, respectively. However, the remarkable enlargement of the vascular compartment was not accompanied by an increase in the number of vessel profiles in a given tissue section, suggesting that angiogenesis and the production of new capillaries or microvessels is not a primary factor in reshaping the vascular network.

In experiments that evaluated the mechanisms involved in CH-induced tissue remodeling in the carotid body, we examined the effects of suramin, a well-known purinergic P2 receptor antagonist, which is also a nonspecific blocker of growth factor receptors. This drug was of particular interest because it had previously been shown to inhibit CH-induced changes in the vascular bed of the lung (17). In our studies, adult male rats received twice-weekly suramin injections (15 mg/kg, i.p.) or vehicle during 14 days of hypobaric hypoxia (380 torr). Age-matched male rats similarly treated with suramin or vehicle served as normoxic controls. Photomicrographs in Figure 1 show that in the normal carotid body (Fig. 1A), chemosensory lobules innervated by the sensory terminals of chemoreceptor axons

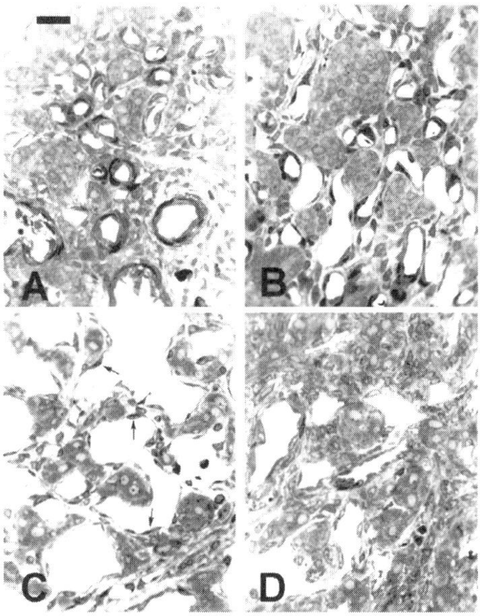

Figure 1 Effect of suramin on CH-induced morphological remodeling in rat carotid body. Adult male rats received twice-weekly suramin injections (15 mg/kg, i.p.) or vehicle during 14 days of hypobaric hypoxia (380 torr). In a vehicle-treated normoxic carotid body (A), chemosensory lobules are comprised of small ($\leq 10\,\mu m$) ovoid type I cells surrounded by type II cells. The chemosensory parenchymal elements are separated and enveloped by connective tissue containing a complex network of small arteries, veins, sinusoidal veins, and capillaries. Similar features were present in normoxic animals treated with suramin (B). Remarkable growth of blood vessels is apparent following 14 days of CH in vehicle-treated animals (C). A substantial increase in the loose connective tissue stroma surrounds these large (up to 100 μm), thin-walled sinusoids. Vessels are lined with darkly staining, closely spaced endothelial cells some of which contain a prominent nucleus and cell soma protruding into the vessel lumen (arrows). Parenchymal cell lobules are noticeably distorted and are drawn out into elongated aggregates, and linear arrays, which contain some flat rather than round type I cells. Individual type I cells are abnormally large (15–20 μm), with a dilated nucleus and prominent nucleolus resembling that of nerve cells. In CH animals treated with suramin (D), changes in the vasculature were much less severe, and compression and distortion of chemoreceptor cell lobules is not evident. However, the incidence of type II cell is increased, and many type I cells are enlarged and displayed a conspicuous nucleolus. A prominent increase of fibroblasts and other mesodermal elements is evident in the connective tissue stroma.

are comprised of small ($\leq 10\,\mu m$) ovoid type I cells with a prominent nucleus, surrounded by darker type II cells. The chemosensory parenchymal elements are separated and enveloped by connective tissue containing a complex network of small arteries, veins, venous sinusoids, and capillaries whose diameter ranges between 4 and 20 μm in normal vehicle-treated animals (18). Similar features were present in normoxic animals treated with suramin (Fig. 1B). Remarkable growth of blood vessels is apparent following 14 days of CH in vehicle-treated animals. This was accompanied by a substantial increase in the loose connective tissue stroma that surrounds these exceedingly large (up to 100 μm), thin-walled sinusoids (Fig. 1C). Many vessels were lined with darkly staining, closely spaced endothelial cells some of which were conspicuous by a prominent nucleus and a cell soma that protruded into the vessel lumen (arrows). Such extensive remodeling and enlargement of the carotid body vasculature involved distortion of parenchymal cell lobules, which appeared to be drawn out into elongated aggregates sometimes containing flat rather than round type I cells. Thus, linear arrays of type I cells were commonly situated in narrow spaces between the largest vessels, a situation in sharp contrast with normal tissue (compare A vs. C in Fig. 1). Additionally, the individual chemosenory type I cells are abnormally large (15–20 μm), with a dilated nucleus lacking chromatin and prominent nucleolus resembling that of nerve cells. In CH animals treated with suramin (Fig. 1D), changes in the vasculature were much less severe, and compression and distortion of chemoreceptor cell lobules was not evident. However, there was a large increase in the incidence of type II cells, and many type I cells were also enlarged and displayed a conspicuous nucleolus. Thus, suramin appeared to have no effect on the hypoxia-induced hypertrophic changes that are apparent in type I cells of vehicle-treated CH animals. In suramin-treated CH tissue, a prominent increase of fibroblasts and other mesodermal elements was also evident in the connective tissue stroma. Collectively, these data suggest that CH induces extensive cellular hyperplasia and tissue remodeling in the carotid body as a consequence of purinergic mechanisms and/or vasculogenic growth factors and their receptors.

The remarkable vascular growth and tissue remodeling observed in the carotid body is similar to changes induced in the heart and lungs by CH. In these O_2-sensitive tissues chronic exposure to low O_2 elicits vascular wall thickening in the pulmonary circuit and hypertrophy of the right ventricle, changes that accompany pulmonary hypertension (19). These maladaptive adjustments have been attributed in part to the actions of vascular endothelial growth factor (VEGF) and VEGF receptors, which are up-regulated in the lung during CH (20). VEGF is a glycoprotein initially isolated from tumor cells and subsequently characterized as a peptide capable of increasing vascular permeability (21). VEGF is known to be a highly specific and potent endothelial cell mitogen and promoter of angiogenesis. The biological actions of VEGF, including vascular proliferation and angiogenesis, are mediated by Flk-1 (fetal liver kinase), a transmembrane tyrosine kinase receptor protein (21,22).

Our assessment of VEGF and Flk-1 receptor transcript levels in the carotid body indicates that, as in the lung, CH elevates the expression of both. In these studies extracted mRNA was converted to cDNA using conventional methods for reverse transcription, and the resulting cDNA was then used in a quantitative PCR analysis. Competition between target (VEGF or Flk-1) cDNA and an internal standard mimic cDNA was comprised of molecules having identical binding sites for primers, but a slightly different base-pair size, which allowed later separation of target and mimic PCR products on agarose gels. Figure 2, which plots data from multiple reactions of serial dilutions of the mimic molecule on the x-axis (concentration), shows a typical quantitative analysis of VEGF transcript expression. In normoxic carotid bodies the concentration of VEGF cDNA in the initial reaction is estimated at the point where the target-to-mimic ratio (y-axis) equals 1. A similar plot is shown for VEGF cDNA from carotid bodies harvested following 14 days of CH. After normalization of each assay to its respective sample protein content, a comparison of the data indicates that the concentration of VEGF mRNA increased more than 15-fold following the 14-day exposure to hypoxia. Table 1 show results from multiple experiments involving carotid bodies harvested following selected intervening periods of CH exposure. They revealed that VEGF expression was similarly elevated at 3 and 7 days of exposure to CH. In the superior cervical ganglion, a nonchemosensitive tissue containing postganglionic sympathetic neurons, exposure to CH for 14 days did not alter the expression of VEGF (transcript ratio: hypoxia/normoxia = 1.09). Similar quantitative assessments of

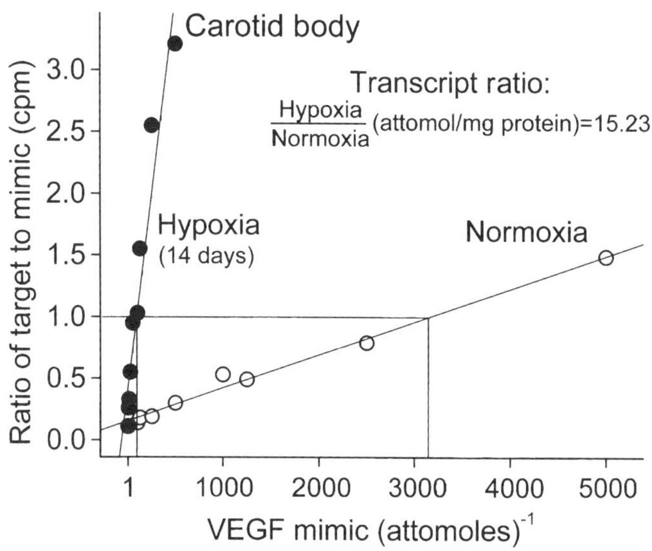

Figure 2 Effect of chronic hypoxia (CH) on vascular endothelial growth factor (VEGF) gene expression in rat carotid body. Data are ratios of target-to-mimic product amounts (y-axis) generated in multiple "competitive" PCR reactions. Open circles (○) = normoxia; closed circles (●) = CH. The ratio of VEGF transcript present in normal versus CH tissue is calculated using data normalized to protein concentration in tissue homogenates. Plots according to least-square best-fit derivations. X-axis is reciprocal of "mimic" molecule concentration.

Table 1 Effect of Chronic Hypoxia on VEGF mRNA Expression in Rat Carotid Body[a]

	Duration of chronic hypoxia (days)		
	3	7	14
Experiment # 1	17.72	8.70	15.23
Experiment # 2	13.48	10.43	9.13

[a] Values are transcript ratios (i.e., CH/normoxia) determined in quantitative RT-PCR assays as described in Figure 2 and text. Experiments 1 and 2 for each duration of CH represent separate groups of normal and hypoxic carotid bodies.

Flk-1 mRNA levels in the carotid body (Table 2) suggest that expression was elevated greater than 10-fold after 3, 7, and 14 days of hypoxic exposure.

Evidence for CH-induced increases in VEGF and Flk-1 was confirmed in immunocytochemical experiments where we observed VEGF expression in normal type I cells and substantially elevated immunostaining intensity after a 14-day exposure to CH. Unlike VEGF, however, immunoreactivity for the Flk-1 receptor was not detectable in the normal carotid body. Prominent Flk-1 immunostaining was present in type I cells after 3 days of CH, and further increases in the receptor protein were evident after 7 and 14 days of hypoxic exposure. VEGF and Flk-1 immunoreactivity were not present in other cell types in the carotid body; surprisingly, however, Flk-1 expression was expressed in a few large blood vessels outside the carotid body, but never in sinusoidal or capillary endothelial cells or other vascular elements within the parenchyma of this chemosensory organ.

The absence of Flk-1 receptor on endothelial cells in normoxic and CH carotid body indicates that VEGF is not directly involved in the extensive vascular reshaping induced by CH. The importance of VEGF for embryonic angiogenesis and organ development is well established (21,23), and VEGF mRNA remains detectable in numerous adult tissues and is expressed at high levels in lung alveolar cells (24). Evidence supporting VEGF involvement in vascular remodeling in adult tissue

Table 2 Effect of Chronic Hypoxia on Flk-1 Receptor mRNA Expression in Rat Carotid Body[a]

	Duration of chronic hypoxia (days)		
	3	7	14
Experiment # 1	9.41	7.04	10.14
Experiment # 2	13.84	14.26	8.33

[a] Values are transcript ratios (i.e., CH/normoxia) determined in quantitative RT-PCR assays as described in Figure 2 and text. Experiments 1 and 2 for each duration of CH represent separate groups of normal and hypoxic carotid bodies.

includes demonstrations that VEGF and Flk-1 receptor mRNA levels are particularly high in ischemic and necrotic areas of tumors undergoing neovascularization (25). However, up-regulation of VEGF and its receptors has not been universally observed in tissues undergoing vasacular readjustments. In a recent study, Deindl et al. (26) found that VEGF, flt-1, and Flk-1 are not up-regulated following the induction of collateral artery growth via femoral artery occlusion. These authors concluded that "under nonischemic conditions, arteriogenesis is neither associated with nor inducible by increased levels of VEGF, and that VEGF is not a natural agent to induce angiogenesis in vivo." Thus the involvement of VEGF in experimentally induced microvascular changes appears to be dependent on specific experimental/pathological conditions. The absence of Flk-1 receptor expression on carotid body endothelial cells correlates with the absence of angiogenesis, as evidenced by the failure of CH to significantly increase the number of small blood vessels observed in tissue sections.

A functional role for VEGF in the carotid body may be indicated by recent reports showing that in addition to regulating vascular remodeling, VEGF can act as a neurotrophic agent via autocrine and/or paracrine mechanisms. Importantly, Flk-1 receptor has been demonstrated in mitotically active cells in mammary carcinoma and regenerating peripheral nerve (27,28). In the latter tissue Flk-1 expression has been identified in proliferating Schwann cells, which are similar to the type II cells of the carotid body. It is noteworthy that the initial elevation of Flk-1 receptor in the carotid body, at day 3 of CH, correlates with the appearance of conspicuous mitotic figures in some cell lobules and is consistent with the possible involvement of VEGF in proliferation of type I as well as type II cells. Moreover, a significant, but poorly understood, aspect of CH-induced remodeling in the carotid body is the rearrangement of chemosenory cells and the redistribution of lobules associated with prolonged exposure to hypoxia. In this regard, recent functional studies of both vascular smooth muscle cells and breast carcinoma cells expressing Flk-1 receptors have shown that exposure to VEGF-165 (10–100 ng/mL) stimulates cell migration on fibronectin substrates (29,30).

III. Adaptation Mediated by the Autocrine/Paracrine Action of Excitatory and Inhibitory Peptides on Chemosensory Type I Cells

Current views suggest that hypoxic chemotransduction in type I cells involves a cascade of events including membrane depolarization, Ca^{2+} influx, and the release of multiple biogenic amine and neuropeptide neurotransmitters that excite synaptic terminals of the CSN (10). Previous efforts to explain the CH-induced increase in chemosensitivity have focused primarily on alterations in neurotransmitter actions (reviewed in Ref. 11). These efforts have identified important changes in the synthesis, storage, and turnover of the numerous endogenous neuroactive agents present in type I cells (e.g., dopamine, norepinephrine, acetylcholine, serotonin, and substance P), but attempts to demonstrate direct involvement of particular neurotransmitters and/or their receptors in altered chemosensitivity have produced negative results and/or conflicting data (e.g., see Refs. 31–33). Recent studies in our laboratory have focused on the roles of two unique peptides in type I cell adaptation to CH, namely endothelin (ET) and atrial natriuretic peptide (ANP). These peptides, like VEGF, have been shown to mediate and regulate the effects of CH on lung and heart.

A. Endothelin

ET is a potent 21-amino-acid vasoconstrictor peptide that was first identified in porcine aortic endothelial cells (see Refs. 34,35 for review); later studies showed three isoforms of ET (ET-1, ET-2, and ET-3). ET and ET-like immunoreactivity have been found in many tissues such as heart and kidney, and on a variety of cells, including airway epithelial cells, lung vascular endothelial cells, and selected neurons in the central and peripheral nervous systems. The actions of ETs are mediated by specific ET_A, ET_B, and ET_C receptors (34), each of which has different affinities for the three ET isoforms. ET receptors are G-protein-coupled to phospholipase C (PLC) and phospholipase A_2 (PLA_2), as well as to adenylate cyclase (AC) (35). The vasoconstrictor effects of ET-1 in the lung occur via its actions on ET_A receptors, which activate PLC (35). ET_A receptor mRNA is highly expressed in vascular smooth muscle cells in the lung (36). Rats exposed to CH display increased steady-state levels of mRNA for ET-1 and ET_A receptors in lung (37), but not in other major organs (e.g., liver, spleen, kidney), suggesting that this response to CH is limited to specific O_2-sensitive tissues. Importantly, the administration of ET_A-receptor antagonists during CH prevents lung tissue remodeling and right ventricular hypertrophy (38–40).

McQueen and his colleagues were the first to demonstrate that intravenous injection of ET peptide elevates respiratory minute volume and elicits CSN excitation in rat carotid body (41). Of particular interest was the observation that these effects were blocked by the specific ET_A-receptor antagonist FR139317. Autoradiographic studies using ^{125}I-ET peptides further demonstrated specific ET binding sites both in carotid body lobules and in surrounding microvascular elements (41,42). Chen et al. reported that ET peptides potentiated hypoxia-evoked nerve activity when applied to rat and rabbit carotid body/CSN preparations superfused in vitro (43,44), where the potent vascular effects of ET-1 are circumvented. This effect of ET-1 is blocked by the highly specific ET_A receptor antagonist BQ-123, but not by the ET_B-receptor antagonist IRL 1038. The cellular mechanisms underlying ET action were indicated in other studies that have shown that incubation of intact rat carotid body in ET-1 increases cyclic AMP (cAMP) levels in type I cells (43). Previous studies have also shown that elevated cAMP levels potentiate neurotransmitter release and stimulus evoked CSN activity (45), suggesting that ET enhances voltage-dependent Ca^{2+} current and intracellular Ca^{2+} levels in type I cells via cAMP. This hypothesis was confirmed in dissociated type I cells from rabbit (44), where ET-1 enhances voltage-gated Ca^{2+} current (Fig. 3) and the hypoxia-evoked intracellular Ca^{2+} response (Fig. 4). Thus, the pharmacological

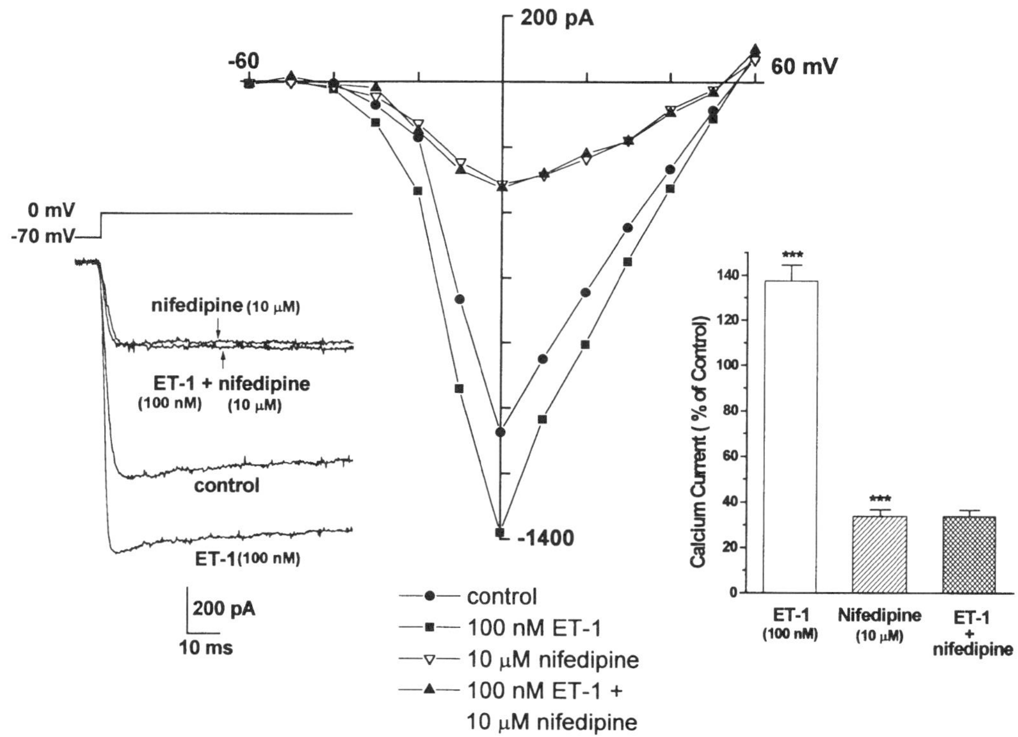

Figure 3 Voltage-sensitive (L-type) Ca^{2+} currents in rabbit type I cells are potentiated in the presence of ET-1. The left panel shows current records obtained in four different experimental conditions: (1) control, in bath solution equilibrated with air; (2) in the presence of 100 nM ET-1; (3) in bath solution containing L-type-specific Ca^{2+}-channel blocker nifedipine (10 μM); and (4) in solution containing 100 nM ET-1 plus 10 μM nifedipine. The representative I–V relationship (center panel) shows the potentiation of inward current by ET-1 and the lack of this effect in the presence of 100 nM nifedipine. Summary data from 10 cells (right panel) indicate significant ($p < 0.001$) elevation of inward current with ET-1 and substantial inhibition in the presence of nifedipine ($p < 0.001$). (From Ref. 44.)

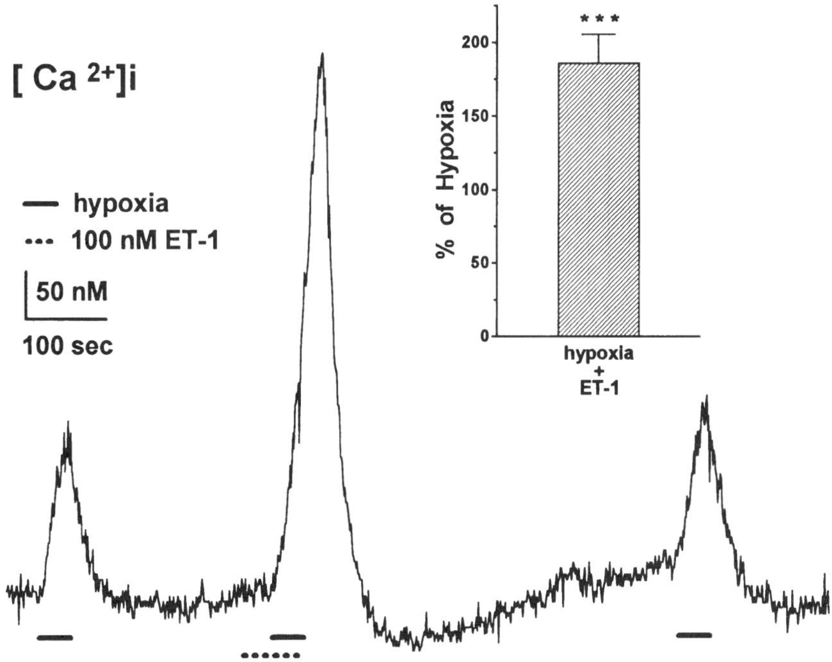

Figure 4 ET-1 potentiates the increase in intracellular Ca^{2+} evoked by hypoxia. Measurements were made on freshly dissociated rabbit type I cells. Basal $[Ca^{2+}]_i$ levels (57.7 ± 5.9 nm; $X \pm SEM$) were determined in superfusate equilibrated with air ($PO_2 = 128$ torr); hypoxic solutions were equilibrated with air and contained 500 µM Na dithionite ($PO_2 = 31$–33 torr). ET-1 (100 nM) did not alter basal $[Ca^{2+}]_i$ levels. Inset summarizes data from six cells; $p < 0.001$ versus control responses (ANOVA for repeated measurements). (From Ref. 44.)

effects of ET appear to involve cAMP- and Ca^{2+}-dependent mechanisms in type I cells, factors that likely potentiate the release of excitatory neurotransmitters during hypoxia (43,44).

In a recent immunocytochemical study of ET in the carotid body we further demonstrated that ET and ET_A receptor are expressed by type I cells. Moreover, staining for ET and ET_A receptor was incrementally increased in type I cells following 3, 7, and 14 days of CH (9). Peptide and receptor up-regulation were confirmed in quantitative RT-PCR assays that showed that 14 days of CH elicited >100-fold and 15-fold increases, respectively, in ET and ET_A-receptor transcripts, suggesting that ET may play an important role in CH-induced adaptation. In physiological studies of rats, in vitro recording of carotid sinus nerve activity following in vivo exposure to CH (380 torr) for 1–16 days demonstrated time-dependent increases in both basal chemoreceptor activity and responses evoked by acute hypoxia. A physiological role for endogenous ET was indicated in experiments using BQ-123. Figure 5 shows an example of the effect of BQ-123 in a normoxic

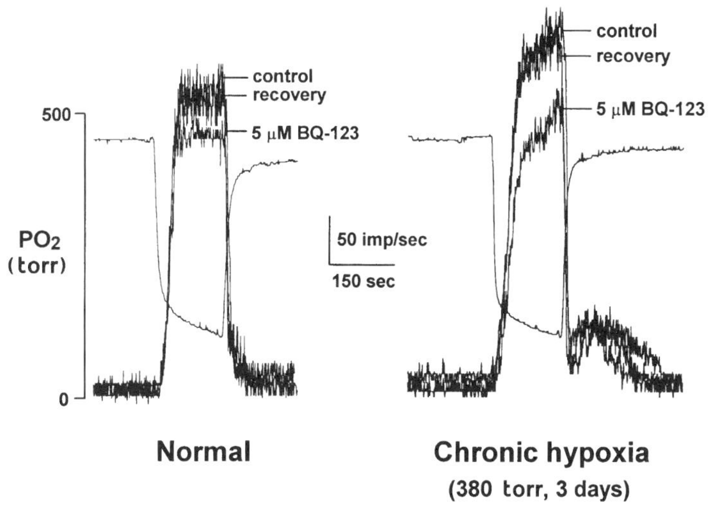

Figure 5 Effect of chronic hypoxia on the sensitivity of chemoreceptor nerve discharge to specific ET_A-receptor antagonist, BQ-123. Left panel shows three superimposed traces of integrated carotid sinus nerve activity in normal rat carotid body; separate trace indicates changes in bath PO_2. Basal and hypoxia-stimulated nerve activity are minimally altered in the presence of 5 μM BQ-123. After 3 days of chronic hypoxia (380 torr; right panel), nerve activity evoked by acute hypoxia is substantially reduced in the presence of the drug. (From Ref. 9.)

versus a CH preparation. In multiple experiments of this type, the averaged CSN discharge evoked by a standard hypoxic stimulus applied to normal carotid bodies was inhibited by 11% in the presence of BQ-123 (5 μM). However, after 3 days of CH the drug diminished the hypoxia-evoked discharge by 20% ($p < 0.01$). This inhibitory effect progressed to 45% at day 9 of CH and to nearly 50% following 12, 14, and 16 days of CH. Furthermore, in the presence of BQ-123 the magnitude of the activity evoked by hypoxia did not differ in normal versus CH preparations, indicating that the increased hypoxic chemosensitivity was mediated by endogenous ET acting on an increasing number of ET_A receptors. However, the increase in basal chemoreceptor discharge was only partially blocked by the ET_A-receptor antagonist, suggesting that changes in the resting nerve activity are in part independent of ET. Collectively, these data suggest that ET and ET_A autoreceptors on oxygen-sensitive type I cells play a critical role in CH-induced increased ehemosensitivity in the rat carotid body.

B. Atrial Natriuretic Peptide

ANP is a 28-amino-acid peptide discovered in cardiomyocytes where heart-wall distension, local hypoxia, and/or calcitonin gene-related peptide (CGRP) causes its release (46,47). In addition to its natriuretic and diuretic properties, ANP is also a potent vasodilator and an effective antimitogenic agent (48,49). Three distinct ANP receptors have been described: ANP-A and ANP-B receptors activate a particulate form of guanylate cyclase (GC) and likely mediate most of the biological effects of ANP via cyclic GMP (cGMP) (50); ANP-C receptors are not linked to GC and function primarily as clearance receptors (50), but nevertheless exhibit the antimitotic effects of A and B receptors (49). All three ANP receptor subtypes are expressed in cultured vascular endothelial cells and smooth muscle cells; however, in the rat, the ANP-C receptor accounts for the overwhelming majority of ANP binding sites in most tissues including lung (48,51). ANP relaxes preconstricted pulmonary arteries in vitro and blunts the vasoconstriction evoked by acute hypoxia in vitro (52). Plasma levels of ANP increase in response to acute hypoxia and CH, as does ANP synthesis in the right heart (53,54). An important role for ANP in CH has been demonstrated in experiments where continuous infusion of ANP during CH, or the administration of agents that block its degradation, attenuates the development of pulmonary hypertension and right ventricular hypertrophy (55–57).

In rat and cat carotid body, immunocytochemical studies revealed the presence of ANP in virtually all type I cells (58,59). Furthermore, in preliminary experiments of the pig carotid body virtually all type I cells were immunopositive utilizing an antibody directed against guanylate-cyclase-coupled-ANP receptors (60). Moreover, brief incubations in submicromolar concentrations of ANP or the ANP analog atriopeptin III (APIII) resulted in elevated cGMP immunoreactivity in type I cells in cat and rat carotid body (61,62). Radioimmunoassay techniques confirmed a dose-response relationship between the increase in tissue cGMP levels and the concentration of APIII in the incubation media (62,63). Electrophysiological recordings in rabbit and cat chemoreceptors showed that submicromolar concentrations APIII potently inhibit hypoxia-evoked CSN activity (58,63). APIII did not alter basal chemoreceptor activity, but it delayed the onset and reduced the peak activity evoked by hypoxia. Importantly, similar chemoreceptor inhibition resulted when carotid bodies were incubated in cell-permeant analogs of cGMP (63). In recent experiments we demonstrated that APIII prevents the hypoxia-induced depression of K^+ currents in type I cells, and that these effects are reversed in the presence of a specific antagonist of protein kinase G, thus further establishing a link between ANP-mediated inhibition and the production of cGMP in type I cells (64). These experiments also demonstrated the involvement of protein phosphatase 2A in a signaling mechanism that enhances K^+ current, while depressing the activity of voltage-gated Ca^{2+} channels via the removal of strategic phosphate groups from channel proteins (64).

In an effort to further elucidate the physiological role of ANP we initiated gene expression studies of the peptide and the A-type receptor in normoxic versus CH carotid body. Results from RT-PCR assays suggest significant upregulation of ANP and ANP-A receptor genes following 14 days of CH in rat chemosensory tissue (65,66). These indications of increased activity of the peptide suggest that CH may involve enhancement of inhibitory mechanisms in the carotid body. Interestingly, Lahiri and his colleagues earlier demonstrated the emergence of chemoreceptor inhibition involving an efferent neural pathway in cats exposed to CH (67). To examine the hypothesis that ANP mediates increased inhibition following CH we

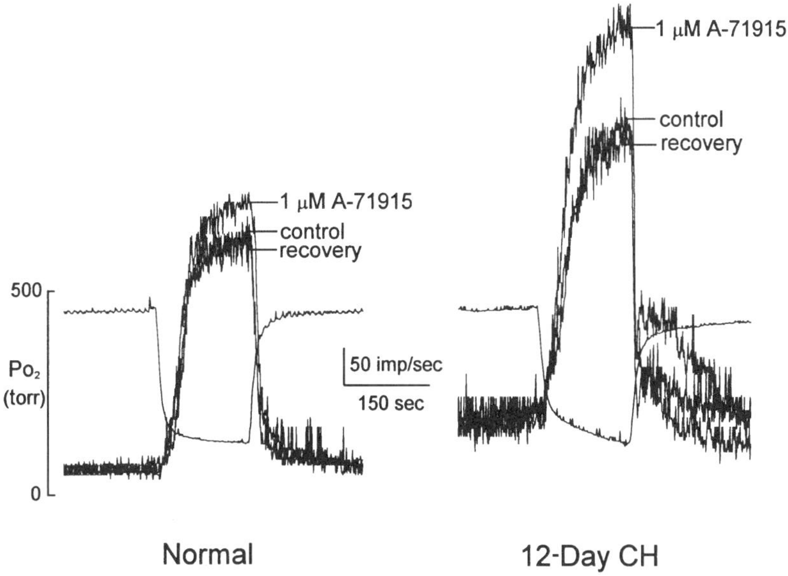

Figure 6 Effect of chronic hypoxia on the sensitivity of rat chemoreceptor nerve discharge to specific ANP-A receptor antagonist A-71915. Left panel shows three superimposed traces of integrated carotid sinus nerve activity in normal rat carotid body; separate trace indicates changes in bath PO_2. Basal and hypoxia-stimulated nerve activity are minimally altered in the presence of 1 μM A-71915. After 12 days of chronic hypoxia (380 torr; right panel), nerve activity evoked by acute hypoxia is substantially increased in the presence of the drug.

have tested the effects of a specific ANP receptor antagonist, A-71915 (1 μM), on CSN activity evoked by an acute hypoxic challenge in vitro. Preliminary results, shown in Figure 6, indicate that blocking ANP-A receptors in the normal carotid body modestly elevates CSN activity evoked by a moderate hypoxic stimulus. However, following 12 days of CH (380 torr), the response to acute hypoxia is enhanced to a substantially greater degree in the presence of the drug. Along with our studies of increased involvement of ET, these findings suggest that CH induces the upregulation of competing excitatory and inhibitory mechanisms in type I cells. Although the functional significance of such complex regulatory mechanisms is unknown, our data indicate that such autocrine/paracrine feedback control of type I cell activity is of utmost importance during the adaptation to chronic hypoxic stimulation.

IV. Adaptation of Chemoreceptor Neurons During Chronic Hypoxia

Important signaling elements in the chemoreflex pathway are chemoafferent neurons with cell bodies in the petrosal ganglia. In the past these elements have been largely ignored as possible adaptive components in VAH owing to the prevailing concept that primary sensory neurons act as simple relays between peripheral sensory organs and the CNS. Recent experimental findings now support the view that certain classes of somatic sensory ganglion cells are able to adapt to altered physiological stimuli; such changes are particularly apparent in the small neurons responding to pain and inflammation. In models of chronic inflammation significant alterations in neuropeptide synthesis, storage, and release occur both in the somata of small primary sensory neurons and in their peripheral and central projections (68,69). Similar changes have also been observed among autonomic afferents. For example, within 24 hr of exposure of vagal nerve terminals in the respiratory tract of guinea pig to an allergen there is a 3–5-fold elevation in the levels of SP, neurokinin A, and CGRP immunoreactivity in the lungs, together with a 10-fold increase in nodose ganglion neurons expressing SP (70,71). Moreover, such rapid changes can be mimicked by tracheal injection of nerve growth factor (NGF), an agent known to induce the preprotachykinin gene (for SP) in other primary sensory systems (71). These and numerous other recent investigations have shown that at least certain primary sensory neurons are capable of dynamic phenotypic adjustments in response to chronically altered physiological conditions.

Chemoafferent neurons that innervate the carotid body, like cutaneous and vagal sensory ganglion cells, contain a variety of neuroactive substances. Presently available evidence indicates the existence of at least two distinct subpopulations of neurons. One group of CSN fibers contains TH, synthesizes DA, and terminates on type I cells with calyciform endings containing well-developed synaptic specializations (72,73); the cell bodies of these neurons are clustered at the distal pole of the ganglion (73). Ishizawa et al. (74) showed that TH-positive neurons in the PG also express the α7-subunit of the nicotinic acetylcholine receptor in a recent double-label immunocytochemical study. A second subpopulation of CSN fibers, immunopositive for SP and CGRP, is comprised of neurons that cluster at the proximal end of the PG (73,75). These neurons contain nitric oxide synthase (NOS), the synthetic enzyme for nitric oxide (NO) (76). SP/CGRP/NOS fibers enter the lobules of chemoreceptor type I cells, but rarely form synaptic appositions. Instead they terminate near type I cells in association with processes of glial-like type II cells (77–79). These neurochemically distinct subgroups of CSN of neurons and their afferent axons are undoubtedly chemosensory in nature because electrophysiological recordings from single units have shown that both Aδ- and C-fibers in the CSN respond to hypoxia (80). Differences in the central sites of termination are also known to exist. Retrograde labeling studies show that the axons of chemoafferent neurons terminate either in the commissural or media subnuclei of the caudal NTS (81–83) and likewise contain either TH or CGRP/NOS (83–85).

A. Effect of Hypoxia on Chemosensory Neuron Phenotype

Evidence that hypoxia-induced adaptation of ventilatory control directly involves peripheral sensory neurons has been provided by Prabhakar et al. (86,87), who showed that 12–24 hr of hypobaric hypoxia elevates the mRNA transcript for the neuronal form of NOS (nNOS) more than 10-fold in nodose ganglia cells

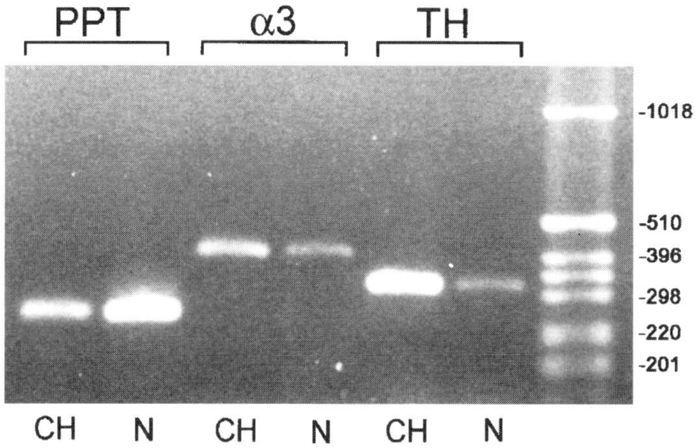

Figure 7 RT-PCR products electrophoresed on an agarose gel. From left to right, each of the three pairs of lanes represents, respectively, mRNA levels for: (1) prepro-tachykinin (PPT); (2) the α3 nicotinic receptor subunit, both in the PG; and (3) TH in the carotid body. Transcript (mRNA) levels for chronic hypoxia (CH; 9–14 days at 380 torr) versus normoxia (N) are represented by the left versus right lanes, respectively, for each pair. Right lane shows DNA molecular-weight markers and indicated base pair sizes.

innervating aortic chemo- and baroreceptors, while the transcript for the endotheial form of NOS was not altered by hypoxia. Also, nNOS protein and enzyme activity nearly doubles following 24 hr of hypoxia. We have used RT-PCR in similar studies to assess changes in mRNA levels in the PG. The genes evaluated include tyrosine hydroxylase (TH), substance P (SP; prepro-tachykinin gene), the α3- and α7-subunits of the nicotinic cholinergic receptor, and the dopaminergic D_2-receptor subtype. Figure 7 shows PCR products electrophoresed on an agarose gel, with the extreme right lane demonstrating the separation of DNA molecular weight markers ("DNA ladder"). From left to right, each of the three pairs of lanes represents, respectively, mRNA levels for: (1), preprotachykinin (PPT); (2) the α3 nicotinic receptor subunit, both in the PG; and (3) TH in the carotid body. Transcript (mRNA) levels for CH versus normoxia are represented by the left versus right lanes, respectively, for each pair. Table 3 summarizes RT-PCR results from preliminary experiments involving the PG, the NG, and the carotid body. The data were confirmed in three separate PCR trials from two batches of cDNA. In the PG, CH (9–14 days of hypobaric hypoxia at 380 torr) elevated the expression of TH and the α3- and α7-subunits of the nicotinic cholinergic receptor, whereas this hypoxic exposure depressed mRNA levels for PPT and the D_2-dopaminergic receptor. Similar changes occurred in the NG, except that CH elevated, rather than depressed, PPT mRNA levels. Parallel experiments with the carotid body confirmed the results of previous studies showing that CH up-regulates the TH gene. In addition, our data indicate increased levels of the α3- and α7-subunits in this tissue following CH. These findings are also in agreement with previous studies by Gauda and Gerfen

Table 3 Effect of Chronic Hypoxia on Phenotypic Expression in Chemosensory Pathways

	Petrosal ganglion		Nodose ganglion		Carotid body	
	N	CH	N	CH	N	CH
Tyrosine hydroxylase	+	+++	+	++	+	+++
Substance P	++	+	+	++	L.C.	L.C.
α3 Nicotinic receptor	+	++	N.A.	N.A.	+	+++
α7 Nicotinic receptor	+	++	+	++	+	++
Dopaminergic D_2 receptor	++	+	++	+	N.A.	N.A.

N: normoxia; CH: chronic hypoxia, 9–14 days at 380 torr; N.A.: not available; L.C.: weak signal; apparent low copy number.

(88), who demonstrated very low levels of PPT expression in the carotid body, and moreover, that these levels do not appear to change substantially following CH.

To further assess the expression of receptors on PG neurons, we developed an in vitro preparation for electrophysiological analysis. The rat PG is positioned in a conventional flow chamber along with attached CSN and lingual nerve (LN) branches of the glossopharyngeal (IXth) nerve for recording of antidromic activity evoked by the introduction of selected neuroactive agents. Utilizing a similar technique Alcayaga et al. (89) found that ACh applied to the cat PG elicits an increased discharge in the CSN, but not in the LN. In normal rat preparations superfused in vitro, a 100-μg bolus of ACh delivered upstream of the PG elicited responses in both the CSN and the LN (Fig. 8). The ACh-evoked activity was markedly depressed by the nicotinic antagonist mecamylamine (100 μM). After 10–14 days of CH (Fig. 9) we found responses elicited by ACh in the CSN are substantially elevated, in accord with our RT-PCR data. Interestingly, CH did not significantly increase the ACh-evoked activity in the LN, suggesting that CH-induced changes in cholinergic receptors are specific for a subset of petrosal neurons whose axons project into the CSN. We assume receptor expression is confined to cell soma in these preparations; nevertheless, the data support the notion that ACh receptors on axon terminals might be involved in the intact carotid body.

B. Chemosensory Synaptic Plasticity Induced by Chronic Hypoxia

In view of the remarkable changes in cholinergic receptor expression in PG neurons exposed to CH, we performed in vitro carotid body–CSN studies of ACh-mediated chemotransmission (90). In superfused preparations, ACh applied to the carotid body elicited larger CSN responses following CH, in accord with the observed increased expression of nicotinic cholinergic receptors in CH-PG neurons. In both normal and CH preparations, the nicotinic antagonist mecamylamine (100 μM) blocked ACh-evoked responses. The responses to an acute hypoxic challenge (superfusate equilibrated with 20% O_2) were likewise inhibited by mecamylamine (100 μM) and involved a reduction of ~80% of CSN activity. Surprisingly, however,

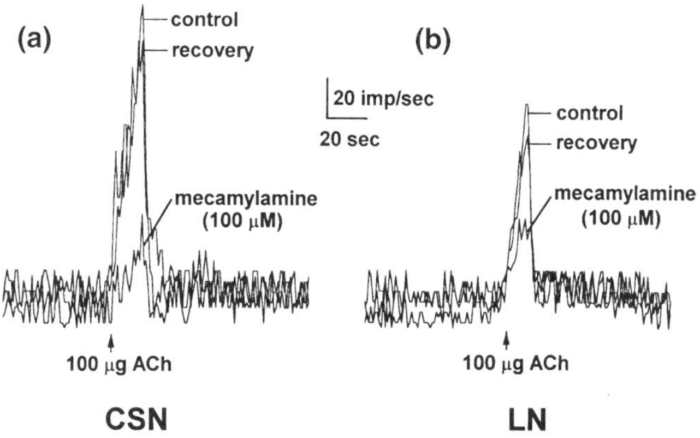

Figure 8 Integrated antidromic neural activity in the carotid sinus nerve (CSN) and lingual nerve (LN) branches of the glossopharyngeal nerve (IXth n.). Nerve preparation was superfused in vitro. Superimposed traces show that application of a 100-μg bolus (arrows) of ACh to the petrosal ganglion evokes a substantial increase in the CSN and LN nerve discharge, which is inhibited in the presence of 100 μM mecamylamine, a nicotinic receptor antagonist.

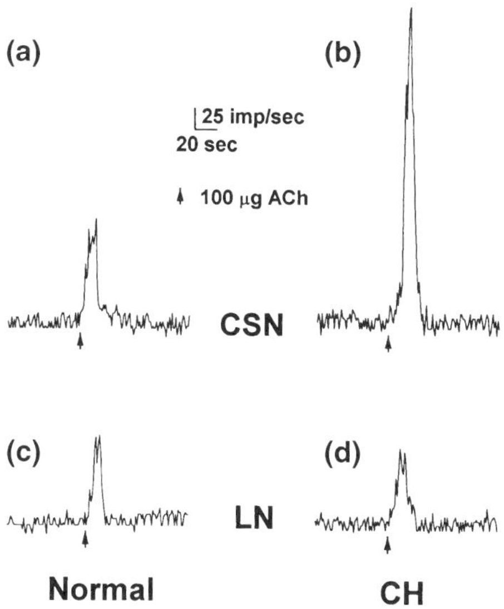

Figure 9 Integrated antidromic neural activity in the carotid sinus nerve (CSN) and lingual nerve (LN) branches of the glossopharyngeal nerve (IXth n.). Nerve preparation was superfused in vitro. Response evoked in the CSN by the application of a 100-μg bolus of ACh (arrows) to the petrosal ganglion is significantly enhanced after a 14-day in vivo exposure to hypobaric (380 torr) hypoxia (CH: chronic hypoxia). Prior exposure to hypoxia does not affect the response to ACh in LN.

after 10–14 days of CH, neither 100 μM nor 500 μM mecamylamine altered activity evoked by an acute hypoxic challenge despite the fact that the latter concentration is far in excess of that required to block most of the hypoxic response in normal animals.

The precise function of ACh in sensory transmission between type I cells and chemoafferent terminals in the carotid body has been a source of controversy for over 30 years. This is due in large part to the fact that cholinergic antagonists, even at high concentrations, fail to fully block CSN activity in response to acute hypoxia or other natural stimuli (see Ref. 91). Recent compelling studies by Fitzgerald, Shirahata (92–94), and their colleagues indicate that ACh acts an excitatory neurotransmitter between type I cells and chemoafferent nerve fibers. These investigators have demonstrated the presence of nicotinic cholinergic receptors on chemoafferent nerve terminals, along with evidence of hypoxia-evoked release of ACh from the carotid body, and they have presented new pharmacological data using highly specific agonists and antagonists. In addition, seminal studies by Zhang et al. (95) involving cocultured rat type I cells and petrosal chemoafferent neurons indicate that hypoxia-evoked synaptic activity is mediated by both ACh and adenosine triphosphate (ATP), acting at nicotinic and purinergic receptors, respectively. The existence of purinergic receptors was confirmed in immunofluorescence studies that localized $P2X_2$ receptor protein on PG neurons and chemoafferent terminals in the carotid body. In our recent in vitro studies, muscarinic (atropine; 10 μM) and α-4-nicotinic (methyllycaconitine; 50 nM) receptor antagonists blocked ~50% of CSN activity evoked by hypoxia in normal carotid body, but after 10–14 days of CH these drugs, like mecamylamine, were completely ineffective against an acute hypoxic challenge. Receptor saturating concentrations of the P2-receptor antagonist suramin (100 μM) blocked 74% of hypoxia-evoked CSN activity in normal carotid body, whereas following CH, only 52% of evoked activity was suramin-sensitive (96). Collectively these findings suggest that cholinergic/purinergic synaptic coupling in the carotid body is a complex process that is highly modified following CH.

In addition to purinergic and cholinergic mechanisms, numerous previous studies also implicate dopaminergic and tachykininergic receptors in chemotransmission. However, our recent experiments that examined CH-induced synaptic plasticity in the rat carotid body demonstrated that tachykininergic (spantide 1; 5 μM) and dopaminergic D_2 (domperidone; 100 nM) antagonists did not alter hypoxia-evoked chemoreceptor activity in either normal or CH preparations (96).

Electrical coupling has been proposed by Eyzaguirre and his colleagues (91,97) as an alternative to chemical synaptic transmission between type I cells and chemoafferent nerve terminals. A variety of electrophysiological techniques demonstrate electrical coupling among type I cells, and their results suggest that the functional properties of the electrical junctions are modulated by intracellular levels of Ca^{2+} and cyclic AMP, as well as by classical chemoreceptor stimulants, such as acidity and hypoxia (98–101). These data appear to conflict with early ultrastructural studies utilizing conventional fixation procedures that found few gap junctions among type I or other cells in the chemosensory lobules (102,103). However, a more recent freeze-substitution ultrastructural study, designed to preserve intracellular relationships, not only found gap junctions between adjacent type I cells, but for the first time identified gap junctions between type I cells and afferent nerve terminals (104). Based on these findings, Eyzaguirre hypothesized that synaptic transmission between type I cells and chemosensory nerve terminals may involve parallel chemical and electrical signaling mechanisms (91,97). Zhang et al. recentiy observed that certain small spontaneous and hypoxia-evoked synaptic potentials remain unblocked in the presence of mecamylamine plus suramin and speculated that these resistant minipotentials could be explained by direct electrical coupling (95).

Abudara et al. have suggested that sustained elevations of cAMP levels during CH may promote the up-regulation of gap-junction-forming connexins (Cxs) in type I cells (105). To follow up this hypothesis we recently studied the effects of 14-day CH exposure on the expression of Cx43 in type I cells and PG neurons utilizing immunocytochemical techniques (106). As is shown in Figure 10, diffuse and granular Cx43-like immunoreactivity was observed in the cytoplasm of type I cells in the normal rat carotid body. In addition, dense punctate spots of the immunoreactive product were present at the margins of these cells (Fig. 10A) and near the borders of chemosensory cell lobules. On the other hand, Cx43-like immunoreactivity was not detectable in petrosal ganglion neurons from normal animals (Fig. 10C). After 2-week exposure to hypobaric (380 torr) hypoxia, Cx43 immunostaining was substantially enhanced in type I cells (Fig. 10B); moreover, CH elicited the expression of Cx43-like immunoreactivity in a subset of afferent neurons distributed throughout the PG (Fig. 10D). Quantitative RT-PCR studies confirmed that CH elicits a substantial increase in Cx43 mRNA levels in the carotid body, along with a marked elevation of Cx43 expression in the PG. Although the functional importance of Cxs in the carotid body is presently unknown, the presence of Cx43 in both type I cells and chemoafferent neurons is consistent with the proposal of Eyzaguirre and Abudara that electrical coupling provides an important and potentially "fail-safe" contribution to synaptic signaling (91). Thus, enhanced development of gap junctions at synapses could both potentiate chemotransmission and ensure chemoreceptor input to the central nervous system during CH.

V. Concluding Remarks

CH-induced enlargement of the carotid body and profound alteration of its vascualar structure were first documented some two decades ago (see Ref. 102). Furthermore, indications that CH results in elevated chemosentivity were demonstrated by elegant nerve recording experiments in 1987/1988 (6,8,107). Yet despite these early reports, adaptation of the carotid body and other O_2-sensitive tissues remains a poorly understood process. Initial efforts to identify the cellular mechanisms underlying these interesting, and clinically important, phenomena were hampered by the lack of basic information of chemotransduction, as well as a failure to appreciate the complexity of chemotransmission in the carotid body. Although still controversial, multiple O_2-sensitive processes in type I cells have been demonstrated in the last

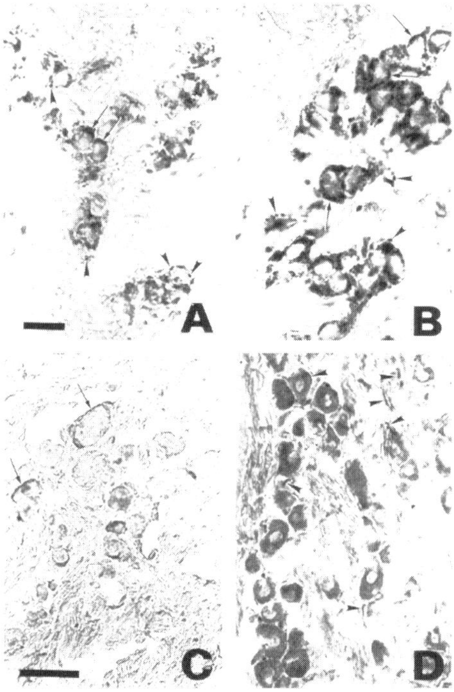

Figure 10 Connexin43 (Cx43) immunocytochemical staining in rat carotid body (A,B) and petrosal ganglion (C,D). In normal carotid body (A), immunostaining appeared as a mixture of diffuse and granular reaction product, located in the cytoplasm of type I cells. Dark punctate spots (arrowheads) and crescentic forms (arrows) were common near the perimeter of type I cells. After chronic hypobaric hypoxia (14 days at 380 torr) (B), the level of Cx43-like immunoreactivity was enhanced in type I cells, and the punctate spots (arrowheads) and crescents (arrows) remained prominent around some type I cells. In the normal petrosal ganglion (C), Cx43 immunoreactivity was not detectable in sensory neurons, but it was present in slender processes of Schwann cells enveloping large neurons (arrows). After chronic hypoxia (D), many neurons throughout the petrosal ganglia were immunopositive for Cx43. Intense staining was noted in axons and in processes emerging from the soma of some cells (arrowheads). Scale bars equal 20 μm for A and B and 50 μm for C and D. (From Ref. 106.)

decade via the application of sophisticated electrophysiological and molecular biological techniques. These methods have also profoundly influenced the discovery of a variety of novel neuroactive agents and neurotransmitter mechanisms between type I cells and chemoafferent nerve terminals. This cascade of new information supports the view that each and every cell type in the carotid body makes a substantial contribution to the changes induced by CH. However, owing to the extreme complexity of the processes involved in each cell type and the multiplicity of interactions between them, we anticipate that the formulation of novel hypotheses and the generation of integrative physiological data will be required to elucidate adaptive mechanisms at the cellular level. Thus, a fundamental conclusion of the studies reviewed here is that adaptation involves multiple tissue components that may respond independently to chronic stimulation. Yet together these diverse elements comprise a highly integrated functional unit. Thus a full understanding of the adaptive process must combine information from studies of CH-induced changes in the carotid body vasculature, O_2-sensitive type I cells, and chemoafferent neurons.

Acknowledgment

This work was supported by USPHS grant NS 12636 and NS 07938.

References

1. Bisgard GE. The role of arterial chemoreceptors in ventilatory acclimatization to hypoxia. In: O'Regan RG, Nolan P, McQueen DS, Paterson DJ, eds. Advances in Experimental Medicine and Biology. Vol. 360, Arterial Chemoreceptors: Cell to System. New York: Plenum Press, 1994:109–122.
2. Olson EB Jr, Dempsey JA. Rat as a model for humanlike ventilatory adaptation to chronic hypoxia. J Appl Physiol Respir Environ Exercise Physiol 1978; 44(5):763–769.
3. Powell FL, Huey KA, Dwinell MR. Central nervous system mechanisms of ventilatory acclimatization to hypoxia. Respir Physiol 2000; 121(2–3):223–236.
4. Dwinell MR, Powell FL. Chronic hypoxia enhances the phrenic nerve response to arterial chemoreceptor stimulation in anesthetized rats. J Appl Physiol 1999; 87(2):817–823.
5. Bisgard GE, Engwall MJ, Weizhen N, Nielsen AM, Vidruk EH. The effect of prolonged stimulation on afferent activity of the goat carotid body. In: Acker H, Trzebski A, O'Regan RG, eds. Chemoreceptors and Chemoreceptor Reflexes. New York: Plenum Press, 1990:165–170.
6. Vizek M, Pickett CK, Well JV. Increased carotid body hypoxic sensitivity during acclimatization to hypobaric hypoxia. J Appl Physiol 1987; 63:2403–2410.
7. Bisgard GE. Increase in carotid body sensitivity during sustained hypoxia. Biol Signals 1995; 4:292–297.
8. Nielsen AM, Bisgard GE, Vidruk EH. Carotid chemoreceptor activity during acute and sustained hypoxia in goats. J Appl Physiol 1988; 65(4):1796–1802.
9. Chen J, He L, Dinger B, Stensaas L, Fidone S. The role of endothelin and endothelin A-type receptor in physiological adaptation of the carotid body during chronic hypoxia. Am J Physiol Lung Cell Mol Physiol 2002; 282:L1314–L1323.
10. Fidone SJ, Gonzalez C, Almaraz L, Dinger B. Cellular mechanisms of peripheral chemoreceptor function. In: Crystal RG, West JB, et al., eds. The Lung: Scientific Foundations. Philadelphia: Lippincott-Raven Publishers, 1997:1725–1746.
11. Bisgard GE. Carotid body mechanisms in acclimatization to hypoxia. Respir Physiol 2000; 121:237–246.

11a. Lahiri S. Plasticity and multiplicity in the mechanisms of oxygen sensing. In: Lahiri S, Prabhakar NR, Forster RE, eds. Advances in Experimental Medicine and Biology: Oxygen Sensing: Molecule to Man. New York: Kluwer Academic/Plenum Publishers, 2000:13–23.

12. Laidler P, Kay JM. A quantitative study of some ultrastructural features of the type I cells in the carotid bodies of rats living at a simulated altitude of 4300 metres. J Neurocytol 1978; 7:183–192.
13. Hellstrom S, Pequignot J-M. Morphometric studies on intact and sympathectomised carotid bodies of long-term hypoxic rats: a light and electron microscopial study. In: Pallot DJ, ed. The Peripheral Arterial Chemoreceptors. London: Croom Helm, 1982:293–301.
14. Laidler P, Kay JM. A quantitative morphological study of the carotid bodies of rats living at a simulated altitude of 4300 metres. J Pathol 1975; 117:183–191.
15. Pequignot JM, Hellstrom S. Intact and sympathectomized carotid bodies of long-term hypoxic rats. Virchows Arch 1983; 400:235–243.
16. Chen J, Dinger B, Stensaas L, Fidone S. Involvement of vascular endothelial growth factor (VEGF) in carotid body vascular remodeling induced by chronic hypoxia. FASEB J 2001; 15:A153.
17. Voelkel NF, Hoeper M, Maloney J, Tuder RM. Vascular endothelial growth factor in pulmonary hypertension. Ann NY Acad Sci 1996; 796:186–193.
18. McDonald DM. A morphometric analysis of blood vessels and perivascular nerves in the rat carotid body. J Neurocytol 1983; 12:155–199.
19. Rabinovitch M, Gamble W, Nadas AS, Miettinen OS, Reid L. Rat pulmonary circulation after chronic hypoxia: hemodynamic and structural features. Am J Physiol 1979; 236(6):H818–H827.
20. Tuder RM, Flook BE, Voelkel NF. Increased gene expression for VEGF and the VEGF receptors KDR/Flk and Flt in lungs exposed to acute or to chronic hypoxia. J Clin Invest 1995; 95:1798–1807.
21. Thomas KA. Vascular endothelial growth factor, a potent and selective angiogenic agent. J Biol Chem 1996; 271(2):603–606.
22. Ferrara N. Role of vascular endothelial growth factor in the regulation of angiogenesis. Kidney Int 1999; 56:794–814.
23. Neufeld G, Tessler S, Gitay-Goren H, Cohen T, Levi B-Z. Vascular endothelial growth factor and its receptors. Prog Growth Factor Res 1994; 5:89–97.
24. Monacci WT, Merrill MJ, Oldfield EH. Expression of vascular permeability factor/vascular endothelial growth factor in normal rat tissues. Am J Physiol (Cell Physiol) 1993; 264(33):C995–C1002.
25. Minchenko A, Bauer T, Salceda S, Caro J. Hypoxic stimulation of vascular endothelial growth factor expression in vitro and in vivo. Lab Invest 1994; 71(3):374–379.
26. Deindl E, Buschmann I, Hoefer IE, Podzuweit T, Boengler K, Vogel S, et al. Role of ischemia and of hypoxia-inducible genes in arteriogenesis after femoral artery occlusion in the rabbit. Circ Res 2001; 89:779–786.
27. Xie B, Tam NNC, Tsao SW, Wong YC. Co-expression of vascular endothelial growth factor (VEGF) and its receptors (flk-1 and flt-1) in hormone-induced mammary cancer in the Noble rat. Br J Cancer 1999; 81:1335–1343.
28. Sondell M, Lundborg G, Kanje M. Vascular endothelial growth factor stimulates Schwann cell invasion and neovascularization of acellular nerve grafts. Brain Res 1999; 846:219–228.
29. Price DJ, Miralem T, Jiang S, Steinberg R, Avraham N. Role of vascular endothelial growth factor in the stimulation of cellular invasion and signaling of breast cancer cells. Cell Growth Different 2001; 12:128–138.

30. Ishida A, Murray J, Saito Y, Kanthou C, Benzakour O, Shibuya M, et al. Expression of vascular endothelial growth factor receptors in smooth muscle cells. J Cell Physiol 2001; 188:359–368.
31. Janssen PL, Dwinell MR, Pizarro J, Bisgard GE. Intracarotid dopamine infusion does not prevent acclimatization to hypoxia. Respir Physiol 1998; 111:33–43.
32. Janssen PL, O'Halloran KD, Pizarro J, Dwinell MR, Bisgard GE. Carotid body dopaminergic mechanisms are functional after acclimatization to hypoxia in goats. Respir Physiol 1998; 111:25–32.
33. Tatsumi K, Pickett CK, Weil JV. Possible role of dopamine in ventilatory acclimatization to high altitude. Respir Physiol 1995; 99:63–73.
34. Simonson MS, Dunn MJ. Cellular signaling by peptides of the endothelin gene family. FASEB J 1990; 4:2989–3000.
35. Rubanyi GM, Polokoff MA. Endothelins: Molecular biology, biochemistry, pharmacology, physiology and pathophysiology. Pharmacol Rev 1994; 46:325–415.
36. Arai H, Hori S, Aramori I, Ohkubo H, Nakanishi S. Cloning and expression of cDNA encoding an endothelin receptor. Nature 1990; 348:730–732.
37. Huaibin L, Chen S-J, Chen Y-F, Meng QC, Durand J, Oparil S, et al. Enhanced endothelin-1 and endothelin receptor gene expression in chronic hypoxia. J Appl Physiol 1994; 77(3):1451–1459.
38. Bonvallet ST, Zamora MR, Hasunuma K, Sato K, Hanasato N, Anderson D, et al. BQ123, and ET_A-receptor antagonist, atenuates hypoxic pulmonary hypertension in rats. Am J Physiol 1994; 266(35):H1327–H1331.
39. Chen S-J, Chen Y-F, Meng QC, Durand J, Dicarlo VS, Oparil S. Endothelin-receptor antagonist bosentan prevents and reverses hypoxic pulmonary hypertension in rats. J Appl Physiol 1995; 79(6):2122–2131.
40. Eddahibi S, Raffestin B, Clozel M, Levame M, Adnot S. Protection from pulmonary hypertension with an orally active endothelin receptor antagonist in hypoxic rats. Am J Physiol 1995; 268(37):H828–H835.
41. McQueen DS, Dashwood MR, Cobb VJ, Bond SM, Man CG, Spyer KM. Endothelin and rat carotid body: autoradiographic and functional pharmacological studies. J Auton Nerv Syst 1995; 53(2–3):115–125.
42. Spyer KM, McQueen DS, Dashwood MR, Sykes RM, Daly Md, Muddle JR. Localisation of [^{125}I]endothelin binding sites in the region of the carotid bifurcation and brain stem of the cat: possible baro and chemoreceptor involvement. J Cardiovasc Pharmacol 1991; 17(Suppl. 7):S385–389.
43. He L, Chen J, Dinger B, Stensaas L, Fidone S. Endothelin modulates chemoreceptor cell function in mammalian carotid body. In: Zapata P, Eyzaguirre C, Torrance RW, eds. Frontiers in Arterial Chemoreception. New York: Plenum Press, 1996:305–311.
44. Chen J, He L, Dinger B, Fidone S. Cellular mechanisms involved in rabbit carotid body excitation elicited by endothelin peptides. Respir Physiol 2000; 121:13–23.
45. Wang W J, Cheng G-F, Yoshizaki K, Dinger B, Fidone S. The role of cyclic AMP in chemoreception in the rabbit carotid body. Brain Res 1991; 540:96–104.
46. Baertschi AJ, Hausmaninger C, Walsh RS, Mentzer RM, Wyatt DA, Pence RA. Hypoxia-induced release of atrial natriuretic factor (ANF) from the isolated rat and rabbit heart. Biochem Biophys Res Commun 1986; 140:427–433.
47. Yamamoto A, Kimura S, Hasui K, Fujisawa Y, Tamaki T, Fukui K et al. Calcitonin gene-related peptide (CGRP) stimulates the release of atrial natriurectic peptide (ANP) from isolated rat atria. Biochem Biophys Res Commun 1988; 155:1452–1458.

48. Suga S, Nakao K, Hosoda KJ, Mukoyama M, Ogawa Y, Shirakami G et al. Receptor selectivity of natriuretic peptide family, atrial natriuretic peptide, brain natriuretic peptide, and C-type natriuretic peptide. Endocrinology 1992; 130:229–239.
49. Cahill PA, Hassid A. Differential antimitogenic effectiveness of atrial natriuretic peptides in primary versus subcultured rat aortic smooth muscle cells: relationship to expression of ANF-C receptors. J Cell Physiol 1993; 154:28–38.
50. Maack T, Suzuki M, Almeida FA, Nussenzveig D, Scarborough RM, McEnroe GA et al. Physiological role of silent receptors of atrial natriuretic factor. Science 1989; 238:675–678.
51. Klinger JR, Moalli R, Warburton RR, Wrenn DS, Hill NS. C-receptor ligand blocks pulmonary clearance of atrial natriuretic peptide in isolated rat lungs. Proc Soc Exp Biol Med 1992; 201:154–158.
52. Jin H, Yang R-H, Chen YF, Jackson RM, Itoh H, Mukoyama M et al. Atrial natriuretic peptide in acute hypoxia-induced pulmonary hypertension in rats. J Appl Physiol 1991; 71:807–814.
53. Klinger JR, Petit RD, Warburton RR, Wrenn DS, Arnal F, Hill NS. Neutral endopeptidase inhibition attenuates development of hypoxic pulmonary hypertension in rats. J Appl Physiol 1993; 75(4):1615–1623.
54. Stockman PT, Will DH, Sides SD, Brunnert SR, Wilner GD, Leathy KM et al. Reversible induction of right ventricular atriopeptin synthesis in hypertrophy due to hypoxia. Circ Res 1988; 63:201–213.
55. Thompson JS, Sheedy W, Morice AH. Neutral endopeptidase (NEP) inhibition in rats with established pulmonary hypertension secondary to chronic hypoxia. Br J Pharmacol 1994; 113:1121–1126.
56. Thompson JS, Sheedy W, Morice AH. Effects of the neutral endopeptidase inhibitor, SCH 42495, on the cardiovascular remodelling secondary to chronic hypoxia in rats. Clin Sci 1994; 87:109–114.
57. Jin H, Yang R-H, Chen Y-F, Jackson RM, Oparil S. Atrial natriuretic peptide attenuates the development of pulmonary hypertension in rats adapted to chronic hypoxia. J Clin Invest 1990; 85:115–120.
58. Wang Z-Z, He L, Stensaas LJ, Dinger BG, Fidone SJ. Localization and *in vitro* actions of atrial natriuretic peptide in the cat carotid body. J Appl Physiol 1991; 70(2):942–946.
59. Stensaas LJ, Wang Z-Z, Dinger B, Fidone S. Alteration of atrial natriuretic peptide (ANP) immunostaining in the rat carotid body evoked by hypoxia. Soc Neurosci Abstr 1991; 17:118.
60. Wang Z-Z, Hirose S, Dinger B, Fidone S, Stensaas LJ. Localization of atrial natriuretic peptide receptors in the carotid body. Soc Neurosci Abstr 1991; 17:118.
61. Wang Z-Z, Stensaas LJ, de Vente J, Dinger B, Fidone SJ. Immunocytochemical localization of cAMP and cGMP in cells of the rat carotid body following natural and pharmacological stimulation. Histochemistry 1991; 96:523–530.
62. Wang Z-Z, Stensaas LJ, Wang W-J, Dinger B, de Vente J, Fidone SJ. Atrial natriuretic peptide increases cyclic guanosine monophosphate immunoreactivity in the carotid body. Neuroscience 1992; 49(2):479–486.
63. Wang W-J, He L, Chen J, Dinger B, Fidone S. Mechanisms underlying chemoreceptor inhibition induced by atrial natriuretic peptide in rabbit carotid body. J Physiol 1993; 460:427–441.
64. He L, Dinger B, Fidone S. Cellular mechanisms involved in carotid body inhibition produced by atrial natriuretic peptide. Am J Physiol Cell Physiol 2000; 278:C845–C852.

65. Chen J, He L, Dinger B, Fidone S. Dynamics of neuroactive and vasoregulatory agents in the carotid body. Proceedings of the XIVth International Symposium on Arterial Chemoreception, Philadelphia, 1999.
66. He L, Chen J, Dinger B, Fidone S. The involvement of multiple vasoactive agents in altered rat carotid body chemosensitivity induced by chronic hypoxia. Soc Neurosci Abstr 1998; 24:375.
67. Lahiri S, Smatresk N, Pokorski M, Barnard P, Mokashi A. Efferent inhibition of carotid body chemoreception in chronically hypoxic cats. Am J Physiol 1983; 245:R678–R683.
68. Calza L, Pozza M, Zanni M, Manzini CU, Manzini E, Hokfelt T. Peptide plasticity in primary sensory neurons and spinal cord during adjuvant-induced arthritis in the rat: An immunocytochemical and *in situ* hybridization study. Neuroscience 1998; 82(2):575–589.
69. Galeazza MT, Garry MG, Yost HJ, Strait KA, Hargreaves KM, Seybold VS. Plasticity in the synthesis and storage of substance P and calcitonin gene-related peptide in primary afferent neurons during peripheral inflammation. Neuroscience 1995; 66(2):443–458.
70. Fischer A, McGregor GP, Saria A, Philippin B, Kummer W. Induction of tachykinin gene and peptide expression in guinea pig nodose primary afferent neurons by allergic airway inflammation. J Clin Invest 1996; 98:2284–2291.
71. Undem BJ, Hunter DD, Liu M, Haak-Frendscho M, Oakragly A, Fischer A. Allergen-induced sensory neuroplasticity in airways. Int Arch Allergy Immunol 1999; 118:150–153.
72. Almaraz L, Wang Z-Z, Stensaas LJ, Fidone SJ. Release of dopamine from carotid sinus nerve fibers innervating type I cells in the cat carotid body. Biol Signals 1993; 2:16–26.
73. Finley JCW, Polak J, Katz DM. Transmitter diversity in carotid body afferent neurons: dopaminergic and peptidergic phenotypes. Neuroscience 1992; 51:973–987.
74. Ishizawa Y, Fitzgerald RS, Shirahata M, Schofield B. Localization of nicotinic acetylcholine receptors in cat carotid body and petrosal ganglion. Adv Exp Med Biol 1996; 410:253–256.
75. Kummer W. Retrograde neuronal labeling and double-staining immunohistochemistry of tachykinin and calcitonin gene-related peptide-immunoreactive pathways in the carotid sinus nerve of the guinea pig. J Auton Nerv Syst 1988; 23:131–141.
76. Wang Z-Z, Stensaas LJ, Dinger BG, Fidone SJ. Nitric oxide mediates chemoreceptor inhibition in the cat carotid body. Neuroscience 1995; 65:217–229.
77. Kondo H, Yamamoto M. Occurrence, ontogeny, ultrastructure and some plasticity of CGRP (calcitonin gene-related peptide)-immunoreactive nerves in the carotid body of rats. Brain Res 1988; 473:283–293.
78. Kummer W, Fischer A, Heym C. Ultrastructure of calcitonin gene-related peptide- and substance P-like immunoreactive nerve fibres in the carotid body and carotid sinus of the guinea pig. Histochemistry 1989; 92:433–439.
79. Torrealba F, Correa R. Ultrastructure of calcitonin gene-related peptide immunoreactive, unmyelinated afferents to the cat carotid body: a case of volume transmission. Neuroscience 1995; 64(3):777–785.
80. Fidone SJ, Sato A. A study of chemoreceptor and baroreceptor A and C-fibres in the cat carotid nerve. J Physiol 1969; 205:527–548.
81. Claps A, Torrealba F. The carotid body connections: a WGA-HRP study in the cat. Brain Res 1988; 455:123–133.

82. Finley JCW, Katz DM. The central organization of carotid body afferent projections to the brainstem of the rat. Brain Res 1992; 572:108–116.
83. Massari VJ, Shirahata M, Johnson TA, Gatti PJ. Carotid sinus nerve terminals which are tyrosine hydroxylase immunoreactive are found in the commissural nucleus of the tractus solitarius. J Neurocytol 1996; 25(3):197–208.
84. Haxhiu MA, Chang CH, Dreshaj IA, Erokwu B, Prabhakar NR, Cherniack NS. Nitric oxide and ventilatory response to hypoxia. Respir Physiol 1995; 101(3):257–266.
85. Torrealba F. Calcitonin gene-related peptide immunoreactivity in the nucleus of the tractus solitarius and the carotid receptors of the cat originates from peripheral afferents. Neuroscience 1992; 47(1):165–173.
86. Prabhakar N, Rao S, Premkumar D, Pieramici SF, Kumar GK, Kalari RK. Regulation of neuronal nitric oxide synthase gene expression by hypoxia: role of nitric oxide in respiratory adaptation to low PO_2. Adv Exp Med Biol 1996; 410:345–348.
87. Prabhakar NR, Pieramici SF, Premkumar DR, Kumar GK, Kalaria RN. Activation of nitric oxide synthase gene expression by hypoxia in central and peripheral neurons. Brain Res Mol Brain Res 1996; 43(1–2):341–346.
88. Gauda EB, Gerfen CR. Expression and localization of enkephalin, substance P and substance P receptor genes in the rat carotid body. In: Zapata P, Eyzaguirre C, Torrance RW, eds. Frontiers in Arterial Chemoreception. New York: Plenum Press, 1996:313–323.
89. Alcayaga J, Barrios M, Bustos F, Miranda G, Molina MJ, Iturriaga R. Modulatory effect of nitric oxide on acetylcholine-induced activation of cat petrosal ganglion neurons in vitro. Brain Res 1999; 825:194–198.
90. He L, Chen J, Dinger B, Fidone S. Altered cholinergic chemotransmission in rat carotid body following chronic hypoxia. FASEB J 2000; 14:A393.
91. Eyzaguirre C, Abudara V. Carotid body glomus cells: chemical secretion and transmission (modulation?) across cell-nerve ending junctions. Respir Physiol 1999; 115:135–149.
92. Fitzgerald RS, Shirahata M, Ide T. Further cholinergic aspects of carotid body chemotransduction of hypoxia in cats. J Appl Physiol 1997; 82(3):819–827.
93. Fitzgerald RS, Shirahata M. Release of acetylcholine from the in vitro cat carotid body. Adv Exp Med Biol 1996; 410:227–232.
94. Shirahata M, Ishizawa Y, Rudisill M, Schofield B, Fitzgerald RS. Presence of nicotinic acetylcholine receptors in cat carotid body afferent system. Brain Res 1998; 814:213–217.
95. Zhang M, Zhong H, Vollmer C, Nurse CA. Co-release of ATP and ACh mediates hypoxic signalling at rat carotid body chemoreceptors. J Physiol 2000; 525.1:143–458.
96. He L, Chen J, Dinger B, Fidone S. Chemical synaptic transmission in the chronically hypoxic rat carotid body. Soc Neurosci Abstr 2001; 26:282.
97. Eyzaguirre C. Carotid body gap junctions: secretion of transmitters and possible electric coupling between glomus cells and nerve terminals. Adv Exp Med Biol 2000; 475:349–357.
98. Abudara V, Eyzaguirre C. Electrical coupling between cultured glomus cells of the rat carotid body: observations with current and voltage clamping. Brain Res 1994; 664:257–265.
99. Abudara V, Eyzaguirre C. Modulation of junctional conductance between rat carotid body glomus cells by hypoxia, cAMP and acidity. Brain Res 1998; 792:114–125.

100. Abudara V, Eyzaguirre C. Effects of calcium on the electric coupling of carotid body glomus cells. Brain Res 1996; 725:125–131.
101. Monti-Bloch L, Abudara V, Eyzaguirre C. Electrical communication between glomus cells of the rat carotid body. Brain Res 1993; 622:119–131.
102. McDonald DM. Peripheral chemoreceptors: structure-function relationships of the carotid body. In: Hornbein TF, ed. Regulation of Breathing. Part I. New York-Basel: Marcel Dekker, 1981:105–319.
103. McDonald DM, Mitchell RA. The innervation of glomus cells, ganglion cells and blood vessels in the rat carotid body: a quantitative ultrastructural analysis. J Neurocytol 1975; 4:177–230.
104. Kondo H, Iwasa H. Re-examination of the carotid body ultrastructure with special attention to intercellular membrane appositions. In: Zapata P, Eyzaguirre C, Torrance RW, eds. Frontiers in Arterial Chemoreception. New York: Plenum Press, 1996:45–50.
105. Abudara V, Garces G, Saez JC. Cells of the carotid body express connexin43 which is up-regulated by cAMP. Brain Res 1999; 849:25–33.
106. Chen J, He L, Dinger B, Stensaas L, Fidone S. Chronic hypoxia upregulates connexin43 expression in rat carotid body and petrosal ganglion. J Appl Physiol 2002. In press.
107. Barnard P, Andronikou S, Pokorski M, Smatresk N, Mokashi A, Lahiri S. Time-dependent effect of hypoxia on carotid body chemosensory function. J Appl Physiol 1987; 63(2):685–691.

26

Neurochemical Processes Involved in Acclimatization to Long-Term Hypoxia

VINCENT JOSEPH

Laval University Schoool of Medicine
Quebec, Ontario, Canada

JEAN-MARC PEQUIGNOT

Université Claude Bernard Lyon I
Lyon, France

I. Introduction

Acclimatization to long-term hypoxia is a process that takes place following ascent to high altitude and that allows a gradual improvement of the ability to tolerate the hypoxic environment and perform physical or mental tasks under optimal conditions. An important component of this process is the hypoxic ventilatory acclimatization (HVA) that develops over several days depending on the altitude reached. Acclimatization to hypoxia reveals a striking plasticity of the chemoreflex, which takes place within the first days of exposure and can be prolonged for weeks, months, or years. The initial rise in ventilation in response to low arterial PO_2 is followed by a later ventilatory reduction (hypoxic ventilatory roll-off) within the first minutes of exposure (1). Extending the exposure to hypoxia for several hours to months elicits a secondary time-dependent increase in ventilatory output despite a continuously increasing arterial PO_2 (2). The mechanisms responsible for HVA are not well understood and have been examined in recent reviews (3–7). There is clear-cut evidence that the peripheral arterial chemoreceptors play a major role in initiating the ventilatory acclimatization to hypoxia, but this does not preclude a role for central structures involved in the translation of chemosensory inputs and that can modulate the integration of carotid chemoafferent inputs. Early and recent studies have shown that the ventilatory plasticity induced by sustained hypoxia is associated with changes in the morphology and phenotype of the carotid chemoreceptors, increases in neurotransmitter biosynthesis and release, modulation of receptor expression in the carotid body, and increased firing rate of the carotid chemoafferent neurons. More recent studies demonstrated that the neuroplasticity also takes place during long-term hypoxia in restricted areas of the central nervous system that have been implicated in respiratory and sympathetic responses to hypoxia.

This chapter deals with the neurochemical plasticity induced by long-term hypoxia in the carotid body (with special emphasis on the dopaminergic pathways) and in brainstem structures involved in translation of the peripheral chemosensory inputs, and their possible role in triggering or modulating HVA.

II. The Carotid Bodies During High-Altitude Acclimatization

The ventilatory acclimatization to hypoxia is almost exclusively dependent on the carotid bodies as demonstrated by two studies carried out on sheep showing that HVA is not achieved after chemodenervation (8), but can be induced by perfusion of the carotid bodies with hypoxic normocapnic blood (9). The drop of arterial oxygen tension consecutive to hypoxic exposure induces a cascade of events—which is still not fully understood—leading to the associated anatomical, neurochemical, and functional changes within the carotid bodies.

A. Anatomical and Neurochemical Changes

Some of the most important characteristics of chronically hypoxemic carotid bodies are a marked hypertrophy and enhanced content, turnover, and synthesis of catecholamines (dopamine and norepinephrine) (10a,10b). In our own studies done in adult male rats permanently living at high altitude (La Paz, Bolivia, 3600 m), the content of norepinephrine and dopamine were respectively 24 and 43 times higher in the carotid bodies compared to age-matched rats at sea level, while the activity of tyrosine hydroxylase (the rate-limiting enzyme for catecholamines synthesis) was almost six times higher in high-altitude rats (10). The components of carotid body hypertrophy under chronic hypoxia are an increased number and volume of chemosensitive cells (11–13), an increased number of fibroblasts located in the conjunctive walls surrounding the carotid body (14), and an increase of blood vessels caused by both vasodilatation and ingrowth of new blood vessels (15). Additionally, glomic cells are submitted to ultrastructural changes in chronic hypoxia revealed by examining hypoxic carotid bodies by electron microscopy. These changes include an increased volume density of mitochondria and enlargement of dense-core vesicles that store dopamine and norepinephrine (11). There is also a gradual transformation of the morphological appearance of glomic cells, which shifts from a bimodal pattern (with both small- and large-vesicles cells) to a unimodal pattern exhibiting mainly large-vesicle cells (11). These changes are accompanied by a global phenotypic transformation with a very large increase in the number of noradrenergic cells (16).

Increased or decreased cell number in a given organ is related to an interplay of factors that regulate cell cycle and the balance between cellular survival and apoptosis. Accordingly, the increased cell number in the carotid bodies under chronic hypoxia may be related to changes affecting the activity of specific growth factors and/or changes in the activity of apoptosis-promoting factors. To date very few studies have assessed these points, but it has been shown that the increase in vascularity of carotid body blood vessels under chronic hypoxia is dependent on an up-regulation of the vascular endothelial growth factor (17). The drastic changes observed in the carotid bodies under chronic hypoxemia appears a promising tool to explore some aspects of O_2-dependent tissue growth, which is a point of broad interest in cancer medicine and developmental biology (18).

Chemosensitive cells (also termed type I, or glomic, cells) within the carotid bodies are directly sensitive to the oxygen tension as in vitro study showed a three- to fourfold increase of glomic cells volume following chronic hypoxic exposure (19). Furthermore, the increased synthesis and content of catecholamine by glomic cells are clearly related to a marked effect of low oxygen tension on the synthesis and stability of tyrosine hydroxylase mRNA (20). If in vitro models are very important to understand the mechanisms of carotid body changes in chronic hypoxia, they may differ from in vivo processes. In vivo sustained hypoxia induces an elevation of catecholamine synthesis that largely overruns the utilization rate and produces a marked increase of catecholamine stores within the glomic cells. This is in striking contrast to in vitro studies showing increased release of dopamine without significant increased dopamine stores within cultured glomic cells after 3 weeks of hypoxia (21).

B. Functional Changes

The functional changes within the carotid bodies are tightly connected to the anatomical and neurochemical changes occurring under chronic hypoxia. If some results show decreased carotid body hypoxic sensitivity after chronic hypoxia (22), it is now well established that successful ventilatory acclimatization relies on an increased sensitivity of carotid body chemoreceptors. Actually, numerous experimental approaches, ranging from hypoxic ventilatory response in awake animals (10,23) to in vitro measurements of ion channel function on glomic cells (24), are consistent with increased carotid body sensitivity following chronic hypoxia.

The Dopaminergic Hypothesis

Among the numerous neuroactive factors synthesized by carotid body glomic cells, the role of dopamine has been particularly studied during chronic hypoxia. Dopamine is found at high concentration and has been recognized as a potent inhibitory neuromodulator of carotid body chemotransduction both under acute hypoxia and after HVA (25–31). This point deserves some explanation as it may appear a mind-puzzling paradox: from the consistent body of literature concerning this aspect, one may reach the conclusions that: (1) carotid body dopamine inhibits ventilation, (2) ventilation progressively increases during a stay at altitude, and finally, (3) carotid body dopamine content and utilization progressively increase during a stay at altitude. If the inhibitory effect of dopamine on carotid body chemoresponsiveness is clearly established, dopamine may also have a marked stimulatory effect (32). For example dopamine appears to be a very important factor of HVA: in transgenic mice lacking the dopaminergic D2 receptors, HVA cannot be achieved as in wild-type control mice (33). These dual actions of carotid body dopamine are considered to depend on the subsets of dopaminergic receptors present on the pre- and postsynaptic elements in the carotid bodies: according to the chemotransduction model of Gonzalez, when released from carotid body type I cells under hypoxemic stimulation, dopamine has two complementary actions: (1) it depolarizes carotid sinus nerve endings by acting through low-affinity excitatory postsynaptic D2 receptors, and (2) it binds to high-affinity presynaptic D2 receptors that exert a negative feedback on the release of dopamine, thus indirectly decreasing the binding of dopamine to postsynaptic excitatory receptors (25). This last effect appears to be predominant as dopamine is mainly considered an inhibitor of carotid body chemoresponsiveness. Furthermore, postsynaptic D1 receptors that are expressed in the carotid sinus nerve endings of both cats and rabbits (34) are probably involved in depolarization of nerve endings following dopamine release from glomic cells.

In this model of dopaminergic chemotransduction, the apparent aforementioned paradox may be resolved if one considers that the enhanced catecholamine content and general carotid body hyperplasia under chronic hypoxia are associated with an enhanced number of functional chemoreceptor units (composed of at least one glomic cell and one sensitive nerve ending forming a synapse) within the carotid body (Fig. 1). In this pattern, the overall activity of the carotid sinus nerve (which conveys the chemosensitive information from the carotid body to the brainstem respiratory centers) may be enhanced as well as dopamine content.

In addition to these anatomical changes, neurochemical adjustments may be involved in more discrete changes associated with HVA; for example, the inhibitory effect of dopamine is markedly reduced following 3 days of hypoxia in cats (35). However, this effect is reduced if hypoxia is prolonged: in two other studies the inhibitory effect of dopamine was transiently reduced in rats following two days of hypoxia, but was restored following 8 days of hypoxia (13,30). This transient desensitization is associated with a marked depletion of carotid body dopamine content, due to enhanced release during the first days of hypoxia (13), rapidly counterbalanced by increased tyrosine hydroxylase activity. Changes in the dopaminergic transmission pathway under chronic hypoxia also affect the expression of D2-receptor mRNA within the carotid body, which shows an early peak of increased expression following 12 hr of hypoxia and a late increase after 7 days of sustained hypoxia (36). In vitro experiments have shown that nicotinic mechanisms participate to a large extent in the increased dopamine release induced by chronic hypoxia, indicating critical interactions between acetylcholine and dopamine within glomic cells (37).

Recent studies by Kline et al. (38), using mice partially deficient in the transcription factor hypoxia inducible factor-1α (HIF-1α) demonstrated the

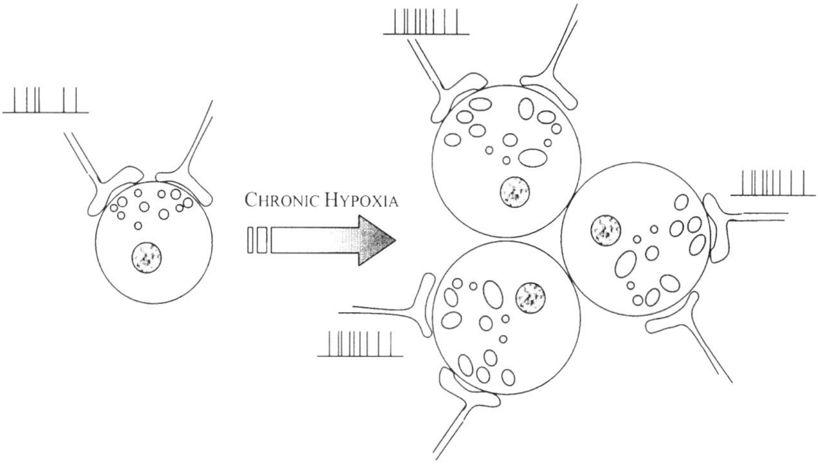

CB IN NORMOXIA	CB IN CHRONIC HYPOXIA
• Normal-sized cells with small and large vesicles.	• Enlarged cells with larger vesicles.
• Low stocks and turnover of catecholamines.	• High stocks and turnover of catecholamines.
• Few or no dividing cells.	• New cells appeared after intense cell division during the first days of hypoxic exposure.

Figure 1 Schematic drawing showing a hypothetical model explaining the changes of CB functions during chronic hypoxia: in normoxia, a fixed number of chemoreceptor units (glomic cell and carotid sinus nerve terminal) sense arterial O_2 and carotid sinus nerve activity is low. After chronic hypoxia each unit increases its firing rate due to constitutive changes and more units sense arterial O_2; this results in increased carotid sinus nerve activity.

implication of HIF-1α in carotid body functional responses to hypoxia. In vitro, the partial deficiency of HIF-1α resulted in a marked decrease of carotid body hypoxic chemosensitivity, but in vivo hypoxic ventilatory response was not impaired, owing to an increased utilization of chemoreceptors other than the carotid bodies in these transgenic mice. These studies also showed enhanced expression of HIF-1α in glomic cells after 45 min of hypoxia in wild-type mice. Following chronic hypoxia, transgenic mice failed to show enhanced hypoxic ventilatory response as in wild-type controls, suggesting impaired ventilatory acclimatization. However, the relative influence of the distinct peripherals and/or central processes was not elucidated in this study (38).

C. Other Mechanisms

A large variety of neuroactive chemicals have been identified in carotid body, which may all be involved in the changes observed during long-term hypoxia. Norepinephrine (NE) is found at high concentrations in glomic cells and sympathetic nerve endings in the carotid body. NE content and the number of noradrenergic cells increase during long-term hypoxia (16), thus a potential role for NE may be hypothesized during HVA. However, experimental data based on sympathetic denervation, $\alpha 2$- or β-receptors blockade, failed to show any noradrenergic component for both the time course and magnitude of HVA (39–41). Nevertheless, the question of the implication of the shift from dopaminergic to noradrenergic phenotype (see above) during HVA remains open. On the other hand, noradrenergic (42) or dopaminergic (43) sources outside the carotid body may also be involved in functional carotid body output during HVA. One study suggested a close link between ventilatory and sympathetic activation during hypoxia, this linkage being brought about by a non-β-adrenergic mechanism (42).

The vasoactive peptide endothelin, localized in glomic cells, enhances the release of catecholamine from hypoxic carotid body in vitro, and its content is increased in glomic cells during chronic hypoxia (44). After chronic hypoxia, the carotid body becomes more sensitive to the endothelin antagonist BQ-123 (45). These findings suggest that endothelin is involved in the carotid body mechanisms mediating HVA. Changes that are not consistent with increased carotid body hypoxic sensitivity during chronic hypoxia have also been reported, including decrease of the excitatory substance P immunoreactivity in glomic cells and increased inhibitory effect of nitric oxide (see Ref. 6 for review).

III. Importance of Acclimatization to Long-Term Hypoxia at Altitude and Hormonal Control of Dopaminergic Metabolism in the Carotid Bodies

A proper acclimatization to hypoxia is very important and determines the overall ability to cope with hypoxia. In climbers at very high altitude, the maximum altitude that an individual can reach is clearly related to the hypoxic ventilatory response measured at sea level and to the extent of the ventilatory acclimatization during ascension (46,46a). At lower altitude, a proper ventilatory acclimatization is also a determinant mechanism for the overall hypoxic acclimatization. Our own observations (unpublished) in rats permanently living at 3600 m above sea level (Bolivian Institute for Altitude Biology—IBBA, La Paz, Bolivia) revealed that following surgical chemodenervation, the hematocrit rose from a baseline level of around 50% to more than 70% within a few weeks and numerous animals died during this period.

In human populations, it is estimated that 10–15% of young and more than 30% of aged (above 50 years) high-altitude permanent residents (>2500 m) are at risk to develop chronic mountain sickness (47,48). Some of the major features of chronic mountain sickness are chronic alveolar hypoventilation associated with low arterial oxygen saturation and low hypoxic ventilatory responsiveness that lead to elevated hematocrit and hemoglobin concentration (49,50). This process, also termed "high altitude deacclimatization," is clearly gender-dependent, with men being much more sensitive than women, at least until menopause, which is accompanied by marked deacclimatization in women (51,52). Gonadal steroids are critically involved in this gender discrepancy (52,53) and progesterone associated with estrogens are efficient in the treatment of chronic mountain sickness by stimulating resting minute ventilation (54). Recent findings from our laboratory showed that ovarian steroids stimulate breathing by reducing the inhibitory dopaminergic drive in the carotid bodies (31). Interestingly, this study revealed, once more, the duality of dopamine actions in the carotid bodies: while domperidone (a specific peripheral D2 antagonist) enhanced resting minute ventilation in ovariectomized females, idomperidone inhibits resting minute ventilation in females following 10 daily injections of ovarian steroids (see Fig. 2). This demonstrates that ovarian steroids are able to modify the dopaminergic component of carotid body chemoresponsiveness that is determinant for high-altitude rats (10,31). Furthermore, these results showed that the resting level of minute ventilation, under strictly normoxic conditions, is submitted to a determinant peripheral dopaminergic drive, and that individual differences in the control of breathing may be related to individual differences in the carotid body dopaminergic drive (Fig. 2). Thus changes in carotid body dopaminergic metabolism may explain the deacclimatization

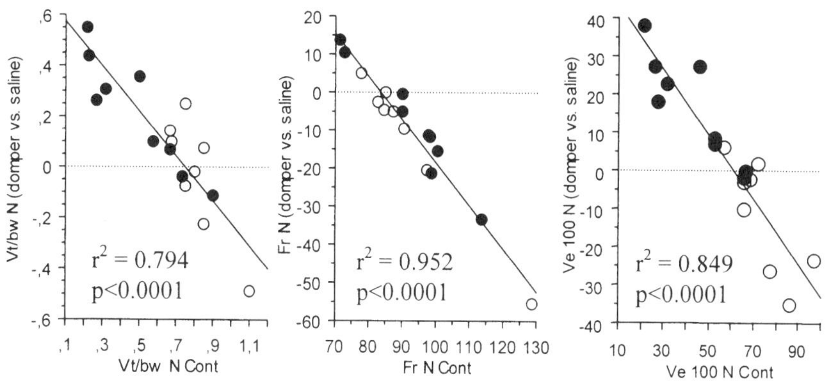

Figure 2 Correlation between individual values of tidal volume (Vt/bw, mL/100 g), respiratory frequency (Fr, breaths/minute), and minute ventilation (Ve 100, mL/min/100 g) measured by whole-body plethysmography after saline control injection (N, x-axis) and the net effect (= value after domperidone injection – value after saline control injection; y-axis) of a single domperidone injection (1 mg/kg, 48 hr after saline injection). All measurements have been done in normoxia, before (●) and after (○) 10 days of hormonal (progesterone + estradiol) treatment in ovariectomized females. For each graph the points above the horizontal dotted line indicate stimulation of the corresponding parameter and points below the line indicate inhibition. Regression lines and regression statistics (r^2 and p) are shown for each graph.

syndrome observed following menopause in permanent-high-altitude women (51,52). Furthermore, individual factors affecting carotid body dopaminergic metabolism may explain individual variability in the amplitude of HVA.

IV. Neurochemical Acclimatization to Hypoxia in the Central Nervous System

Although the neurochemical and morphological plasticity of the peripheral arterial chemoreceptors certainly plays a prominent role in increasing sensitivity of the carotid body to hypoxia, the adaptive response to hypoxia may involve further alternative mechanisms (4,6,7,55). Evidence is accumulating to suggest that, under long-term hypoxia, the central translation of peripheral chemoreceptor inputs may be modulated by the plasticity of brainstem cardiorespiratory structures. Systemic hypoxia can induce c-Fos expression in the dorsomedial and ventrolateral medulla oblongata including the nucleus tractus solitarius (NTS), ventrolateral medulla, area postrema, as well as in the caudal raphe nuclei (nucleus raphe pallidus and nucleus raphe magnus), or in the pons medulla including the lateral parabrachial nucleus, the Kölliker-Fuse nucleus, the A5 area, and the locus ceruleus (56). This widespread cellular activation emphasizes the diversity of central structures activated by systemic and/or central hypoxemia.

Neurophysiological and neuroanatomical studies have selectively labeled the central projections of chemosensitive fibers caudally to the obex within the NTS in cats and rats, whereas barosensory fibers project rostral to the obex (57–59). The NTS, which contains the premotor neurons of the dorsal respiratory group (60), is now recognized as the major projection site of chemosensory fibers. Neurons of caudal NTS with carotid chemoreceptor inputs project onto the ventrolateral medulla (61) that contains the premotor neurons of the ventral respiratory group (60).

A. Neurotransmitters and Neuromodulators

Excitatory Amino Acid: Glutamate

Under acute hypoxia carotid sinus nerve terminals within the NTS release glutamate, which in turn stimulates breathing (62–65). It has been shown that hypoxia induces c-Fos protein expression in NMDA-glutamate-receptor-labeled neurons within the NTS, an effect that was attenuated by MK-801, a specific antagonist of NMDA-glutamate receptors (62,66). Thus, NMDA-glutamate receptors are critical for the central synaptic relays leading to an appropriate ventilatory response during acute hypoxia. Considered with the pivotal role that glutamate plays in the acute hypoxic ventilatory response, change in the expression of NMDA receptors appears as a potential candidate to contribute to time-dependent changes in the hypoxic ventilatory response during chronic hypoxia. The future directions should test the hypothesis that synaptic plasticity of excitatory amino acid neurotransmission is part of the central mechanisms of HVA.

Monoamines

Many of the neurons in the central nervous system expressing the c-Fos protein under hypoxia are serotonergic or catecholaminergic (56). Both electrical stimulation of the carotid sinus nerve and hypoxic stimulation of the carotid bodies induce Fos-like immunoreactivity in serotonergic neurons of the caudal raphe nuclei and the parapyramidal cell group that play a role in modulating ongoing cardiorespiratory activity. The same study showed a colocalization of the c-Fos protein and tyrosine hydroxylase in catecholaminergic neurons located in several brainstem areas involved in cardiorespiratory functions, i.e, the NTS area, the dorsal vagal complex, the ventrolateral medulla oblongata, the locus ceruleus, and the A5 noradrenergic cell group in pons (56).

Serotonin

Serotonin is known to be an important neuromodulator of respiratory output (67). Long-term facilitation of respiratory motor output following episodic hypoxia is dependent on activation of serotonin receptors (68–70). Long-term hypoxia changes the central metabolism of serotonin in discrete brain regions involved in cardiorespiratory control, including the raphe nuclei, the NTS area, the locus ceruleus, and the ventrolateral medulla (71), and it has been hypothesized that serotonin is a neural component necessary for facilitation of ventilation associated with acclimatization to hypoxia. However, the available evidence based on pharmacological manipulations fails to show the role of serotonin in the control of HVA. Indeed, systemic depletion of serotonin stores by *p*-chlorophenylalanine in rats did not alter the time-dependent increase in ventilatory output associated with HVA (72). Similar results were obtained after blockade of serotonin receptors by methysergide, although methysergide was able to enhance the ventilatory response to acute hypoxia (73).

Dopamine

Both medullary catecholaminergic neurons and, to a lesser extent, the carotid sinus nerve endings (74,75) contribute to the pool of tyrosine hydroxylase protein present in caudal NTS. Catecholamines released from both compartments, i.e., dopamine from the chemosensory nerve endings and norepinephrine from A2 cells, respectively, have been implicated in the functional responses to hypoxia.

Dopaminergic neurons and dopamine D2 receptors have been localized in the NTS (74,76). In response to hypoxia, dopamine is released from dopaminergic neurons in the NTS (77). Central blockade of dopamine D2 receptors decreases ventilation for a given level of arterial chemoreceptor activity (78), implying that central dopamine D2 receptors facilitate the hypoxic ventilatory response. Thus, an enhanced effect of central dopamine on ventilation during chronic exposure to hypoxia may contribute to the increased central gain of the hypoxic ventilatory response. This point has been demonstrated by recent studies in rats exposed to 0, 2, and 8 days of hypoxia: following blockade of dopamine D2 receptors in the central nervous system, minute ventilation decreased significantly more after 8 days of hypoxia than in normoxia, but did not change significantly after 2 days of hypoxia

(30). Thus, a transitory insensitivity to dopamine appeared during the first days of hypoxia, and central dopamine contributes to ventilatory acclimatization between 2 and 8 days of hypoxia. The level of mRNA encoding dopamine D2 receptors was initially increased in the caudal NTS within the first 12 hr of hypoxia and then decreased below control levels within the next 7 days (36). These time-dependent changes in dopamine D2 receptors could also contribute to HVA.

Norepinephrine

There is growing evidence that the medullary catecholaminergic areas participate in the control of cardiorespiratory functions under normoxia or hypoxia (60,79–81) and can display a functional neuroplasticity under environmental challenges such as hypoxia (82,83). Caudally the NTS contains the A2 noradrenergic cell group and rostrally the C2 adrenergic cell group, although there is a midmedullary zone where the groups overlap (74). The A2 cell group is adjacent to the dorsal respiratory premotor neurons and a role for A2 cells has been attributed in neuromodulatory control of the central respiratory output. Respiratory neurons possess adrenergic receptors (84) and receive close appositions from tyrosine hydroxylase–immunoreactive neurons (85). Central catecholaminergic neurons tonically modulate the brainstem respiratory rhythm generator (86) and this effect is mediated by $\alpha 2$-adrenergic receptors (84,87,88). Discrete lesions of the caudal NTS in the commissural nucleus (caudal part of the A2 cell group), which is the primary site of projection of carotid chemoafferent fibers, attenuated the ventilatory response to hypoxia whereas lesions rostral to the obex were ineffective (89). Consistent with the finding of Housley and Sinclair (89), we described a functional segregation of the A2 catecholaminergic neurons since long-term hypoxic rats display activation of A2 neurons located caudal to the obex. The changes include increased turnover of norepinephrine and increased activity and expression of tyrosine hydroxylase protein and mRNA (82,83,90). Tyrosine hydroxylase activity in A2 neurons increased gradually after several days of sustained hypoxic exposure to reach the maximal level after 10–14 days in rats (90). The changes in tyrosine hydroxylase protein amount within the NTS were significantly correlated with concomitant increases in ventilatory output after 2 weeks of hypoxia and a possible link between HVA and stimulation of A2 neurons has been suggested (90).

Systemic hypoxia stimulates A1/C1 catecholamine cells in caudal ventrolateral medulla (91). Consistent with this finding, increased levels of tyrosine hydroxylase mRNA have been found in the caudal ventrolateral medulla of rats subjected to long-term hypoxia (82). Neuroanatomical studies showed close interrelationships between noradrenergic neurons in the ventrolateral medulla and respiratory premotoneurons (92,93). From all the available evidence, it is tempting to speculate that the changes in neuronal noradrenergic function might act as a fine tuning of the increasing ventilatory output in long-term hypoxic rats: a short-term low noradrenergic activity in medullary areas implicated in respiratory control could be a factor favoring the gradual increase in hypoxic ventilatory response, in opposition with the delayed high noradrenergic neuronal activity, which limits the magnitude of HVA, and contribute to the ventilatory steady-state reached after completion of HVA (Fig. 3).

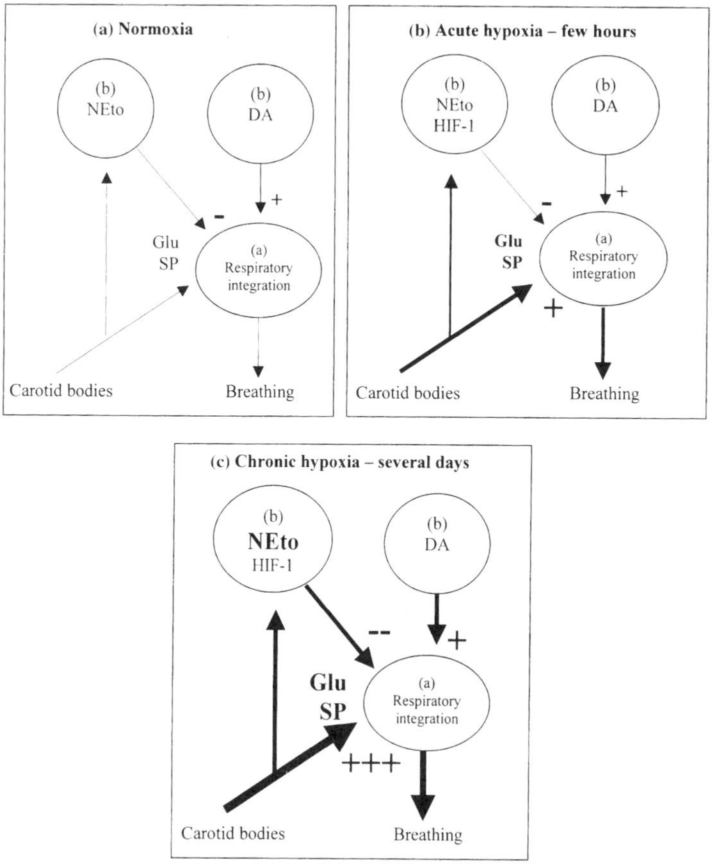

Figure 3 Hypothetical model of a putative catecholaminergic modulation of ventilation following acute (b) and chronic hypoxia (c) compared to resting conditions (normoxia—a). Carotid bodies send axonal projections to distinct sets of neurons including neurons directly involved in respiratory rhythm generation and modulation (a) and catecholaminergic neurons (b). In normoxia the level of peripheral stimulation is low (a). During acute hypoxic exposure (b), carotid terminals release glutamate (Glu—see text) and substance P (SP—see text) in target neurons, which stimulate breathing. HIF-1 protein level rapidly increases in noradrenergic neurons, probably as a result of direct neural O_2 sensing (see text). HIF-1 stimulates the expression of tyrosine hydroxylase (TH) mRNA, among a variety of HIF-1 responsive genes. After several days of hypoxic exposure (c), extensive transformation of carotid bodies is completed, and the release of glutamate and substance P is sustained at a high level. Norepinephrine turnover is increased owing to concomitant effects of transsynaptic activation and increased mRNA and protein expression. This high level of noradreneric activity exerts an inhibitory tone on breathing. At the same time, the stimulatory effect of dopamine is reinforced (see text). Alternatively, changes in receptor density at synapses (not represented) may also participate in subtle changes during HVA. Relative thickness of the arrows represents the level of activation of the corresponding pathway. Abbreviations: DA: dopamine; NEto: norepinephrine turnover; Glu: glutamate; SP: substance P; HIF-1: hypoxia-inducible factor 1.

Neuropeptides: Neuropeptide Y, Substance P

Neuropeptide Y is colocalized with catecholamines in the noradrenergic A1 cell group and in the locus ceruleus (94,95). Neuropeptide Y microinjections in the ventrolateral medulla induce hypertensive and bradycardic responses (96) and long-term hypoxia enhances the neuropeptide Y content in the rat ventrolateral medulla. This may be linked to the excitation of a bulbospinal neuropeptide Y sympathoexcitatory system mediating the hypoxia-induced sympathoadrenal activation (97).

Several anatomical, neurochemical, and pharmacological studies have identified substance P as a putative neurotransmitter in the acute hypoxic chemotransduction pathway. Substance P–like immunoreactive fibers have been identified in the carotid sinus nerve and a high density of substance P–immunoreactive fibers and cell bodies are closely located to respiratory neurons (98). Microinjection of substance P into the NTS induces increases in tidal volume and respiratory frequency (99). Acute hypoxia increases the release of substance P in NTS in cats and rabbits and this effect is not apparent after bilateral chemodenervation (100,101). Substance P directly applied to the medulla oblongata of rabbits causes an increase in both tidal volume and ventilatory frequency, while a substance P antagonist applied on the ventrolateral surface of medulla oblongata of rabbit pups inhibits the ventilatory response to acute hypoxia (102). On the other hand, long-term hypoxia failed to induce any change in substance P content of the NTS, providing no evidence to support a significant role for substance P in acclimatization processes to hypoxia (97).

B. Putative Mechanisms Involved in Control of Neuronal Acclimatization to Hypoxia

The activity of most central neurons is depressed by hypoxia, an effect that provides some protection of neuronal integrity by reducing the O_2 uptake (103). There is, however, increasing evidence that central neurons might develop appropriate metabolic responses to hypoxia. Using cytochrome oxidase histochemistry as an indicator of changes in energy demand in response to long-term hypoxia, LaManna et al. (104) found overall decreased cytochrome oxidase activity within the hypoxic rat brain. Brainstem regions associated with respiratory control centers, however, maintain relatively preserved cytochrome oxidase activity. It has been proposed that these responsive neurons may be part of a widespread central O_2-chemosensing complex associated with cardiorespiratory control (105).

Chemosensory mechanisms in the NTS of rat can develop following irreversible disruption of the chemoafferent pathway. Ventilatory acclimatization to hypoxia was altered but still persisted in carotid sinus nerve–transected rats (106). Since the caudal NTS is a major site of integration for peripheral chemosensory inputs, it is worth noting that the hypoxia-induced increase in tyrosine hydroxylase (TH) mRNA in caudal NTS was not impaired by carotid chemodenervation (106). On the other hand, transsynaptic activation by carotid body inputs is a key element in the changes of norepinephrine turnover during long-term hypoxia since carotid body denervation can prevent them (107). The reasons for this discrepency are not fully understood, but the fact that TH mRNA in NTS is still increased by hypoxia after chemodenervation, is a strong argument in favor of central oxygen sensing.

In vivo and in vitro studies demonstrate that direct central oxygen sensing may elicit excitatory and inhibitory effects on ventilation (108–114). The influence of humoral factors derived from deoxygenated blood on respiratory centers has been demonstrated by a recent study: Lipton and colleagues (115) provided evidence that S-nitrosothiols (complexes of NO bound to a thiol group in cysteine, released from deoxyhemoglobin) participate in respiratory response to hypoxia by acting on neurons in the NTS. In addition to this nongenomic effect of hypoxia on NTS neurons, modulation of gene expression by increased level of the hypoxia-inducible factor-1 (HIF-1) contributes to induction of adaptive responses to hypoxia. HIF-1 is responsible for the activation of a number of genes that contribute to the metabolic or hematological adjustments required by the reduction in oxygen availability (116–118). An oxygen-responsive fragment containing AP1, AP2, and HIF-1 cis-regulatory elements has been identified within the tyrosine hydroxylase gene promoter, which may regulate the transcription of tyrosine hydroxylase gene (119,120). Pascual and colleagues (121) showed that moderate hypoxia (10% O_2) sustained for several hours can induce HIF-1α protein expression selectively in neurons located in the caudal NTS and in the ventrolateral medulla. A subset of A2C2 and A1C1 catecholaminergic neurons colocalized tyrosine hydroxylase and HIF-1α protein, suggesting that HIF-1 may participate in the control of hypoxia-induced central expression of tyrosine hydroxylase. Therefore, neurons in the NTS also appear as a target for direct modulatory effects of hypoxia. Under chronic hypoxia, Chavez et al. (122) described a sustained elevation of HIF-1α, which persisted 2 weeks before returning to baseline, in neurons and endothelial cells of rat brain. This enhanced HIF-1 expression was associated with increased VEGF and glucose transporter GLUT-1 expression, thus showing central vascular and metabolic remodelling under chronic hypoxia. The functional implication in terms of ventilatory acclimatization is not described yet.

V. Conclusion

Exposure to long-term hypoxia elicits a number of changes in neurotransmitters present in the carotid body and brainstem cardiorespiratory areas. It is meaningful that the changes in the activity of neurotransmission were detected at different levels of the metabolic pattern, i.e., genic and protein expression, neurotransmitter turnover, receptor expression, and are associated with cellular plasticity of the glomic cells and of neurons involved in integration of chemoafferent inputs. From a cellular point of view, hypoxia may be considered a stress factor that disturbs a preestablished steady state and progressively leads to subsequent reorganization of cellular systems. For integrative physiology, the level of ventilatory output after acclimatization to long-term hypoxia is determined by the balance between a variety of excitatory and inhibitory processes that take place both in carotid bodies and in central areas (see also Fig. 3). Future studies should provide more data on the molecular and genetic basis of the mechanisms responsible for this plasticity and on the role of neurotransmitters in HVA. Promising perspectives include investigations on the plasticity of excitatory amino acid receptors during long-term exposure to hypoxia and on the role of blood NO derivates and transcriptional factors as signaling factors potentially able to trigger the neuronal plasticity associated to HVA and its potential physiopathological implications.

References

1. Hoop B, Beagle JL, Maher TJ, Kazemi H. Brainstem amino acid neurotransmitters and hypoxic ventilatory response. Respir Physiol 1999; 118:117–129.
2. Weil J. Control of ventilation in chronic hypoxia: role of peripheral chemoreceptors. In: Lahiri S, Cherniak N, Fitzgerald R, eds. Response and Adaptation to Hypoxia: Organ to Organelle. New York: Oxford University Press, 1991:122–132.
3. Bisgard GE. Increase in carotid body sensitivity during sustained hypoxia. Biol Signals 1995; 4:292–297.
4. Powell FL, Milsom WK, Mitchell GS. Time domains of the hypoxic ventilatory response. Respir Physiol 1998; 112:123–134.
5. Lahiri S, Rozanov C, Cherniack NS. Altered structure and function of the carotid body at high altitude and associated chemoreflexes. High Alt Med Biol 2000; 1:63–74.
6. Bisgard GE. Carotid body mechanisms in acclimatization to hypoxia. Respir Physiol 2000; 121:237–246.
7. Powell FL, Huey KA, Dwinell MR. Central nervous system mechanisms of ventilatory acclimatization to hypoxia. Respir Physiol 2000; 121:223–236.
8. Smith C, Bisgard G, Nielsen A, Daristotle L, Kressin N, Forster H, Dempsey J. Carotid bodies are requiered for ventilatory acclimatization to chronic hypoxia. J Appl Physiol 1986; 60:1003–1010.
9. Busch MA, Bisgard GE, Forster HV. Ventilatory acclimatization to hypoxia is not dependant on arterial hypoxemia. J Appl Physiol 1985; 58:1874–1880.
10. Joseph V, Soliz J, Pequignot J, Sempore B, Cottet-Emard JM, Dalmaz Y, Favier R, Spielvogel H, Pequignot JM. Gender differentiation of the chemoreflex during growth at high altitude: functional and neurochemical studies. Am J Physiol Regul Integr Comp Physiol 2000; 278:R806–R816.
10a. McGregor KH, Gil J, Lahiri S. A morphometric study of the carotid body in chronically hypoxic rats. J Appl Physiol 1984; 57:1430–1438.
10b. Hanbauer I, Karoum F, Hellstrom S, Lahiri S. Effects of hypoxia lasting up to one month on the catecholamine content in rat carotid body. Neuroscience 1981; 6:81–86.
11. Pequignot JM, Hellstrom S, Johansson C. Intact and sympathectomized carotid bodies of long-term hypoxic rats: a morphometric ultrastructural study. J Neurocytol 1984; 13:481–493.
12. Pallot D, Bee D, Barer G, Jacob S. Some effects of chronic stimulation on the rat carotid body. In: Eyzaguirre C, Fidone S, Fitzgerald R, Lahiri S, McDonald D, eds. Arterial Chemoreception. New-York: Springer-Verlag, 1990:293–301.
13. Bee D, Pallot DJ. Acute hypoxic ventilation, carotid body cell division, and dopamine content during early hypoxia in rats. J Appl Physiol 1995; 79:1504–1511.

14. Laidler P, Kay J. A quantitative morphological study of the carotid bodies of rats living at a stimulated altitude of 4300 m. J Pathol 1975; 117:183–191.
15. Pequignot JM, Hellstrom S. Intact and sympathectomized carotid bodies of long-term hypoxic rats. A morphometric light microscopical study. Virchows Arch A Pathol Anat Histopathol 1983; 400:235–243.
16. Verna A, Schamel A, Pequignot JM. Long term hypoxia increases the number of norepinephrine-containing glomus cells in the rat carotid body: a correlative immunohistochemical and biochemical study. J Auton Nerv Syst 1993; 44:171–177.
17. Chen J, Major J, Dinger B, Stensaas L, Fidone S. Inhibition of hypoxia-induced angiogenesis in the rat carotid body by a growth factor-receptor blocker. FASEB J 1998; 12:A168.
18. Maltepe E, Simon M. Oxygen, genes, and development: An analysis of the role of hypoxic gene regulation during murin vascular development. J Mol Med 1998; 76:391–401.
19. Mills L, Nurse C. Chronic hypoxia in vitro increases volume of dissociated carotid body chemoreceptor. Neuroreport 1993; 4:619–622.
20. Czyzyk-Krzeska M, Bayliss D, Lawson E, Millhorn D. Regulation of tyrosine hydroxylase gene expression in the rat carotid body by hypoxia. J Neurochem 1992; 58:1538–1546.
21. Jackson A, Nurse C. Dopaminergic properties of cultured rat carotid body chemoreceptors grown in normoxic and hypoxic environments. J Neurochem 1997; 69:645–654.
22. Tatsumi K, Pickett CK, Weil JV. Decreased carotid body hypoxic sensitivity in chronic hypoxia: role of dopamine. Respir Physiol 1995; 101:47–57.
23. Vizek M, Pickett CK, Weil JV. Increased carotid body hypoxic sensitivity during acclimatization to hypobaric hypoxia. J Appl Physiol 1987; 63:2403–2410.
24. Hempleman SC. Increased calcium current in carotid body glomus cells following in vivo acclimatization to chronic hypoxia. J Neurophysiol 1996; 76:1880–1886.
25. Gonzalez C, Almaraz L, Obeso A, Rigual R. Carotid body chemoreceptors: from natural stimuli to sensory discharges. Physiol Rev 1994; 74:829–898.
26. Gonzalez C, Lopez-Lopez JR, Obeso A, Perez-Garcia MT, Rocher A. Cellular mechanisms of chemoreception in the carotid body. Respir Physiol 1995; 102:137–147.
27. Iturriaga R, Larrain C, Zapata P. Effects of dopaminergic blockade upon carotid chemosensory activity and its hypoxia-induced excitation. Brain Res 1994; 663:145–154.
28. Zapata P, Torrealba F. Blockade of dopamine-induced chemosensory inhibition by domperidone. Neurosci Lett 1984; 51:359–364.
29. Joseph V, Dalmaz Y, Cottet-Emard JM, Pequignot JM. Dexamethasone's influence on tyrosine hydroxylase activity in the chemoreflex pathway and on the hypoxic ventilatory response. Pflügers Arch 1998; 435:834–839.
30. Huey KA, Brown IP, Jordan MC, Powell FL. Changes in dopamine D(2)-receptor modulation of the hypoxic ventilatory response with chronic hypoxia. Respir Physiol 2000; 123:177–187.
31. Joseph V, Soliz J, Soria R, Pequignot J, Favier R, Spielvogel H, Pequignot JM. Dopaminergic metabolism in carotid bodies and high altitude acclimatization in female rats. Am J Physiol Regul Integr Comp Physiol 2002; 282:R765–R773.

32. Fidone S, Gonzalez C, Obeso A, Gomez-Nino A, Dinger B. Biogenic amine and neuropeptide transmitters in carotid body chemotransmission. In: Sutton J, Coates G, Remmers J, eds. Hypoxia: The Adaptations. Philadelphia: BC Dekker, 1990:116–126.
33. Huey KA, Low MJ, Kelly MA, Juarez R, Szewczak JM, Powell FL. Ventilatory responses to acute and chronic hypoxia in mice: effects of dopamine D2 receptors. J Appl Physiol 2000; 89:1142–1150.
34. Bairam A, Frenette J, Dauphin C, Carroll JL, Khandjian EW. Expression of dopamine D1-receptor mRNA in the carotid body of adult rabbits, cats and rats. Neurosci Res 1998; 31:147–154.
35. Tatsumi K, Pickett CK, Weil JV. Possible role of dopamine in ventilatory acclimatization to high altitude. Respir Physiol 1995; 99:63–73.
36. Huey KA, Powell FL. Time-dependent changes in dopamine D(2)-receptor mRNA in the arterial chemoreflex pathway with chronic hypoxia. Brain Res Mol Brain Res 2000; 75:264–270.
37. Jackson A, Nurse CA. Role of acetylcholine receptors and dopamine transporter in regulation of extracellular dopamine in rat carotid body cultures grown in chronic hypoxia or nicotine. J Neurochem 1998; 70:653–662.
38. Kline DD, Peng YJ, Manalo DJ, Semenza GL, Prabhakar NR. Defective carotid body function and impaired ventilatory responses to chronic hypoxia in mice partially deficient for hypoxia-inducible factor 1alpha. Proc Natl Acad Sci USA 2002; 99:821–826.
39. Clar C, Dorrington KL, Robbins PA. Ventilatory effects of 8 h of isocapnic hypoxia with and without beta-blockade in humans. J Appl Physiol 1999; 86:1897–1904.
40. Ryan ML, Hedrick MS, Pizarro J, Bisgard GE. Carotid body noradrenergic sensitivity in ventilatory acclimatization to hypoxia. Respir Physiol 1993; 92:77–90.
41. Ryan ML, Hedrick MS, Pizarro J, Bisgard GE. Effects of carotid body sympathetic denervation on ventilatory acclimatization to hypoxia in the goat. Respir Physiol 1995; 99:215–224.
42. Asano K, Mazzeo RS, McCullough RE, Wolfel EE, Reeves JT. Relation of sympathetic activation to ventilation in man at 4300 m altitude. Aviat Space Environ Med 1997; 68:104–110.
43. Serebrovskaya TV, Karaban IN, Kolesnikova EE, Mishunina TM, Swanson RJ, Beloshitsky PV, Ilyin VN, Krasuk AN, Safronova OS, Kuzminskaya LA. Geriatric men at altitude: hypoxic ventilatory sensitivity and blood dopamine changes. Respiration 2000; 67:253–260.
44. He L, Chen J, Dinger B, Stensaas L, Fidone S. Endothelin modulates chemoreceptor cell function in mammalian carotid body. Adv Exp Med Biol 1996; 410:305–311.
45. Dinger B, Chen J, He L, Fidone S. Endothelin is involved in the increased chemosensitivity of the chronically hypoxic carotid body. FASEB J 1999; 13.
46. Schoene RB, Lahiri S, Hackett PH, Peters RM, Jr., Milledge JS, Pizzo CJ, Sarnquist FH, Boyer SJ, Graber DJ, Maret KH, West JB. Relationship of hypoxic ventilatory response to exercise performance on Mount Everest. J Appl Physiol 1984; 56:1478–1483.
46a. Lahiri S, Delaney RG, Brody JS, Simpser M, Velasquez T, Motoyama EK, Polgar C. Relative role of environmental and genetic factors in respiratory adaptation to high altitude. Nature 1976; 261:133–135.
47. Monge-C C, Arregui A, León-Velarde F. Pathophysiology and epidemiology of chronic mountain sickness. Int J Sports Med 1992; 13:S79–S81.

48. León-Velarde F, Arregui A, Monge-C C, Ruiz Y, Ruiz H. Aging at high altitudes and the risk of Chronic Mountain Sickness. J Wild Med 1993; 4:183–188.
49. Severinghaus JW, Bainton GR, Carcelen A. Respiratory insensitivity to hypoxia in chronically hypoxic man. Respir Physiol 1966; 1:308–334.
50. Kryger M, McCullough RE, Doekel R, Collins DD, Weil JV, Grover RF. Excessive polycythemia of high altitude: Role of ventilatory drive and lung disease. Am Rev Respir Dis 1978; 118:659–666.
51. León-Velarde F, Ramos MA, Hernández JA, De Idiáquez D, Muñoz LS, Gaffo A, Córdova S, Durand D, Monge-C C. The role of menopause in the development of chronic mountain sickness. Am J Physiol 1997; 272:R90–R94.
52. León-Velarde F, Rivera-Chira M, Tapia R, Huicho L, Monge-C C. Relationship of ovarian hormones to hypoxemia in women residents of 4300 m. Am J Physiol 2001; 280:R488–R493.
53. Ou LC, Sardella GL, Leiter JC, Brinck-Johnsen, Smith RP. Role of sex hormones in the development of chronic mountain sickness in rats. J Appl Physiol 1994; 77:427–433.
54. Kryger M, McCullough RE, Collins DD, Scoggin CH, Weil JV, Grover RF. Treatment of excessive polycythemia of high altitude with respiratory stimulant drugs. Am Rev Respir Dis 1978; 117:455–464.
55. Bisgard GE, Neubauer JA. peripheral and central effects of hypoxia. In: Dempsey JA, Pack AI, eds. Regulation of Breathing. Vol. 79. New-York: Marcel Dekker, 1995:617–668.
56. Erickson JT, Millhorn DE. Hypoxia and electrical stimulation of the carotid sinus nerve induce Fos-like immunoreactivity within catecholaminergic and serotoninergic neurons of the rat brainstem. J Comp Neurol 1994; 348:161–182.
57. Donoghue S, Felder R, Jordan D, Spyer K. The central projections of carotid body baroreceptors and chemoreceptors in the cat: a neurophysiological study. J Physiol (Lond) 1984; 347:397–409.
58. Finley J, Katz D. The central organization of carotid-body afferents-projections to the brainstem of the rat. Brain Res 1992; 572:108–116.
59. Housley GD, Martin-Body RL, Dawson NJ, Sinclair JD. Brain stem projections of the glossopharyngeal nerve and its carotid sinus nerve branch in the rat. Neuroscience 1987; 22:237–250.
60. Bianchi AL, Denavit-Saubié M, Champagnat J. Central control of breathing in mammals: neuronal circuitry, membrane properties and neurotransmitters. Physiol Rev 1995; 75:1–45.
61. Koshiya N, Guyenet PG. NTS neurons with carotid chemoreceptor inputs arborize in the rostral ventrolateral medulla. Am J Physiol 1996; 270:R1273–R1278.
62. Gozal D, Xue YD, Simakajornboon N. Hypoxia induces c-Fos protein expression in NMDA but not AMPA glutamate receptor labeled neurons within the nucleus tractus solitarii of the conscious rat. Neurosci Lett 1999; 262:93–96.
63. Mizusawa A, Ogawa H, Kikuchi Y, Hida W, Kurosawa H, Okabe S, Takishima T, Shirato K. In vivo release of glutamate in nucleus tractus solitarii of the rat during hypoxia. J Physiol (Lond) 1994; 478:55–65.
64. Ohtake PJ, Torres JE, Gozal YM, Graff GR, Gozal D. NMDA receptors mediate peripheral chemoreceptor afferent input in the conscious rat. J Appl Physiol 1998; 84:853–861.

65. Vardhan A, Kachroo A, Sapru HN. Excitatory amino acid receptors in commissural nucleus of the NTS mediate carotid chemoreceptor responses. Am J Physiol 1993; 264:R41–R50.
66. Haxhiu MA, Strohl KP, Cherniack NS. The N-methyl-D-aspartate receptor pathway is involved in hypoxia-induced c-Fos protein expression in the rat nucleus of the solitary tract. J Auton Nerv Syst 1995; 55:65–68.
67. McCrimmon DR, Mitchell GS, Dekin MS. Glutamate, GABA and serotonin in ventilatory control. In: Dempsey JA, Pack AI, eds. Regulation of Breathing. Vol. 79. New-York: Marcel Dekker, 1995:151–218.
68. Bach KB, Lutcavage ME, Mitchell GS. Serotonin is necessary for short-term modulation of the exercise ventilatory response. Respir Physiol 1993; 91:57–70.
69. Bach KB, Mitchell GS. Hypoxia-induced long-term facilitation of respiratory activity is serotonin dependent. Respir Physiol 1996; 104:251–260.
70. Millhorn DE, Eldridge FL, Waldrop TG. Prolonged stimulation of respiration by endogenous central serotonin. Respir Physiol 1980; 42:171–188.
71. Poncet L, Denoroy L, Dalmaz Y, Pequignot JM. Activity of tryptophan hydroxylase and content of indolamines in discrete brain regions after a long-term hypoxic exposure in the rat. Brain Res 1997; 765:122–128.
72. Olson EB, Jr. Ventilatory adaptation to hypoxia occurs in serotonin-depleted rats. Respir Physiol 1987; 69:227–235.
73. Herman JK, O'Halloran KD, Mitchell GS, Bisgard GE. Methysergide augments the acute, but not the sustained, hypoxic ventilatory response in goats. Respir Physiol 1999; 118:25–37.
74. Kalia M, Fuxe K, Goldstein M. Rat medulla oblongata. II. Dopaminergic, noradrenergic (A1 and A2) and adrenergic neurons, nerve fibers, and presumptive terminal processes. J Comp Neurol 1985; 233:308–332.
75. Massari VJ, Shirahata M, Johnson TA, Gatti PJ. Carotid sinus nerve terminals which are tyrosine hydroxylase immunoreactive are found in the commissural nucleus of the tractus solitarius. J Neurocytol 1996; 25:197–208.
76. Yokoyama C, Okamura H, Nakajima T, Taguchi J, Ibata Y. Autoradiographic distribution of [3H]YM-09151-2, a high-affinity and selective antagonist ligand for the dopamine D2 receptor group, in the rat brain and spinal cord. J Comp Neurol 1994; 344:121–136.
77. Goiny M, Lagercrantz H, Srinivasan M, Ungerstedt U, Yamamoto Y. Hypoxia-mediated in vivo release of dopamine in nucleus tractus solitarii of rabbits. J Appl Physiol 1991; 70:2395–2400.
78. Smatresk NJ, Pokorski M, Lahiri S. Opposing effects of dopamine receptor blockade on ventilation and carotid chemoreceptor activity. J Appl Physiol 1983; 54:1567–1573.
79. Coles SK, Dick TE. Neurones in the ventrolateral pons are required for post-hypoxic frequency decline in rats. J Physiol 1996; 497:79–94.
80. Guyenet PG, Koshiya N, Huangfu D, Verberne AJM, Riley TA. Central respiratory control of A5 and A6 pontine noradrenergic neurons. Am J Physiol 1993; 264:R1035–R1044.
81. Guyenet PG. Neural structures that mediate sympathoexcitation during hypoxia. Respir Physiol 2000; 121:147–162.
82. Dumas S, Pequignot JM, Ghilini G, Mallet J, Denavit-Saubie M. Plasticity of tyrosine hydroxylase gene expression in the rats nucleus tractus solitarius after ventilatory acclimatazation to hypoxia. Brain Res Mol Brain Res 1996; 40:188–194.

83. Soulier V, Gestreau C, Borghini N, Dalmaz Y, Cottet-Emard JM, Pequignot JM. Peripheral chemosensitivity and central integration: neuroplasticity of catecholaminergic cells under hypoxia. Comp Biochem Physiol A Physiol 1997; 118:1–7.
84. Pérez H, Ruiz S, Laurido C, Hernandez A. Locus coeruleus-mediated inhibition of chemosensory responses in the rat nucleus tractus solitarius is mediated by alpha2-adrenoreceptors. Neurosci Lett 1998; 249:37–40.
85. Sun QJ, Pilowsky P, Minson J, Arnolda L, Chalmers J, Llewellyn-Smith IJ. Close appositions between tyrosine hydroxylase immunoreactive boutons and respiratory neurons in the rat ventrolateral medulla. J Comp Neurol 1994; 340:1–10.
86. Errchidi SG, Monteau R, Hilaire G. Noradrenergic modulation of the medullary respiratory rythm generator in the newborn rat: an in vitro study. J Physiol (Lond) 1991; 443:477–498.
87. Champagnat J, Denavit-Saubie M, Moyanova S, Rondouin G. Catecholaminergic depressant effects on bulbar repiratory mechanisms. Brain Res 1979; 160:57–68.
88. Denavit-Saubié M, Champagnat J, Rondouin G. Central effect of adrenergic compounds on cat respiratory mechanisms. In: Fuxe K, Goldstein B, Höckfelt B, Höckfelt T, eds. Central Adrenaline Neurons: Basic Aspects and Their Role in Cardiovascular Functions. Oxford: Pergamon, 1980:277–287.
89. Housley GD, Sinclair JD. Localization by kainic lesions of neurons transmitting the carotid chemoreceptor stimulus for respiration in rat. J Physiol (Lond) 1988; 406:99–114.
90. Schmitt P, Soulier V, Pequignot JM, Pujol JF, Denavit-Saubie M. Ventilatory acclimatization to chronic hypoxia: relationship to noradrenaline metabolism in the rat solitary complex. J Physiol (Lond) 1994; 477:331–337.
91. Smith DW, Buller KM, Day TA. Role of ventrolateral medulla catecholamine cells in hypothalamic neuroendocrine cell responses to systemic hypoxia. J Neurosci 1995; 15:7979–7988.
92. Ellenberger HH, Feldman JL, Zhan WZ. Subnuclear organization of the lateral tegmental field of the rat. II: Catecholamine neurons and ventral respiratory group. J Comp Neurol 1990; 294:212–222.
93. Pilowsky PM, Jiang C, Lipsky J. An intracellular study of respiratory neurons in the rostral ventrolateral medulla of the rat and their relationship to catecholamine-containing neurons. J Comp Neurol 1990; 301:604–617.
94. Everitt BJ, Hokfelt T, Terenius L, Tatemoto K, Mutt V, Goldstein M. Differential coexistence of neuropeptide Y (NPY)-like immunoreactivity with catecholamines in the central nervous system of the rat. Neuroscience 1984; 11:443–462.
95. Harfstrand A, Fuxe K, Terenius L, Kalia M. Neuropeptide Y-immunoreactive perikarya and nerve terminals in the rat medulla oblongata: relationship to cytoarchitecture and catecholaminergic cell groups. J Comp Neurol 1987; 260:20–35.
96. Tseng CJ, Mosqueda-Garcia R, Appalsamy M, Robertson D. Cardiovascular effects of neuropeptide Y in rat brainstem nuclei. Circ Res 1989; 64:55–61.
97. Poncet L, Denoroy L, Dalmaz Y, Pequignot JM, Jouvet M. Alteration in central and peripheral substance P- and neuropeptide Y-like immunoreactivity after chronic hypoxia in the rat. Brain Res 1996; 733:64–72.
98. Leibstein AG, Dermietzel R, Willenberg IM, Pauschert R. Mapping of different neuropeptides in the lower brainstem of the rat: with special reference to the ventral surface. J Auton Nerv Syst 1985; 14:299–313.

99. Chen ZB, Hedner J, Hedner T. Local effects of substance P on respiratory regulation in the rat medulla oblongata. J Appl Physiol 1990; 68:693–699.
100. Lindefors N, Yamamoto Y, Pantaleo T, Lagercrantz H, Brodin E, Ungerstedt U. In vivo release of substance P in the nucleus tractus solitarii increases during hypoxia. Neurosci Lett 1986; 69:94–97.
101. Srinivasan M, Goiny M, Pantaleo T, Lagercrantz H, Brodin E, Runold M, Yamamoto Y. Enhanced in vivo release of substance P in the nucleus tractus solitarii during hypoxia in the rabbit: role of peripheral input. Brain Res 1991; 546:211–216.
102. Yamamoto Y, Lagercrantz H. Some effects of substance P on central respiratory control in rabbit pups. Acta Physiol Scand 1985; 124:449–455.
103. Neubauer JA, Melton JE, Edelman NH. Modulation of respiration during brain hypoxia. J Appl Physiol 1990; 68:441–451.
104. LaManna JC, Kutina-Nelson KL, Hritz MA, Huang Z, Wong-Riley MT. Decreased rat brain cytochrome oxidase activity after prolonged hypoxia. Brain Res 1996; 720:1–6.
105. Sun MK, Reis DJ. Central neural mechanisms mediating excitation of sympathetic neurons by hypoxia. Prog Neurobiol 1994; 44:197–219.
106. Roux JC, Pequignot JM, Dumas S, Pascual O, Ghilini G, Pequignot J, Mallet J, Denavit-Saubie M. O2-sensing after carotid chemodenervation: hypoxic ventilatory responsiveness and upregulation of tyrosine hydroxylase mRNA in brainstem catecholaminergic cells. Eur J Neurosci 2000; 12:3181–3190.
107. Soulier V, Cottet-Emard JM, Pequignot J, Hanchin F, Peyrin L, Pequignot JM. Differential effects of long-term hypoxia on norepinephrine turnover in brain stem cell groups. J Appl Physiol 1992; 73:1810–1814.
108. Horn EM, Waldrop TG. Suprapontine control of respiration. Respir Physiol 1998; 114:201–211.
109. Kawai Y, Qi J, Comer AM, Gibbons H, Win J, Lipski J. Effects of cyanide and hypoxia on membrane currents in neurones acutely dissociated from the rostral ventrolateral medulla of the rat. Brain Res 1999; 830:246–257.
110. Nolan PC, Waldrop TG. In vivo and in vitro responses of neurons in the ventrolateral medulla to hypoxia. Brain Res 1993; 630:101–114.
111. Nolan PC, Waldrop TG. In vitro responses of VLM neurons to hypoxia after normobaric hypoxic acclimatization. Respir Physiol 1996; 105:23–33.
112. Nolan PC, Waldrop TG. Ventrolateral medullary neurons show age-dependent depolarizations to hypoxia in vitro. Brain Res Dev Brain Res 1996; 91:111–120.
113. Reis DJ, Golanov EV, Ruggiero DA, Sun MK. Sympatho-excitatory neurons of the rostral ventrolateral medulla are oxygen sensors and essential elements in the tonic and reflex control of the systemic and cerebral circulations. J Hypertens Suppl 1994; 12:S159–S180.
114. Solomon IC, Edelman NH, Neubauer JA. Pre-Botzinger complex functions as a central hypoxia chemosensor for respiration in vivo. J Neurophysiol 2000; 83:2854–2868.
115. Lipton AJ, Johnson MA, Macdonald T, Lieberman MW, Gozal D, Gaston B. S-nitrosothiols signal the ventilatory response to hypoxia. Nature 2001; 413:171–174.
116. Fandrey J. Hypoxia-inducible gene expression. Respir Physiol 1995; 101:1–10.
117. Semenza GL. Hypoxia-inducible factor 1 and the molecular physiology of oxygen homeostasis. J Lab Clin Med 1998; 131:207–214.
118. Wenger RH, Gassmann M. Oxygen(es) and the hypoxia-inducible factor-1. Biol Chem 1997; 378:609–616.

119. Czyzyk-krzeska M. Molecular aspects of oxygen sensing in physiological adaptation to hypoxia. Respir Physiol 1997; 110:99–111.
120. Mishra RR, Adhikary G, Simonson MS, Cherniack NS, Prabhakar NR. Role of c-fos in hypoxia-induced AP-1 cis-element activity and tyrosine hydroxylase gene expression. Brain Res Mol Brain Res 1998; 59:74–83.
121. Pascual O, Denavit-Saubie M, Dumas S, Kietzmann T, Ghilini G, Mallet J, Pequignot JM. selective cardiorespiratory and catecholaminergic areas express the hypoxia-inducible factor-1α (HIF-1α) under *in vivo* hypoxia in rat brainstem. Eur J Neurosci 2001; 14:1981–1991.
122. Chavez JC, Agani F, Pichiule P, LaManna JC. Expression of hypoxia- inducible factor-1alpha in the brain of rats during chronic hypoxia. J Appl Physiol 2000; 89:1937–1942.

27

Biology of Reactive Oxygen Species

Their Role in Oxygen Chemoreception in the Carotid Body

CONSTANCIO GONZALEZ, MARIA TERESA AGAPITO, ASUNCIÓN ROCHER, GLORIA SANZ ALFAYATE, and ANA OBESO

University of Valladolid
Valladolid, Spain

I. Introduction

Reactive oxygen species (ROS) include a large number of oxygen-containing molecules having in common a greater reactivity than molecular O_2. Free radicals are a subgroup of ROS that, in addition to the great reactivity, have one or more unpaired electrons. In the context of this chapter the most important ROS are H_2O_2 (hydrogen peroxide), $O_2^{\cdot -}$ (superoxide radical), and $OH^{\cdot}$ (hydroxyl radical).

In vertebrates, ROS have been considered metabolic by-products, triggers or effectors of cellular damage, except in phagocytic cells where the damage produced by ROS is aimed to kill the phagocyted-invading organisms or cancerous cells. However, recently it has been suggested that ROS production in specific reactions and/or subcellular compartments can be regulated; these ROS would be able to act as second messengers and regulate cellular processes. Yet, we want to emphasize with Halliwell and Gutteridge (1) that basically all data supporting a role for ROS as messengers or regulators have been generated in in vitro systems with established cell lines. The metabolic peculiarities of tumoral cells and their ability to support insulting conditions (e.g., high levels of ROS) require caution in considering the physiological significance of the observed phenomena in in vivo systems.

II. Origin of ROS in Vertebrates

In vertebrates, the genesis of most free radicals occurs at reactions involving the transference of electrons, and among them the most important are those reactions in which a single electron is transferred to a molecule of O_2 to form the superoxide radical, $O_2^{\cdot-}$. In nonphagocytic cells, the main source of $O_2^{\cdot-}$ is mitochondrial electron transport chain where univalent reduction of O_2 may take place in some steps of the overall transport system. In the two sites where $O_2^{\cdot-}$ can be produced (complex I and ubiquinone pool) there are electron shuttle molecules (flavin adenine mononucleotide and ubiquinone) that can accept one or two electrons. When they accept one electron, they become free radicals themselves capable of passing the electron to molecular O_2. It is estimated that 1–3% of the total O_2 consumed by mitochondria at normal PO_2 is converted to $O_2^{\cdot-}$. Mitochondria have an enzyme system to eliminate $O_2^{\cdot-}$, thereby preventing its reaction with mitochondrial constituents that will deteriorate the energy-generating mitochondrial functions (i.e., the Krebs cycle and the electron transport chain). This enzyme is the cyanide-resistant manganese superoxide dismutase (SOD), which transforms $O_2^{\cdot-}$ into H_2O_2 that diffuses through the mitochondrial membranes and is less reactive than $O_2^{\cdot-}$. Classically, mitochondrial ROS production has been considered a by-product in the process of O_2 reduction, and thereby the rate of $O_2^{\cdot-}$ production has been considered to run parallel to O_2 consumption (2); good evidence for that is obtained in exercising muscle both in vitro and in vivo (3). Alternatively, at any rate of electron transport (e.g., at a normal rate of O_2 consumption), the production of $O_2^{\cdot-}$ would increase with increasing cellular PO_2, i.e., the leakage of single electrons to partially oxidize O_2 would be favored by the greater PO_2 in the mitochondrial milieu. However, in recent years Chandel and co-workers have reported that hypoxia (<35 mmHg) increases (500–1000%) ROS production in isolated cardiomyocytes and other cell types (4,5), claiming that mitochondrial ROS are the triggers for hypoxia-induced transcription. In other words, Chandel et al. changed the status of mitochondrial ROS from by-products to second messengers controlling the expression of genes as important as erythropoietin (see Ref. 5). This is surprising because the high affinity of cytochrome oxidase for O_2 tends to assure a normal flow of electrons to molecular O_2 and to prevent their leakage at the quinone level until the PO_2 reaches very low levels (6). Interestingly, in one of the latest papers of the Chandel group (7) it has been found that pulmonary artery smooth muscle cells, which respond to hypoxia with contraction (i.e., they require an increased O_2 consumption to support the contraction), generate only 30% more ROS than normoxic cells when exposed to hypoxia of identical intensity of that used with cardiomyocytes in previous studies.

Quantitatively, the transport of O_2 by erythrocytes also is an important source of $O_2^{\cdot-}$. Although essentially all O_2 binding to hemoglobin does not imply the oxidation of Fe^{2+} in the heme moiety, there is some delocalization of an electron when O_2 is bound, and thereby the appearance of an intermediate structure that can be written as a resonant "equilibrium":

$$\text{Heme-Fe}^{2+} - O_2 \leftrightarrow \text{hemo-Fe}^{3+} - O_2^{\cdot-}$$

Usually, the Fe^{2+} of heme group keeps the delocalized electron and hemoglobin releases O_2, but occasionally (3% of hemoglobin/day) the electron is kept by O_2 and then hemoglobin releases $O_2^{\cdot-}$, the Fe^{2+} is transformed into Fe^{3+}, and methemoglobin is formed. Methemoglobin reductase reduces methemoglobin to hemoglobin and recovers its O_2-transporting function. In addition, to dispose $O_2^{\cdot-}$, erythrocytes have a high concentration of copper-zinc SOD to transform the $O_2^{\cdot-}$ into H_2O_2 and high levels of glutathione peroxidase, which transforms H_2O_2 into water ($2GSH + H_2O_2 \rightarrow GSSG + 2H_2O$), and of catalase, which transforms H_2O_2 into water plus O_2 ($2H_2O_2 \rightarrow 2H_2O + O_2$). The oxidized glutathione (GSSG) is converted to reduced glutathione (GSH) by the enzyme glutathione reductase, which uses as reducing agent NADPH generated in the powerful pentose phosphate pathway. These are the most important mechanisms used not only by erythrocytes but by all body cells to dispose of ROS.

The smooth endoplasmic reticulum contains enzymes involved in reactions of oxidation and hydroxylation of many types of molecules (alcohols, barbiturates, antibiotics, steroid hormone synthesis and metabolism, etc.). These enzymes use cytochrome P450 and b_5 as electron transport system and can also generate $O_2^{\cdot-}$ and/or H_2O_2.

Phagocytes possess an enzymatic complex, the NADPH oxidase, that catalyzes one electron reduction of O_2 to form $O_2^{\cdot-}$ using NADPH as electron donor: $NADPH + 2O_2 \rightarrow NADP^+ + H^+ + 2\ O_2^{\cdot-}$. NADPH oxidase is a highly regulated enzymatic complex composed of a number cytosolic proteins, namely, p67*phox*, p47*phox*, p40*phox*, and Rac1/Rac2, and two membrane proteins, gp91*phox* and p22*phox*, which, together, form cytochrome b_{558}. When the enzymatic complex is activated, the cytosolic components migrate to the membranes and associate with the membrane components to assemble the catalytically active oxidase. Cytochrome b_{558}, which contains both a flavin and two heme groups and a consensus sequence to bind NADPH, is responsible for electron transfer from NADPH to oxygen, but 67*phox* also possesses an NADPH-binding site. P47*phox* is phosphorylated via protein kinase C and seems to initiate the enzyme assembly, while p67*phox* would act as an activator of the catalysis by cytochrome b_{558}. The function of Rac and p40*phox* is not well defined, but all cytoplasmic components are important as negative dominants of each of them produces partial or total loss of enzymatic activity (8). Several cell types were found to contain one or another subunits of NADPH oxidase, but lacked the gp91-*phox* subunit, being difficult to relate the production of ROS in these cells with the plasma-membrane-linked NADPH oxidase. However, a search for homologs of gp91-*phox* resulted in the description of several homologs whose distribution is only partially understood (9–12). These homologs have several interesting properties: in addition to their domains to bind the heme and the nucleotides, some of them exhibit constitutive activity and are regulated by intracellular messengers including intracellular Ca^{2+}. Then, the ROS generated by these enzymes could be important in cell signaling for two reasons: first, extracellular signals acting on their receptors can readily activate the genesis of ROS via intracellular Ca^{2+}, and second, the ROS generated could be released intracellularly (contrary to phagocytes) in potentially restricted compartments (8).

There are other potential sources of free radicals: (1) Phospholipase A_2 is a ubiquitous enzyme that releases arachidonic acid in the cells. Arachidonic acid serves as a substrate for cyclooxygenase and lipooxygenase and is the precursor for the major eicosanoids. Several intermediates in the biosynthetic pathways are free radicals that can transfer an electron to O_2 and form $O_2^{\cdot -}$. (2) In situations of tissue injury such as the ischemia-reperfusion process, xanthine dehydrogenase, which oxidizes hypoxanthine and xanthine to uric acid using NAD^+ as acceptor of electrons, is converted into xanthine oxidase by oxidative or proteolytic processes, this new enzyme being capable of transferring the electrons from the purines directly to O_2 to form $O_2^{\cdot -}$. (3) There are autooxidation reactions in which several biologically important molecules, such as tyrosine, catecholamines, and tryptophan, slowly reduce O_2 to form $O_2^{\cdot -}$. $O_2^{\cdot -}$ in turn is capable of further oxidizing those molecules and a cycle of unwanted reactions might be started with potentially deleterious effects to the cells. (4) The cleavage of the chemical bonds of water by high-energy radiation (water radiolysis) might constitute an important source of free radicals according to the reaction:

$$H_2O \rightarrow H^{\cdot} \text{ (hydrogen radical)} + OH^{\cdot} \text{ (hydroxyl radical)}$$

Although ultraviolet (UV) light does not have enough energy to produce radiolysis of water, UV light, and even visible light in the presence of some sensitizer molecules, can produce homolytic scission of hydrogen peroxide in the skin to generate hydroxyl radicals ($H_2O_2 \rightarrow 2\ OH^{\cdot}$). This can be an important source of ROS at high altitude owing to the high levels of UV radiation. Additionally, violet-blue light can cause photoreduction of flavins leading to activation of flavin-containing oxidases in every cellular compartment and production of H_2O_2; the same light can, in turn, produce homolytic scission of H_2O_2 to generate $OH^{\cdot}$. These effects of light also are important methodologically as the measured rates of ROS production by cells in fluorescence imaging experiments can greatly depend on the intensity and duration of the illumination pulses and on the sampling rate (13).

III. General Reactivity of ROS Including Disposal Reactions

In preceding paragraphs we have seen that the most important primary ROS produced in vertebrate cells is $O_2^{\cdot -}$ (oxygen transport, respiratory chain, NADPH oxidase, cytochrome P450–dependent oxidations, autooxidations), although $OH^{\cdot}$ and H_2O_2 can also be primary ROS. These primary ROS can undergo a large variety of reactions with other molecules to give rise to new ROS or to species that are oxidized or reduced. New ROS formed are capable of further reactions originating chain reactions until one of the formed radicals reacts with a molecule (scavenger) to generate an unreactive compound. This widespread reactivity is the basis for the toxic effects of free radicals. For example, in their reactions with membrane lipids, $OH^{\cdot}$ abstract an atom of hydrogen from polyunsaturated fatty acids forming H_2O

and transforming the fatty acid in a carbon-centered radical capable of reacting with molecular O_2 to form a peroxyl radical (R-COO$^\cdot$-COOH), which in turn is capable of abstracting another hydrogen atom from a nearby polyunsaturated fatty acid spreading out the alteration of lipids. When a peroxyl radical attacks another double bond in the same molecule, a cyclic compound is formed that eventually breaks down to generate malonaldehyde and other by-products, whose levels are measured to define the magnitude of lipid peroxidation in certain circumstances. These alterations in the lipids change many basic properties of the cell membrane systems; in addition, peroxyl radicals, $OH^\cdot$ and carbon-centered radicals, attack membrane proteins to generate all kinds of distorted molecules (breakdown, intermolecular or intramolecular cross-linking of proteins with the formation of aggregates, etc.), which implies the loss of the membrane function and its eventual destruction. The same type of cross-linking and breakdown reactions can be made by $OH^\cdot$ in the DNA molecules and between DNA and nuclear proteins leading to alterations in the repair of DNA or in its replication or transcription. In sum, $OH^\cdot$ is so reactive that it will react in the very same place where it is produced and disappear immediately, but at the same time new reactive molecules would appear that would spread the damaging effects.

The $O_2^{\cdot -}$ is much less reactive than $OH^\cdot$. It can be protonized to form $HO_2^\cdot$ (hydroperoxyl radical), which has a greater reducing activity and which, in addition, is freely permeable to membranes owing to its lack of charge. Although the concentration of $HO_2^\cdot$ is 1/100–400 that of $O_2^{\cdot -}$, its greater reactivity and permeability make $HO_2^\cdot$ an important free radical. In addition, $HO_2^\cdot$ is involved in the spontaneous dismutation of $O_2^{\cdot -}$ to H_2O_2 and molecular O_2 according to the equations: $H^+ + O_2^{\cdot -} \rightarrow HO_2^\cdot$, $HO_2^\cdot + HO_2^{\cdot -} \rightarrow H_2O_2 + O_2$. Superoxide radical can donate one electron to transition metals transforming Fe^{3+}, Cu^{2+}, and Mn^{3+} into Fe^{2+}, Cu^{+}, and Mn^{2+}, respectively, and then it acts as a reducing agent, but it can also act as an oxidizing agent reversing the reaction. The enzymes responsible for the cellular elimination of $O_2^{\cdot -}$, the SOD (cytosolic or copper-zinc SOD and mitochondrial or manganese SOD), are metalloproteins that through cycles of reduction and oxidation of their transition metals greatly accelerate the disappearance of $O_2^{\cdot -}$. The SOD catalyzed reaction is:

$$\text{Enz-oxidized metal} + O_2^{\cdot -} \longrightarrow \text{Enz-reduced metal} + O_2$$
$$\text{Enz-reduced metal} + O_2^{\cdot -} + 2H^+ \longrightarrow \text{Enz-oxidized metal} + H_2O_2$$

Other important reactions of $O_2^{\cdot -}$ include oxidation of ascorbate and the catechol ring to form ascorbate and diphenol radicals and H_2O_2. Summarizing, we can state that most of $O_2^{\cdot -}$ ends up as H_2O_2 owing to spontaneous and enzyme-catalyzed dismutation, implying that H_2O_2 is the final common path of most free radicals in the organism.

However, H_2O_2 is also harmful to the cells. The toxicity of H_2O_2 would explain the existence of specific enzymatic systems to destroy it, including catalases ($2H_2O_2 \rightarrow 2H_2O + O_2$) and peroxidases, the most important of which is glutathione peroxidase ($H_2O_2 + 2GSH \rightarrow GSSG + 2H_2O$; GSSG is back-reduced to GSH by the action of the NADPH-based glutathione reductase). The toxicity of H_2O_2 is not exerted directly, unless it is applied at high concentrations ($>10\,\mu M$), but by the $OH^{\cdot}$ it can generate upon reaction with transition metals by the multistep Fenton reaction ($Fe^{2+} + H_2O_2 \rightarrow \quad \rightarrow \quad \rightarrow Fe^{3+} + OH^{\cdot} + OH^{-}$) or upon homolytic fission by the action of ultraviolet light. We have already examined the effects of $OH^{\cdot}$.

IV. Signaling by ROS

A. Via Irreversible Reactions

Many of the reactions of ROS described are widespread within every cell compartment, irreversible, and damaging for the target molecules. However, it is conceivable that ROS can signal through functionally irreversible reactions. For example, ROS can oxidize transcription factors, which upon oxidation would be degraded and thereby ROS reactions would stop the function of the transcribed gene. Along these lines, it is known that iron regulatory protein-2 (a transcription factor controlling the expression of several genes involved in iron metabolism) binds iron in relation to cellular levels of iron in such a manner that the binding will be maximal in iron-replete cells. Bound iron, in turn, promotes the oxidation of the regulatory protein, and the degree of oxidation, in turn, directs ubiquitination and proteosome degradation (14). It seems established that the iron-binding domain in iron regulatory protein-2, which contains five cysteines, circumscribes the oxidative damage into those residues because the $OH^{\cdot}$ produced by the Fenton reaction with the bound iron react in the same place where they are produced and disappear as such $OH^{\cdot}$. In the same schema, it has been proposed that in normoxia the prevailing high levels of ROS would oxidize critical cysteine residues of hypoxia-inducible factor-1α (HIF-1α) rendering it suitable for ubiquitination and degradation in the proteosome; in hypoxia the decrease in ROS levels would favor the reduced status of HIF-1α, its accumulation, heterodimerization with HIF-1β, binding to DNA and activation of the transcription of hypoxia-sensitive genes (15). Contrary to the case with iron regulatory protein-2, in HIF-1α there is not an obvious mechanism to circumscribe the oxidative damage to HIF-1α itself. It might be argued that a source of ROS funnels its products to the cell compartment where HIF-1α is located, or alternatively that the cysteine residues in HIF-1α susceptible of oxidation are especially sensitive to ROS (see below). This ROS-mediated mechanism for HIF-1α stabilization is in dispute (5,16,17).

Signaling via irreversible reactions such as H_2O_2-dependent cross-linkage of tyrosine residues has also been described. When the eggs of the sea urchin are fertilized, they form an envelope to block new spermatozoa from entering. The envelope is formed by cross-linkage of tyrosine residues of extracellular matrix proteins. The formation of dityrosine bridges is catalyzed by a peroxidase released by the egg that uses H_2O_2 generated by a NADPH oxidase complex (18). Protection from the potential deleterious effects of this oxidative mechanism is afforded by ovothiol, an intracellular antioxidant that can be regenerated by GSH. High-molecular-weight oxidases containing two domains, one with homology to peroxidase and the other to gp91*phox* have been cloned in human, *Drosophila*, and *C. elegans* and termed duox. Human duox ($p138^{Tox}$) is the enzyme catalyzing the H_2O_2-dependent condensation of iodinated tyrosyl residues, which supports thyroid hormone biosynthesis (9).

B. Via Reversible Reactions

In their review Thannickal and Famburg (19) describe protein dimerization by intermolecular disulfide linkage as another redox-dependent signaling mechanism. In this type of signaling the reactions are reversible and dependent on the prevailing redox environment of the cell (20). For example, it is known that some protein kinases are formed by two subunits that possess two (or more) cysteine residues and are inactive in their reduced dissociated form. An increase in ROS levels would lead to the formation of disulfide bonds among the monomers to form an active dimer. The same mechanism controls the activity of glutamate cysteine ligase (glutamyl-cysteine synthetase), the limiting enzyme in GSH synthesis. The enzyme, in both vertebrates and invertebrates, is formed by two subunits, a catalytic subunit and a regulatory subunit, which contain cysteine residues, which are oxidized when the levels of ROS are high and the two subunits form a complex via disulfide bonds. The complex exhibits a much lower Km for glutamate and a much higher Ki for GSH accelerating the synthesis of GSH; when the levels of ROS are low (and GSH increase), the subunit would tend to dissociate with the subsequent increase in the Km for glutamate and decrease in the Ki for GSH and slowing in the rate of GSH synthesis (21). A comparable, ROS-dependent regulation might occur in the interaction between calmodulin and plasma membrane Ca^{2+}-dependent ATPase: H_2O_2 can oxidize reversibly precise methionine residues of the ATPase to methionine sulfoxide, decreasing the ability of calmodulin to activate the ATPase (see Ref. 22). In another model of control of the activity by the redox environment the enzyme (or growth factor) is bound to a regulatory subunit (e.g., thioredoxin) that has two or more sulfhydryl groups. The enzyme in the complex is inactive. The increase in ROS levels would cause the intramolecular oxidation of the —SH groups of the regulatory subunit to form an intramolecular disulfide bond, producing the dissociation of the complex and the enzyme becoming active. In all these models, it is not obvious how the compartimentalization of the ROS signals required to gain specificity takes place (see below).

Additional redox signaling reactions include intramolecular modifications of proteins by oxidation of cysteine (and methionine) residues and intramolecular disulfide bonding in monomeric proteins. Some of these reactions are readily reversible and therefore they could potentially be the most relevant for hypoxic transduction in the carotid body chemoreceptor cells. The reversibility of these reactions makes them most suitable to switch on and off the responses to hypoxia in the CB, which are fast in onset and termination (23,24). Therefore, we will describe in some detail the reactions as well as the potential mechanisms to circumscribe them to the molecular effectors of the hypoxic response.

Sulfhydryl groups in the proteins can be present as such sulfhydryls (Prot-SH), as disulfides (Prot-S-S-Prot), or as mixed disulfides with GSH (Prot-S-SG) or other molecules containing sulfhydryl groups such as thioredoxin, cysteine, or γ-glutamyl-cysteine. The predominant mixed disulfides would be with GSH owing to its much higher concentration. Prot-SH thiolation and dethiolation with GSH (i.e., glutathiolation) should be considered a dynamic and reversible physiological process that should reflect the redox environment of the cell in such a manner that in response to an increase in ROS levels, the glutathiolation of proteins would predominate (20). Two important reactions for glutahiolation are:

1. Thiol disulfide exchange:

Increase in ROS levels $\rightarrow$ increase in H_2O_2 (spontaneous and catalyzed dismutation)

$H_2O_2 + 2GSH \longrightarrow GSSG + 2H_2O$ (glutathione peroxidase)

Prot-SH + GSSG $\longrightarrow$ Prot-S-SG + GSH (thioltransferase)

The reaction of glutathiolation can proceed to form intra- or intermolecular disulfide bonding:

Prot-S-SG + Prot-SH $\longrightarrow$ Prot-S-S-Prot + GSH

In this exchange it is important to note that GSSG disappears and GSH is recovered totally or partially, and therefore the GSH/GSSG ratio (which is the main determinant of the reduction potential of the cell) tends to be maintained in spite of the increase in ROS levels. When the production of ROS returns to normal levels, the thioltransferase reaction is reversed and the GSSG generated is reduced to 2GSH by action of glutathione reductase.

2. Two-electron oxidation of the protein thiol to sulfenic acid and ulterior reaction with GSH. With a series of intermediates the overall reaction for this process is (25):

Prot-S^- + H^+ + $O_2^{\cdot -}$ $\longrightarrow$ Prot-$SO^{\cdot}$ + OH^- $\rightarrow$

$\rightarrow$ Prot-SOH + GSH $\longrightarrow$ Prot-S-SG + H_2O

The pK_a of cysteine is around 8.5 implying that in the normal cell environment less than 10% of the Prot-SH groups are in their dissociated form: Prot-S^- + H^+. However, basic amino acids can enhance the dissociation of sulfhydryl groups in such a manner that in the intimate environment of a protein, sulfhydryl groups can have pK_a as low as 3.5. As it is the thiolate anion (Prot-S^-) that controls the oxidation of thiols, it is obvious that the thiols with lower pK_a can represent specific targets for ROS (20). In other words, there can be domains or motifs within the structure of the protein that favor their glutathiolation as is the case for their nitrosylation (26). These mechanisms would circumscribe the modification of proteins produced by ROS to specific thiol residues, and thereby the apparent pleiotropic and delocalized reactivity of ROS can gain selectivity.

V. ROS Reactions and Hypoxic Transduction in the Carotid Body Chemoreceptors

The molecular machinery involved in the transduction of the hypoxic stimulus into a neurosecretory response in chemoreceptor cells of the carotid body (CB) is not completely known, but current models (e.g., 23,24,27) include the minimal molecular steps and molecular elements shown in the Table 1.

All molecular elements involved in low PO_2 transduction should possess in their structure cysteine (or methionine) residues susceptible of redox modification, and in the case of the putative hemoproteic O_2 sensor, the presence of iron would provide an additional mechanism for redox regulation as the reduction or oxidation of iron would modify the ability of the sensor to react with O_2. In fact, there is an ample body of evidence showing that oxidation and reduction of sulfhydryl groups with sulfhydryl reagents and H_2O_2 alters the gating kinetics and/or amplitudes of K^+ currents in many preparations including chemoreceptor cells (e.g., 27–33; see Ref. 34). Similarly, Ca^{2+} currents can be altered by sulfhydryl reagents and H_2O_2 (35). However, the direction of the modifications observed varies from study to study; e.g., reducing agents in some cases activate the currents while others inhibit them, reflecting probably different degrees of oxidation or reduction of a given ion channels in a specific preparation or different ratios of cysteine/methionine residues among different channels (25).

There are many questions that we can raise regarding the significance of ROS reactions in the hypoxic transduction in the chemoreceptor cells of the CB. In the context of this chapter, we probably should ask first if there is any known mechanism funneling ROS to the transduction machinery and/or if there are cysteine residues in the transducing proteins that circumscribe the reactions of ROS to the molecular machinery involved in the transduction. Alternatively, it might be asked if in any of the proteins involved in the hypoxic transduction glutathiolation or deglutathiolation has been described after physiological hypoxic stimulation. So far none of these questions have answers derived from experimental evidence as is the case in other

Table 1 Steps in the Transduction of Hypoxic Stimulus in Chemoreceptor Cells and Molecular Elements Involved

Oxygen transduction steps	O_2 transduction molecular elements
Low PO_2	
↓	
O_2 detection	Hemoproteic O_2 sensor
↓	
Cell depolarization	K^+ channels regulatory subunits
↓	
Calcium entry	Voltage-operated calcium channels
↓	
Release of neurotransmitters	Exocytotic machinery

Table 2 Criteria for Considering ROS Messengers or Mediators in Physiological Cellular Responses

1.	ROS should increase or decrease during stimulation and a given relationship should exist between the intensity of the stimulus and the change in ROS level.
2.	Cell mechanisms responsible for the change in ROS level should be identified.
3.	The molecular targets for ROS should be identified. The change produced in the target by ROS should be defined. The mechanisms of target/effector coupling should also be defined.
4.	The cellular responses elicited by the effector activated by ROS should be defined.
5.	The mechanism(s) responsible for the inactivation of the ROS signal should be defined.
6a.	Manipulation of ROS levels by alteration of glutathione levels or by inhibition or activation of ROS scavenging enzymes should produce the predicted change in the cell response.
6b.	Overexpression and elimination of the enzymes scavenging ROS should produce the predicted change in the cell responses.

systems (22,25,26,36). Lacking direct evidence we must base our suggestions and proposals on indirect observations. In a recent article on this topic (34), we have proposed a set of criteria (Table 2) that will help us to examine the available literature.

1. In previous paragraphs we have discussed the opposing views on whether hypoxia increases or decreases ROS production and levels. We have also discussed the apparent direct relationship between metabolic rate and production of ROS at normal PO_2, exemplifying this relationship by an increase in ROS production in contracting muscle in vitro or in the intact animals during exercise. For an important range of hypoxia (arterial PO_2 between 75 and 40 mmHg or even below) the CB is functioning in a situation comparable to that of exercise: an increased metabolic rate (37,38) and a PO_2 high enough to support the normal functioning of cytochrome oxidase. It should be considered that the high blood flow of the CB that increases during hypoxia tends to assure an adequate supply of O_2 in spite of hypoxia. We will assume that the increased electron flow through the respiratory chain of chemoreceptor cells will proceed until arterial PO_2 drops to $\approx$20 mmHg: up to those levels, hypoxia produces a sustained increase in the chemoreceptor activity (24).

With those facts in mind, it should be stated that there are no direct measurements of ROS in chemoreceptor cells. Cross et al. (39) reported a parallel increase in NAD(P)H fluorescence and chemoreceptor discharge in entire CBs during stepwise decreases in PO_2; it was concluded that hypoxia would decrease the production of ROS. Recently, we have measured in calf carotid bodies (40) levels and GSH/GSSG ratios, which is the most significant index of the redox status of the cells (20), and found that hypoxia does not alter the levels or ratios of GSH, suggesting that hypoxia does not alter significantly the levels of ROS. Additional experiments using *N*-acetylcysteine, which scavenges ROS and increases GSH levels and GSH/GSSG ratio (40), did not alter the response of chemoreceptor cells to hypoxia, high external K^+, or the Ca^{2+} ionophore inomycin; these finding

Table 3 Lack of Significant Effect of *N*-Acetyl Cysteine (NAC) on the Release of ^{3}H-Catecholamines Elicited by Hypoxia and High Extracellular K^+

	Calf		Rabbit	
	S_2/S_1 control	S_2/S_1 NAC-treated	S_2/S_1 control	S_2/S_1 NAC-treated
Hypoxia	0.73 ± 0.12 ($n = 10$)	0.72 ± 0.12 ($n = 10$)	0.52 ± 0.06 ($n = 12$)	0.61 ± 0.10 ($n = 12$)
High K_e^+	0.54 ± 0.02 ($n = 10$)	0.51 ± 0.04 ($n = 10$)	0.64 ± 0.07 ($n = 7$)	0.77 ± 0.08 ($n = 7$)

Each CB from both species was stimulated twice. In control CBs both stimuli consisted in a period of 10 min of incubation in a solution made hypoxic ($PO_2 \approx 46$ mmHg) or containing 35 mM K^+. In the experimental NAC-treated CBs, the organs were similarly stimulated but in the 40-min period prior to the second stimulation and the 10-min period of the second stimulation the incubating solutions contained 2 mM NAC. S_2/S_1 represents the quotient of the evoked release in the second presentation of the stimulus to that obtained in the first presentation.

suggested that ROS levels are not critical to control chemoreceptor cell responses to hypoxia (Table 3).

2. There is immunocytochemical evidence for the presence of all the subunits of the neutrophil NADPH oxidase in the CB (41), although more recent studies have shown that most of the intraglomic immunostaining is localized in macrophages infiltrating the CB tissue (42). Diphenyleneiodonium (DPI; an inhibitor of NADPH oxidase) does activate chemoreceptor cells promoting release of catecholamines in intact rat and rabbit CB (43), suggesting that inhibition of the oxidase (and thereby of the genesis of ROS) would mimic hypoxia, as proposed by Cross et al. (39). However DPI does not affect the response to hypoxia and it does not occlude the response to hypoxia in either species, and two additional inhibitors of the oxidase do not modify the release of catecholamines in normoxic or in hypoxic conditions suggesting that DPI activation of chemoreceptor cells was not related to its ability to inhibit NADPH oxidase (43). In the same line are the results of Roy et al. (44) showing that knockout mice for gp91*phox* exhibit normal chemoreceptor responses to hypoxia evaluated either at the chemoreceptor cell or at the ventilation level. These findings are in line with the conclusions reached in the previous paragraph that suggest that ROS are not critical messengers for chemoreceptor cell functioning. Of course, the possibility exists that chemoreceptor cells express one of the recently cloned NADPH oxidases and that the increase in intracellular Ca^{2+} that is known to occur in chemoreceptor cells during hypoxia activates the enzyme (see above). If this were the case, then ROS could act as modulators, not triggers, of the chemoreceptor cell responses to hypoxia. The potential role of mitochondria as sources of ROS has been discussed, but again, the fact that cyanide or antimycin A, which increase mitochondrial ROS, as well as rotenone, which decreases them (4), are powerful chemostimulant agents (for references see Refs. 24,34) would suggest that ROS of

mitochondrial origin are not critical messengers in the genesis of the hypoxic responses in chemoreceptor cells.

3. The putative molecular elements involved in the different steps of the hypoxic transduction in chemoreceptor cells have been presented in Table 1. Data obtained by Ganfornina and López-Barneo (45) and Riesco-Fagundo et al. (27) in *inside-out* patches of chemoreceptor cell membranes showing that hypoxia decreases the opening probability (Po) of K^+ channels provide evidence that the putative hemoproteic O_2-sensor as well as the K^+ channels can respond to PO_2 changes without the concurrence of ROS. This is so because O_2, at the working pressures (10–150 mmHg), cannot oxidize or reduce thiol or other functional groups in proteins (except possibly for some autoxidation of tyrosine and tryptophan; see Ref. 1). Then, to involve ROS in the genesis of these observations it is necessary to postulate that a redox system capable of transferring the electrons back and forth to the channel protein is present in the isolated membrane patches (see Ref. 36). However, this seems unlikely because CO is able to prevent or to reverse the action of hypoxia on whole K^+ currents in rabbit chemoreceptor cells (46) and in rat membrane isolated patches (27), and there are no known mechanisms by which the highly unreactive CO can substitute O_2 to increase the rate of ROS production.

These considerations, however, do not exclude the possibility that ROS can modulate the responses. For example, in *inside-out* patches of rabbit chemoreceptor cells GSH decreased the Po of the O_2-sensitive K^+ channel in a reversible manner (29), although in *inside-out* patches of rat chemoreceptor cells dithithreitol increased the Po of the O_2-sensitive channel in a reversible manner (27). As stated previously, the disparity of the results might well be due to different degrees of reduction of the K^+ channels, but again, the physiological significance of these findings with reducing agents is uncertain as increases in the GSH/GSSG ratio from 30 to 60 in intact CB tissue do not affect the sensitivity of chemoreceptor cells to hypoxia (40). The effects of hypoxia and reducing agents on the Ca^{2+} channel in general and in chemoreceptor cells in particular are similarly disparate (see Ref. 47 vs. Refs. 35 and 48).

4. Although we can define the cellular responses elicited by hypoxia in chemoreceptor cells, we have serious difficulties in defining the cellular effects produced by the ROS level changes occurring during hypoxia because we do not know what those changes are. And even if we knew the changes in ROS levels produced during hypoxia, we do not know if they are the cause or consequence of activation of the effector and cellular responses.

In chemoreceptor cells hypoxia reduces the Po of the O_2-sensitive K^+ channels and produces cell depolarization, activation of voltage-operated Ca^{2+} channels, and release of neurotransmitters (see Table 1). In the preceding criterion we have seen that sulfhydryl reagents alter the Po or amplitude of the currents, and therefore it might be proposed that ROS, by altering the redox environment of the cells, can produce similar alterations. Since hypoxia reduces the Po of O_2-sensitive K^+ channels, and thereby the amplitude of the whole current, if ROS were to increase those parameters it might be suggested that hypoxia decreases ROS production: supporting these view are, among other, the data of Vega-Saenz de Miera and Rudy (28), Fu et al. (49), and Barlow et al. (50). However, if it is found that ROS mimic hypoxia on K^+ channels, the findings would favor the notion that hypoxia increases ROS production (33,51). Direct application of H_2O_2 to the intact CB does not support a critical role for ROS in O_2 chemoreception (52).

5. Since we do not know if hypoxia increases or decreases ROS production, we do not know if we should be looking for mechanisms that scavenge ROS or if, on the contrary, we should be looking for mechanisms that generate ROS. In the first case (hypoxia increases ROS production), we should be talking of the general defense mechanisms against ROS discussed earlier, and in situations of chronic hypoxia, we should be looking for induction of ROS scavenging enzymes (i.e., SOD, GSH peroxidase, etc.). In the second case (hypoxia decreases ROS), we should be asking for decreases in the activity of ROS-generating systems, and in situations of chronic hypoxia, we should look for down-regulation of the scavenging enzymes (1). Some studies suggest that chronic hypoxia might decrease ROS production.

6. Interpretation of the data resulting from experimental maneuvers aimed at manipulating ROS levels or the redox environment of the cells is not always straightforward. For example, ebselen, which is a selenium-containing mimic of GSH peroxidase, decomposes peroxides in the presence of GSH, with the net result of a decrease in H_2O_2 levels and a concomitant decrease in GSH levels and GSH/GSSG ratio. The redox environment produced by ebselen in the cells is singular, and different from that encountered physiologically. Naturally, in living cells, a decrease in the GSH levels and GSH/GSSG is produced by an increase of H_2O_2 (or ROS levels in general). The interpretation of the data obtained by the use of nitro blue tetrazolium is similarly complex: this agent is frequently used as an $O_2^{\cdot -}$ scavenger, but on reaction with the $O_2^{\cdot -}$ it is transformed to tetrazolinyl radical, which can react with O_2 to generate $O_2^{\cdot -}$ (1). Thus, the ability of nitro blue tetrazolium to scavenge $O_2^{\cdot -}$ will depend on the prevailing PO_2. Contrary to those situations, *N*-acetylcysteine, which is a precursor of GSH and direct scavenger of ROS, would reasonably mimic a situation in which there is a decreased production of ROS and would oppose a situation of increased rate of ROS production. Diamide is an agent that reacts with GSH (and other compounds having sulfhydryl groups) and oxidizes it to GSSG and thereby it would mimic a situation in which there is an increased production of ROS: it will cause a decrease in GSH/GSSG ratio and therefore the cellular power for scavenging of ROS leading to an increase in ROS levels. These few examples represent a word of caution when selecting the appropriate pharmacological tools to establish the significance of ROS in oxygen chemoreception in the CB.

There are only a few studies of the CB aiming to alter the elimination of ROS by the cells. Osanai et al. (52) found that aminotriazole, an inhibitor of catalase that should produce an increase in H_2O_2 levels and thereby a tendency for GSH and GSH/GSSG to decrease, does not affect the frequency of discharge in the carotid sinus nerve elicited by hypoxia. We have already commented on recent data from our laboratory (40) showing that *N*-acetylcysteine increased absolute GSH levels and GSH/GSSG ratio and yet it did not alter the basal normoxic or the hypoxia-induced release of catecholamines from chemoreceptor cells.

VI. Conclusion

From their classical consideration as potentially toxic by-products of oxidative metabolism, ROS have gained the status of potential second messengers during the last decade. Among the many cellular processes that can generate ROS, the mitochondrial electron transport chain is, quantitatively speaking, the most important, but the ROS produced by the several isoforms of NADPH oxidases present in different cell types might represent a signal more compartimentalized with the capability of action on a restricted set of molecular targets. Specific sequences in the proteins may be critical to achieve further specificity in signaling. The reactions of glutathiolation/deglutathiolation of proteins, which are readily reversible, could be well suited to mediate or to modulate the responses of the CB to hypoxia.

To examine the potential ROS in the transduction of the hypoxic stimulus in the CB chemoreceptor cells, we propose a set of criteria that ROS should satisfy to be considered true messengers. Unfortunately, none of the criteria are univocally satisfied. Thus, it is not certain if hypoxia increases or decreases ROS production in the CB, and therefore we do not know if mitochondria or NADPH oxidase is activated or inhibited during *physiological* stimulation of the CB. Assuming that the targets for ROS are the molecular elements involved in the cascade of hypoxia transduction, there is no information on the modifications produced in those targets by alterations in ROS levels. The changes in the functioning of chemoreceptor cells elicited by exogenous alteration of ROS levels are not consistent, the inconsistencies being probably due to uncontrolled differences in the experimental maneuvers. The manipulation of ROS scavenging systems does not produce significant changes in the CB responses to hypoxia. Further work is needed to accept or to deny a physiological role for ROS in the chemoreception of oxygen.

Acknowledgments

This work has been supported by grant BFI2001-1713 from the Ministry of Science and Technology of Spain.

References

1. Halliwell B, Gutteridge JMC. Free Radicals in Biology and Medicine. (3rd ed). Oxford: Oxford University Press, 1999.
2. Chance B, Sies H, Boveris A. Hydroperoxide metabolism in mammalian organs. Physiol Rev 1979; 59:527–605.
3. Sen CK, Packer L. Thiol homeostasis and supplements in physical exercise. Am J Clin Nutr 2000; 72:653S–669S.
4. Duranteau J, Chandel NS, Kulisz A, Shao Z, Schumacker PT. Intracellular signaling by reactive oxygen species during hypoxia in cardiomyocytes. J Biol Chem 1998; 273:11619–11624.

5. Chandel NS, Schumacker PT. Cellular oxygen sensing by mitochondria: old questions, new insight. J Appl Physiol 2000; 88:1880–1889.
6. Gnaiger E, Mendez G, Hand SC. High phosphorylation efficiency and depression of uncoupled respiration in mitochondria under hypoxia. Proc Natl Acad Sci USA 2000; 97:11080–11085.
7. Waypa GB, Chandel NS, Schumacker PT. Model for hypoxic pulmonary vasoconstriction involving mitochondrial oxygen sensing. Circ Res 2001; 88:1259–1266.
8. Griendling KK, Sorescu D, Ushio-Fukai M. NAD(P)H oxidase: role in cardiovascular biology and disease. Circ Res 2000; 86:494–501.
9. Lambeth JD, Cheng G, Arnold RS, Edens WA. Novel homologs of gp91phox. Trends Biochem Sci 2000; 25:459–461.
10. Geiszt M, Kopp JB, Varnai P, Leto TL. Identification of renox, an NAD(P)H oxidase in kidney. Proc Natl Acad Sci USA 2000; 97:8010–8014.
11. Banfi B, Molnar G, Maturana A, Steger K, Hegedus B, Demaurex N, Krause KH. A Ca^{2+}-activated NADPH oxidase in testis, spleen, and lymph nodes. J Biol Chem 2001; 276:37594–37601.
12. Cheng G, Cao Z, Xu X, van Meir EG, Lambeth JD. Homologs of gp91phox: cloning and tissue expression of Nox3, Nox4, and Nox5. Gene 2001; 269:131–140.
13. Hockberger PE, Skimina TA, Centonze VE, Lavin C, Chu S, Dadras S, Reddy JK, White JG. Activation of flavin-containing oxidases underlies light-induced production of H_2O_2 in mammalian cells. Proc Natl Acad Sci USA 1999; 96:6255–6260.
14. Iwai K, Drake SK, Wehr NB, Weissman AM, LaVaute T, Minato N, Klausner RD, Levine RL, Rouault TA. Iron-dependent oxidation, ubiquitination, and degradation of iron regulatory protein 2: implications for degradation of oxidized proteins. Proc Natl Acad Sci USA 1998; 95:4924–4928.
15. Huang LE, Arany Z, Livingston DM, Bunn HF. Activation of hypoxia-inducible transcription factor depends primarily upon redox-sensitive stabilization of its alpha subunit. J Biol Chem 1996; 271:32253–32259.
16. Srinivas V, Leshchinsky I, Sang N, King MP, Minchenko A, Caro J. Oxygen sensing and HIF-1 activation does not require an active mitochondrial respiratory chain electron-transfer pathway. J Biol Chem 2001; 276:21995–21998.
17. Vaux EC, Metzen E, Yeates KM, Ratcliffe PJ. Regulation of hypoxia-inducible factor is preserved in the absence of a functioning mitochondrial respiratory chain. Blood 2001; 98:296–302.
18. Shapiro BM. The control of oxidant stress at fertilization. Science 1991; 252: 533–536.
19. Thannickal VJ, Fanburg BL. Reactive oxygen species in cell signaling. Am J Physiol Lung Cell Mol Physiol 2000; 279:L1005–L1028.
20. Schafer FQ, Buettner GR. Redox environment of the cell as viewed through the redox state of the glutathione disulfide/glutathione couple. Free Radic Biol Med 2001; 30:1191–1212.
21. Fraser JA, Saunders RD, McLellan LI. *Drosophila melanogaster* glutamate cysteine ligase activity is regulated by a modifier subunit with a similar mechanism of action to the mammalian form. J Biol Chem 2002; 277:1158–1165.
22. Hoshi T, Heinemann SH. Regulation of cell function by methionine oxidation and reduction. J Physiol 2001; 531:1–11.
23. Gonzalez C, Almaraz L, Obeso A, Rigual R. Oxygen and acid chemoreception in the carotid body chemoreceptors. Trends Neurosci 1992; 15:146–153.

24. Gonzalez C, Almaraz L, Obeso A, Rigual R. Carotid body chemoreceptors: from natural stimuli to sensory discharges. Physiol Rev 1994; 74:829–898.
25. Barrett WC, DeGnore JP, Keng YF, Zhang ZY, Yim MB, Chock PB. Roles of superoxide radical anion in signal transduction mediated by reversible regulation of protein-tyrosine phosphatase 1B. J Biol Chem 1999; 274:34543–34536.
26. Stamler JS, Lamas S, Fang FC. Nitrosylation. the prototypic redox-based signaling mechanism. Cell 2001; 106:675–683.
27. Riesco-Fagundo AM, Pérez-García MT, Gonzalez C, López-López JR. Ca^{2+}-dependent K^+ channels are modulated by hypoxia and CO in isolated membrane patches of rat chemoreceptor cells. Circ Res 2001; 89:430–436.
28. Vega-Saenz de Miera E, Rudy B. Modulation of K^+ channels by hydrogen peroxide. Biochem Biophys Res Commun 1992; 186:1681–1687.
29. Lopez-Barneo J, Pardal R, Montoro RJ, Smani T, Garcia-Hirschfeld J, Urena J. K^+ and Ca^{2+} channel activity and cytosolic $[Ca^{2+}]$ in oxygen-sensing tissues. Respir Physiol 1999; 115:215–227.
30. Perez-Garcia MT, Lopez-Lopez JR, Gonzalez C. Kvβ1.2 subunit modulates the sensitivity to hypoxia of cloned voltage-gated K^+ channels in a subfamily-specific manner. J Gen Physiol 1999; 113:897–907.
31. Thuringer D, Findlay I. Contrasting effects of intracellular redox couples on the regulation of maxi-K^+ channels in isolated myocytes from rabbit pulmonary artery. J Physiol 1997; 500:583–592.
32. Soh H, Jung W, Uhm DY, Chung S. Modulation of large conductance calcium-activated potassium channels from rat hippocampal neurons by glutathione. Neurosci Lett 2001; 298:115–118.
33. Tang XD, Daggett H, Hanner M, Garcia ML, McManus OB, Brot N, Weissbach H, Heinemann SH, T Hoshi. Oxidative regulation of large conductance calcium-activated potassium channels. J Gen Physiol 2001; 117:253–274.
34. Gonzalez C, Sanz-Alfayate G, Agapito MT, Gomez-Niño A, Rocher A, Obeso A. Significance of ROS in oxygen sensing in cell systems with sensitivity to physiological hypoxia. Respir Physiol Neurobiol 2002; 132:17–41.
35. Fearon IM, Palmer AC, Balmforth AJ, Ball SG, Varadi G, Peers C. Modulation of recombinant human cardiac L-type Ca^{2+} channel alpha1C subunits by redox agents and hypoxia. J Physiol 1999; 514:629–637.
36. Eu JP, Sun J, Xu L, Stamler JS, Meissner G. The skeletal muscle calcium release channel: coupled O_2 sensor and NO signaling functions. Cell 2000; 102:499–509.
37. Obeso A, Gonzalez C, Rigual R, Dinger B, Fidone S. Effect of low O_2 on glucose uptake in rabbit carotid body. J Appl Physiol 1993; 74:2387–2393.
38. Obeso A, Rocher A, Herreros B, Gonzalez C. Oxygen consumption and energy metabolism of the carotid body. In: Gonzalez C, ed. The Carotid Body Chemoreceptors. New York: Springer-Verlag, 1997:31–45.
39. Cross AR, Henderson L, Jones OTG, Delpiano MA, Hentschel J, Acker H. Involvement on NAD(P)H oxidase as a PO_2 sensor protein in the rat carotid body. Biochem J 1990; 272:743–747.
40. Sanz-Alfayate G, Obeso A, Agapito MT, Gonzalez C. Reduced to oxidized glutathione ratios and oxygen sensing in calf and rabbit carotid body chemoreceptor cells. J Physiol 2001; 537:209–220.
41. Kummer W, Acker H. Immunohistochemical demonstration of four subunits of neutrophil NAD(P)H oxidase in type I cells of carotid body. J Appl Physiol 1995; 78:1904–1909.

42. Dvorakova M, Hohler B, Vollerthun R, Fischbach T, Kummer W. Macrophages: a major source of cytochrome b558 in the rat carotid body. Brain Res 2000; 852:349–354.
43. Obeso A, Gomez-Niño A, Gonzalez C. Inhibition of NADPH oxidase does not interfere with low PO_2 transduction in chemoreceptor cells. Am J Physiol 1999; 276:C593–C601.
44. Roy A, Rozanov C, Mokashi A, Daudu P, Al-mehdi AB, Shams H, Lahiri S. Mice lacking in gp91 phox subunit of NAD(P)H oxidase showed glomus cell $[Ca^{2+}]_i$ and respiratory responses to hypoxia. Brain Res 2000; 872:188–193.
45. Ganfornina MD, López-Barneo J. Single K^+ channels in membrane patches of arterial chemoreceptor cells are modulated by O_2 tension. Proc Natl Acad Sci USA 1991; 88:2927–2930.
46. Lopez-Lopez JR, Gonzalez C. Time course of K^+ current inhibition by low oxygen in chemoreceptors cells of adult rabbit carotid body: effects of carbon monoxide. FEBS Lett 1992; 299:251–254.
47. Montoro RJ, Urena J, Fernandez-Chacon R, Alvarez de Toledo G, Lopez-Barneo J. Oxygen sensing by ion channels and chemotransduction in single glomus cells. J Gen Physiol 1996; 107:133–143.
48. Summers BA, Overholt JL, Prabhakar NR. Augmentation of L-type calcium current by hypoxia in rabbit carotid body glomus cells: evidence for a PKC-sensitive pathway. J Neurophysiol 2000; 84:1636–1644.
49. Fu XW, Nurse CA, Wang YT, Cutz E. Selective modulation of membrane currents by hypoxia in intact airway chemoreceptors from neonatal rabbit. J Physiol 1999; 514:139–150.
50. Barlow RS, El-Mowafy AM, White RE. H_2O_2 opens BK(Ca) channels via the PLA_2-arachidonic acid signaling cascade in coronary artery smooth muscle. Am J Physiol 2000; 279:H475–H483.
51. DiChiara TJ, Reinhart PH. Redox modulation of *hslo* Ca^{2+}-activated K^+ channels. J Neurosci 1997; 17:4942–4955.
52. Osanai S, Mokashi A, Rozanov C, Buerk DG, Lahiri S. Potential role of H_2O_2 in chemoreception in the cat carotid body. J Auton Nerv Syst 1997; 63:39–45.

28

Optical Analysis of the Oxygen-Sensing Signal Pathway

HELMUT ACKER, UTTA BERCHNER-PFANNSCHMIDT, and CHRISTOPH WOTZLAW

Max Planck Institute of Molecular Physiology
University of Essen
Dortmund, Germany

CHRISTINE HUCKSTORF and TINO STRELLER

University of Rostock
Rostock, Germany

I. Tissue Oxygen Sensing

The heterogeneous oxygen partial pressure (PO_2) distribution in mammalian organs ranging from about 0 to 90 torr, under normoxic conditions with an arterial PO_2 of about 100 torr as shown in Figure 1, requires an O_2-sensing signal cascade to adapt cellular functions to PO_2 heterogeneity. Under normoxic conditions O_2 sensing leads to optimization of cell function (normoxic response) as exemplified by enhanced phosphoenol pyruvate carboxykinase (PCK) expression in the periportal liver zone with high PO_2 levels and of enhanced glukokinase (GK) expression in the perivenous liver zone with low PO_2 levels (1). Under arterial hypoxia tissue PO_2 frequency distribution is left-shifted leading to a hypoxic response of O_2 sensing for adaptation of cell function. The response includes enhanced expression of an array of proteins involved in regulation of different functions such as erythropoietin (EPO) in red cell formation, vascular endothelial growth factor (VEGF) in blood vessel formation (2), or lactate dehydrogenase (LDH) in energy metabolism (3). Furthermore, potassium channel gating is altered in carotid body type I cells to release various transmitter-exciting, synaptically connected nerve fibers for nervous regulation of ventilation and blood circulation (4,5), in neuroepithelial bodies (NEB)

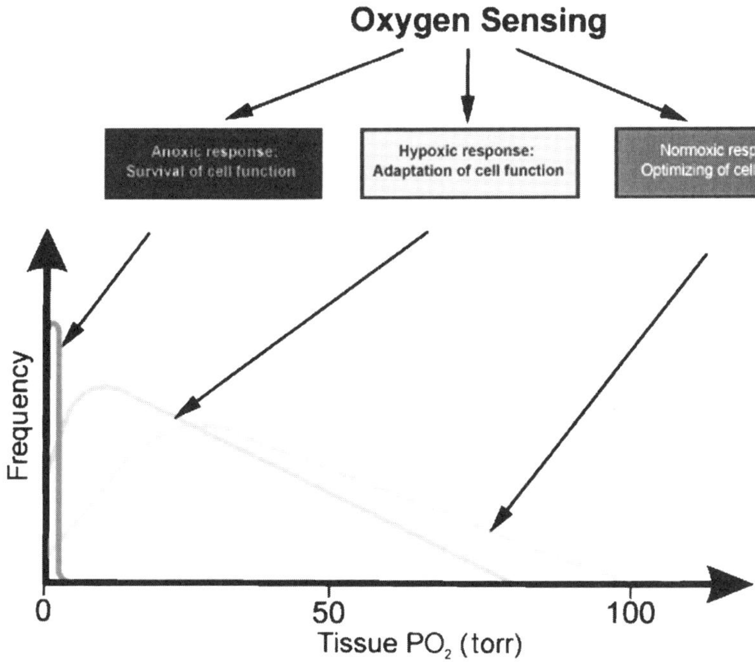

Figure 1 Frequency distribution of oxygen partial pressure in tissue under normoxic, hypoxic, and anoxic conditions of the arterial blood. Oxygen sensing provides for all three conditions a response to trigger different cell functions (70).

to release serotonin controlling the bronchial muscular tone during hypoxia (6–8) as well as in vascular smooth muscle cells leading to peripheral blood vessel dilatation (9) or to lung vessel vasoconstriction (10). The anoxic response supports survival of cell function as it occurs in lower vertebrates by drastically reducing protein synthesis and thereby reducing O_2 consumption enabling life for months under water without ventilation (11). Reduction of protein synthesis was also observed in the peri-infarct area (penumbra) after stroke induced by middle cerebral artery occlusion probably facilitating survival of cell function by means of oxygen sensing (12).

II. Nature of Oxygen Sensor

It is not clear whether these different responses are induced by various O_2-sensing signal cascades. One hypothesis supposes a cascade to consist of mitochondrial or nonmitochondrial heme proteins sensing oxygen with a subsequent second-messenger formation, as for instance reactive oxygen species (ROS), which influence via an iron-mediated Fenton reaction the stability of hypoxia-inducible transcription factor HIF-1α for altering its binding capacity to hypoxic responsive elements of the GK, EPO, VEGF, or LDH gene or change the thiol status of ion channel proteins (2,13–15). Among the nonmitochondrial heme proteins isoforms of the neutrophil NADPH oxidase are discussed as oxygen sensors (6,16–18). Also nitric oxide synthase is assumed to be an oxygen sensor in carotid body tissue and brainstem respiratory center involving PO_2-dependent NO formation with subsequent NO-heme interaction or S-nitrosothiol biochemistry (19–21). Mitochondrial complex II, III, and IV are discussed as oxygen sensors for Hep3B liver tumor cells triggering EPO production (22), for cardiomyocytes regulating hibernation during hypoxia (23), as well as for carotid body tissue-enhancing nervous chemoreceptor discharge in dependence on PO_2 (24–26). Another hypothesis supposes an oxygen-sensing cascade comprising nonheme iron-binding proteins such as HIF prolyl-hydroxylase (HIF-PH) regulating HIF-1α and HIF-2α (ubiquitylation and degradation by the von Hippel–Lindau tumor suppressor (pVHL) E3 ligase complex under normoxia through hydroxylation at proline residues 402 and 564 located in the HIF oxygen-dependent degradation domain (27–30).

III. Aim of the Study

Detection, characteristics, and interaction of three different oxygen sensor proteins will be the main subject of this chapter. Atomic structure, visible light absorption spectra, and catalytic cycle of heme a and heme a3 in mitochondrial cytochrome c oxidase are well described in recent literature (31,32). The question was addressed, therefore, whether the deconvolution method of light absorption spectra as applied by our group to oxygen-sensing HepG2 cells (33,34) or rat carotid body (35) is able to detect different cytochrome c oxidase species probably acting as oxygen sensors. In the light of recent findings on different gp91 subtypes in various tissues (36–38) and gp91 knockout mice (6,10,39,40), the meaning of different NADPH oxidase isoforms as oxygen sensor is also addressed. To complete optical analysis of the oxygen-sensing pathway, intracellular compartments harboring the iron-mediated Fenton reaction involved in ROS degradation, and probably recovering the ferrous iron state of HIF-PH for regulating HIF stabilization and nuclear translocation, are visualized three-dimensionally by means of two-photon confocal laser microscopy. Based on these measurements, two hypotheses will be outlined combining the three oxygen sensors mentioned for ion channel or gene expression regulation.

IV. Oxygen Sensor and Ion Channel Conductivity

A. Cytochrome c Oxidase

Detection of heme proteins by light absorption photometry was carried out on superfused isolated rat carotid body, cervical ganglion, as well as HepG2 cell spheroids (35,41). Differential spectra were obtained by first recording in

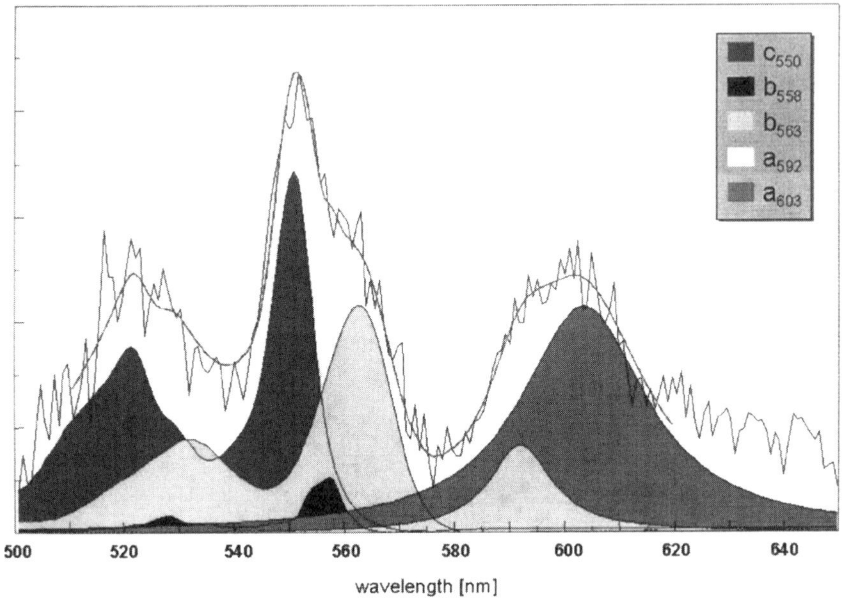

Figure 2 N_2 versus aerobic steady-state spectrum as a mean of six carotid bodies fitted by different mitochondrial and nonmitochondrial cytochrome spectra as indicated. The superposition curve obtained by varying the amplitude of the optical density of five cytochromes closely fits the experimental curve.

superfusion medium equilibrated with 28.8% O_2, 4% CO_2, and 67.2% N_2 (aerobic steady state). These spectra were automatically subtracted from those obtained under reducing conditions, produced by equilibrating the saline with different concentrations of N_2, O_2, CO, or cyanide at constant CO_2 (4%). Since the differential spectra obtained are composed of various mitochondrial and nonmitochondrial heme proteins, differential spectra of various isolated cytochromes within a wavelength range between 510 and 620 nm were used in a deconvolution procedure (33,35). The purpose was to identify peaks and shoulders and to fit the experimental curve to a superposition curve. Figure 2 summarizes a mean N_2 versus aerobic steady-state spectra recorded from six rat carotid bodies 6 min after onset of hypoxia in the superfusion medium (60 torr). The amplitude of the optical density of five isolated cytochrome spectra (nonmitochondrial cytochrome b_{558} and mitochondrial cytochrome b_{563}, cytochrome c_{550}, cytochrome a_{592}, cytochrome $aa3_{603}$) was varied by multiplying with weight factors (33) for deconvolution to fit the experimental curve as close as possible with a superposition curve representing the sum of optical densities of the different cytochromes at each wavelength. Table 1 compares weight factor values to fit N_2 versus aerobic steady-state spectra of different cell types. Cytochrome b_{558} as described for the neutrophil NADPH oxidase contributes significantly in the rat carotid body tissue to the spectrum caused by abundantly present macrophages, which in addition to type I cells contain this nonmitochondrial

Table 1 Cell-Type-Dependent Weight Factors Used to Fit Experimental N_2 Versus Aerobic Steady-State Spectra by Deconvolution in Percent of Total Cytochrome Contribution

	Cytochrome b558M	Cytochrome b563	Cytochrome c550	Cytochrome a592	Cytochrome a(3)598	Cytochrome a601	Cytochrome aa(3)603	Cytochrome aa3(605)
HepG2	2.6	23.1	22		19.1	27		6.2
Carotid body	9.1	22.7	31.3	7.6			29.3	
Cervical ganglion	6.6	19.3	35.4				38.7	

cytochrome (42). Cyanide-insensitive heme a of mitochondrial cytochrome c oxidase dominates by a blue shift peaking at 592 nm. This shift is described as a result of a single site mutation of Arginin 45 hydrogen bonded to the formyl group of heme a to methionine in *Paracoccus denitrificans* (43). As a yield a drastically lowered midpoint potential of heme a was measured and, correspondingly, a dominance of direct Cu_A-heme a3 electron transfer over the normal electron pathway Cu_A-heme a-heme a3. To fit N_2 versus aerobic steady-state spectra of HepG2 cells six cytochromes had to be used (see Table 1). In addition to cytochrome b_{558}, cytochrome b_{563}, and cytochrome c, three spectra of cytochrome c oxidase species have been introduced for this purpose (44). One was obtained by forming a N_2 versus CN^- aerobic steady-state spectrum (45) giving heme a of cytochrome aa3 (32) with a peak at 601 instead of at 605 nm as described (32,46). The second was measured by forming a CN^- aerobic steady-state spectrum versus total cytochrome aa3 peaking at 596 nm instead of at 595 nm as characteristic for the Fe^{+II}-O_2 Cu^{+I} species of cytochrome a3 named compound A (31). The third was the spectrum of total cytochrome aa3 peaking at 605 nm instead of at 604 nm as described (46). In the case of the cervical ganglion cytochrome b_{558}, cytochrome b_{563}, cytochrome c_{550}, and cytochrome $aa3_{603}$ are sufficient to fit the N_2 versus CN^- aerobic steady-state spectrum (see Table 1).

Hypothesis One

Figure 3 depicts as a hypothetical model the possible involvement of cytochrome a_{592} in carotid body oxygen sensing. This unusual cytochrome bypassed from the normal electron pathway as outlined above might underlie a nonlinear reduction already starting at high PO_2, giving an apparent low PO_2 affinity of carotid body cytochrome c oxidase. Assuming cytochrome a_{592} is involved in intracellular calcium regulation altering ion channel conductivity and transmitter release, the nonlinear PO_2—chemoreceptor discharge relationship might be related thereto. Also, other heme ligands such as CN^-, CO, or NO might influence chemoreceptor discharge by binding to cytochrome a3-CuB(20) (26,47) and thus altering the redox

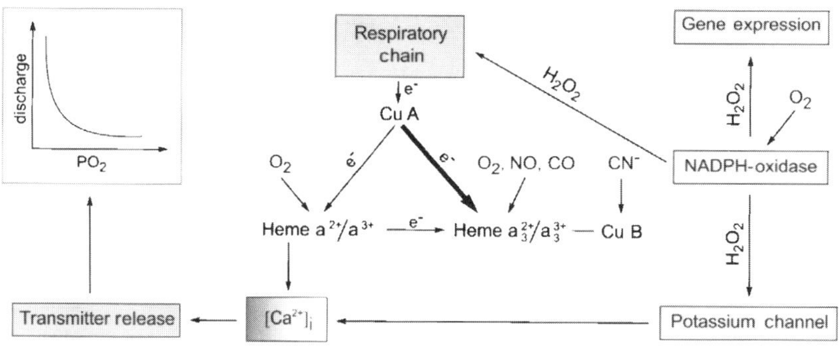

Figure 3 Two-cytochrome model hypothesizing the oxygen-sensing mechanism in rat carotid body tissue.

state of cytochrome a_{592} nonlinearly. This special arrangement of cytochrome c oxidase is unique for the carotid body specialized in oxygen sensing. The other cell types, as shown in Table 1, seem to contain a normal mitochondrial chain that is unlikely to function as oxygen sensor. Defects in mitochondrial complex I, III, and IV do not impair oxygen sensor function in HepG2 and other cell types (48,49).

B. NADPH Oxidase

Figure 3 shows the NADPH oxidase as oxygen sensor presumably regulating ion channel conductivity and gene expression. Different NADPH oxidase isoforms with components (50) p22*phox*, gp91*phox*, p47*phox*, p40*phox*, p67*phox*, and Rac1,2 (51) have to be considered. The isoforms (Nox1-4 and the Duox group) concern especially gp91*phox*, as reviewed by Lambeth et al. (36,52). The neutrophil NADPH oxidase has been also identified in endothelial cells (40) and neuroepithelial bodies (15). Gp91*phox* knockout mice showed an impaired hypoxic ventilatory control in neonatal animals due to a decreased oxygen sensitivity of NEB potassium channel conductivity (6,53) having oxygen sensor function of the NADPH oxidase in this cell type and related strains (7,54). On the other hand, gp91*phox* knockout mice showed no impaired oxygen-sensing function of pulmonary vasculature smooth muscle cells (10) or carotid body hypoxic drive (39), contrasting findings describing an impaired sensing function of these organs under diphenyleniodonium (DPI) or AEBSF application as specific NADPH oxidase inhibitors (17,35,55,56) as well as identification of the NADPH oxidase in these cells by histology, photometry, and molecular biology (56–58). We assume, therefore, as shown in Figure 3, an unidentified NADPH oxidase isoform inhibiting potassium channel conductivity in carotid body type I cells under hypoxia. Oxygen sensing in carotid body tissue is therefore hypothesized to be based on an interaction between cytochrome a_{592} as a cytosolic and NADPH oxidase as a plasma membrane oxygen sensor.

V. Oxygen Sensor and Gene Expression

A. ROS

The potential effect of PO_2-dependent H_2O_2 production with respect to O_2-dependent gene expression, as shown in Figure 3, was demonstrated for EPO production in HepG2 cells. H_2O_2-production rate was positively correlated with rising pericellular PO_2 (49,59). Under normoxia, EPO production is low owing to high H_2O_2 levels. Under hypoxia, however, when H_2O_2 production decreases, full expression of the gene is permitted. Indeed, the external application of H_2O_2 to HepG2 cells cultured under hypoxia depressed the hypoxia-stimulated EPO production and mimicked high PO_2 values (60). These H_2O_2-mediated effects were antagonized by Fe^{3+} chelator desferrioxamine (DFO) and $OH^{\cdot}$ scavengers dimethylthiourea or tetramethylthiourea (D/TMTU) indicating the involvement of an iron mediated Fenton reaction (61) degrading H_2O_2. The same reaction seems to be also involved in the reciprocal modulation of the glucagon-activated PCK gene and the insulin-activated GK gene by O_2 in primary hepatocytes (1,62).

B. Localization of Cellular Fenton Reaction

The Fenton reaction has two kinetics for degrading H_2O_2:

$$H_2O_2 + Fe^{2+} \rightarrow Fe^{3+} + OH^- + OH^{\cdot}$$

$$H_2O_2 + Fe^{3+} \rightarrow Fe^{2+} + OOH^- + H^+$$

Fluorescent rhodamine 123 converted from nonfluorescent dihydrorhodamine 123 by scavenging $OH^{\cdot}$ generated from H_2O_2 in a cellular Fe^{2+}-dependent Fenton reaction (33) could be localized as distinct "hot spots" in the endoplasmatic reticulum of HepG2 cells using two-photon laser scanning microscopy (2P-CLSM), as shown in Figure 4. A mode-locked Ti : Sapphire laser was applied pumped by a high-power 532-nm laser resulting in a pulse width of about 120 fs (10^{-15} sec) at a repetition rate of 76 MHz and an average power of about 48 mW at 850 nm, which diminished to 4 mW in front of the $60\times/1{,}2$NA water objective used for cell visualization. For 3-D reconstruction series of up to 64 optical z-sections (average section distance 300 nm) were recorded with an x–y resolution of 512×512 pixels. For deconvolution of data sets 3-D point spread functions of the 2P-CLSM were measured and calculation of isosurface and data visualization was performed on a Unix workstation (63). In contrast to one-photon systems, two-photon excitation offers many advantages such as inherent optical slicing, low out-of-focus bleaching, less photo damage, no oxygen radical formation, and deeper penetration of the specimen. For identifying endoplasmatic reticulum HepG2 cells were transiently

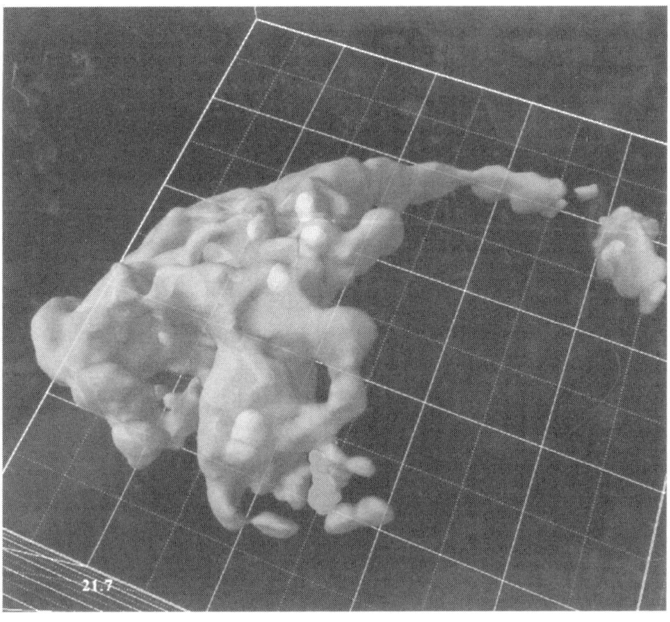

Figure 4 Three-dimensional in vivo 2P-CLSM illustration of Fe^{2+}-mediated Fenton reaction converting nonfluorescent dihydrorhodamine 123 into fluorescent rhodamine 123 at hot spots (light gray) inside the perinuclear endoplasmatic reticulum (dark gray) of one HepG2 cell.

transfected with gene constructs in which expression of enhanced cyan fluorescent protein (ECFP, emission at 470 nm) was directed by the endoplasmic reticulum targeting sequence of calreticulin and the KDEL retrieval sequence. Figure 4 shows in gray the 3D extensions of perinuclear endoplasmatic reticulum in one HepG2 cell kept alive under normoxic tissue culture conditions in an observation chamber mounted on the microscope stage. Within the first 10 min after application of dihydrorhodamine 123, several distinct "hot spots" within the endoplasmatic reticulum show rhodamine 123 fluorescence (recorded at 620 nm to avoid cross-talk with ECFP fluorescence), indicating a localized Fe^{2+}-mediated Fenton reaction.

C. HIF Stabilization and Nuclear Translocation

The HIF DNA-binding complex is a heterodimer of α- and β-subunits. In oxygenated cells, the α-subunits are unstable, being targeted for proteosomal destruction by specific degradation domains. This process is dependent on the von Hippel–Lindau tumor suppressor protein (pVHL), which serves as the recognition component of a ubiquitin ligase that promotes ubiquitin-dependent proteolysis of HIF-1α. In hypoxic cells, HIF-1α degradation is suppressed leading to nuclear translocation, dimerization, and transcriptional activation of target genes (64).

2P-CLSM was applied to visualize three-dimensionally HIF-1α trafficking after cytoplasmatic stabilization into the nucleus of HepG2 cells stimulated by $CoCl_2$ (100 μM, 4 hr). After stimulation, cells were fixed and immunostained with fluorescent antibodies raised against HIF-1α and ECFP targeting the endoplasmatic reticulum. Figure 5 shows two HepG2 cells at low and high magnification. The cell

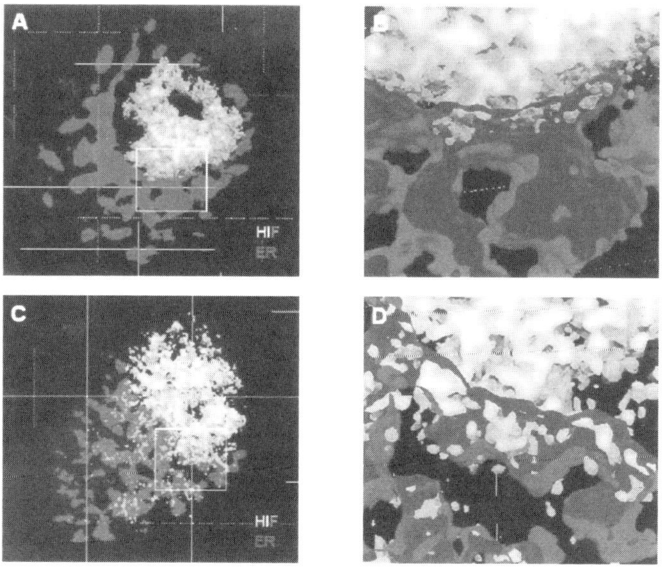

Figure 5 Three-dimensional 2P-CLSM visualization of HIF-1α nuclear translocation at lower (A,C) and higher (B,D) magnification of two HepG2 cells incubated with 100 μM $COCl_2$ for 4 hr (HIF-1α: light gray; endoplasmatic reticulum: dark gray).

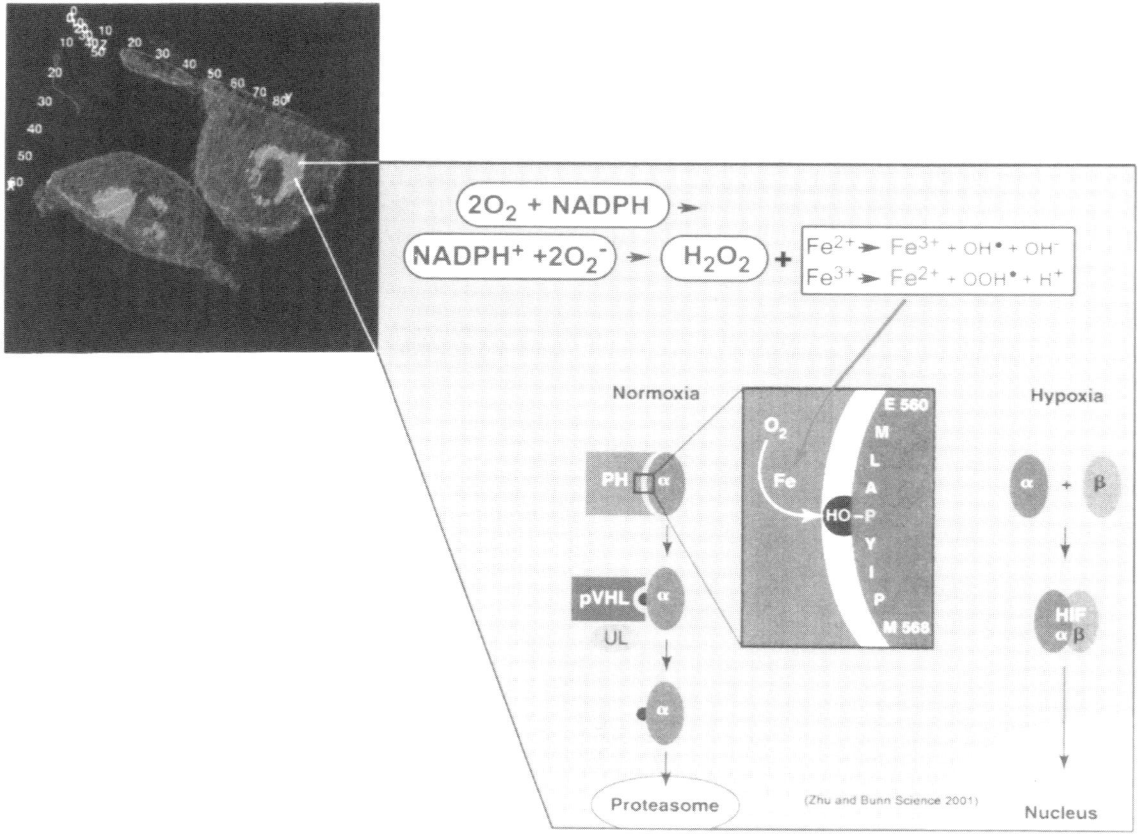

Figure 6 Model of HIF-PH-mediated HIF-1α degradation (59) due to PH-mediated HIF-1α hydroxylation at P564 with subsequent pv HL-mediated ubiquitination. H_2O_2 degradation mediated by Fenton reaction is supposed to be required for recovering Fe^{2+} under normoxic conditions. (Adapted from Refs. 62,65.)

in Figure 5A,B has a complete nuclear translocation of HIF-1α (light gray) whereas the second cell (Fig. 5C,D) shows two events: stabilization of HIF-1α in the perinuclear endoplasmatic reticulum (dark gray) and trafficking to as well as accumulation of HIF-1α in the nucleus. It is obvious from the high magnification (Fig. 5B,D) that the endoplasmatic reticulum seems to form pores with the outer nuclear membrane used for HIF-1α entry into the nucleus.

Hypothesis Two

Figure 6 shows the assumed meaning of the Fenton reaction for HIF-PH-regulated HIF-1α ubiquitylation and degradation by the von Hippel–Lindau tumor suppressor protein (65). HIF-PH has an absolute requirement for dioxygen as cosubstrate as well as α-oxoglutarate and ferrous iron. With PH-mediated hydroxylation Fe^{3+}-O^- is produced, which inhibits PH as described for collagen synthesis. Therefore, PH gets a high activity under ferric iron–reducing conditions such as vitamin C application or hypoxia during wound healing when ferrous iron is abundant (66). We hypothesize that the F^{3+}-mediated Fenton reaction recovers Fe^{2+} to maintain PH activity and low HIF-1α levels under normoxia. As already outlined for hypothesis one, oxygen sensing might also mean here an interaction between two components: HIF-PH and an NADPH oxidase isoform (18) generating H_2O_2. An assumed low-PO_2 affinity of both systems impairs their activity under hypoxia. The high Fe^{2+} level, however, keeps the HIF-PH alert to react immediately to reoxygenation with a subsequent cycling of the iron charge state during Fenton reaction. It should be mentioned, however, that HIF-1α stabilization is induced also by other means as MAP kinases for regulation of catecholamine synthesis (67) or p53 for regulation of tumor angiogenesis (68). Furthermore, hypoxic gene expression can be up-regulated independent of HIF (69).

VI. Conclusion

Optical methods, as described, are powerful tools to characterize members of the oxygen-sensing pathway leading to HIF-1α stabilization or ion channel gating. Improvement of optical resolution enables visualization of colocalization of cellular compartments with single steps of the signal cascade leading to a better understanding and a more specific experimental or therapeutic interference with oxygen sensing.

References

1. Kietzmann T, Porwol T, Zierold K, Jungermann K, Acker H. Involvement of a local fenton reaction in the reciprocal modulation by O_2 of the glucagon-dependent activation of the phosphoenolpyruvate carboxykinase gene and the insulin-dependent activation of the glucokinase gene in rat hepatocytes. Biochem J 1998; 335(Pt 2):425–432.

2. Bunn HF, Poyton RO. Oxygen sensing and molecular adaptation to hypoxia. Physiol Rev 1996; 76:839–885.
3. Ebert BL, Gleadle JM, O'Rourke JF, Bartlett SM, Poulton J, Ratcliffe PJ. Isoenzyme-specific regulation of genes involved in energy metabolism by hypoxia: similarities with the regulation of erythropoietin. Biochem J 1996; 313:809–814.
4. Lopez-Barneo J, Pardal R, Ortega-Saenz P. Cellular mechanisms of oxygen sensing [Review]. Annu Rev Physiol 2001; 63:259–287.
5. Lahiri S. Historical perspectives of cellular oxygen sensing and responses to hypoxia. J Appl Physiol 2000; 88(4):1467–1473.
6. Fu XW, Wang D, Nurse C, Dinauer MC, Cutz E. NADPH oxidase is an O_2 sensor in airway chemoreceptors: evidence from K^+ current modulation in wild-type and oxidase-deficient mice. Proc Natl Acad Sci USA 2000; 97(8):4374–4379.
7. O'Kelly I, Lewis A, Peers C, Kemp PJ. O_2 sensing by airway chemoreceptor-derived cells. J Biol Chem 2000; 275:7684–7692.
8. Skogvall S, Korsgren M, Grampp W. Evidence that neuroepithelial endocrine cells control the spontaneous tone in guinea pig tracheal preparations. J Appl Physiol 1999; 86(3):789–798.
9. Franco-Obregon A, Ureña J, López-Barneo J. Oxygen-sensitive calcium channels in vascular smooth muscle and their possible role in hypoxic arterial relaxation. Proc Natl Acad Sci USA 1995; 92:4715–4719.
10. Archer SL, Reeve HL, Michelakis E, Puttagunta L, Waite R, Nelson DP et al. O-2 sensing is preserved in mice lacking the gp91 phox subunit of NADPH oxidase. Proc Natl Acad Sci USA 1999; 96:7944–7949.
11. Kwast KE, Hand SC. Acute depression of mitochondrial protein synthesis during anoxia—contributions of oxygen sensing, matrix acidification, and redox state. J Biol Chem 1996; 271(13):7313–7319.
12. Hata R, Maeda K, Hermann D, Mies G, Hossmann KA. Dynamics of regional brain metabolism and gene expression after middle cerebral artery occlusion in mice. J Cereb Blood Flow Metab 2000; 20(2):306–315.
13. Duprat F, Guillemare E, Romey G, Fink M, Lesage F, Lazdunski M, et al. Susceptibility of cloned K^+ channels to reactive oxygen species. Proc Natl Acad Sci USA 1995; 92:11796–11800.
14. Taglialatela M, Castaldo P, Iossa S, Pannaccione A, Fresi A, Ficker E, et al. Regulation of the human ether-a-gogo related gene (herg) k+ channels by reactive oxygen species. Proc Natl Acad Sci USA 1997; 94:11698–11703.
15. Wang DS, Youngson C, Wong V, Yeger H, Dinauer MC, Demiera EVS, et al. Nadph-oxidase and a hydrogen peroxide-sensitive k+ channel mag function as an oxygen sensor complex in airway chemoreceptors and small-cell lung-carcinoma cell-lines. Proc Natl Acad Sci USA 1996; 93:13182–13187.
16. Jones RD, Hancock JT, Morice AH. NADPH oxidase: a universal oxygen sensor? [Review]. Free Rad Biol Med 2000; 29(5):416–424.
17. Cross AR, Henderson L, Jones OT, Delpiano MA, Hentschel J, Acker H. Involvement of an NAD(P)H oxidase as a pO2 sensor protein in the rat carotid body. Biochem J 1990; 272(3):743–747.
18. Zhu H, Qiu H, Yoon HP, Huang S, Bunn HF. Identification of a cytochrome b-type NAD(P)H oxidoreductase ubiquitously expressed in human cells. Proc Natl Acad Sci USA 1999; 96(26):14742–14747.

19. Abu-Soud HM, Rousseau DL, Stuehr DJ. Nitric oxide binding to the heme of neuronal nitric-oxide synthase links its activity to changes in oxygen tension. J Biol Chem 1996; 271(51):32515–32518.
20. Prabhakar NR. Oxygen sensing by the carotid body chemoreceptors [Review]. J Appl Physiol 2000; 88(6):2287–2295.
21. Lipton AJ, Johnson MA, Macdonald T, Liebermann MW, Gozal D, Gaston B. S-Nithrosothiols signal the ventilatory response to hypoxia. Nature 2001; 413:171–174.
22. Chandel NS, Maltepe E, Goldwasser E, Mathieu CE, Simon MC, Schumacker PT. Mitochondrial reactive oxygen species trigger hypoxia-induced transcription. Proc Natl Acad Sci USA 1998; 95:11715–11720.
23. Budinger GRS, Duranteau J, Chandel NS, Schumacker PT. Hibernation during hypoxia in cardiomyocytes—role of mitochondria as the o-2 sensor. J Biol Chem 1998; 273:3320–3326.
24. Baysal BE, Ferrell RE, Willett-Brozick JE, Lawrence EC, Myssiorek D, Bosch A et al. Mutations in SDHD, a mitochondrial complex II gene, in hereditary paraganglioma. Science 2000; 287(5454):848–851.
25. Mills E, Jöbsis FF. Mitochondrial resoiratory chain of carotid body and chemoreceptor response to changes in oxygen tension. J Neurophysiol 1972; 35:405–428.
26. Wilson DF, Mokashi A, Chugh D, Vinogradov S, Osanai S, Lahiri S. The primary oxygen sensor of the cat carotid body is cytochrome a_3 of the mitochondrial respiratory chain. FEBS Lett 1994; 351:370–374.
27. Masson N, Willam C, Maxwell PH, Pugh CW, Ratcliffe PJ. Independent function of two destruction domains in hypoxia-inducible factor-alpha chains activated by prolyl hydroxylation. EMBO J 2001; 20(18):5197–5206.
28. Jaakkola P, Mole DR, Tian YM, Wilson MI, Gielbert J, Gaskell SJ, et al. Targeting of HIF-{alpha} to the von Hippel–Lindau ubiquitylation complex by O_2-regulated prolyl hydroxylation. Science 2001.
29. Epstein AC, Gleadle JM, McNeill LA, Hewitson KS, O'Rourke J, Mole DR, et al. *C. elegans* EGL-9 and mammalian homologs define a family of dioxygenases that regulate HIF by prolyl hydroxylation. Cell 2001; 107(1):43–54.
30. Ivan M, Kondo K, Yang H, Kim W, Valiando J, Ohh M, et al. HIF{alpha} targeted for VHL-mediated destruction by proline hydroxylation: implications for O_2 sensing. Science 2001; 292:464–468.
31. Michel H, Behr J, Harrenga A, Kannt A. Cytochrome c oxidase—structure and spectroscopy [Review]. Annu Rev Biophys Biomol Struct 1998; 27:329–356.
32. Hellwig P, Grzybek S, Behr J, Ludwig B, Michel H, Mntele W. Electrochemical and ultraviolet/visible/infrared spectroscopic analysis of hemea and a_3 redox reactions in the cytochrome c oxidase from *Paracoccus denitrificans*: separation of heme a and a_3 contributions and assignment of vibrational modes. Biochemistry 1999; 38:1685 1694.
33. Porwol T, Ehleben W, Zierold K, Fandrey J, Acker H. The influence of nickel and cobalt on putative members of the oxygen-sensing pathway of erythropoietin-producing HepG2 cells. Eur J Biochem 1998; 256(1):16–23.
34. Ehleben W, Porwol T, Fandrey J, Kummer W, Acker H. Cobalt and desferrioxamine reveal crucial members of the oxygen sensing pathway in HepG2 cells. Kidney Int 1997; 51(2):483–491.
35. Lahiri S, Ehleben W, Acker H. Chemoreceptor discharges and cytochrome redox changes of the rat carotid body: role of heme ligands. Proc Natl Acad Sci USA 1999; 96(16):9427–9432.

36. Lambeth JD, Cheng G, Arnold RS, Edens WA. Novel homologs of gp91phox. Trends Biochem Sci 2000; 25(10):459–461.
37. Geiszt M, Kopp JB, Varnai P, Leto TL. Identification of renox, an NAD(P)H oxidase in kidney. Proc Natl Acad Sci USA 2000; 97(14):8010–8014.
38. De D, X, Wang D, Many MC, Costagliola S, Libert F, Vassart G, et al. Cloning of two human thyroid cDNAs encoding new members of the NADPH oxidase family. J Biol Chem 2000; 275(30):23227–23233.
39. Roy A, Rozanov C, Mokashi A, Daudu P, Al mehdi AB, Shams H, et al. Mice lacking in gp91 phox subunit of NAD(P)H oxidase showed glomus cell [Ca2+](i) and respiratory responses to hypoxia. Brain Res 2000; 872(1–2):188–193.
40. Görlach A, Brandes RP, Nguyen K, Amidi M, Dehghani F, Busse R. A gp91phox containing NADPH oxidase selectively expressed in endothelial cells is a major source of oxygen radical generation in the arterial wall [see comments]. Circ Res 2000; 87(1):26–32.
41. Görlach A, Holtermann G, Jelkmann W, Hancock JT, Jones SA, Jones OT, et al. Photometric characteristics of haem proteins in erythropoietin- producing hepatoma cells (HepG2). Biochem J 1993; 290(Pt 3):771–776.
42. Dvorakova M, Höhler B, Vollerthun R, Fischbach T, Kummer W. Macrophages: a major source of cytochrome b558 in the rat carotid body. Brain Res 2000; 852:349–354.
43. Kannt A, Pfitzner U, Ruitenberg M, Hellwig P, Ludwig B, Mantele W, et al. Mutation of Arg-54 strongly influences heme composition and rate and directionality of electron transfer in Paracoccus denitrificans cytochrome c oxidase. J Biol Chem 1999; 274(53):37974–37981.
44. Porwol T, Ehleben W, Brand V, Acker H. Tissue oxygen sensor function of NADPH oxidase isoforms, an unusual cytochrome aa3 and reactive oxygen species. Respir Physiol 2001; 128:331–348.
45. Ehleben W, Bolling B, Merten E, Porwol T, Strohmaier AR, Acker H. Cytochromes and oxygen radicals as putative members of the oxygen sensing pathway. Respir Physiol 1998; 114(1):25–36.
46. Liao GL, Palmer G. The reduced minus oxidized difference spectra of cytochromes a and a_3. Biochim Biophys Acta Mol Cell Res 1996; 1274:109–111.
47. Hill BC. The pathway of CO binding to cytochrome c oxidase: can the gateway be closed? FEBS Lett 1993; 354:284–288.
48. Srinivas V, Leshchinsky I, Sang N, King MP, Minchenko A, Caro J. Oxygen sensing and HIF-1 activation does not require an active mitochondrial respiratory chain electron-transfer pathway. J Biol Chem 2001; 276(25):21995–21998.
49. Vaux EC, Metzen E, Yeates KM, Ratcliffe PJ. Regulation of hypoxia-inducible factor is preserved in the absence of a functioning mitochondrial respiratory chain. Blood 2001; 98(2):296–302.
50. Babior BM. NADPH Oxidase: an update. Blood 1999; 93(3):1464–1476.
51. Cool RH, Merten E, Theiss Ch, Acker H. Rac1, and not Rac2, is involved in the regulation of the intracellular hydrogen peroxide level in HepG2 cells. Biochem J 1998; 332:5–8.
52. Cheng GJ, Cao ZH, Xu XX, Van Meir EG, Lambeth JD. Homologs of gp91phox: cloning and tissue expression of Nox3, Nox4, and Nox5. Gene 2001; 269(1–2):131–140.
53. Kazemian P, Stephenson R, Yeger H, Cutz E. Respiratory control in neonatal mice with NADPH oxidase deficiency. Respir Physiol 2001; 126(2):89–101.

54. O'Kelly I, Peers C, Kemp PJ. NADPH oxidase does not account fully for O_2-sensing in model airway chemoreceptor cells. Biochem Biophys Res Commun 2001; 283(5):1131–1134.
55. Grimminger F, Weissmann N, Spriestersbach R, Becker E, Rosseau S, Seeger W. NADPH oxidase-inhibitors on hypoxic vasoconstriction in buffer-perfused rabbit lungs. Am J Physiol Lung Cell and Mol Physiol 1995; 12:L 747-L 752.
56. Weissmann N, Tadic A, Hanze J, Rose F, Winterhalder S, Nollen M, et al. Hypoxic vasoconstriction in intact lungs: a role for NADPH oxidase-derived H(2)O(2)? [In Process Citation]. Am J Physiol Lung Cell Mol Physiol 2000; 279(4):L683–L690.
57. Kummer W, Acker H. Immunohistochemical demonstration of four subunits of neutrophil NAD(P)H oxidase in type I cells of carotid body. J Appl Physiol 1995; 78(5):1904–1909.
58. Marshall C, Mamary AJ, Verhoeven AJ, Marshall BE. Pulmonary-artery nadph-oxidase is activated in hypoxic pulmonary vasoconstriction. Am J Respir Cell Mol Biol 1996; 15:633–644.
59. Canbolat O, Fandrey J, Jelkmann W. Effects of modulators of the production and degradation of hydrogen-peroxide on erythropoietin synthesis. Respir Physiol 1998; 114:175–183.
60. Tabata M, Tarumoto T, Ohmine K, Furukawa Y, Hatake K, Ozawa K, et al. Stimulation of GATA-2 as a mechanism of hydrogen peroxide suppression in hypoxia-induced erythropoietin gene expression. J Cell Physiol 2001; 186(2):260–267.
61. Fandrey J, Frede S, Ehleben W, Porwol T, Acker H, Jelkmann W. Cobalt chloride and desferrioxamine antagonize the inhibition of erythropoietin production by reactive oxygen species. Kidney Int 1997; 51(2):492–496.
62. Kietzmann T, Fandrey J, Acker H. Oxygen radicals as messengers in oxygen-dependent gene expression. News in Physiological Science 2000; 15:202–208.
63. Strohmaier A, Porwol T, Acker H, Spiess E. Three-dimensional organization of microtubules in tumor cells studied by confocal laser scanning microscopy and computer-assisted deconvolution and image reconstruction. Cells Tissues Organs 2000; 167(1):1–8.
64. Wenger RH. Mammalian oxygen sensing, signalling and gene regulation. J Exp Biol 2000; 203(Pt 8):1253–1263.
65. Zhu H, Bunn HF. How do cells sense oxygen? Science 2001.
66. Takahashi Y, Takahashi S, Shiga Y, Yoshimi T, Miura T. Hypoxic induction of prolyl 4-hydroxylase α(I) in cultured cells. J Biol Chem 2000; 275:14139–14146.
67. Conrad PW, Conforti L, Kobayashi S, Beitner-Johnson D, Rust RT, Yuan Y, et al. The molecular basis of O2-sensing and hypoxia tolerance in pheochromocytoma cells. Comp Biochem Physiol B Biochem Mol Biol 2001; 128(2):187–204.
68. Ravi R, Mookerjee B, Bhujwalla ZM, Sutter CH, Artemov D, Zeng Q, et al. Regulation of tumor angiogenesis by p53-induced degradation of hypoxia-inducible factor 1alpha. Genes Dev 2000; 14(1):34–44.
69. Dong Z, Venkatachalam MA, Wang J, Patel Y, Saikumar P, Semenza GL, et al. Up-regulation of apoptosis inhibitory protein IAP-2 by hypoxia. Hif-1-independent mechanisms. J Biol Chem 2001; 276(22):18702–18709.
70. Porwol T, Ehleben W, Brand V, Acker H. Tissue oxygen sensor function of NADPH oxidase isoforms, an unusual cytochrome aa3 and reactive oxygen species. Respir Physiol 2001; 128: 331–348.

29

Mitochondria as Vascular Oxygen Sensors

The Redox Hypothesis of Vascular O_2 Sensing

ZLATKO I. POZEG, BERNARD THÉBAUD, BENJAMIN D. TYRRELL, EVANGELOS D. MICHELAKIS, and STEPHEN L. ARCHER

University of Alberta
Edmonton, Alberta, Canada

I. Introduction

Virtually all cells of aerobic organisms are sensitive to hypoxia, in the sense that, if severe and prolonged, hypoxia causes cell damage or death. However, certain cardiovascular and pulmonary cells display intrinsic, adaptive responses that are critical for survival of aerobic organisms during hypoxia. The function of these specialized, O_2-sensitive tissues is evident in response to environmental hypoxia, cardiopulmonary disease, and during birth. A review of O_2 sensing in a variety of mammalian O_2-sensitive tissues reveals widely conserved mechanisms that are important in the acute adaptive response to changes in PO_2.

Aerobic organisms require O_2 for the production of adenosine triphosphate (ATP). Most of this high-energy phosphate is produced within the mitochondria by oxidative phosphorylation, with the remainder generated in the cytosol by glycolysis (Fig. 1). The products of glycolysis, fatty acid β-oxidation, and protein deamination enter the mitochondria where, after conversion to Acyl Co A, they enter Krebs' cycle. This oxidative phosphorylation, in addition to generating ATP, produces reducing equivalents (e.g., nicotinamide and flavin adenine dinucleotides, NADH, $FADH_2$) that have high (negative) redox potential and thus a tendency to donate electrons (Fig. 1). These electron donors enter the electron transport chain (ETC) in

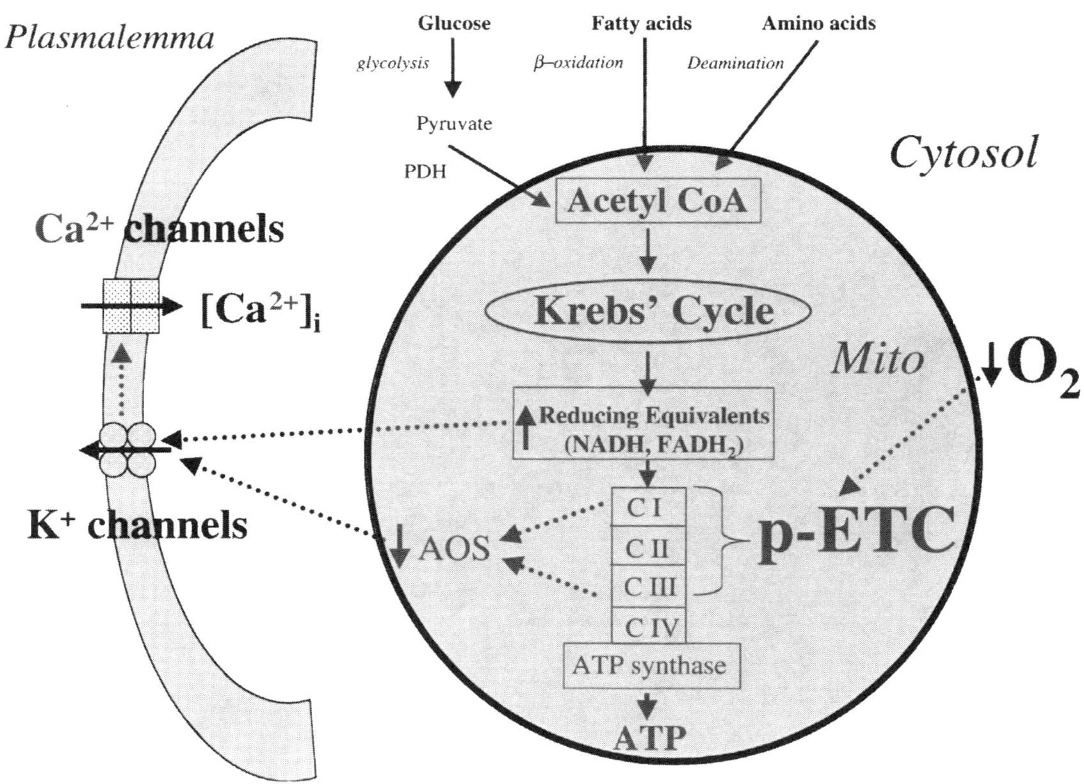

Figure 1 Mitochondria affect ion channel function through the production of diffusible reducing equivalents and AOS in response to hypoxia. p-ETC, proximal electron transport chain; PDH, pyruvate dehydrogenase.

the mitochondrial inner membrane, and their electrons flow through the ETC complexes until they react with O_2, in the distal ETC (complex IV). The electrochemical energy stored during the ETC electron flow (discussed in detail later) is used by ATP synthase to generate ATP. In the course of this redox cascade two important events occur, in addition to ATP synthesis, that are relevant to O_2 sensing. First, some single-electron reductions occur resulting in the formation of activated O_2 species (AOS), particularly in the proximal part of the ETC. Second, if acceptance of electrons is inhibited there may be upstream accumulation of electron donors. Both AOS and the electron donors can act as signaling molecules in the O_2-sensing pathway. During hypoxia, where the availability of the final electron acceptor, oxygen, is limited, the flow of electrons is compromised. This results in an increase in reducing equivalents and a concomitant decrease in AOS generation. Alterations in these substances, some of which are freely diffusible, can alter the function of cytosolic or plasmalemmal targets, initiating the cellular response to hypoxia, potentially upstream of ATP depletion. This feedback mechanism is therefore a very attractive means of linking hypoxia and metabolism with cellular function and could serve as an O_2-sensing mechanism. There has even been a recent report linking mitochondrial dysfunction to a form of pulmonary hypertension that afflicts poultry (1).

O_2-sensing systems consist of a *sensor* that alters the production of a *mediator* in response to changes in PO_2. The mediator, in turn, alters the function of one or more *effectors*, which ultimately mediate the physiological response of the system. Teleologically it is optimal that the sensor monitors a variable that is rapidly modified by mild hypoxia, before ATP levels decline or tissue damage occurs. An array of O_2-sensing systems has evolved to maintain the PO_2 within a tight physiological range and thus stabilize ATP production and promote survival during the periodic exposures to hypoxia that occurs in most aerobic lives. Often the sensor, mediator, and effector are linked in a functional unit. This chapter will focus on a sensing unit in which the sensor is the proximal portion of the mitochondrial ETC, the mediators are ETC-derived AOS, and the effectors are redox-sensitive plasmalemmal potassium channels (K^+ channels). Although it is unlikely that there is a single, universal O_2 sensor, there is evidence suggesting that the proximal mitochondrial ETC is involved in O_2 sensing in several tissues and species. The proposed central role of the mitochondria in many O_2-sensing systems comes at a time when the participation of these organelles in controlling apoptosis and cytosolic Ca^{2+}, $[Ca^{2+}]_i$, is increasingly recognized (2).

II. Comparative Physiology of O_2 Sensing

There are species, developmental, and interorgan differences in metabolic needs and aerobic metabolism that are reflected in species and tissue-specific variations in O_2-sensing strategies. However, different species and tissues, faced with similar challenges (e.g., environmental hypoxia) frequently share mechanisms of O_2 sensing. A form of hypoxic pulmonary vasoconstriction (HPV), for example, is found in hagfish (3), trout (4), bullfrogs (5), and most mammals, including humans (6–8). In HPV, a sensor in the smooth muscle cells of small pulmonary arteries (PASMC) detects local decreases in the alveolar O_2 eliciting vasoconstriction. This is possible because the resistance PAs are thin walled and are surrounded by alveoli. HPV shifts blood to better-oxygenated areas of the lungs, thus optimizing ventilation/perfusion matching and improving systemic oxygenation. The carotid body (CB) is a highly innervated, vascular organ in the carotid artery bifurcation that responds to decreases in arterial PO_2. The O_2 sensor in the carotid body lies within the type 1 or *glomus* cells. As PO_2 declines, the type 1 cells proportionately increase their rate of afferent discharge to the brainstem via cranial nerve IX, which then stimulates ventilation. In adult mammals, PASMCs and the type 1 cells of the CB are two of the best-characterized O_2-sensing cells. Both these sensors respond, within seconds, to physiological levels of hypoxia (rather than severe hypoxia or anoxia). Although these fine-tuned systems have a different output, contraction of the PASMC versus increased nerve discharge for the CB, they may have similar sensors (mitochondria), mediators (AOS and other redox mediators), and effectors (K^+ channels). The response of both organs to hypoxia optimizes tissue O_2 delivery and ATP production.

Additional O_2 sensors are active in the fetus. In the fetus, the ductus arteriosus (DA) is tonically relaxed by its hypoxic environment. This allows venous blood to bypass the nonventilated lungs, facilitating flow to the fetus's oxygenating organ, the placenta. However, as the PO_2 rises following birth, the DA SMC contract, forcing venous blood to pass through the ventilated lungs where it is oxygenated. It is remarkable that this vessel has the exact opposite response to hypoxia as the adjacent PA. The ductal sensor may appear paradoxical, responding to high, rather than low, PO_2, but like all other sensors, the DA's response increases the organism's O_2 supply (by diverting blood flow to the newly expanded lungs).

In addition, the ADMC in the fetal adrenal glands sense the transient hypoxia that normally occurs during labor and secrete catecholamines, thus preparing the fetus for the stress of labor (9). The role of the neuroepithelial body (NEB) is less clear. NEB cells are aggregated in clusters within the airway. These cells are activated by hypoxia to secrete their vasoactive and bronchospastic peptides, such as bombesin and serotonin. Hypoxic secretion of these spasmogenic peptides may regulate airway reactivity and tone during hypoxia, particularly in the immature lung (10,11).

It is striking that in all of these O_2-sensitive tissues, the effectors appear to be K^+ channels (Fig. 2). Isolated PASMCs, ADMCs (or the pheochromocytoma cells, PC12, used as models for ADMCs), and cells from the CB and NEB respond to hypoxia with a decrease in the outward K^+ current (Fig. 2). This results in depolarization of the plasmalemmal membrane, increased opening of voltage-gated Ca^{2+} channels, influx of Ca^{2+}, and an increase in $[Ca^{2+}]_i$. In the case of PASMC, this results in the activation of actin-myosin and contraction. In CB glomus cells, it results in release of dopamine from vesicles and increased neural transmission to the brainstem via the IX cranial nerve. Similarly, in ADM and NEB cells K^+ current inhibition, membrane depolarization, and increases in $[Ca^{2+}]_i$ cause the release of

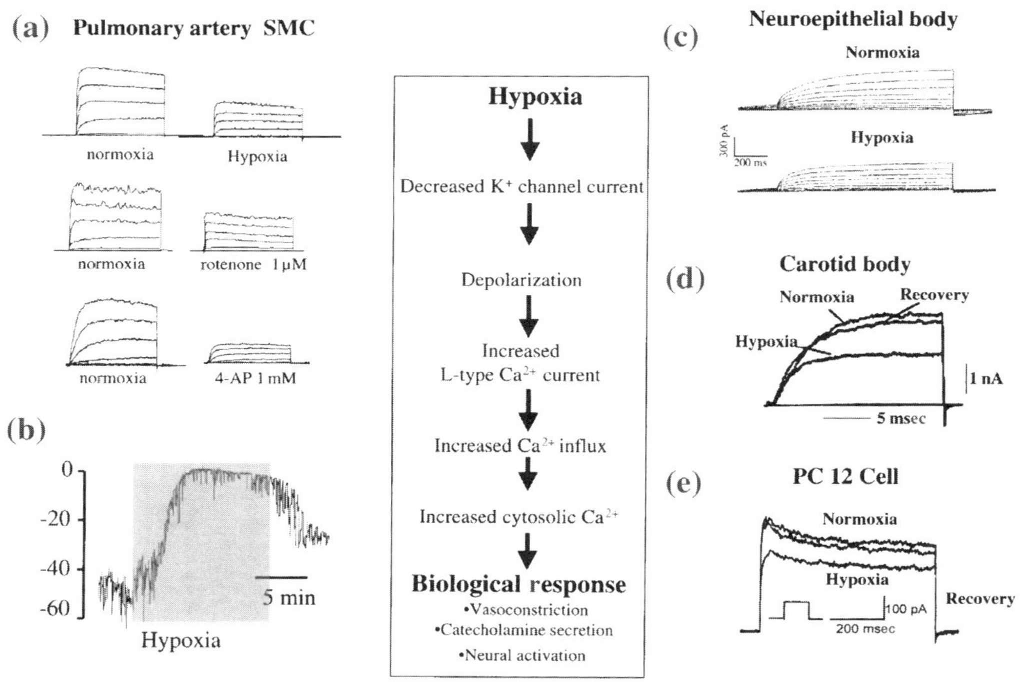

Figure 2 Hypoxia inhibits K^+ channels in several O_2-sensitive tissues. (a) Whole-cell patch-clamp experiments show K^+ current inhibition in response to hypoxia in rat resistance PASMCs. The complex I ETC blocker rotenone and the Kv blocker 4-aminopyridine (4-AP) also inhibit K^+ current. (b) Using the current clamp mode, resistance PASMCs depolarize in response to hypoxia. (From Ref. 29.) (c,d,e) Hypoxia inhibits outward K^+ current in isolated cells from the NEB, CB, and PC-12 cells. (From Refs. 39, 105, 106.) Inset: Proposed mechanism for HPV, featuring the effector portion of the pathway.

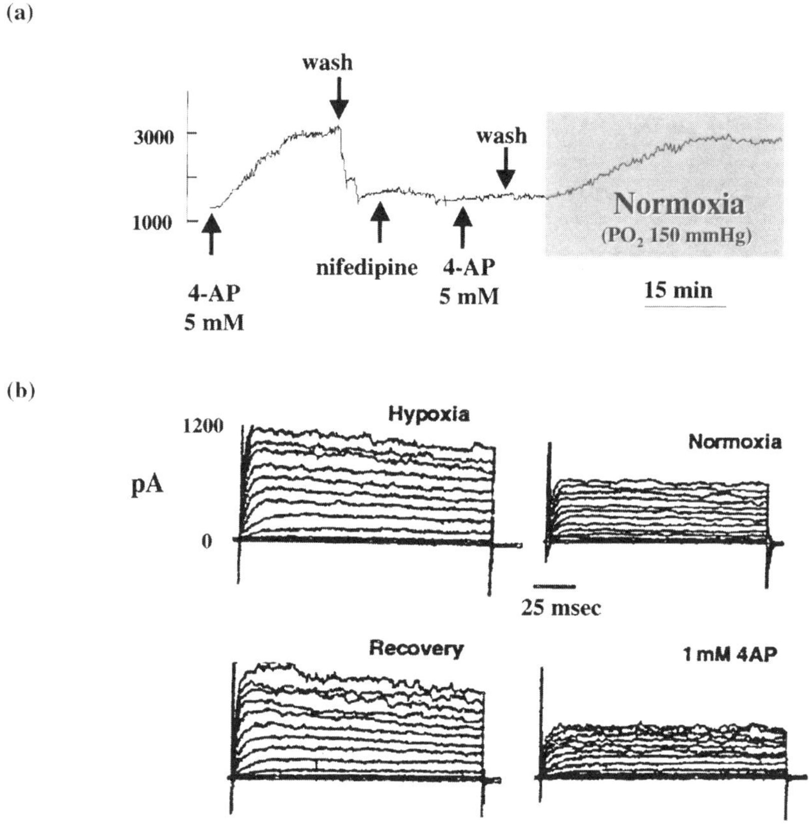

Figure 3 K^+ channels are the effectors in the O_2 constriction of the DA. (a) A freshly isolated human DA ring, studied in hypoxia to mimic in utero conditions, constricts to 4-aminopyridine and normoxia. The L-type Ca^{2+} channel blocker nifedipine inhibits the constriction to 4-aminopyridine suggesting that constriction occurred because of 4-aminopyridine-induced depolarization, increased opening of the L-type Ca^{2+} channels, and an influx of Ca^{2+}. (From Ref. 107.) (b) Patch-clamp studies of isolated rabbit DASMC show an O_2-induced inhibition of K^+ current, similar to that achieved by 4-aminopyridine. (From Ref. 12.)

catecholamines and serotonin, respectively. Unlike the other O_2 sensors, the DASMCs are activated by increases in PO_2 that accompany the initiation of respiration at birth, rather than by hypoxia. Increases in PO_2 cause acute K^+ current inhibition, SMC membrane depolarization, and contraction (Fig. 3) (12,13).

The similarity in the effector mechanism in several of these systems raises the possibility for similarities in the sensor mechanisms. As we discuss below, there is now evidence supporting a role of mitochondria and redox signals in the best-studied O_2-sensitive tissues (e.g., the PASMC, DASMC, and the CB type 1 cell). The following sections will discuss the evidence that K^+ channels, redox mediators, and mitochondrial sensors are involved in O_2 sensing. The discussion will focus mostly on HPV and the PASMC O_2-sensing system.

III. HPV and the Role of Ca^{2+} and K^+ Channels as Effectors

HPV is defined as a rapid (seconds) (14), fully reversible, increase in pulmonary vascular resistance that occurs in response to alveolar hypoxia. Hypoxic contraction occurs in salt-perfused isolated lungs, PA rings, as well as isolated PASMCs (Fig. 4). Although HPV is evident in medium-sized PAs and even pulmonary veins, it is strongest in small, distal PAs (Fig. 4) (15). HPV occurs when the FiO_2 falls below a relatively modest threshold, being routinely elicited at a FiO_2 of 12.5% in humans (16). The role of endothelium in HPV has largely been studied in endothelium-

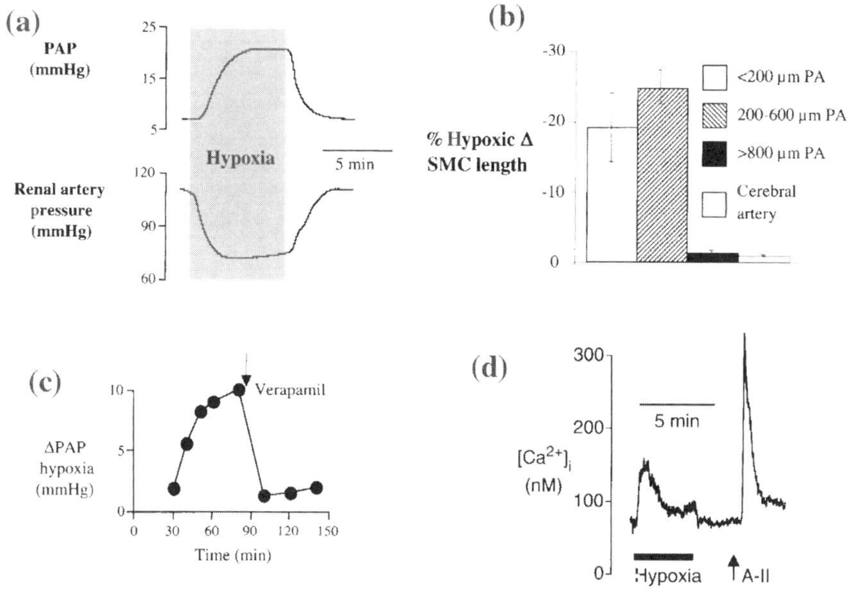

Figure 4 HPV is intrinsic to the resistance PA. (a) When rat lungs and kidneys are perfused in series, physiological hypoxia (induced by ventilating the lungs with hypoxic gas) causes pulmonary vasoconstriction and renal vasodilatation. This occurs in the presence of NO synthase and prostaglandin synthesis blockers, suggesting it is intrinsic to the SMC. (From Ref. 108.) (b) Isolated SMCs from small and medium-sized PAs constrict to hypoxia, whereas SMC from large PAs and systemic arteries do not. (From Ref. 109.) (c) The L-type Ca^{2+} channel blocker verapamil reverses HPV, suggesting that the opening of voltage-gated Ca^{2+} channels is required for HPV. (From Ref. 19.) (d) Intracellular Ca^{2+} levels increase in Fura-2-loaded PASMC in response to superfusion with a hypoxic solution or Angiotensin II. (From Ref. 36.)

denuded ring models. In these models, some investigators report that HPV has two distinct phases, an acute, transient, endothelium-independent phase (phase I), lasting minutes, followed by a sustained endothelium-dependent phase II (17). However, the denuded PA rings usually require priming with a vasoconstrictor, such as phenylephrine, to elicit HPV, raising questions as to the validity of this model for authentic HPV. Furthermore, in vivo there is no obvious physiological correlate of phase 1 and 2 and HPV onsets rapidly and is sustained. Although denudation does generally diminish HPV, it is difficult to distinguish how the altered vascular reactivity relates to the intended loss of the endothelium versus the nonspecific trauma to the vessel wall.

The initial studies suggesting that SMC ion channels in the pulmonary circulation were important to HPV derived from pharmacological experiments in isolated lungs and anesthetized animals, each demonstrating that numerous, pharmacologically distinct, L-type Ca^{2+} channel blockers inhibit HPV (Fig. 4) (18–22). Ca^{2+} channel blockers have similar inhibitory effects on human HPV in vivo (23). Despite one report to the contrary (24), Ca^{2+} channel blockers usually inhibit HPV in PA rings (25). Conversely, Ca^{2+} channel agonists, such as BAY K 8644, enhance HPV (26,27).

The link between the K^+ and Ca^{2+} channels is through plasmalemmal membrane potential. The open-state probability of voltage-gated, L-type, Ca^{2+} channels increases as the membrane depolarizes, resulting in increased Ca^{2+} current, elevation of $[Ca^{2+}]_i$ and activation of the contractile apparatus. In the adult pulmonary circulation the voltage-gated K^+ channels (Kv) control membrane potential and inhibiting these channels, using 4-aminopyridine, causes pulmonary vasoconstriction (28,29). In the past decade, the patch-clamp technique has been used to directly confirm that hypoxia inhibits whole-cell K^+ current (I_K) and depolarizes PASMCs (Fig. 2a,b) (30). This finding has been confirmed by other groups in other species (rats and humans) (31–33). It appears that inhibition of I_K is the result of hypoxic inhibition of a family of 4-aminopyridine-sensitive K^+ channels (29,34,35). Candidate O_2-sensitive Kv channels include Kv1.5 and Kv2.1 (Fig. 5) (36). Indeed, mice lacking the gene for Kv1.5 have diminished HPV and display less O_2-sensitive Kv current (Fig. 5a,b) (37). Other candidate O_2-sensitive channels in PASMC include heterotetramers of Kv1.2/1.5 and Kv2.1/9.3 (32), as well as homotetramers of Kv3.1b (38). In other O_2-sensitive tissues, such as the PC12 cell, other types of K^+ channels are thought to be O_2-sensitive, including Kv1.2 (PC-12 cell) (39) and the calcium-sensitive K^+ channels, K_{Ca} (40), or leak channels (41) in the type 1 CB cells. The predominant type of O_2-sensitive K^+ channel can change within a single tissue during normal maturation (42) or with disease such as chronic hypoxic pulmonary hypertension (43). In the fetal PASMC, the large conductance channel, BK_{Ca}, is the predominant O_2-sensitive fetal channel and it appears to be activated in response to O_2-induced increases in nitric oxide synthesis (44). How does PO_2 alter the function of K^+ channels? To answer this question it is helpful to briefly review K^+ channel structure.

K^+ channels are tetrameric transmembrane proteins consisting of four α-subunits. The gating and kinetics of K^+ channels can be altered by their association

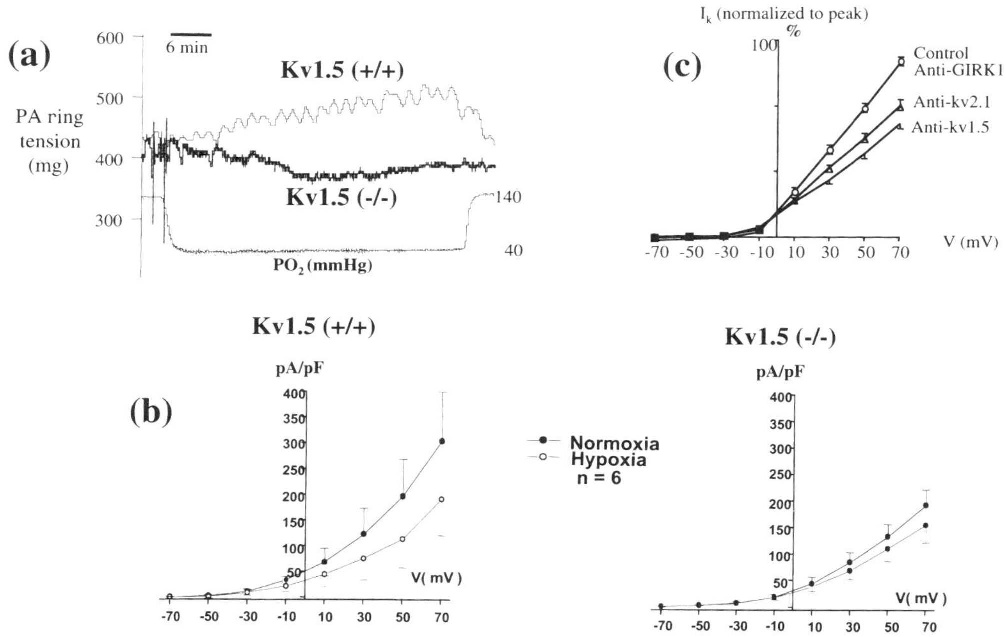

Figure 5 The role of Kv1.5 in PASMC. (a) Resistance PA rings from mice lacking Kv1.5 have diminished HPV compared to wild-type mice. HPV was elicited from these rings without priming with a vasoconstrictor. (From Ref. 37.) (b) PASMCs from mice lacking Kv1.5 have decreased O_2-sensitive current compared to wild-type mice, as shown in these I-V curves from whole-cell patch-clamp experiments. (From Ref. 37.) (c) Antibodies against Kv1.5 and Kv2.1 but not against other K^+ channels (such as the Kir channel GIRK) inhibit K^+ current and depolarize PASMC (not shown) when applied intracellularly via that patch-clamp pipette. These immunoelectrophysiology experiments suggest that Kv1.5 and Kv2.1 are active and important for the control of K^+ conductance and membrane potential and thus might regulate PA tone. (From Ref. 36.)

with four auxiliary β-subunits. The channel pore, with its signature K^+ recognition sequence, is formed by coordinated alignment of key amino acids in the S5–6 region of each of the α-subunits. Some (but not all) K^+ channels are well suited to redox regulation, as occurs in O_2 sensing, by virtue of possessing key cysteine and methionine groups. Reduction or oxidation of these residues by a redox mediator, such as an AOS, could cause conformational changes in the channel and thus alter pore function (45). Certain K^+ channels, including Kv1.5, respond to reduction and oxidation by changing their gating and open-state probability. In the PASMC, oxidants increase I_K (e.g., H_2O_2, diamide, oxidized glutathione) whereas reducing agents and agents that facilitate electron shuttling inhibit I_K (e.g., reduced glutathione and duroquinone, the synthetic coenzyme Q mimetic) (46). Concordant with their electrophysiological effects, oxidants dilate the pulmonary circulation (mimicking O_2) while reducing agents mimic hypoxia (Fig. 6). Hypoxia and redox agents may alter the function of ion channels, specifically K^+ channels, directly (47) or by modulating levels of a diffusible redox mediator, such as an AOS. The

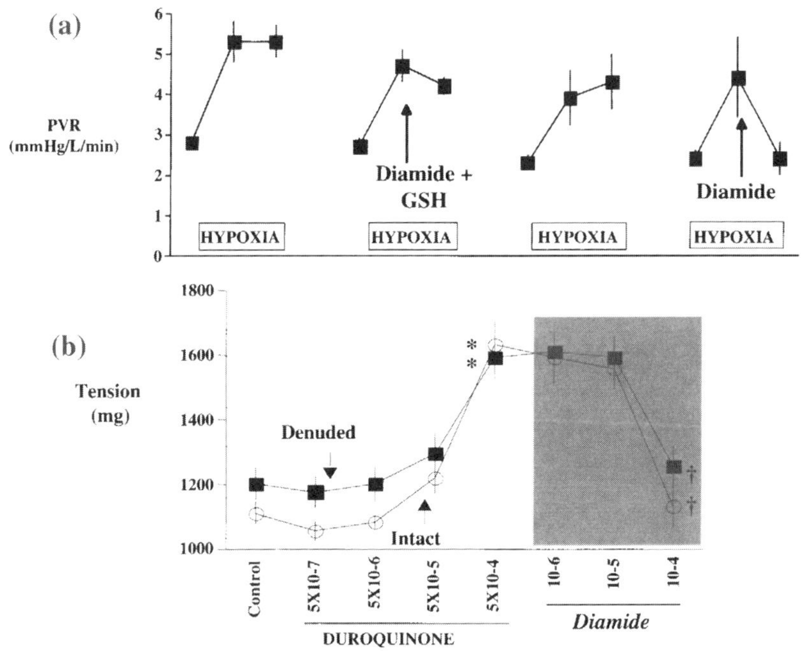

Figure 6 Redox control of PA tone. (a) The sulfhydryl oxidant diamide reverses HPV in anesthetized dogs. Preincubation of diamide with reduced glutathione prevents this effect. (From Ref. 110.) (b) Endothelium-denuded PA rings constrict in response to the electron-shuttling agent duroquinone and this is reversed by diamide. These data suggest that reducing agents (and, by extension, hypoxia) may cause PA constriction via a redox mechanism that involves a shift to a more reduced state. The fact that this occurs in denuded PAs suggests that this process is endothelium independent.

existence of redox-sensitive K^+ channels does not establish the question whether the electrophysiological effects of O_2 are direct or mediated via a remote sensor that produces a mediator, which ultimately alters channel activity.

IV. Redox Physiology and the Role of AOS as Mediators in HPV

AOS were originally thought to be too toxic to serve a physiological role, but like nitric oxide, in low amounts these species are important signaling molecules. Although AOS can be produced by xanthine oxidase, cyclooxygenase, and nitric oxide synthase, the PO_2-responsive production of AOS primarily occurs in the mitochondrial electron transport chain (ETC) and several vascular oxidases, including nicotinamide adenine dinucleotide phosphate NAD(P)H and novel vascular oxidases (NOX). The most common species is superoxide radical, produced by the mitochondrial ETC and vascular oxidases. Superoxide radical is very unstable and has a short diffusion radius, making it a poor mediator; however, it is rapidly converted to the more stable and diffusible H_2O_2 by MnSOD in the mitochondria and Cu/Zn SOD in the cytoplasm. H_2O_2 production is thereby still linked to PO_2 and is an attractive mediator. By oxidizing sulfhydryl and metal groups, AOS can alter the function of numerous target proteins, transcription factors, and genes. The production of AOS is balanced with their removal by antioxidant systems (dismutases, catalase, and sulfhydryl antioxidants systems, such as glutathione). This balance determines the redox status of the cell. The amount of AOS produced by a stimulus as well as their half-life depends on the redox potential of their surrounding environment

Redox potential is a measure of the relative tendency of a substance to acquire or donate electrons. In the case of mitochondrial ETC, electrons flow from electron donors, NADH and FADH, to distal electron recipients because of the dinucleotide's more negative reduction-oxidation potential (E_0) (Fig. 7). The more negative E_0 the more likely a substance is to donate an electron, while more positive E_0 predicts a diathesis to accept an electron. In the mitochondrial ETC, electrons flow down a potential gradient of redox potential ranging from -0.35, for NADH/NAD+, to $+0.82$, for O_2/H_2O (Fig. 7). Physiological generation of AOS occurs during normal electron shuttling by cytochromes within the ETC. Although most of the O_2 used by the mitochondrial ETC is eventually reduced to H_2O, $\sim$2% is incompletely reduced and yields radicals, particularly superoxide radical, $O_2^{\dot{-}}$, owing to single-electron reductions (see review in Ref. 48).

Most radical production in the mitochondria occurs within complex I (49,50) and III (51,52) (Fig. 7). Although superoxide radical itself is unable to leave the mitochondria, there is release of superoxide-derived, H_2O_2 by mitochondria during respiration (53). It is noteworthy that the rate of H_2O_2 released by liver mitochondria, measured in the mouse, rat, guinea pig, rabbit, pig, and cow (and in flight muscle mitochondria of *Drosophila melanogaster*), is inversely related to the life span potential of the species, a reminder both that redox signaling is also important

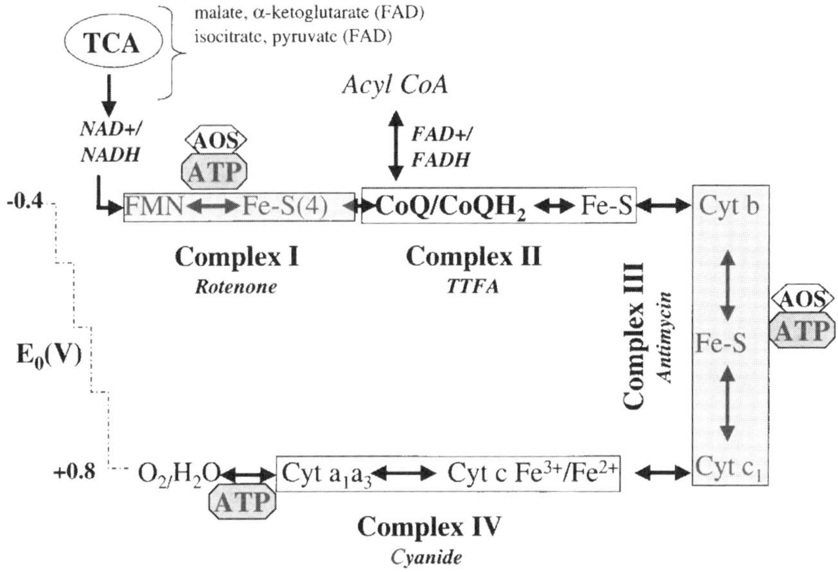

Figure 7 Schematic illustrating sites of AOS generation in the mitochondrial ETC. Electrons flow down the ETC driven by a difference in the redox potential between ~ -0.4 in complex I and $+0.8$ in complex IV. As the electrons flow toward their final target, oxygen, they release electrochemical energy, which is used by the ATP synthase to synthesize ATP. Most of the "drop" in redox potential occurs in complex I and III, sites that are responsible for most of the mitochondrial AOS production and ATP synthesis. Rotenone blocks complex I; thenoyltrifluoroacetone (TTFA) blocks complex II (87). Both agents inhibit electron entry to ubiquinone. Antimycin and cyanide are blockers of complex III and IV, respectively.

in apoptosis (54) and that there is likely mitochondrial heterogeneity among species and tissues.

AOS, whether diffusing out of mitochondria or produced by membrane-bound oxidases, can affect target proteins that control vascular tone, including Kv channels. AOS are attractive candidate O_2-sensitive mediators for several reasons (55,56). Their production varies rapidly in response to changes in PO_2, over the physiological range (20–100 mmHg). AOS can alter the reduction-oxidation of sulfhydryl groups of certain amino acids (e.g., cysteine), thereby altering the three-dimensional conformation and function of K^+ channel channels and other proteins (45). A number of other important signaling pathways, such as kinases and phosphatases, are redox-sensitive, and these redox signaling pathways are likely important in regulating cell proliferation, vascular wall remodeling, and ionic remodeling that occur during the chronic responses to hypoxia (57–60).

The redox hypothesis of HPV proposes that there is a tonic, basal, normoxic production of AOS in the PASMCs. These AOS are most likely derived from their mitochondria, although contributions from other sources, such as plasmalemmal NADPH oxidase, cannot be excluded. This normoxic generation of AOS establishes a basal, physiological level of K^+ channel oxidation, thereby setting open-state probability and membrane potential. This relative normoxic membrane hyperpolarization favors the L-type Ca^{2+} channel staying in the closed state and maintains the pulmonary circulation's low resting tone (one-fourth the blood pressure of the systemic circulation with the same flow). Thus, AOS are a signal for "normoxic vasodilatation." During alveolar hypoxia, production of AOS falls (Fig. 8) (61–63), withdrawing this tonic oxidant signal and reducing key amino acids, thereby inhibiting K^+ channels. It is likely that the link between the PASMC mitochondria and vascular tone is the dynamic modulation of basal AOS production by the mitochondrial ETC in response to changes in PO_2. However, it is uncertain whether the diminution of AOS production directly reduces K^+ channels, promoting their closure, or is a marker for diminished electron transport and the resulting cytosolic accumulation of reduced forms of other redox couples, such as NADP/NADPH or GSSG/GSH, that could also alter ion channel function (46). The concept that AOS, when produced in small, regulated amounts, can serve as physiological signals, rather than mediators of toxic injury, was controversial when initially proposed (64) and remains the subject of debate. However, the established role of another radical, nitric oxide, in diverse physiological mechanisms is a relevant precedent. It is important to reinforce that the redox hypothesis envisions small amounts of AOS, levels that are vasodilatory and do not cause capillary leak in the pulmonary

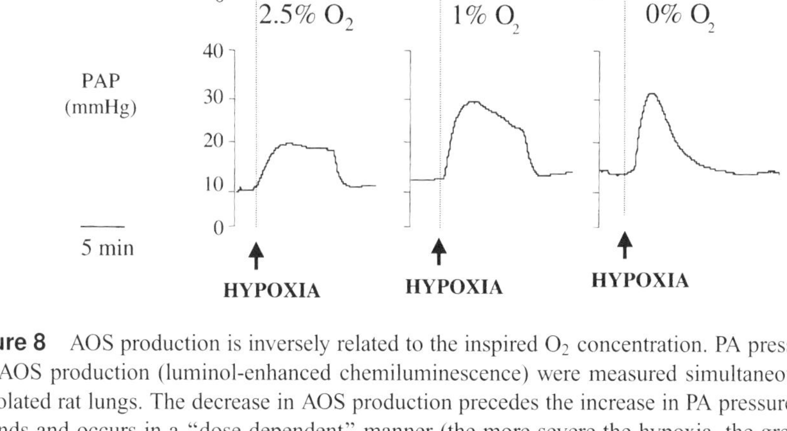

Figure 8 AOS production is inversely related to the inspired O_2 concentration. PA pressure and AOS production (luminol-enhanced chemiluminescence) were measured simultaneously in isolated rat lungs. The decrease in AOS production precedes the increase in PA pressure by seconds and occurs in a "dose-dependent" manner (the more severe the hypoxia, the greater the fall in AOS). Similar findings are made when AOS are recorded from isolated PA rings (rather than whole lungs, data not shown). (From Ref. 111.)

circulation, whereas it is clear that larger amounts of AOS cause pulmonary vasoconstriction and edema (65,66).

Two major sensors have been proposed that could act as physiological sources of the AOS and fulfill the sensor role, namely, nicotinamide adenine dinucleotide phosphate (NADPH oxidase) (67,68) and the mitochondrial ETC (35). What is the evidence for each potential source of signaling AOS?

V. Mitochondria as Vascular O_2 Sensors

Although mitochondria have been dismissed as poor candidates for O_2 sensing because the Km of their cytochromes has been said to be too low for modulation by physiological levels of hypoxia, intact cells and mitochondria exhibit a reversible inhibition of respiration during exposure to hypoxia as a result of inhibition of the catalytic activity of cytochrome c oxidase (48). Furthermore, the mitochondria's role as the predominant site for O_2 consumption and ATP synthesis suggests that they might be ideal sites for an O_2 sensor. In 1981 Rounds and McMurtry noted that inhibitors of the mitochondrial ETC caused pulmonary vasoconstriction (69), although this was attributed to changes in ATP production. While anoxia causes ATP depletion, this results in pulmonary vasodilatation, not constriction. In fact, in mild to moderate hypoxia, ATP and phosphocreatine levels are preserved (70). Furthermore, HPV is not altered by experimentally depleting ATP and phosphocreatine levels (71). Indeed, the lung is quite resistant to hypoxia, teleologically reflecting its role as a supplier, rather than a consumer, of O_2. In the isolated rat lung, ATP and ATP/ADP ratios are preserved after an hour's exposure to an alveolar PO_2 of 7 mmHg or CO/O_2 ratios of 10/1 (72). This reflects the lung's very low O_2 consumption (relative to most organs) and also indicates the high avidity of the lung mitochondria's ETC for O_2.

The O_2-sensing function of PASMC mitochondria appears to relate to the redox cascade within ETC (Fig. 7). Electrons from reduced nicotinamide adenine dinucleotide (NADH) and flavin adenine dinucleotide ($FADH_2$), are transferred, down a redox gradient, along a chain of ubiquinones and cytochromes and are ultimately transferred to oxygen, forming water (Fig. 7). The free energy produced in the transfer of electrons is used to pump protons from the mitochondrial matrix to the intermembrane space. This produces an electrochemical gradient across the inner mitochondrial membrane, which is utilized by ATP synthase to phosphorylate ADP to ATP, an essential substrate for all cellular energy-requiring functions (Fig. 1). The redox potential is the driving force for these reduction-oxidation reactions, which drives both ATP synthesis and AOS generation.

The mitochondrial ETC consists of four huge, multicomponent complexes whose function is to transfer electrons from reducing equivalents such as NADH and $FADH_2$, which are produced during glycolysis and the Krebs cycle (Fig. 1). Complex I of the ETC is an NADH dehydrogenase (Fig. 7). Here, NADH is reduced to NAD^+ as it transfers its two electrons to ubiquinone, which is oxidized to ubiquinol. Rotenone is a specific inhibitor of complex I. Complex II is a flavoprotein dehydrogenase, e.g., succinate dehydrogenase. It is at this site that succinate, an exogenous substrate, transfers electrons to FAD^+ to form $FADH_2$. $FADH_2$ is reduced as ubiquinone is oxidized to ubiquinol. Complex III of the ETC is cytochrome c reductase. This complex transfers electrons from ubiquinol to cytochrome c. There is an intermediate step involving the cytochrome bc1 complex. In the bc1 complex, ubiquinol becomes oxidized to ubisemiquinone as it passes an electron to the Rieske iron-sulfur center. Ubisemiquinone is the only reduced electron carrier in complex III capable of reducing O_2 to superoxide radical (51). This is a critical step in electron transfer as it represents a balance between the availability of O_2 and the size of the ubisemiquinone pool (51). Myxothiazol is an inhibitor of the electron transfer to the Rieske iron-sulfur center of the bc1 complex. Antimycin inhibits the transfer of electrons from ubisemiquinone to ubiquinone. Complex IV is the cytochrome oxidase, composed of two cytochromes, a and a_3. This complex accepts electrons from reduced cytochrome c and passes them to O_2 (Fig. 7).

Inhibitors that block the ETC upstream of ubisemiquinone, such as rotenone, myxothiazol, and amytal, tend to inhibit the formation of ubisemiquinone and thereby reduce the formation of AOS (Fig. 7). Conversely, ETC inhibitors that act downstream of that site, such as cyanide and azide, increase AOS production by increasing the production of ubisemiquinone (48). Our data are consistent with hypoxia, rotenone, and antimycin acting upstream and inhibiting AOS formation (Fig. 9 and Fig. 10) (35). Alterations in the production of AOS by the ETC could alter the cellular redox state and hence serve as an oxygen sensor. The superoxide radical generated at complexes I and III is dismutated by mitochondrial-specific manganese superoxide dismutase (MnSOD)-generating H_2O_2 (73). H_2O_2 is an ideal candidate as a signaling molecule connecting the mitochondrial sensor to the K^+ channel effector because it has a wide diffusion radius and has the ability to alter the function of K^+ channels (Fig. 1) (74) and important enzymes such as guanylate cyclase (75,76). Hypoxia and metabolic inhibitors also shift the balance of cytosolic redox couples, such as NADPH/NADP or GSSG/GSH, to a more reduced ratio (Fig. 1).

The evidence suggesting that the mitochondrial ETC could be important in O_2 sensing is based in part on the concordant effects of certain ETC inhibitors and hypoxia. Inhibitors of complex I (rotenone) and complex III (antimycin), but not inhibitors of complex IV (cyanide), mimic hypoxia's effect in the pulmonary circulation (Fig. 7) (35,77). Inhibitors of the mitochondrial ETC mimic hypoxia's effects on the carotid body (e.g., increase sinus nerve activity) (78) and PA (cause vasoconstriction) (35,69) (Fig. 9). The proximal ETC inhibitors also relax the DA (unpublished data). Like hypoxia, rotenone and antimycin decrease I_K in PASMCs (Fig. 2a) and decrease the production of AOS (Fig. 9, Fig. 10). The mitochondrial uncoupler carbonyl cyanide *p*-trifluoromethoxyphenyl-hydrazone (FCCP) also decreases PASMC Kv current (79). Buckler and Vaughan-Jones made similar observations in the carotid body, noting that both mitochondrial uncouplers, FCCP and 2,4-dinitrophenol (DNP), excite the carotid body by inhibiting a background K^+ conductance and inducing a small inward current. This leads to membrane depolarization of the type 1 cell and voltage-gated Ca^{2+} entry (80). As would be

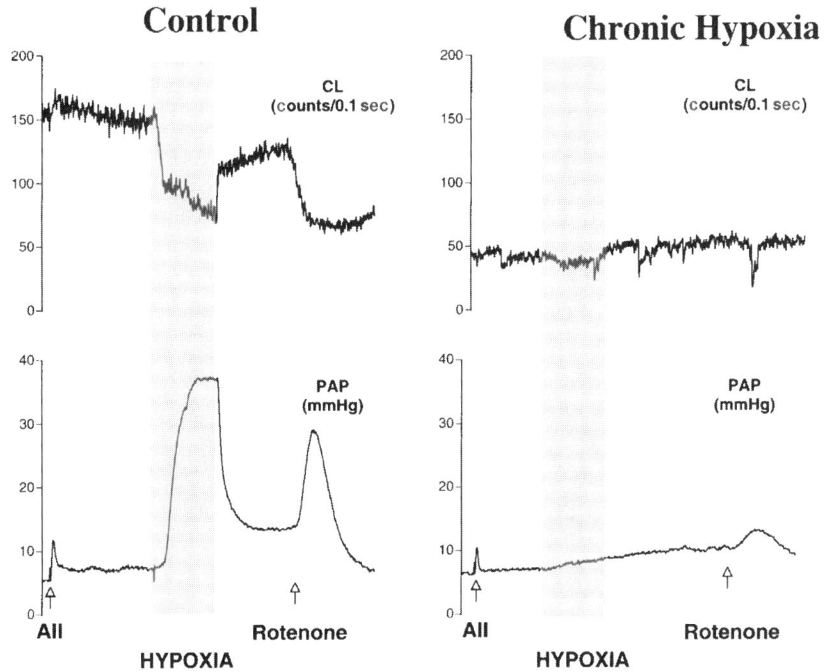

Figure 9 Chronic hypoxia reveals similarities between acute hypoxia and rotenone. Simultaneous luminol chemiluminescence and PA pressure measurements in a normoxic isolated rat lung (left) show that rotenone constricts the PA and decreases AOS production in a manner similar to hypoxia. In lungs from a rat exposed to chronic hypoxia (0.45 atmospheres for ~ 3 weeks) HPV is suppressed, while angiotensin II constriction is unchanged. The fact that constriction to rotenone and hypoxia are concordantly suppressed suggests the two share a common, redox mechanism. (From Ref. 43.)

predicted, the proximal ETC inhibitors increase I_K in the normoxic DASMC (unpublished data).

The study of changes in redox chemistry and vascular reactivity in chronic hypoxia provides further circumstantial evidence that the mitochondrial ETC participates in O_2 sensing. It is well established that chronic exposure to hypoxia paradoxically and selectively impairs acute HPV (19). Constriction to phenylephrine, angiotensin II, and 4-aminopyridine is preserved or enhanced (Fig. 9) (81). This loss of HPV persists for several days after return to a normoxic environment (43,82) and is associated with a decrease in normoxic AOS production in the lung (Fig. 9) and an increase in normoxic levels of reduced glutathione (Fig. 11). Thus chronic hypoxia creates a reduced state in which the acute reduction normally caused by hypoxia (Fig. 8) is no longer sensed as being "hypoxic" and indeed the dynamic decrease in AOS with acute hypoxia no longer occurs (Fig. 9). In chronic hypoxia,

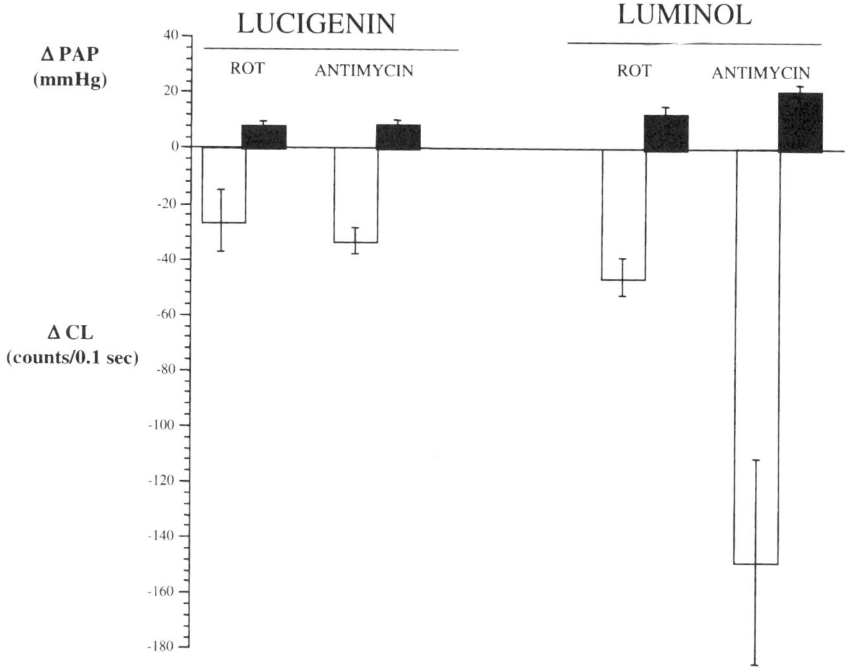

Figure 10 The effects of ETC inhibitors and hypoxia on AOS generation and pulmonary vascular resistance are concordant. In the Krebs-albumin-perfused isolated rat lungs, PA pressure reflects pulmonary vascular resistance because flow is held constant. Chemiluminescence, measured by two different methods (luminol and lucigenin enhancement), is decreased (white bars) whereas PA pressure is increased (black bars) in response to both rotenone and antimycin. The complex IV inhibitor cyanide does not lower AOS generation (not shown). (From Ref. 35.)

O_2-sensitive K^+ current is suppressed, much as occurs in PASMC with administration of reduced glutathione (Fig. 11b) (43).

VI. Controversies in O_2 Sensing

Although there is growing agreement that mitochondria or a vascular oxidase are redox O_2 sensors, there is debate as to whether hypoxia and ETC inhibitors decrease (61,63,67) or increase (77,83) AOS production. For example, Waypa et al. concur with our finding that hypoxia is sensed by inhibiting complex III but find that complex inhibition increased AOS and speculate that this superoxide production initiated Ca^{2+} release and vasoconstriction (77). It is possible that much of the controversy results from the use of 2′,7′-dichlorodihydrofluorescein diacetate (DCF) to measure AOS. Although none of the techniques for AOS measurement (chemiluminescence, spin trapping, fluorescent probes) is perfect, DCF is sub-

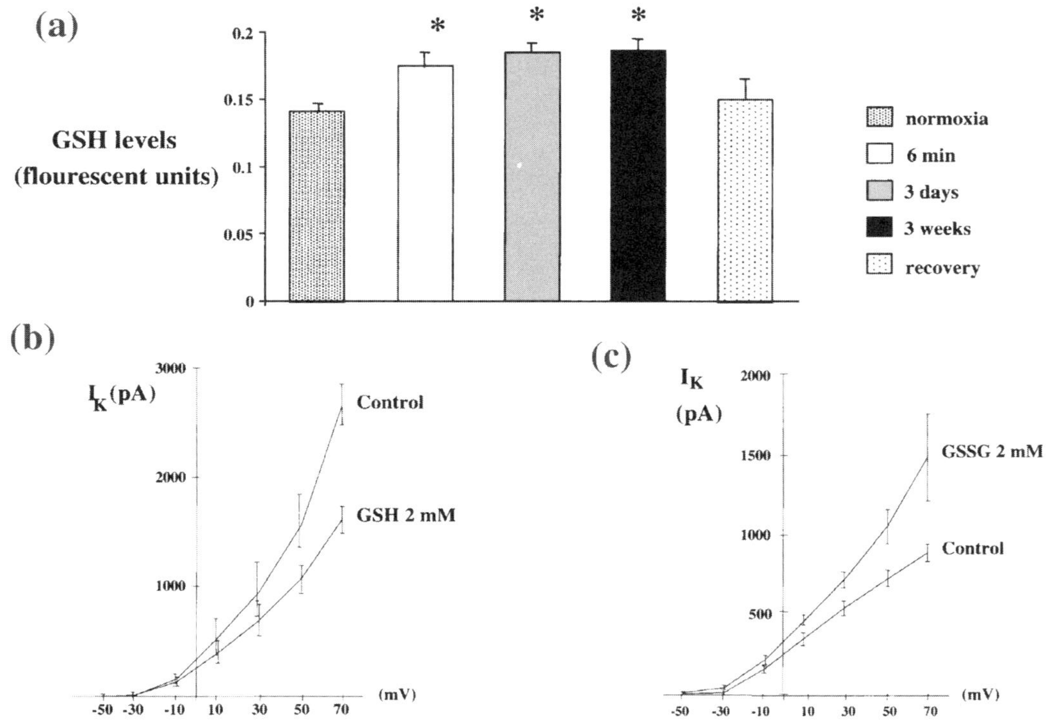

Figure 11 Hypoxia increases levels of reduced glutathione, a potential K^+-channel inhibitor. (a) GSH levels (measured in whole-lung homogenates) increase rapidly after the onset of hypoxia in the rat lung and remain elevated for the duration of hypoxia. (From Ref. 43.) (b,c) IV curves from whole-cell patch-clamp experiments show that reduced glutathione (GSH) inhibits whereas its oxidized form (GSSG) activates PASMC I_K. This suggests that an increase in the GSH/GSSH ratio induced by hypoxia in the cytoplasm might, together with the decrease in AOS production, lead to PASMC depolarization and HPV. (From Ref. 74.)

optimal for measuring O_2-based radicals. DCF readily detects nitric oxide (even more efficiently than it detects AOS) (84). In addition, DCF can itself increase generation of H_2O_2, a species it is often used to measure (85). Rota et al. evaluated the sensitivity and specificity of DCF using several techniques, including electron spin resonance, and found that DCF catalytically stimulates the formation of AOS in a manner that is dependent on, and affected by, various biochemical reducing agents. They concluded, "DCF cannot be used conclusively to measure superoxide or hydrogen peroxide formation in cells undergoing oxidative stress" (86). In addition, some investigators have shown that rotenone decreases AOS production using DCF (87). Certainly, the results of chemiluminescence measurements (using either luminol or lucigenin, Fig. 10) consistently show that hypoxia and mitochondrial ETC inhibitors rapidly decrease AOS production (35,62,63). Although lucigenin and luminol have their pitfalls (lack of specificity for superoxide versus H_2O_2), they remain sensitive and useful probes for measuring *oxygen*-derived free radicals (as opposed to nitrogen-based radicals). Lucigenin is insensitive to nitric oxide and luminol detects NO only at very alkalotic pH (88).

Another controversy in the role of the mitochondria as possible O_2 sensor is whether it is a diffusible AOS or the accumulation on the cytosol of reducing equivalents that elicits HPV. Inhibition of mitochondrial electron transport occurs in the hypoxic state, and this leads to the accumulation of reducing agents, thereby increasing the overall ratio of NADPH/NADP and GSH/GSSG (Fig. 11). This accumulation of electron donors in the cytosol occurs in a dose-dependent fashion. NADPH and GSH may also then act to inhibit K^+ channels (46,89,90) (Fig. 11). This cytosolic change in redox status accompanies the change in AOS production and could itself contribute to the sensor function (64). The relative importance of reduced redox couples versus reduced levels of AOS in the redox control of vascular tone is difficult to define, in part because of the interconnected nature of redox couples, such that a change in AOS alters levels of reducing equivalents and vice versa.

VII. NADPH Oxidase as an Alternative Redox O_2 Sensor

The postulated role of NADPH oxidase in the redox hypothesis is much the same as that of the mitochondria—sources of mediator AOS. A priori, there is little to favor one redox sensor over another, and indeed, there is no reason precluding involvement of both pathways. NADPH oxidase is a flavocytochrome found in phagocytes, type 1 cells of the CB, NEB cells, PASMCs, and endothelial cells. The flavocytochrome contains two membrane-bound subunits, the 91-kDa-membrane subunit of the phagocyte oxidase (gp91phox) and p22phox and two cytosolic proteins, p47phox and p67phox. This plasmalemmal enzyme shuttles electrons to oxygen, yielding superoxide radical in the process. NADPH oxidase (or a variant that preferentially uses NADH as a substrate) produces AOS in proportion to the PO_2 (67,91). This constitutes a potential redox signal (67,91,92) and the hypothesis has been put forward that this oxidase may be an oxygen sensor (93,94). In HPV, it is hypothesized that hypoxia reduces AOS production by the NADPH oxidase system (91), thereby producing a more reduced redox state. As with the mitochondrial sensor theory, the reduced redox state created by hypoxic inhibition of NADPH oxidase could inhibit a redox-gated K^+ channel, resulting in membrane depolarization with subsequent Ca^{2+} entry into the cell and vasoconstriction (35). However, much like the controversy regarding the effects of ETC inhibitors on AOS generation, one group has reported that hypoxia increases AOS production by a gp91phox containing NADPH or NADH oxidase (95). Much of the evidence favoring NADPH oxidase as an O_2 sensor relates to a relatively nonspecific pharmacological probe, diphenyleneiodonium (DPI). DPI is an inhibitor of the NADPH oxidase and does, in several respects, mimic hypoxia. DPI reduces AOS production in the PA during normoxia and inhibits I_K in PASMCs (96). After causing slight vasoconstriction, DPI decreases subsequent pressor responses to hypoxia (68,96), as occurs with mitochondrial ETC inhibitors (35). Unfortunately, DPI nonspecifically inhibits flavoprotein-containing enzymes, including NOS and complex 1 of the mitochondrial ETC (97). Thus, DPI is not a good or a very specific probe for determining the source of AOS in complex environments, such as the PASMC, in which there could be several sources.

The development of mice deficient in NADPH oxidase activity by mutation of the X-linked gene for the 91-kDa-membrane subunit of the phagocyte oxidase (gp91phox) (Fig. 12), provided an opportunity to study the role of NADPH oxidase in HPV. Most O_2-responsive cell types contain a similar form of the oxidase, containing the gp91phox component (98). In these mice, loss of a functional NADPH oxidase dramatically lowers AOS production (whether measured using luminol, lucigenin, or unenhanced (Fig. 13a) (99). However, HPV and the O_2-sensitive portion of I_K are preserved in mice lacking a functional gp91 phox (Fig. 13b,d). This suggests that gp91 phox NADPH oxidases present in PASMCs and PA endothelium (Fig. 12) are not required for O_2 sensing in HPV (99). Moreover, rotenone constriction is preserved or enhanced in these NADPH-oxidase-deficient mice, consistent with the mitochondrial O_2-sensor hypothesis (Fig. 13c). Preserved O_2 sensing has been reported in the type 1 cell of the carotid body from gp91 phox knockout mice (100), although the O_2 sensing appears to be impaired in their NEBs (101). These findings argue against the classic NADPH oxidase system as an O_2 sensor in HPV, although they do not exclude a role for the novel oxidases or a role for NADPH oxidase in certain O_2-sensitive tissues. It is certainly plausible that several O_2-sensing mechanisms can operate simultaneously in some tissues, as has been recently suggested by O'Kelly et al., who found evidence for both NADPH oxidase-sensitive and -insensitive O_2-sensing mechanisms in NEBs (102).

Vascular SMCs also contain gp91phox homologs, called NOX, which preferentially utilize NADPH versus NADH as substrate. NOXs are also inhibited by DPI and are potentially important sources of AOS production in systemic vascular SMCs (103,104). However, to date, most studies of NOX have been performed in systemic arteries and have measured AOS production in response to vasoconstrictors (e.g., angiotensin II) or mitogens (e.g., platelet-derived growth factor). There is little evidence that NOX are involved in PASMC AOS production or HPV. It is

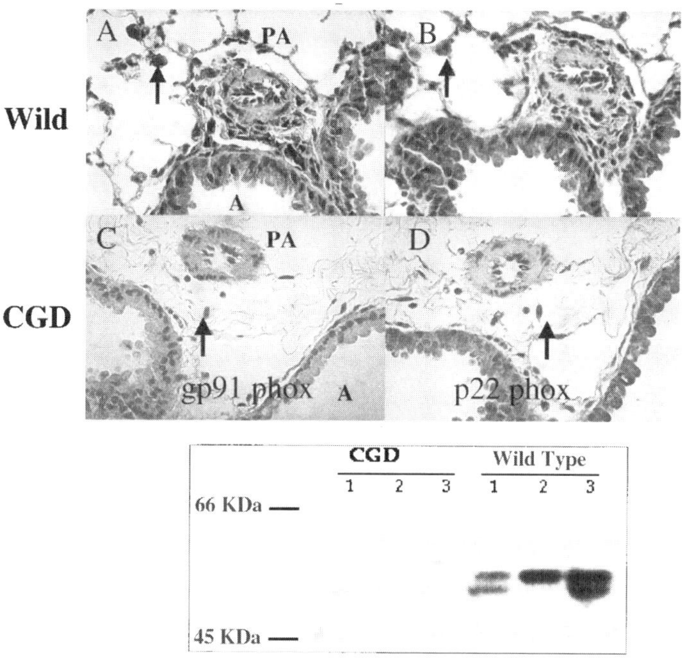

Figure 12 CGD mice lack the gp91 phox subunit of NADPH oxidase. (a) Immunohistochemistry (above) and immunoblotting (below, note the lower molecular weight is common in mice) experiments show that CGD mice lack the gp91 phox subunit of the enzyme, compared to wild mice (darker stain). Expression of p22 phox is similar in the CGD versus the wild mice. Note that in the CGD mice, the subunit is absent from the PA and alveolar (A) epithelial cells as well as from phagocytes (arrows). (From Ref. 99.)

noteworthy that angiotensin II does not cause an acute change in AOS production in mouse lungs, at doses that cause vasoconstriction (99). Perhaps the predominant source of AOS differs between PAs, where AOS may be signaling molecules serving to optimize O_2 uptake from the environment, versus systemic arteries, where AOS may be involved in the pathogenesis of atherosclerosis and systemic hypertension (both of which are rare in the pulmonary circulation).

VIII. Summary

We propose the following model for O_2 sensing, as outlined in Figure 14. In normoxia, the proximal mitochondrial ETC produces AOS, some of which (such as H_2O_2), can diffuse to the plasmalemma, oxidizing and thus opening K^+ channels. This favors normoxic membrane hyperpolarization and vasodilatation. Vascular oxidases contribute to the AOS production, and their role, relative to mitochondria, as sources of O_2-sensitive AOS may vary among tissues. Hypoxia is sensed in the

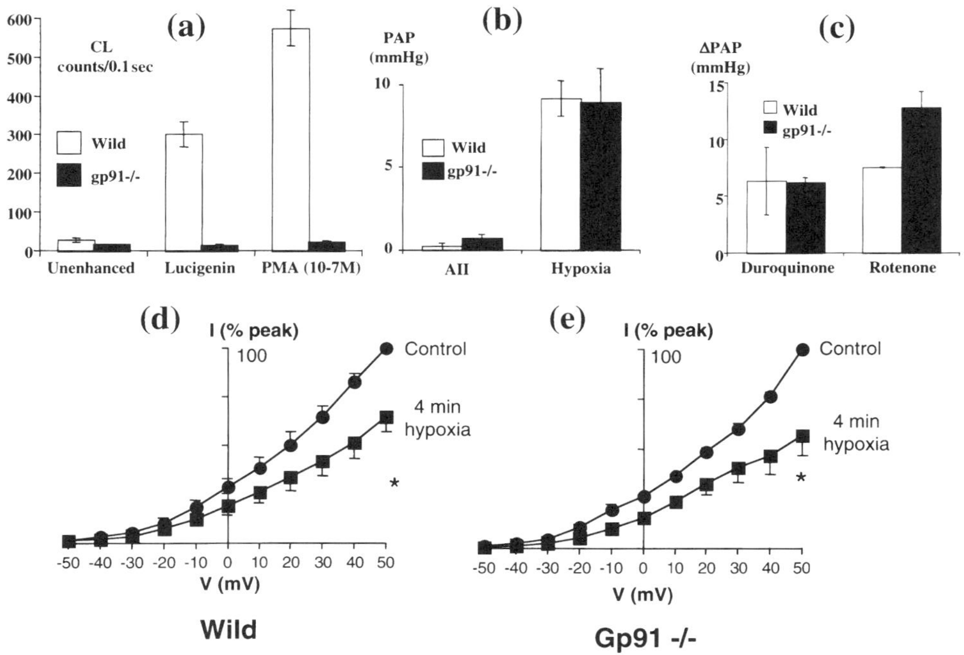

Figure 13 O_2 sensing is preserved in CGD mice lacking the gp91 phox subunit of NADPH oxidase. (a) The CGD mice lacking the gp91 phox subunit of NADPH oxidase have greatly decreased chemiluminescence, compared to wild mice. This is true for unenhanced chemiluminescence, luminol-enhanced chemiluminescence, and the difference persists following phorbol acetate myristate (PMA) stimulation of AOS generation. (b) Despite this, isolated perfused lungs from CGD mice have an identical contractile response to angiotensin II and hypoxia, compared to the wild-type mice. (c) The PAs from CGD mice constrict more to rotenone than those from wild-type mice. This is perhaps because the ETC-derived AOS are now more important in the control of tone, since the NADPH-derived AOS are missing. (d) The CGD and wild-type mice have similar hypoxia-sensitive K^+ currents, consistent with the similarity of their hypoxic pressor responses. (From Ref. 99.)

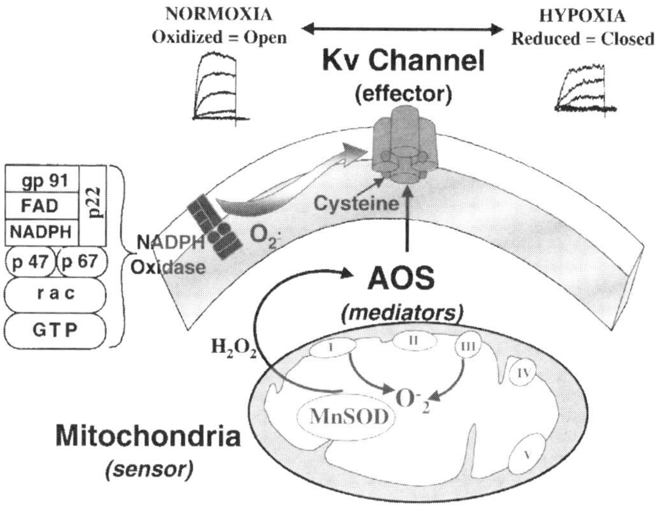

Figure 14 A proposed mechanism for O_2 sensing. The redox theory of HPV suggests that AOS are produced in proportion to PO_2 by O_2 sensors, likely complexes I and III of the mitochondrial ETC, and possibly also cytochrome-based vascular oxidases. The changes in AOS production alter the gating and open probability of the effectors of HPV, the O_2- and redox-sensitive Kv channels, such as Kv1.5 and Kv2.1.

proximal ETC, and perhaps vascular oxidases, resulting in a decrease in AOS production. This decrease in the levels of AOS, as well as the resulting increase in reducing equivalents in the cytoplasm, promotes a reduced state that causes K^+-channel inhibition, membrane depolarization, and vasoconstriction. In other words, the decrease of the production of a tonically produced vasodilator (such as H_2O_2) by hypoxia results in HPV. There is now significant evidence that supports this or a similar mechanism in the PASMC and the CB. In the case of the DA, increasing PO_2 also increases in AOS generation, including H_2O_2; however, in the DA this increased oxidant signal acts in reverse, inhibiting K^+ channels (13). Possible diversity in the nature of K^+ channels and their response to redox modulation in this vessel versus the PA, such as the type of Kv channel heterotetramer expression and the presence or not of β-subunits, might explain this intriguing difference.

Abbreviations

Adrenomedullary cells (ADMC); ductus arteriosus smooth muscle cells (DASMC); neuroepithelial body (NEB); pulmonary artery smooth muscle cells (PASMC); hypoxic pulmonary vasoconstriction (HPV); nicotinamide adenine dinucleotide phosphate [NAD(P)H]; activated oxygen species (AOS); inspired O_2 concentration (FiO_2); voltage-gated K+ channels (Kv); smooth muscle cells (SMCs); dipheny-leneiodonium (DPI); 91-kilodalton-containing phagocyte oxidase (gp91phox); novel vascular oxidase (NOX); 2′,7′-dichlorofluorescin diacetate (DCF); electron transport chain (ETC); adenosine triphosphate (ATP); whole-cell K+ current (I_K); voltage-sensitive K+ current (Kv); pheochromocytoma cells (PC-12); reduction-oxidation potential (E_0); *p*-trifluoromethoxyphenyl-hydrazone (FCCP); 2,4-dinitrophenol (DNP).

Acknowledgments

Both E.D. Michelakis and S.L. Archer are supported by the Alberta Heritage Foundation for Medical Research, the Canadian Foundation for Innovation, the Heart and Stroke Foundation of Canada, and the Canadian Institutes for Health Research.

References

1. Iqbal M, Cawthon D, Wideman RF Jr, Bottje WG. Lung mitochondrial dysfunction in pulmonary hypertension syndrome. I. Site-specific defects in the electron transport chain. Poult Sci 2001; 80:485–495.
2. Duchen MR. Contributions of mitochondria to animal physiology: from homeostatic sensor to calcium signalling and cell death. J Physiol 1999; 516:1–17.
3. Olson KR, Russell MJ, Forster ME. Hypoxic vasoconstriction of cyclostome systemic vessels: the antecedent of hypoxic pulmonary vasoconstriction? Am J Physiol Regul Integr Comp Physiol 2001; 280:R198–R206.
4. Smith MP, Russell MJ, Wincko JT, Olson KR. Effects of hypoxia on isolated vessels and perfused gills of rainbow trout. Comp Biochem Physiol A Mol Integr Physiol 2001; 130:171–181.
5. Shelton G. The effects of lung ventilation on blood flow to the lungs and body of the amphibian, *Xenopus laevis*. Respir Physiol 1970; 9:183–196.
6. Hambraeus-Jonzon K, Bindslev L, Mellgard AJ, Hedenstierna G. Hypoxic pulmonary vasoconstriction in human lungs: a stimulus-response study. Anesthesiology 1997; 86:308–315.
7. Carlsson AJ, Bindslev L, Santesson J, Gottlieb I, Hedenstierna G. Hypoxic pulmonary vasoconstriction in the human lung: the effect of prolonged unilateral hypoxic challenge during anaesthesia. Acta Anaesthesiol Scand 1985; 29:346–351.
8. Robin ED, Theodore J, Burke CM, Oesterle SN, Fowler MB, Jamieson SW, Baldwin JC, Morris AJ, Hunt SA, Vankessel A. Hypoxic pulmonary vasoconstriction persists in the human transplanted lung. Clin Sci (Lond) 1987; 72:283–287.
9. Slotkin TA, Seidler FJ. Adrenomedullary catecholamine release in the fetus and newborn: secretory mechanisms and their role in stress and survival. J Dev Physiol 1988; 10:1–16.
10. Johnson DE, Georgieff MK. Pulmonary neuroendocrine cells: their secretory products and their potential roles in health and chronic lung disease in infancy. Am Rev Respir Dis 1989; 140:1807–1812.
11. Fu XW, Wang D, Pan J, Farragher SM, Wong V, Cutz E. Neuroepithelial bodies in mammalian lung express functional serotonin type 3 receptor. Am J Physiol Lung Cell Mol Physiol 2001; 281:L931–L940.

12. Tristani-Firouzi M, Reeve HL, Tolarova S, Weir EK, Archer SL. Oxygen-induced constriction of the rabbit ductus arteriosus occurs via inhibition of a 4-aminopyridine-sensitive potassium channel. J Clin Invest 1996; 98:1959–1965.
13. Reeve HL, Tolarova S, Nelson DP, Archer S, Weir EK. Redox control of oxygen sensing in the rabbit ductus arteriosus. J Physiol 2001; 533:253–261.
14. Jensen K, Mico A, Czartolomna J, Latham L, Voelkel N. Rapid onset of hypoxic vasoconstriction in isolated lungs. J Appl Physiol 1992; 72:2018–2023.
15. Madden J, Vadula M, Kurup V. Effects of hypoxia and other vasoactive agents on pulmonary and cerebral artery smooth muscle cells. Am J Physiol 1992; 263:L384–L393.
16. Naeije R, Melot C, Mols P, Hallemans R. Effects of vasodilators on hypoxic pulmonary vasoconstriction in normal man. Chest 1982; 82:404–410.
17. Ward JP, Robertson TP. The role of the endothelium in hypoxic pulmonary vasoconstriction. Exp Physiol 1995; 80:793–801.
18. Kjaeve J, Bjertnaes LJ. Interaction of verapamil and halogenated inhalation anesthetics on hypoxic pulmonary vasoconstriction. Acta Anaesthesiol Scand 1989; 33:193–198.
19. McMurtry I, Davidson B, Reeves J, Grover R. Inhibition of hypoxic pulmonary vasoconstriction by calcium antagonists in isolated rat lungs. Circ Res 1976; 38:99–104.
20. Tucker A, McMurtry IF, Grover RF, Reeves JT. Attenuation of hypoxic pulmonary vasoconstriction by verapamil in intact dogs. Proc Soc Exp Biol Med 1976; 151:611–614.
21. Kennedy T, Summer W. Inhibition of hypoxic pulmonary vasoconstriction by nifedipine. Am J Cardiol 1982; 50:864–868.
22. Archer S, Yankovich R, Chesler E, Weir E. Comparative effects of nisoldipine, nifedipine and bepridil on experimental pulmonary hypertension. J Pharmacol Exp Ther 1985; 233:12–17.
23. Melot C, Naeije R, Hallemans R, Lejeune P, Mols P. Hypoxic pulmonary vasoconstriction and pulmonary gas exchange in normal man. Respir Physiol 1987; 68:11–27.
24. Robertson TP, Hague D, Aaronson PI, Ward JP. Voltage-independent calcium entry in hypoxic pulmonary vasoconstriction of intrapulmonary arteries of the rat. J Physiol 2000; 525 Pt 3:669–680.
25. Zhang F, Woodmansey PA, Morice AH. Acute hypoxic vasoconstriction in isolated rat small and large pulmonary arteries. Physiol Res 1995; 44:7–18.
26. McMurtry I. Bay K8644 potentiates and A23187 inhibits hypoxic vasoconstriction in rat lungs. Am J Physiol 1985; 249:H741–H746.
27. Tolins M, Weir E, Chesler E, Nelson D, From A. Pulmonary vascular tone is increased by a voltage-dependent calcium channel potentiator. J Appl Physiol 1986; 60:942–948.
28. Hasunuma K, Rodman D, McMurtry I. Effects of K^+ channel blockers on vascular tone in the perfused rat lung. Am Rev Respir Dis 1991; 144:884–887.
29. Archer SL, Huang JMC, Reeve HL, Hampl V, Tolarova S, Michelakis E, Weir EK. Differential distribution of electrophysiologically distinct myocytes in conduit and resistance arteries determines their response to nitric oxide and hypoxia. Circ Res 1996; 78:431–442.
30. Post J, Hume J, Archer S, Weir E. Direct role for potassium channel inhibition in hypoxic pulmonary vasoconstriction. Am J Physiol 1992; 262:C882–C890.
31. Yuan X-J, Goldman W, Tod M, Rubin L, Blaustein M. Hypoxia reduces potassium currents in cultured rat pulmonary but not mesenteric arterial myocytes. Am J Physiol 1993; 264:L116–L123.

32. Hulme JT, Coppock EA, Felipe A, Martens JR, Tamkun MM. Oxygen sensitivity of cloned voltage gated K^+ channels expressed in the pulmonary vasculature. Circ Res 1999; 85:489–497.
33. Osipenko ON, Evans AM, Gurney AM. Regulation of the resting potential of rabbit pulmonary artery myocytes by a low threshold, O_2-sensing potassium current. Br J Pharmacol 1997; 120:1461–1470.
34. Yuan X-J. Voltage gated K^+ currents regulate resting membrane potential and $[Ca^{2+}]_i$ in pulmonary artery myocytes. Circ Res 1995; 77:370–378.
35. Archer SL, Huang J, Henry T, Peterson D, Weir EK. A redox based oxygen sensor in rat pulmonary vasculature. Circ Res 1993; 73:1100–1112.
36. Archer SL, Souil E, Dinh-Xuan AT, Schremmer B, Mercier JC, El Yaagoubi A, Nguyen-Huu L, Reeve HL, Hampl V. Molecular identification of the role of voltage-gated K^+ channels, Kv1.5 and Kv2.1, in hypoxic pulmonary vasoconstriction and control of resting membrane potential in rat pulmonary artery myocytes. J Clin Invest 1998; 101:2319–2330.
37. Archer SL, London B, Hampl V, Wu X, Nsair A, Puttagunta L, Hashimoto K, Waite RE, Michelakis ED. Impairment of hypoxic pulmonary vasoconstriction in mice lacking the voltage-gated potassium channel Kv1.5. FASEB J 2001; 15:1801–1803.
38. Osipenko ON, Tate RJ, Gurney AM. Potential role for kv3.1b channels as oxygen sensors. Circ Res 2000; 86:534–540.
39. Conforti L, Millhorn DE. Selective inhibition of a slow-inactivating voltage-dependent K^+ channel in rat PC12 cells by hypoxia. J Physiol (Lond) 1997; 502:293–305.
40. Peers C. Hypoxic suppression of K^+ currents in type I carotid body cells: selective effect on the Ca^{2+}-activated K^+ current. Neurosci Lett 1990; 119:253–256.
41. Buckler K. A novel oxygen-sensitive potassium current in rat carotid body type I cells. J Physiol 1997; 498:649–662.
42. Reeve H, Weir E, SLA, Cornfield D. A maturational shift in pulmonary K^+ channels, from Ca^{2+} sensitive to voltage dependent. Am J Physiol 1998; 275(6 Pt 1: L1019–L1025.
43. Reeve HL, Michelakis E, Nelson DP, Weir EK, Archer SL. Alterations in a redox oxygen sensing mechanism in chronic hypoxia. J Appl Physiol 2001; 90:2249–2256.
44. Cornfield DN, Reeve HL, Tolarova S, Weir EK, Archer SL. Oxygen causes fetal pulmonary vasodilation through activation of a calcium-dependent potassium channel. Proc Natl Acad Sci USA 1996; 93:8089–8094.
45. Ruppersberg J, Stocker M, Pongs O, Heinemann S, Frank R, Koenen M. Regulation of fast inactivation of cloned mammalian 1K(A) channels by cysteine oxidation. Nature 1991; 352:711–714.
46. Reeve HL, Weir EK, Nelson DP, Peterson DA, Archer SL. Opposing effects of oxidants and antioxidants on K^+ channel activity and tone in vascular tissue. Exp Physiol 1995; 80:825–834.
47. Jiang C, Haddad GG. A direct mechanism for sensing low oxygen levels by central neurons. Proc Natl Acad Sci USA 1994; 91:7198–7201.
48. Chandel NS, Schumacker PT. Cellular oxygen sensing by mitochondria: old questions, new insight. J Appl Physiol 2000; 88:1880–1889.
49. Ide T, Tsutsui H, Kinugawa S, et al. Mitochondrial electron transport complex I is a potential source of oxygen free radicals in the failing myocardium. Circ Res 1999; 85:357–363.

50. Barja G, Herrero A. Localization at complex I and mechanism of the higher free radical production of brain nonsynaptic mitochondria in the short-lived rat than in the longevous pigeon. J Bioenerg Biomembr 1998; 30:235–243.
51. Turrens JF, Alexandre A, Lehninger AL. Ubisemiquinone is the electron donor for superoxide formation by complex III of heart mitochondria. Arch Biochem Biophys 1985; 237:408–414.
52. Herrero A, Barja G. Sites and mechanisms responsible for the low rate of free radical production of heart mitochondria in the long-lived pigeon. Mech Ageing Dev 1997; 98:95–111.
53. Herrero A, Barja G. H_2O_2 production of heart mitochondria and aging rate are slower in canaries and parakeets than in mice: sites of free radical generation and mechanisms involved. Mech Ageing Dev 1998; 103:133–146.
54. Sohal RS, Svensson I, Brunk UT. Hydrogen peroxide production by liver mitochondria in different species. Mech Ageing Dev 1990; 53:209–215.
55. Acker H, Xue D. Mechanisms of O_2 sensing in the carotid body in comparison with other O_2-sensing cells. News in Physiological Science 1995; 10:211–216.
56. Archer SL, Weir EK, Reeve HL, Michelakis E. Molecular identification of O_2 sensors and O_2-sensitive potassium channels in the pulmonary circulation. Adv Exp Med Biol 2000; 475:219–240.
57. Griendling KK, Sorescu D, Lassegue B, Ushio-Fukai M. Modulation of protein kinase activity and gene expression by reactive oxygen species and their role in vascular physiology and pathophysiology. Arterioscler Thromb Vasc Biol 2000; 20:2175–2183.
58. Abe J, Takahashi M, Ishida M, Lee JD, Berk BC. c-Src is required for oxidative stress-mediated activation of big mitogen-activated protein kinase 1. J Biol Chem 1997; 272:20389–20394.
59. Seko Y, Takahashi N, Tobe K, Kadowaki T, Yazaki Y. Hypoxia and hypoxia/reoxygenation activate p65PAK, p38 mitogen-activated protein kinase (MAPK), and stress-activated protein kinase (SAPK) in cultured rat cardiac myocytes. Biochem Biophys Res Commun 1997; 239:840–844.
60. Nakamura K, Hori T, Sato N, Sugie K, Kawakami T, Yodoi J. Redox regulation of a src family protein tyrosine kinase p56lck in T cells. Oncogene 1993; 8:3133–3139.
61. Freeman BA, Crapo JD. Hyperoxia increases oxygen radical production in rat lungs and lung mitochondria. J Biol Chem 1981; 256:10986–10992.
62. Paky A, Michael J, Burke-Wolin T, Wolin M. Endogenous production of superoxide by rabbit lungs: effects of hypoxia or metabolic inhibitors. J Appl Physiol 1993; 74:2868–2874.
63. Archer SL, Nelson DP, Weir EK. Simultaneous measurement of oxygen radicals and pulmonary vascular reactivity in the isolated rat lung. J Appl Physiol 1989; 67:1903–1911.
64. Archer S, Will J, Weir E. Redox status in the control of pulmonary vascular tone. Herz 1986; 11:127–141.
65. Archer SL, Peterson D, Nelson DP, DeMaster EG, Kelly B, Eaton JW, Weir EK. Oxygen radicals and antioxidant enzymes alter pulmonary vascular reactivity in the rat lung. J Appl Physiol 1989; 66:102–111.
66. Burghuber O, Mathies MM, McMurtry IF, Reeves JT, Voelkel NF. Lung edemea due to hydrogen peroxide is independent of cyclooxygenase products. J Appl Physiol 1984; 56:900–905.

67. Mohazzab-H K, Fayngersh R, Kaminski P. Potential role of NADPH oxidoreductase-derived reactive O_2 species in calf pulmonary arterial PO_2-elicited responses. Am J Physiol 1995; 269:L637-L644.
68. Thomas H III, Carson R, Fried E, Novitch R. Inhibition of hypoxic pulmonary vasoconstriction by diphenyleneiodonium. Biochem Pharmacol 1991; 42:R9–R12.
69. Rounds S, McMurtry I. Inhibitors of oxidative ATP production cause transient vasoconstriction and block subsequent pressor responses in rat lungs. Circ Res 1981; 48:393–400.
70. Buescher P, Perse D, Pillai R, Litt M, Mitchell M, Sylvester JT. Energy state and vasomotor tone in hypoxic pig lungs. J Appl Physiol 1991; 70:1874–1881.
71. Archer SL, Nelson D, Zimmer S, From AHL, Weir EK. Hypoxic pulmonary vasoconstriction is unaltered by creatine depletion induced by dietary β-guanidino propionic acid. Life Sci 1989; 45:1081–1088.
72. Fisher AB, Dodia C. Lung as a model for evaluation of critical intracellular PO_2 and PCO_2. Am J Physiol 1981; 241:E47–E50.
73. Boveris A, Chance B. The mitochondrial generation of hydrogen peroxide. Biochem J 1973; 134:707–716.
74. Weir EK, Archer SL. The mechanism of acute hypoxic pulmonary vasoconstriction: the tale of two channels. FASEB J 1995; 9:183–189.
75. Wolin M, Burke T. Hydrogen peroxide elicits activation of bovine pulmonary arterial soluble guanylate cyclase by a mechanism associated with its metabolism by catalase. Biochem Biophys Res Commun 1987; 143:20–25.
76. Wolin MS. Interactions of oxidants with vascular signaling systems. Arterioscler Thromb Vasc Biol 2000; 20:1430–1442.
77. Waypa GB, Chandel NS, Schumacker PT. Model for hypoxic pulmonary vasoconstriction involving mitochondrial oxygen sensing. Circ Res 2001; 88:1259–1266.
78. Mulligan E, Lahiri S, Storey BT. Carotid body O_2 chemoreception and mitochondrial oxidative phosphorylation. J Appl Physiol 1981; 51:438–446.
79. Yuan XJ, Sugiyama T, Goldman WF, Rubin LJ, Blaustein MP. A mitochondrial uncoupler increases KCa currents but decreases KV currents in pulmonary artery myocytes. Am J Physiol 1996; 270:C321–C331.
80. Buckler KJ, Vaughan-Jones RD. Effects of mitochondrial uncouplers on intracellular calcium, pH and membrane potential in rat carotid body type I cells. J Physiol 1998; 513:819–833.
81. Isaacson T, Hampl V, Weir E, Nelson D, Archer S. Increased endothelium-derived nitric oxide in hypertensive pulmonary circulation of chronically hypoxic rats. J Appl Physiol 1994; 76:933–940.
82. McMurtry IF, Petrun MD, Reeves JT. Lungs from chronically hypoxic rats have decreased pressor response to acute hypoxia. Am J Physiol 1978; 235:H104–H109.
83. Leach RM, Hill HM, Snetkov VA, Robertson TP, Ward JP. Divergent roles of glycolysis and the mitochondrial electron transport chain in hypoxic pulmonary vasoconstriction of the rat: identity of the hypoxic sensor. J Physiol 2001; 536:211–224.
84. Rao KM, Padmanabhan J, Kilby DL, Cohen HJ, Currie MS, Weinberg JB. Flow cytometric analysis of nitric oxide production in human neutrophils using dichlorofluorescein diacetate in the presence of a calmodulin inhibitor. J Leukoc Biol 1992; 51:496–500.

85. Rota C, Chignell CF, Mason RP. Evidence for free radical formation during the oxidation of 2′-7′-dichlorofluorescein to the fluorescent dye 2′-7′-dichlorofluorescein by horseradish peroxidase: possible implications for oxidative stress measurements. Free Radic Biol Med 1999; 27:873–881.
86. Rota C, Fann YC, Mason RP. Phenoxyl free radical formation during the oxidation of the fluorescent dye 2′7′-dichlorofluorescein by horseradish peroxidase: possible consequences for oxidative stress measurements. J Biol Chem 1999; 274:28161–28168.
87. Corda S, Laplace C, Vicaut E, Duranteau J. Rapid reactive oxygen species production by mitochondria in endothelial cells exposed to tumor necrosis factor-alpha is mediated by ceramide. Am J Respir Cell Mol Biol 2001; 24:762–768.
88. Kikuchi K, Nagano T, Hayakawa H, Hirata Y, Hirobe M. Detection of nitric oxide production from a perfused organ by a luminol-H_2O_2 system. Anal Chem 1993; 65:1794–1799.
89. Park M, Lee S, Lee S, Ho W, Earm Y. Different modulation of Ca-activated K channels by the intracellular redox potential in pulmonary and ear arterial smooth muscle cells of the rabbit. Eur J Physiol 1995; 430:308–314.
90. Lee S, Park M, So I, Earm Y. NADH and NAD modulates Ca^{2+}-activated K^+ channels in small pulmonary arterial smooth muscle cells of the rabbit. Pflügers Arch 1994; 427:378–380.
91. Mohazzab-H K, Wolin M. Properties of a superoxide anion-generating microsomal NADH oxidoreductase, a potential pulmonary artery PO_2 sensor. Am J Physiol 1994; 267:L823–L831.
92. Mohazzab KM, Wolin MS. Oxidant signalling and vascular oxygen sensing: role of H_2O_2 in responses of the bovine pulmonary artery to changes in PO_2. Adv Exp Med Biol 2000; 475:249–258.
93. Acker H, Dufau E, Huber J, Sylvester D. Indications to an NADPH oxidase as a possible pO_2 sensor in the rat carotid body. FEBS Lett 1989; 256:75–78.
94. Cross A, Henderson L, Jones O, Delpiano M, Hentschel J, Acker H. Involvement of an NAD(P)H oxidase as a pO_2 sensor protein in the rat carotid body. Biochem J 1990; 272:743–747.
95. Marshall C, Mamary AJ, Verhoeven AJ, Marshall BE. Pulmonary artery NADPH-oxidase is activated in hypoxic pulmonary vasoconstriction. Am J Respir Cell Mol Biol 1996; 15:633–644.
96. Weir EK, Wyatt CN, Reeve J, Huang J, Archer SL, Peers C. Diphenyleneiodonium inhibits both potassium and calcium currents in isolated pulmonary artery smooth muscle cells. J Appl Physiol 1994; 76:2611–2615.
97. Gatley S, Sherratt H. The effects of diphenyleneiodonium on mitochondrial reactions. Biochem J 1976; 158:307–315.
98. Youngson C, Nurse C, Yeger H, Curnutte JT, Vollmer C, Wong V, Cutz E. Immunocytochemical localization of O_2-sensing protein (NADPH oxidase) in chemoreceptor cells. Micro Res Tech 1997; 37:101–106.
99. Archer SL, Reeve HL, Michelakis ED, Puttagunta L, Waite R, Nelson DP, Dinauer MC, Weir EK. Oxygen sensing is preserved in mice lacking the gp91 fox subunit of NADPH oxidase. Proc Natl Acad Sci, USA 1999; 96:7944–7949.
100. Roy A, Rozanov C, Mokashi A, Daudu P, Al-mehdi AB, Shams H, Lahiri S. Mice lacking in gp91 phox subunit of NAD(P)H oxidase showed glomus cell [Ca(2+)](i) and respiratory responses to hypoxia. Brain Res 2000; 872:188–193.

101. Kazemian P, Stephenson R, Yeger H, Cutz E. Respiratory control in neonatal mice with NADPH oxidase deficiency. Respir Physiol 2001; 126:89–101.
102. O'Kelly I, Peers C, Kemp PJ. NADPH oxidase does not account fully for O_2-sensing in model airway chemoreceptor cells. Biochem Biophys Res Commun 2001; 283:1131–1134.
103. Sorescu D, Somers MJ, Lassegue B, Grant S, Harrison DG, Griendling KK. Electron spin resonance characterization of the NAD(P)H oxidase in vascular smooth muscle cells. Free Radic Biol Med 2001; 30:603–612.
104. Lassegue B, Sorescu D, Szocs K, Yin Q, Akers M, Zhang Y, Grant SL, Lambeth JD, Griendling KK. Novel gp91(phox) homologues in vascular smooth muscle cells : nox1 mediates angiotensin II-induced superoxide formation and redox- sensitive signaling pathways. Circ Res 2001; 88:888–894.
105. Lopez-Barneo J, Lopez-Lopez J, Urena J, Gonzalez C. Chemotransduction in the carotid body: K^+ current modulated by PO_2 in type I chemoreceptor cells. Science 1988; 242:580–582.
106. Youngson C, Nurse C, Yeger H, Cutz E. Oxygen sensing in airway chemoreceptors. Nature 1993; 365:153–155.
107. Michelakis E, Rebeyka I, Bateson J, Olley P, Puttagunta L, Archer S. Voltage-gated potassium channels in human ductus arteriosus. Lancet 2000; 356:134–137.
108. Hampl V, Weir EK, Archer SL. Endothelium-derived nitric oxide is less important for basal tone regulation in the pulmonary than the renal circulation of the adult rat. J Vasc Med Biol 1994; 5:22–30.
109. Madden J, Dawson C, Harder D. Hypoxia-induced activation in small isolated pulmonary arteries from the cat. J Appl Physiol 1985; 59:113–118.
110. Weir E, Eaton J, Chesler E. Redox status and pulmonary vascular reactivity. Chest 1985; 88:249S–252S.
111. Archer S, Eaton J, Nelson D, Kelly P, Peterson D, Weir E. Oxygen radicals and antioxidant enzymes alter pulmonary vascular reactivity in the rat lung. J Appl Physiol 1988; 66:102–111.

30

Roles for NAD(P)H Oxidases as Vascular Oxygen Sensors and Their Influence on Oxidant-Regulated Signaling Mechanisms

MICHAEL S. WOLIN, SACHIN A. GUPTE, and RICHARD A. OECKLER

New York Medical College
Valhalla, New York, U.S.A.

I. Introduction

This chapter will focus on examining how the influence of O_2 tension on the activity of NAD(P)H oxidases can function as a PO_2 sensor in vascular tissue through controlling the production of specific reactive O_2 species (ROS) that regulate signaling systems. This field began with the initial suggestions that ROS could function as mediators in the pulmonary arterial response to hypoxia (1–4) and that a phagocytic cell-like NAD(P)H oxidase could function as the PO_2 sensor in the carotid body (5). While emphasis will be placed on mechanisms through which PO_2 regulates vascular contractile function, some of the processes controlled by prolonged changes in ROS-dependent signaling systems may also influence mechanisms involved in vascular homeostasis, growth, and remodeling. The signaling systems most sensitive to changes in the basal production of individual ROS caused by localized changes in PO_2 are likely to be important participants in responses that are observed. Since physiological changes in PO_2 also influence the function of various other O_2-metabolism-dependent enzymes through mechanisms that do not involve ROS, such as cytochrome oxidase, nitric oxide synthase (NOS), cytochrome P450 (P450), and cyclooxygenase (COX), processes regulated by metabolic effects or nitric oxide (NO) and eicosanoid mediators derived from these systems also need to be considered in evaluating the integrated actions of PO_2 on vascular function.

II. NAD(P)H Oxidases

Multiple oxidoreductase enzymes or electron transport chains have NAD(P)H oxidase activity under certain conditions. These include phagocytic cell-like NAD(P)H oxidases, and other systems including the mitochondrial electron transport chain, NOS, COX, and xanthine dehydrogenase/oxidase (XO), which can be modified generally under pathophysiological conditions to become significant vascular sources of ROS (6). The factors that influence ROS production by each of these oxidases will be considered to help define conditions where they could potentially function as PO_2 sensors.

A. Phagocytic Cell-Like NAD(P)H Oxidases

Cells present in the vessel wall such as endothelium (7–9) and vascular smooth muscle (10–12) appear to contain NAD(P)H oxidases that seem to be similar to the NADPH oxidase present in phagocytic cells. The phagocytic cell NADPH oxidase is thought to have an electron transport chain containing a flavocytochrome b_{558} that appears to generate superoxide anion as a result of O_2 reacting with the ferrous (Fe^{2+}) form of this cytochrome (13). While membrane-bound gp91phox and p22phox subunits of this oxidase appear to contain the key components of this electron transport chain, this system does not produce superoxide until cell stimulation-associated signaling activates the binding of cytosolic p40phox, p47phox, and p67phox subunits. The G-protein Rac and phosphorylation of the p47phox subunit appear to have important roles in activation of this oxidase. Vascular cells appear to have similar oxidases that seem to have a basal superoxide-generating activity under physiological conditions (7–12). While the p22phox subunit also appears to be an essential component of the vascular oxidase, subunits homologous to gp91phox termed Nox-1 and Nox-4 seem to replace this subunit in vascular smooth muscle (14). However, subunits very similar to gp91phox seem be present in endothelium (7,8). There is evidence that Rac-1, p47phox, and p67phox may have important roles in activation of vascular forms of the oxidase by angiotensin II and thrombin receptors (15,16). Angiotensin II also appears to increase the expression of the p22phox and Nox-1 subunits (12,14). The influence of lactate on basal superoxide production and PO_2-elicited contractions to hypoxia in endothelium-removed bovine calf pulmonary arteries and the high levels of a flavoprotein-containing NADH oxidase activity present in this tissue resulted in the proposal that an oxidase whose activity is controlled by the redox status of cytosolic NAD(H) functions as a PO_2 sensor in this vascular segment (10,17). Additional studies modulating the generation of hydrogen peroxide (H_2O_2) and its metabolism were most consistent in supporting a role for removal of a H_2O_2-elicited vasodilator mechanism in the observed contractile response to hypoxia (18). Based on evidence that mice lacking the gp91phox subunit show relatively normal pulmonary arterial responses to hypoxia (19), if this response is dependent on changes in ROS, it is possible that a Nox-containing NAD(P)H oxidase functions as the PO_2 sensor.

B. Other NAD(P)H Oxidases

Any oxidase whose activity is regulated by physiological changes in PO_2 has the potential to function as an O_2 sensor for signaling mechanisms influenced by ROS derived from the oxidase. In addition to the flavocytochrome b_{558}-type oxidases, the mitochondrial electron transport chain, NOS, COX, P450, and XO are systems with NAD(P)H oxidase activity that is often increased under specific pathophysiological conditions to levels where these oxidases can become significant vascular sources of ROS (6). The mitochondrial electron transport chain produces superoxide when sites in the regions of the NADH dehydrogenase and coenzyme Q systems are reduced (20). These sites of ROS generation are being considered in one of the actively investigated hypotheses for the PO_2 sensor that mediates the pulmonary artery response to hypoxia (21,22). In addition, there is general agreement that high rates of mitochondrial oxidant production appear to be an important component signaling mechanisms involved in apoptosis (23). A deficit in the availability of L-arginine and/or the tetrahydrobioptrin cofactor for NOS results in an increase in the oxidase activity of this enzyme associated with a loss of NO biosynthesis from L-arginine (24,25). Thus, the NADPH oxidase activity of NOS can become a major source of superoxide-derived ROS in vascular cells with high levels of this enzyme, such as the endothelium, and this has been observed in several important vascular diseases including diabetes and hypertension (25,26). Cytochrome P450 is an additional enzyme present in the vascular endothelium that has NAD(P)H oxidase activity (27). When COX is synthesizing prostaglandin H_2 from arachidonic acid, it has a peroxidase activity with NAD(P)H-like oxidase activity because it produces NAD(P)·, an unstable radical that is thought to rapidly react with O_2 to form superoxide (28). It appears that conditions that produce stimulation of the NAD(P)H oxidase activities of NOS, P450, and COX in endothelium can result in the release of vasodilator levels of H_2O_2 (24,25,27–29). Xanthine dehydrogenase is an endothelial cell enzyme that may have some oxidase activity under basal physiological conditions. However, the oxidase activity of this enzyme is increased by processes involving thiol oxidation and by limited proteolysis (30). Exposure of tissues to either ischemia and reperfusion or hypoxia and reoxygenation appears to generally cause marked increases in endothelial xanthine oxidase activity, associated with increased availability of its substrates NADH and/or hypoxanthine as a result of hypoxia and the depletion of tissue ATP associated with increased release of adenosine and its metabolism to hypoxanthine (31). While ROS produced by these oxidases appear to be involved in the activation of signaling mechanisms associated with alterations in vascular function and injury (see Fig. 2), minimal consideration has been given to how ROS and responses originating from these oxidases could be influenced by changes in PO_2.

III. How Are Vascular Signaling Mechanisms Regulated by Oxidant Species?

Vascular tissue is exposed to multiple stimuli that influence the activities of oxidases that produce ROS and metabolic conditions that influence the control of redox systems that influence both the metabolism of individual oxidant species and/or the expression of signaling regulated by these species. Changes in PO_2 can influence both the release of vasoactive mediators from the endothelium and processes contained within vascular smooth muscle through changes in ROS. These interactions create an environment for the existence of a wealth of PO_2 sensing and signaling mechanisms in the vessel wall.

A. Influence of PO_2 on Signaling Mechanisms Regulating the Release of Endothelium-Derived Mediators

Endothelium contains multiple oxidases and signaling systems controlled by ROS, as shown in Figure 1. It appears that phagocytic cell-like NAD(P)H oxidase(s) seem to be an important basal source of ROS under physiological conditions, whereas activation of the endothelium and pathophysiological conditions increase the oxidase activities of many of the NAD(P)H oxidases including NOS, COX, P450, and XO. It appears that stimulation of the release of prostaglandins and NO are some of the more readily observed actions of H_2O_2 (32–34). Peroxide seems to stimulate

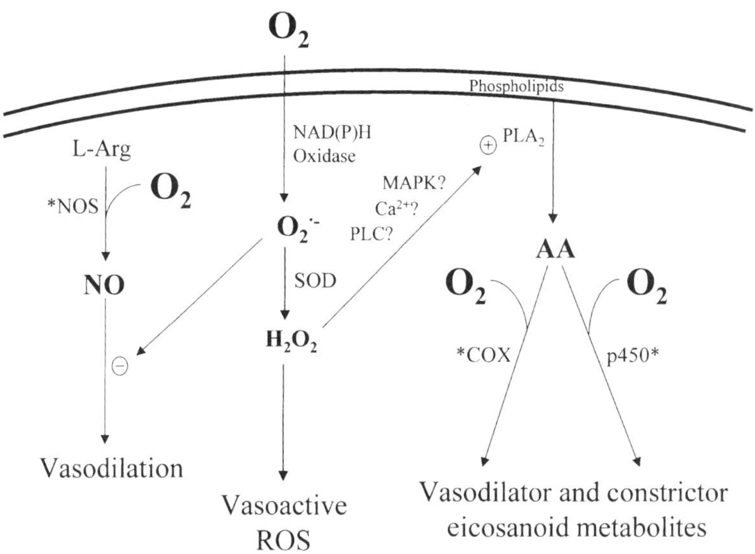

Figure 1 O_2-sensing mechanisms involving endothelial mediators, including potential roles for ROS. * Potential oxidases; PLA_2, phospholipase A_2; MAPK, mitogen-activated protein kinase; PLC, phospholipase C; AA, arachidonic acid; COX, cyclooxygenase; SOD, superoxide dismutase; NOS, nitric oxide synthase.

phospholipases that release arachidonic acid, possibly through mechanisms involving increases in intracellular calcium, and the activities of protein kinase C (PKC) and mitogen-activated protein (MAP) kinases (35–38). The metabolism of this fatty acid by COX into prostaglandins is activated by peroxides as a result of an oxidation of the heme of this enzyme into its catalytically active ferryl form (39). There is evidence for peroxide causing increases in both the NO-generating activity of NOS and the expression of this enzyme (32,40). A peroxide-induced increase in endothelial cell Ca^{2+} and activation of protein kinase B (PKB) may partially explain the acute stimulation of NOS activity (41). Activation of prostaglandin-mediated responses is one of the most readily observed reactions elicited when microvascular preparations are exposed to peroxide (42,43). While arachidonic acid–derived eicosanoid mediators including prostaglandins and metabolites of P450 are thought to be important mediators of many of the microvascular responses to changes in PO_2 (44–46), the role of ROS such as peroxide in these mechanisms remains to be investigated. Elucidating the role of ROS in the actions of PO_2 on the release of endothelium-derived vasoactive mediators such as NO and eicosanoids is a difficult question to address because NOX, COX, and P450 are enzymes that require O_2 for biosynthesis of the mediators they produce, have activities that are stimulated by ROS, and are systems capable of directly generating ROS. In addition, based on studies in a variety of different endothelium-containing preparations, NOS, COX, and P450 have the capacity to produce endothelium-derived vasodilator levels of ROS such as H_2O_2 (25,27,29).

B. Influence of PO_2 on Signaling Mechanisms That Regulate Vascular Smooth Muscle Function

Changes in PO_2 appear to have direct effects on force generation by vascular smooth muscle, and evidence exists that ROS-dependent mechanisms may mediate some of the responses that are observed. The contraction of pulmonary arteries to hypoxia and responses of vascular preparations to posthypoxic reoxygenation are some of conditions for which there is significant evidence for a role for ROS (18,21). The most sensitive signaling mechanisms activated by localized changes in ROS are usually linked to the metabolism of these species.

Role of Systems Linked to the Metabolism of ROS

Proteins with high rates of reaction with individual ROS should have the greatest potential for sensing the species involved, if the reaction produces a means of interacting with a cellular signaling system. As considered in previous reviews (18,47), the metabolism of peroxide by enzymes including heme peroxidases, catalase, and glutathione peroxidase has interactions with some of the most sensitive cellular systems regulated by ROS. As discussed earlier, the metabolism of peroxide by the peroxidase reaction of COX oxidizes the heme of this enzyme into the ferryl form needed for the biosynthesis of prostaglandins. Studies in vascular tissue and with purified enzymes have identified a mechanism of stimulation of cGMP production by activation of the soluble form of guanylate cyclase (sGC) resulting from the metabolism of peroxide by catalase, which is associated with formation of the oxidized compound I intermediate of catalase (1,4,48). There is evidence that this mechanism participates in PO_2-elicited responses such as posthypoxic reoxygenation in endothelium-removed bovine pulmonary and coronary arteries and human placental vessels (48–50). The activity of sGC is regulated by multiple additional mechanisms involving ROS and NO-derived species that could be regulated by PO_2. Some of these processes include the stimulation of sGC by NO and heme oxygenase-derived CO, and loss of this activation by CO and NO as a result of oxidation of the heme of sGC, the extremely efficient inactivation of NO through its reaction with superoxide anion, or the inhibition of H_2O_2-catalase dependent stimulation of sGC by NO and/or superoxide (4,17,51,52). The metabolism of peroxide by glutathione peroxidase produces oxidized glutathione (GSSG), and the formation of GSSG can be important in signaling due to its use for S-glutathionation or S-thiolation of key sites on proteins regulated by thiol redox mechanisms. There is evidence that many proteins, such as potassium and calcium channels, thought to be important in PO_2-elicited responses (21,53), can be regulated through S-thiolation or other thiol redox-associated changes (54,55). Thus, multiple signaling mechanisms linked to the metabolism of peroxide appear associated with PO_2-elicited responses in vascular tissue, as summarized in Figure 2.

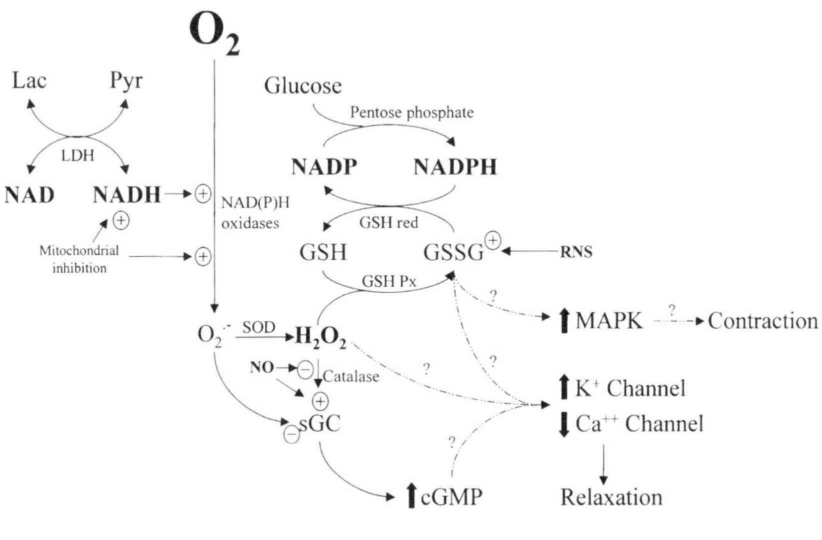

Figure 2 Potential relationships between NAD(P)H-derived ROS, redox systems, and signaling mechanisms regulated by PO_2 in vascular smooth muscle. LDH, lactate dehydrogenase; GSH red, glutathione reductase; GSH Px, glutathione peroxidase; RNS, reactive nitrogen species; SOD, superoxide dismutases; sGC, soluble guanylate cyclase; MAPK, mitogen-activated protein kinases.

Signaling Systems Regulated by Reactions with ROS

The chemical reaction of certain ROS with key protein sites potentially linked to the control of signaling systems is another process potentially involved in oxidant signaling and PO_2 sensing. Certain protein thiol groups are activated by the environment in which they are located, often to make them more acidic, and this can result in selective interactions with ROS and peroxynitrite. Peroxynitrite is the initial product of the extremely efficient reaction of NO with superoxide, and this species oxidizes, nitrosates, and nitrates thiols (6,56). An example of this is the extreme susceptibility of a key active site thiol on tyrosine phosphatases to modification by ROS and peroxynitrite (57,58). Although physiological conditions may result in a basal level of activity for certain tyrosine kinases, some of the members of this family of kinases are regulated by self phosphorylation reactions. Thus, inhibition of tyrosine phosphatases can be associated with the activation of signaling mechanisms dependent on tyrosine phosphorylation reactions. Some of the most sensitive systems to stimulation by ROS include some members of the PKB, PKC, and MAP kinase families (6,59). The extracellular regulated kinase (ERK) and p38 MAP kinase and PKC systems have mechanisms of promoting the development of force (60–62). The prolonged stimulation of these protein kinases is often associated with alterations in gene expression and mechanisms that control vascular cell growth and protect against apoptosis (59,63).

Many vascular potassium and calcium channels and ATP-dependent reuptake systems have reactive thiols and appear to controlled by thiol redox-linked mechanisms (55,64–66). These channels could potentially be regulated by ROS directly interacting with channel thiols (54,65,66), by PO_2 and/or ROS altering these channels through changes in cellular redox systems that control thiol redox (2,21), or by redox-sensitive signaling systems involving mediators such as cGMP (18) (see Fig. 2). There is also evidence that potassium channels have NAD(P)H oxido-reductase activity (67) and that they are directly sensitive to changes in PO_2 (68). While pulmonary arterial smooth muscle has been shown to possess voltage-regulated potassium and calcium channels that appear to contribute to functional responses regulated by changes in PO_2, there is much debate regarding the actual mechanisms that are involved in controlling the activity of these channels.

Signaling Systems Regulated by Changes in Cellular Redox Control Mechanisms

Certain cellular redox systems such as cytosolic NADP(H), NAD(H), and glutathione (GSH/GSSG) and related mitochondrial systems (including the electron transport) have important roles in the control of signaling mechanisms regulated by ROS (see Fig. 2). It is thought that the pentose phosphate pathway of glucose metabolism is the primary source of cytosolic NADPH in vascular tissue (69). NADPH is an important cofactor for the regeneration of GSH from GSSG, a process needed for the metabolism of peroxide, and for reduction of protein thiols modified by oxidation or S-thiolation, for methemoprotein reductase needed to keep the heme of sGC in the ferrous form that appears to be required for stimulation by NO, and for the biosynthesis of NO by NOS and superoxide by NADPH oxidases (6,70,71). The redox status of cytosolic NAD(H) is thought to be primarily controlled by the balance by NADH removal by mitochondrial shuttle mechanisms and the lactate dehydrogenase reaction, and its biosynthesis by glyceraldehyde-3-phosphate dehydrogenase in glycolysis. NADH is a cofactor for NADH oxidases and for the reduction of oxidized protein thiols (6). The inhibition of mitochondrial respiration by physiological levels of NO may also be an important factor in the control of signaling by ROS through its effects on energy metabolism, redox systems, and the production of ROS (72,73). Thus, cellular redox systems have the potential for multiple interactions with ROS-associated signaling because they influence both the levels of specific ROS and the expression of some of the signaling mechanisms regulated by ROS. Change in PO_2 is known to have major effects on cellular redox systems. However, the impact of these changes on the function of most redox-controlled signaling mechanisms remains to be investigated.

IV. Roles for NAD(P)H Oxidases and Oxidant Signaling Mechanisms in the Regulation of Vascular Responses by PO_2

The availability of selective probes for mitochondrial function and endothelium-derived mediators produced by COX, NOS, and P450 has helped to both establish roles for products of these systems in PO_2-elicited vascular responses and to define the importance of responses mediated by other mechanisms. NAD(P)H oxidases and signaling mechanisms controlled by ROS generated by these enzymes have been given active consideration for mediating certain PO_2-elicited responses present in vascular smooth muscle. The concentration dependence of these oxidases for O_2 as a substrate for the production of ROS over the physiological range permits these oxidases to function as a PO_2 sensor, and the signaling mechanisms influenced by these changes mediate the physiological responses that are observed. Studies suggesting NAD(P)H oxidases could function as the PO_2 sensor in the carotid body (5) resulted in consideration of these oxidases as a PO_2 sensor in pulmonary arteries (10,19,74) because of previous suggestions that changes in ROS could be mediating responses of these vessels to changes in PO_2 (1–4). However, identifying the oxidases and roles for the signaling mechanisms involved has not been easy because of the complexity of systems that exist. NAD(P)H oxidases including the Nox-type systems, mitochondrial electron transport-linked sites, and oxidoreductases present in voltage-regulated potassium channels are currently being given the most consideration as the pulmonary artery PO_2 sensor for the process of hypoxic vasoconstriction. Redox-regulated signaling mechanisms, including potassium (19) and calcium channels (53), sGC (18), and processes resulting from the activation of tyrosine kinases such as PKC and MAP kinase mechanisms (36,75), are currently being given the most consideration for the signal transduction linked to this PO_2 sensor in pulmonary arteries (see Fig. 2). Since these systems appear to be many of the most sensitive mechanisms to regulation by ROS such as H_2O_2, it is possible that changes in multiple redox-regulated processes may mediate the pulmonary arterial contractile response to hypoxia.

Oxygen-sensing mechanisms sensitive to physiological levels of O_2 in the systemic vasculature have been an area of major interest to cardiovascular physiologists because of the important role of PO_2 in the control of organ blood flow and vascular contractile function. While roles and mechanisms related to certain tissue metabolites such as adenosine that accumulate during exposure to hypoxia in the control of microvascular blood flow have been extensively investigated (76), other metabolites such as lactate, initially proposed in 1880 to be a metabolic regulator of blood flow (77), have only recently become of interest (18,78,79). There is rather strong evidence that exposure of vascular preparations to ischemia-reperfusion or hypoxia-reoxygenation activates ROS production by multiple oxidases and responses linked to these species. The recently identified role of what appears to be a ROS and cGMP-mediated relaxation to lactate in skeletal muscle arterioles (78), coupled with evidence that lactate and pyruvate influence many physiological mechanisms that control microcirculatory blood flow (79), is consistent with a role for a relaxing mechanism mediated by ROS such as H_2O_2 and the redox status cytosolic NADH, respectively, in one of the important metabolic mechanisms regulating microcirculatory blood flow. If this is true, it is likely that NADH oxidase functions as an important microcirculatory PO_2 and metabolic sensor.

V. Concluding Remarks

The multiple NAD(P)H oxidases available for the production of ROS and the diversity of signaling mechanisms linked to the generation of specific ROS permits these systems to potentially have multiple roles in vascular PO_2 sensing mechanisms under both physiological and pathophysiological conditions. It appears that NAD(P)H oxidase(s) containing a b_{558}-type flavocytochrome with homologs of the phagocytic oxidase gp91phox subunit are designed for PO_2 sensing in vascular tissue under physiological conditions, because the unstimulated form of this system seems to have a basal activity that generates ROS in amounts that are close to the range needed for activation of the most sensitive signaling mechanisms controlled by ROS such as H_2O_2. The absence of a consensus on a common vascular PO_2-sensing mechanism is likely to originate from the diversity of ROS-dependent and -independent mechanisms available, and much additional investigation is needed to define the importance of each of the PO_2-sensing mechanisms that has been identified or proposed. Thus, the controversy, in this area, will continue to exist.

Acknowledgments

Recent studies from the authors' laboratory have been funded by USPHS grants HL31069, HL43023, and HL66331.

References

1. Burke-Wolin T, Wolin MS. H_2O_2 and cGMP may function as an O_2 sensor in the pulmonary artery. J Appl Physiol 1989; 66:167–170.
2. Archer SL, Will JA, Weir EK. Redox status in the control of pulmonary vascular tone. Herz 1986; 11:127–141.
3. Archer SL, Nelson DP, Weir EK. Simultaneous measurement of O_2 radicals and pulmonary vascular reactivity in rat lung. J Appl Physiol 1989; 67:1903–1911.
4. Burke TM, Wolin MS. Hydrogen peroxide elicits pulmonary arterial relaxation and guanylate cyclase activation. Am J Physiol 1987; 252:H721–H732.
5. Acker H, Dufau E, Huber J, Sylvester D. Indications to an NADPH oxidase as a possible pO_2 sensor in the rat carotid body. FEBS Lett 1989; 256:75–78.
6. Wolin MS. Interactions of oxidants with vascular signaling systems. Arterioscler Thromb Vasc Biol 2000; 20:1430–1442.
7. Gorlach A, Brandes RP, Nguyen K, Amidi M, Dehghani F, Busse R. A $gp91_{phox}$ containing NADPH oxidase selectively expressed in endothelial cells is a major source of oxygen radical generation in the arterial wall. Circ Res 2000; 87:26–32.
8. Jones SA, O'Donnell VB, Wood JD, Broughton JP, Hughes EJ, Jones OT. Expression of phagocyte NADPH oxidase components in human endothelial cells. Am J Physiol 1996; 271:H1626–H1634.
9. Mohazzab KM, Kaminski PM, Wolin MS. NADH oxidoreductase is a major source of superoxide anion in bovine coronary artery endothelium. Am J Physiol 1994; 266:H2568–H2572.
10. Mohazzab KM, Wolin MS. Properties of a superoxide anion-generating microsomal NADH oxidoreductase, a potential pulmonary artery PO_2 sensor. Am J Physiol 1994; 267:L823–L831.
11. Griendling KK, Minieri CA, Ollerenshaw JD, Alexander RW. Angiotensin II stimulates NADH and NADPH oxidase activity in cultured vascular smooth muscle cells. Circ Res 1994; 74:1141–1148.
12. Griendling KK, Sorescu D, Ushio-Fukai M. NAD(P)H oxidase: role in cardiovascular biology and disease. Circ Res 2000; 86:494–501.
13. Babior BM. NADPH oxidase: an update. Blood 1999; 93:1464–1476.
14. Lassegue B, Sorescu D, Szocs K, Yin Q, Akers M, Zhang Y, Grant SL, Lambeth JD, Griendling KK. Novel $gp91_{phox}$ homologues in vascular smooth muscle cells: nox1 mediates angiotensin II-induced superoxide formation and redox-sensitive signaling pathways. Circ Res 2001; 88:888–894.
15. Patterson C, Ruef J, Madamanchi NR, Barry-Lane P, Hu Z, Horaist C, Ballinger CA, Brasier AR, Bode C, Runge MS. Stimulation of a vascular smooth muscle cell NAD(P)H oxidase by thrombin. Evidence that $p47_{phox}$ may participate in forming this oxidase in vitro and in vivo. J Biol Chem 1999; 274:19814–19822.
16. Rey FE, Cifuentes ME, Kiarash A, Quinn MT, Pagano PJ. Novel competitive inhibitor of NAD(P)H oxidase assembly attenuates vascular O_2^- and systolic blood pressure in mice. Circ Res 2001; 89:408–414.
17. Omar HA, Mohazzab KM, Mortelliti MP, Wolin MS. O_2-dependent modulation of calf pulmonary artery tone by lactate: potential role of H_2O_2 and cGMP. Am J Physiol 1993; 264:L141–L145.

18. Wolin MS, Burke-Wolin TM, Mohazzab-H KM. Roles for NAD(P)H oxidases and reactive oxygen species in vascular oxygen sensing mechanisms. Respir Physiol 1999; 115:229–238.
19. Archer SL, Reeve HL, Michelakis E, Puttagunta L, Waite R, Nelson DP, Dinauer MC, Weir EK. O_2 sensing is preserved in mice lacking the $gp91_{phox}$ subunit of NADPH oxidase. Proc Natl Acad Sci USA 1999; 96:7944–7949.
20. Turrens JF, Boveris A. Generation of superoxide anion by the NADH dehydrogenase of bovine heart mitochondria. Biochem J 1980; 191:421–427.
21. Archer SL, Weir EK, Reeve HL, Michelakis E. Molecular identification of O_2 sensors and O_2-sensitive potassium channels in the pulmonary circulation. Adv Exp Med Biol 2000; 475:219–240.
22. Waypa GB, Chandel NS, Schumacker PT. Model for hypoxic pulmonary vasoconstriction involving mitochondrial oxygen sensing. Circ Res 2001; 88:1259–1266.
23. Cadenas E, Davies KJ. Mitochondrial free radical generation, oxidative stress, and aging. Free Radic Biol Med 2000; 29:222–230.
24. Cosentino F, Barker JE, Brand MP, Heales SJ, Werner ER, Tippins JR, West N, Channon KM, Volpe M, Luscher TF. Reactive oxygen species mediate endothelium-dependent relaxations in tetrahydrobiopterin-deficient mice. Arterioscler Thromb Vasc Biol 2001; 21:496–502.
25. Katusic ZS. Vascular endothelial dysfunction: does tetrahydrobiopterin play a role? Am J Physiol Heart Circ Physiol 2001; 281:H981–H986.
26. Harrison DG. Cellular and molecular mechanisms of endothelial cell dysfunction. J Clin Invest 1997; 100:2153–2157.
27. Fleming I, Michaelis UR, Bredenkotter D, Fisslthaler B, Dehghani F, Brandes RP, Busse R. Endothelium-derived hyperpolarizing factor synthase (cytochrome P450 2C9) is a functionally significant source of reactive oxygen species in coronary arteries. Circ Res 2001; 88:44–51.
28. Kukreja RC, Kontos HA, Hess ML, Ellis EF. PGH synthase and lipoxygenase generate superoxide in the presence of NADH or NADPH. Circ Res 1986; 59:612–619.
29. Sobey CG, Heistad DD, Faraci FM. Potassium channels mediate dilatation of cerebral arterioles in response to arachidonate. Am J Physiol 1998; 275:H1606–H1612.
30. Granger DN. Role of xanthine oxidase and granulocytes in ischemia-reperfusion injury. Am J Physiol 1988; 255:H1269–H1275.
31. Xia Y, Khatchikian G, Zweier JL. Adenosine deaminase inhibition prevents free radical-mediated injury in the postischemic heart. J Biol Chem 1996; 271:10096–10102.
32. Shimizu S, Saitoh Y, Yamamoto T, Momose K. Stimulation by hydrogen peroxide of L-arginine metabolism to L-citrulline coupled with nitric oxide synthesis in cultured endothelial cells. Res Commun Chem Pathol Pharmacol 1994; 84:315–329.
33. Taylor L, Menconi MJ, Polgar P. The participation of hydroperoxides and oxygen radicals in the control of prostaglandin synthesis. J Biol Chem 1983; 258:6855–6857.
34. Callahan KS, Garcia JG. Oxidant exposure stimulates cultured coronary artery endothelial cells to release 15-HETE: differential effects on PGI_2 and 15-HETE synthesis. J Lab Clin Med 1994; 124:569–578.
35. Chakraborti S, Chakraborti T. Down-regulation of protein kinase C attenuates the oxidant hydrogen peroxide-mediated activation of phospholipase A_2 in pulmonary vascular smooth muscle cells. Cell Signal 1995; 7:75–83.

36. Zhang J, Jin N, Liu Y, Rhoades RA. Hydrogen peroxide stimulates extracellular signal-regulated protein kinases in pulmonary arterial smooth muscle cells. Am J Respir Cell Mol Biol 1998; 19:324–332.
37. Boyer CS, Bannenberg GL, Neve EP, Ryrfeldt A, Moldeus P. Evidence for the activation of the signal-responsive phospholipase A_2 by exogenous hydrogen peroxide. Biochem Pharmacol 1995; 50:753–761.
38. Rao GN, Runge MS, Alexander RW. Hydrogen peroxide activation of cytosolic phospholipase A_2 in vascular smooth muscle cells. Biochim Biophys Acta 1995; 1265:67–72.
39. Bakovic M, Dunford HB. Intimate relation between cyclooxygenase and peroxidase activities of prostaglandin H synthase. Peroxidase reaction of ferulic acid and its influence on the reaction of arachidonic acid. Biochemistry 1994; 33:6475–6482.
40. Drummond GR, Cai H, Davis ME, Ramasamy S, Harrison DG. Transcriptional and posttranscriptional regulation of endothelial nitric oxide synthase expression by hydrogen peroxide. Circ Res 2000; 86:347–354.
41. Fulton D, Gratton JP, McCabe TJ, Fontana J, Fujio Y, Walsh K, Franke TF, Papapetropoulos A, Sessa WC. Regulation of endothelium-derived nitric oxide production by the protein kinase Akt. Nature 1999; 399:597–601.
42. Wolin MS, Rodenburg JM, Messina EJ, Kaley G. Oxygen metabolites and vasodilator mechanisms in rat cremasteric arterioles. Am J Physiol 1987; 252:H1159–H1163.
43. Gurtner GH, Burke-Wolin T. Interactions of oxidant stress and vascular reactivity. Am J Physiol 1991; 260:L207–L211.
44. Messina EJ, Sun D, Koller A, Wolin MS, Kaley G. Increases in oxygen tension evoke arteriolar constriction by inhibiting endothelial prostaglandin synthesis. Microvasc Res 1994; 48:151–160.
45. Wolin MS, Rodenburg JM, Messina EJ, Kaley G. Similarities in the pharmacological modulation of reactive hyperemia and vasodilation to hydrogen peroxide in rat skeletal muscle arterioles: effects of probes for endothelium-derived mediators. J Pharmacol Exp Ther 1990; 253:508–512.
46. Harder DR, Narayanan J, Birks EK, Liard JF, Imig JD, Lombard JH, Lange AR, Roman RJ. Identification of a putative microvascular oxygen sensor. Circ Res 1996; 79:54–61.
47. Wolin MS. Reactive oxygen species and vascular signal transduction mechanisms. Microcirculation 1996; 3:1–17.
48. Burke-Wolin TM, Wolin MS. Inhibition of cGMP-associated pulmonary arterial relaxation to H_2O_2 and O_2 by ethanol. Am J Physiol 1990; 258:H1267–H1273.
49. Figueroa R, Martinez E, Fayngersh RP, Tejani N, Mohazzab HK, Wolin MS. Alterations in relaxation to lactate and H_2O_2 in human placental vessels from gestational diabetic pregnancies. Am J Physiol Heart Circ Physiol 2000; 278:H706–H713.
50. Mohazzab HK, Kaminski PM, Fayngersh RP, Wolin MS. Oxygen-elicited responses in calf coronary arteries: role of H_2O_2 production via NADH-derived superoxide. Am J Physiol 1996; 270:H1044–H1053.
51. Dierks EA, Burstyn JN. The deactivation of soluble guanylyl cyclase by redox-active agents. Arch Biochem Biophys 1998; 351:1–7.
52. Zhao Y, Brandish PE, DiValentin M, Schelvis JP, Babcock GT, Marletta MA. Inhibition of soluble guanylate cyclase by ODQ. Biochemistry 2000; 39:10848–10854.
53. Madden JA, Dawson CA, Harder DR. Hypoxia-induced activation in small isolated pulmonary arteries from the cat. J Appl Physiol 1985; 59:113–118.

54. Iesaki T, Wolin MS. Thiol oxidation activates a novel redox-regulated coronary vasodilator mechanism involving inhibition of Ca^{2+} influx. Arterioscler Thromb Vasc Biol 2000; 20:2359–2365.
55. Weir EK, Archer SL. The mechanism of acute hypoxic pulmonary vasoconstriction: the tale of two channels. FASEB J 1995; 9:183–189.
56. Wink DA, Cook JA, Pacelli R, Liebmann J, Krishna MC, Mitchell JB. Nitric oxide (NO) protects against cellular damage by reactive oxygen species. Toxicol Lett 1995; 82-83:221–226.
57. Natarajan V, Scribner WM, al-Hassani M, Vepa S. Reactive oxygen species signaling through regulation of protein tyrosine phosphorylation in endothelial cells. Environ Health Perspect 1998; 106 (Suppl 5):1205–1212.
58. Takakura K, Beckman JS, MacMillan-Crow LA, Crow JP. Rapid and irreversible inactivation of protein tyrosine phosphatases PTP1B, CD45, and LAR by peroxynitrite. Arch Biochem Biophys 1999; 369:197–207.
59. Griendling KK, Sorescu D, Lassegue B, Ushio-Fukai M. Modulation of protein kinase activity and gene expression by reactive oxygen species and their role in vascular physiology and pathophysiology. Arterioscler Thromb Vasc Biol 2000; 20:2175–2183.
60. Leinweber BD, Leavis PC, Grabarek Z, Wang CL, Morgan KG. Extracellular regulated kinase (ERK) interaction with actin and the calponin homology (CH) domain of actin-binding proteins. Biochem J 1999; 344 (Pt 1):117–123.
61. Yamboliev IA, Hedges JC, Mutnick JL, Adam LP, Gerthoffer WT. Evidence for modulation of smooth muscle force by the p38 MAP kinase/HSP27 pathway. Am J Physiol Heart Circ Physiol 2000; 278:H1899–H1907.
62. Ruzycky AL, Morgan KG. Involvement of the protein kinase C system in calcium-force relationships in ferret aorta. Br J Pharmacol 1989; 97:391–400.
63. Irani K. Oxidant signaling in vascular cell growth, death, and survival: a review of the roles of reactive oxygen species in smooth muscle and endothelial cell mitogenic and apoptotic signaling. Circ Res 2000; 87:179–183.
64. Grigoriev SM, Skarga YY, Mironova GD, Marinov BS. Regulation of mitochondrial KATP channel by redox agents. Biochim Biophys Acta 1999; 1410:91–96.
65. Elliott SJ, Meszaros JG, Schilling WP. Effect of oxidant stress on calcium signaling in vascular endothelial cells. Free Radic Biol Med 1992; 13:635–650.
66. Grover AK, Xu A, Samson SE, Narayanan N. Sarcoplasmic reticulum Ca^{2+} pump in pig coronary artery smooth muscle is regulated by a novel pathway. Am J Physiol 1996; 271:C181–C187.
67. McCormack T, McCormack K. Shaker K^+ channel beta subunits belong to an NAD(P)H-dependent oxidoreductase superfamily. Cell 1994; 79:1133–1135.
68. Jiang C, Haddad GG. A direct mechanism for sensing low oxygen levels by central neurons. Proc Natl Acad Sci USA 1994; 91:7198–7201.
69. Asahina T, Kashiwagi A, Nishio Y, Ikebuchi M, Harada N, Tanaka Y, Takagi Y, Saeki Y, Kikkawa R, Shigeta Y. Impaired activation of glucose oxidation and NADPH supply in human endothelial cells exposed to H_2O_2 in high-glucose medium. Diabetes 1995; 44:520–526.
70. Chance B, Sies H, Boveris A. Hydroperoxide metabolism in mammalian organs. Physiol Rev 1979; 59:527–605.
71. Gupte SA, Rupawalla T, Phillibert D Jr, Wolin MS. NADPH and heme redox modulate pulmonary artery relaxation and guanylate cyclase activation by NO. Am J Physiol 1999; 277:L1124–L1132.

72. Wolin MS, Hintze TH, Shen W, Mohazzab-H KM, Xie YW. Involvement of reactive oxygen and nitrogen species in signalling mechanisms that control tissue respiration in muscle. Biochem Soc Trans 1997; 25:934–939.
73. Boveris A, Costa LE, Poderoso JJ, Carreras MC, Cadenas E. Regulation of mitochondrial respiration by oxygen and nitric oxide. Ann NY Acad Sci 2000; 899:121–135.
74. Grimminger F, Weissmann N, Spriestersbach R, Becker E, Rosseau S, Seeger W. Effects of NADPH oxidase inhibitors on hypoxic vasoconstriction in buffer-perfused rabbit lungs. Am J Physiol 1995; 268:L747–L752.
75. Jin N, Packer CS, Rhoades RA. Pulmonary arterial hypoxic contraction: signal transduction. Am J Physiol 1992; 263:L73–78.
76. Berne RM. The role of adenosine in the regulation of coronary blood flow. Circ Res 1980; 47:807–813.
77. Gaskell W. On the tonicity of the heart and blood vessels. J Physiol 1880; 3:48–75.
78. Chen YL, Wolin MS, Messina EJ. Evidence for cGMP mediation of skeletal muscle arteriolar dilation to lactate. J Appl Physiol 1996; 81:349–354.
79. Ido Y, Chang K, Woolsey TA, Williamson JR. NADH: sensor of blood flow need in brain, muscle, and other tissues. FASEB J 2001; 15:1419–1421.

31

Oxygen Sensing in Pulmonary Neuroepithelial Bodies and Related Tumor Cell Model

ERNEST CUTZ, XIAO WEN FU, and HERMAN YEGER

University of Toronto and
The Hospital for Sick Children
Toronto, Ontario, Canada

CHRIS PEERS and PAUL J. KEMP

University of Leeds
Leeds, England

I. Introduction

Innervated clusters of amine- and peptide-containing cells widely distributed within the airway epithelium of mammalian lungs, termed neuroepithelial bodies (NEB), were described over 30 years ago (1). Although much progress has been made since then in defining the structural aspects of NEB, their precise function remains unknown. The idea that NEB may represent airway sensors monitoring changes in airway gas concentration has been proposed by early neuroanatomists (2). The first experimental evidence indicating that NEB may indeed represent hypoxia-sensitive airway chemoreceptors has been reported by Lauweryns et al. (3,4). However, the underlying mechanisms and the effects on lung function remained undefined. Further advances in characterizing functional aspects of NEB cells have occurred with development of suitable in vitro models to overcome the problem of their small numbers and widespread distribution within anatomically complex lung parenchyma (5,6). Using cultures of NEB cells isolated from near-term fetal rabbit lungs, amine (serotonin, 5-HT) release was observed after exposure to different levels of hypoxia confirming their intrinsic O_2 sensitivity (6). The same model was subsequently used for electrophysiological characterization of ionic currents and the O_2-sensing mechanism (7). These latter studies have strengthened the view that NEB represent airway O_2 sensors analogous to carotid body (CB) glomus cells, well-defined arterial chemoreceptors (8). Development of additional experimental models, such as fresh lung slice preparations, has allowed direct electrophysiological recordings from NEB cells in their natural environment (9). The identification of O_2-sensing properties in small-cell lung carcinoma (SCLC) cell line H-146, a tumor counterpart of NEB in normal lungs, has been instrumental in studies of molecular mechanism involved in O_2 sensitivity (10). There is also new information provided by anatomical studies on the distribution and innervation of NEB (11). The principal aim of this chapter is to review recent progress in the cell and molecular biology of NEB using various experimental models and approaches, focusing on their potential role as airway chemoreceptors.

II. Structure, Distribution, and Development

The various morphological aspects of pulmonary neuroendocrine cell (PNEC) system, comprised of solitary cells and innervated cell clusters, NEB have been subject of several recent reviews (12–15). This section will focus on morphological features of NEB indicative of their chemoreceptor nature and their potential role as airway O_2 sensors.

A. General Morphology and Ultrastructure

Typical pulmonary NEB are composed of 5–15 innervated cell clusters distributed throughout the epithelium of intrapulmonary airways (Fig. 1). At the ultrastructural level, NEB exhibit many features consistent with a chemoreceptor organ, including: (1) the presence of organelles involved in the synthesis and storage of amine/peptide neurotransmitters and neuromodulators, (2) direct exposure to the airway lumen, and (3) afferent sensory innervation. Classic NEBs form distinct ovoid corpuscles with lateral cell membranes interdigitating with neighboring cells via gap junctions and desmosomes. At the apical membrane, tight junctions (zonula occludens) are present and short microvilli are covered by glycocalyx. It is presumed that the O_2 sensor is located at this microvillous membrane since it is in direct contact with intraluminal air. Studies using scanning electron microscopy have shown that the apical surface of NEB exhibit species and developmental variations in respect to their exposure to the airway lumen (16,17).

The nuclei of NEB cells are usually elongated, often with an indented nuclear membrane and evenly distributed chromatin. The most characteristic cytoplasmic feature is the presence of numerous dense core vesicles (DCV), the storage site of amine and peptides. Other cytoplasmic organelles include small mitochondria, rough endoplasmic reticulum, ribosomes, microtubules, and bundles of microfilaments. A well-developed Golgi complex shows a variety of associated vesicles related to the formation of DCV. In the rabbit fetus and neonate two morphological types of nerve endings are found: the afferent-like nerve fibers containing numerous small mitochondria representing sensory nerve fibers, and efferent-like (motor) nerve fibers containing agranular vesicles, microtubules, and sparse mitochondria. Serial section

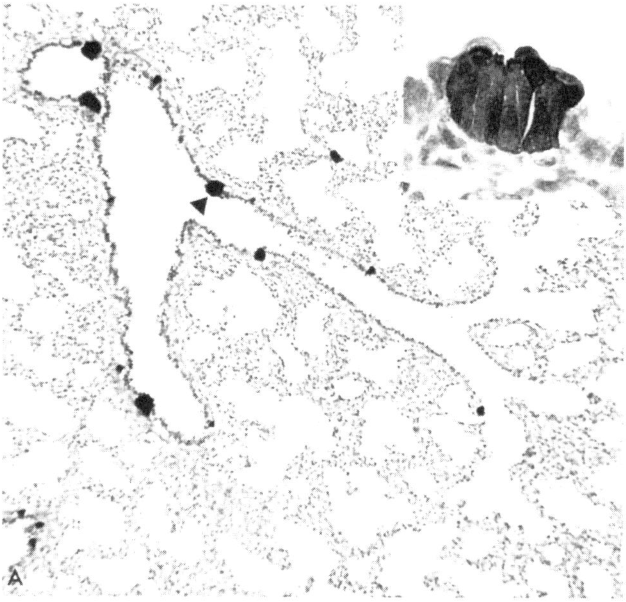

Figure 1 Serotonin immunoreactive NEBs in rabbit fetal lung (26-day gestation). NEB are located within the mucosa of airways, particularly near bifurcations (arrowhead). Immunoperoxidase method for 5-HT (×100). Insert: Higher magnification of NEB with apical surface in contact with airway lumen (×250). (From Ref. 7.)

studies have shown that these two types of nerve endings may be in continuity, an arrangement consistent with local modulation via axon reflex (18).

B. Amine/Peptides and Neuroendocrine Markers

The various immunohistochemical and molecular markers of NEB cells reported in mammalian lungs are summarized in Table 1. Only those markers relevant to the role of NEB as O_2 sensor will be reviewed.

Serotonin (5-Hydroxytryptamine, 5-HT)

It is now well established that NEB cells synthesize and release 5-HT. By immunohistochemical methods, 5-HT has been identified in NEB cells in lungs of human and several animal species (2,12,13) (Fig. 1). Significant amounts of 5-HT were detected by HPLC in extracts of NEB cell cultures isolated from fetal rabbit lung (6). Immunoreactivity for tryptophan hydroxylase (TH), a rate-limiting enzyme in 5-HT synthesis, as well as aromatic amino acid decarboxylase (AADC) has been reported in NEB of human, rat, and mouse lung (19). The expression of mRNA for both TH and AADC has been identified in extracts of whole rabbit fetal lung and cell cultures (6). With single-cell RT-PCR, expression of mRNA encoding TH has been

Table 1 Immunohistochemical and Molecular Markers of NEB Cells in Mammalian Lungs[a]

Amine	Neuroendocrine/neurosecretion markers
Serotonin (5-hydroxytryptamine, 5-HT) (12,13,24)	Neuron-specific enolase (NSE) (38)
Amine-metabolizing enzymes	Protein gene product 9.5 (PGP9.5) (39,40)
Tryptophan hydroxylase (6,19)	Chromogranin A (41)
Aromatic amino acid decarboxylase (6,19)	NCAM/MOC-1 (43,44)
Regulatory peptides	Leu 7/NHK (42)
Bombesin/gastrin-releasing peptide (GRP) (21–25)	Goα (45)
Calcitonin (26)	Synaptophysin (46)
Calcitonin gene–related peptide (CGRP) (28–30)	Synaptic vesicle protein 2 (SV_2) (47)
Cholecystokinin (CCK) (31)	Synaptobrevin (47)
Substance P (32)	Synaptogamin (47)
Somatostatin (34)	SNAP-25 (47)
Endothelin (33)	Calbindin D28K (48)
Peptide YY (35)	
Helodermin (36)	
Pituitary adenyl cyclase–activating protein (37)	
Transcription factors	Functional proteins
Mash-1 (63–65)	NADPH oxidase ($gp91^{phox}$, $p22^{phox}$, $p47^{phox}$, rac2) (74)

[a] There is marked species variation in the expression of different markers. However, the majority have been identified in NEB cells of human lung and/or experimental animals (i.e., rabbit, rat, hamster, mouse).

confirmed in NEB cells microdissected from rabbit neonatal lung (20). The precise role of 5-HT in NEB cells or in lung physiology has not been fully characterized.

Bombesin/Gastrin-Releasing Peptide (GRP)

A tetradecapeptide, bombesin, originally isolated from frog skin was the first NEB cell peptide identified in human lung (21). This amphibian peptide is structurally related to a 27-amino-acid mammalian peptide, gastrin-releasing peptide (GRP), with similar biological and immunohistochemical properties (22). High expression of bombesin/GRP is found in human NEB cells, particularly during the fetal/neonatal period (23), whereas in other species minimal or no bombesin/GRP immunoreactivity is observed (24).

The precise role of bombesin/GRP in the lung remains unknown. Since bombesin/GRP exhibits growth factor-like properties and appears to be developmentally regulated, a role during lung morphogenesis and/or differentiation has been proposed (22,24). A GRP-preferring receptor mRNA expression has been localized in airway epithelial cells, NEB cells, and adjacent mesenchyma in human and rabbit fetal/neonatal lungs (24). A possible link between the O_2-sensing function of NEB, the relatively hypoxic intrauterine environment, and lung morphogenesis mediated via bombesin/GRP peptides has recently been proposed (25).

Calcitonin and Calcitonin Gene–Related Peptide (CGRP)

Immunoreactivity for calcitonin and calcitonin gene–related peptide (CGRP) has been localized in NEB cells of human as well as several animal species (26). Both peptides are derived from a single calcitonin gene by alternate RNA processing (27). Expression of CGRP is particularly prominent in NEB cells and submucosal nerves in rodent lungs (28). The precise function of calcitonin or CGRP in the lung is unknown. Involvement of CGRP in modulation of hypoxia-induced pulmonary vasoconstriction has been postulated (29,30).

Other Peptides

Additional regulatory peptides identified by immunohistochemistry in NEB cells of human and animal lungs include cholecystokinin (31), substance P (32), endothelin (33), and somatostatin (34). A minor subpopulation of PNEC or NEB was found to express peptide YY (35), helodermin (36), and pituitary adenyl cyclase–activating protein (37). The function of these peptides in O_2 sensing by NEB cells or in lung physiology in general has not been defined.

Neuroendocrine/Neurosecretion Markers

Immunoreactivity for neuron-specific enolase (NSE), a cytosolic enzyme with glycolytic activity, has been demonstrated in PNEC during early stages of lung development, preceding expression of amine and peptides (38). Another functional neuroendocrine marker, protein gene product 9.5 (PGP9.5), was found widely distributed in peripheral nervous tissue including PNEC and NEB cells as well as intrapulmonary nerves and ganglia (39). In early fetal lungs, PGP9.5 is expressed in primitive airway epithelium diffusely and later becomes restricted to neural components and PNEC/NEB cells (40).

Chromogranin A belongs to a family of acidic secretory proteins associated with dense core granules in a variety of endocrine and neuroendocrine cells (41). A lymphocyte surface marker, Leu7/NHK, cross-reacts with related epitope on the plasma membrane of PNEC and NEB (42). The other NEB cell surface markers include neural adhesion molecule (NCAM) and related epitope MOC-1, originally identified in the small cell lung carcinoma cell line (43,44).

Immunoreactivity for G protein (Go_{α}) was identified in PNEC/NEB of hamster lungs (45). The various proteins involved in the function and trafficking of DCV have also been identified in PNEC/NEB cells using immunohistochemistry, including: synaptophysin (46), synaptic vesicle protein 2 (SV2), synaptobrevin, synaptogamin, and SNAP-25 (47). Calcium-binding protein, calbindin-D28k, immunoreactivity was detected in PNEC/NEB cells by Ito and colleagues (48).

C. Innervation

The nature and origin of intraepithelial nerve endings in contact with NEB cells have been extensively investigated in rabbit neonatal lungs using electron microscopy and various vagotomy procedures (49,50). The key ultrastructural findings indicated that the predominant or type 1 nerve endings with afferent-like morphology represent sensory fibers of vagal origin with the cell bodies residing in the nodose ganglion (49,50). The less frequent, type 2 (efferent-like) nerve endings were interpreted as axon collaterals, which could modulate secretory activity of NEB cells. The suggestion that afferent nerve endings in contact with NEB cells are involved in hypoxia signal transduction was supported by ultrastructural changes induced by acute hypoxia (51).

Recently, the use of various neuroendocrine/neural markers and confocal microscopy has allowed more detailed analysis of NEB innervation (Fig. 2). In the adult Wistar rat, three different sensory fiber populations innervating NEB have been identified. The predominant subpopulation was represented by vagally derived afferents confirmed by retrograde labeling with DiI and positive immunoreactivity for calbindin-D28k or P2X3 purinoreceptors (10,52). The second population derived from the spinal ganglia was immunoreactive for CGRP, while the third component included nNOS-immunoreactive nerve endings originating from intrinsic peribron-

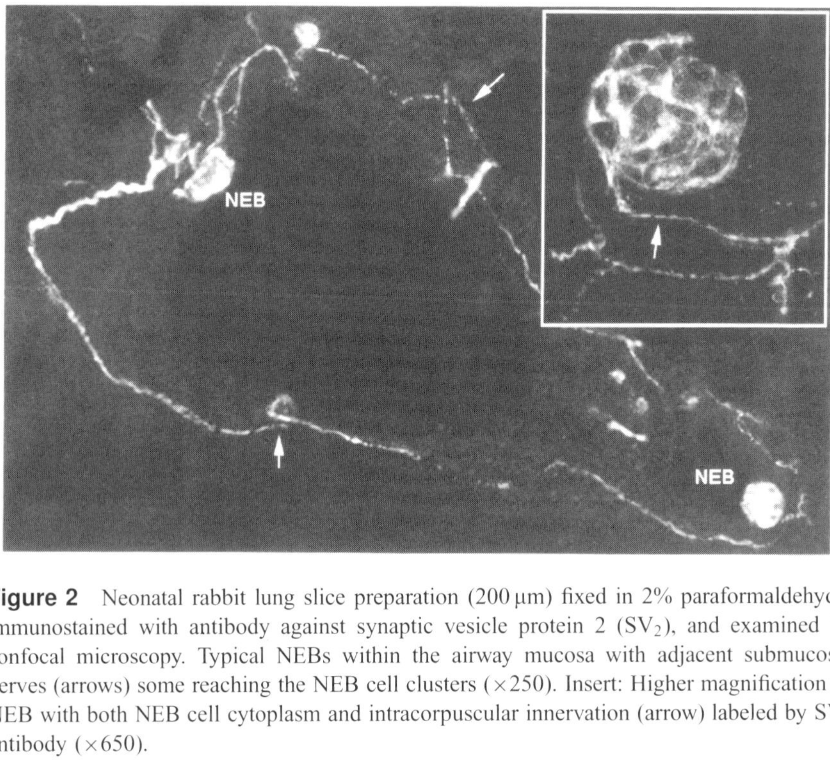

Figure 2 Neonatal rabbit lung slice preparation (200 μm) fixed in 2% paraformaldehyde, immunostained with antibody against synaptic vesicle protein 2 (SV_2), and examined by confocal microscopy. Typical NEBs within the airway mucosa with adjacent submucosal nerves (arrows) some reaching the NEB cell clusters (×250). Insert: Higher magnification of NEB with both NEB cell cytoplasm and intracorpuscular innervation (arrow) labeled by SV_2 antibody (×650).

chial ganglia (53). In many instances, CGRP and nNOS-immunoreactive nerve fibers colocalized in the same NEB. Complex innervation of NEB is reminiscent of the carotid body (CB) where the various components of innervation have been well characterized (54). By analogy with CB, it has been proposed that the nitriergic component of NEB innervation may be inhibitory to the sensory discharges generated by airway hypoxia. However, confirmation by pharmacological and physiological approaches is lacking.

D. Distribution, Frequency, and Development

Quantitative studies on the distribution and frequency of NEB in rabbit fetal and postnatal lungs revealed striking developmental changes with prominence of NEB during the neonatal period and a decline postnatally (55). The frequency of NEB was highest at airway bifurcations with up to 60% located on or near airway branch points. A similar pattern of distribution and frequency of NEB was observed in human lungs with peaking at birth and decline postnatally (23,56). Quantitation of PNEC/NEB in lungs of humans between 2 and 95 years of age revealed that the relative numbers of these cells did not change over time, suggesting that the total number of PNEC/NEB may remain constant, but their density changes owing to "dilutional" effects of postnatal lung growth (57).

Recent studies using a combination of immunohistochemistry, electron microscopy, and stereological methods reexamined the question of NEB frequency and developmental changes. Van Lommel and Lauweryns (58) confirmed the highest density of NEB occurring at birth, particularly in species with relatively immature lungs. Postnatally the density of NEB decreased owing to expansion and growth of the alveoli, but the number of NEB cell nuclei and cytoplasmic DCV remained unchanged. On the other hand, there was an increase in the number of afferent-like nerve terminals and a reduction in efferent-like nerve endings, suggesting postnatal maturation of NEB innervation. Furthermore, in neonatal rabbits the innervation of NEB was found to be more mature compared to that of CB from the same animals (59). These findings suggested that NEB are most active in neonatal lungs and therefore may complement the chemoreceptor function of CB, particularly during the early phase of postnatal development when CB function is still immature (60). This chemoreceptor function of NEB could be critical for species whose lungs are not fully mature at birth with less efficient gas exchange and may require local mechanisms to detect changes in pO_2 to prevent hypoxemia (58,59).

At present the predominant view is that NEB cells originate from foregut endoderm, as has been demonstrated for the endocrine cells of the gastrointestinal (GI) tract and islets of Langerhans (61). In early fetal lungs, cells exhibiting neuroendocrine features (precursors of PNEC/NEB) are the first cell type to differentiate within primitive airway epithelium (56). Ito et al. (62) have shown that PNEC/NEB develop in culture from early fetal airway epithelium without mesenchyma, confirming their epithelial origin.

Although the precise mechanisms controlling neuroendocrine differentiation are not fully understood, recent studies have identified several neurogenic genes and related transcription factors involved in the ontogeny of PNEC/NEB cells. Proneural bHLH (basic-helix-loop-helix) transcription factors appear to play a significant role in cell fate determination of the airway epithelium (63–65). First shown to be critical for *Drosophila* neurosensory development, bHLH transcription factors of the achaete-scute family have mammalian homologs with similar functional characteristics. The human achaete-scute homology 1 (HASH-1) gene is 95% homologous to the murine homolog (MASH-1) (66). HASH-1 was found to be selectively expressed in fetal PNEC, and disruption of the ASH-1 gene in mice led to failure of PNEC differentiation in the lung while GI endocrine cells were present [63]. Interestingly, lung morphogenesis did not appear to be affected by the loss of ASH-1 expression although these mice died soon after birth. Studies on the roles of MASH-1 and related genes were extended in knockout mice models by Ito and co-workers, who showed enhanced PNEC differentiation in hairy-enhancer-split (HES-1)-deficient mice and absence of PNEC in MASH-1-deficient mice (65). Additional studies demonstrated that MASH-1 and HES-1 likely act in concert with Notch/Notch ligand signaling pathways [64].

The developmental pattern including the spatial and temporal relationship of PNEC/NEB to the developing bronchial tree has been investigated by Sorokin et al. (67). They described centrifugal waves of PNEC/NEB differentiation progressing from trachea to intrapulmonary airways. Based on developmental changes in NEB cell differentiation and maturation of its innervation, Van Lommel and Lauweryns proposed a dual role for NEBs as having a local paracine or endocrine role during intrauterine life and postnatally that of chemoreceptors, when neural connections have fully matured (58).

III. Oxygen-Sensing Mechanisms

A. Experimental Models

The investigation of O_2-sensing mechanisms in NEB cells became possible with the development of suitable in vitro models and by devising means to identify NEB in a living state using neutral red for patch-clamp analysis. Our initial model used cultures of intact NEB isolated from near-term rabbit fetal lungs (7). This method, while highly reproducible, is time consuming and may induce potential artifacts related to enzymatic treatment and cell culture. Because of these limitations we have developed an alternate model using fresh lung slice preparations (9). The advantages of this model include: (1) patch-clamp recordings are made directly from NEB in their "natural" environment; (2) this preparation may be suitable for investigation of synaptic connections between NEB cells and their innervation; and (3) the technique is easily applied to different animal species.

A third alternative has been the use of human tumor cell lines related to PNEC/NEB, namely, small-cell lung carcinoma (SCLC). Although more than 60 SCLC cell lines have been described and partially characterized (68,69), relatively few are being used as models for PNEC/NEB cell function. More detailed analysis of SCLC cell lines revealed that most retained multiple phenotypic and biochemical characteristics of their counterpart in normal lung (69). The finding that SCLC cell lines express HASH-1, a transcription factor involved in early PNEC differentiation, suggests that these tumors are derived from PNEC/NEB progenitors (70). This is also supported by findings that SCLC cell line H-146 expresses many features of the O_2-sensing mechanism described in native NEB (11). The advantage of tumor cell lines is their easy propagation providing sufficient cell numbers for studies using biochemical and molecular biological approaches (see below).

B. Ionic Currents and O_2-Sensing Properties of NEB Cells in Fetal/Neonatal Lungs

Our initial observations that NEB cells exhibit membrane properties of excitable cells were made on cultures of NEB isolated from near-term rabbit fetal lung. Using whole-cell patch-clamp configuration, depolarizing voltage steps from holding potential of −60 to +20 mV activated both a fast transient inward current and a prolonged outward current (7). Further characterization of these currents revealed that NEB cells from fetal rabbit lung express voltage-activated K^+, Na^+, and Ca^{2+} currents. Upon exposure to hypoxia (pO_2 25–30 mmHg) there was reversible reduction (25–30%) in outward K^+ current while inward Na^+ and Ca^{2+} currents were unaffected by hypoxia. In current clamp mode, closure of K^+ channels by hypoxia produced an increase in the spontaneous firing frequency and slope of the depolarization pacemaker potential in NEB cells.

Passive Membrane Properties

Generally, the electrophysiology findings from the NEB culture model were reproduced in the lung slice model using tissues from different species during the late-fetal or neonatal period (Fig. 3). For example, the mean resting potential of NEB cells was found to be -51.2 ± 1.6 mV in neonatal rabbit, -51.9 ± 1.9 mV in neonatal mouse, and -50.4 ± 2.9 mV in neonatal hamster, almost identical values for the three species. In neonatal rabbit, the resting input resistance and capacitance were 1.08 ± 0.06 GΩ and 2.42 ± 0.17 pF, respectively, and the membrane time constant was 2.5 sec. Assuming NEB cells are spherical, and specific membrane capacitance is 1 μF cm^{-2}, the diameter of a single NEB cell in situ was estimated at ~8.7 μm (9). In NEB cell culture, whole-cell voltage-clamp recordings indicated input resistances of 2.1 GΩ and capacitance values of 6.1 pF suggesting larger cell size (~10 μm) compared with lung slices, perhaps related to the culture conditions. This could also explain the more prominent inward and outward currents recorded from NEB cells maintained in culture compared with those obtained in situ from the lung slices.

K^+ Currents

Following a depolarizing pulse to +30 mV from a holding potential of −60 mV, NEB cells in neonatal rabbit lung slices generated an outward current that was blocked (~50%) by 20 mM tetraethylammonium (TEA) or 2 mM 4-aminopyridine (4-AP). The activation threshold of this K^+ current was around −45 mV. The on-

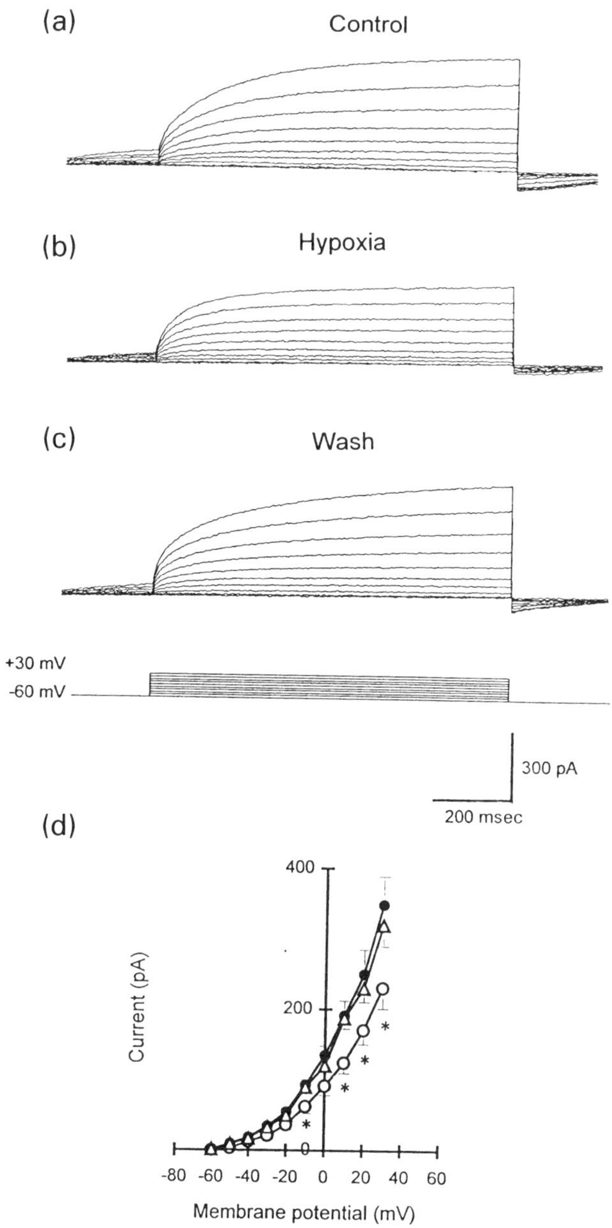

Figure 3 Effect of hypoxia on K^+ current in neonatal rabbit NEB cells in lung slice preparation. (a) Outward K^+ current evoked by depolarizing steps from -60 to $+30\,mV$ in control normoxic Krebs solution. (b) Outward current evoked by same voltage steps as in (a) was reduced by hypoxia. (c) Washout of the hypoxic solution caused a recovery of the outward K^+ current. (d) I-V relationships for the current in control solution (●) and in hypoxic solution (○) are plotted together with recovery K^+ current (Δ). Holding potential was $-60\,mV$; * significant difference from control ($p < 0.05$). Data represent means ± SEM for a sample of 10 cells. (From Ref. 9.)

set of this current was relatively slow and inactivation was not apparent during 600-msec pulses. The current amplitudes at a test potential of +30 mV were 422.5 ± 9.6 pA in the range of 100–700 pA. The mean density of the K^+ channel currents in NEB cells was 91 pA pF^{-1}. With respect to the contribution of K^+ to the membrane potential of NEB cells, we have recently shown in amperometric experiments that high K^+ (50 mM) in extracellular medium, induced serotonin (5-HT) release from the NEB cells as a result of membrane potential depolarization (71).

Characterization of O_2-Sensitive K^+ Channel

The outward K^+ currents were reversibly inhibited (~40%) by hypoxia (pO_2 ~20 mmHg) in both NEB culture and lung slices (Fig. 3). The O_2-sensitive, voltage-gated K^+ current was blocked (40%) by TEA or 4-AP. Among the O_2-sensitive K^+ currents, two types of K^+ current have been observed, a Ca^{2+}-independent ($I_{K(V)}$) and Ca^{2+}-dependent ($I_{K(Ca)}$) component. Of the O_2-sensitive K^+ current $I_{K(Ca)}$ represented ~55% and $I_{K(V)}$ ~45%, suggesting that both components were equally suppressed by hypoxia. There was no additional inhibition K^+ current in NEB cells in lung slices, with either TEA or 4-AP plus hypoxia, suggesting that TEA or 4-AP-sensitive K^+ current corresponds to O_2-sensitive K^+ current (9).

O_2 Sensor and Modulation of O_2-Sensitive K^+ Channel

We have previously reported that mRNAs for both the hydrogen peroxide (H_2O_2)-sensitive voltage-gated K^+ channel subunit (KH_2O_2) KV3.3a and membrane components of NADPH oxidase (gp^{91phox} and p^{22phox}) are coexpressed in the NEB cells of fetal rabbit and neonatal human lungs as well as related SCLC cell lines (72). Using whole-cell voltage clamp we have shown that the K^+ current of fetal rabbit NEB cells in culture exhibited inactivating properties similar to KV3.3a transcripts expressed in the *Xenopus oocyte* model (73). Exposure of NEB cells to H_2O_2, the dismuted by-product of the oxidase, under normoxia resulted in an increase of the outward K^+ current indicating that H_2O_2 could be the transmitter modulating to the O_2-sensitive K^+ channel (76) (Fig. 4). Additional evidence in support of NADPH oxidase as an O_2-sensing protein in NEB cells included: (1) demonstration of expression of important proteins ($gp91^{phox}$, $p22^{phox}$, $p47^{phox}$, rac2) in NEB cells that together constitute the multimeric NADPH oxidase complex (74); (2) hypoxia reduced rhodamine 123 fluorescence in cultured NEB cells (indicative of reduced free radical formation); (3) NADPH oxidase blocker diphenylene iodonium (DPI) blocks hypoxic K^+ channel inhibition (7,9,72) (Fig. 5). This hypothesis was further strengthened by experiments on NEB cells from wild-type (WT) and oxidase-deficient (OD) mice (75). In WT NEB cells, hypoxia ($pO_2 = 15$–20 mmHg) suppressed (~46%) K^+ currents of both Ca^{2+}-dependent and Ca^{2+}-independent components. In contrast, hypoxia had no effect on K^+ current in NEB cells from OD mice, even though both K^+ current components were expressed. These studies provided strong evidence in support of the "membrane model" of the O_2-sensing mechanism; i.e., the K^+ channel is closely associated with an O_2-sensing NADPH oxidase, and the interaction occurs via a membrane-delimited pathway (72,76).

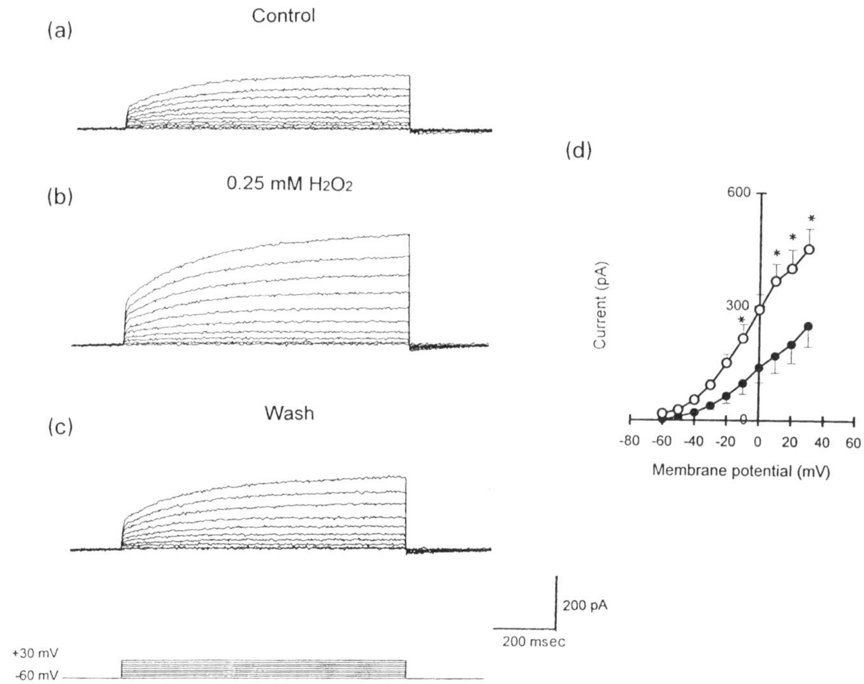

Figure 4 Effects of H_2O_2 on K^+ current. (a) Outward K^+ current evoked by depolarizing steps from −60 to +30 mV in control Krebs solution. (b) The K^+ outward currents were increased significantly ($*p < 0.05$, $n = 4$) after perfusing with 0.25 mM H_2O_2. (c) Washout of the H_2O_2 caused a recovery of the outward K^+ current. (d) I-V relationships for a sample of four cells were plotted under control conditions (●) and after perfusing with 0.25 mM H_2O_2 (○). Data are shown as ±SEM. (From Ref. 9.)

It is now clear that at least in pulmonary NEB cells, NADPH oxidase is the predominant O_2 sensor that regulates Kv3.3 function via a redox modulation of its fast amino-terminal-dependent inactivation. Recent studies on the structural organization of the cytoplasmic region of Kv channels, particularly the T1-domain segment, suggest that it is not directly attached to the membrane-associated channel protein but instead hangs as a "gondola" below a dirigible (77). This "hanging gondola" model predicts that the cytoplasmic pore access occurs through four "widows" formed by "cables" linking T1 to the S1 transmembrane helix. Based on these data and recent information on the interaction of α-β-subunits of the Kv channels, Patel and Honore (78) proposed a new model for O_2-sensor K^+-channel complex in pulmonary NEB (Fig. 6). This model incorporates parts of our earlier schema where H_2O_2-generating NADPH oxidase was closely associated with the O_2-sensitive K^+ channel and its gating involved critical cysteine residues with a "ball and chain" mechanism occluding the inner pore of the channel (72).

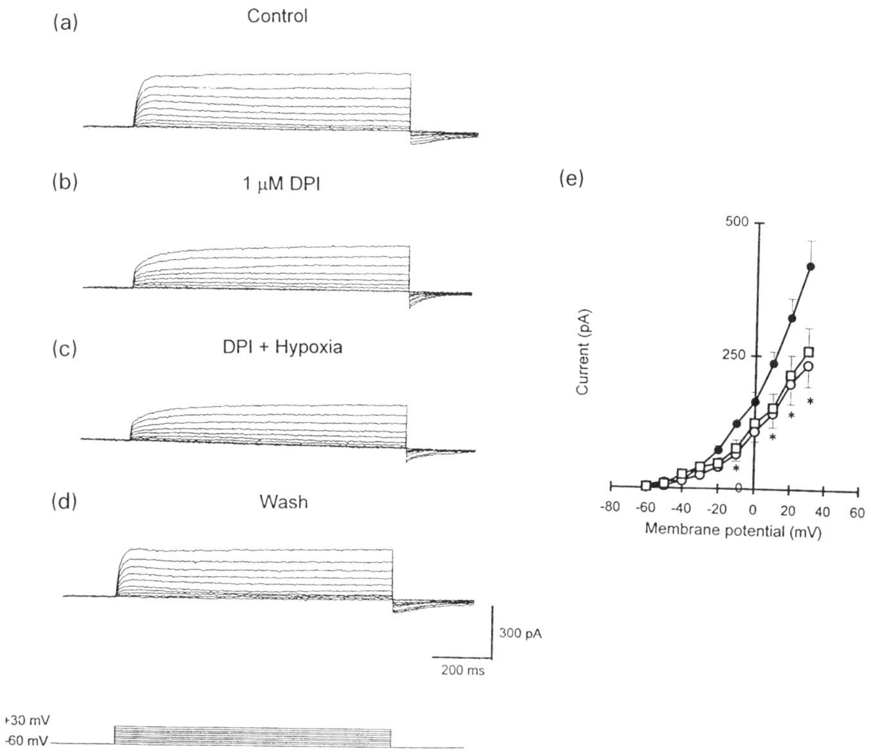

Figure 5 Effect of DPI on K^+ current. (a) Outward K^+ current evoked by depolarizing steps from -60 to $+30$ mV in control Krebs solution. (b) Outward current was reduced after bath application of 1 μM DPI. (c) K^+ current in (b) was not altered by further exposure to both DPI and hypoxia. (d) Washout of the DPI + hypoxic solution caused a recovery of the outward K^+ current. (e) I-V relationships obtained from six cells under control conditions (●), after perfusing the cells with 1 μM DPI + hypoxic solution (○). Recorded currents in DPI were significantly ($*p < 0.05$, $n = 6$) reduced relative to control; there was no significant difference between currents recorded in DPI and DPI + hypoxia. Data are shown as means ± SEM. (From Ref. 9.)

Na^+ Currents

The Na^+ current of NEB cells in rabbit neonatal lung slices exhibits fast activation and inactivation and is selectively carried by Na^+ ions, since it is completely blocked by nanomolar concentrations of tetrodotoxin. The activation threshold of I_{Na} was around -50 mV and reached maximal amplitude at -10 mV; the mean peak current was 42.1 ± 5.4 pA with the range of 15–70 pA. The Na^+ current density was 18 ± 3.9 pA pF^{-1} and hypoxia failed to modify the amplitude of the Na^+ current (9). In cultured NEB cells, Na^+ current was activated at depolarizing voltages between -20 and -30 mV; its peak was 282 pA and appeared to have both a TTX-sensitive and a TTX-insensitive component. In the current clamp recording mode, cultured

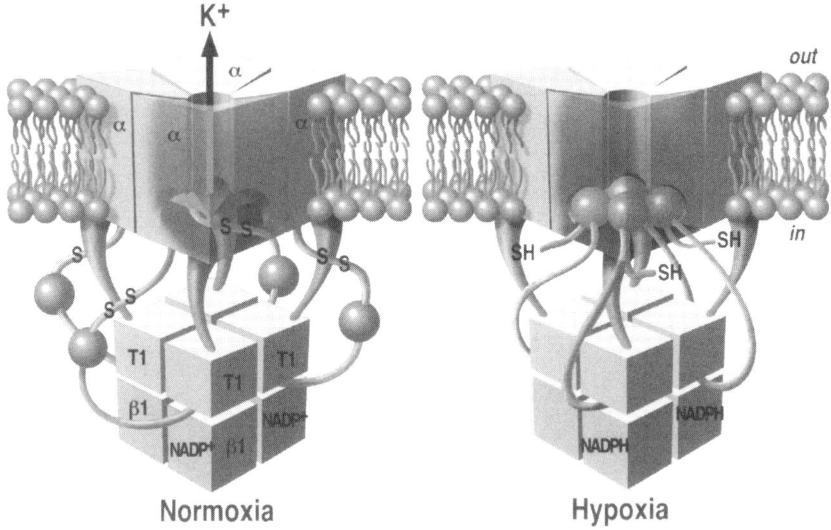

Figure 6 Molecular model of O_2 sensor complex in NEB cell. Shown is the α-β potassium (K^+) channel complex with tetramer of α-subunits forming the ionic pore. β-subunits interact with the assembly domains T1 in the cytosol. A positively charged amino terminal ball domain of the α-subunit (and possibly of the β-subunit) underlies fast inactivation. The reduced nicotinamide adenine dinucleotide phosphate (NADPH) oxidase complex is shown to be associated with this K^+-channel complex in NEB cells. Reactive oxygen intermediates produced by the NADPH oxidase modulate the inactivation process by oxidizing specific cysteine residues in the amino terminus, forming disulfide bridges with other cysteines located in the channel and thus immobilizing the inactivation balls. (From Ref. 78, Courtesy of Dr. Honore.)

NEB cells exhibited Na^+ potentials, sometimes as trains of spikes (7). Using amperometry to record directly 5-HT secretion from NEB cells in lung slices, the hypoxia-evoked 5-HT secretion was abolished in the presence of 0.1 μM TTX, a blocker of the Na^+-dependent action potential. This confirms that the hypoxic chemotransduction steps in NEBs ultimately lead to increase in frequency of cell firing (7,71).

Ca^{2+} Currents

In NEB cells from neonatal rabbit lung slices, Ca^{2+} currents have a threshold at around −48 mV, reach a peak at 0 mV, and are blocked by 100 μM Cd^{2+}. The amplitudes of the Ca^{2+} current at 0 mV ranged between 65 and 117 pA when 10 mM Ba^{2+} was used as the charge carrier, with a mean of 85 ± 9.5 pA. The current density was 36 pA pF^{-1}. Perfusion of the cells with hypoxia solution had no significant effects on the amplitude of this current (9). Using amperometry method, hypoxia-induced 5-HT release was not affected by 1 μM ω-cgTx (ω-conotoxin GVIA toxin), a specific N-type Ca^{2+} channel blocker. In contrast, 2 μM nifedipine, a specific L-type Ca^{2+} channel blocker, inhibited (90%) hypoxia-induced 5-HT release, indicating that hypoxia enhances Ca^{2+} entry into NEB cells and triggers exocytosis, primarily through L-type $Ca2^{+}$ channels (71).

Chloride Currents

Although Cl^- currents may play a role in regulation of intracellular pH and membrane potential, their role in O_2-sensing cells has not been extensively studied. Stea and Nurse (79) have demonstrated large conductance voltage-independent Cl^- channel in CB glomus cells of newborn rat. However, there are no electrophysiological studies on the effects of hypoxia on Cl^- channel activity in O_2-sensing cells. We have recently reported expression of cystic fibrosis transmembrane regulator (CFTR) and Cl^- conductances in NEB cells of neonatal rabbit lung using molecular biology and electrophysiology approaches (20). CFTR is a Cl^--channel regulated by phosphorylation and the presence of nucleotides, is localized to the apical plasma membrane of epithelial cells, including those in the lung, and participates in vectorial Cl^- secretion (80). Mutations in the CFTR gene cause cystic fibrosis, a disease characterized by reduced anion-driven fluid secretion leading to alterations in periciliary fluid, increased thickened mucus, bacterial colonization with chronic infection, and destructive lung disease (81). Since NEB are an integral part of the mucociliary epithelium, they may be involved in the pathophysiology of CF lung disease. The specific role of CFTR and Cl^- conductances in NEB cell O_2-sensing and/or neurosecretory functions is currently under investigation.

C. O_2-Sensing Properties of the SCLC Cell Line H146

Although much progress has been achieved employing NEB cells both in culture and in situ, the use of molecular biological tools as methods by which to identify specific components of the acute O_2-sensing signal transduction cascade in native tissue is inherently problematic. Furthermore, the study of chronic modulation of acute O_2 sensing by pathologically and environmentally relevant stimuli is difficult in native systems where transcription and translation are not constant over time in culture. One way in which to address such shortcomings has been to establish an immortalized model of NEB cells, and one particular SCLC cell line, H146, has proved particularly useful in this regard (see Ref. 82 for recent review).

The Role of NADPH Oxidase as O_2-Sensing Protein

In both rodent NEBs (7,9) and H146 cells (11,83,84), hypoxia causes K^+-channel inhibition and cell depolarization (Fig. 7) and the key upstream event in airway O_2 sensing is modulation of cellular redox potential via substrate-limited generation of reactive oxygen species (ROS) by NADPH oxidase (72,75,83,84). The NADPH oxidase model predicts that the tonic and sequential generation of superoxide and

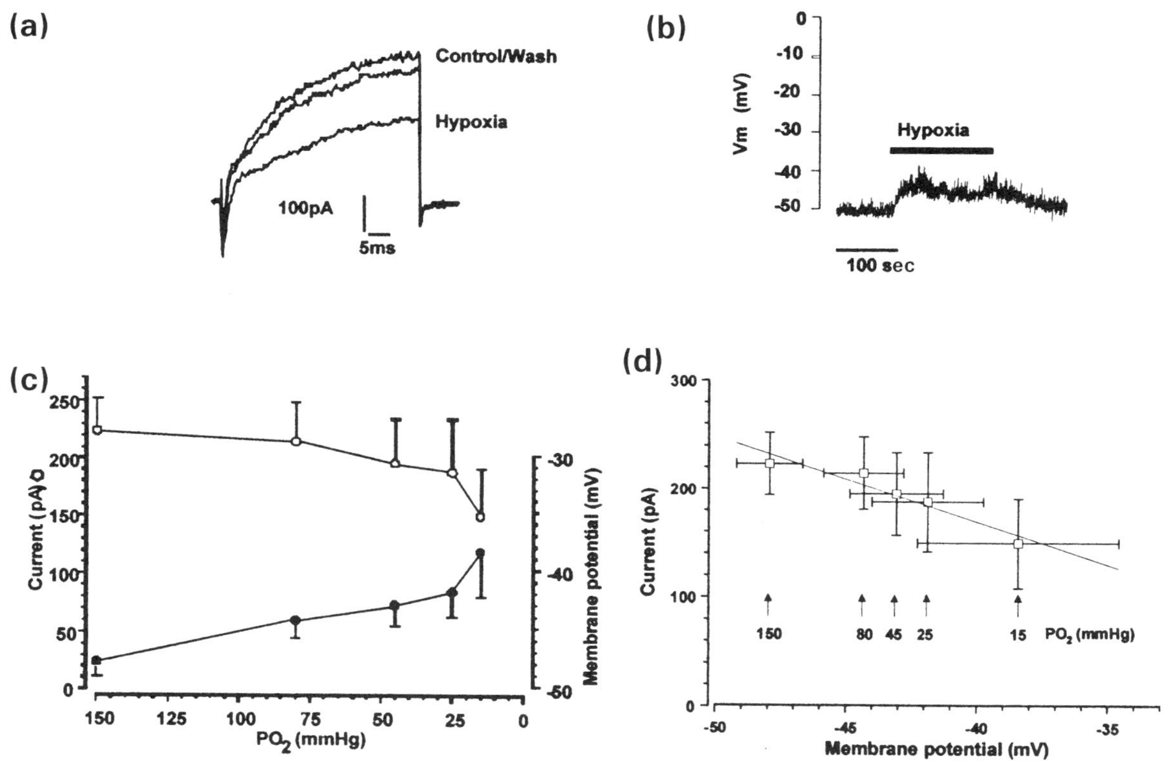

Figure 7 O_2 sensitivity of H146 cells. (a) Exemplar whole-cell current traces before (control), during (hypoxia), and following (wash) bath perfusion with hypoxic solution ($pO_2 \sim 30$ mmHg) when employing a voltage-step protocol from −70 mV to 0 mV. (b) A typical depolarization response to acute hypoxia in current clamp mode. (c) Mean current amplitudes (at 0 mV) (○) and membrane potentials (●) during graded hypoxic challenge as shown on the x-axis. (d) Correlation between membrane potential and K^+ current magnitude at each grade of hypoxia.

H_2O_2 from environmental O_2 under normoxic conditions should promote K^+-channel activity and that any reduction in H_2O_2 levels, duc to reduced pO_2 would, therefore, lead to K^+-channel closure and cell depolarization.

The similarities between native NEB cell and H146 cell O_2 signal transduction are striking. Thus, during hypoxia, depression of H146 cellular H_2O_2 levels can be clearly demonstrated using the fluorescent dye, 2′,7′-dichlorodihydrofluorescein diacetate (83). Central to the NADPH oxidase model is the idea that the primary effector (a K^+-channel protein) must be able to respond to H_2O_2. In H146 cells, the O_2-sensitive K^+ current is 4-aminopyridine-insensitive (11) and activation of this 4-AP-resistant K^+ current by H_2O_2 is observed during hypoxic challenge (83); like native NEB cells, this response is transient. The key support for NADPH as the primary sensor in H146 cells comes from experiments that exploited the fact that NADPH oxidase activity can be up-regulated by protein kinase C (PKC)-dependent phosphorylation of two components of the enzyme complex, $p67^{phox}$ and $p47^{phox}$ (85). Treatment of H146 cells with the phorbol ester, 12-*O*-tetradecanoylphorbol-13 acetate (TPA), activates PKC, which increases the turnover of NADPH oxidase via translocation of the $p47^{phox}$ and $p67^{phox}$ subunits to the plasma membrane. Such activation results in augmented NADPH-oxidase-dependent H_2O_2 production, which has little effect on K^+-channel activity in normoxia since the channels appear maximally active. However, in hypoxia, PKC activation results in a dramatic amelioration of the K^+-current depression simply because the cellular H_2O_2 levels cannot be reduced to a level normally associated with hypoxia in the absence of PKC activation (Fig. 8) (83). In other words, PKC activation results in maintenance of a relatively oxidized K^+-channel environment even during hypoxic challenge, a process that is consistent with a PKC-dependent increase in O_2 affinity.

The observation that K^+ currents recorded from NEB cells of OD mouse lung are insensitive to acute hypoxia (75) is at slight odds with data from H146 cells that suggest that human SCLC cells express a second, but converging, O_2-sensing system (84). Thus, in the H-146 cell model maximal inhibitory concentrations of either of two structurally unrelated inhibitors of NADPH oxidase activity, diphenylene iodonium (DPI) and phenylarsine oxide (PAO), are unable to abolish completely the hypoxic depression of K^+ currents (84). Whether the quantitative contributions of this oxidase versus an additional sensory pathway (which may be different in neonatal rabbit NEB cells and the human tumor cell model) are a result of species differences or cell type remains to be investigated.

It is clear that the NEB cell primary O_2 sensor is NADPH oxidase, while in the other O_2-sensing cells (notably, the pulmonary vasculature and carotid body) it is not. However, one controversial mechanism that has been implicated in a variety of O_2-sensitive tissues (86), including pulmonary arteriolar smooth muscle cells (87), is modulation of mitochondrial production of reactive oxygen species. However, in H146 cells at least, mitochondrial involvement is at best limited since inhibition of the mitochondrial respiratory chain by myxothiazol as well as ρ^0 cells (cells genetically manipulated to be free of mitochondria) does not demonstrate significant perturbation in their acute hypoxic response (Fig. 9a–d).

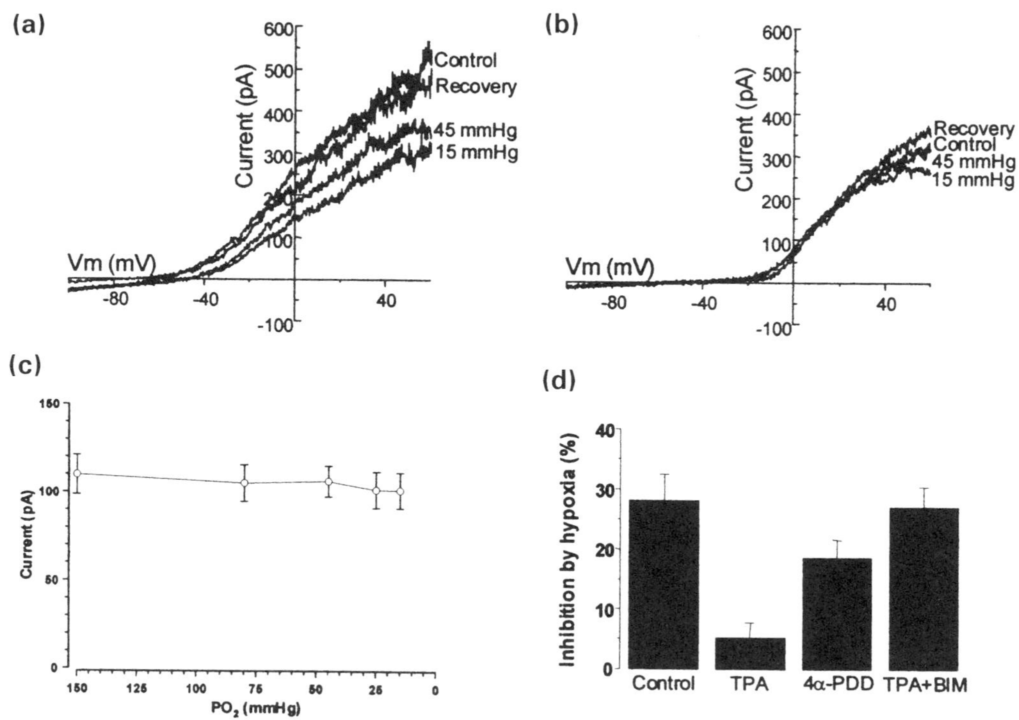

Figure 8 Hypoxia signal transduction in H146 cells. (a) Exemplar I-V relationships during normoxia and two grades of hypoxia (45 mmHg and 15 mmHg) in untreated and (b) TPA pretreated cells. Currents were recorded during 1-sec ramp depolarizations from a holding potential of −70 mV. (c) Plots of mean outward K^+ current recorded from TPA-treated cells during graded reduction in pO_2 from 150 mmHg to 15 mmHg. (d) Mean current inhibition by hypoxia at 0 mV with no bath additions (control) and following pretreatment with the agents shown beneath each bar. (From Ref. 83.)

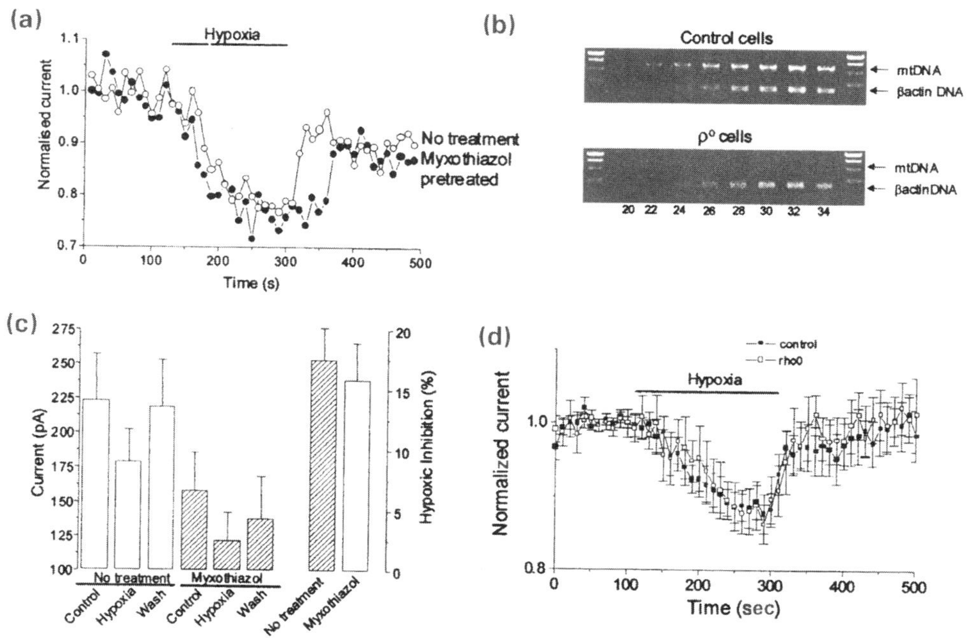

Figure 9 Lack of effects of hypoxia in H146 cells treated with mitochondrial respiratory chain inhibitor and in mitochondria-free cells. (a) Time courses of normalized K^+-current depression by acute hypoxia (30 mmHg) in typical H146 cells following preincubation with vehicle (0.01% DMSO) (no treatment) or myxothiazol (100 nM), a potent inhibitor of mitochondrial respiratory chain at cytochrome b-c. The duration of application of hypoxia is indicated by the horizontal bar. (b) The effects of hypoxia on mean current in untreated (open bars) and myxothiazol-treated (hatched bars) H146 cells. Mean normalized hypoxic inhibitions in the two groups of cells are shown by the bars on the right. (c) Exemplar semiquantitative RT-PCR reactions of control (upper gel) and cells (lower gel) using primer pairs directed against mitochondrial DNA (mtDNA) (upper band) and βactin DNA (lower band). Right lane in both gels shows DNA size markers of 150, 300, 500, and 1000 bp. Numbers to the bottom of the lower gel indicate the number of cycles and apply to both gels. (d) Mean time courses of hypoxia-evoked inhibition of K^+ currents recorded from control (solid symbols) and cells (open symbols). The period of perfusion with hypoxic (30 mmHg) solution is indicated by the horizontal bar.

Tandem P-Domain K^+ Channels and Their Role in O_2 Sensing

The amenability of H146 cells to electrophysiological study and molecular intervention has provided insights into the physiological importance and molecular identity of the O_2-sensitive K^+ channel. Thus, even in the presence of a maximally effective concentration of 4-AP, hypoxia causes membrane depolarization in H146 cells, demonstrating that the O_2-sensitive K^+ channels influence cell membrane potential. The pharmacological profile of this channel, coupled with its ability to govern membrane potential, suggests that it belongs to the newly emerging tandem P-domain K^+-channel (K_{2P}) family (88). To address this in more detail, the H146 model has allowed rapid screening for the presence of mRNA encoding eight separate members of this family (89), an approach that would be technically very difficult in native NEB cells. Employing primer pairs directed against the published sequences of human K_{2P} channels (TWIK1 and 2, TASK1, 2, and 3, TREK1 and 2, and TRAAK), only hTWIK1 and hTRAK were not amplified from DNase-treated, reverse-transcribed H146 cell mRNA (90). Further pharmacological profiling of the O_2-sensitive K^+-current relative resistance to tetraethylammonium (11,89) arachidonate—and halothane sensitivity and dithiothreitol resistance (89)—together with our observations of Zn^{2+} and 4-AP resistance (11) suggest that TASK3 underlies the O_2-sensitive K^+ current. Most convincingly, using a molecular abrogation approach, we found that an antisense probe directed against TASK3 and TASK1 (TASK1 involvement was already dismissed owing to the Zn^{2+} resistance of the O_2-sensitive current) was able to abolish almost completely TASK transcription/translation and this protein knockdown correlated with ablation of the hypoxic K^+-channel inhibition in these cells; missense and lipofectamine-only treated cells retained their ability to respond to hypoxia (Fig. 10) (89). Thus, the O_2-sensitive channel in this immortalized model of NEB cells is the K_{2P} channel, TASK3 (89). Whether this channel is expressed in native NEB cells is still under investigation.

There is also evidence for the involvement of K_{2P} channels in O_2 signal transduction in other chemosensory tissues. Thus, TASK1 has been implicated indirectly in carotid body glomus cell O_2 sensitivity (90) and suggested to be involved in maintenance of resting membrane potential in isolated pulmonary artery smooth muscle cells (91). These are exciting electrophysiological observations in native tissues. They now have a molecular and functional correlate as evidenced by our recent recombinant studies demonstrating directly that human TASK1, when expressed in HEK293 cells (92), demonstrates both the expected pH sensitivity and, most importantly, is reversibly and rapidly inhibited, in a pH-sensitive manner, as O_2 availability is reduced (92).

D. Modulation of O_2 Sensing by Chronic Hypoxia

Healthy individuals undergo adaptive responses to prolonged hypoxia when acclimatizing to high altitude. Most recently, evidence has emerged that even short-duration exposures to very mild hypoxia, such as those commonly associated with the reduction in cabin pressure during commercial airline flight, can induce an acclimatization response (93). Individuals with a variety of cardiorespiratory

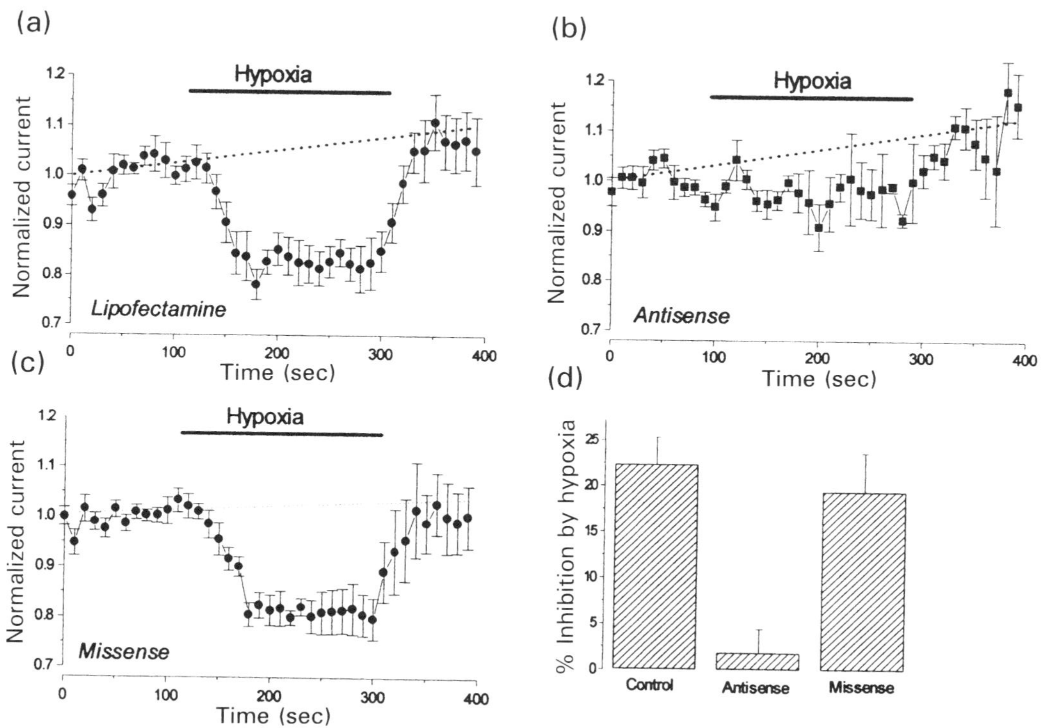

Figure 10 Molecular identity of K^+ channel in H146 cells. (a) Mean time courses of the effect of acute hypoxia (applied for the periods indicated by the horizontal bar) on cells treated only with lipofectamine, (b) on cells treated with missense oligodeoxynucleotide, and (c) on cells treated with antisense oligodeoxynucleotide directed against hTASK3. (d) A bar graph plots the mean hypoxic inhibition in lipofectamine-only treated, hTASK antisense-, or missense oligodeoxynucleotides-treated cells, as indicated. (From Ref. 89.)

diseases that cause a prolonged reduction of arterial pO_2 (e.g., chronic obstructive pulmonary disease, apnea of sleep) will also undergo adaptive responses. Such changes have been demonstrated in animal models; for example, selective suppression of voltage-gated K^+-channel expression has been documented in pulmonary vascular smooth muscle cells in chronic hypoxia, an effect that may account for pulmonary hypertension associated with chronic hypoxia (94). Such an effect may be offset, or its severity lessened, by resetting of certain pulmonary hypoxic reflexes—reduction in the acute O_2 sensitivity of NEBs may, therefore, lessen the deleterious effect of pulmonary smooth muscle hypoxic vasoconstriction by reducing the secretion of 5-HT into the pulmonary vasculature. Similarly, altered expression of ion channels in carotid body glomus cells has been reported when these cells are kept under hypoxic conditions, either in vivo (95) or in vitro (96), suggesting this is a widespread phenomenon among O_2-sensing cells. Cellular responses to chronic hypoxia have not been investigated in NEBs, but NEB cell hyperplasia is associated with apnea of prematurity and SIDS (97), which suggests strongly that they are likely to undergo some form of remodeling as an adaptive response to chronic changes of ambient pO_2. If chronic hypoxia were able to modulate acute O_2 sensitivity of airway chemoreception, there are a number of points along the transduction pathway that could be remodeled; these include the O_2 sensor (NADPH oxidase), the K^+ channel effector (TASK3), and the Ca^{2+} influx pathways (principally voltage-gated Ca^{2+} channels). In H146 cells, we have evidence for remodeling of both K^+- and Ca^{2+}-channel functional expression. H146 cells cultured in chronic hypoxia (71 mmHg for 24 hr) display: (1) decreased K^+-current density (98); (2) reduced K^+-current sensitivity to graded, acute hypoxia (98); (3) reduced voltage-gated Ca^{2+} entry (99); and (4) differential regulation of Ca^{2+}-entry pathways (99).

Thus, employing competitive polymerase chain reaction assays, we have shown that the down-regulation of both acute O_2 sensitivity (Fig. 11a) and K^+-current density (Fig. 11b) is not associated with reduced steady-state mRNA levels of either hTASK1 or TASK3, as assessed using competitive PCR (98). This is a surprising result, but suggests that chronic hypoxia is principally a posttranscriptional regulatory signal. In addition to a decrease in K^+ channel sensitivity, Ca^{2+} influx is significantly attenuated (Fig. 11c). This reduction is associated with a switch from Ca^{2+} influx through L-type Ca^{2+} channels to influx principally through N-type Ca^{2+} channels (Fig. 11d). Thus, the overall effect of chronic hypoxia is to reduce the cells' potential to release vasoactive amines during acute hypoxic insult, a response that could be an important protective mechanism during periods of prolonged hypoxia such as occur in chronic lung disease.

IV. Mechanisms of Chemotransduction, Pre- and Postsynaptic Receptors

The candidate neurotransmitters that mediate fast chemosensory transmission of hypoxia signal from NEB cells via vagal sensory afferents include 5-HT, acetylcholine (ACh), and adenosine triphosphate (ATP) as recently proposed for CB (100). Although it is well established that NEB synthesize and release 5-HT, the evidence for the role of ACh and ATP is at present tentative. The evidence for a cholinergic mechanism in NEB includes: (1) demonstration of acetylcholinesterase activity (101); (2) the presence of small clear vesicles in efferent-like nerve endings by EM; and (3) immunohistochemical demonstration of VChat in NEB cells (unpublished observation). The possible involvement of ATP and purinergic receptors is based on histochemical demonstration of quinicrine fluorescence in NEB cells and $P2/_{X3}$ immunoreactivity of NEB nerve endings (52).

The evidence for the role of 5-HT as a transmitter of hypoxia stimulus in NEB cells was suggested by earlier in vivo and in vitro studies (3,4,6). Recently, using amperometry with carbon fiber electrodes we have confirmed hypoxia induced 5-HT release from intact NEB in newborn rabbit lung slices (71). Although under normoxia (pO_2 ~155 mmHg) there was no detectable 5-HT release, hypoxia elicited a dose-dependent response in a pO_2 range from mild to severe hypoxia (pO_2 95–18 mmHg) (Fig. 12a,b). It is of interest that pO_2 in tracheal air is ~149 mmHg whereas in intrapulmonary airways and alveoli pO_2 is ~104 mmHg. Since NEB are distributed only within intrapulmonary airways up to the bronchiole-alveolar junction (Fig. 1), these physiological pO_2 values (~104 mmHg) would be expected to invoke "spontaneous" (basal) 5-HT release. Thus 5-HT in NEB cells fulfills the basic requirements for a function as a neurotransmitter, i.e., demonstration of spontaneous and physiological stimulus-evoked release (102). It would be expected that hypoxia-induced release of 5-HT from NEB cells would lead to activation of post- and presynaptic (auto) receptors. It has been shown previously that in the rabbit, activation of 5-HT receptors results in depolarization of both nodose neurons and isolated vagus nerve (103). Furthermore, the electrophysiological properties of 5-HT receptors on rabbit nodose neurons have been well characterized (104).

A. 5-HT-3 Receptors

Recently we reported expression and functional characterization of 5-HT3-receptor (5-HT3-R), a ligand-gated ion channel in NEB cells (105). Using nonisotopic in situ hybridization, expression of 5-HT3-R mRNA was localized in the cytoplasm of NEB cells of different mammals. By means of dual immunohistochemistry for 5HT and 5-HT3-R the two epitopes were colocalized in NEB cells, with 5-HT3-R immunoreactivity localized to the plasma membrane The electrophysiological and pharmacological properties of 5-HT3-R were studied in NEB from neonatal hamster lung slices. The application of 5-HT or its agonist (2-methyl-5-HT) induced inward currents in a concentration-dependent manner. The 5-HT-induced current was blocked by a specific 5-HT3-R blocker (ICS-205-930) whereas blockers of other 5-HT receptors had no significant effect. We have also demonstrated that the 5-HT3-R blocker inhibited hypoxia-evoked 5-HT secretion, suggesting that 5-HT3-R in NEB cells functions as an autoreceptor with positive feedback modulation of hypoxia signaling (105).

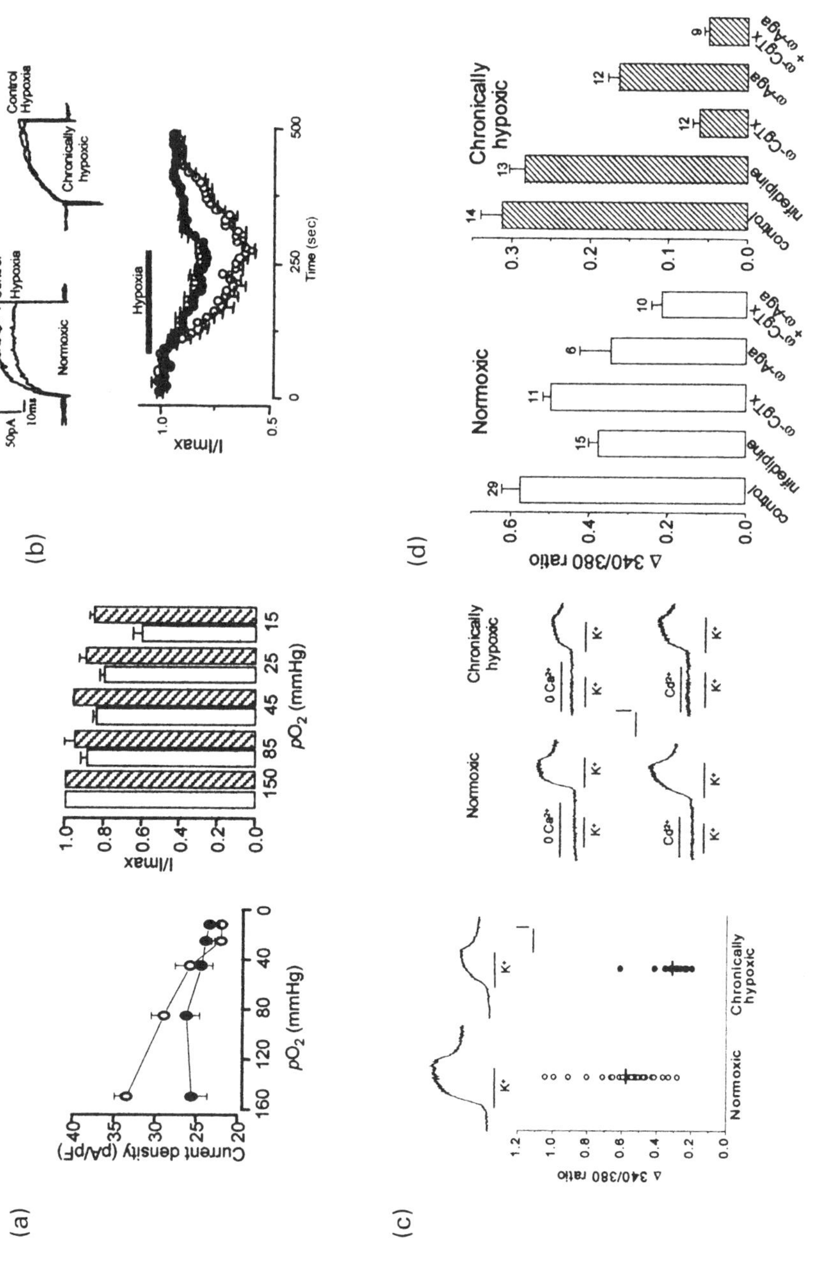
(a)
Current density (pA/pF)
pO2 (mmHg)
I/Imax
(b)
Control
Hypoxia
Normoxic
Chronically hypoxic
Time (sec)
(c)
Δ 340/380 ratio
Normoxic
Chronically hypoxic
(d)
Normoxic
Chronically hypoxic
control
nifedipine
ω-CgTx
ω-Aga
ω-CgTx + ω-Aga

Figure 11 Effects of chronic hypoxia on regulation of H146 K^+ currents, K^+-current O_2 sensitivity, and Ca^{2+}-influx pathways. (a) The left panel shows the effect of acute, graded hypoxic challenges upon K^+ current density in H146 cells maintained at 37°C for 24 hr in normoxia (142 mmHg) (open symbols) or chronic hypoxia (71 mmHg) (solid symbols). The right panel shows the mean normalized current responses to graded hypoxia in normoxic (open bars) and chronically hypoxic (hatched bars) cells. (b) Top panels show exemplar currents recorded at 0 mV (from a holding potential of −70 mV) before (control) and during (hypoxia) acute hypoxic challenge in cells maintained for 24 hr in normoxia (top left) or chronic hypoxia (top right). The lower panel shows the mean time courses of the acute hypoxic inhibition of K^+ currents in normoxic (open symbols) and chronically hypoxic (solid symbols) cells. (c) Upper-right panel shows example recordings of $[Ca^{2+}]_i$ rises in response to cell depolarization evoked by application of 50 mM K^+ (K^+ applied for the periods indicated by the horizontal bars). Recordings were made from normoxic and chronically hypoxic cells (as indicated below bar graph). Also plotted (lower-left panel) are peak rises of $[Ca^{2+}]_i$ evoked by 50 mM K^+ in all normoxic cells (open symbols) and all chronically hypoxic cells (solid symbols). Right panels show typical effects of Ca^{2+}-free perfusate on K^+-evoked rises of $[Ca^{2+}]_i$ in normoxic and chronically hypoxic (top right) cells, as indicated. Also shown are typical effects of 200 μM Cd^{2+} on K^+-evoked rises of $[Ca^{2+}]_i$ in normoxic and chronically hypoxic cells (lower right), as indicated. For all examples, the first application of 50 mM K^+ was during perfusion with Ca^{2+}-free or Cd^{2+}-containing perfusate and the second application was following reperfusion with normal bath solution (as indicated). Scale bars: vertical, 0.2 ratio units; horizontal, 30 sec. These apply to all traces. (d) Mean rises of $[Ca^{2+}]_i$ recorded in normoxic cells (open bars, left panel) and chronically hypoxic cells (hatched bars, right panel) in response to application of perfusate containing 50 mM K^+. Recordings were made in the absence of Ca^{2+}-channel blockers (control), or in the presence of 2 μM nifedipine, or after preincubation with 1 μM ω-conotoxin GVIA or 200 nM ω-agatoxin GIVA, or both toxins, as indicated. Each bar represents the mean response (with vertical SEM bar), taken from the number of recordings shown above each bar.

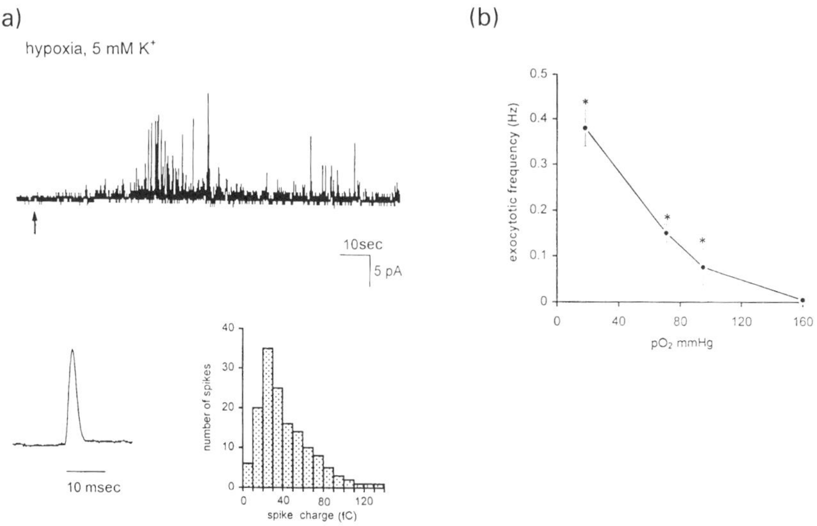

Figure 12 Amperometric recording of hypoxia-induced 5-HT release from NEB cell in neonatal rabbit lung slice preparation. (a) Example of exocytosis with 5-HT release induced by perfusion with a hypoxia solution (pO_2 ~18 mmHg). At the point indicated by the arrow, the perfusate was changed to hypoxia solution. Large spike-like exocytic events are shown at an expanded time base (bottom left). Also shown is frequency distribution of the charge of secretory events evoked by hypoxia in six different cells (bottom right). (b) Graded response to hypoxia with four different pO_2 levels on secretory response from NEB cells. Each point represents the mean values of between 8 and 18 cells (mean ± SEM). (From Ref. 71.)

B. Nicotinic Acetylcholine (nACh) Receptor

There is a great deal of data linking smoking to SCLC in humans (106) and on effects of nicotine on metabolic and proliferative activity of PNEC/NEB in experimental animals or in vitro models (107). We have recently studied the expression and functional characterization of ACh receptors in NEB cells using a combination of in situ hybridization, immunohistochemistry, and electrophysiology (108). The expression of mRNA as well as immunoreactivity for the β2-subunit of nACh receptor was localized in NEB cells of hamster neonatal lung. In whole-cell patch-clamp study, both nicotine and ACh induced transient inward currents that were concentration and voltage dependent (Fig. 13A). Under current clamp, nicotine (50 μM) induced rapid depolarization followed by repolarization (Fig. 13B). The effects of nicotine-induced current were abolished by mecamylamine (50 μM), an ACh-receptor blocker. No significant change in ACh-induced current was observed under either mild (pO_2 ~ 75 mmHg) or severe (pO_2 ~20 mmHg) hypoxia (Fig. 13A). This finding is in contrast to that reported for CB glomus cells, where even a slight decrease in pO_2 enhanced significantly ACh-induced inward current (109). Our findings indicate that the nACh receptor is a predominant cholinergic receptor in

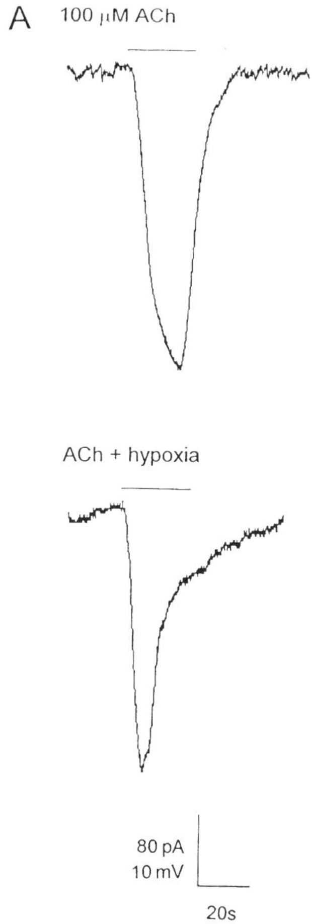

Figure 13 (A) Nicotinic ACh receptor in NEB cells of neonatal hamster lung. ACh-induced inward current under normoxia and hypoxia conditions ($pO_2 = 20$ mmHg. Holding potential was −60 mV). (B) Application of nicotine evoked an inward current (a). Holding potential was −60 mV. Effects of holding potential on inward currents evoked by 50 μM nicotine. Each plotted point is the mean peak inward current amplitude taken from between five and eight cells at each holding potential. (c) Nicotine evoked a membrane potential depolarization. (d) The peak currents evoked at each concentration are expressed relative to the peak current evoked by 50 mM nicotine and plotted against the log [nicotine]; mean response taken from five to eight cells. The experimental data were fitted by the Hill equation with a Hill coefficient of 0.9 and $EC_{50} = 4$ μM.

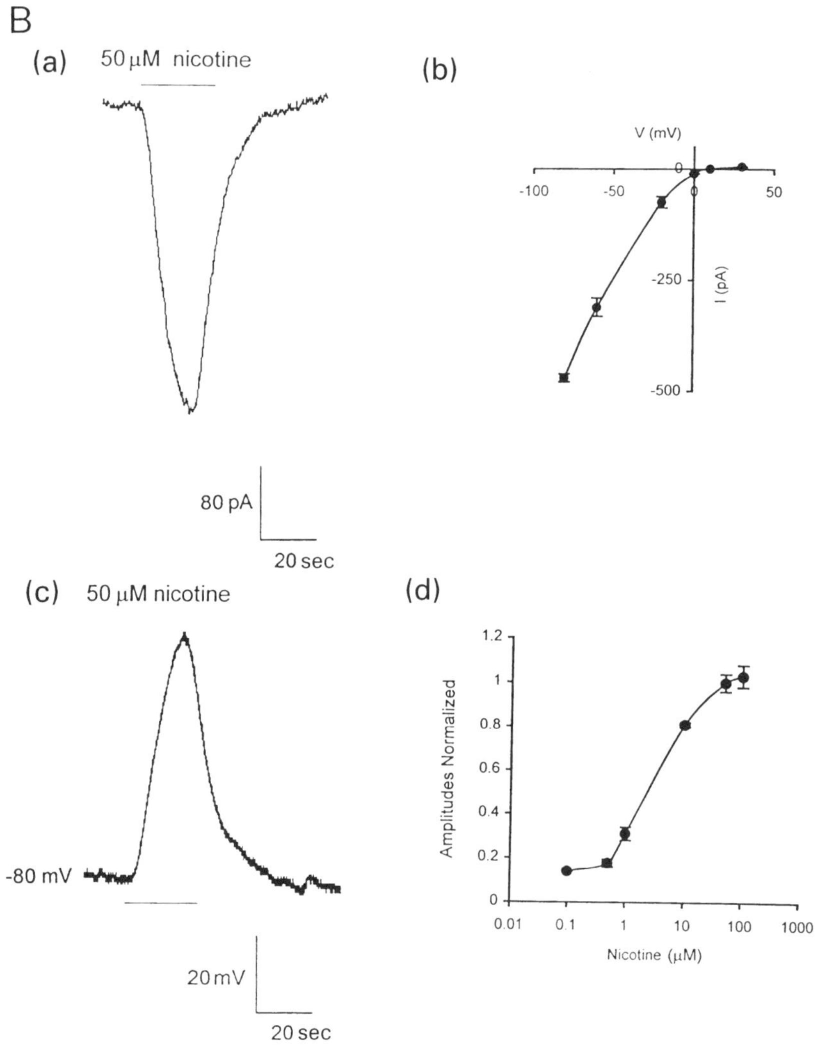

Figure 13 Continued.

NEB cells and that its activation could lead to excitation of NEB cells causing membrane depolarization and opening of voltage-gated Ca^{2+} channels followed by neurotransmitter release, a fundamental step in the chemotransduction process.

V. Summary and Conclusions

Our updated working model for mechanism of chemotransduction in NEB cells is shown in Figure 14. Generally, this model is similar to other O_2-sensitive

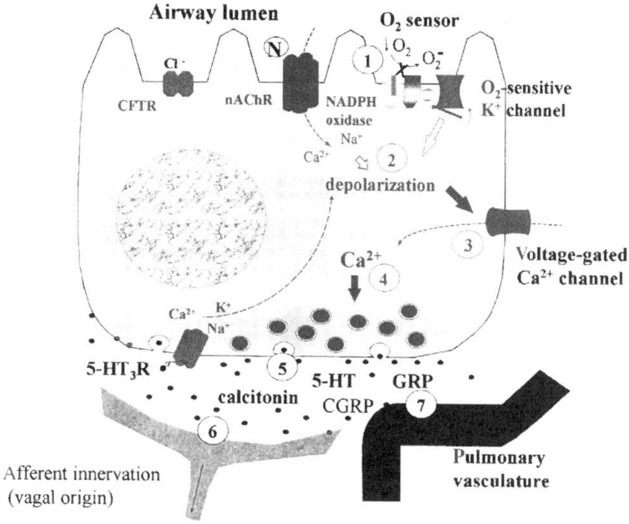

Figure 14 Proposed mechanism of chemotransduction of acute hypoxia in pulmonary NEB cells. (1) Hypoxia leads to decreased NADPH oxidase activity with reduced reactive oxygen species (ROS) generation (i.e., O_2^- and H_2O_2). This in turn alters the redox status of the O_2-sensitive K^+ channel causing its inactivation or closure (Fig. 6). In native NEB cells the predominant O_2-sensitive K^+ channel is represented by a voltage-activated Shaw-related KV3.3a (72). In H146 cells an additional TASK-like channel with O_2 sensitivity may represent an NADPH oxidase-independent pathway (84). The convergence of these diverse pathways in NEB cells has been proposed (82). (2) Closure of the O_2-sensitive K^+ channel leads to membrane depolarization allowing the NEB cell to reach threshold faster and thus increasing its spiking frequency. (3) The increase in spike activity causes waves of depolarization to spread over the cell membrane, activating voltage-gated Ca^{2+} channels (4) and leading to Ca^{2+} influx followed by exocytosis with neurotransmitter release (5). The various amine and neuropeptide transmitters packaged in DCV (i.e., 5-HT, GGRP, calcitonin, GRP, etc.) (Table 1) are likely coreleased and may reach their respective target(s). Binding of 5-HT to vagal sensory afferents in contact with NEB cells relays information to the brainstem respiratory center to affect ventilation (6). In addition, 5-HT and/or CGRP and other peptides may also affect the pulmonary vasculature (7) or exhibit local paracrine activity. NEB cell function may be further modulated by activation of neurotransmitter receptors (autoreceptors) expressed on NEB cell membrane. For example 5-HT_3 receptors, a ligand-gated ion channel, may provide a positive feedback, amplifying the hypoxia signal (106). On the other hand, nACh receptor does not appear to be directly involved in O_2 sensing by NEB cells. However, exposure to nicotine (as during smoking) is facilitated by a direct airway contact of NEB cell apical membrane, and thus activation of an nACh receptor may indirectly affect NEB function. Expression of CFTR and Cl conductance in NEB cells is of interest as this membrane channel may affect both O_2 sensing and neurosecretory function (20). (Modified from Refs. 15 and 25.)

neurosecretory cells (110) with some exceptions. For example, there is now strong evidence for direct participation of NADPH oxidase in O_2 sensing in NEB cells and the related H146 cell line (72,83), in contrast to other O_2-sensing cells (i.e., carotid body, pulmonary artery, smooth muscle cells) where alternate mechanisms have been proposed (110). This diversity and complexity of O_2-sensing mechanisms may reflect local modification and adaptation to the level and range of hypoxia, developmental stage, or coupling to different effector mechanisms. In spite of significant recent advances in cellular and molecular biology of NEB cells, their precise function in the lung remains to be defined. Based on the many similarities with arterial chemoreceptors, a role for NEB as airway O_2 sensors affecting respiratory control, particularly during the neonatal period, has been proposed. At present there is only circumstantial evidence indicating possible interaction between these two chemoreceptor systems, complementing rather than duplicating their function. For example, in central hypoventilation syndrome, there is hyperplasia of NEB perhaps to compensate for hypoplastic CB (111). Furthermore, we have observed recently that neonatal OD mice whose NEB cells are unresponsive to hypoxia exhibit an abnormal pattern of respiration not found in adult OD mice (112,113). It is hoped that our expanding knowledge base on the cellular and molecular mechanisms of NEB cells will facilitate future experiments to elucidate their physiological functions as well as involvement in the pathophysiology of various disease processes.

Acknowledgments

The authors thank Jie Pan for providing the original photograph for Figure 2. We also thank Mike Starr for preparation of figures and diagram (Fig. 14). This work was supported by grants from the Canadian Institutes for Health Research (MOP 12742 and MGP 15270) to EC and HY, and The Wellcome Trust and British Heart Foundation to PJK and CP.

References

1. Lauweryns JM, Cokelaere M, Theunuyck P. Neuroepithelial bodies in mammalian respiratory mucosa of various mammals: light optical, histochemical and ultrastructural investigation. Z Zellforsch 1972; 135:569–592.
2. Sorokin SP, Hoyt RF. Neuroepithelial bodies and solitary small-granule cells. In: Massaro D, ed. Lung Cell Biology. New York: Marcel Dekker, 1989:191–344.
3. Lauweryns JM, Cokelaere M. Hypoxia sensitive neuroepithelial bodies, intrapulmonary secretory neuroreceptors undulated by the CNS. Z Zellforsch 1973; 145:521–540.
4. Lauweryns JM, Cokelaere M, Lerut T. Cross-circulation studies on the influence of hypoxia and hypoxaemia on neuroepithelial bodies in young rabbits. Cell Tissue Res 1978; 193:373–386.
5. Cutz E, Yeger H, Wong V, Bienkowski E, Chan W. In vitro characteristics of pulmonary neuroendocrine cells from rabbit fetal lung. I. Effects of culture media and nerve growth factor. Lab Invest 1985; 53:672–683.

6. Cutz E, Speirs V, Yeger H, Newman C, Wang D, Perrin DG. Cell biology of pulmonary neuroepithelial bodies—validation of an in vitro model. Anat Rec 1993; 236:41–52.
7. Youngson C, Nurse CA, Yeger H, Cutz E. Oxygen sensing in airway chemoreceptors. Nature 1993; 365:153–155.
8. Gonzalez C, Almaraz L, Obeso A, Rigual R. Carotid body chemoreceptors: from natural stimuli to sensory discharges. Physiol Rev 1994; 74:829–898.
9. Fu XW, Nurse CA, Wang YT, Cutz E. Selective modulation of membrane currents by hypoxia in intact airway chemoreceptors from neonatal rabbit. J Physiol 1999; 514:139–150.
10. Van Lommel A, Lauweryns JM, Berthoud HR. Pulmonary neuroepithelial bodies are innervated by vagal afferent nerves: an investigation with in vitro antergrade DiI tracing and confocal microscopy. Anat Embryol 1998; 197:325–330.
11. O'Kelly I, Stephens RH, Peers C, Kemp PJ. Potential identification of the O_2-sensitive K^+ current in human neuroepithelial body-derived cell line. Am J Physiol 1999; 276:L96–L104.
12. Adriansen D, Scheuermann DW. Neuroendocrine cells and nerves of the lung. Anat Rec 1993; 236:70–85.
13. Cutz E. Cellular and Molecular Biology of Airway Chemoreceptors. Austin, TX: Landes Bioscience and Chapman & Hall, 1997:1–33.
14. Van Lommel A, Bolle T, Fannes W, Lauweryns JM. The pulmonary neuroendocrine system: the past decade. Arch Histol Cytol 1999; 62:1–16.
15. Cutz E, Jackson A. Neuroepithelial bodies as airway oxygen sensors. Respir Physiol 1999; 115:201–214.
16. Cutz E, Chan W, Sonstegard K. Identification of neuroepithelial bodies in rabbit fetal lungs by scanning electron microscopy: a correlative light, transmission and scanning electron microscopic study. Anat Rec 1978; 192:459–466.
17. Hung KS, Chapman AL, Mestermacher MA. Scanning electron microscopy of bronchiolar neuroepithelial bodies in neonatal mouse lungs. Anat Rec 1979; 193:913–926.
18. Lauweryns JM, Van Lommel A, Dom RJ. Innervation of rabbit intrapulmonary neuroepithelial bodies: quantitative and qualitative ultrastructural study after vagotomy. J Neurol Sci 1985; 65:81–92.
19. Lauweryns JM, Van Ranst L. Immunohistochemical localization of aromatic-L-amino acid decarboxylase in human, rat and mouse broncho-pulmonary and gastro-intestinal endocrine cells. J Histochem Cytochem 1988; 36:1181–1186.
20. Yeger H, Pan J, Fu XW, Bear C, Cutz E. Expressions of CFTR and Cl-conductances in pulmonary neuroepithelial bodies. Am J Physiol 2001; 281:L713–L721.
21. Wharton J, Polak JM, Bloom SR, Ghatei MA, Solcia E, Brown MR, Pearse AGE. Bombesin-like immunoreactivity in the lung. Nature 1978; 273:769–770.
22. Sunday ME, Kaplan LM, Motoyama E. Gastrin-releasing peptide (mammalian bombesin) gene expression in health and disease. Lab Invest 1988; 59:5–24.
23. Cutz E, Gillan JE, Track NS. Pulmonary endocrine cells in developing human lung and during neonatal adaptation. In: Becker KL, Gazdar AF, eds. The Endocrine Lung in Health and Disease, Philadelphia: Saunders, 1984:210–231.
24. Wang D, Yeger H, Cutz E. Expression of gastrin-releasing peptide receptor gene in developing lung. Am J Respir Mol Biol 1996; 14:409–416.
25. Sunday ME, Cutz E. Role of neuroendocrine cells in fetal and postnatal lung. In: Meldelson CR, ed. Endocrinology of the Lung. Totowa, NJ: Humana Press, 2000:299–336.

26. Tsutsumi Y. Immunohistochemical analysis of calcitonin and calcitonin-gene related peptide in human lung. Hum Pathol 1989; 20:869–902.
27. Rosenfeld MG, Mernod JJ, Aurora SG. Production of a novel neuropeptide encoded by the calcitonin gene via tissue-specific RNA processing. Nature 1983; 304:129–135.
28. Cadieux A, Springall DR, Mulderry PK, Rodrigo J, Ghatei MA, Therengi G, Bloom SR, Polak JM. Occurrence, distribution and ontogeny of CGRP immunoreactivity in the rat lower respiratory tract: effects of capsaicin treatment and surgical denervations. Neuroscience 1986; 19:605–627.
29. Roncalli M, Springall DR, Maggioni M, Moradogli-Haftvani RJD, Winter L, Zhao G, Coggi G, Polak JM. Early changes in the calcitonin gene-related peptide (CGRP) content of pulmonary endocrine cells concomitant with vascular remodelling in the hypoxic rat. Am J Respir Cell Mol Biol 1993; 9:467–474.
30. Tjen-A-Looi S, Krainczi M, Ekman R, Keith I. Sensory CGRP depletion by capsaicin exacerbates hypoxia-induced pulmonary hypertension in rats. Regul Pept 1998; 74:1–10.
31. Wang YY, Cutz E. Localization of cholecystokinian-like peptide in neuroendocrine cells of mammalian lungs: a light and electron microscopic immunohistochemical study. Anat Rec 1993; 236:198–205.
32. Gallego R, Garcia-Caballero T, Roson E, Beiras A. Neuroendocrine cells of human lung express substance P-like immunoreactivity. Acta Anat 1990; 139:278–282.
33. Giaid A, Polak JM, Gaitonde V, Hamid QA, Moscoso G, Legon S, Uwanogho D, Roncalli M, Shinmi O, Sawamura T, Kimura S, Yanagisawa M, Masaki T, Springall DR. Distribution of endothelian-like immunoreactivity and mRNA in the developing and adult human lung. Am J Respir Cell Mol Biol 1991; 4:50–58.
34. Polak JM, Becker KL, Cutz E, Gail DB, Goniakowska-Witalinska L, Gosney JR, Lauweryns JM, Linnoila I, McDowell EM, Miller YE, Scheuermann DW, Springall DR, Sunday ME, Zacconi G. Lung endocrine cell markers, peptides and amine. Anat Rec 1993; 236:169–171.
35. Keith IM, Ekman R. PYY-like material and its spatial relationship with NPY, CGRP and 5-HT in the lung of the Syrian golden hamster. Cell Tissue Res 1990; 262:543–550.
36. Luts A, Uddman R, Absood A, Hakanson R, Sundler F. Chemical coding of endocrine cells of the airways: presence of helodermin-like peptides. Cell Tissue Res 1991; 265:425–433.
37. Uddman R, Luts A, Arimura A, Sundler F. Pituitary adenylate cyclase-activating peptide (PACAP) a new vasoactive intestinal peptide (VIP)-like peptide in the respiratory tract. Cell Tissue Res 1991; 265:197–201.
38. Wharton J, Polak JM, Cole GA, Marangos PJ, Pearse AGE. Neuron specific enolase as an immunohistochemical marker for diffuse neuroendocrine system in human fetal lung. J Histochem Cytochem 1981; 29:1359–1364.
39. Lauweryns JM, Van Ranst L. Protein gene product 9.5 expression in lungs of human and other mammals. Immunohistochemical detection in neuroepithelial bodies, neuroendocrine cells and nerves. Neurosci Lett 1988; 85:311–316.
40. Haley K, Drazen J, Osathanondh R, Sunday ME. Comparison of ontogeny of protein gene product 9.5, chromogranin A, and proliferating cell nuclear antigen in human fetal lung. Microsc Res Tech 1997; 37:62–68.
41. Lloyd RV, Incangelo A, Eiden LE, Cano M, Jin L, Grimes M. Chromogranin A and B messenger ribonucleic acids in pituitary and other normal and neoplastic human endocrine tissues. Lab Invest 1989; 60:548–565.

42. Lauweryns JM, Van Ranst L. Leu-7 like immunoreactivity in human, monkey and pig bronchopulmonary neuroepithelial bodies and neuroendocrine cells. J Histochem Cytochem 1987; 35:687–691.
43. Jin L, Hemperly JJ, Lloyd RV. Expression of neural cell adhesion molecule in normal and neoplastic human neuroendocrine tissues. Am J Pathol 1991; 138:961–969.
44. DeLeij H, Poppema S, Klein-Nulend J, ter Haar A, Schevander E, Ebbens T, Postumus F, The TM. Neuroendocrine differentiation antigen of human lung carcinoma and Kultchitzky cells. Cancer Res 1985; 45:2192–2200.
45. Ito T, Udaka N, Kawano N, Kitamura H. Ontogeny of pulmonary neuroendocrine cells which express alpha subunit of guanine-binding protein G_{0_α}. Histochem Cell Biol 1999; 111:389–395.
46. Lee I, Gould VE, Moll R, Wiedemann B, Frank WW. Synaptophysin expressed in the bronchopulmonary tract: neuroendocrine cells, neuroepithelial bodies and neuroendocrine neoplasms. Differentiation 1987; 34:115–125.
47. Ito T. Differentiation and proliferation of pulmonary neuroendocrine cells. Prog Histochem Cytochem 1999; 4:247–322.
48. Ito T, Udaka N, Inayama Y, Kitamura H, Kanisawa M. Hamster pulmonary endocrine cells with positive immunostain for calbindin D28K. Histochem Cell Biol 1998; 109:67–73.
49. Lauweryns JM, Van Lommel AT, Dom JR. Innervation of rabbit intrapulmonary neuroepithelial bodies: Quantitative and qualitative ultrastructural study after vagotomy. J Neurol Sci 1985; 67:81–92.
50. Lauweryns JM, Van Lommel AT. Effects of various vagotomy procedures on the reaction to hypoxia of rabbit neuroepithelial bodies: modulation by intrapulmonary axon reflexes? Exp Lung Res 1986; 11:319–339.
51. Lauweryns JM, Van Lommel A. Morphometric analysis of hypoxia induced synaptic activity in pulmonary neuroepithelial bodies. Cell Tissue Res 1982; 226:201–214.
52. Brouns I, Andriansen D, Burnstock G, Timmermans JP. Intraepithelial vagal sensory nerve terminals in rat pulmonary neuroepithelial bodies express $P2X_3$ receptors. Am J Respir Cell Mol Biol 2000; 23:52–61.
53. Brouns I, Van Genechten J, Scheuermann DW, Timmermans JP, Adriansen D. Neuroepithelial bodies: the link between neuronal nitric oxide and sensitivity to airway hypoxia? J Comp Neurol 2002; 449:343–354.
54. McDonald DM. Structure–function relationships of chemoreceptive nerves in the carotid body. Annu Rev Respir Dis 1997; 115:193–207.
55. Cho T, Chan W, Cutz E. Distribution and frequency of neuroepithelial bodies in postnatal rabbit lung: quantitative study with monoclonal antibody against serotonin. Cell Tissue Res 1989; 255:353–362.
56. Cutz E, Gillan JE, Bryan AC. Neuroendocrine cells in the developing human lung: morphologic and functional considerations. Pediatr Pulmonol 1985; 1:S21–S29.
57. Gosney JR. Neuroendocrine cell populations in postnatal human lungs: minimal variation from childhood to old age. Anat Rec 1993; 236:177–180.
58. Van Lommel A, Lauweryns JM. Postnatal development of the pulmonary neuroepithelial bodies in various animal species. J Auton Nerv Syst 1997; 65: 17–24.
59. Bolle T, Lauweryns JM, Van Lommel A. Postnatal maturation of neuroepithelial bodies and carotid body innervation: a quantitative investigation in the rabbit. J Neurocytol 2000; 29:241–248.

60. Donnelly DF. Developmental aspects of oxygen sensing by the carotid body. J Appl Physiol 2000; 88:2296–2301.
61. Sindhu GS. The endodermal origin of digestive and respiratory APUD cells. Am J Pathol 1979; 96:5–20.
62. Ito T, Nogawa H, Udaka N, Kitamura H, Kanisawa M. Development of pulmonary neuroendocrine cells of fetal hamster in explant culture. Lab Invest 1997; 77:449–457.
63. Borges M, Linnoila RI, van de Velde HJ, Chen H, Nelkin BD, Mabry M, Baylin SB, Ball DW. An achaete-scute homologue essential for neuroendocrine differentiation in the lung. Nature 1997; 386:852–855.
64. Ito T, Udaka N, Yazawa T, Okudela K, Hayashi H, Sudo T, Guillemot F, Kageyama R, Kitamura H. Basic helix-loop-helix transcription factors regulate neuroendocrine differentiation of fetal pulmonary epithelium. Development 2000; 127:3913–3921.
65. Ito T, Udaka N, Ikeda M, Yazawa T, Kageyama R, Kitamura H. Significance of proneural basic helix-loop-helix transcription factors in neuroendocrine differentiation of fetal lung epithelial cells and lung carcinoma cells. Histol Histopathol 2001; 16:335–343.
66. Ball DW, Azzoli CG, Baylin SB, Chi D, Sou S, Donis-Keller H, Cumaraswamy A, Borges M, Nelkin BD. Identification of human achaete-scute homolog highly expressed in neuroendocrine tumors. Proc Natl Acad Sci USA 1993; 90:6548–6552.
67. Sorokin SP, Hoyt R, Shaffer MJ. Ontogeny of neuro-epithelial bodies: correlations with mitogenesis and innervation. Microsc Res Tech 1997; 37:43–61.
68. Carney DN, Gazdar AF, Bepler G, Guccion JG, Marangos PJ, Moody TW, Zweig MH, Minna JD. Establishment and identification of small cell lung cancer cell lines having classic and variant features. Cancer Res 1985; 45:2913–2923.
69. Gazdar AF, Carney DN, Nau MM, Minna JD. Characterization of variant subclasses of cell lines derived from small cell lung cancer having distinctive biochemical, morphological and growth properties. Cancer Res 1985; 45:2924–2930.
70. Chen H, Biel MA, Borges MW, Thiagalingam A, Nelkin BD, Baylin SB, Ball DW. Tissue-specific expression of human achaete-scute homologue-1 in neuroendocrine tumors: transcriptional regulation by dual inhibitory regions. Cell Growth Differ 1997; 8:677–686.
71. Fu XW, Nurse CA, Wong V, Cutz E. Hypoxia-induced secretion of serotonin from intact pulmonary neuroepithelial bodies in neonatal rabbit. J Physiol 2002; 539:503–510.
72. Wang D, Youngson C, Wong V, Yeger H, Dinauer M, Vega-Saenz de Miera E, Rudy B, Cutz E. NADPH-oxidase and a hydrogen peroxide-sensitive K^+ channel may function as an oxygen sensor complex in airway chemoreceptors and small cell lung carcinoma cell lines. Proc Natl Acad Sci USA 1996; 93:13182–13187.
73. Vega-Saenz de Miera E, Rudy B. Modulation of K^+ channels by hydrogen peroxide. Biochem Biophys Res Commun 1992; 186:1681–1687.
74. Youngson C, Nurse CA, Yeger H, Curnutte JT, Cutz E. Immunohistochemical localization of O_2 sensing protein (NADPH) oxidase in chemoreceptor cells. Microsc Res Tech 1997; 37:101–106.
75. Fu XW, Wang D, Nurse CA, Dinauer MC, Cutz E. NADPH Oxidase is an O_2 sensor in airway chemoreceptors: evidence from K^+ current modulation in wild-type and oxidase-deficient mice. Proc Natl Acad Sci USA 2000; 97:4374–4379.
76. Lopez-Barneo J. Oxygen-sensing by ion channels and the regulation of cellular functions. Trends Neurosci 1996; 19:435–440.

77. Kobertz WR, Williams C, Miller E. Hanging gondola structure of the T_1 domain of a voltage-gated K^+ channel. Biochemistry 2000; 39:10347–10352.
78. Patel AJ, Honore E. Molecular physiology of oxygen-sensitive potassium channels. Eur Respir J 2001; 18:221–227.
79. Stea A, Nurse CA. Chloride channels in cultured glomus cells of the rat carotid body. Am J Physiol 1989; 275:C174–C181.
80. Schwiebert EM, Benos JD, Egan ME, Stutts J, Guggino WB. CFTR is a conductance regulator as well as a chloride channel. Physiol Rev 1999; 79:S145–S166.
81. Pilewski JM, Frizzell RA. Role of CFTR in airway disease. Physiol Rev 1999; 79:S215–S255.
82. Peers C, Kemp PJ. Acute oxygen sensing: diverse but convergent mechanisms in airway and arterial chemoreceptors. Respir Res 2001; 2:145–149.
83. O'Kelly I, Lewis A, Peers C, Kemp PJ. O_2 sensing by airway chemoreceptor-derived cells: protein kinase C activation reveals functional evidence for involvement of NADPH oxidase. J Biol Chem 2000; 275:7684–7692.
84. O'Kelly I, Peers C, Kemp PJ. NADPH oxidase does not account fully for O_2 sensing in model airway chemoreceptor cells. Biochem Biophys Res Commun 2001; 283:1131–1134.
85. Tauber AI. Protein kinase C and the activation of the human neutrophil NADPH-oxidase. Blood 1987; 69:711–720.
86. Chandel NS, Schumaker PT. Cellular oxygen sensing by mitochondria: old questions, new insight. J Appl Physiol 2000; 88:1880–1889.
87. Waypa GB, Chandel NS, Schumaker PT. Model for hypoxic pulmonary vasoconstriction involving mitochondrial oxygen sensing. Circ Res 2001; 88:1259–1266.
88. Lesage F, Lazdunski M. Molecular and functional properties of two-pore-domain potassium channels. Am J Renal Physiol 2000; 279(5):F793–801.
89. Hartness ME, Lewis A, Searle GJ, O'Kelly I, Peers C, Kemp PJ. Combined antisense and pharmacological approaches implicate hTASK as an airway O_2 sensing K^+ channel. J Biol Chem 2001; 276:26499–26508.
90. Buckler KJ, Williams BA, Honore E. An oxygen-, acid- and anaesthetic-sensitive TASK-like background potassium channel in rat arterial chemoreceptor cells. J Physiol 2000; 525:135–142.
91. Gurney AM, Osipenko ON, MacMillan D, Kempsill FEJ. Potassium channels underlying the resting membrane potential of pulmonary artery smooth muscle cells. Clin Exp Pharmacol Physiol 2002; 29(4):330–333.
92. Lewis A, Hartness ME, Chapman CG, Fearon IM, Meadows HJ, Peers C, Kemp PJ. Recombinant nTASK1 is an O_2-sensitive K^+ channel. Biochem Biophys Res Commun 2001; 285:1290–1294.
93. Fatemian M, Kim DY, Poulin MJ, Robbins PA. Very mild exposure to hypoxia for 8 h can induce ventilatory acclimatization in humans. Pflügers Arch 2001; 441:840–843.
94. Smirnov SV, Robertson TP, Ward JP, Aaronson PI. Chronic hypoxia is associated with reduced delayed rectifier K^+ current in rat pulmonary artery muscle cells. Am J Physiol 1994; 266:H365–H370.
95. Wyatt CN, Wright C, Bee D, Peers C. O_2-sensitive K^+ currents in carotid-body chemoreceptor cells from normoxic and chronically hypoxic rats and their roles in hypoxic chemotransduction. Proc Natl Acad Sci USA 1995; 92:295–299.

96. Stea A, Jackson A, Macintyre L, Nurse CA. Long-term modulation of inward currents in O_2 chemoreceptors by chronic hypoxia and cyclic AMP in vitro. J Neurosci 1995; 15:2192–2202.
97. Perrin DG, MacDonald TJ, Cutz E. Hyperplasia of bombesin-immunoreactive pulmonary neuroendocrine cells and neuroepithelial bodies in sudden infant death syndrome. Pediatr Pathol 1991; 11:431–447.
98. Kemp PJ, Searle GJ, O'Kelly I, Peers C. Dynamic down-regulation of acute O_2-sensing in a model of airway neuroepithelial bodies by chronic hypoxia is independent of K^+ channel expression. FASEB J 2001 (Abstr).
99. Colebrooke RL, Smith IF, Kemp PJ, Peers C. Calcium channel remodelling by chronic hypoxia in a human airway chemoreceptor cell line. Neurosci Lett 2002; 18(2):69–72.
100. Nurse CA, Zhang M. Acetylcholine contributes to hypoxic chemotransmission in co-cultures of rat type 1 cells and petrosal neurons. Respir Physiol 1999; 155:189–199.
101. Cutz E. Neuroendocrine cells of the lung. An overview of morphologic characteristics and development. Exp Lung Res 1982; 3:185–208.
102. Katz B. The Release of Neural Transmitter Substances. Springfield, IL: Charles C Thomas, 1969.
103. Higashi H, Nishi S. 5-Hydroxytryptamine receptors on visceral primary afferent neurones of rabbit nodose ganglia. J Physiol 1982; 323:543–567.
104. Peters JA, Malone HM, Lambert JJ. An electrophysiological investigation of the properties of 5-HT3 receptors on the rabbit nodose ganglion neurons in culture. Br J Pharmacol 1993; 110:665–676.
105. Fu XW, Wang D, Pan J, Farragher SM, Wong V, Cutz E. Neuroepithelial bodies in mammalian lung express functional serotonin type 3 receptor. Am J Physiol Lung Cell Mol Physiol 2001; 281:L931–L940.
106. Aguayo SM. Pulmonary neuroendocrine cells in tobacco-related lung disorders. Anat Rec 1993; 236:122–127.
107. Nylen ES, Becker KL, Snider RH, Tabassian AR, Cassidy MM, Linnoila I. Cholinergic-nicotinic control of growth and secretion of cultured pulmonary neuroendocrine cells. Anat Rec 1993; 236:129–135.
108. Fu XW, Jackson A, Wang D, Cutz E. Expression of functional nicotinic acetylcholine receptor in pulmonary neuroepithelial bodies. Abs. http://www.neuro.affil.edu/post/AAA.html.
109. Shirahata M, Sham JSK. Roles of ion channels in carotid body chemotransmission of acute hypoxia. Jpn J Physiol 1999; 49:213–228.
110. Lopez-Barneo J, Pardal R, Ortega-Saenz P. Cellular mechanisms of oxygen sensing. Annu Rev Physiol 2001; 63:259–287.
111. Cutz E, Ma TKF, Perrin DG, Moore AM, Becker LE. Peripheral chemoreceptors in congenital central hypoventilation syndrome. Am J Respir Crit Care Med 1997; 155:358–363.
112. Kazemian P, Stephenson R, Yeger H, Cutz E. Respiratory control in neonatal mice with NADPH oxidase deficiency. Respir Physiol 2001; 126:89–101.
113. Roy A, Rozanov C, Mokashi A, Daudu P, Al-Mehdi A, Shams H, Lahiri S. Mice lacking in $gp91^{phox}$ subunit of NAD(P)H oxidase showed glomus cell $[Ca^{2+}](i)$ and respiratory responses to hypoxia. Brain Res 2000; 872:188–193.

32

Oxygen Sensing by Neonatal Adrenal Chromaffin Cells

A Role for Mitochondria?

COLIN A. NURSE and IAN M. FEARON

McMaster University
Hamilton, Ontario, Canada

ADELE JACKSON

Ottawa Health Research Institute
Ottawa, Ontario, Canada

ROGER J. THOMPSON

University of Colorado Health Sciences Center
Denver, Colorado, U.S.A.

I. Introduction

The ability to sense and make appropriate adjustments to O_2 availability is critical for normal physiological and pathological processes encountered by aerobic organisms. Therefore, the mechanisms underlying cellular responses to O_2 deficiency are both biologically important and clinically relevant. Though it is recognized that probably all cells have some ability to sense O_2, as will be discussed elsewhere in this volume, a major focus over the last ~15 years has been on the more specialized O_2-sensing cells, including the carotid body chemoreceptors (1–3), pulmonary neuroepithelial bodies (4,5), and pulmonary vascular smooth muscle cells (6,7). In these cell types, the critical issues that are still controversial are the molecular identity and location of the O_2 sensor, as well as the signaling pathways that couple the sensor to the cellular response. Whereas a neutrophil-like, membrane-associated NADPH oxidase appears to fulfill the role for the O_2 sensor in pulmonary neuroepithelial bodies, based on use of an oxidase-deficient mouse model (8), the same cannot be said for the pulmonary vasculature (9) or carotid body receptors (10,11). Recent studies on pulmonary vasculature strongly favor a role for mitochondria, and heme-protein(s) in the electron transport chain, as the initial sites of oxygen sensing (6,7,12).

More recently the adrenal medulla, known more for its role in mediating the "fight or flight" response during stress, has become an attractive model for studying mechanisms of O_2 sensing and is the particular focus of this review. It has been known for some time that catecholamine (CA) release from adrenomedullary chromaffin cells (AMC) plays a critical role in the ability of animals and humans to survive stressors associated with delivery and the transition to extrauterine life (13,14). This CA release is vital for the proper cardiovascular, respiratory, and metabolic responses to the natural stressors at birth, e.g., hypoxia or asphyxia. These responses include regulation of cardiac function via stimulation of α-adrenergic receptors (prior to the subsequent switch over to the adult β-adrenergic receptors) and initiation of lung respiration (15). The latter involves the stimulation of surfactant secretion via β_2-receptors, and transformation of the physiological properties of the lung epithelium from a state of net fluid secretion to one of net fluid absorption. In the fetus, at least in sheep, the effects of hypoxia on adrenal medullary cells result in a redirection of blood to critical organs such as the heart and brain, and this response is crucial to fetal survival (16,17).

All the above physiological responses to hypoxia take place before a functional innervation of the adrenal gland. For example, in some species including rat and human, sympathetic innervation of the adrenal medulla is immature or absent in the *neonate*, yet the animal can still elicit the vital CA surge in response to hypoxic challenge (13). Seidler and Slotkin (13) showed that in the newborn rat, acute hypoxia reduces adrenal CA through a "nonneurogenic" mechanism, which is lost postnatally along a time course that roughly coincides with the maturation of the sympathetic innervation of the adrenal medulla. Furthermore, denervation experiments demonstrated that removal of the splanchnic innervation in mature or adult animals resulted in a gradual reappearance of this nonneurogenic mechanism (15). This correlation between lack of functional adrenal innervation and the presence of an apparent direct hypoxia-sensing mechanism that promotes CA secretion from chromaffin cells is also seen in fetal sheep (18,19). The mechanisms responsible for the direct nonneurogenic response of the adrenal medulla to hypoxia have only recently been investigated. We discuss below some of the progress made in this area, and highlight some the similarities between these mechanisms and those described in other O_2-sensing cells, especially carotid body chemoreceptors, which share a similar developmental origin with adrenal chromaffin cells.

II. O_2-Sensitive Voltage-Gated K^+ Channels in Chromaffin Cells

Studies of the nonneurogenic mechanisms underlying hypoxia-induced CA secretion from perfused fetal sheep adrenal glands revealed the requirement for entry of extracellular calcium through voltage-gated Ca^{2+} channels (19). This CA release was abolished by removal of extracellular Ca^{2+} from the adrenal perfusate and was significantly inhibited by the L-type Ca^{2+} channel blocker nifedipine (19). These data suggest that membrane depolarization is involved in the nonneurogenic adrenal CA response to hypoxia. A general mechanism by which this can occur is via hypoxic inhibition of K^+ channels, as demonstrated in other amine-secreting O_2-sensitive cells in carotid (20,21) and neuroepithelial (4,5,8) bodies. Indeed, we obtained direct evidence for this mechanism in short-term cultures of *neonatal* rat AMC, in which hypoxia caused a reversible inhibition of voltage-dependent outward K^+ current (22). A typical example, using ramp depolarizations, is illustrated in Figure 1a, where hypoxia inhibited the outward K^+ current at more positive potentials. The hypoxia-sensitive current in this cell is shown as a "difference" current (IKO_2) in Figure 1b, obtained by subtracting the current in hypoxia (h) from the control (c) current in normoxia (Fig. 1a). In Figure 1b, the reversal potential of IKO_2 is near the K^+ equilibrium potential, confirming selectivity of the underlying channels for K^+ ions. In similar neonatal rat AMC cultures, ~1 hr exposure to hypoxia (5–10% O_2) was previously shown to stimulate CA secretion via a nifedipine-sensitive mechanism (22), as was previously observed in the intact hypoxia-perfused fetal sheep adrenals (19). Brief exposures (a few minutes) to severe hypoxia have also been reported to stimulate an increase in intracellular Ca^{2+} and CA secretion in freshly isolated *neonatal* rat AMC, as detected by fluorimetric and voltammetric techniques, respectively (23).

A detailed analysis of the hypoxia-sensitive, voltage-dependent K^+ current in neonatal rat AMC (i.e., IKO_2 in Fig. 1b) indicated it consisted of a complex modulation of three different O_2-sensitive K^+ currents (24). Anoxia (24) or hypoxia (22; unpublished observations) *inhibited* both a large-conductance, iberiotoxin-sensitive Ca^{2+}-dependent K^+ current ($IBKO_2$) and a Ca^{2+}-independent delayed

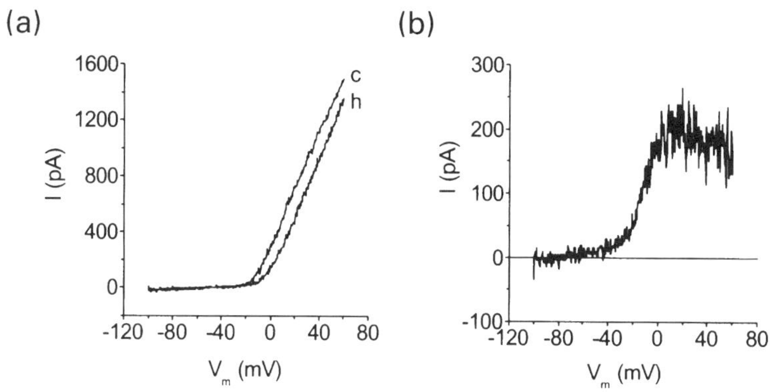

Figure 1 Effects of hypoxia on whole-cell K^+ current in neonatal rat chromaffin cells. (a) Ramp I-V relation showing suppression of outward K^+ current by hypoxia (h; PO_2~10 torr) at positive potentials; normoxic I-V relation is indicated by control (c) trace. In (b) the hypoxia-sensitive component is shown as a "difference" current (trace c–trace h). In (b) note reversal of this hypoxia-sensitive current near the K^+ equilibrium potential ($E_K \simeq 83$ mV). Voltage-dependent Na^+ currents were blocked with tetrodotoxin in this experiment.

rectifier-type K^+ current ($IKVO_2$), but *activated* a glibenclamide- and Ca^{2+}-sensitive K^+ current (IK_{ATP}). Estimates of the relative contributions of these currents to the net IKO_2 revealed that $IBKO_2$ was the largest contributor (24). For a voltage step to + 30 mV during anoxia, the current density of the glibenclamide-sensitive (i.e., anoxia-activated) component IK_{ATP} was ~25 pA/pF compared to ~10 pA/pF for the anoxia-inhibited $IKVO_2$ component. Therefore, for a net IKO_2 of ~40 pA/pF observed at this step potential (Fig. 1b; see also Fig. 3, Ref. 24), the inhibitory contribution of $IBKO_2$ during anoxia must be ~55 pA/pF; this value is close to the magnitude of the iberiotoxin-sensitive component of outward current (~48%) in these cells under normoxic conditions (24). Taken together, these data suggest that hypoxic inhibition of Ca^{2+}-dependent BK currents is likely to be a major contributor to the low PO_2-dependent CA secretion in *neonatal* rat chromaffin cells. Direct confirmation of this point is considered later.

III. Does Mitochondrial Inhibition Mimic Hypoxic Regulation of O_2-Sensitive Voltage-Dependent K^+ Currents in Neonatal Chromaffin Cells?

In prototypic chemoreceptors (type 1 cells) in the mammalian carotid body (CB), mitochondria have long been considered the initial site of O_2 sensing (25–27). Mitochondrial oxidative phosphorylation via the electron transport chain (ETC) has frequently been considered a key element in the O_2 signaling pathway, and the heme-containing protein cytochrome a_3 has been proposed as the primary oxygen sensor in cat CB (27). The idea that the mitochondrial ETC may also act as the site of O_2 sensing in neonatal rat AMC was initially proposed by Mojet et al. (23). These workers found that inhibition of mitochondrial respiration in these cells by either cyanide (CN; a complex IV inhibitor) or rotenone (a complex I inhibitor) mimicked the effects of severe hypoxia by reversibly increasing both $[Ca^{2+}]_i$ and catecholamine secretion. However, these latter two cellular responses are downstream events in the signaling pathway that could occur through alteration in intracellular Ca^{2+} stores, without involvement of plasma membrane K^+ channels. Therefore, the question arose whether or not these mitochondrial inhibitors mimic the early events of hypoxia, e.g., inhibition of outward K^+ current. As shown in Figure 2a,b using perforated-patch whole-cell recordings, inhibition of mitochondrial respiration and ATP production in neonatal AMC by either CN (5 μM–5 mM) or the mitochondrial uncoupler 2,4-dinitrophenol (DNP; 2.5 μM–2.5 mM) had no effect on outward K^+ current at any test potential; this was the case even after perfusion of the drugs for ~15 min. Therefore, these inhibitors failed to mimic this early effect of hypoxia, though hypoxia was still effective in suppressing outward K^+ current in the *same* cells (not shown). In contrast to these results, the complex I inhibitor rotenone (0.5–7 μM) did indeed mimic hypoxia in causing a dose-dependent inhibition of outward K^+ current, as shown in Figure 2c, and the effects of the two stimuli were nonadditive (unpublished observations). These data support a mitochondrial origin of the primary O_2 sensor in neonatal AMC, but indicate it is

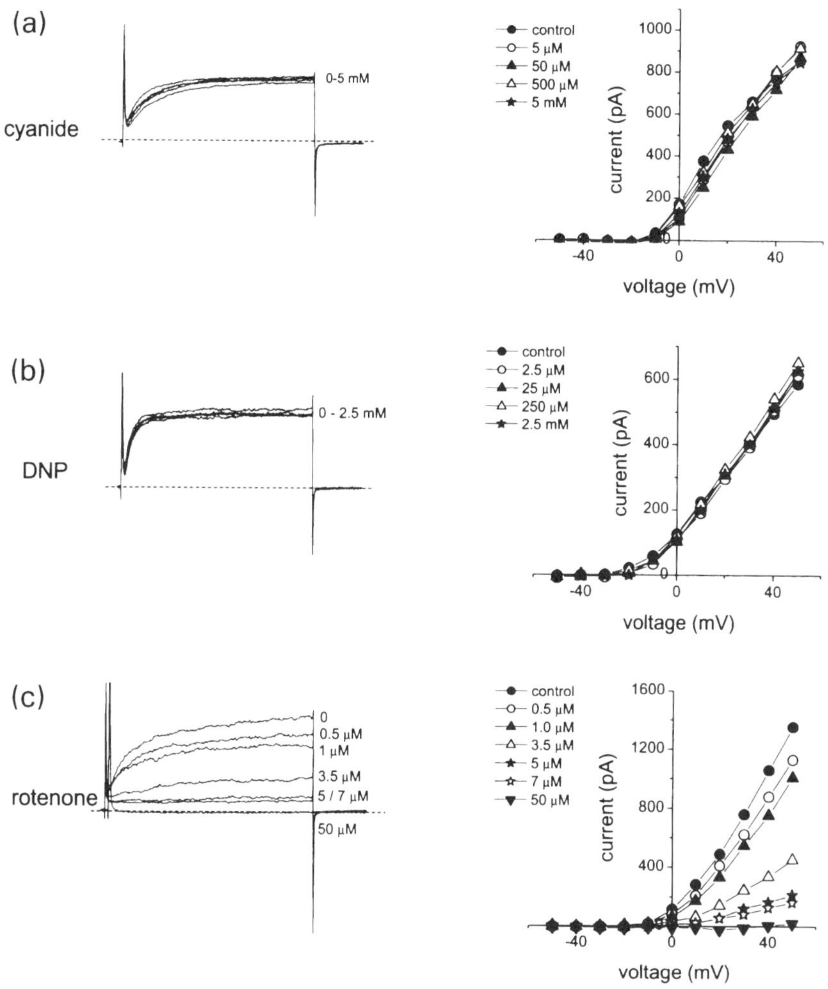

Figure 2 Effects of inhibitors of the mitochondrial electron transport chain on outward K^+ current in neonatal rat chromaffin cells. Representative traces for a voltage step to +30 mV are shown on the left, for a variety of inhibitor concentrations as indicated in (a) cyanide, (b) 2,4-dinitrophenol (DNP), and (c) rotenone. The corresponding I-V relation for each cell is shown on the right. Note that both cyanide and DNP had no effect, whereas rotenone caused a dose-dependent inhibition of outward K^+ current.

unlikely to be at the complex IV (i.e., cytochrome a_3) site. Moreover, they point to a nonequivalence in the action of proximal versus distal inhibitors of the mitochondrial ETC, as observed in other O_2-sensing cells (6).

How can the data of Mojet et al. (23), suggesting that the effects of CN and severe hypoxia share a common mechanism, be reconciled with our routine observations that CN (unlike hypoxia) had no effect on outward K^+ current in neonatal AMC? As discussed above, and as will be argued later, we propose that inhibition of this current (and particularly the Ca^{2+}-dependent BK component) is an important step in hypoxia-evoked CA secretion from these cells. In general, the effects of CN on membrane properties of O_2-chemoreceptive cells appear complex, though it is known to be a potent CB stimulus. In rat CB type 1 cells CN inhibited outward K^+ current (28,29), but surprisingly, this occurred regardless of whether ATP was included in the intracellular pipette solution (28). In contrast, CN enhanced whole-cell K^+ currents in rabbit CB type 1 cells, by an effect that appeared to be mediated via release of Ca^{2+} from intracellular stores (30,31). The paradox relating to the stimulatory effects of CN on secretion (23), but without any apparent effect on outward K^+ current in neonatal AMC (Fig. 2a), may reflect its high affinity for heme-containing proteins other than those contained in the electron transport chain. Interestingly, the failure of CN to mimic hypoxia in the present study suggests there are alternative pathways for electron flow from the ETC to molecular O_2, other than the well-characterized route through cytochrome a_3; these pathways could be important for the hypoxic regulation of K^+ channels in neonatal AMC. Indeed, different subunits of the ETC have "leak" pathways for generating reactive oxygen species (ROS), which are potential signaling molecules during hypoxia-sensing (12), and these sites vary between tissue and metabolic state (32).

IV. Does Mitochondrial Inhibition Mimic the Hypoxia-Induced Receptor Potential in Neonatal Chromaffin Cells?

In addition to the three O_2-sensitive K^+ currents described in our previous study on neonatal rat AMC (24) and reviewed above, the possible occurrence of a fourth was revealed during current-clamp recordings. Both anoxia (24) and hypoxia (Fig. 3a) evoked membrane depolarization or a receptor potential in single quiescent AMC. Though this current was not initially identified (24), all or part of it persisted during anoxia in the presence of cadmium, 4-aminopyridine, TEA, or charybdotoxin, suggesting the involvement of a background O_2-sensitive current. Clues to its identity were suggested in subsequent studies on *adult* rat (33) and fetal sheep (17) chromaffin cells, in which hypoxia inhibited an apamin-sensitive, small-conductance Ca^{2+}-dependent K^+ current (SK). In the latter studies, both hypoxia and apamin blocked a K^+ current that contributed to the resting potential and led to membrane depolarization. Moreover, in fetal sheep chromaffin cells, the depolarization due to hypoxia and apamin produced a rise in intracellular calcium that was nonadditive when the two stimuli were applied together (17), suggesting convergence onto a common pathway. In our initial attempts using apamin as a probe for SK channels in *neonatal* rat AMC, we obtained variable and inconclusive results, in part because the effects of apamin were slow in onset and poorly reversible. More recently, we tested the effects of bicuculline (100 μM), a more rapid and readily reversible blocker of

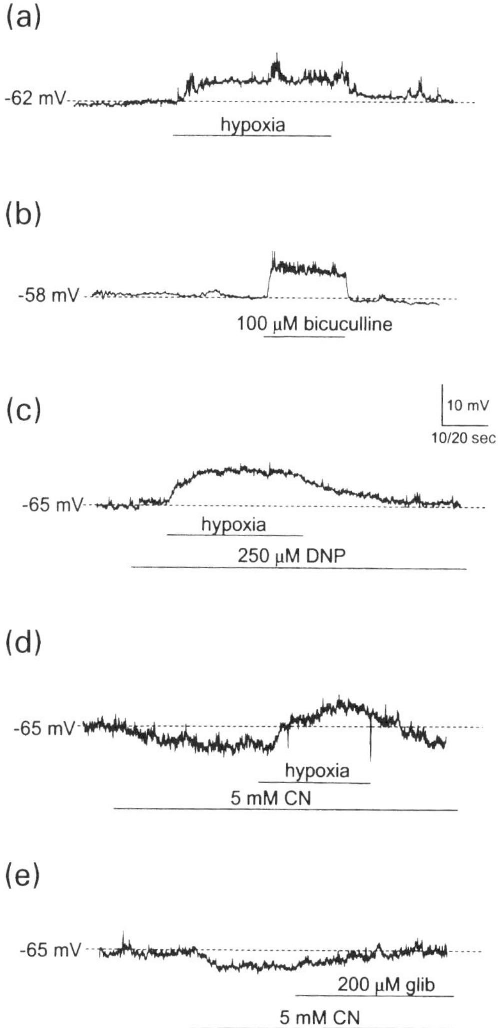

Figure 3 Effects of various stimuli and mitochondrial inhibitors on the resting membrane potential of quiescent chromaffin cells. A typical example of hypoxia-induced membrane depolarization is shown in (a), using nystatin perforated-patch whole-cell recording. In (b), bicuculline (100 μM), a reversible inhibitor of small-conductance Ca^{2+}-dependent K^+ channels (SK), also caused membrane depolarization similar to hypoxia. The mitochondrial inhibitors 2,4-dinitrophenol (DNP) and cyanide (CN) did not mimic the hypoxia-induced membrane depolarization seen in (c) and (d), respectively; in fact, in (d), CN caused membrane hyperpolarization, though in most cases no change in membrane potential was observed. Both DNP and CN were usually without effect even after perfusing the drug for >10 min. In (e), the hyperpolarizing effect of CN was reversed in the presence of 200 μM glibenclamide, a blocker of K_{ATP} channels.

SK channels (34), and found that it depolarized neonatal rat AMC by comparable amounts to that seen with hypoxia (Fig. 3a,b). These preliminary data suggest that SK channels are open at the resting potential of *neonatal* rat AMC and are therefore candidates for mediating the receptor potential or membrane depolarization following closure by hypoxia.

Does mitochondrial inhibition mimic the hypoxia-induced receptor potential in neonatal rat AMC? As illustrated in Figure 3c–e, neither DNP (250 μM) nor CN (5 mM) induced membrane depolarization in these cells, though hypoxia did. This was the case even when application of either drug lasted for >10 min. In most cases these mitochondrial inhibitors caused no significant change in resting potential, though occasionally CN caused membrane hyperpolarization, as indicated in Figure 3d; since hypoxia depolarized this cell, the O_2 sensor is unlikely to involve the CN-binding site or complex IV of the mitochondrial ETC. Though not studied systematically, it is possible that the hyperpolarizing effect of CN in such cells was mediated via activation of K_{ATP} channels, since the effect was reversed by 200 μM glibenclamide (Fig. 3e), a blocker of K_{ATP} channels. Interestingly, as observed in voltage clamp studies (Fig. 2c), the complex I inhibitor rotenone mimicked the hypoxia-induced membrane depolarization in neonatal AMC, suggesting that the more proximal mitochondrial ETC could be involved in O_2 sensing (unpublished observations).

How do these data on the effects of mitochondrial inhibitors on the receptor potential in neonatal AMC compare with those obtained in O_2-sensitive CB type 1 cells? In type 1 cells CN caused membrane hyperpolarization (30) or depolarization (35), whereas in *adult* AMC a hyperpolarization has been observed (36). Additionally, CN increased cytoplasmic $[Ca^{2+}]$ and/or catecholamine secretion in neonatal rat type 1 cells (35), neonatal AMC (23), and PC 12 (37) cells. Interestingly, in neonatal rat CB type 1 cells, the mitochondrial uncoupler DNP (another strong CB stimulus) mimicked hypoxia in causing membrane depolarization and voltage-gated Ca^{2+} entry, apparently via inhibition of an O_2-sensitive background K^+ conductance (35). Indeed, in the latter study the possibility was raised that the ADP/ATP ratio may modulate the background K^+ conductance in these cells. In contrast to neonatal rat type 1 cells, the failure of DNP to mimic hypoxia and evoke membrane depolarization in neonatal rat AMC (Fig. 3b) may relate to the presence of different types of O_2-sensitive background K^+ channels, i.e., TASK-1-like K^+ channels in type 1 cells (38) versus Ca^{2+}-dependent SK channels in AMC (17,33). Notably, the complex I inhibitor rotenone mimicked hypoxia in both cell types, raising the possibility that a common mitochondrial sensor or signaling pathway could lead to regulation of membrane ion channels via different end products, e.g., ATP/ADP or ROS levels. Conceivably, it was the decrease in ATP/ADP ratio during CN that led to its hyperpolarizing effect in Figure 3d, probably owing to activation of K_{ATP} channels as ATP levels decreased. Consistent with this proposal, the K_{ATP} channel blocker glibenclamide reversed the hyperpolarizing effect of CN on membrane potential in this cell (Fig. 3e).

V. Proposed Physiological Roles of O_2-Sensitive K^+ Currents in Neonatal Chromaffin Cells

In the foregoing sections we reviewed the evidence favoring the presence of as many as four different O_2-sensitive K^+ currents in neonatal rat AMC. What are their respective physiological roles? We propose that inhibition of Ca^{2+}-dependent BK channels in these cells plays a major role in evoking "nonneurogenic" CA secretion from the adrenal medulla during neonatal stress induced by hypoxia or asphyxia. Direct evidence in support of this idea is considered in the next section. Inhibition of the BK current would lead to action potential broadening, increased voltage-gated Ca^{2+} entry, and enhanced secretion. Indeed, the generally accepted role of BK channels in chromaffin cells is to modulate action potential repolarization (39), as well as their ability to fire repetitively (40). In this regard, we have actually recorded this expected broadening of the action potential during hypoxia in spontaneously active neonatal AMC (41; unpublished observations). Interestingly, spontaneous activity was frequently recorded in neonatal AMC that were members of a cell cluster (Fig. 4a,c), a characteristic arrangement of chromaffin cells in situ, whereas single isolated cells were frequently quiescent (24; Fig. 4a,b). The occurrence of spontaneous activity ensures that the voltage-dependent BK channels are frequently open in normoxia, and therefore available for closure and promoting action potential broadening during hypoxia. On the other hand, the occurrence of quiescent cells (Fig. 4b) suggests a potential role for the O_2-sensitive, small-conductance Ca-dependent SK channels. It may well be that inhibition of these background or "resting" SK channels by hypoxia, and the resulting membrane depolarization, play an important role in quiescent cells insofar as the probability of cell firing and CA secretion are increased. In the case of cell clusters, which show a high degree of spontaneous activity, hypoxic inhibition of SK channels would appear to increase the safety factor for spike generation, rather than being a major *direct* contributor to CA secretion. Ca^{2+}-dependent SK channels have been assigned major roles in setting the resting membrane potential, generation of slow afterhyperpolarizations, and spike frequency adaptation in other preparations (34). However, in spontaneously active neurons, the effect of SK channels on firing frequency was not easily predicted and appeared to depend on interactions with other currents present (34); for example, closure of such channels could result in a depolarization sufficiently large to reduce firing frequency via inactivation of voltage-dependent Na^+ channels.

Two other O_2-sensitive K^+ currents were observed in neonatal AMC, a delayed-rectifier-type K^+ current ($IKVO_2$) that was inhibited by hypoxia, and a glibenclamide- and Ca^{2+}-sensitive K^+ current (IK_{ATP}) that was *activated* by hypoxia (24). Each of these currents may be viewed as canceling the effects of the other, though the extent to which this occurs is likely to depend on the PO_2 level. Under anoxic conditions, the relative contribution of IK_{ATP} was ~25 pA/pF compared to ~10 pA/pF for $IKVO_2$ (step to +30 mV; 24). Since IK_{ATP} activation is enhanced by lower intracellular ATP and higher intracellular Ca^{2+} levels, its contribution may fall off more steeply at a higher PO_2. In general, activation of IK_{ATP} during hypoxia may play a protective role during severe hypoxia by aiding membrane repolarization and

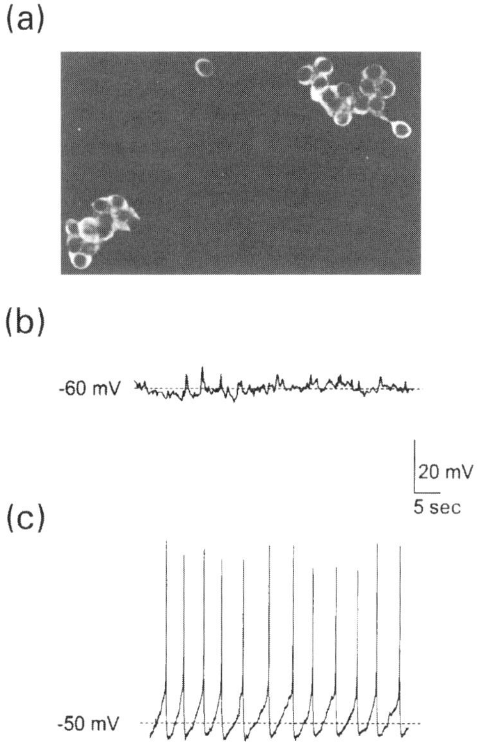

Figure 4 Properties of single versus clustered neonatal rat chromaffin cells. (a) Tyrosine hydroxylase (TH)-positive immunofluorescence of chromaffin cells in culture; note the presence of small clusters and occasional single isolated cells. Perforated-patch recordings of membrane potential revealed that single cells tended to be quiescent (b), whereas cells in clusters tended to fire spontaneous action potentials at room temperature (c).

thereby blunting the increase in intracellular calcium. The resulting preservation of CA stores might allow the chromaffin cells to maintain an influence on the cardiovascular and respiratory systems during asphyxia or severe hypoxic episodes in the perinatal period. The inhibition of $IKVO_2$ during hypoxia would be expected to slow membrane repolarization and promote voltage-gated Ca^{2+} entry and CA secretion, but its contribution is significantly less than that of BK current inhibition (24).

VI. Role of O_2-Sensitive BK Channels in Catecholamine Secretion from Neonatal Chromaffin Cells

As discussed above, inhibition of Ca^{2+}-dependent BK channels appears to be an important step in hypoxia sensing by neonatal AMC. Exposure of such cultures to hypoxia (5 and 10% O_2) for ~1 hr has previously been shown to stimulate CA secretion via entry of extracellular Ca^{2+} through L-type Ca^{2+} channels (22). Since the majority of AMC in these cultures are present in clusters (Fig. 4a), in which spontaneous spiking is common (Fig. 4c), we presumed that hypoxia-induced CA secretion was due primarily to inhibition of BK channels, expected to be frequently open in such spiking AMC. If so, direct inhibition of BK channels under normoxic conditions with the specific blocker iberiotoxin (IbTx) should stimulate CA secretion similar to hypoxia. As illustrated in Figure 5a, this was indeed the case since incubation of AMC cultures in 200 nM IbTX for ~1 hr resulted in a significant increase (~ 3×) in both epinephrine (E) and norepinephrine (NE) release. Moreover, the ratio of E : NE in the bathing fluid was similar for both hypoxia and IbTX (Fig. 5b), consistent with release originating from the same vesicular pool for these two stimuli. In contrast, when high extracellular K^+ was used as a stimulus, CA secretion was still enhanced (22), but the E : NE ratio was significantly reduced (Fig. 5b). These data suggest that the sustained membrane depolarization due to high

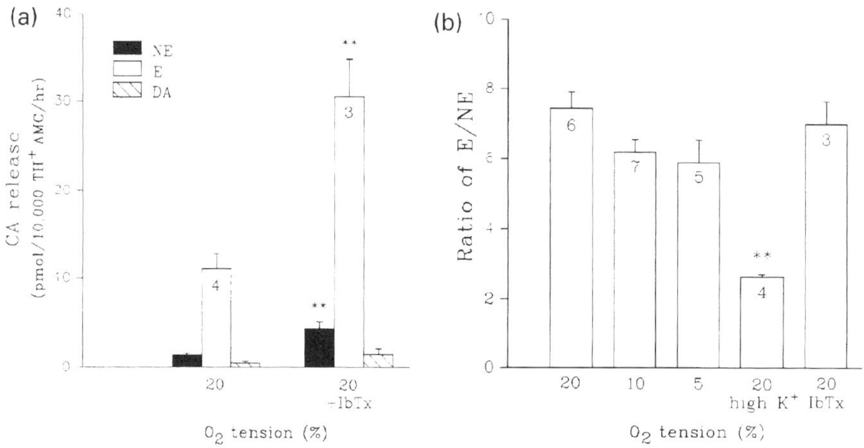

Figure 5 Effects of depolarizing stimuli on catecholamine release from cultured, neonatal rat chromaffin cells, measured by high-performance liquid chromatography (HPLC). The stimulatory effects of iberiotoxin (IbTx; 200 nM), a selective blocker of Ca^{2+}-dependent BK channels, on epinephrine (E), norepinephrine (NE), and dopamine (DA) release are shown in (a), under normoxic conditions (20% O_2). Release was normalized to the number of TH-positive chromaffin cells present in the culture at the end of the experiment (see Fig. 4a). The normalized release of E and NE was significantly ($**p < 0.01$) greater following 1 hr exposure to IbTX, relative to untreated controls. In (b), the ratio of E : NE release is compared following exposure of the cultures to normoxia (20% O_2), hypoxia (10 and 5% O_2), high extracellular K^+ (30 mM), and IbTX (200 nM) for 1 hr. This ratio was similar for all treatments except high K^+, where it was significantly reduced ($**p < 0.01$). Note that blockade of BK channels with IbTx mimicked hypoxia in stimulating CA release (a), and in the same relative proportions (b); this is consistent with inhibition of BK channels acting as an important mediator of hypoxic chemosensitivity in these cells. Bars represent mean ± SEM for the number of cultures indicated in (a) and (b).

K^+ does not mimic the same secretory pattern induced by hypoxia, whereas inhibition of BK channels does. They further caution against a total reliance on generalized amine secretion as a measure of the true hypoxic response, and emphasize that the pattern of stimulation (phasic vs. tonic) may be an important determinant of the proportion of the individual CA secreted. Taken together, these data strongly support an important role for the inhibition of BK channels in the hypoxic chemosensitivity of neonatal rat chromaffin cells.

VII. How Does Hypoxia Regulate BK Channels in Neonatal Chromaffin Cells?

The mechanisms by which hypoxia regulates BK and other O_2-sensitive K^+ channels in neonatal chromaffin cells are not completely understood. These mechanisms are also relevant to *neonatal* rat CB type 1 cells (42) and neocortical neurons (43) where hypoxic regulation of BK channels did not occur in excised patches, presumably owing to loss of cytoplasmic factors. This view was recently challenged in *adult* rat type 1 cells, where hypoxic regulation of BK channels was observed in inside-out patches and was proposed to occur by a membrane-delimited mechanism (44). While the reason for this discrepancy is unknown and may be age-related, the possibility that intracellular organelles (e.g., mitochondria) might remain attached to membrane patches after excision cannot be excluded (45).

In neonatal rat chromaffin cells, a role for the involvement of the more proximal, but not distal, mitochondrial electron transport chain in O_2 sensing was discussed above. Since the inhibitors of mitochondrial ATP production, cyanide and DNP, failed to mimic several of the effects of hypoxia on AMC, it is unlikely that a change in ATP/ADP ratio is the important physiological signal mediating O_2 sensing and CA secretion. However, the fall in ATP during hypoxia is likely to play a role in the *activation* of O_2-sensitive K_{ATP} channels, but the effect is to inhibit, not augment, secretion. Since the complex I inhibitor rotenone mimicked several of the effects of hypoxia, it must act via pathways other than those leading to inhibition of ATP production. One route by which this can occur is via alteration in reactive oxygen species (ROS) generation, though there is a major controversy as to whether hypoxia *increases* or *decreases* ROS in various cell types (7,12). We propose that in neonatal AMC hypoxia causes a *decrease* in ROS levels, which in turn acts as an internal signal ultimately leading to regulation of K^+ channels and CA secretion. In recent studies we have actually measured such a decrease in ROS levels during hypoxia (41) and have found that scavengers of ROS mimic the hypoxic inhibition of K^+ current in neonatal AMC (unpublished observations). Further studies at the single-channel level are required to elucidate the signaling pathway.

VIII. Future Directions and Considerations

In earlier studies (22,23), hypoxic sensitivity was observed in neonatal (P1–2) but not juvenile (P13–20) rat AMC, consistent with a loss of the nonneurogenic

O_2-sensing mechanism with development of a mature adrenal innervation (13,15). It would be of considerable interest to unravel the mechanisms leading to this loss of O_2 sensitivity and the likely role of sympathetic innervation, notwithstanding the observation that some *adult* rat chromaffin cells may possess O_2-sensitive currents after acute isolation (33) or short-term culture (46). These studies, as well as others (17,19,47), support our original observations that O_2 sensing in isolated chromaffin cells can occur at moderate PO_2 levels (22), well above 5 torr, the critical limit suggested by the experiments of Mojet et al. (23). Interestingly, in *adult* guinea pig chromaffin cells cyanide mimicked the effects of mild hypoxia ($PO_2 \sim 58$ torr) in inducing membrane depolarization and CA secretion (47). However, since cyanide did not mimic hypoxia in our studies on *neonatal* rat chromaffin cells, the mechanisms underlying this species difference also need clarification. In the study of Mojet et al. (23), severe hypoxia was achieved by bubbling N_2 in the presence of Na dithionite. Though Na dithionite was used in our subsequent characterization of O_2-sensitive K^+ currents in neonatal AMC (24), the same currents appeared to be modulated by hypoxia, in the absence of dithionite. Recent studies, however, caution against the use of dithionite in studies of O_2 sensitivity in secretory cells. First, Carpenter et al. (48) reported that dithionite evoked catecholamine secretion from carotid body type 1 cells and PC 12 cells by activating an artifactual Ca^{2+} influx pathway. Second, dithionite is known to be a generator of ROS (49), which would clearly complicate analysis of hypoxia-sensing mechanisms, especially if mitochondrial ROS generation is implicated as an intermediary signal. These considerations preclude a clear interpretation of secretion studies using dithionite and emphasize the need to avoid its use in future studies of O_2-sensing mechanisms in general. Our evidence favors a *decrease* in ROS as the signal linking hypoxia to K^+-channel modulation in neonatal rat chromaffin cells (41). Therefore, it would be of interest to investigate how ROS regulate single hypoxia-sensitive K^+ channels in inside-out membrane patches derived from these cells.

Acknowledgments

This work was supported by operating grants from the Heart and Stroke Foundation of Ontario and the Natural Sciences and Engineering Research Council of Canada to CAN. In addition, AJ and RJT were recipients of a Heart and Stroke Foundation of Ontario Research Traineeship. IMF was supported by a Wellcome International Prize Travelling Research Fellowship (Ref: 06154/B/00/Z).

References

1. Gonzalez C, Almaraz L, Obeso A, Rigual R. Carotid body chemoreceptors: from natural stimuli to sensory discharges. Physiol Rev 1994; 4:829-897.
2. Peers C. Oxygen-sensitive ion channels. Trends Pharmacol Sci 1997; 18:405–408.
3. Prabhakar NR. Oxygen sensing by the carotid body chemoreceptors. J Appl Physiol 2000; 88:2287–2295.

4. Youngson C, Nurse C, Yeger H, Cutz E. Oxygen sensing by airway chemoreceptors. Nature 1993; 365:153–155.
5. Cutz E, Jackson A. Neuroepithelial bodies as airway oxygen sensors. Respir Physiol 1999; 115:201–214.
6. Archer SL, Huang J, Henry T, Peterson D, Weir EK. A redox based oxygen sensor in rat pulmonary vasculature. Circ Res 1993; 73:1100–1112.
7. Waypa GB, Chandel NS, Schumacker PT. Model for hypoxic pulmonary vasoconstriction involving mitochondrial oxygen sensing. Circ Res 2001; 88:1259–1266.
8. Fu XW, Wang D, Nurse CA, Dinauer MC, Cutz E. NADPH oxidase is an O_2 sensor in airway chemoreceptors: evidence from K^+ current modulation in wild-type and oxidase-deficient mice. Proc Natl Acad Sci USA 2000; 97:4374–4379.
9. Archer SL, Reeve HL, Michelakis E, Puttagunta L, Waite R, Nelson DP, Dinauer MC, Weir EK. O_2 sensing is preserved in mice lacking the gp91 phox subunit of NADPH oxidase. Proc Natl Acad Sci USA 1999; 96:7944–7949.
10. Roy A, Rozanov C, Mokashi A, Daudu P, Al-Mehdi AB, Shams H, Lahiri S. Mice lacking in gp91 phox subunit of NAD(P)H oxidase showed glomus cell $[Ca^{2+}]_i$ and respiratory responses to hypoxia. Brain Res 2000; 872:188–193.
11. He L, Chen J, Dinger B, Sanders K, Sundar K, Hoidal J, Fidone S. Characteristics of carotid body chemosensitivity in NADPH oxidase-deficient mice. Am J Physiol 2002; 282:C27-C33.
12. Chandel NS, Schumacker PT. Cellular oxygen sensing by mitochondria: old questions, new insight. J Appl Physiol 2000; 88:1880–1889.
13. Seidler FJ, Slotkin TA. Adrenomedullary function in the neonatal rat: responses to acute hypoxia. J Physiol (Lond) 1985; 385:1–16.
14. Lagercrantz H, Slotkin TA. The stress of being born. Sci Am 1986; 254:100–107.
15. Slotkin TA, Seidler FJ. Adrenomedullary catecholamine release in the fetus and newborn: secretory mechanisms and their role in stress and survival. J Dev Physiol 1988; 10:1–16.
16. Phillippe, M. Fetal catecholamines. Am J Obstet Gynecol 1983; 146:840–855.
17. Keating D, Rychkov GY, Roberts ML. Oxygen sensitivity in the sheep adrenal medulla: role of SK channels. Am J Physiol 2001; 281:C1434–C1441.
18. Cheung CY. Fetal adrenal medulla catecholamine response to hypoxia—direct and neural components. Am J Physiol 1990; 258:R1340–R1346.
19. Adams MB, Simonetta G, McMillen IC. The non-neurogenic catecholamine response of the fetal adrenal to hypoxia is dependant on activation of voltage sensitive Ca^{2+}-channels. Dev Brain Res 1996; 94:182–189.
20. Lopez-Barneo J. Oxygen-sensing by ion channels and the regulation of cellular functions. Trends in Neurosciences 1996; 19:435–440.
21. Jackson A, Nurse C. Dopaminergic properties of cultured rat carotid body chemoreceptors grown in normoxic and hypoxic environments. J Neurochem 1997; 69:645–654.
22. Thompson RJ, Jackson A, Nurse CA. Developmental loss of hypoxic chemosensitivity in rat adrenomedullary chromaffin cells. J Physiol (Lond) 1997; 498:503–510.
23. Mojet MH, Mills E, Duchen MR. Hypoxia-induced catecholamine secretion in isolated newborn rat adrenal chromaffin cells is mimicked by inhibition of mitochondrial respiration. J Physiol (Lond) 1997; 504:175–189.
24. Thompson RJ, Nurse CA. Anoxia differentially modulates multiple K^+ currents in neonatal rat adrenal chromaffin cells. J Physiol (Lond) 1998; 512:421–434.

25. Mills E, Jobsis FF. Mitochondrial respiratory chain of carotid body and chemoreceptor response to changes in oxygen tension. J Neurophysiol 1972; 35:405–428.
26. Duchen MR, Biscoe TJ. Mitochondrial function in type I cells isolated from rabbit arterial chemoreceptors. J Physiol (Lond) 1992; 450:13–31.
27. Wilson DF, Mokashi A, Chugh D, Vinogradov S, Osanai S, Lahiri S. The primary oxygen sensor of the cat carotid body is cytochrome a_3 of the mitochondrial respiratory chain. FEBS Lett 1994; 351:370–374.
28. Peers C, O'Donnell J. Potassium currents recorded in type 1 carotid body cells from the neonatal rat and their modulation by chemoexcitatory agents. Brain Res 1990; 522:259–266.
29. Donnelly D.F. Response to cyanide of two types of glomoid cells in mature rat carotid body. Brain Res 1993; 630:157–168.
30. Biscoe TJ, Duchen MR. Electrophysiological responses of dissociated type 1 cells of the rabbit carotid body to cyanide. J Physiol (Lond) 1989; 413:447–468.
31. Biscoe TJ, Duchen MR, Eisner DA, O'Neill SC, Valdeolmillos M. Measurements of intracellular Ca^{2+} in dissociated type 1 cells of the rabbit carotid body. J Physiol (Lond) 1989; 416:421–434.
32. Barja G. Mitochondrial oxygen radical generation and leak: sites of production in states 4 and 3, organ specificity, and relation to aging and longevity. J Bioenerg Biomemb 1999; 31:347–366.
33. Lee J, Lim W, Eun S-Y, Kim SJ, Kim J. Inhibition of apamin-sensitive K^+ current by hypoxia in adult rat adrenal chromaffin cells. Pflügers Arch Eur J Physiol 2000; 439:700–704.
34. Johansson S, Druzin, A., Haage D, Wang M-D. The functional role of a bicuculline-sensitive Ca^{2+}-activated K^+ current in rat medial preoptic neurons. J Physiol (Lond) 2001; 532:625–635.
35. Buckler KJ, Vaughan-Jones RD. Effect of mitochondrial uncouplers on intracellular calcium, pH and membrane potential in rat carotid body type 1 cells. J Physiol (Lond) 1998; 5133:819–933.
36. Latha MV, Borowitz JL, Yim GKW, Kanthasamy A, Isom GE. Plasma membrane hyperpolarization by cyanide in chromaffin cells: role of potassium channels. Arch Toxicol 1994; 68:370–374.
37. Taylor SC, Shaw SM, Peers C. Mitochondrial inhibitors evoke catecholamine release from pheochromocytoma cells. Biochem Biophys Res Commun 2000; 273:17–21.
38. Buckler KJ, Williams B, Honore E. An oxygen-, acid- and anaesthetic-sensitive TASK-like background potassium channel in rat arterial chemoreceptor cells. J Physiol (Lond) 2000; 525:135–142.
39. Pancrazio JJ, Johnson PA, Lynch C III. A major role for calcium-dependent potassium current in action potential repolarization in adrenal chromaffin cells. Brain Res 1994; 668:246–251.
40. Solaro CR, Prakriya M, Ding JP, Lingle CJ. Inactivating and noninactivating Ca^{2+}- and voltage-dependent K^+ current in rat adrenal chromaffin cells. J Neurosci 1995; 15:6110–6123.
41. Thompson RJ. Mechanisms of O_2-chemosensing in developing chromaffin cells. Ph.D. dissertation, McMaster University, Hamilton, Ontario, 2000.
42. Wyatt CN, Peers C. Ca^{2+}-activated K^+ channels in isolated type 1 cells of the neonatal rat carotid body. J Physiol (Lond) 1995; 483:559–565.

43. Liu H, Moczydlowski E, Haddad GG. O_2 deprivation inhibits Ca^{2+}-activated K^+ channels via cytosolic factors in mice neocortical neurons. J Clin Invest 1999; 104:577–588.
44. Riesco-Fagundo AM, Perez-Garcia MT, Gonzalez C, Lopez-Lopez JR. O_2 modulates large-conductance Ca^{2+}-dependent K^+ channels of rat chemoreceptor cells by a membrane-restricted and CO-sensitive mechanism. Circ Res 2001; 89:430–436.
45. Rustenbeck I, Dickel C, Herrmann C, Grimmsmann T. Mitochondria present in excised patches from pancreatic B-cells may form microcompartments with ATP-dependent potassium channels. Biosci Rep 1999; 19:89–98.
46. Mochisuki-Oda N, Takeuchi Y, Matsumura K, Oosawa Y, Watanabe Y. Hypoxia-induced catecholamine release and intracellular Ca^{2+} increase via suppression of K^+ channels in cultured rat adrenal chromaffin cells. J Neurochem 1997; 69:377–387.
47. Inoue M, Fujishiro N, Imanaga I. Hypoxia and cyanide induce depolarization and catecholamine release in dispersed guinea-pig chromaffin cells. J Physiol (Lond) 1998; 507:807–818.
48. Carpenter E, Hatton CJ, Peers C. Effects of hypoxia and dithionite on catecholamine release from isolated type 1 cells of the rat carotid body. J Physiol (Lond) 2000; 523:719–729.
49. Archer SL, Hampl V, Nelson DP, Sidney E, Peterson, DA and Weir EK. Dithionite increases radical formation and decreases vasoconstriction in the lung: evidence that dithionite does not mimic alveolar hypoxia. Circ Res 1995; 77:174–181.

33

O_2 Sensing in Neurons

Evidence and Requirement for a Multitude of Sensors

GABRIEL G. HADDAD

Albert Einstein College of Medicine
and Children's Hospital at Montefiore
New York, New York, U.S.A.

I. Introduction

A large body of experimental work has been performed in a variety of cell types, tissues, and organs studying how tissues and cells sense and respond to O_2 deprivation. O_2 deprivation in these studies has varied in severity and in duration. As a result of this large amount of work over the past two decades, there has been a plethora of ideas and theories regarding O_2 sensing and response of cells to hypoxia. Every tissue that has been studied senses O_2 deprivation and responds in a specific way, and neuronal tissue is no exception. Furthermore, neurons show a variety of responses depending on the severity, and duration of the hypoxia, the age of the animal, and the type and region of the neuron, to name just a few variables.

We and others have focused on the study of nerve and glial cells during O_2 deprivation and have concentrated on their response, from the initial stage, i.e., from sensing, to the stage of injury and irreversible damage and death. In certain instances, when the duration and severity of hypoxia are not too overwhelming, we have evidence that these nerve cells may adapt and possibly survive over prolonged periods of hypoxia. These studies have also put in perspective the various responses of nerve cells to hypoxia since these responses have unmasked the cascades of events that take place inside and outside cells when tissues are exposed to lack of O_2. In particular, different cascades of events can occur depending on several factors. It is these cascades that I will detail in this chapter. The aim in this chapter therefore is to highlight observations that will lend credence to the idea that, depending on the acuteness, severity, and duration of hypoxia, there are a number of potential O_2 sensors in nerve cells. Toward that aim, I will detail results and data regarding, on the one hand, ionic flux and the rapidity with which it occurs in nerve cells, and their control mechanisms. On the other hand, I will detail some newer observations regarding gene activation during O_2 deprivation that occurs in nerve cells and other cell types and that can shed light on mechanisms of O_2 sensing over prolonged O_2 deprivation. Such contrast between short- and long-term hypoxia will therefore emphasize the different sensing mechanisms and the potential multitude of O_2 sensors.

II. Mechanisms of Ionic Flux

A. General

To study the neuronal response to hypoxia with maturation, hypoglossal neurons from adult and neonatal rat (3–7, 14–16, 21, and 28 days) brain stem slices were subjected to O_2 deprivation (1). All neurons depolarized and showed no evidence of hyperpolarization at any time during the hypoxic period. The magnitude of this depolarization was much larger in adult hypoglossal neurons (mean = 32.0 mV) than in young neonatal neurons (mean = 10.4–11.2 mV). The changes in the adult started to occur usually within 20–30 sec of the institution of hypoxia. In the early phase of hypoxia, adult neurons increased peak and steady-state spike frequency to induced current injections. Later, both spike frequencies decreased, and in one-half of adult neurons, there was a depolarization block. Input resistance (R_N) of most adult neurons increased during hypoxia (R_N = 180% of control after 5 min). Though neonatal neurons increased firing frequency, none had depolarization block and there was no increase in R_N. To determine whether the hypoxia-induced depolarization was synaptically mediated, tetrodotoxin (TTX), tetraethylammonium (TEA), apamin, high Mg^{2+}/low Ca^{2+} solutions, and intracellular ethyleneglycol-bis-(aminoethylether) *NNN″N″*· tetracetic acid (EGTA) were used, and these did not reduce the magnitude of depolarization in hypoglossal neurons of 4-week-old and adult rats. Strophanthidin application depolarized hypoglossal neurons but decreased R_N.

The hypoxic response in these slices was also assessed in both newborn and adult slices by examining extracellular K^+ concentrations using ion-selective electrodes to measure extracellular fluid K^+ (K_o^+). This analysis revealed a major increase in K_o^+ (mean = 3.2 mM) in the adult hypoglossal area but not in the newborn tissue (mean = 0.65 mM). This probably reflects a difference in the amount of K^+ efflux between neonatal and adult hypoglossal neurons. Shrinkage in the extracellular compartment in the adult may also account for some of the difference. Under more profound O_2 deprivation, K_o^+ can reach very high levels, sometimes as high as 100 mM.

These results suggest that neonatal neuronal tissue is much less sensitive and more tolerant to hypoxia than the adult, with the inherent cellular properties being maintained in newborn but not in adult neurons. The reason(s) for the difference in K_o^+ is not known but could be due, to some extent, to different rates of intracellular ATP depletion, maturational aspects of K^+ efflux mechanisms, and failure of the Na^+-K^+ pump. How these changes in sensing and in response to O_2 lack are transduced in the newborn versus adult neurons is not well understood. However, it should be clear that under these hypoxic conditions, the overall electrophysiological alterations occurred very quickly after the start of hypoxia. Hence, since the electrophysiological responses to hypoxia do occur very fast, the sensing mechanisms that lead to these responses have to be very fast also, on the order of seconds.

B. Potassium Ionic Flux and Its Regulation

Since K^+ homeostasis has been shown to change in a major way in brain tissue during hypoxia (2), the regulation of this efflux becomes important to understand. Indeed, K^+ channel modulation has been shown to be an integral and important cellular response to O_2 deprivation. Part of this modulation occurs as a result of changes in concentrations of several cytosolic factors such as ATP and Ca^{2+}. We and others have shown that a number of cytosolic factors change during hypoxia and that these, in turn, modulate a number of ion channels, including K^+ channels. Hence, these studies demonstrate that the function of membrane proteins such as K^+ channels is linked to K^+ efflux during hypoxia. It is unknown, however, whether there are mechanisms other than those originating from changes in the cytosol. To test the hypothesis that membrane-delimited mechanisms participate in the O_2-sensing process and are involved in the modulation of K^+ channel activity in central neurons, we performed experiments using patch-clamp techniques and dissociated cells from the rat neocortex and substantia nigra (3,4). Whole-cell outward currents were studied in voltage-clamp mode using Na^+-free or low-Na^+ (5 mM, with 1 μM TTX) extracellular medium. O_2 deprivation produced a biphasic response in current amplitude, i.e., an initial transient increase followed by a pronounced decrease in outward currents. The reduction in outward currents was a reversible process since perfusion with a medium of $PO_2 > 100$ mmHg (1 mmHg = 133 Pa) led to a complete recovery. In cell-free excised membrane patches, we found that a specific K^+ current (large conductance, inhibited by micromolar concentrations of ATP and activated by Ca^{2+}) was reversibly inhibited by lack of O_2. This was characterized by a marked decrease in channel open-state probability and a slight reduction in unitary conductance. The magnitude of channel inhibition by O_2 deprivation was closely dependent on O_2 tension. The PO_2 level for 50% channel inhibition was about 10 mmHg, with little or no inhibition at $PO_2 > 20$ mmHg. Our data showed that O_2 deprivation had no effect on another K^+ current characterized by a much smaller conductance and Ca^{2+} independence. This provides evidence for the selective nature of the hypoxia-induced inhibition of some K^+ channels and therefore provides the first evidence for regulation of K^+ channel activity by O_2 deprivation in cell-free excised patches from central neurons (3,4). How this inhibition occurs is not well understood.

By the use of specific agents that chelate metal in metal-containing O_2-sensing centers, including heme, nonheme iron, copper, and flavin, we also demonstrated that iron-center but not copper-center blockers inhibited the channel in excised patches in a similar fashion as low PO_2. These results strongly suggested that K^+-channel activity is modulated during O_2 deprivation by iron-containing proteins, thus providing evidence for an O_2-sensing mechanism in neuronal membranes. It is important to mention that this sensing mechanism does not depend on cytosolic changes and therefore seems to be different since it is localized in neuronal plasma membranes.

C. Sodium Ionic Flux and Its Regulation

Other important ions whose concentrations change markedly during hypoxia are Na^+ and Cl^-. Indeed, we have shown that anoxia induces a drop in extracellular Na^+ in the brain slice and that removal of extracellular Na^+ prevents the anoxia-induced morphological changes in dissociated hippocampal neurons (5–7). To understand the mechanisms that sense O_2 and lead to acute neuronal swelling during anoxia, we studied the ionic movements of Cl^- and Na^+ during O_2 deprivation in the hypoglossal (XII) neurons of rat brain slices using double-barrelled ion-selective microelectrodes. Baseline extracellular Cl^- and Na^+ activities ($[Cl^-]_o$), and $[Na^+]_o$ were measured in both adult and neonatal brain slices (8,9). During a period of anoxia (4 min), $[Na^+]_o$ decreased by about 40 mM in adult slices while $[Na^+]_o$ did not show any significant change in the neonatal slice. Although anoxia induced a significant decrease of $[Cl^-]_o$ in both adult and neonate, $[Cl^-]_o$ dropped seven times more in the adult than in the neonate. Furthermore, intracellular Cl^- activity ($[Cl^-]_i$) was studied in adult hypoglossal cells. All showed an increase in $[Cl^-]_i$ with O_2 deprivation. We also hypothesized that intracellular Na^+ increases during anoxia. To study intracellular Na^+ (Na_i^+) in isolated neurons, we used the fluorophore Sodium Green in freshly dissociated rat neurons and SBFI in cultured cortical neurons. We found that 10 min of anoxia caused an increase in Na_i^+, with a latency of about 2 min. In these neurons, fluorescence increased by an average of about 20%. We conclude that during anoxia: (1) intracellular $[Cl^-]$ and $[Na^+]$ increase in the adult and this occurs most likely because of entry of extracellular ions into the cytosol, and (2) there is a major maturational difference in mechanisms regulating Cl^- and Na^+ homeostasis between newborn and adult brain tissue. We speculate that these mechanisms may account, at least partly, for the relative insensitivity and tolerance to anoxia in the newborn.

Although it is well known that Na_i^+ increases during hypoxia, the sensors for this ionic alteration are not well documented. It is possible that exchangers/transporters and channels are somehow located within the cascade of events that lead to this increase. However, if one examines the regulation of the voltage-sensitive Na^+ channels during hypoxia, one would be surprised to see the results obtained, keeping in mind the increase in intracellular Na^+ (9). Indeed, hippocampal neurons respond to acute oxygen deprivation with an *inhibition* of whole-cell Na^+ current (I_{Na}) (10–12) (Fig. 1). Kinases can modulate I_{Na} and kinases are activated during

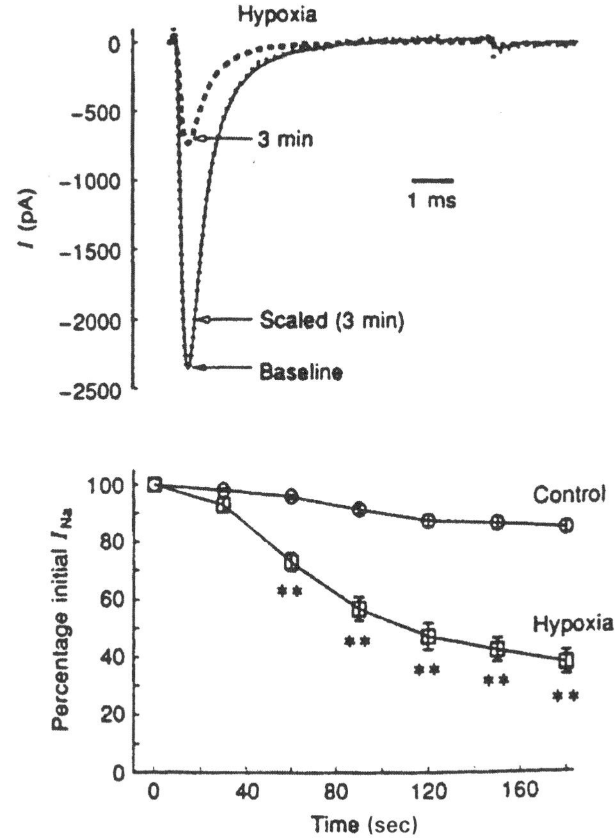

Figure 1 Top: Whole-cell inward Na^+ current (I_{Na}) plotted under baseline (21% O_2, solid line) and after 3 min of hypoxia (dashed line). A scaled inward current is shown (dotted line). Note that the decrease in current is substantial after 3 min and that the kinetics of the inward current did not change in hypoxia. Bottom: Time course of the percentage of initial I_{Na} plotted as a function of time. Note the decrease or inhibition of I_{Na} with hypoxia that starts seconds after the initiation of hypoxia.

hypoxia. Therefore, we hypothesized that kinase activation may play a role in the hypoxia-induced inhibition of I_{Na}. Single-electrode patch-clamp techniques were used in dissociated hippocampal CA1 neurons from the rat. I_{Na} was recorded at baseline, during exposure to kinase activators (with and without kinase inhibitors), and during acute hypoxia (with and without kinase inhibitors). Hypoxia (3 min) reduced I_{Na} to about 40% of initial values, and shifted the steady-state inactivation in the negative direction (10–12). Hypoxia produced no effect on activation or fast inactivation.

Protein kinase A (PKA) activation with adenosine 3′,5′-cyclic adenosine monophosphate, N^6, O_2-dibutyryl, sodium salt (db-cAMP) resulted in a reduction of

I_{Na} to 63% without an effect on activation or steady-state inactivation. I_{Na} was also reduced by activation of protein kinase C (PKC) with phorbol 12-myristate 13-acetate (PMA) to 40% or 1-oleoyl-2-acetyl-sn-glycerol (OAG) to 46%. In addition, steady-state inactivation was shifted in the negative direction by PKC activation. Neither the activation curve nor the kinetics of fast inactivation was altered by PKC activation. The response to PKA activation was blocked by the PKA inhibitor *N*-[2-(*p*-bromocinnamylamino)ethyl]-5-isoquinolinesulphonamide (H-89) and by PKA inhibitory peptide PKA_{5-24} (PKA_i). PKC activation was blocked by the kinase inhibitor 1-(5-isoquinolinesulphonyl)-2-methylpiperazine (H-7), by the PKC inhibitor calphostin C, and by the inhibitory peptide PKC_{19-31} (PKC_i). The hypoxia-induced inhibition of I_{Na} and shift in steady-state inactivation were greatly attenuated by H-7, calphostin C, or PKC_i, but not by H-89 or PKA_i (12). We conclude that hypoxia activates PKC in rat CA1 neurons, and that PKC activation leads to the hypoxia-induced inhibition of I_{Na}. These data would therefore indicate that kinases can modulate I_{Na} and can induce an inhibition, very much like that during hypoxia. How kinases are activated during hypoxia, what is upstream of them is not clear at present. However, they seem to be in the sensing cascade that affects ionic flux, especially Na^+.

From the above analysis, we should highlight a few issues pertaining to O_2 sensing. (1) In the studies detailed above, sensing is very rapid; whatever the sensor is, the reactions that lead to the electrophysiological responses observed using our techniques and approaches must be very quick, on the order of seconds. (2) Most of the alterations are not genetically mediated and no gene expression is presumably altered in this short period.

III. Mechanisms of Response: Longer-Term O_2 Deprivation

While the response to O_2 deprivation over a short (e.g., minutes), acute period is characterized mostly by ionic homeostatic alterations, as detailed above, the response to longer-term (e.g., hours, days, months) hypoxia is manifested by alterations in gene expression, protein levels, and a variety of other changes inside and outside cells. Gene transcription and translation of both plasma and mitochondrial membrane proteins in vertebrates and invertebrates have been studied in our laboratory and others. Below are a few examples that illustrate the idea that a number of proteins change their expression and that these have a major impact on excitability and metabolism in nerve cells. These changes do not take place except after hours and days of O_2 deprivation.

A. Plasma Membrane Proteins

We have shown previously that chronic hypoxia can regulate the expression of membrane proteins. Since there are virtually no glucose stores in the brain and glucose transport can be rate-limiting during stress, the role of glucose transporters becomes crucial for cell survival under stress. We therefore asked whether glucose transporter 1 (GT1), which is expressed in a variety of cells in the brain, especially in the microvessels (for glucose transport from blood vessels to brain), change in response to chronic hypoxia (13). Because major developmental changes occur in the rat central nervous system (CNS) early in life, we studied brain GT1 at different ages. We tested this question in fetuses (from embryonic day 10 to birth), developing rats (from birth to 30 days old), or adult rats (from 90 to 120 days old), and exposed rats to hypoxia (fractional inspired O_2 9%). Our data show that (1) GT1 mRNA level was much lower in the newborn than in the adult and increased with age, (2) chronic hypoxia caused a decrease of ~65% in GT1 mRNA in adult brain but induced an increase in fetal (more than 50%) and developing (~80%) rats, and (3) the response of the housekeeping gene (glyceraldehyde 3-phosphate dehydrogenase) was not similar to that of GT1, suggesting that the changes of GT1 mRNA with hypoxia are specific to glucose transporters. We believe that the increase in GT1 mRNA in brain tissue in response to long-term O_2 deprivation is a reflection of differential reliance on various intermediary metabolic pathways in the immature versus the mature rat brain and that hypoxia induces genetic changes involved in the regulation of glycolysis and glycolytic enzymes (13).

Since we have previously observed that prolonged O_2 deprivation alters membrane protein expression and membrane excitability in the CNS, we studied the effect of prolonged O_2 deprivation on rat cortical and hippocampal neuronal Na^+ channels during postnatal development (14,15). Using electrophysiological recordings in brain slices, RNA analysis (northern and slot blots), and saxitoxin (a specific ligand for Na^+ channels) binding autoradiography, we showed that (1) baseline membrane properties of cortical and hippocampal CA1 neurons from rats chronically exposed to hypoxia were not substantially different from those of naïve neurons, (2) acute stress (e.g., hypoxia) elicited a markedly exaggerated response in the exposed neurons as compared to naïve ones, (3) chronic hypoxia increased Na^+ channel mRNA and saxitoxin-binding density in the cortex and hippocampus as compared to control ones, and (4) the enhanced neuronal response to acute hypoxia in the exposed cortical and CA1 neurons was considerably attenuated by applying TTX, a voltage-sensitive Na^+ channel blocker, in a dose-dependent manner. We conclude that prolonged O_2 deprivation can lead to major electrophysiological disturbances, especially when exposed neurons are stressed acutely, which renders the chronically exposed neurons more vulnerable to subsequent microenvironmental stress. We suggest that this Na^+-channel-related overexcitability is likely to constitute an example of how cells try to sense and adapt to long-term O_2 deprivation. How these nerve cells sense hypoxia and how this sensing is "translated" into alterations of membrane protein expression, including Na^+ channels, is unknown at present.

Since the response to long-term hypoxia in terms of Na^+ channel density may depend on the age of exposure, we also asked whether there is a difference between the adult and prenatal brains regarding Na^+ channel expression with hypoxic exposure. Our results show that Na^+ channel messenger RNA and saxitoxin-binding density decreased after prolonged hypoxia in adult brain homogenates, and this was in sharp contrast to the changes observed in fetal brains, which tended to increase Na^+-channel messenger RNA and protein after hypoxia (14,15). We speculate that the down-regulation of Na^+ channels during long-term hypoxia in mature brains is an adaptive cellular response, aimed at minimizing the mismatch between energy supply and demand, since maintenance of Na^+ gradients is a major energy-requiring process. However, the prenatal brain does not seem to depend on this adaptive mechanism in response to hypoxic stress. Here again, the sensing mechanisms are not quite understood, but it is clear that the response to hypoxia depends on age and it is possible that sensing mechanisms evolve with age in a way to change responses.

B. Mitochondrial and Other Proteins

During long-term hypoxia, not only are plasma membrane proteins affected but also mitochondrial membrane proteins. For example, we have evidence from work that we and others have performed that proteins such as bcl2 are up-regulated (16). In addition, proteins that are involved in the apoptotic pathways are also involved whether they lead to apoptosis or are antiapoptotic. In some of our studies, for instance, we hypothesized that (1) hypoxia alone can induce neuronal apoptosis, (2) hypoxic severity alters the time course of neuronal apoptosis, (3) hypoxia increases neuronal p53, and this increase in p53 is critical for neuronal apoptosis (17). Embryonic neocortical neurons cultured for 7–10 days were studied with O_2 levels set at 0.1%, 1%, and 3% O_2. Under all hypoxic conditions, observed changes in cellular morphology and DNA fragmentation, detected by the TUNEL method and gel electrophoresis, were consistent with apoptosis. These alterations were seen after a shorter period with increasing hypoxic severity. Immunoblot analysis revealed an increase in p53 protein in hypoxia-exposed neurons. Analysis of immunofluorescence-stained neurons revealed increases in p53 with increased duration and severity of hypoxia (Fig. 2). Antisense oligonucleotides for p53 significantly increased the number of surviving neurons during hypoxic exposure (17). Therefore, we conclude that hypoxia-induced neuronal apoptosis is, at least in part, a p53-dependent process whose time course is influenced by hypoxic severity and duration. We also speculate that some of the O_2-sensing mechanisms may be different depending on whether or not apoptotic pathways are triggered.

In a separate series of experiments, we wished to examine some of the adaptive changes that can take place in organisms that clearly tolerate hypoxia. We took advantage of the *Drosophila melanogaster*'s extraordinary resistance to anoxia to study the molecular mechanisms underlying this phenomenon of tolerance. We analyzed mRNA expression of heat shock proteins (HSP) (HSP26 and HSP70), ubiquitins (UB) (UB3 and UB4), cytochrome oxidase I (COXI), and superoxide dismutase (SOD) using slot blot analysis (18). The expression of HSP genes, especially HSP70, was remarkably up-regulated (up to a thousandfold) while those of UB4 and COXI were down-regulated (10–60%) in response to the anoxic stress. The expression of UB3 gene was up-regulated by $1.5\times$ and the expression of SOD gene was not significantly affected. The expression of HSP genes peaked by 15 min into anoxia and then declined but stayed above baseline (18). From these results, we conclude that (1) anoxia differentially regulates gene expression and (2) the up-regulation of HSP70 and down-regulation of UB4 by anoxia are consistent with the

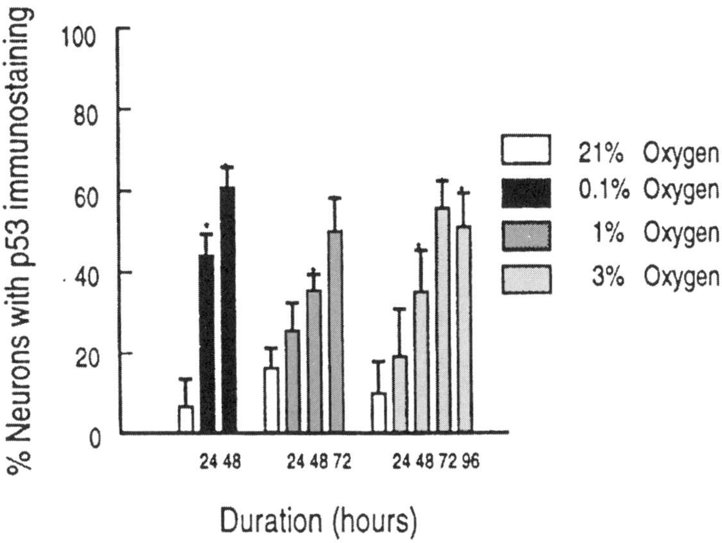

Figure 2 Effect of hypoxia on cellular p53. Relative cellular fluorescence intensity of p53 immunofluorescence-stained neurons was determined by computer-assisted calculation of the ratio of the mean pixel intensity of experimental over control (21% O_2). Note that altering either the severity or duration increased the percentage of neurons showing p53.

idea that *D. melanogaster* resist anoxia, at least in part, by protecting proteins against degradation.

These studies also indicate that a number of proteins change their expression, most likely through either transcriptional regulation or RNA stability. How O_2-sensing mechanisms operate to increase or decrease transcription or alter RNA stability may also be different than, say, those sensing mechanisms that alter cell functions and manifest themselves via very quick electrophysiological responses.

C. Observations Recently Derived from Newer Approaches

Knowing that there are vast differences in the animal kingdom between O_2 tolerant and O_2-sensitive organisms, we thought that it would be beneficial to compare and contrast the responses to hypoxia and the sensing mechanisms to O_2 deprivation between those animals that "sense" and those that are much less sensitive to lack of O_2. Since we were convinced that a large component that could explain the differences in O_2 sensitivity is genetic in nature, we considered a number of potential genetic models for use. It was also clear to us that the freshwater turtle, which was considered an anoxia-tolerant organism par excellence, was not an optimal organism if cutting-edge techniques were to be employed. At that time we discovered the *D. melanogaster* and its resistance to lack of O_2 (19,20). When *Drosophila* fruit flies are subjected to extremely low oxygen concentrations ($\leq 0.01\%$) and their physiological and behavioral responses are studied before, during, and after anoxia, they show a very interesting phenomenon. When they are first exposed to anoxia (complete lack of O_2), *Drosophila* lose coordination, fall, and become motionless after about 30–60 sec in anoxia. However, they tolerate a complete N_2 atmosphere for several hours following which they seem to totally recover without apparent injury. Of interest, mean O_2 consumption per gram of *Drosophila* tissue was substantially reduced in low O_2 concentrations (20% of control) (19,20).

It is interesting to note that O_2 sensing and the resistance to anoxia can be manifested differently in different organisms. For example, *Drosophila* definitely senses the lack of O_2, as it responds quickly in a similar way to mammals; these flies develop anoxic stupor when the O_2 level is very low. In addition, they show a physiological response that is commensurate with this behavioral response. For example, extracellular recordings from flight muscles in response to electrical stimulation reveal that, within a very short period (i.e., seconds), muscle evoked potentials (EP) are totally silenced with anoxia (19,20). Furthermore, there is a complete recovery of muscle EP with reoxygenation, after a latency that is proportional to the anoxic period. This stereotypical response in flies is very different from that in turtles. Turtles do not seem to lose neuronal activity even after very prolonged anoxia (2), extending to hours. Hence, flies, unlike turtles, seem to have different strategies for O_2 sensing as well as for survival under very low O_2 conditions. From the point of view of sensing, flies seem to have mechanisms that render them similar to mammals since they certainly sense the low O_2 condition and respond to it like mammals. However, the recovery from prolonged anoxia in flies seem to be possible, while, in general, this is not the case in mammalian organisms and tissues.

Although some of the genetic models, including *Drosophila*, have been in use for many decades, the conservation of complements of genes with evolution, from prokaryotes to eukaryotes, and from yeast, *Drosophila*, *Caenorhabditis elegans*, zebra fish to man, has become more appreciated in the past decade. With this important discovery of conservation, these models have become even more useful in trying to solve problems relevant to human physiology, biology, and disease. Genes responsible for functions as varied as circadian rhythms, alcohol intoxication, development of tracheal buds, heart chambers, and central nervous system have all been cloned first in model systems (19,21) and then studied in mammals as well as humans. Genes responsible for programmed cell death were first cloned in *c. elegans* (21) and their homologs were found in mammals and humans afterward. The whole field of molecular physiology and, in particular, ion channels received a major boost after the cloning of the K^+ channel from the fruit fly (22).

If our interest is in O_2 sensing and response to low O_2, how does a genetic model, such as *Drosophila*, help us? Since *Drosophila* survives long periods of anoxia, we took advantage of this situation and specifically asked what is the *Drosophila* endowed with genetically to be able to sense but resist anoxia for many hours? To answer this question, we have used a few strategies and obtained very interesting results. Below is a brief outline of the genetic approach that we have used and a summary of the results obtained.

Assuming that there are genes in the *Drosophila* that sense the lack of O_2 and are adaptive and protect cells and tissues from anoxic injury, the main idea is to mutagenize the fly genome and develop a mutagenesis screen to identify flies that have lost these genes, i.e., flies that have loss-of-function mutations. Then these mutations are mapped in detail using marker chromosomes and small deficiencies and then cloned. The genes responsible for the abnormal phenotype are then ascertained in various ways, including the use of transgenic techniques and rescue experiments. There are at least two main advantages with this approach: (1) we start with a phenotype that is useful and is relevant to the question asked; (2) there is no bias in terms of genes and molecules that will be discovered. We therefore mutagenized (x-ray, 4000 rad) wild-type males and crossed them to attached-X females [*c*(1)*ywf*]. By so doing, irradiated males will transmit their mutated X-chromosome to the male offspring (very different from the usual situation in which the male offspring inherits the X-chromosome from the female parent). Therefore, by testing the first male-generation progeny, we could test for phenotype of mutations, irrespective of whether the mutations are recessive or dominant since, like humans and mammals, male flies have one X-chromosome.

In a specialized apparatus, more than 22,000 flies, carrying mutagenized chromosomes, have been screened so far in our laboratory. Since we had already studied in detail the wild-type responses and developed distribution profiles (histograms) about the recovery from anoxia, a threshold (close to the 96th percentile of the wild-type distribution) was used to identify and isolate mutants. We have so far identified 10 mutants that have loss-of function mutations and eight complementation groups (8 mutations have two weaker alleles isolated as well) and the mutants have profoundly altered distribution of recovery times after reoxygenation. The marked delay in recovery after anoxia displayed by these mutant flies suggested that they were much more sensitive to a lack of O_2 (19). To further our understanding of these mutations, we directly examined their effect on CNS function. Identified neurons that can be studied electrophysiologically in *Drosophila* are those of the Giant Fiber system. During reoxygenation, the wild-type flies started to respond by firing evoked potentials (neurons in CNS are stimulated and muscle action potentials recorded across several synapses) after 2 min into recovery. However, flies with mutations had a much longer latency time to firing of the first evoked potential, with some mutant flies requiring up to 25–30 min for the first evoked response (19)!

Mapping of the induced mutations was performed with X-chromosomal markers and complementation tests. Cloning of the first mutation revealed an interesting gene at the basis of the phenotype of anoxia resistance in that particular line of flies. It turned out to be a double-stranded pre-mRNA editase that edits adenosine into inosine and often changes amino acid residues and protein structure and function (23). The targets of this gene are not all delineated but a number of them are related to membrane channels and receptors (23). Hence, this particular gene is responsible for O_2 sensing and tolerance of *Drosophila* to hypoxia. Clearly, this is only one gene that has a major impact and it is only one of potentially many—as our mutagenesis screen would indicate at present.

The current genetic screen has not been saturated. However, we had to screen more than 22,000 mutagenized flies to get the mutations obtained. This suggested that there is a limited number of genes in *Drosophila* that may be involved in O_2 sensitivity and response to hypoxia. Because these mutations profoundly disrupted the recovery from anoxia, we believe that this approach can be used effectively to dissect the genetic basis of resistance to anoxia and to O_2 sensing and help delineate the genes responsible for sensing, responsiveness, and protection against low O_2.

In summary, responses to O_2 deprivation can manifest itself in many ways depending on the tools that are used, on the duration and severity of O_2 deprivation, on tissues and cell types, and on ontogeny and phylogeny of organisms. The sensing mechanisms are bound to be varied and different in various conditions and in various organisms, as we illustrated above. Some of the mechanisms are in membranes, some in the cytosol; others respond very fast, and yet others take a much longer period to take place. Hence, in my opinion, there has to be a multitude of sensing mechanisms for various circumstances and for various functions. Although not irrevocable, the evidence is strong to start considering the existence of multiple O_2 sensors and widening our search.

References

1. Haddad GG, Donnelly DF. O_2 deprivation induces a major depolarization in brain stem neurons in the adult but not in the neonatal rat. J Physiol 1990; 429:411–425.
2. Haddad GG, Jiang C. O_2 deprivation in the central nervous system: on mechanisms of neuronal response, differential sensitivity and injury. Prog Neurobiol 1993; 40(3):277–318.
3. Jiang C, Haddad GG. A direct mechanism for sensing low oxygen levels by central neurons. Proc Natl Acad Sci USA 1994; 91:7198–7201.
4. Jiang C, Haddad GG. Oxygen deprivation inhibits a K^+ channel independently of cytosolic factors in rat central neurons. J Physiol 1994; 481.1:15–26.
5. Haddad GG, Jiang C. Mechanisms of anoxia-induced depolarization in brainstem neurons: in vitro current and voltage clamp studies in the adult rat. Brain Res 1993; 625:261–268.
6. Kwei S, Jiang C, Haddad GG. Acute anoxia-induced alterations in MAP2 immunoreactivity and neuronal morphology in rat hippocampus. Brain Res 1993; 620:203–210.
7. Fung M-L, Haddad GG. Anoxia-induced depolarization in CA1 hippocampal neurons: role of Na^+-dependent mechanisms. Brain Res 1997; 762:97–102.
8. Jiang C, Agulian S, Haddad GG. Cl^- and Na^+ homeostasis during anoxia in rat hypoglossal neurons: intracellular and extracellular in vitro studies. J Physiol 1992; 448:697–708.
9. Friedman JE, Haddad GG. Anoxia induces an increase in intracellular sodium in rat central neurons in vitro. Brain Res 1994; 663:329–334.
10. Lung M-L, Croning MDR, Haddad GG. Sodium homeostasis in rat hippocampal slices during oxygen and glucose deprivation: role of voltage-sensitive sodium channels. Neurosci Letters 1999; 275:41–44.

11. Cummins TR, Donnelly DF, Haddad GG. Effect of metabolic inhibition on the excitability of isolated hippocampal CA1 neurons: developmental aspects. J Neurophysiol 1991; 66(5):1471–1482.
12. O'Reilly JP, Cummins TR, Haddad GG. Oxygen deprivation inhibits Na^+ current in rat hippocampal neurons via protein kinase C. J Physiol 1997; 503.3:479–488.
13. Xia Y, Warshaw JB, Haddad GG. Chronic hypoxia causes opposite effects on glucose transporter 1 mRNA in mature versus immature rat brain. Brain Res 1995; 675:224–230.
14. Xia Y, Haddad GG. Effect of prolonged O_2 deprivation of Na^+ channels: differential regulation in adult versus fetal rat brain. Neuroscience 1999; 94(4):1231–1243.
15. Xia Y, Fung M-L, O'Reilly JP, Haddad GG. Increased neuronal excitability after long-term O_2 deprivation is mediated mainly by sodium channels. Mol Brain Res 2000; 76:211–219.
16. Banasiak KJ, Cronin T, Haddad GG. bcl-2 Prolongs neuronal survival during hypoxia-induced apoptosis. Mol Brain Res 1999; 72:214–225.
17. Banasiak KJ, Haddad GG. Hypoxia-induced apoptosis: effect of hypoxic severity and role of p53 in neuronal cell death. Brain Res 1998; 797:295–304.
18. Ma E, Haddad GG. Anoxia regulates gene expression in the central nervous system of *Drosophila melanogaster*. Mol Brain Res 1997; 46:325–328.
19. Haddad GG, Sun Y-A, Wyman RJ, Xu T. Genetic basis of tolerance to O_2 deprivation in *Drosophila melanogaster*. Proc Natl Acad Sci USA 1997; 94:10809–10812.
20. Krishnan SN, Sun YA, Mohsenin A, Wyman RJ, Haddad GG. Behavioral and electrophysiological responses of *Drosophila melanogaster* to prolonged periods of anoxia. J Insect Physiol 1997; 43:203–210.
21. Haddad GG. Enhancing our understanding of the molecular responses to hypoxia in mammals using *Drosophila melanogaster*. J Appl Physiol 2000; 88:1481–1487.
22. Salkoff L, Baker K, Butler A, Covarrubias M, Pak MD, Wei A. An essential set of K^+ channels conserved in flies, mice and humans. Trends Neurosci 1992; 15(5):161–166.
23. Ma E, Gu XQ, Wu X, Xu T, Haddad GG. Mutation in pre-mRNA adenosine deaminase markedly attenuates neuronal tolerance to O_2 deprivation in *Drosophila melanogaster*. J Clin Invest 2001; 107(6):685–693.

34

Oxygen Sensitivity of Central Cardiorespiratory Regions

JUDITH A. NEUBAUER, JAGADEESHAN SUNDERRAM, NICOLA RITUCCI, and DOMINIC D'AGOSTINO

UMDNJ–Robert Wood Johnson Medical School
New Brunswick, New Jersey, U.S.A.

I. Introduction

It has been known for more than a century that respiration is stimulated by hypoxia. In 1868 Pflüger (1) demonstrated that ventilation increased when dogs inhaled nitrogen. Prior to identification of the hypoxia-sensitive carotid chemoreceptors in 1930 (2), the most commonly held theory was that the respiratory centers themselves were sensitive to hypoxia, presumably by an indirect effect on "central acidity" (2a,2b,3,4). Identification of the carotid and aortic chemoreceptors in 1930 by Heymans and colleagues (2) and the demonstration that, in the absence of intact carotid sinus and aortic depressor nerves, inhalation of nitrogen resulted in only a slight stimulation of respiration represented a major breakthrough in understanding the reflex stimulation of respiration by chemical stimuli, and effectively ended the search for a central oxygen sensor. Interest in identifying central sites of oxygen chemosensitivity shifted toward elucidating the mechanism of oxygen transduction by the carotid body chemoreceptors.

II. Hypoxic Modulation of Central Neural Networks

While reflex stimulation of respiration during hypoxia is ascribed to the arterial chemoreceptors, hypoxia still affects the central structures responsible for processing this afferent information. Since the effect of hypoxia on the central nervous system (CNS) is primarily to reduce excitability of neurons, hypoxic modulation of the central respiratory network tends to attenuate the reflex stimulation from peripheral sources and promote respiratory depression. The depressant effect of central hypoxia has been observed in several situations. For example, the central depressant effect of hypoxia has been used to explain the biphasic nature of the respiratory response to hypoxia in both the neonatal and adult mammal (4a–7), and respiratory depression with hypoxia is a consistent finding in anesthetized chemodenervated animals, with the magnitude of depression directly correlated to the degree of brain hypoxia (8,9). Thus, in general, brain hypoxia causes a reduction in neuronal excitability presumably through mechanisms that may very well conserve energy during the stress of reduced availability of oxygen. However, when brain oxygenation becomes severely compromised, respiratory depression is replaced with a pronounced respiratory excitation in the form of gasping. This gasping pattern of respiration, characterized as a short-duration, high-amplitude burst of activity associated exclusively with inspiratory discharge (10,11), is not terminal gasping but appears to be a functionally important reflex response to severe oxygen distress. Gasping has been found to be a highly effective gas exchange pattern and, in conjunction with enhanced central sympathetic activation, promotes autoresuscitation (12). In fact, failure to gasp has been proposed as a potential cause of sudden infant death syndrome (13) suggesting that the underlying mechanisms responsible for this reflex stimulation of respiratory and sympathetic outputs act as a backup "chemosensor" for these central neural networks.

III. Hypoxic Excitation

The excitation of respiration during severe hypoxia at levels of hypoxia associated with diminished activation of peripheral chemoreceptors suggests that hypoxia is directly exciting some site within the respiratory network to elicit the gasp-like pattern of respiratory activity. Studies investigating the location of this "gasping" center have been largely the work of St John and his associates. Early pontomedullary transection studies pointed to the location of the gasping center as medullary (10). More recent studies by these investigators have localized the medullary site to the lateral tegmental field by showing that electrical stimulation of this region can evoke premature gasps while lesioning this region can eliminate gasping (14,15).

Interestingly, this region of the rostral ventrolateral medulla is in very close proximity to the pre-Bötzinger complex (perhaps overlapping this region), which has been identified as containing neurons essential for rhythm generation both in vitro in neonatal rat brainstem and transverse medullary slice preparations (16,17) and in vivo in the adult cat (18,19) and rat (20). More recently, we have found that chemical activation of the pre-Bötzinger complex using discrete microinjections of a glutamate analog can elicit "gasping" in a normoxic anesthetized cat (21), thus further localizing the gasping site to the same region implicated in the generation of respiratory rhythm. The pre-Bötzinger complex has been characterized as having a high proportion of inspiratory-related and preinspiratory neurons, presumably the essential elements for respiratory rhythm generation (16,19,22,23), about 15% of which have spinal projections (24). That gasping occurs as rhythmic bursts of augmented inspiratory activity without evidence of patterning effects of reciprocal inhibition from expiratory activity could be envisioned to occur if the locus responsible for the generation of rhythm is common to both gasping and eupnea. In the case of severe hypoxia, gasping would supplant eupnea because of a hypoxic depression of the network with a loss of inhibition concomitant with an excitation of the rhythm-generating (pacemaker) site.

Hypoxic excitation of a respiratory-related region is not entirely unexpected since recent work suggests that there are unique sites within the brain that are directly stimulated by hypoxia. For example, sympathoexcitatory neurons in the rostral ventrolateral medulla, as well as pontine and hypothalamic neurons with projections that can excite or inhibit ventilation, have been shown to increase their activity in response to local hypoxia (25–31). Dillon and Waldrop (26), using whole-cell patch recordings in tissue slices of the caudal hypothalamus, found that neurons depolarized and increased their firing rates when exposed to hypoxia. Direct stimulation of this region of the hypothalamus has been shown to result in increases in respiratory frequency (32), suggesting that direct hypoxic stimulation of these neurons may mediate the tachypneic response to hypoxia in awake chemodenervated animals. Additionally, there is indirect evidence suggesting that brain hypoxia may directly stimulate neurons in supramedullary centers that then act to depress respiration. These studies in neonates (25,27,33) found that hypoxic depression is attenuated in fetal and newborn lambs, rabbits, and rats after a site in the upper lateral pons is transected. Direct hypoxic excitation of neurons has also been found for rostral ventrolateral medullary neurons, including both the C1 sympathoexcitatory and the pre-Bötzinger complex neurons.

While reflex stimulation of sympathetic activity during hypoxia is also ascribed to the arterial chemoreceptors, hypoxia increases sympathetic activity in the absence of peripheral chemoreceptors, suggesting a central site of excitation (34,35). Early studies indicated that cerebral ischemia (36) and systemic hypoxia (31) excite reticulospinal sympathoexcitatory neurons in the C1 region of rostral ventrolateral medulla (just caudal to the pre-Bötzinger complex). These neurons are essential for the tonic generation of vasomotor tone and the reflex regulation of arterial blood pressure (37). Discrete microinjection of NaCN into the C1 region of the rostral ventrolateral medulla has been found to increase tonic sympathetic nerve discharge (28) and to excite the rostral ventrolateral medulla spinal sympathoexcitatory neurons in a dose-dependent, reversible manner (30,31). The response of these sympathoexcitatory neurons to systemic hypoxia appears to be chemically selective, since hypercapnia and acidosis do not modify the discharge rates of these neurons (31).

In addition to the hypoxia-sensitive sympathoexcitatory region, the respiratory-related region located just rostral to the C1 sympathoexcitatory region, the pre-Bötzinger complex, is also directly excited by local hypoxia (38). Analogous to the ability to excite sympathoexcitatory neurons with local hypoxia, chemical activation of the pre-Bötzinger complex using discrete microinjections of a glutamate analog or NaCN can elicit augmented respiratory bursts (gasp or sigh-like) in a normoxic anesthetized cat with little change in blood pressure (21,38). In contrast to the sympathoexcitatory C1 region, the pre-Bötzinger complex does exhibit sensitivity to local acidosis (39). Production of local acidosis by microinjection of acetazolamide or methazolamide into the pre-Bötzinger complex produces an increase in respiratory frequency and, in some instances, the appearance of augmented inspiratory bursts. It is noteworthy that these two chemosensitive regions of the rostral ventrolateral medulla, the C1 sympathoexcitatory and pre-Bötzinger complex, while anatomically discrete and functionally distinct, are juxtaposed to each other, perhaps suggesting that there is some anatomical organization of these oxygen-sensing neurons, sharing a common oxygen-sensing mechanism, but functionally distinct by virtue of their separate cardiorespiratory projections.

As mentioned, the pre-Bötzinger complex has been identified in the in vitro neonatal rat brainstem and slice preparations (16,17,40) and in the adult cat (18,19) as containing neurons essential for rhythm generation. The observation that intense stimulation of the pre-Bötzinger complex can elicit augmented inspiratory bursts has suggested to some that the locus responsible for the generation of rhythm (the pre-Bötzinger complex) is common to both augmented and eupneic breathing patterns. Thus, the respiratory patterns of eupnea and gasping reflect a reconfiguration of the respiratory network, which share the same neural elements responsible for rate ("pacemaker") and pattern formation with different expression of these individual components (41). The work of Smith et al. (16) strongly implicates the pre-Bötzinger complex of the rostral ventrolateral medulla as containing the "pacemaker" neurons that are essential for respiratory rhythm generation. In keeping with the hybrid model for respiratory rhythm generation (i.e., the "pacemaker" is conditional), rhythm generation during eupnea may be primarily generated by reciprocal inhibition of inspiratory and expiratory neurons of the network with some undetermined influence from the "pacemaker" (42,43). In contrast, during severe hypoxia, wherein the network generation of rhythm and pattern fails owing to a loss of inhibition, the pacemaker may become the primary driver of the network, which responds by producing a more "square-wave" inspiratory pattern (gasp-like). Considering the circumstances under which a pacemaker-driven system would be most evident, e.g., during severe hypoxia, it would be appropriate if the pacemaker were not depressed by hypoxia but rather excited by hypoxia.

It is noteworthy that gasps and augmented breaths (sighs) share some characteristics in that both have an onset characterized with a rapid rate of rise and a high-amplitude burst of activity associated exclusively with inspiratory discharge (10,11,44–46). Gasps differ from sighs and other augmented breaths largely in their short inspiratory duration, which may reflect the effect of severe hypoxia on the ability to sustain a prolonged inspiratory burst. Of interest when distinguishing between hypoxia-induced gasps and other augmented inspiratory bursts is that hypoxia is not essential for the generation of augmented breaths. Spontaneous sighs are a common feature of normal respiration occurring during both wakefulness and sleep (47–49). Sighs can be elicited in response to stimulation of pulmonary afferents and modified by afferent input from peripheral chemoreceptors, cutaneous stimuli, and behavioral factors (50–54). During sleep, the majority of sighs are associated with electroencephalographic signs of arousal, starting either before or after a sigh (48,55). However, a potent activator of sighs is hypoxia. Hypoxia increases the frequency of sighs presumably owing to both the peripheral and central effects of hypoxia. The integration of increased peripheral chemoreceptor activity with the central excitation of the pre-Bötzinger complex results in the emergence of a greater number of augmented breaths. Thus, sighs and gasps form a continuum of augmented respiratory output from the occurrence of a few spontaneous sighs during normoxia to a predominant pattern of gasping during severe hypoxia. The emergent properties of the pacemaker oscillator (56) in conjunction with the hypoxic sensitivity of the pre-Bötzinger complex are likely the primary determinants of an augmented breath and its frequency.

IV. Cellular Mechanisms Underlying the Neuronal Responses to Hypoxia

A. Hypoxic Depression of Neural Activity

The cellular events that determine whether hypoxia will excite or depress the activity of brainstem neurons are of great interest. Since a reduction in neuronal excitability is the most commonly observed response to hypoxia, the underlying mechanism of this inhibition has been the subject of intense investigation. Numerous studies have examined the reduction in neuronal excitability in a variety of brain regions including the hippocampus (57–59), cortex (60,61), and hypoglossal nucleus (62) and have found that the depression of neuronal activity is due to both presynaptic and postsynaptic effects of hypoxia. The mechanisms responsible for the reduction in excitability appear to vary somewhat with the severity of the hypoxic stimulus and regional location of the neuron as well as with the age of the animal. Mechanisms that have been reported include presynaptic reductions in neurotransmitter release, reuptake or synthesis that act to reduce the net excitatory input (57,63–65), as well as postsynaptic changes in I_{Ca}^{2+} (61,66), gK^+ (58,59,61,67–70), Na^+ channel inactivation (71,72), and mitochondrial depolarization and Ca^{2+} uptake (73). Results obtained in cultured rostral ventral medullary neurons suggests that hypoxic depression of neuronal activity is intrinsic and likely due to a decreased conductance of a cation channel (74). Hypoxia also promotes the metabolic production of adenosine, GABA, and serotonin (75–77) and lactic acid (78), neuromodulators that reduce the excitability of CNS neurons by both pre- and postsynaptic effects. Taken together, these events can result in a spectrum of responses ranging from impairment of function to complete silencing of neurons. In regard to the respiratory network, CNS hypoxia depresses both inspiratory and expiratory neurons in both the ventral respiratory group and dorsal respiratory group (79), and intracellular recordings from respiratory neurons show a loss of both excitatory and inhibitory postsynaptic potentials suggesting a nonselective depression of synaptic transmission (80).

B. Hypoxic Excitation of Neural Activity

The mechanism(s) by which hypoxia directly excites neurons is unresolved. One possibility is that these neurons, which are acting as oxygen "chemosensors," are excited via cellular responses similar to those in carotid chemosensitive cells. Although a great deal is known about the glomus cell and its response to hypoxia, the mechanism of carotid chemosensitivity is still hotly disputed. One of the proposed mechanisms of hypoxic excitation of the glomus cell is a decrease in the conductance of a K^+ channel (81) that ultimately leads to the opening of voltage-dependent Ca^{2+} channels and a Ca^{2+}-mediated release of an excitatory neurotransmitter (82,83). Some similarity between these peripheral oxygen-sensing cells and central rostral ventral medullary "chemosensitive" neurons may be quite likely. For example, NaCN reversibly decreases the K^+ current of C1 bulbospinal neurons identified using retrograde labeling techniques (84). In addition, Sun and Reis (85), using an in vitro brainstem slice preparation, demonstrated that hypoxia induced an excitation of rostral ventrolateral medullary "pacemaker" neurons and this excitation was associated with an increase in a Ca^{2+} conductance. Studies performed on isolated rostral ventrolateral medullary neurons in cell culture, a preparation that avoids most of the space clamp problems associated with voltage-clamping Ca^{2+} channels in an intact preparation, found support for a hypoxia-induced increase in a Ca^{2+} current by demonstrating that excitation of these neurons is likely due to an increased conductance of a cation channel (74,86). However, in addition to an increased conductance in a Ca^{2+} channel, other ion channel candidates are equally likely, including an increased conductance of a Na^+ channel or a mixed Na^+, K^+ channel (I_h).

Another theory that has been postulated to explain the transduction of the hypoxic signal into an increased excitability of the carotid body is that heme-related proteins are important for oxygen sensing in the carotid body (87). For example, Wilson et al. (87a) found that the photochemical action spectra observed during the reversal of the CO effect on the oxygen sensing of the carotid body was consistent with the heme protein cytochrome c oxidase. Cross et al. (88) found a photometrically measurable heme signal that increased with hypoxia, a response that could be attenuated by inhibiting NADPH oxidase. They suggested that this heme protein may contribute to chemoreception in the carotid body by regulating ion channel conductance through its ability to alter the production of H_2O_2, which changes protein conformation by inducing changes in the GSH/GSSG redox system. However, since cGMP levels decrease in the carotid body during hypoxia (89), another likely possibility is that the regulation of ion channel conductance is due to a heme protein linked to guanylyl cylase. Two such proteins present in the carotid body are nitric oxide synthase (NOS) and heme oxygenase (HO) (90). The ventilatory response to hypoxia has been found to be impaired after inhibition of NOS (91) as well as in mice deficient in NOS-3 (92) or knockout mice lacking the enzyme required to activate S-nitrosoglutathione (93). Within the carotid body, NOS is present only in nerve endings and not glomus cells and nitric oxide appears to be an important neurotransmitter in the efferent inhibition of the carotid body (90,94).

HO is also present in the carotid body in the oxygen-sensing glomus cells and inhibition of HO with metalloporphyrins markedly alters the afferent activity of the carotid body (95).

C. Heme Oxygenase

Heme oxygenase converts heme and O_2 to biliverdin and CO (96,97). Recent interest has focused on CO as a biological messenger owing to its ability to activate guanylyl cyclase as well as having direct effects on ion channels (98). Heme oxygenase exists in three forms: a constituitively expressed form (HO-2), a hypoxia inducible form (HO-1), and an isoform whose sequence is highly homologous to that of HO-2, which has been cloned recently but not characterized fully (HO-3) (97,99). HO-2 is found in discrete neuronal populations throughout the brain closely paralleling the distribution of guanylyl cyclase (100,101) as well as in peripheral autonomic ganglia including the petrosal, superior cervical, nodose, and myenteric ganglia (102). In addition to its proposed role in the O_2-sensing mechanism of carotid body glomus cells (95) and the pulmonary vascular endothelium (102,103), a role for HO-mediated production of CO has been proposed to be important for long-term potentiation in the hippocampus (104,105), regulation of cGMP-gated channels in olfactory receptor neurons (106), and regulation of Na,K ATPase activity in the cerebellum (107). Our studies have found both mRNA for HO-2 and HO-2 immunoreactivity in the rostral ventrolateral medulla (108), suggesting that the activation of HO may be important in the hypoxic sensitivity of sympathoexcitatory and pre-Bötzinger complex neurons in the rostral ventrolateral medulla.

In contrast to HO-2, HO-1 is expressed in high concentrations in the liver and spleen but not in the brain or pulmonary arteries under normoxic conditions (109,110). However, in the brain HO-1 mRNA expression is activated by heat shock or the severe oxidative stress of cerebral ischemia (111), a response that is thought to be protective largely because the heme products act as antioxidants (112). The induction of HO-1 expression by hypoxia is regulated through the activation of another O_2-sensitive factor, the HIF alpha subunits (113). Within the rostral ventrolateral medulla, we have found that exposing rats to chronic hypoxia (10% O_2 for 10 days) induces the expression of HO-1 mRNA in this brain region in a relatively specific manner (108). The fact that HO-1 induction during chronic hypoxia is specific to these cells suggests that it is involved in a function unique to these cells, and it may not be acting merely as a general stress response protein. In the pulmonary artery, hypoxic induction of HO-1 is thought to act in an oxygen-sensing capacity to enhance vasodilation in the face of hypoxia of pulmonary vasoconstriction (103,114). The absence of HO-1 in this setting results in maladaptive responses of the right ventricle and pulmonary arteries during chronic hypoxia. For example, studies performed in HO-1 null mice subjected to hypoxia for 5–7 weeks revealed greater ventricular weights, more dilated right ventricles, right ventricular infarcts, and organized mural thrombi than occurred in wild-type mice. In addition, lipid peroxidation and oxidative damage occurred in the right ventricular cardiomyocytes in HO-1-/-, but not wild-type mice (115). These data demonstrate that mice deficient in HO-1 have maladaptive cardiomyocyte responses to chronic hypoxia. Thus, HO-1 induction may provide an important substrate for the cardiorespiratory adaptations to chronic hypoxia.

As already noted, apart from being induced by hypoxia, induction of HO-1 by agents that cause oxidative stress occurs in human skin fibroblasts (116), glial cells and astrocytes (117–119), the retina (120), and cultured retinoblastoma cells (121). The induction of HO-1 in response to oxidative stress is believed to have a protective role through the antioxidant properties of biliverdin and bilirubin. Of particular interest is the fact that in clinical conditions associated with chronic repeated episodes of hypoxia, such as obstructive sleep apnea syndrome (OSA), the episodic occurrence of hypoxia and reoxygenation provides for a potential scenario of not only hypoxia but also repetitive oxidative stress providing two stimuli for the induction of HO-1.

V. Clinical Relevance

Numerous physiological and clinical conditions are associated with hypoxemia, ranging from acute episodes of hypoxia (asthma) to chronic sustained hypoxia (ascent to high altitude, COPD) or chronic intermittent hypoxia (OSA). During these episodes of hypoxia, activation of both peripheral and central sites likely occurs, with the net response reflecting a coordination of these drives by the pattern-generating network.

Chronic hypoxemia is a physiological consequence of ascent to high altitude and a common clinical consequence of a number of respiratory and cardiovascular disorders including COPD, hypoventilation syndromes, and congestive heart failure. Respiratory and sympathetic adaptations to chronic hypoxia have been well characterized. Chronic hypoxia produces ventilatory acclimatization manifested as a secondary increase in breathing over a period of several days (122). Following acclimatization to hypoxia, there is also an increase in respiratory responses to acute hypoxia and hypercapnia (123–125). Chronic hypoxia also results in a progressive increase in sympathetic activity and blood pressure as demonstrated by an elevation of catecholamines in the plasma and urine with high-altitude ascent (126,127).

Most studies examining the mechanisms responsible for respiratory and sympathetic acclimatization during chronic hypoxia have largely focused on characterizing the cellular and biochemical adaptations of the peripheral chemoreceptors. However, changes in the peripheral chemoreceptors may not be the only site along the neuraxis that undergoes adaptation in response to chronic hypoxia. For example, even in the absence of peripheral chemoreceptors, hypoxia can still elicit changes in respiratory and sympathetic activity, suggesting that the central respiratory and sympathetic neural networks are directly sensitive to hypoxia as well. These changes could reflect a generalized increase in the excitability of the cardiorespiratory network or they could reflect changes within the unique sites of the rostral ventrolateral medulla that function to sense changes in oxygenation, the C1-sympathoexcitatory region, and the pre-Bötzinger complex.

In addition to the potential for adaptations in the central oxygen sensors during chronic sustained hypoxia, it is likely that adaptations occur in these central cardiorespiratory oxygen sensors during chronic intermittent hypoxia as well. Of particular interest is understanding what the impact of chronic intermittent hypoxia is on the cardiorespiratory responses and the particular adaptive or maladaptive changes in the cellular processing of the hypoxic stimulus (128–130). For example, how much of the clinical sequalae associated with OSA can be attributed to the responses to hypoxia?

VI. Obstructive Sleep Apnea Syndrome

OSA is characterized by repetitive occlusions of the upper airway during sleep resulting in significant hypoxemia, hypercapnia, arousals, and sleep fragmentation. Of the many pathophysiological consequences of OSA, it is of interest that hypertension occurs with a high frequency (131). In fact, OSA appears to be an independent risk factor for systemic hypertension (132,133). There are multiple factors that could contribute to the acute increase in blood pressure during an obstructive apneic event (134), including increases in sympathetic tone in response to hypoxia or arousal and mechanical events required to overcome airway obstruction. It is likely that all of these factors contribute to the acute increases in blood pressure during an apneic event. However, several studies have shown a major role for hypoxia in mediating both the acute rise in blood pressure during an apnea as well as for the persistent hypertension associated with repetitive apneas. For example, the degree of desaturation during an apneic event is directly related to the magnitude of the increase in blood pressure associated with apnea (135), and supplemental oxygen greatly reduces the increase in blood pressure in subjects with simulated recurrent apneas (136). Increased sympathetic activity during apnea has been shown to occur in OSA patients (137,138) and is likely a direct consequence of the progressive hypoxemia associated with the apneic events (139–141).

OSA is not only associated with elevation of blood pressure during an apneic event, but repetitive apneas lead to a gradual rise in resting blood pressure. For example, patients with sleep apnea have hypertension and high levels of sympathetic tone while awake (142,143). The mechanism underlying this sustained increase in sympathetic tone is largely undefined. As already pointed out, sleep apnea results in a number of changes (hypoxia, hypercapnia, arousal, and sleep fragmentation) all of which could result in persistent elevation of sympathetic tone and daytime blood pressure. However, recent work in rats (144,145) and dogs (146) has shown that chronic intermittent hypoxia alone can induce a sustained increase in daytime blood pressure due to elevated sympathetic tone (147). The neural site(s) responsible for this adaptive increase in blood pressure is not known; however, it is reasonable to speculate that the sympathoexcitatory region in the rostral ventrolateral medulla is likely a potential site of adaptation.

VII. Conclusion

The search for a central oxygen sensor that mediates cardiorespiratory responses has led to the identification of two critical sites within these neural networks: the C1 sympathoexcitatory region and the respiratory pre-Bötzinger complex. Hypoxic excitation of these sites produces increases in sympathetic tone and augmented breaths (gasps, sighs) that, in the extreme condition, likely play an important role in autoresuscitation. The cellular mechanisms utilized by these hypoxia-sensitive neurons are largely unknown but may parallel those mechanisms used to sense oxygen in other chemosensitive cells such as the carotid body. Since most theories of oxygen sensitivity in the carotid body suggest a major role for a heme protein in the transduction process, it is reasonable to expect a similar process will underlie the oxygen sensitivity of the rostral ventrolateral chemosensors. The observation that heme oxygenase isoforms are expressed constituitively (HO-2) and are inducible (HO-1) in these rostral ventrolateral regions of the medulla makes heme oxygenase a plausible candidate for sensing oxygen in these cardiorespiratory sites. The physiological and clinical relevance of these central oxygen sensors in determining the cardiorespiratory responses to acute and chronic hypoxia has become an area of major scientific interest.

References

1. Pflüger E. Ueber die Ursache der Athembewegungen, sowie der Dyspnoë und Apnoë. Pflügers Arch Gesamte Physiol Meschen Tiere 1868; 1:61–106.
2. Heymans C, Bouckaert JJ, Dautrebande L. Sinus carotidien et réflexes respiratoires. II. Influence respiratoire réflexes de l'acidose, de l'alcalose, de l'anhidride carbonique, de l'ion hydrogène et de l'anoxémie. Sinus carotidiens et échanges respiratoires dans les poumons et au-delà des poumons. Arch Int Pharmacodyn Ther 1930; 39:400–448.

2a. Haldane JS, Priestley JG. The regulation of the lung-ventilation. J Physiol 1905; 32:225–266.

2b. Lahiri S. Historical perspectives of cellular oxygen sensing and responses to hypoxia. J Appl Physiol 2000; 88:1467–1473.

3. Winterstein H. The hydrogen ion theory of the regulation of breathing. Pflügers Arch 1921; 187:293–298.
4. Gesell R. On the chemical regulation of respiration: I. The regulation of respiration with special reference to the metabolism of the respiratory center and the coordination of the dual function of hemoglobin. Am J Physiol 1923; 66:40–47.

4a. Andronikou S, Shirahata M, Mokashi A, Lahiri S. Carotid body chemoreceptor and ventilatory response to sustained hypoxia and hypercapnia in the cat. Respir Physiol 1988; 72:361–374.

5. Easton PA, Slykerman LJ, Anthonisen NR. Ventilatory response to sustained hypoxia in normal adults. J Appl Physiol 1986; 61:906–911.
6. Kagawa S, Stafford MJ, Waggener TB, Severinghaus JW. No effect of naloxone on hypoxia-induced ventilatory depression in adults. J Appl Physiol 1982; 52:1030–1034.
7. Lawson EE, Long WW. Central origin of biphasic breathing pattern during hypoxia in newborns. J Appl Physiol 1983; 55:483–488.

8. Neubauer JA, Melton JE, Edelman NH. Modulation of respiration during brain hypoxia. J Appl Physiol 1990; 68:441–451.
9. Neubauer JA, Santiago TV, Posner MA, Edelman NH. Ventral medullary pH and ventilatory responses to hyperfusion and hypoxia. J Appl Physiol 1985; 58:1659–1668.
10. St John WM, Knuth KV. A characterization of the respiratory pattern of gasping. J Appl Physiol 1981; 50:984–993.
11. Zhou D, Wasicko MJ, Hu J-M, St John WM. Differing activities of medullary respiratory neurons in eupnea and gasping. J Appl Physiol 1991; 70:1265–1270.
12. Guntheroth WG, Kawabori I. Hypoxic apnea and gasping. J Clin Invest 1975; 56:1371–1377.
13. French JW, Morgan BC, Guntheroth WG. Infant monkeys—a model for crib death. Am J Dis Child 1972; 123:480–484.
14. St John WM, Bledsoe TA, Sokol HW. Identification of medullary loci critical for neurogenesis of gasping. J Appl Physiol 1984; 56:1008–1019.
15. St John WM, Bledsoe TA, Tenney SM. Characterization by stimulation of medullary mechanisms underlying gasping neurogenesis. J Appl Physiol 1985; 58:121–128.
16. Smith JC, Ellenberger HH, Ballanyi K, Richter DW, Feldman JL. Pre-Bötzinger complex: a brainstem region that may generate respiratory rhythm in mammals. Sci 1991; 254:726–729.
17. Koshiya N, Smith JC. Neuronal pacemaker for breathing visualized in vitro. Nature 1999; 400:360–363.
18. Connelly CA, Dobbins EG, Feldman JL. Pre-Bötzinger complex in cats: respiratory neuronal discharge patterns. Brain Res 1992; 590:337–340.
19. Schwarzacher SW, Smith JC, Richter DW. Pre-Bötzinger complex in the cat. J Neurophysiol 1995; 73:1452–1461.
20. Gray PA, Janczewski WA, Mellen N, McCrimmon DR, Feldman JL. Normal breathing requires pre-Bötzinger complex neurokinin-1 receptor-expressing neurons. Nature Neurosci 2001; 4:927–930.
21. Solomon IC, Edelman NH, Neubauer JA. Patterns of phrenic motor output evoked by chemical stimulation of neurons located in the pre-Bötzinger complex in vivo. J Neurophysiol 1999; 81:1150–1161.
22. Richter DW, Spyer KM. Studying rhythmogenesis of breathing: comparison of in vivo and in vitro models. Trends Neurosci 2001; 24:464–472.
23. Smith JC, Greer JJ, Liu G, Feldman JL. Neural mechanisms generating respiratory pattern in mammalian brainstem-spinal cord in vitro. I. Spaciotemporal patterns of motor and medullary neuron activity. J Neurophysiol 1990; 64:1149–1169.
24. Ellenberger HH, Feldman JL. Subnuclear organization of the lateral tegmental field of the rat. I. Nucleus and ventral respiratory group. J Comp Neurol 1990; 294:202–211.
25. Dawes GS, Gardner WN, Johnston BM, Walker DW. Breathing in fetal lambs: the effect of brainstem section. J Physiol (Lond) 1983; 335:535–553.
26. Dillon GH, Waldrop TG. In vitro response of caudal hypothalamic neurons to hypoxia and hypercapnia. Neuroscience 1992; 51:941–950.
27. Martin-Body RL, Johnston BM. Central origin of the hypoxic depression of breathing in the newborn. Res Physiol 1988; 71:25–32.
28. Mitra J, Dev NB, Trivedi R, Amini S, Ernsberger P, Cherniack NS. Intramedullary sodium cyanide injection of respiratory and vasomotor responses in cats. Res Physiol 1993; 93:71–82.
29. Nolan PC, Waldrop TG. In vivo and in vitro response of neurons in the ventrolateral medulla to hypoxia. Brain Res 1993; 630:101–114.

30. Sun M-K, Jeske IT, Reis DJ. Cyanide excites medullary sympathoexcitatory neurons in rats. Am J Physiol 1992; 262:R182-R189.
31. Sun M-K, Reis DJ. Hypoxia selectively excites vasomotor neurons of rostral ventrolateral medulla in rats. Am J Physiol 1994; 266:R245-R256.
32. Waldrop TG, Bauer RM, Iwamoto GA. Microinjection of GABA antagonists into the posterior hypothalamus elicits locomotor activity and cardiorespiratory activation. Brain Res 1988; 444:84–94.
33. Okada Y, Kawai A, Mückenhoff K, Scheid P. Role of the pons in hypoxic respiratory depression in the neonatal rat. Res Physiol 1998; 111:55–63.
34. Wasicko MJ, Melton JE, Neubauer JA, Krawciw N, Edelman NH. Cervical sympathetic and phrenic nerve responses to progressive brain hypoxia. J Appl Physiol 1990; 68:53–58.
35. Guyenet PG. Neural structures that mediate sympathoexcitation during hypoxia. Res Physiol 2000; 121:147–162.
36. Guyenet PG, Brown DL. Unit activity in nucleus paragigantocellularis lateralis during cerebral ischemia in the rat. Brain Res 1986; 364:301–314.
37. Ross CA, Ruggiero PA, Park DH, Joh TH, Sved AF, Fernandez-Pardal J, Saacerdra JM, Reis DJ. Tonic vasomotor control by the rostral ventrolateral medulla: effect of electrical or chemical stimulation of the area containing C1 adrenaline neurons on arterial pressure, heart rate, and plasma catecholamines and vasopressin. J Neurosci 1984; 4:474–494.
38. Solomon IC, Edelman NH, Neubauer JA. Pre-Bötzinger complex functions as a central hypoxia chemosensor for respiration in vivo. J Neurophysiol 2000; 83:2854–2868.
39. Solomon IC, Edelman NH, O'Neal MH 3rd. CO_2/H^+ chemoreception in the cat pre-Bötzinger complex in vivo. J Appl Physiol 2000; 88:1996–2007.
40. Johnson SM, Koshiya N, Smith JC. Isolation of the kernel for respiratory rhythm generation in a novel preparation: the pre-Bötzinger complex "island." J Neurophysiol 2001; 85:1772–1776.
41. Lieske SP, Thoby-Brisson M, Telgkamp P, Ramirez JM. Reconfiguration of the neural network controlling multiple breathing patterns: eupnea, sighs and gasps. Nature Neurosci 2000; 3:600–607.
42. Rekling JC, Feldman JL. Pre-Bötzinger complex and pacemaker neurons: Hypothesized site and kernel for respiratory rhythm generation. Annu Rev Physiol 1998; 60:385–405.
43. Smith JC, Butera RJ, Koshiya N, Del Negro C, Wilson CG, Johnson SM. Respiratory rhythm generation in neonatal and adult mammals: the hybrid pacemaker-network model. Respir Physiol 2000; 122:131–147.
44. Orem J, Trotter RH. Medullary respiratory neuronal activity during augmented breaths in intact unanesthetized cats. J Appl Physiol 1993; 74:761–769.
45. Takeda M, Matsumoto S. Discharge patterns of dorsal and ventral respiratory group neurons during spontaneous augmented breaths observed in pentobarbital anesthetized rats. Brain Res 1997; 749:95–100.
46. Takeda M, Matsumoto S. Medullary post-inspiratory neuronal and diaphragm post-inspiratory activity during spontaneous augmented breaths in anesthetized rats. Brain Res 1997; 758:218–222.
47. Issa FG, Porostocky S. Effect of sleep on changes in breathing pattern accompanying sigh breaths. Respir Physiol 1993; 93:175–187.
48. Perez-Padilla R, West P, Kryger MH. Sighs during sleep in adult humans. Sleep 1983; 6:234–243.

49. Thach BT, Taeusch HW Jr. Sighing in newborn human infants: role of inflation-augmenting reflex. J Appl Physiol 1976; 41:502–507.
50. Bartlett D. Origin and regulation of spontaneous deep breaths. Respir Physiol 1971;12:230–238.
51. Bendixen HH, Smith GM, Mead J. Pattern of ventilation in young adults. J Appl Physiol 1964; 19:195–198.
52. Cherniack NS, von Euler C, Glogowska M, Homma I. Characteristics and rate of occurrence of spontaneous and evoked augmented breaths. Acta Physiol Scand 1974; 111:349–360.
53. Glogowska M, Richardson PS, Widdicombe JC, Winning AJ. The role of the vagus nerves, peripheral chemoreceptors and other afferent pathways in the genesis of augmented breaths in cats and rabbits. Respir Physiol 1972; 16:179–196.
54. Shea SA, Horner RL, Banner NR, McKenzie E, Heaton R, Yacoub MH, Guz A. The effect of human heart-lung transplantation upon breathing at rest and during sleep. Respir Physiol 1988; 72:131–150.
55. Wulbrand H, McNamara F, Thach BT. Suppression of σ spindle electroencephalographic activity as a measure of transient arousal after spontaneous and occlusion-evoked sighs and startles. Pediatr Res 1998; 44:767–773.
56. Butera RJ Jr, Rinzel J, Smith JC. Models of respiratory rhythm generation in the pre-Bötzinger complex. II. Populations of coupled pacemaker neurons. J Neurophysiol 1999; 81:398–415.
57. Davis AK, Janigro D, Schwartzkoin PA. Effect of tissue preincubation and hypoxia on CA3 hippocampal neurons in the in vitro slice. Brain Res 1986; 370:44–53.
58. Fujiwara N, Higashi H, Shimoji K, Yoshimura M. Effects of hypoxia on rat hippocampal neurones in vitro. J Physiol (Lond) 1987; 384:131–151.
59. Leblond J, Krnjevic K. Hypoxia changes in hippocampal neurons. J Neurophysiol 1989; 62:1–14.
60. Doll CJ, Hochachka PW, Reiner PB. Effects of anoxia and metabolic arrest on turtle and rat cortical neurons. Am J Physiol 1991; 260:R747-R755.
61. Silver IA, Erecinska M. Intracellular and extracellular changes of [Ca^{2+}] in hypoxia and ischemia in rat brain in vivo. J Gen Physiol 1990; 95:837–866.
62. Haddad GG, Donnelly DF. O_2 deprivation induces a major depolarization in brain stem neurons in the adult but not in the neonatal rat. J Physiol (Lond) 1990; 429:411–428.
63. Erecinska M, Nelson D, Wilson DF, Silver IA. Neurotransmitter amino acids in the CNS. I. regional changes in amino acid levels in rat brain during ischemia and reperfusion. Brain Res 1984; 304:9–22.
64. Gibson GE, Shimada M, Blass JP. Alterations in acetylcholine synthesis and in cyclic nucleotides in mild cerebral hypoxia. J Neurochem 1978; 31:757–760.
65. Young JN, Somjen GG. Suppression of presynaptic calcium currents by hypoxia in hippocampal tissue slices. Brain Res 1992; 573:70–76.
66. Mironov SL, Richter DW. Hypoxic modulation of L-type Ca(2+) channels in inspiratory brainstem neurones: intracellular signalling pathways and metabotropic glutamate receptors. Brain Res 2000; 869:166–77.
67. Hansen AJ, Hounsgaard J, Jahnsen H. Anoxia increases potassium conductance in hippocampal nerve cells. Acta Physiol Scand 1982; 115:301–310.
68. Jiang C, Sigworth FJ, Haddad GG. Oxygen deprivation activates an ATP-inhibitable K^+ channel in substantia nigra neurons. J Neurosci 1994; 14:5590–5602.

69. Murphy KPSJ, Greenfield SA. Neuronal selectivity of ATP-sensitive potassium channels in guinea pig substantia nigra by response to anoxia. J Physiol (Lond) 1992; 453:167–183.
70. Mironov SL, Richter DW. Intracellular signalling pathways modulate K(ATP) channels in inspiratory brainstem neurones and their hypoxic activation: involvement of metabotropic receptors, G-proteins and cytoskeleton. Brain Res 2000; 853:60–67.
71. Cummins TR, Jiang C, Haddad GG. Human neocortical excitability is decreased during anoxia via sodium channel modulation. J Clin Invest 1993; 91:608–615.
72. Mironov SL, Richter DW. Cytoskeleton mediates inhibition of the fast Na^+ current in respiratory brainstem neurons during hypoxia. Eur J Neurosci 1999; 11:1831–1834.
73. Mironov SL, Richter DW. Oscillations and hypoxic changes of mitochondrial variables in neurons of the brainstem respiratory centre of mice. J Physiol 2001; 533:227–236.
74. Mazza E Jr, Edelman NH, Neubauer JA. Hypoxic excitation in neurons cultured from the rostral ventrolateral medulla of the neonatal rat. J Appl Physiol 2000; 88:2319–2329.
75. Winn HR, Rubio R, Berne RM. Brain adenosine concentration during hypoxia in rats. Am J Physiol 1981; 241:H235–H242.
76. Richter DW, Schmidt-Garcon P, Pierrefiche O, Bischoff AM, Lalley PM. Neurotransmitters and neuromodulators controlling the hypoxic respiratory response in anaesthetized cats. J Physiol 1999; 514:567–578.
77. Centonze D, Saulle E, Pisani A, Bernardi G, Calabresi P. Adenosine-mediated inhibition of striatal GABAergic synaptic transmission during in vitro ischaemia. Brain 2001; 124:1855–1865.
78. Neubauer JA, Simone A, Edelman NH. Role of brain lactic acidosis in hypoxic depression of respiration. J Appl Physiol 1988; 65:1324–1331.
79. Bianchi AL, St John WM. Changes in antidromic latencies of medullary respiratory neurons in hypercapnia and hypoxia. J Appl Physiol 1985; 59:1208–1213.
80. Richter DW, Bischoff A, Anders K, Bellingham M, Windhorst U. Response of the medullary respiratory network of the cat to hypoxia. J Physiol (Lond) 1991; 443:231–256.
81. Lopez-Lopez JR, Gonzalez C, Urena J, Lopez-Barneo J. Low pO_2 selectively inhibits K channel activity in chemoreceptor cells of the mammalian carotid body. J Gen Physiol 1989; 93:1001–1015.
82. Biscoe TJ, Duchen MR. Responses of Type I cells dissociated from the rabbit carotid body to hypoxia. J Physiol (Lond) 1990; 428:39–59.
83. Biscoe TJ, Duchen MR, Eisner DA, O'Neill SC, Valeolmillos M. Measurements of intracellular Ca^{2+} in dissociated Type I cells of the rabbit carotid body. J Physiol 1989; 416:421–434.
84. Wang G, Zhou P, Repucci MA, Golanov EV, Reis DJ. Specific actions of cyanide on membrane potential and voltage-gated ion currents in rostral ventrolateral medulla neurons in rat brainstem slices. Neurosci Lett 2001; 309:125–129.
85. Sun M-K, Reis DJ. Hypoxia-activated Ca^{2+} currents in pacemaker neurones of rat rostral ventrolateral medulla in vitro. J Physiol (Lond) 1994; 476:101–116.
86. Kawai Y, Qi J, Comer AM, Gibbons H, Win J, Lipski J. Effects of cyanide and hypoxia on membrane currents in neurones acutely dissociated from the rostral ventrolateral medulla of the rat. Brain Res 1999; 830:246–257.
87. Prabhakar NR, Overholt JL. Cellular mechanisms of oxygen sensing at the carotid body: meme proteins and ion channels. Respir Physiol 2000; 122:209–221.

87a. Wilson DF, Mokashi A, Chugh D, Vinogradov S, Osanai S, Lahiri S. The primary oxygen sensor of the cat carotid body is cytocrome a_3 of the mitochondrial respiratory chain. FEBS Lett 1994; 351:370–374.
88. Cross AR, Henderson L, Jones OTG, Delpiano MA, Hentschel J, Acker H. Involvement of an NAD(P)H oxidase as a pO_2 sensor protein in the rat carotid body. Biochem J 1990; 272:743–747.
89. Wang WJ, Cheng GF, Dinger BG, Fidone SJ. Effects of hypoxia on cyclic nucleotide formation in rabbit carotid body in vitro. Neurosci Lett 1989; 105:164–168.
90. Prabhakar NR. NO and CO as second messengers in oxygen sensing in the carotid body. Respir Physiol 1999; 115:161–168.
91. Haxhiu MA, Chang CH, Dreshaj IA, Erokwu B, Prabhakar NR, Cherniack NS. Nitric oxide and ventilatory response to hypoxia. Respir Physiol 1995; 101:257–266.
92. Kline DD, Yang T, Premkumar DR, Thomas AJ, Prabhakar NR. Blunted respiratory responses to hypoxia in mutant mice deficient in nitric oxide synthase-3. J Appl Physiol 2000; 88:1496–1508.
93. Lipton AJ, Johnson MA, Macdonald T, Lieberman MW, Gozal D, Gaston B. S-nitrosothiols signal the ventilatory response to hypoxia. Nature 2001; 413:171–174.
94. Wang ZZ, Stensaas LJ, Dinger BG, Fidone SJ. Nitric oxide mediates chemoreceptor inhibition in the cat carotid body. Neuroscience 1995; 65:217–229.
95. Prabhakar NR, Dinerman JL, Agani FH, Snyder SH. Carbon monoxide: a role in carotid body chemoreception. Proc Natl Acad Sci USA 1995; 92:1994–1997.
96. Ingi T, Chiang G, Ronnett GV. The regulation of heme turnover and carbon monoxide biosynthesis in cultured primary rat olfactory receptor neurons. J Neurosci 1996; 16:5621–5628.
97. Maines MD. Heme oxygenase: function, multiplicity, regulatory mechanisms, and clinical applications. FASEB J 1988; 2:2557–2568.
98. Wang R, Wu L, Wang Z. The direct effect of carbon monoxide on K_{Ca} channels in vascular smooth muscle cells. Pflügers Arch Eur J Physiol 1997; 434:285–291.
99. McCoubrey WK, Huang TJ, Maines MD. Isolation and characterization of a cDNA from the rat brain that encodes hemoprotein heme oxygenase-3. Eur J Biochem 1997; 247:725–732.
100. Maines MD, Mark JA, Ewing JF. Heme oxygenase, a likely regulator of cGMP production in the brain: incuction in vivo of HO-1 compensates for depression in NO synthase activity. Mol Cell Neurosci 1993; 4:398–405.
101. Verma A, Hirsch DJ, Glatt CE, Ronnett GV, Synder SH. Carbon monoxide: a putative neural messenger. Science 1993; 259:381–384.
102. Zakhary R, Gaine SP, Dinerman JL, Ruat M, Flavahan NA, Snyder SH. Heme oxygenase 2: endothelial and neuronal localization and role in endothelium-dependent relaxation. Proc Natl Acad Sci USA 1996; 93:795–798.
103. Morita T, Perrella MA, Lee M-E, Kourembanas S. Smooth muscle cell-derived carbon monoxide is a regular of vascular cGMP. Proc Natl Acad Sci USA 1995; 92:1475–1479.
104. Stevens CF, Wang Y. Reversal of long-term potentiation by inhibitors of haem oxygenase. Nature 1993; 364:147–148.
105. Zhou M, Small SA, Kandel ER, Hawkins RD. Nitric oxide and carbon monoxide produce activity-dependent long-term synaptic enhancement in hippocampus. Science 1993; 260:1946–1950.

106. Leinders-Zufall T, Shepherd GM, Zufall F. Regulation of cyclic nucleotide-gated channels and membrane excitability in olfactory receptor cells by carbon monoxide. J Neurophysiol 1995; 74:1498–1508.
107. Nathanson JA, Scavone C, Scanlon C, McKee M. The cellular Na^+ pump as a site of action for carbon monoxide and glutamate: a mechanism for long-term modulation of cellular activity. Neuron 1995; 14:781–794.
108. Mazza E, Thakker-Varia S, Tozzi CA, Neubauer JA. Expression of heme oxygenase in the oxygen-sensing regions of the rostral ventrolateral medulla. J Appl Physiol 2001; 91:379–385.
109. Cruse I, Maines MD. Evidence suggesting that the two forms of heme oxygenase are products of different genes. J Biol Chem 1988; 263:3348–3353.
110. Sun Y, Rotenberg MO, Maines MD. Developmental expression of heme oxygenase isozymes in rat brain. J Biol Chem 1990; 265:8212–8217.
111. Paschen W, Uto A, Djuricic B, Schmitt J. Hemeoxygenase expression after reversible ischemia of rat brain. Neurosci Lett 1994; 180:5–8.
112. Stocker R, Yamamoto Y, McDonagh AF, Glazer AN, Ames BN. Bilirubin is an antioxidant of possible physiological importance. Science 1987; 235:1043–1046.
113. Ivan M, Kondo K, Yang H, Kim W, Valiando J, Ohh M, Salic A, Asara JM, Lane WS, Kaelin WG Jr. HIF[alpha] targeted for VHL-mediated destruction by proline hydroxylation: implications for O_2 sensing. Science 2001; 292:464–468.
114. Morita T, Mitsialis SA, Koike H, Liu Y, Kourembanas S. Carbon monoxide controls the proliferation of hypoxic vascular smooth muscle cells. J Biol Chem 1997; 272:32804–32809.
115. Yet S-F, Perrella M, Layne MD, Hseih C-M, Maemura K, Kobzik L, Wiesel P, Christou H, Kourembanas S, Lee M-E. Hypoxia induces severe right ventricular dilatation and infarction in heme oxygenase-1 null mice. J Clin Invest 1999; 103:R23–R29.
116. Keyse SM, Tyrrell RM. Heme oxygenase is the major 32-kDa stress protein induced in skin fibroblasts by UVA radiation, hydrogen peroxide, and sodium arsenite. Proc Natl Acad Sci USA 1989; 86:99–103.
117. Ewing JF, Maines MD. Rapid induction of heme oxygenase-1 mRNA and protein by hyperthermia in rat brain: heme oxygenase-2 is not a heat shock protein. Proc Natl Acad Sci USA 1991; 88:5354–5358.
118. Ewing JF, Haber SN, Maines MD. Normal and heat-induced patterns of expression of heme oxygenase–1 in rat brain: hyperthermia causes rapid induction of mRNA and protein. J Neurochem 1992; 58:4023–4029.
119. Matz P, Turner C, Weinstein PR, Massa SM, Panter SS, Sharp FR. Heme oxygenase-1 induction in glia throughout rat brain following experimental subarachnoid hemorrhage. Brain Res 1991; 713:211–222.
120. Kutty RK, Kutty G, Wiggert B, Chader GJ, Darrow RM, Organisciak DT. Induction of heme oxygenase-1 in the retina by intense visible light: suppression by the antioxidant dimethylthiolurea. Proc Natl Acad Sci USA 1995; 92:1171–1181.
121. Kutty G, Hayden B, Osawa Y, Wiggert B, Chader GJ, Kutty RK. Heme oxygenase: expression in human retina and modulation by stress agents in human retinoblastoma cell model system. Curr Eye Res 1992; 11:153–160.
122. Dempsey JA, Forster HV. Mediations of ventilatory adaptations. Physiol Rev 1982; 62:262–346.
123. Aaron EA, Powell FL. Effect of chronic hypoxia on hypoxic ventilatory response in awake rats. J Appl Physiol 1993; 74:1635–1640.

124. Engwall MJA, Bisgard GE. Ventilatory responses to chemoreceptor stimulation after hypoxic acclimatization in awakc goats. J Appl Physiol 1990; 69:1236–1243.
125. Forster HV, Dempsey JA, Birnbaum ML, Reddam WG, Thoden J, Grover RE, Rankin J. Effect of chronic exposure to hypoxia on ventilatory responses to CO_2 and hypoxia. J Appl Physiol 1971; 31:586–592.
126. Cunningham WL, Becker EJ, Kreuzer F. Catecholamines in plasma and urine at high altitude. J Appl Physiol 1965; 20:607–610.
127. Sharma SC, Hoon RS, Balasubramanian V, Chadha KS. Urinary catecholeamine excretion in temporary residents of high altitude. J Appl Physiol 1978; 44:725–727.
128. Neubauer JA. Invited Review: Physiological and pathophysiological responses to intermittent hypoxia. J Appl Physiol 2001; 90:1593–1599.
129. Gozal E, Gozal, D. Respiratory plasticity following intermittent hypoxia: developmental interactions. J Appl Physiol 2001; 90:1995–1999.
130. Mitchell GS, Baker TL, Nanda SA, Fuller DD, Zabka AG, Hodgeman BA, Bavis RW, Mack KJ, Olson EB Jr. Invited review: Intermittent hypoxia and respiratory plasticity. J Appl Physiol 2001; 90:2466–2475.
131. Fletcher E.C. The relationship between systemic hypertension and obstructive sleep apnea: facts and theory. Am J Med 1995; 98:118–128.
132. Kiselk J, Clark M, Pera V, Rosenberg C, Redline S. The association between hypertension and sleep apnea in obese patients. Chest 1993; 104:775–780.
133. Young T, Peppard P, Palti M, Mae Hla K, Finn L, Morgan B, Skatrud J. Population-based study of sleep-disordered breathing as a risk factor for hypertension. Arch Intern Med 1997; 157:1746–1752.
134. Fletcher EC. Sympathetic activity and blood pressure in the sleep apnea syndrome. Respiration 1997; 64:22–28.
135. Shepard JW Jr. Gas exchange and hemodynamics during sleep. Med Clin North Am 1985; 69:1234–1263.
136. Van Der Aardweg JG, Karemaker JM. Repetitive apneas induce periodic hypertension in normal subjects through hypoxia. J Appl Physiol 1992; 72:821–827.
137. Hedner, J, H Ejnall, J Sellgren, T Hedner and G Wallin. High muscle nerve sympathetic activity in the sleep apnea syndrome. J Hypertens 1988; 6:S529–S531.
138. Leuenberger UA, Mawji Z, Waravdekar NV, Zwillich CW. Sympathetic neural responses to transient hypoxia are increased in obstructive sleep apnea. Circulation 1996; 94(S1):I544.
139. Cantu RC, Nahas CG, Manger WC. Effect of hypercapnic acidosis and of hypoxia on adrenal catecholamine output of the spinal dog. Proc Soc Exp Biol Med 1966; 122:433–437.
140. Heistad DD, Abboud FM. Circulatory adjustments to hypoxia. Circulation 1988; 61:463–470.
141. Somers VK, Mark AL, Zavala DC, Abboud FM. Influence of ventilation and hypocapnia on sympathetic nerve responses to hypoxia. J Appl Physiol 1989; 67:2095–2100.
142. Carlson JT, Hedner J, Elam M, Ejnell H, Sellgren J, Wallin BG. Augmented resting sympathetic activity in awake patients with obstructive sleep apnea. Chest 1993; 103:1763–1768.
143. Somers VK, Dyken ME, Clary MP, Abboud FM. Sympathetic neural mechanisms in obstructive sleep apnea. J Clin Invest 1995; 96:1897–1904.

144. Fletcher EC, Bao G, Li R. Renin activity and blood pressure in response to chronic episodic hypoxia. Hypertension 1999; 34:309–314.
145. Fletcher EC, Lesske J, Qian W, Miller CC III, Unger T. Repetitive episodic hypoxia causes elevation of blood pressure in rats. Hypertension 1992; 19:555–561.
146. Brooks D, Horner RL, Kozar LG, Render-Teixeir CL, Phillipson EA. Obstructive sleep apnea as a cause of systemic hypertension. Evidence from a canine model. J Clin Invest 1997; 99:106–109.
147. Fletcher EC, Lesske J, Culman J, Miller CC, Unger T. Sympathetic denervation blocks blood pressure elevation in episodic hypoxia. Hypertension 1992; 20:612–619.

35

Oxygen Sensing by the Brainstem in Respiratory Control

IRENE C. SOLOMON and NORMAN H. EDELMAN

State University of New York at Stony Brook
Stony Brook, New York, U.S.A.

I. Introduction

That hypoxia modulates breathing has long been recognized. Prior to the discovery of the function of the arterial chemoreceptors in 1930 (1), the locus of this phenomenon had been assigned to the central nervous system (CNS). With that discovery, however, full attention was paid to the transducer role of the carotid bodies in the ventilatory response to hypoxia, and only in the last two decades or so has attention returned to the consideration that the CNS may serve as the initial transducer as well as an important modulator in the increasingly large array of recognized ventilatory responses to acute and more prolonged hypoxia.

Both depressant and excitatory effects of hypoxia have been assigned to the loci primarily in the CNS and for each, both direct and indirect mechanisms have been proposed. In this review, direct mechanisms refer to a proposed chemoreceptor-like function in specific loci in the CNS. Indirect mechanisms refer to a variety of phenomena such as disinhibition of diencephalic facilitation of respiratory activity by the depressant effect of hypoxia on the cortex or the increase of levels of both excitatory and inhibitory neuromodulators in the nucleus tractus solitarius (NTS) associated with stimulation of the peripheral chemoreceptors.

It is the thesis of this review that for the most part, both the excitatory and depressant ventilatory effects of CNS hypoxia are specific adaptive processes, phylogenetically and perhaps ontologically conserved, rather than nonspecific events brought about by insufficient substrate availability for neuronal energy metabolism.

II. Depressant Effects of CNS Hypoxia

A. Manifestations of CNS-Mediated Respiratory Depressant Effects of Hypoxia

Conceptually, the most straightforward way to demonstrate the ventilatory effects of brain hypoxia is to expose peripherally chemodenervated animals to a hypoxic environment. The results of such studies, however, are not uniform with the major confounding variable being arousal state. In awake, unanaesthetized, chemodenervated animals both excitation and depression have been reported (as reviewed in Ref. 2). The response most often observed consists of tachypnea with variable effects on tidal volume and little or no change in alveolar ventilation. In one study, the tachypnea was related to an increase in systemic metabolic rate and reversed by adrenergic blockade (3).

In anesthetized, chemodenervated animals, the response to systemic hypoxia is highly predictable. Progressive CNS hypoxia first reduces the amplitude of inspiratory (e.g., phrenic) and expiratory (e.g., triangularis sterni) nerve discharges followed by a decrease of burst frequency (4,5). The hypoglossal nerve discharge seems more sensitive to the depressant effects of hypoxia than the phrenic nerve discharge (6), and expiratory nerve discharges are more readily depressed than inspiratory nerve discharges (5,7).

Within the brainstem, there is a decrease in the discharge rate of respiratory neurons located in the dorsal respiratory group, the ventral respiratory group, and the Bötzinger complex (8,9). Intracellular recordings from respiratory neurons have demonstrated a loss of EPSPs and IPSPs, indicating that both excitatory and inhibitory presynaptic inputs are reduced by CNS hypoxia (9). In addition, there is a modulation in the phase relationships of inspiratory-related neuronal activities, as demonstrated by an apparent phase shift of late-inspiratory neurons, suggesting that they either fire earlier in the respiratory cycle or silence before the early-inspiratory neurons (10). In our model of progressive brain hypoxia in the deafferented cat, phrenic silence follows the reduction in inspiratory neural output and usually occurs when the arterial O_2 content is reduced by $\sim$50% (4). It is important to note that, at this point, the capacity for neuronal excitation remains intact as evidenced by stimulation of phrenic nerve activity in response to CO_2 (4) and carotid sinus nerve stimulation (11), enhanced tonic discharge in the preganglionic cervical sympathetic nerve (although the respiratory phasic discharge is silenced with phrenic silence) (12,13), and the fact that progression to more severe hypoxia elicits excitation of respiratory output in the form of gasping (12–16).

In the "pregasping" phase of progressive brain hypoxia, however, some other respiratory excitatory phenomena are diminished. For example, we have shown that the phenomenon of poststimulus excitation ("afterdischarge") is blunted by progressive CNS hypoxia. In these experiments using the deafferented animal model, before hypoxia, stimulation of the central end of the cut carotid sinus nerve evoked augmentation of the amplitude of phrenic nerve bursts that did not return to baseline immediately upon cessation of stimulation, but declined exponentially with a time constant of approximately 60 sec (11). Progressive brain hypoxia, produced by inhalation of 0.5% carbon monoxide, blunted this phenomenon in a graded fashion such that there was a direct relationship between arterial O_2 content and the time constant of the decay of phrenic nerve activity (11). Since poststimulus excitation has been proposed to play a role in stabilization of respiratory output (i.e., prevention or reduction of apnea) in the face of intermittent hypoxia, its loss may contribute to cyclic breathing with prolonged apneas in settings such as the obstructive sleep apnea syndrome.

Associated with poststimulus excitation of phrenic nerve amplitude is a decline below baseline of respiratory frequency. Recent studies suggest that posthypoxic frequency decline, which is seen in the carotid sinus intact animal, may be mediated in the ventrolateral pons (17,18), and that the phenomenon appears to require NMDA receptor activation (19).

A much-studied respiratory depressant effect of hypoxia that is either directly or indirectly mediated by the CNS has been termed "hypoxic ventilatory decline." That the ventilatory decline with arterial hypoxia in carotid sinus nerve (CSN) denervated anesthetized cats with maintained arterial P_{CO_2} could be reversed by electrical stimulation of the cut CSN suggests that the CSN input maintained ventilation with concomitant central nervous system (CNS) hypoxic depression (19a). This evidence also suggests that hypoxic CNS inhibition is an active process (see later, Refs. 2,4,11). In adult humans and some animals, when isocapnic hypoxia is produced in a square-wave fashion, the initial enhancement of respiratory output is followed, in about 5–7 min, by a decline to approximately 50% of the peak level (20–22). The phenomenon is best observed in the unanaesthetized state, but may be seen in anesthetized animals as well. It is far more prominent in the early neonate (1–5 days in humans) where the decline in respiratory output may be to baseline or below the initial level of respiratory output. Recent studies in the neonate have demonstrated a significant genetic variation to expression of this phenomenon (23).

B. Mechanisms of CNS-Modulated Respiratory Depressant Effects of Hypoxia

There is strong evidence that all of the phenomena described above are not related to inadequate metabolic substrate for neuronal activation, but may be considered to be "active" inhibitory processes (2). The evidence includes the facts that the phenomena take place at levels of oxygenation well above those associated with energy depletion (24), responses to CO_2 and carotid sinus nerve stimulation remain intact (4,11), and in the progressive brain hypoxia model, more severe hypoxia results in a shift from phrenic silence to respiratory excitation that takes the form of gasping (12,13,16). In addition, the concept of "active inhibition" is buttressed by the finding of increased antidromic latencies in most medullary respiratory-modulated neurons during mild to moderate hypoxia, indicating that these neurons are hyperpolarized (25).

The source of the hyperpolarization almost definitely includes either a global or regionally specific alteration in the balance of inhibitory and excitatory neurotransmitters/neuromodulators. In the progressive brain hypoxia model, good evidence exists for an important role played by increased activity of GABA, acting at the $GABA_A$ receptor (26). In the neonate, there is good evidence for significant mediation of the depressant effect by adenosine (27,28) and elaboration of endogenous opiates (29). In the adult, however, less strong evidence exists for roles played by adenosine, opiates, serotonin, and acidosis due to increased lactate elaboration (as reviewed in Ref. 2). Intermediate molecular events in hypoxic ventilatory depression continue to be discovered. Recently, Gozal and colleagues (30) have shown that release of platelet-derived growth factor-BB and activation of its receptor are critical components of hypoxic ventilatory decline.

It is important to note that modulation of brain neuroeffectors by systemic hypoxia need not be a manifestation of a direct effect of hypoxia on brain metabolism; indirect effects may be important in both depressant and excitatory phenomena. Thus, in adult awake cats and rats, the phenomenon of hypoxic ventilatory decline is abolished by carotid body denervation, implying that stimulation of respiration by the carotid bodies is necessary for the subsequent depression to take place (31,32). Stimulation of the carotid bodies has been shown to release both excitatory and inhibitory neuromodulators in the NTS. Among the latter is dopamine, and to that point, the dopamine antagonist haloperidol has been shown to block hypoxic ventilatory decline (33). At the cellular level, the initial response to a reduction in oxygen is a transient intracellular alkalinization followed by acidosis (34). Potassium conductance rises (35) resulting in an increase in membrane hyperpolarization and a reduction in neuronal excitability (35,36). Further, there is a large negative shift in the ATP-dependent inactivation of the voltage-dependent sodium current (36). These events are readily reversible upon reoxygenation.

The picture is somewhat different in the neonate where transection studies have repeatedly suggested various pontine (37) and suprapontine sites as loci for the inhibitory effects upon ventilation of systemic hypoxia. A recent study implicates the parafasicular nuclear region of the thalamus (38). Supramedullary mediation of hypoxic depression has also been shown in the amphibian brain (39).

As pointed out in a previous review (2), the adaptive value of an "active" process of CNS-mediated hypoxic depression of breathing would seem to be to conserve energy. This is well known in a variety of lower (than mammals) species, especially those known to be unusually tolerant of hypoxic environments. A good example may be seen in freshwater turtles, which can survive in virtually O_2-free cold water for prolonged periods by suppressing motor and neuronal activity. Interestingly, this species seems to accomplish this using a mechanism observed in mammals, increased GABAergic inhibition of neuronal activity (40).

In mammals, the adaptive value of hypoxic depression is most clearly apparent in the fetus. In the uterine environment a respiratory response to hypoxia would not increase O_2 delivery to tissues but would increase O_2 consumption. However, the peripheral chemoreceptors must have some degree of functional development in utero to deal with hypoxic challenges in the immediate postnatal period. A good solution to this problem is a well-developed central mechanism for hypoxic depression. This probably accounts for the greater-than-adult hypoxic ventilatory depression observed in the early neonate as development of the full adult ventilatory response to hypoxia involves both a decline of hypoxic ventilatory depression and full maturation of the peripheral hypoxic chemoreflex.

III. Excitatory Effects of CNS Hypoxia

A. Manifestations of Centrally Mediated Hypoxic Respiratory Excitation

In the intact animal, there are a number of phenomena that are thought to represent either direct or indirect CNS-mediated hypoxic excitation of respiratory output. It has been shown in several species, although not in humans, that there is a full recovery of the ventilatory response to acute hypoxia following peripheral chemodenervation in about 21–90 days (41–46). This does not appear to reflect return of peripheral chemoreceptor function but, rather, a considerable reorganization of the central hypoxia chemoreflex pathways (46).

As noted above, in the unanesthetized animal, acute systemic hypoxia immediately following chemodenervation or inhalation of carbon monoxide (to produce systemic hypoxia without stimulating arterial chemoreceptors) elicits a tachpneic response. Dillon and Waldrop (47), using whole-cell patch recording in tissue slices, have shown that neurons in the caudal hypothalamus depolarize and increase firing rates when exposed to hypoxia. In addition, direct stimulation of this region in the intact animal elicits tachypnea. They propose that this site is the locus of the tachypneic response to hypoxia in the acutely denervated animal.

In recent years much attention has been paid to the modification of CNS processing of chemoreception (central "plasticity") brought about by repeated short bouts of hypoxia (48–53). Not all findings have been consistent, perhaps owing to the use of different experimental paradigms and because the developmental phase of the experimental animal seems to be an important variable (for recent review see Ref. 54).

In adult humans and animals, chronic intermittent hypoxia has a facilatory effect manifest by enhanced responses to acute hypoxia (52,53), which appears to involve a serotonin-dependent mechanism (53). Both facilatory and inhibitory effects have been observed in neonatal animals (48,50,51). In addition, state of arousal has been shown to be a factor in the response to repetitive hypoxia in the neonate. In newborn lambs, for example, repetitive hypoxia rapidly became ineffective as a stimulus during active sleep but retained its responses during quiet sleep (49).

In the developing rat, the facilatory effect of intermittent hypoxia has been shown to be manifest as a reduction in hypoxic ventilatory decline observed with continuous hypoxia. This phenomenon was abolished by administrations of 7-nitroindazole, an antagonist of neuronal nitric oxide (NO) synthase, leading to the conclusion that this manifestation of the facilatory effect of chronic intermittent hypoxia is mediated by enhanced expression of neuronal NO (50). On the other hand, mutant mice deficient in neuronal nitric oxide synthase manifest augmented hypoxic responses (55). Whether the locus of this modulation is peripheral or central is unclear.

Probably the most extensively studied example of a CNS-mediated facilatory effect of continuous prolonged hypoxia is the acclimatization to high altitudes and its analogs produced using various experimental paradigms. The basic observation is that following the acute response to hypoxia, there is a gradual increase in ventilation over a period of hours to days. The more severe the hypoxia, the longer the time needed to reach a new plateau of ventilation. In general, mammals other than humans require a shorter time to manifest full acclimatization.

Earlier studies focused on the role of the central CO_2/H^+ chemoreceptors while recent ones have focused on heightened responsiveness to hypoxia by the carotid bodies. However, there are a variety of studies that show that either continuous hypoxia or continuous stimulation of the carotid sinus nerves for durations on the order of magnitude of hours modifies the CNS milieu in a facilatory fashion. This has been termed "long-term facilitation" of respiratory output. Forster, Dempsey, and co-workers proposed a central facilitation of respiratory drive as the mechanism for acclimatization to altitude based on a series of observations that showed heightened responses to a wide variety of respiratory stimuli (reviewed in Ref. 56). Several workers have proposed a change in brain monoamine metabolism (57–60). Millhorn et al. (57,58) showed that prolonged carotid sinus stimulation produced a lasting (beyond the stimulus) hyperventilation in anesthetized or decerebrate cats that could be blocked by serotonin antagonism. On the other hand, long-term facilitation could not be blocked by serotonin antagonism in the rat (60).

B. Centrally Mediated Sympathetic Excitation

It is our view that the most convincing examples of direct, i.e., chemoreceptor-like, hypoxic excitation of motor output mediated by discrete sites in the CNS are excitation of sympathetic discharge and gasping brought about by severe brain hypoxia.

Following chemodenervation in the anesthetized animal, systemic hypoxia produces a powerful increase in sympathetic discharge and arterial blood pressure (61–63). That this is mediated in the brainstem is indicated by a similar effect of the injection of hypoxic saline or cyanide (NaCN) in the right vertebral artery of the anesthetized deafferented cat (64). This response has been shown to be mediated by reticulospinal vasomotor neurons located in the RVLM (63,65,66) that appear to be essential for the generation and reflex regulations of arterial blood pressure (67).

There is substantial evidence from studies in the intact animal that these neurons are directly stimulated by hypoxia. Sun and colleagues have shown that microinjection of cyanide into the RVLM of rats evokes a pressor response (68). The RVLM reticulospinal sympathoexcitatory vasomotor neurons, many of which exhibit pacemaker-like activity, are rapidly and reversibly excited in a dose-dependent manner when the cyanide is delivered by either microinjection or microiontophoresis (63,66,68,69). This excitation is not altered by blockade of ionotropic excitatory amino acid (EAA) receptors in this region (70). Further, during hypoxic excitation of these RVLM reticulospinal sympathoexcitatory vasomotor neurons, their response to baroreceptor stimulation is preserved, suggesting that the hypoxic response is not mediated by a reduction in GABAergic neurotransmission or GABA receptor desensitization (68). Taken together, these observations provide evidence suggesting that RVLM reticulospinal sympathoexcitatory vasomotor neurons are directly excited by hypoxia (i.e., they are hypoxia-chemosensitive).

Further studies by Sun and Reis with the in vitro medullary slice preparation provide additional evidence for a direct excitatory effect of hypoxia on these neurons (69). In this reduced preparation, the RVLM sympathoexcitatory neurons retain excitability by hypoxia and focal NaCN. When synaptic transmission is blocked by application of tetrodotoxin (TTX), both hypoxia and NaCN continue to elicit neuronal membrane depolarization while the membrane response is abolished by application of Co^{2+}, a nonselective Ca^{2+}-channel blocker (68–70). These observations argue for intrinsic hypoxic chemosensitivity of these sympathoexcitatory neurons of the RVLM.

C. Gasping—Manifestation and Sites of Origin

Gasping is a highly stereotypical respiratory pattern that is ordinarily associated with severe hypoxia or massive supramedullary tissue damage. In the adult, it is mostly an agonal form of breathing while in the neonate it has been shown to be an effective autoresuscitative mechanism following respiratory arrest associated with severe hypoxia or asphyxia. During gasping, phrenic nerve discharge is characterized by a high-amplitude, abrupt-onset (i.e., rapid rate of rise), short-duration burst of activity (15). Single neuron recordings show synchronization of many medullary bulbospinal inspiratory-modulated neurons, and the discharge of expiratory-modulated neurons is severely reduced or abolished (71,72). This pattern of respiratory motor output is not unique to the severely hypoxic brain, and has been produced in the classic studies of Lumsden (73) as well as by St. John and Knuth (15) by pontomedullary transection as well as in more recent studies by chemical perturbations in the pre-Bötzinger complex (pre-BötC) (13,74,75; see below). All studies indicate that medullary neurons are sufficient for expressing gasping behavior.

Two regions in the rostral medulla have been identified as playing a key role either in the genesis or in the expression of gasping behavior of the respiratory system. The first was identified by St. John and colleagues in cats (76,77) and rats (78) and is located in the lateral segmental field (LTF). Electrical stimulation in this area during gasping produced by pontomedullary dissociation results in a premature gasp. Further, obliteration of the LTF eliminates gasping without alteration of the eupneic pattern of respiratory output. The second site that appears to play a critical role in the origin of gasping has been identified as the pre-BötC. This region was originally identified as both necessary and sufficient for respiratory rhythmogenesis in elegant studies by Feldman, Smith, and colleagues involving serial rostral-to-caudal and caudal-to-rostral sectioning of the neonatal rat brainstem (79). An analogous region, from both functional and anatomical points of view, has been established in the adult cat (74,80–82).

Evidence for the key role of the pre-BötC in the genesis of gasping has come primarily from our laboratory. We have demonstrated that both focal activation of ionotropic EAA receptors and focal hypoxia produced by microinjection of NaCN elicit gasp-like bursts in the whole phrenic nerve neurogram of the anesthetized, ventilated, hyperoxic cat (13,74,75). In addition, Huang et al. (83) have shown that obliteration of a region of the ventrolateral medulla that includes the pre-BötC and pre-I region (which is located rostral to the pre-BötC) in decerebrate neonatal rats abolished gasping elicited by severe hypoxia, but was ineffective in altering the eupneic pattern of breathing. In our studies in the adult cat, however, more discrete chemical lesioning of the pre-BötC abolished both eupneic respiratory output and hypoxia-induced gasping (13,84,85; see below). Recently, Gray and colleagues (86) have shown that near-total destruction of the neurokinin-1-expressing neurons located in the pre-BötC results in severe disruption but not obliteration of breathing. Finally, Lieske and colleagues (87) have recently demonstrated that in medullary slices that contain the pre-BötC, anoxia shifts inspiratory-modulated neuronal activity in this region as well as the eupneic pattern of hypoglossal motor activity to gasp-synchronous or gasp-like discharges, confirming in the reduced preparation that the primary locus for respiratory rhythm generation is common to both eupnea and gasping.

D. Gasping Mechanisms

We believe that there is strong evidence for both disinhibition and excitation playing key roles in the origin of gasping, and that they probably operate synergistically. There is much evidence for an important role for disinhibition. These are the classic pontomedullary transection experiments of Lumsden (73) as well as the analogous studies of St. John and Knuth (15), who demonstrated that transection, cooling, and radiofrequency lesioning at the level of the pontomedullary junction all produce a shift from eupneic pattern to a gasp-like pattern of phrenic nerve discharge. Most recently, we have proposed that the gasping "generator" that resides in the pre-BötC is ordinarily under strong GABAergic synaptic inhibition from more rostral influences in the brainstem (13). To test this hypothesis, we carried out focal microinjections of the $GABA_A$ receptor antagonist bicuculline methiodide (BIC) into the pre-BötC of the anesthetized, paralyzed, ventilated, deafferented cat. Unilateral microinjection of BIC into the pre-BötC evoked several responses in phrenic motor output, including high-amplitude, rapid-rate-of-rise, short-inspiratory-duration (i.e., gasp-like) bursts. In addition, we observed gasp-like bursts interposed between eupneic bursts and gasp-like bursts appended to the end of a eupneic burst or an "augmented burst" pattern of discharge. These findings suggest that during eupneic breathing, some population of neurons within the pre- BötC that are capable of producing gasp-like behavior are under $GABA_A$-mediated synaptic inhibition. Whether or not these are the pacemaker neurons posited to exist in the pre-BötC by Smith and Feldman and colleagues (79,88–92) remains to be determined.

Studies from other laboratories have also explored the role of the pre-BötC in the genesis of gasping. Huang et al. (83) have demonstrated that lesioning the pre-BötC in decerebrate neonatal rats with kainic acid abolished the gasping response to severe hypoxia but left the eupneic pattern of respiratory output intact. In contrast,

Ramirez et al. (82) produced bilateral synaptic blockade within the pre-BötC in pentobarbitone-anesthetized cats by injection of TTX and abolished eupneic output without eliminating the gasping response to severe hypoxia or asphyxia. It is important to note in this regard that Sun and Reis have shown that hypoxic excitation of reticulospinal sympathoexcitatory vasomotor neurons seems to be independent of synaptic transmission, or, at least, TTX-sensitive mechanisms (63,69,70; described above). If similar mechanisms exist in the pre-BötC, the inability of TTX to block hypoxia-induced gasping is not unexpected.

Our own preliminary studies suggest that the pre-BötC is necessary for generation of both eupneic and gasping patterns in the anesthetized cat. We have examined phrenic nerve discharge in response to both unilateral and bilateral microinjection of kainic acid or ibotenic acid into the pre-BötC under hyperoxic normocapnic and severe hypoxic conditions (13,84,85). An example of results may be found in Figure 1. We found that both unilateral and bilateral lesions of the pre-BötC eliminated the gasp-like response to severe brain hypoxia. Interestingly, in the preliminary studies, both the unilateral and bilateral lesions eliminated eupneic respiratory output as well, suggesting that a critical mass of neurons from this region is necessary for the expression of rhythmic respiratory behavior. Our studies also showed (see below) that microinjection of BIC into the pre-BötC reduced the degree of hypoxia necessary to elicit gasping in the progressive brain hypoxia model (13; see below). Taken together, these findings suggest a synergistic role for brain hypoxia and $GABA_A$-mediated disinhibition in the production of gasping.

In the next section, we posit that the pre-BötC transduces hypoxia in a manner analogous to the carotid bodies. However, before that we must consider other potential mechanisms by which hypoxia may elicit excitation in a neural network. There are several. For example, reduced ATP availability may limit the ATP-dependent phosphorylation required to maintain functional integrity of GABA receptors (93–95) and thereby result in disinhibition. In addition, in some regions of the brain, hypoxia increases the release of glutamate (96,97), which may potently excite *N*-methyl-D-aspartate receptors (98,99) enhancing Ca^{2+} influx (98,100–102). Glutamate, moreover, has been shown to inhibit presynaptic release of GABA and also cause postsynaptic inhibition of GABA-mediated synaptic potentials via cyclic GMP-dependent processes (103,104). Further, hypoxia may induce glutamate-mediated activation of L-type Ca^{2+} channels and concomitant stimulation of metabotropic glutamate receptors of the mGLUR1/5 type (105). Finally, in neonatal rats, nitric oxide production has been proposed to play an important facilitory role in the genesis of gasping (106).

E. The Chemoreceptor-like Characteristics of the Pre-BötC

Here, we argue that the pre-BötC has the capacity to manifest a chemoreceptor-like function (i.e., act as a primary hypoxic transducer), and that this capacity is most readily apparent when the neurons in this region are released from strong GABAergic inhibition. First, we (and others) have shown that the pre-BötC is a key site of excitatory input to the respiratory motor output. Microinjection of the

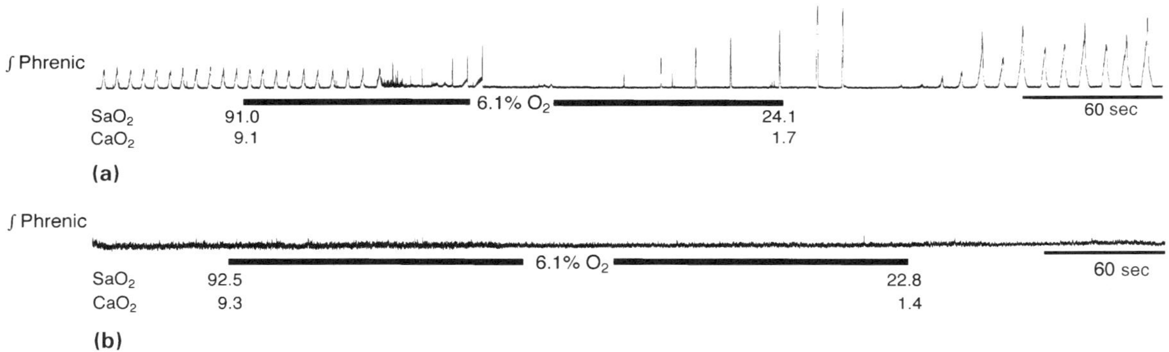

Figure 1 Effect of progressive brain hypoxia on phrenic nerve discharge before and after selective lesion of cell bodies in the pre-BötC. Cell body destruction was produced by unilateral microinjection of kainic acid (KA; 4.69 mM; 20 nl). (a) Before unilateral KA lesion of pre-BötC, isocapnic progressive brain hypoxia depressed phrenic nerve activity to the level of apnea, followed by gasping. Unilateral microinjection of KA into the pre-BötC (i.e., lesion of cell bodies) produced a transient excitation of phrenic nerve discharge (not shown), which was followed by apnea. (b) Following unilateral KA lesion of pre-BötC, isocapnic progressive brain hypoxia was ineffective in eliciting gasping or any other form of respiratory rhythmic activity. Similar effects were observed following bilateral KA lesion. Traces in both (a) and (b) show integrated phrenic nerve activity. SaO_2, arterial O_2 saturation (%); CaO_2, arterial O_2 content (vol %). (From Ref. 13.)

excitatory amino acid DL-homocysteic acid (DLH) elicited a variety of phrenic nerve outputs including tonic discharge, gasp-like patterns, augmented burst patterns, and enhanced eupneic patterns (74). Second, in 49 of 81 sites in which DLH elicited excitation, microinjection of NaCN elicited excitation as well (75). Two such responses are shown in Figure 2. In the first, microinjection of NaCN rapidly produced rapid-onset, high-amplitude, short-duration bursts typical of gasping behavior. It should be noted that the respiratory pattern in Figure 2 differs from gasping produced by severe systemic hypoxia in that the expiratory time relative to that in eupnea is decreased while with systemic hypoxia it is usually increased. This difference may represent a depressant effect of hypoxia at sites other than the pre-BötC during systemic hypoxia. The second pattern elicited by focal NaCN-induced hypoxia in the pre-BötC was an enhancement of the eupneic pattern. This implies that chemoreceptor-like activation of the pre-BötC may result not only in gasping, but stimulation of eupneic activity as well.

The most compelling evidence that the pre-BötC has the capacity to play a hypoxic chemoreceptor-like role in respiratory control comes from studies in which we examined the effects of systemic hypoxia on respiratory output before and after

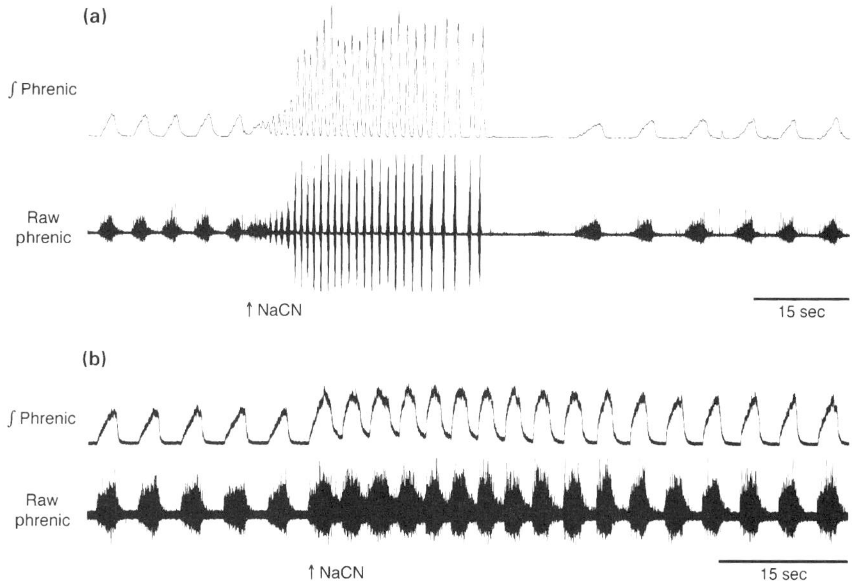

Figure 2 Examples of respiratory excitation evoked by NaCN microinjection into the pre-BötC. Traces from top to bottom in both (a) and (b): integrated phrenic nerve activity and raw phrenic nerve activity. Unilateral microinjection of NaCN (1 mM; 21 nl) into the pre-BötC produced either (a) a rapid series of high-amplitude, rapid-rate-of-rise, short-duration bursts or (b) an increase in frequency of phrenic bursts accompanied by an increase in the amplitude of integrated phrenic nerve discharge. See text for description of other excitatory patterns elicited by microinjection of NaCN into the pre-BötC. (From Refs. 13 and 75.)

focal microinjection of BIC into the pre-BötC in anesthetized, deafferented cats (13). One such experiment is shown in Figure 3. Prior to microinjection of BIC, the characteristic sequence of responses to progressive brain hypoxia in the anesthetized, paralyzed, ventilated, deafferented cat was observed: that is, a gradual decline in phrenic amplitude to phrenic silence, followed by gasp-like output. After unilateral microinjection of BIC, the baseline pattern manifested augmented bursts. Under these conditions, when progressive hypoxia was applied, gasp-like activity became quickly apparent. Most interestingly, eupneic phrenic activity was enhanced as well. The eupneic activity was abolished by progressive hypoxia while gasp-like activity persisted. Taken together, these observations suggests that the disinhibited pre-BötC may transduce a hypoxic signal in the physiological range of hypoxia. When thus stimulated by less than severe hypoxia, both augmentation of eupnea and elicitation of gasp-like behavior may be observed. With more severe systemic hypoxia, the eupneic output of the respiratory controller is abolished, most probably by depressant mechanisms not residing in the pre-BötC. Gasping, however, is resistant to the depressant effects of systemic hypoxia and persists. How gasping remains intact when eupnea is suppressed by hypoxia is unclear. One possibility is that gasping is minimally dependent upon neurotransmitter-mediated synaptic transmission, and therefore, is insensitive to GABAergic inhibition. It is interesting in this regard to recall that Sun and Reis (63,69,70) have shown that hypoxia transduction by the sympathoexcitatory vasomotor neurons of the ventrolateral medulla is not abolished by TTX-mediated inhibition of synaptic transmission. In addition, recent data from our laboratory have shown that both the pre-BötC and regions of the central CO_2/H^+ chemoreceptor are rich in connexins, a marker for electrogenic (i.e., nonneurotransmitter mediated) cell-to-cell neural transmission (107,108).

Other laboratories have provided support for the idea that neurons in the pre-BötC are excited by hypoxia. Ramirez et al. (109) examined the hypoxic responses of pre-BötC neurons in in vitro transverse medullary slices from neonatal mice. They found a biphasic response consisting of an initial excitation of inspiratory motor output in response to anoxia. This augmentation was characterized by an increase in the frequency of rhythmic discharges in pre-BötC neurons and an increase in frequency of burst discharges on the hypoglossal nerve. Further, in a subsequent study, they demonstrated that some pre-BötC pacemaker neurons continue to burst rhythmically during anoxia (110). Taken together, the observations from both in vivo and in vitro studies suggest that respiratory excitation in response to severe brain hypoxia results, at least in part, from direct hypoxic excitation (i.e., hypoxic chemosensitivity) of pre-BötC neurons, suggesting that these cells may function as central oxygen detectors in the control of breathing.

IV. Summary

From this brief review, it is apparent that brain hypoxia most likely modulates respiratory output in an anatomically and metabolically diverse manner. In addition, a variety of circumstances, including state of arousal, stage of development, and even heredity, impact upon these responses.

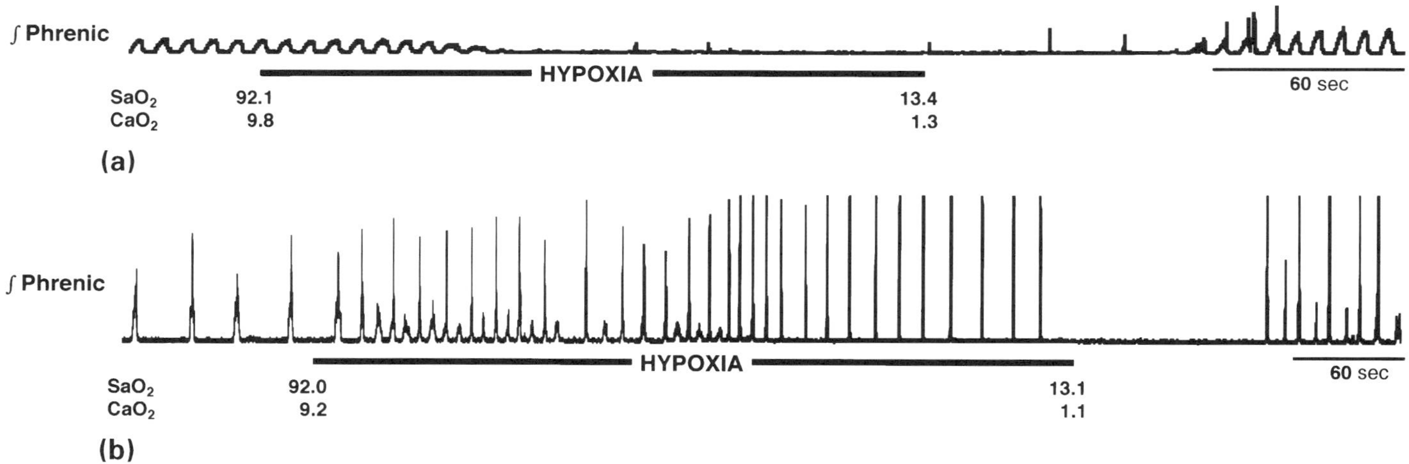

Figure 3 Effect of progressive brain hypoxia on phrenic nerve discharge before and after $GABA_A$ receptor antagonism in the pre-BötC. $GABA_A$ receptor antagonism was produced by unilateral microinjection of bicuculline methiodide (BIC; 2 mM; 20 nl). (a) Before blockade of $GABA_A$ receptors in the pre-BötC, progressive brain hypoxia depressed phrenic nerve activity to the level of apnea, followed by gasping. Unilateral microinjection of BIC into the pre-BötC produced augmented bursts in the phrenic neurogram. (b) Following $GABA_A$ receptor antagonism, progressive brain hypoxia produced respiratory excitation. This respiratory excitation consisted of elicitation of low-amplitude, eupneic-like bursts in the phrenic neurogram and an increase in the frequency of the BIC-induced augmented bursts. As hypoxia progressed, these low-amplitude, eupneic-like bursts were depressed, leaving only augmented and gasp-like phrenic bursts. Traces in both (a) and (b) show integrated phrenic nerve activity. SaO_2, arterial O_2 saturation (%); CaO_2, arterial O_2 content (vol %). (From Ref. 13.)

Depressant effects (other than those from neuronal failure due to energy depletion) may well reflect a residue of mechanisms that have evolved to deal with the peculiar challenge of the perinatal period. Nonspecific effects of brain hypoxia have been described, but more recent work points to direct excitation of medullary neurons by hypoxia as seen in both the sympathoexcitatory neurons of the RVLM and the putative pacemaker neurons of the pre-BötC. In our view, the role of the excitation of the pre-BötC in the genesis of gasping has been established and the possibility that hypoxia may modulate eupneic breathing has been raised.

Acknowledgments

The authors' research was supported by the National Institutes of Health (ICS, R01-HL63175; NHE, R01-HL16022), American Heart Association, New Jersey Affiliate (ICS, Grant-in-Aid #96-G-04), and American Lung Association (ICS, RG-008-N). ICS was an Edward Livingston Trudeau Scholar from the American Lung Association.

References

1. Neymans C, Bouckaert JJ, Dautrebande L. Sinus carotidien et réflexes respiratoires. II. Influences respiratoires réflexes de l'acidose, de l'anoxémie, de l'anhydride carbonique, de respiratoires dans les poumons et au dela des poumons. Arch Int Pharmacodyn 1930; 39:400–448.
2. Neubauer JA, Melton JE, Edelman NH. Modulation of respiration during brain hypoxia. J Appl Physiol 1990; 68:441–451.
3. Santiago TV, Edelman NH. Mechanism of the ventilatory response to carbon monoxide. J Clin Invest 1976; 57:977–986.
4. Melton JE, Neubauer JA, Edelman NH. CO_2 sensitivity of cat phrenic neurogram during hypoxic respiratory depression. J Appl Physiol 1988; 65:736–743.
5. Oyer-Chae L, Melton JE, Neubauer JA, Edelman NH. Triangularis sterni and phrenic nerve responses to progressive brain hypoxia. J Appl Physiol 1992; 72:1522–1528.
6. Wasicko MJ, Neubauer JA, Melton JE, Harangozo AM, Edelman NH. The effect of progressive brain hypoxia on respiratory activity of the hypoglossal nerve. Fed Proc 1987; 46:1418.
7. Fregosi RF, Knuth SL, Ward DK, Bartlett JD. Hypoxia inhibits abdominal expiratory activity. J Appl Physiol 1987; 63:211–220.
8. St John WM, Wang SC. Response of medullary respiratory neurons to hypercapnia and isocapnic hypoxia. J Appl Physiol 1977; 43:812–821.
9. Richter DW, Bischoff A, Anders K, Bellingham M, Windhorst U. Response of the medullary respiratory network of the cat to hypoxia. J Physiol 1991; 443:231–256.
10. England SJ, Melton JE, Douse MA, Duffin J. Activity of respiratory neurons during hypoxia in the chemodenervated cat. J Appl Physiol 1995; 78:856–861.
11. Melton JE, Yu QP, Neubauer JA, Edelman NH. Modulation of respiratory responses to carotid sinus nerve stimulation by brain hypoxia. J Appl Physiol 1992; 73:2166–2171.

12. Wasicko MJ, Melton JE, Neubauer JA, Krawciw N, Edelman NH. Cervical sympathetic and phrenic nerve responses to progressive brain hypoxia. J Appl Physiol 1990; 68:53–58.
13. Solomon IC. Excitation of phrenic and sympathetic output during acute hypoxia: contribution of medullary oxygen detectors. Respir Physiol 2000; 121:101–117.
14. Guntheroth WG, Kawabori I. Hypoxic apnea and gasping. J Clin Invest 1975; 56:1371–1377.
15. St John WM, Knuth KV. A characterization of the respiratory pattern of gasping. J Appl Physiol 1981; 50:984–993.
16. Oyer-Chae L, Melton JE, Neubauer JA, Edelman NH. Phrenic and sympathetic nerve responses to glutamergic blockade during normoxia and hypoxia. J Appl Physiol 1993; 74:1954–1963.
17. Coles SK, Dick TE. Neurones in the ventrolateral pons are required for post-hypoxic frequency decline in rats. J Physiol 1996; 497:79–94.
18. Dick TE, Coles SK. Ventrolateral pons mediates short-term depression of respiratory frequency after brief hypoxia. Respir Physiol 2000; 121:87–100.
19. Coles SK, Ernsberger P, Dick TE. A role for NMDA receptors in posthypoxic frequency decline in the rat. Am J Physiol 1998; 274:R1546–R1555.

19a. Lahiri, S. Depressant effect of acute and chronic hypoxia on ventilation. In: Paintal AS, ed. Morphology and Mechanisms of Chemoreceptors. Delhi: V.P. Chest Institute, University of Delhi, 1976:138–146.

20. Easton PA, Slykerman LJ, Anthonisen NR. Ventilatory response to sustained hypoxia in normal adults. J Appl Physiol 1986; 61:906–911.
21. Vizek M, Pickett CK, Weil JV. Biphasic ventilatory response of dult cats to sustained hypoxia has central origin. J Appl Physiol 1987; 63:1658–1664.
22. Vizek M, Bonora M. Diaphragmatic activity during biphasic ventilatory response to hypoxia in rats. Respir Physiol 1998; 111:153–162.
23. Bissonnette JM, Knopp SJ. Developmental changes in the hypoxic ventilatory response in C57BL/6 mice. Respir Physiol 2001; 128:179–186.
24. Siesjö BK, Nilsson L. The influence of arterial hypoxemia upon labile phosphates and upon extracellular and intracellular lactate and pyruvate concentration in the rat brain. Scand J Clin Lab Invest 1971; 27:83–96.
25. Bianchi AL, St John WM. Changes in antidromic latencies of medullary respiratory neurons in hypercapnia and hypoxia. J Appl Physiol 1985; 59:1208–1213.
26. Melton JE, Neubauer JA, Edelman NH. GABA antagonism reverses hypoxic respiratory depression in the cat. J Appl Physiol 1990; 69:1296–1301.
27. Yan S, Laferriere A, Zhang C, Moss IR. Microdialyzed adenosine in nucleus tractus solitarii and ventilatory response to hypoxia in piglets. J Appl Physiol 1995; 79:405–410.
28. Chau A, Koos BJ. Metabolic and cardiorespiratory responses to hypoxia in fetal sheep: adenosine receptor blockade. Am J Physiol 1999; 276:R1805–R1811.
29. De Boeck C, Van Reempts P, Rigatto H, Chernick V. Naloxone reduces decrease in ventilation induced by hypoxia in newborn infants. J Appl Physiol 1984; 56:1507–1511.
30. Gozal D, Simakajornboon N, Czapla MA, Xue YD, Gozal E, Vlasic V, Lasky JA, Liu JY. Brainstem activation of platelet-derived growth factor-beta receptor modulates the late phase of the hypoxic ventilatory response. J Neurochem 2000; 74:310–319.

31. Long WQ, Giesbrecht GG, Anthonisen NR. Ventilatory response to moderate hypoxia in awake chemodenervated cats. J Appl Physiol 1993; 74:805–810.
32. Tabata M, Kurosawa H, Kikuchi Y, Hida W, Ogawa H, Okabe S, Tun Y, Hattori T, Shirato K. Role of GABA within the nucleus tractus solitarii in the hypoxic ventilatory decline of awake rats. Am J Physiol Regul Integr Comp Physiol 2001; 281:R1411–R1419.
33. Tatsumi K, Pickett CK, Weil JV. Effects of haloperidol and domperidone on ventilatory roll off during sustained hypoxia in cats. J Appl Physiol 1992; 72:1945–1952.
34. Silver IA, Erecinska M. Intracellular and extracellular changes of Ca^{2+} in hypoxia and ischemia in rat brain in vivo. J Gen Physiol 1990; 95:837–866.
35. Leblond J, Krnjevic K. Hypoxic changes in hippocampal neurons. J Neurophysiol 1989; 62:1–14.
36. Cummins TR, Jiang C, Haddad GG. Human neocortical excitability is decreased during anoxia via sodium channel modulation. J Clin Invest 1993; 91:608–615.
37. Okada Y, Kawai A, Mückenhoff K, Scheid P. Role of the pons in hypoxic respiratory depression in the neonatal rat. Respir Physiol 1998; 111:55–63.
38. Koos BJ, Chau A, Matsuura M, Punla O, Kruger L. Thalamic locus mediates hypoxic inhibition of breathing in fetal sheep. J Neurophysiol 1998; 79:2383–2393.
39. Gargaglioni LH, Branco LG. Role of nucleus isthmi in the ventilatory response to hypoxia of Bufo paracnemis. Respir Physiol 2000; 119:31–39.
40. Lutz PL, Edwards R, McMahon PM. Gamma-Aminobutyric acid concentrations are maintained in anoxic turtle brain. Am J Physiol 1985; 249:R372–R374.
41. Bisgard GE, Forster HV, Klein JP. Recovery of peripheral chemoreceptor function after denervation in ponies. J Appl Physiol 1980; 49:964–970.
42. Smith PG, Mills E. Restoration of reflex ventilatory response to hypoxia after removal of carotid bodies in the cat. Neuroscience 1980; 5:573–580.
43. Martin-Body RL, Robson GJ, Sinclair JD. Restoration of hypoxic respiratory responses in the awake rat after carotid body denervation by sinus nerve section. J Physiol 1986; 380:61–73.
44. Pan LG, Forster HV, Martino P, Strecker PJ, Beales J, Serra A, Lowry TF, Forster MM, Forster AL. Important role of carotid afferents in control of breathing. J Appl Physiol 1998; 85:1299–1306.
45. Forster HV, Pan LG, Lowry TF, Serra A, Wenninger J, Martino P. Important role of carotid chemoreceptor afferents in control of breathing of adult and neonatal mammals. Respir Physiol 2000; 119:199–208.
46. Roux JC, Peyronnet J, Pascual O, Dalmaz Y, Pequignot JM. Ventilatory and central neurochemical reorganisation of O_2 chemoreflex after carotid sinus nerve transection in rat. J Physiol (Lond) 2000; 522(Pt 3):493–501.
47. Dillon GH, Waldrop TG. In vitro responses of caudal hypothalamic neurons to hypoxia and hypercapnia. Neuroscience 1992; 51:941–950.
48. Johnston RV, Grant DA, Wilkinson MH, Walker AM. Repetitive hypoxia rapidly depresses arousal from active sleep in newborn lambs. J Physiol (Lond) 1998; 510:651–659.
49. Johnston RV, Grant DA, Wilkinson MH, Walker AM. Repetitive hypoxia rapidly depresses cardio-respiratory responses during active sleep but not quiet sleep in the newborn lamb. J Physiol (Lond) 1999; 519:571–579.
50. Gozal D, Gozal E. Episodic hypoxia enhances late hypoxic ventilation in developing rat: putative role of neuronal NO synthase. Am J Physiol 1999; 276:R17–R22.

51. Miller MJ, Haxhiu MA, Haxhiu-Poskurica B, Dreshaj IA, Difiore JM, Martin RJ. Recurrent hypoxic exposure and reflex responses during development in the piglet. Respir Physiol 2000; 123:51–61.
52. Mahamed S, Duffin J. Repeated hypoxic exposures change respiratory chemoreflex control in humans. J Physiol 2001; 534:595–603.
53. Ling L, Fuller DD, Bach KB, Kinkead R, Olson EB, Jr, Mitchell GS. Chronic intermittent hypoxia elicits serotonin-dependent plasticity in the central neural control of breathing. J Neurosci 2001; 21:5381–5388.
54. Gozal E, Gozal D. Respiratory plasticity following intermittent hypoxia: developmental interactions. J Appl Physiol 2001; 90:1995–1999.
55. Kline DD, Yang T, Huang PL, Prabhakar NR. Altered respiratory responses to hypoxia in mutant mice deficient in neuronal nitric oxide synthase. J Physiol 1998; 511:273–287.
56. Dempsey JA, Forster HV. Mediation of ventilatory adaptations. Physiol Rev 1982; 62:262–346.
57. Millhorn DE, Eldridge FL, Waldrop TG. Prolonged stimulation of respiration by endogenous central serotonin. Respir Physiol 1980; 42:171–188.
58. Millhorn DE, Eldridge FL, Waldrop TG. Prolonged stimulation of respiration by a new central neural mechanism. Respir Physiol 1980; 41:87–103.
59. Olson EB, Jr, Vidruk EH, McCrimmon DR, Dempsey JA. Monoamine neurotransmitter metabolism during acclimatization to hypoxia in rats. Respir Physiol 1983; 54:79–96.
60. Olson EB, Jr. Ventilatory adaptation to hypoxia occurs in serotonin- depleted rats. Respir Physiol 1987; 69:227–235.
61. Rohlicek CV, Polosa C. Hypoxic responses of sympathetic preganglionic neurons in sino-aortic-denervated cats. Am J Physiol 1983; 244:H681–H686.
62. Heistad DD, Abboud FM. Circulatory adjustments to hypoxia. Circulation 1988; 61:463–470.
63. Sun MK, Reis DJ. Central neural mechanisms mediating excitation of sympathetic neurons by hypoxia. Prog Neurobiol 1994; 44:197–219.
64. Mitra J, Dev NB, Romaniuk JR, Trivedi R, Prabhakar NR, Cherniack NS. Cardiorespiratory changes induced by vertebral artery injection of sodium cyanide in cats. Respir Physiol 1992; 87:49–61.
65. Sun MK, Spyer KM. Responses of rostroventrolateral medulla spinal vasomotor neurones to chemoreceptor stimulation in rats. J Auton Nerv Syst 1991; 33:79–84.
66. Sun MK, Reis DJ. Hypoxia selectively excites vasomotor neurons of rostral ventrolateral medulla in rats. Am J Physiol 1994; 266:R245–R256.
67. Ross CA, Ruggiero PA, Park DH, Joh TH, Sved AF, Fernandez-Pardal J, Saacerdra JM, Reis DJ. Tonic vasomotor control by the rostral ventrolateral medulla: Effect of electrical or chemical stimulation of the area containing C1 adrenaline neurons on arterial pressure, heart rate, and plasma catecholamines and vasopressin. J Neurosci 1984; 4:474–494.
68. Sun MK, Jeske IT, Reis DJ. Cyanide excites medullary sympathoexcitatory neurons in rats. Am J Physiol 1992; 262:R182–R189.
69. Sun MK, Reis DJ. Hypoxia-activated Ca^{2+} currents in pacemaker neurones of rat rostral ventrolateral medulla in vitro. J Physiol 1994; 476:101–116.
70. Sun MK, Reis DJ. Hypoxic excitation of medullary vasomotor neurons in rats are not mediated by glutamate or nitric oxide. Neurosci Lett 1993; 157:219–222.

71. Zhou D, Wasicko MJ, Hu JM, St John WM. Differing activities of medullary respiratory neurons in eupnea and gasping. J Appl Physiol 1991; 70:1265–1270.
72. Tomori Z, Fung ML, Donic V, Donicova V, St John WM. Power spectral analysis of respiratory responses to pharyngeal stimulation in cats: comparisons with eupnoea and gasping. J Physiol 1995; 485:551–559.
73. Lumsden T. Observation on the respiratory centres in the cat. J Physiol (Lond) 1923; 57:153–160.
74. Solomon IC, Edelman NH, Neubauer JA. Patterns of phrenic motor output evoked by chemical stimulation of neurons located in the pre-Bötzinger complex in vivo. J Neurophysiol 1999; 81:1150–1161.
75. Solomon IC, Edelman NH, Neubauer JA. Pre-Bötzinger complex functions as a central hypoxia chemosensor for respiration in vivo. J Neurophysiol 2000; 83:2854–2868.
76. St John WM, Bledsoe TA, Sokol HW. Identification of medullary loci critical for neurogenesis of gasping. J Appl Physiol 1984; 56:1008–1019.
77. St John WM, Bledsoe TA, Tenney SM. Characterization by stimulation of medullary mechanisms underlying gasping neurogenesis. J Appl Physiol 1985; 58:121–128.
78. Fung ML, Wang W, St John WM. Medullary loci critical for expression of gasping in adult rats. J Physiol 1994; 480:597–611.
79. Smith JC, Ellenberger HH, Ballanyi K, Richter DW, Feldman JL. Pre-Bötzinger complex: a brainstem region that may generate respiratory rhythm in mammals. Science 1991; 254:726–729.
80. Connelly CA, Dobbins EG, Feldman JL. Pre-Bötzinger complex in cats: respiratory neuronal discharge patterns. Brain Res 1992; 590:337–340.
81. Schwarzacher SW, Smith JC, Richter DW. Pre-Bötzinger complex in the cat. J Neurophysiol 1995; 73:1452–1461.
82. Ramirez JM, Schwarzacher SW, Pierrefiche O, Olivera BM, Richter DW. Selective lesioning of the cat pre-Bötzinger complex in vivo eliminates breathing but not gasping. J Physiol 1998; 507:895–907.
83. Huang Q, Zhou D, St John WM. Lesions of regions for in vitro ventilatory genesis eliminate gasping but not eupnea. Respir Physiol 1997; 107:111–123.
84. Solomon IC, Neubauer JA, Edelman NH. Role of the pre-Bötzinger complex in the production of hypoxia-induced gasping. FASEB J 1998; 12:A497.
85. Solomon IC, Edelman NH. Effects of unilateral and bilateral pre-Bötzinger complex lesions on eupneic breathing in vivo. FASEB J 2000; 14:A642.
86. Gray PA, Janczewski WA, Mellen N, McCrimmon DR, Feldman JL. Normal breathing requires pre-Bötzinger complex neurokinin-1 receptor-expressing neurons. Nat Neurosci 2001; 4:927–930.
87. Lieske SP, Thoby-Brisson M, Telgkamp P, Ramirez JM. Reconfiguration of the neural network controlling multiple breathing patterns: eupnea, sighs and gasps [see comment]. Nat Neurosci 2000; 3:600–607.
88. Smith JC, Funk GD, Johnson SM, Feldman JL. Cellular mechanisms generating respiratory rhythm: insights from in vitro and computational studies. In: Trouth CO, Millis R, Kiwull-Schone H, Schlaefke M, eds. Ventral Brainstem Mechanisms and Control of Respiration and Blood Pressure. New York: Marcel Dekker, 1995:463–496.
89. Smith JC, Butera RJ, Koshiya N, Del Negro C, Wilson CG, Johnson SM. Respiratory rhythm generation in neonatal and adult mammals: the hybrid pacemaker-network model. Respir Physiol 2000; 122:131–147.

90. Feldman JL, Smith JC. Neural control of respiratory pattern in mammals: an overview. In: Dempsey JA, Pack AI, eds. Regulation of Breathing. New York: Marcel Dekker, 1995:39–69.
91. Rekling JC, Feldman JL. Pre-Bötzinger complex and pacemaker neurons: hypothesized site and kernel for respiratory rhythm generation. Annu Rev Physiol 1998; 60:385–405.
92. Koshiya N, Smith JC. Neuronal pacemaker for breathing visualized in vitro. Nature 1999; 400:360–363.
93. Stelzer A, Kay AR, Wong RKS. GABA-receptor functions in hippocampal cells is maintained by phosphorylation factors. Science 1988; 241:339–341.
94. Chen QX, Stelzer A, Kay AR, Wong RKS. $GABA_A$ receptor function is regulated by phosphorylation in acutely dissociated guinea-pig hippocampal neurons. J Physiol (Lond) 1990; 420:207–221.
95. Shirasaki T, Aibara K, Akaike N. Direct modulation of $GABA_A$ receptor by intracellular ATP in dissociated nucleus tractus solitarii neurones of rat. J Physiol (Lond) 1992; 449:551–572.
96. Hagberg H, Lehmann A, M. S, Nystrom B, Jacobsen I, Hamberger I. Ischemia-induced shift of inhibitory and excitatory amino acids from intra- to extracellular compartments. J Cereb Blood Flow Metab 1985; 5:413–419.
97. Rothman SM, Olney JW. Glutamate and the pathophysiology of hypoxic—ischemic brain damage. Ann Neurol 1986; 19:105–111.
98. Rothman S. Synaptic release of excitatory amino acid neurotransmitter mediates anoxic neuronal death. J Neurosci 1984; 7:1884–1891.
99. Goldberg MP, Weiss JH, Pham P, Choi DW. *N*-methyl-D-aspartate receptors mediate hypoxic neuronal injury in cortical culture. J Pharmacol Exp Ther 1987; 243:784–791.
100. Eyzaguirre C, Koyano H. Effects of some pharmacological agents on chemoreceptor discharges. J Physiol (Lond) 1965; 178:410–437.
101. MacDermott AB, Mayer ML, Westbrook GL, Smith SJ, Barker JL. NMDA-receptor activation increases cytosolic calcium concentration in cultured spinal cord neurons. Nature 1986; 321:519–522.
102. Zorumski CF, Thid LL. Properties of vertebrate glutamate receptors: calcium mobilization and desensitization. Prog Neurobiol 1992; 39:295–336.
103. Glaum SR, Miller RJ. Activation of metabotropic glutamate receptors produces reciprocal regulation of ionotropic glutamate and GABA responses in the nucleus of the tractus solitarius of the rat. J Neurosci 1993; 13:1636–1641.
104. Stefani A, Pisani A, Mercuri NB, Bernardi G, Calabresi P. Activation of metabotropic glutamate receptors inhibits calcium currents and GABA-mediated synaptic potentials in striatal neurons. J Neurosci 1994; 14:6734–6743.
105. Mironov SL, Richter DW. Hypoxic modulation of L-type Ca^{2+} channels in inspiratory brainstem neurones: intracellular signalling pathways and metabotropic glutamate receptors. Brain Res 2000; 869:166–177.
106. Gozal D, Torres JE, Gozal E, Nuckton TJ, Dixon MK, Gozal YM, Hornby PJ. Nitric oxide modulates anoxia-induced gasping in the developing rat. Biol Neonate 1998; 73:264–274.
107. Solomon IC, Halat TJ, El-Maghrabi R, O'Neal MH, 3rd. Differential expression of connexin26 and connexin32 in the pre-Bötzinger complex of neonatal and adult rat. J Comp Neurol 2001; 440:12–19.

108. Solomon IC, Halat TJ, El-Maghrabi MR, O'Neal MH, 3rd. Localization of connexin26 and connexin32 in putative CO_2-chemosensitive brainstem regions in rat. Respir Physiol 2001; 129:101–121.
109. Ramirez JM, Quellmalz UJ, Wilken B, Richter DW. The hypoxic response of neurones within the in vitro mammalian respiratory network. J Physiol 1998; 507:571–582.
110. Thoby-Brisson M, Telgkamp P, Ramirez JM. The role of the hyperpolarization-activated current in modulating rhythmic activity in the isolated respiratory network of mice. J Neurosci 2000; 20:2994–3005.

36

Petrosal Ganglion Responses In Vitro

From Entire Ganglion to Single Cell

JULIO ALCAYAGA

University of Chile
Santiago, Chile

RODRIGO VARAS and RODRIGO ITURRIAGA

Catholic University of Chile
Santiago, Chile

I. Introduction

The petrosal ganglion (PG) is the main sensory ganglion of the glossopharyngeal nerve, projecting peripherally to the innermost part of the oral cavity and to the carotid bifurcation, through the glossopharyngeal branch and the carotid sinus nerve (CSN), respectively. The ganglia contain a heterogeneous population (1) of pseudomonopolar neurons, completely encircled by satellite (glial) cells (2) that project peripherally with myelinated or unmyelinated fibers (3–5). According to the sensory modality they serve, or the associated transduction mechanisms, PG neurons can be divided into two categories. A vast population of these neurons are mechanoreceptors that innervate the pharynx and the carotid sinus, while the remaining neurons innervate chemoreceptors of the tongue and the carotid body. PG mechanoreceptor neurons are probably directly activated by physical deformation of their terminals, as the soma of nodose ganglia baroreceptor neurons (6) somatosensory neurons from dorsal root ganglia (7). On the other hand, chemosensory neurons are synaptically activated by neurotransmitters released from a sensory cell in the carotid body (type I, glomus) or in the taste buds.

II. Electrical Properties

Intracellular and whole-cell patch-clamp recordings have shown that PG neurons can be roughly classified in two major groups according to their electrical characteristics. A population of neurons responds to electrically imposed depolarization with a fast action potential, followed by a brief afterhyperpolarization (Fig. 1a). A second population of neurons presents an inflection (hump) in the repolarizing phase of the spike and a longer-lasting afterhyperpolarization (Fig. 1b) (8–11). Similarly, in rat PG neurons in culture two populations have been described, differing in their Na^+ voltage-dependent currents (12). Although this classification reflects only intrinsical electrical properties, electrophysiological recordings of identified PG neurons show that some relationship exists between the electrical properties and the sensory modality served by the neuron.

A. Neurons Projecting Through the Carotid Sinus Nerve (CSN)

Baroreceptor Neurons

Intracellular recordings from identified cat PG baroreceptor neurons, with myelinated axons, show that they respond to electrically imposed short depolariza-

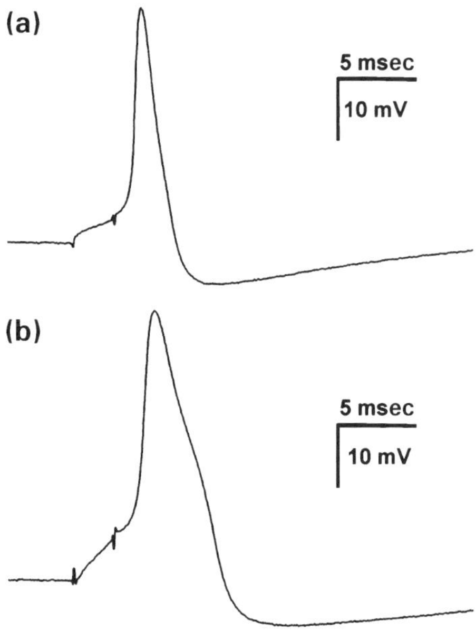

Figure 1 Intracellularly recorded action potentials, evoked by brief pulses, from cat petrosal ganglion neurons in tissue culture. (a) A neuron that responds with a brief action potential with a smooth repolarizing phase. (b) A neuron responding with an action potential presenting a clear hump, or inflection, in the repolarizing phase.

tion with a spike followed by a brief afterhyperpolarization. When these neurons are depolarized by long-lasting pulses, they present spiking activity whose frequency of discharge depends on the magnitude of the depolarization. Because they comprise the fastest-conducting neurons and because of the brief duration of the action potential these neurons have been called F-type (8). A second population of myelinated baroreceptor cells, called S-type because of its slower conduction velocity, responds tonically to long-lasting pulses, but the action potentials show a hump in the repolarizing phase of the spike and a longer-duration afterhyperpolarization. Unmyelinated baroreceptor neurons in the same preparation respond as F-type or S-type neurons, although the magnitude of the hump was smaller in these unmyelinated neurons (8). Thus, baroreceptor neurons appear to be a relatively heterogeneous population, comprising spiking neurons that generate fast or humped action potentials.

Chemosensory Neurons

Identified cat PG neurons with myelinated fibers projecting to the carotid body are characterized by slower conduction velocities than F-type neurons, presenting a prominent hump in the repolarizing phase of the spike and a long-lasting afterhyperpolarization (8). Furthermore, most of these neurons respond to long depolarizing pulses with one or two spikes (93%) or as many as four spikes (7%), independently of the duration or amplitude of the applied depolarizing pulse. Because of their prominent hump, these neurons are termed H-type (8). Unmyelinated chemosensory neurons in the cat present characteristics similar to those of F-type and S-type neurons, although the hump was less prominent than in S-type neurons (8). In the rat PG most chemosensory neurons project to the carotid body with unmyelinated fibers (5), presenting only fast-action potentials, and discharging tonically during long-lasting depolarizing pulses (13). These data indicate that most myelinated chemosensory afferents in the cat present humped action potentials with phasic activity, but unmyelinated chemosensory units respond tonically, irrespective of the form of their action potentials. Thus, phasic or tonic activity of chemosensory neurons projecting to the cat carotid body depends on the myelination of their axons. However, because myelination of the axons depends on neuronal size (14,15), the spiking capacity of these neurons would depend on their size as well. Similarly, the voltage-dependent K^+ currents appear to be related to the size of the chemosensory neuron (16). In the rat, where most chemosensory afferents are unmyelinated (5), only fast-spiking neurons project to the carotid body (13).

B. Neurons Projecting Through the Glossopharyngeal Branch

Recordings have also been performed from cat myelinated PG neurons projecting peripherally through the glossopharyngeal nerve beyond the apparent origin of the carotid sinus nerve, the so-called glossopharyngeal branch (GPB). These recordings indicate that fast, humped-nonspiking and humped-spiking action-potential-generating neurons, similar to F-, H-, and S-type neurons, are present within this population (9). However, the sensory modality served by each type of neuron was not determined (9). Despite these similarities, all these neurons present higher thresholds than their equivalent types projecting through the carotid nerve (9). Patch-clamp recordings from isolated rat PG neurons that project to the tongue, giving rise to both somatosensory and taste buds innervation, indicate that they also conform a heterogeneous population in which fast, humped-nonspiking and humped-spiking action-potential-generating neurons are almost equally represented (11).

III. Responses to Neurotransmitters

Transmitter release from receptor cells is a key step in the generation of afferent activity in both gustatory and arterial chemoreceptor afferent pathways. Thus, great attention has been placed on the responses evoked on PG neurons by application of putative chemoreceptor transmitters, both in the whole ganglion and in isolated neurons in vitro.

A. Acetylcholine (ACh)

Application of ACh to the cat PG in vitro induces a brief, dose-related increase in frequency discharge of the neurons projecting through the CSN (Fig. 2a,b), having little or no effect on the neurons projecting peripherally through the GPB (17,18). The responses present a high degree of temporal desensitization, are mimicked by nicotine, and are reversibly blocked by the nicotinic receptor blocker hexamethonium and mechamylamine (17,18). The presence of neostigmine, an inhibitor of the acetylcholinesterase, has little effect on the time course of the response, but increases the sensitivity to ACh (19). In cultured cat PG neurons, application of ACh induces depolarization and spike generation when the threshold is reached (Fig. 2c), in both fast-type and humped-type neurons (20). Patch-clamp recordings show that ACh induces depolarization and an inward current in about 70% of rat PG neurons (Fig. 2d), an effect that is mimicked by nicotine and blocked by hexamethonium (21,22). Similarly, in cocultures of rat carotid body cells with PG neurons, hexamethonium abolishes the basal spiking activity and the hypoxia-induced increase in spike discharge (21,23). In cat PG neurons in culture ACh has been reported to induce inward, outward, and biphasic currents, paralleled by membrane potential changes. Inward currents are blocked by specific $\alpha 4$ or $\alpha 7$ ACh receptor subunit antagonists, while outward currents are sometimes blocked by atropine (24). Immunocytochemical studies have revealed the presence of both $\alpha 4$ and $\alpha 7$ subunits of the nicotinic ACh receptor (24–26).

These data taken together suggest that PG neurons projecting through the CSN are endowed with nicotinic ACh receptors, and that ACh released from the glomus cells may increase the chemosensory afferent activity in the CSN. Further experiments are needed to confirm the presence of muscarinic receptors in PG neurons. However, about 70% of the cultured neurons respond to ACh (20,22), a percentage that exceeds the expected population of PG neurons projecting through the CSN (27). In the rat, about 50% of PG neurons projecting to the tongue responds to ACh with depolarization (11), indicating that—at least in the rat—this population

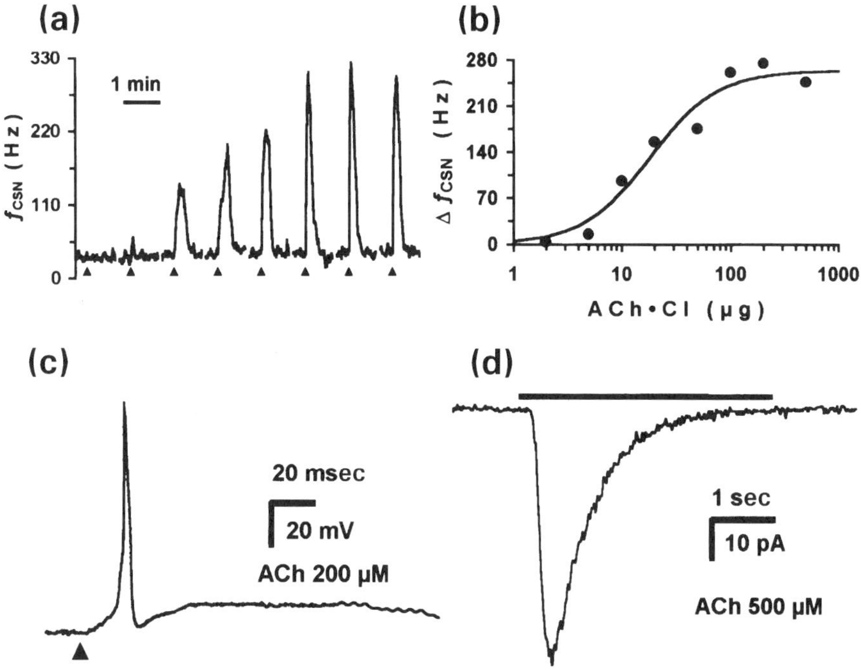

Figure 2 ACh-induced responses in the cat petrosal ganglion and in isolated neurons in culture. (a) Increases in the carotid sinus nerve frequency discharge (f_{CSN}) evoked by the application of increasing ACh doses (2–500 μg) to the ganglion (arrowhead). (b) Dose-response relationship for the significant increases in f_{CSN} (Δf_{CSN}) observed in (a). (c) Depolarization and firing of multiple action potentials, recorded with intracellular microelectrode, elicited by application of an ACh (200 μM) bolus (arrowhead). (d) Inwardly directed inactivating current, recorded in whole-cell voltage-clamp configuration at $V_m = -60$ mV, induced by a 4-sec ACh (500 μM) pulse (continuous line).

is also endowed with ACh receptors. Thus, ACh-induced responses are not only confined to PG neurons projecting through the CSN, but species differences can account for the responses evoked on isolated rat PG neurons projecting to the tongue through the GPB. Similarly, the high incidence of ACh-induced responses recorded in cultured PG neurons suggests that culture conditions could select special subsets of neurons or change the expression of receptors in the soma of the neurons.

B. Adenosine 5′-Triphosphate (ATP)

ATP applied to the cat PG in vitro increases briefly and in a dose-dependent manner the frequency of discharges in both the CSN and the GPB (Fig. 3a,b). However, responses in the CSN present lower threshold and larger amplitudes than those evoked in the GPB (18,28). This response shows little temporal desensitization; it is marginally mimicked by adenosine 5′-monophosphate, and an antagonist of

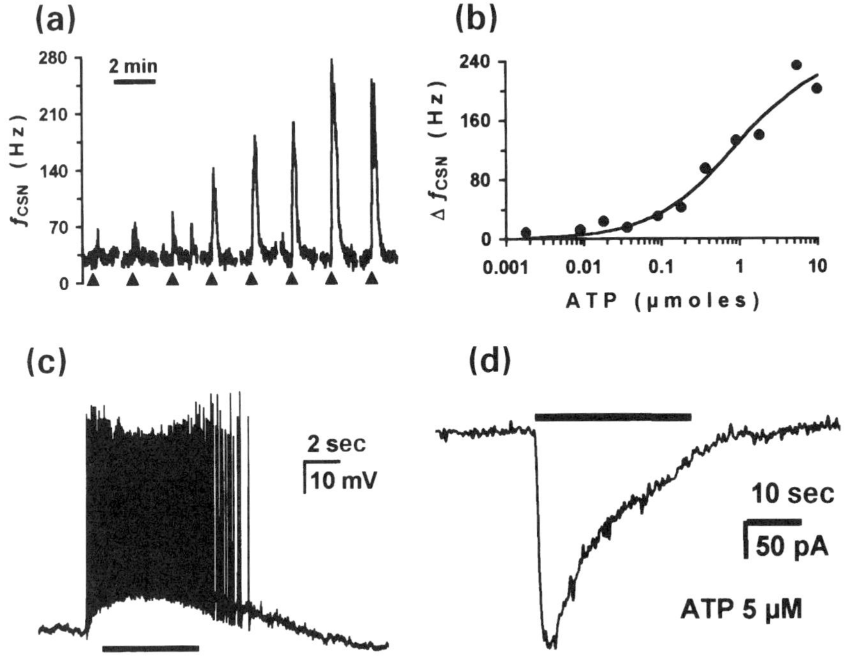

Figure 3 ATP-induced responses in the cat petrosal ganglion and in isolated neurons in culture. (a) Increases in the carotid sinus nerve frequency discharge (f_{CSN}) evoked by the application of increasing ATP doses (0.5–10 μ moles) to the ganglion (arrowhead). (b) Dose-response relationship for the significant increases in f_{CSN} (Δf_{CSN}) observed in (a). (c) Depolarization and firing of multiple action potentials recorded in whole-cell current-clamp by application of 5-sec ATP (5 μM) pulse (continuous line). (d) Inwardly directed inactivating current, recorded in whole-cell voltage-clamp configuration at $V_m = -60$ mV, induced by a 30-sec ATP (5 μM) pulse (continuous line).

metabotropic (G-protein coupled) nucleotide receptors (18,28) does not modify it. Similarly, in rat PG neurons in culture, ATP induces a dose-dependent depolarization, which is blocked by suramin (29). Voltage-clamp recordings show that ATP induces a fast, partially inactivating current, which effect is mimicked by α,β–methylene ATP and blocked by suramin, in a dose-dependent manner (29). At resting membrane potential ($V_m = -60$ mV), ATP induces an inward current, which reverses in direction at membrane potentials above zero (29). The pharmacological properties of ATP-induced responses in PG neurons suggest that these neurons express ionotropic ATP receptors (P2X). Immunostaining of PG ganglia with antibodies against the $P2X_2$ subunit indicates that this subunit is present in the majority of PG neurons and in its peripheral processes within the carotid body (29). It is noteworthy that in cocultures of carotid body and PG neurons, basal activity of spontaneously active neurons, as well as hypoxia-induced responses, is reduced by suramin (29). Moreover, similar effects are obtained with blockade by hexamethonium, while the application of both suramin and hexamethonium produces a total blockade of both basal and hypoxia-induced activity (29). These data suggest that ATP, either alone or combined with other transmitters, may participate in the generation of afferent chemosensory activity. On the other hand, release of ATP within a sensory ganglion has been reported (30), suggesting a possible role for ATP in intraganglionar communication.

C. Dopamine (DA)

Application of DA to the isolated PG in vitro has no direct effect on the activity recorded from both the CSN and the GPB (19,31). Similarly, DA receptor blockade in cocultures of rat carotid body cells and PG neurons affects neither the basal activity of spontaneously active neurons nor the hypoxia-induced responses (23). However, when DA is applied prior to ACh it produces a dose-related modification of the responses induced by the latter. Thus, for a given ACh dose, the lowest DA dose potentiates the response, while the largest dose inhibits the response (19,31). The inhibitory effect of DA on ACh-induced responses is partly reversed by the D2 receptor antagonist spiperone (31). The presence of dopaminergic neurons (32) as well as mRNA for D2 receptors (33) has been shown in a population of PG neurons. These data suggest that DA may act as a modulator of afferent activity in the terminals of PG neurons, and if released at this terminal, it could modulate both the receptor cell and the terminal.

D. Serotonin (5-HT)

Serotonin shows no effect on the activity of PG afferent nerves when applied to isolated cat ganglia in vitro (17) and does not modify the ACh-induced responses. Conversely, in rat PG neurons in culture, 5-HT induces a depolarization and the generation of action potentials in about 50% of the recorded cells (34) and in over 60% of the PG neurons projecting to the tongue (11). Serotonin induces a dose-dependent, inwardly directed current at the resting membrane potential ($V_m = -60\,\text{mV}$) (11,34), an effect that is mimicked by 2-methyl-serotonin and blocked by MDL-72222, specific 5-HT_3 receptor agonist and antagonist, respectively (34). Immunocytochemical and RT-PCR experiments also show the presence of 5-HT_{5a} receptors in the rat PG (35). Serotonin-induced responses present a threshold concentration at about 1 μM, saturate at about 100 μM, and have a half-effective dose (EC_{50}) around 4 μM (11,34). These data indicate that 5-HT receptors are present in rat PG neurons, and that they are expressed in a substantial number of those neurons projecting to the tongue (11).

E. δ-Aminobutyric Acid (GABA)

On the entire PG in vitro, GABA has no effect on the activities recorded from the CSN and the GPB (17); it also shows no effect on ACh-induced responses. Over 80% of rat cultured PG projecting to the tongue responds to GABA, 43% of them with hyperpolarizations and the remaining percentage with depolarizations, but all the responses are accompanied by a reduction of input resistance (11). The responses to GABA are mimicked by muscimol, a $GABA_A$ receptor agonist, but not by baclofen, a $GABA_B$ receptor agonist (11). These data suggest that GABA, probably through $GABA_A$ receptors, may participate in the peripheral activation of PG neurons projecting to the tongue. As inhibitory responses could be masked in the whole-nerve recordings, we cannot rule out the actions of GABA in the PG neurons that project to the carotid body.

F. Substance P (SP)

In tissue cultures, rat PG neurons that project to the tongue respond to a single SP concentration (1 μM) mostly with a depolarization, although hyperpolarizing responses also can be recorded. The depolarizing responses were also accompanied by the generation of action potentials (11). These data indicate that SP can depolarize PG neurons projecting to the tongue, but more experiments are needed to establish the nature of this response and whether this response can also be elicited on PG neurons projecting through the CSN.

G. Nitric Oxide (NO)

Sodium nitroprusside (SNP), a NO generator, applied to the isolated PG in vitro has no effect on the ongoing activity recorded from the peripheral branches of the glossopharyngeal nerve (36). Nevertheless, SNP modifies the ACh-induced responses, almost doubling the EC_{50} from 23.3 μg to 40.0 μg and reducing the slope (Hill coefficient) from 1.4 to 0.7 (36). The changes induced by SNP reduced the sensibility of the preparation to ACh, but increased the response evoked by the lowest ACh dose. Conversely, the blockade of the NO synthase (NOS) with L-NAME slightly increased the sensitivity of the ACh-induced responses (36). NOS has been detected in a population of small PG neurons (15–20 μm) located near the central end of the ganglion (37–39), containing also SP, but segregated from tyrosine hydroxylase–positive neurons (37). It has been postulated that these NOS-containing neurons would participate in an inhibitory pathway in the carotid body, in which NO would depress the oxygen-induced responses by acting on the glomus (receptor) cells (40). Our results suggest that NO may also act at the postsynaptic level, modifying ACh-induced responses in PG neurons.

IV. Responses to Chemosensory Natural Stimuli

The responses of PG neurons to natural stimuli have not been investigated extensively. Thus, although an important part of the ganglionar population corresponds to mechanoreceptors, little is known about their responses to mechanical deformation, as in baroreceptors of the nodose ganglia (6) or somatosensory afferents of the dorsal root ganglia (7). On the other hand, the responses of PG neurons to carotid body natural stimuli have been studied mostly in tissue culture, contrasting the responses evoked from neurons in pure ganglia cultures with those elicited in cocultures of PG neurons with carotid body cells. A complementary approach has been developed studying the responses evoked on the peripheral nerves by application of natural carotid body stimuli to the PG in vitro.

A. Hypoxic and Metabolic Hypoxia

Hypoxic hypoxia, reducing media PO_2 to about 10–12 mmHg, has no efffect on the discharges recorded from the CSN or the GPB, even though the hypoxic challenge lasted for up to 60 min (41). During hypoxic perfusion, ACh-induced responses remain largely unmodifed, suggesting that the neuronal responsiveness is maintained under this condition (18). In pure ganglia cultures, hypoxic hypoxia (PO_2 about 5–25 mmHg) has no effect on membrane potential or voltage-activated currents (23). Conversely, in cocultures of PG and carotid body cells, hypoxic hypoxia induces a depolarization and generation of action potentials in neurons located near carotid body cell clusters (23,29). This hypoxia-induced response is partially blocked by hexamethonium and suramin and totally blocked by the combination of both blockers (23,29). These results indicate that PG neurons are not sensitive to hypoxic hypoxia, and that responses to hypoxia are developed only when the neurons are in synaptic contact with the carotid body receptor cells (23,29).

Metabolic hypoxia induced by metabolic blockers (sodium cyanide, NaCN; sodium azide, NaN_3) and uncouplers (2,4-dinitrophenol, 2,4DNP; antimycin A; oligomycin) has been normally used to evoke responses in the carotid body. Although application of 2,4-DNP or antimycin A to the PG in vitro has no major effects, NaCN or NaN_3 produces a dose-dependent increase in the activity recorded from both the CSN and the GPB (18,41). However, the magnitude and the duration of the responses induced by cyanide in the CSN far exceed those evoked in the GPB, and even those evoked by NaN_3 in the same CSN (41). Thus, cyanide-induced responses in the CSN lasted, at least, 10–20 times more than the ones elicited by NaN_3. This suggests that a population of PG neurons that project through the CSN are largely insensitive to hypoxic hypoxia, but can be differentially activated by cyanide, in a way that does not resemble other forms of metabolic inhibition (18,41). Intraganglionar injections of cyanide to the PG in situ produce a dose-dependent pressor response, suggesting a direct activation of PG neurons by cyanide (42). The cyanide-induced responses in the CSN are reduced in amplitude and duration, in a dose-dependent manner, when DA precedes cyanide application (41). Moreover, application of DA during the development of the cyanide-induced response produces a rapid reduction of the increased frequency discharge, returning the activity toward basal levels within 30 sec (41).

B. Acidification

Acidification of the medium perfusing the PG in vitro has little or no effect on the activities recorded from the CSN or the GPB. Recordings from cultured cat PG neurons show that local acidification, achieved by pressure pulse ejection of acidified (pH 6.5) medium from a micropipette, has no effect on the discharge of PG neurons. However, in most neurons (81%), acidification blocked the action potentials evoked by a slightly suprathreshold depolarizing current pulse (10). Variable effects on the membrane potential that never surpassed 5 mV accompanied the blockade of action potentials (10). Similar effects of pH on visceral sensory neurons have been described (43) and appear to depend on the effects of H^+ on voltage-dependent channels (44). Thus, pH has no major effect on the PG neurons in culture. In cocultures of cat PG neurons and carotid body cells, most cells respond as previously described. However, in about 10% of the cases, acidification produces a depolarization and, when the threshold is reached, the firing of action potentials occurs (10). The magnitude of the depolarization and the number of action potentials fired are related to the duration of the acidic pulse, suggesting that the degree of acidification depends on the mixture of the acidic stimuli with the normal medium. Consecutive acid pulses maintain a steady depolarizing level, without further depolarizing the neuron (10). These data indicate that PG neurons are insensitive to acidification, and that acid-induced responses are present in a population of PG neurons only when carotid body cells are present. The exact mechanism by which this response develops in cocultures has not been studied, but the involvement of synaptic contacts between the glomus cells and PG neurons appears as the most probable mechanism (23,29,43).

V. Conclusions

The PG is constituted by pseudomonopolar sensory neurons that project to the carotid bifurcation, the pharynx, and the tongue, where two populations of neurons can be distinguished: (1) mecanoreceptor neurons, and (2) chemosensory neurons. According to the waveform of the action potential, two types of neurons can be recognized: (1) neurons with fast, monotonic, spikes, and (2) neurons with humped spikes. Large neurons with fast-action potentials fire tonically, while large neurons with humped-action potentials discharge phasically. On the other hand, small neurons fire tonically, irrespective of the waveform of their action potentials. Although evidence exists to support a relationship between the sensory modality and the form of the action potential for myelinated fibers projecting through the CSN, no absolute correlation between the sensory modality and the electrical properties of the neurons in the ganglion has been found.

Acetylcholine, acting on nicotinic ACh receptors, selectively activates the cat PG neurons projecting through the CSN, but a broader population in the case of rat PG neurons. However, ACh-responsive neurons in culture appear to comprise a larger population, suggesting the possibility of neuronal selection by tissue culture conditions. The activation of P2X receptors by ATP induces responses in both the CSN and the GPB, and in a large population of neurons in culture. There is evidence of the participation of both ACh and ATP in the generation of sensory activity in reconstituted chemoreceptors in tissue culture. Evidence points to the fact that other transmitters, such as SP, GABA, and 5-HT, may also activate subsets of PG neurons and may be involved in the generation of afferent activity, although species differences have been reported.

Natural stimuli of the carotid body (low PO_2, acidification) have no effect on PG neurons, but stimuli-related responses can be recorded only in cocultures of carotid body and PG cells. The reconstitution of synaptic contacts between carotid body cells and PG neurons appears to be necessary for the generation of chemosensory activity.

Metabolic blockade by cyanide and azide produces brief activation of PG neurons projecting through the GPB, while cyanide produces a specific, long-lasting activation of PG neurons projecting through the CSN. The mechanism of cyanide-induced activation needs further clarification.

Acknowledgments

We thank Mrs. Carolina Larraín for her assistance in the preparation of the manuscript. This work was supported in part by grants 197-1013 and 199-0030 from FONDECYT (National Fund for Scientific and Technological Development), Chile.

References

1. Alcayaga J, Arroyo J, Font MI, Gutierrez O. The petrosal ganglion of the adult cat: neuronal count, sectional area, and their respective distributions. Biol Res 1996; 29:189–194.
2. Stensaas LJ, Fidone SJ. Ultrastructural study of cat petrosal ganglia: a search for autonomic ganglion cells. Brain Res 1977; 124:28–39.
3. Eyzaguirre C, Uchizono K. Observations on the fibre content of nerves reaching the carotid body of the cat. J Physiol (Lond) 1961; 159:268–281.
4. Fidone SJ, Sato A. A study of chemoreceptor and baroreceptor A and C-fibers in the cat carotid nerve. J Physiol (Lond) 1969; 205:527–548.
5. McDonald D. Morphology of the carotid sinus nerve. II. Number and size of axons. J Neurocytol 1983; 12:373–392.
6. Kraske S, Cunningham JT, Hajduczok G, Chapleau MW, Abboud FM, Wachtel RE. Mechanosensitive ion channels in putative aortic baroreceptor neurons. Am J Physiol 1998; 275:H1497–H1501.
7. McCarter GC, Reichling DB, Levine JD. Mechanical transduction by the rat dorsal root ganglion neurons in vitro. Neurosci Lett 1999; 273:179–182.
8. Belmonte C, Gallego R. Membrane properties of cat sensory neurons with chemoreceptor and baroreceptor endings. J Physiol (Lond) 1983; 342:603–614.
9. Morales A, Ivorra I, Gallego R. Membrane properties of glossopharyngeal sensory neurons in the petrosal ganglion of the cat. Brain Res 1987; 401:340–346.
10. Alcayaga J, Arroyo J. Responses of cat petrosal ganglion neurons are modified by the presence of carotid body cells in tissue cultures. Adv Exp Med Biol 1996; 410:195–201.
11. Koga T, Bradley RM. Biophysical properties and responses to neurotransmitters of petrosal and geniculate ganglion neurons innervating the tongue. J Neurophysiol 2000; 84:1404–1413.
12. Stea A, Nurse CA. Whole-cell currents in two subpopulations of cultured rat petrosal neurons with different tetrodotoxin sensitivities. Neuroscience 1992; 47:727–736.

13. Donnelly DF. Developmental changes in membrane properties of chemoreceptor afferent neurons of the rat petrosal ganglia. J Neurophysiol 1999; 82:209–215.
14. Harper AA, Lawson SN. Conduction velocity is related to morphological cell type in rat dorsal root ganglion neurons. J Physiol (Lond) 1985; 259:31–46.
15. Lee KH, Chung K, Chung JM, Coggeshall RE. Correlation of cell body size, axon size, and signal conduction velocity for individually labelled dorsal root ganglion cells in the cat. J Comp Neurol 1986; 243:335–346.
16. Andrews EM, Kunze DL. Voltage-gated K^+ channels in chemoreceptor sensory neurons of rat petrosal ganglion. Brain Res 2001; 897:199–203.
17. Alcayaga J, Iturriaga R, Varas R, Arroyo J, Zapata P. Selective activation of carotid nerve fibers by acetylcholine applied to the cat petrosal ganglion in vitro. Brain Res 1998; 786:47–54.
18. Alcayaga J, Varas R, Arroyo J, Iturriaga R, Zapata P. Responses of petrosal ganglion neurons in vitro to hypoxic stimuli and putative transmitters. Adv Exp Med Biol 2000; 475:389–396.
19. Zapata P, Larran C, Iturriaga R, Alcayaga J, Eyzaguirre C. Interactions between acetylcholine and dopamine in chemoreception. Adv Exp Med Biol 2000; 475:495–506.
20. Varas R, Alcayaga J, Zapata P. Acetylcholine sensitivity in primary sensory neurons dissociated from the cat petrosal ganglion. Brain Res 2000; 882:201–205.
21. Zhong H, Nurse CA. Co-cultures of rat petrosal neurons and carotid body type 1 cells: a model for studying chemosensory mechanisms. Adv Exp Med Biol 1996; 410:189–193.
22. Zhong H, Nurse CA. Nicotinic acetylcholine sensitivity of rat petrosal sensory neurons in dissociated cell culture. Brain Res 1997; 766:153–161.
23. Zhong H, Zhang M, Nurse CA. Synapse formation and hypoxic signalling in cocultures of rat petrosal neurons and carotid body type 1 cells. J Physiol (Lond) 1997; 503:599–612.
24. Shirahata M, Ishizawa Y, Rudisill M, Sham JS, Schofield B, Fitzgerald RS. Acetylcholine sensitivity of cat petrosal ganglion neurons. Adv Exp Med Biol 2000; 475:377–387.
25. Ishizawa Y, Fitzgerald RS, Shirahata M, Schofield B. Localization of nicotinic acetylcholine receptors in cat carotid body and petrosal ganglion. Adv Exp Med Biol 1996; 410:253–256.
26. Shirahata M, Ishizawa Y, Rudisill M, Sham JS, Schofield B, Fitzgerald RS. Presence of nicotinic acetylcholine receptors in cat carotid body afferent system. Brain Res 1998; 814:213–217.
27. Eyzaguirre C, Zapata P. Perspectives in carotid body research. J Appl Physiol 1984; 57:931–957.
28. Alcayaga J, Cerpa V, Retamal M, Arroyo J, Iturriaga R, Zapata P. Adenosine triphosphate-induced peripheral nerve discharges generated from the petrosal ganglion in vitro. Neurosci Lett 2000; 282:185–188.
29. Zhang M, Zhong H, Vollmer C, Nurse CA. Co-release of ATP and ACh mediates hypoxic signalling at rat carotid body chemoreceptors. J Physiol (Lond) 2000; 525:143–158.
30. Matsuka Y, Neubert JK, Maidment NT, Spigelman I. Concurrent release of ATP and substance P within Guinea pig trigeminal ganglia in vivo. Brain Res 2001; 915:248–255.
31. Alcayaga J, Varas R, Arroyo J, Iturriaga R, Zapata P. Dopamine modulates carotid nerve responses induced by acetylcholine on the cat petrosal ganglion in vitro. Brain Res 1999; 831:97–103.

32. Katz DM, Black IB. Expression and regulation of catecholaminergic traits in primary sensory neurons: relationship to target innervation in vivo. J Neurosci 1986; 6:983–989.
33. Czyzyk-Krzeska MF, Lawson EE, Millhorn DE. Expression of D_2 dopamine receptor mRNA in the arterial chemoreceptor afferent pathway. J Auton Nerv Syst 1992; 41:31–40.
34. Zhong H, Zhang M, Nurse CA. Electrophysiological characterization of 5-HT receptors on rat petrosal neurons in dissociated cell culture. Brain Res 1999; 816:544–553.
35. Wang Z-Y, Keith IM, Beckman MJ, Bronfield MS, Vidruk EH, Bisgard GE. 5-HT5a receptors in the carotid body chemoreception pathway of rat. Neurosci Lett 2000; 278:9–12.
36. Alcayaga J, Barrios M, Bustos F, Miranda G, Molina MJ, Iturriaga R. Modulatory effect of nitric oxide on acetylcholine-induced activation of cat petrosal ganglion neurons in vitro. Brain Res 1999; 825:194–198.
37. Wang Z-Z, Bredt DS, Fidone SJ, Stensaas LJ. Neurons synthesizing nitric oxide innervate the mammalian carotid body. J Comp Neurol 1993; 336:419–432.
38. Wang Z-Z, Stensaas LJ, Bredt DS, Dinger BG, Fidone SJ. Localization and actions of nitric oxide in the cat carotid body. Neuroscience 1994; 60:275–286.
39. Wang Z-Z, Stensaas LJ, Bredt DS, Dinger BG, Fidone SJ. Mechanisms of carotid body inhibition. Adv Exp Med Biol 1994; 360:229–235.
40. Wang Z-Z, Stensaas LJ, Dinger BG, Fidone SJ. Nitric oxide mediates chemoreceptor inhibition in the cat carotid body. Neuroscience 1995; 65:217–229.
41. Alcayaga J, Varas R, Arroyo J, Iturriaga R, Zapata P. Responses to hypoxia of petrosal ganglia in vitro. Brain Res 1999; 845:28–34.
42. Sun M-K, Reis DJ. Dopamine or transmitter release from rat carotid body may not be essential to hypoxic chemoreception. Am J Physiol 1994; 267:R1632–R1639.
43. Alcayaga J, Eyzaguirre C. Electrophysiological evidence for the reconstitution of chemosensory units in co-cultures of carotid body and nodose ganglion neurons. Brain Res 1990; 534:324–328.
44. Hille B. Charges and potentials at the nerve surface: divalent ions and pH. J Gen Physiol 1968; 51:221–236.

37

Comparative Aspects of O_2 Chemoreception
Anatomy, Physiology, and Environmental Adaptations

MARK L. BURLESON

The University of Texas at Arlington
Arlington, Texas, U.S.A.

WILLIAM K. MILSOM

University of British Columbia
Vancouver, British Columbia, Canada

I. Introduction

While all cells are sensitive to hypoxia to various extents, there is a special subset whose physiological function is to detect changing levels of O_2 in either the environment or tissues and elicit compensatory reactions to maintain O_2 uptake. These are the O_2-sensitive chemoreceptors that mediate cardiovascular and ventilatory reflex responses to acute hypoxia and also play a role in long-term physiological changes in response to chronic hypoxia. Most of what we know about O_2 chemoreception comes from studies on mammals where these cells are concentrated in the aortic, carotid, and pulmonary neuroepithelial bodies. It is these cells that have been examined in the greatest detail and are the subject of the bulk of this book. This chapter will focus on the O_2-sensitive chemoreceptors in animals other than mammals.

Disproportionately more is known about O_2 chemoreception in mammals despite the fact that they make up only about 10% of known vertebrate species. There seems to be a general phylogenetic trend with O_2-sensitive chemoreceptors having a more diffuse, internal/external distribution in lower vertebrates to the more localized interior receptor groups in aortic and carotid bodies of mammals. This concentration of chemoreceptor loci follows the phylogenetic trend seen in the internalization and reduction of gill arches.

II. Anatomy

A. Location and Innervation of O_2 Chemoreceptors

The exact anatomical sites of peripheral O_2-sensitive chemoreceptors in fish remain unknown. Recent studies provide strong evidence that in fish, O_2-sensitive chemoreceptors are found on all gills innervated by cranial nerves IX and X, in the spiracle and pseudobranch in species that have these innervated by cranial nerves VII and IX (13,66), as well as within the walls of the orobranchial cavity innervated by cranial nerves V and VII (16) (Fig. 1). It is beginning to appear, however, that the exact distribution of receptors is highly variable from species to species and from one physiological process to another. Thus, the distribution of receptors associated with the hypoxic bradycardia seen in fish is restricted to the first gill arch in many species (trout, cod, traira; 27,81,84), but can be found throughout the first three pairs of gill arches in others (catfish; 12) and within the orobranchial cavity in the elasmobranch fishes (dogfish; 16). The receptors associated with the hypoxic ventilatory response, on the other hand, are only confined to the gill arches in a few species (catfish; 12) while in many fishes, gill denervation fails to eliminate the

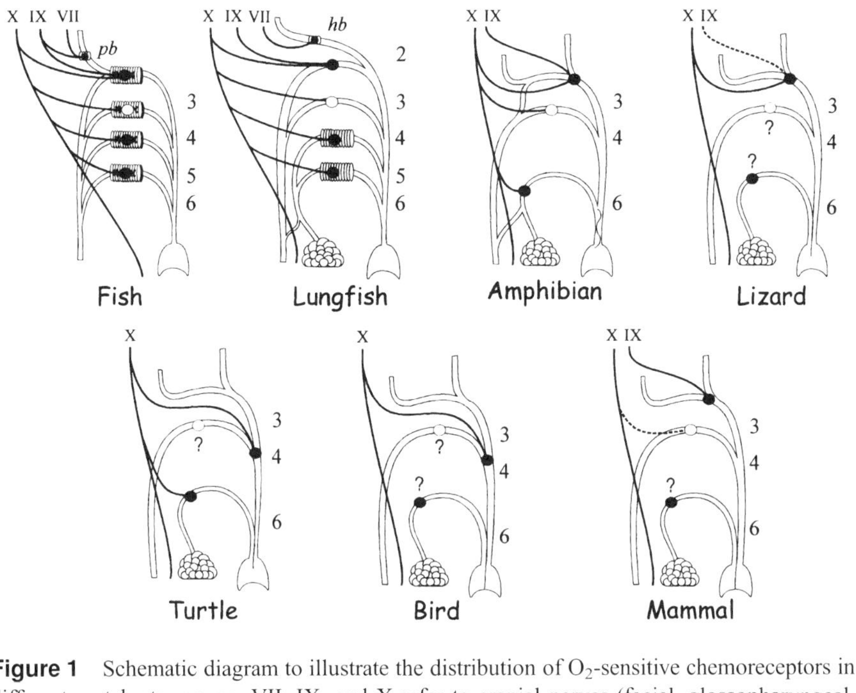

Figure 1 Schematic diagram to illustrate the distribution of O_2-sensitive chemoreceptors in different vertebrate groups. VII, IX, and X refer to cranial nerves (facial, glossopharyngeal, and vagus nerves) while 2, 3, 4, 5, and 6 refer to arteries supplying the respective embryonic gill arches. While O_2 chemoreceptors are found in the aortic bodies of mammals, they do not appear to serve a respiratory role. See text for further details.

hypoxic ventilatory response (sea raven, tench, tambaqui; 77,84) and receptors have been localized to the orobranchial cavity (tambaqui; 67).

Air breathing has arisen in fishes many times utilizing various parts of the buccal and pharyngeal cavities as well as the alimentary canal. The most-derived forms make use of either air-breathing organs or true lungs (33). O_2-sensitive chemoreceptors in these fishes regulate not only changes in heart rate and gill ventilation but also ventilation of air-breathing organs. In the actinopterygian fishes the gar and bowfin, they are found diffusely distributed throughout the gills and pseudobranch innervated by cranial nerves VII, IX, and X (60,80). In the dipnoan lungfishes, they are also distributed throughout the anterior hemibranch and all gill arches innervated by cranial nerves VII, IX, and X (Fig. 1).

In amphibians, the homologs of these sites, except perhaps the orobranchial cavity, remain active as O_2-sensing sites. Amphibian tadpoles begin life as skin and gill breathers and during late metamorphosis both gills and lungs are present. The first extant gill arch (embryonic arch 3) becomes the carotid labyrinth (58) and is innervated by cranial nerve IX. Studies indicate that the initial rapid response to environmental hypoxia arises exclusively from this arch while the subsequent slow component of the hypoxic ventilatory response arises from receptors at other sites (44,45). In amphibians, the aortic arch is a homolog of the 4th embryonic gill arch innervated by the Xth cranial nerve and the pulmocutaneous artery is a homolog of the 6th embryonic gill arch innervated by cranial nerve X. In adults, all appear to contain O_2 chemoreceptors that participate in respiratory reflexes (Fig. 1) (40,41,87,93).

In lizards, indirect evidence suggests that O_2 receptors are located where the internal carotid artery arises from the common carotid artery innervated by the superior laryngeal branch of cranial nerve X and perhaps IX (Fig. 1) (47,64). In birds and turtles, on the other hand, chemoreceptive tissue is not present at the extant carotid bifurcation, but during embryological development the aortic arches retreat backward and the carotid bifurcation remains close to the heart. The external carotids then atrophy and the internal carotids (now called common carotids) divide, secondarily, in the head region (47). As a result the site homologous to the carotid bifurcation in amphibia and mammals is now part of the aortic arch. In birds and turtles, the largest aggregations of chemoreceptive tissue are found in the central cardiovascular area and are innervated by one or more branches of the vagus (Fig. 1) (1,41,43). Given the weight of the evidence (innervation aside), these receptors are believed to be homologous to the carotid chemoreceptors of other vertebrates.

In mammals the carotid bodies are situated at the bifurcations of the common carotid arteries into their internal and external branches and are innervated by the sinus nerve, a branch of cranial nerve IX (Fig. 1). The aortic bodies in mammals, on the other hand, are located in the region of the aortic arch and the roots of the major arteries of the thorax. Their afferent fibers run in the aortic nerve, a branch of the vagus (Fig. 1). They appear to make little contribution to the resting ventilatory drive in eucapnic normoxia and may not contribute to the hypoxic ventilatory response in many species. It would appear that they participate almost exclusively in cardiovascular reflexes in this group (see Ref. 25 for review). Glomus tissue that may be homologous to the aortic chemoreceptors of mammals has also been described within the aortic walls of birds (48,71). Oxygen chemoreceptors have also been reported to occur on the pulmonary arteries of turtles, birds, and mammals but only those in turtles have been reported to give rise to ventilatory reflexes (42,43).

This information is summarized in Figure 1 and reveals two significant trends. The first is that all O_2 chemoreceptors involved in respiratory control are situated peripherally; there is no strong evidence yet to argue for the presence of central O_2 chemoreceptors involved in respiration. The second is that there has been a reduction in the distribution of O_2 chemoreceptors from multiple, dispersed sites in fish and amphibia toward a single dominant receptor site in birds and mammals. While still highly speculative, this latter trend would appear to be correlated with the transition from aquatic respiration and bimodal breathing in animals with intracardiac shunts (where the ability to sense O_2 at multiple sites would be an advantage) to strictly air breathing in animals with no intracardiac shunts. In amphibia and reptiles with well-developed intracardiac shunts, increases in arterial O_2 transport can be achieved by both reducing the degree of cardiac shunt and increasing ventilation. Having multiple O_2 receptors at appropriate sites may allow these species to make appropriate changes that most effectively increase O_2 transport under different conditions (see below). In birds and mammals, respiratory control is much less flexible (one exchange site that must receive the total cardiac output) and it would appear that under these conditions efficient control can be achieved with chemoreceptive information from a single source. Thus, O_2 receptors associated with homologous structures to the first gill arch of fish appear to become the primary respiratory O_2 chemoreceptor while those associated with structures homologous to the remaining gill arches take on a secondary role and/or become more involved in cardiovascular reflexes.

B. Orientation: Hypoxia Versus Hypoxemia

Based on reflex studies, the hypoxic bradycardia exhibited by most teleost fish appears to be triggered by activation of externally oriented receptors that monitor aquatic O_2 levels. These receptors are found largely on the first gill arch but have been reported on other gill arches in some species (see Refs. 13,65 for reviews). Not all fish follow this pattern, however. In the gar and the tambaqui they appear to be sensitive to both internal and external changes in P_{O_2} (79,85) while in the neotropical fish the traira they appear to only monitor the P_{O_2} of the blood (84). By contrast, the O_2-sensitive receptors instrumental in producing the increases in ventilation frequency and amplitude in most teleost fish appear to monitor both the blood and the water (13,67,84,85).

The only direct evidence for the presence of O_2-sensitive chemoreceptors in fish comes from three series of studies in trout and tuna (specifically, the pseudobranch and first gill arch) (14,52–54,62). The studies on both tuna and trout noted afferent activity arising from fibers that responded only, or preferentially, to changes in external (water) O_2 stimulus levels and others that responded only, or preferentially, to changes in internal (perfusate/blood) O_2 stimulus levels (Fig. 2)

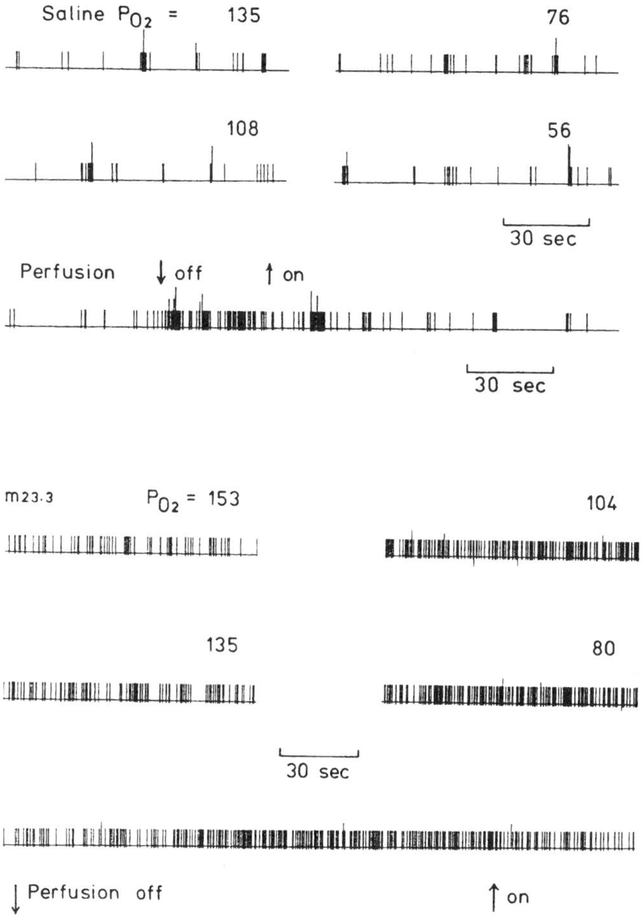

Figure 2 ENG of activity from oxygen chemoreceptors in the first gill arch of skipjack tuna illustrating the effect of interrupting perfusion on afferent activity. The fiber depicted in the top panel is not sensitive to changes in the P_{O_2} of the bathing solution but does respond to changes in perfusion. The fiber illustrated in the bottom panel responds to changes in the P_{O_2} of the bathing solution but is not very sensitive to changes in perfusion. (Reproduced from Ref. 62.)

(14,62). In these studies, some fibers were found that were sensitive to changes in O_2 partial pressure in both the water bathing the gills and the fluid perfusing the gills. These data support the reflex studies and suggest that O_2 chemoreceptors can be rather diffusely distributed throughout the gill.

O_2-sensitive chemoreceptors in exclusively air-breathing species all appear to be designed to monitor only the internal environment. All fish and amphibian larvae studied to date appear to have the ability to distinguish between changes in O_2 levels in water versus blood. The intraspecies differences that are seen in interbranchial distribution (internal versus external orientation) are even harder to explain than their intrabranchial distribution and do not support the earlier hypothesis that hypoxia-tolerant species are more sensitive to arterial hypoxemia while less tolerant species are more sensitive to aquatic hypoxia. In the case of those species with both gills and lungs, however, this appears to allow them to make appropriate adjustments in the relative use of the different respiratory pumps and gas exchange sites to meet environmental and physiological demands.

C. Histology

Although it has become clear that all cells can respond to hypoxia, only glomus (type 1 cells) associated with the vasculature and neuroepithelial cells associated with respiratory epithelia are generally accepted as specific O_2-sensitive chemoreceptors (17). Glomus and neuroepithelial cells are similar and have the same embryological origin, the neural crest (56,86). Both glomus and neuroepithelial cells are characterized by large spherical nuclei, numerous mitochondria, well-developed Golgi apparatus, and dense core vesicles containing various neurochemicals. Both cell types have cytoplasmic processes that make contact with nerves, capillaries, and other glomus and neuroepithelial cells (see reviews in Refs. 59,88). Some glomus and neuroepithelial cells do not appear to be innervated, yet others are innervated by both afferent and efferent nerves. The presence of gap junctions indicates that there is direct communication between glomus cells in a cluster (23) suggesting that they may function together as a unit. Sustentacular cells (type II cells) are closely associated with glomus and neuroepithelial cells, and although they do not appear to play a direct role in O_2 chemoreception, some studies suggest that they may be necessary for normal chemoreceptor function (22).

The microscopic anatomy of glomus cells in vertebrates has been described in the aortic and carotid bodies of mammals (59,88) and birds (48,68,71) and the carotid labyrinth of amphibians (40,49). Although there is reflex and neurophysiological evidence for O_2-sensitive chemoreceptors in the gills of larval amphibia (44,45,83), they have not been examined microscopically. Reptiles do not appear to have distinct carotid bodies or labyrinths. However, clusters of glomus cells have been identified microscopically in the central cardiovascular region, in connective tissue around the pulmonary, carotid, and aortic arches in turtles and lizards (4,42,75).

Neuroepithelial cells are histologically very similar to glomus cells and have been found in the lungs and/or respiratory passages of all vertebrates examined (86). Although the hypoxic reflexes that may be mediated by these receptors are not well defined, they are considered O_2-sensitive chemoreceptors (17). It has been suggested that pulmonary neuroepithelial cells may be important for O_2 chemoreception in very young animals born with immature lungs (86). These cells have also been described in the gills, respiratory swim bladders, and lungs of approximately 10 different species of fishes (55,97–99).

Neuroepithelial cells in the gills and airways of non-mammalian vertebrates may occur singly or in innervated clusters called neuroepithelial bodies (see Refs.

86,98 for reviews). In many cases neuroepithelial bodies are found in close association with capillaries (30). Neuroepithelial cells have also been classified as either "open"- or "closed"-type cells and have been described in the gills and air-breathing organs of fish and in amphibian lungs (30,97). Open cells are found in the outer epithelial layers and make contact with the mucosal surface of the gas exchange organs (Fig. 3). This location puts them in a good position to monitor the ventilatory flow, be it water or air. Open-type cells are polarized. The apical membrane in contact with the respiratory passages have cilia or microvilli (Fig. 3) (30). Dense core vesicles are concentrated in the basal portion of the cell where they come into synaptic contact with nerves (86). Closed-type cells are located on the basal lamina separated from the ventilatory flow, often by several layers of cells.

Branchial neuroepithelial cells are currently the best candidates for the O_2 chemoreceptor cells that mediate the reflex responses to hypoxia in fish. In the gills, these cells are located in the primary epithelium of the gill filaments and lie on the basal lamina between the inhalant water flowing over the gill and blood flow through the gill in an ideal anatomical location to function as O_2 sensors. Branchial neuroepithelial cells share many of the characteristic anatomical features of mammalian O_2 sensors (glomus and pulmonary neuroepithelial cells). Like mammalian cells, branchial neuroepithelial cells are innervated and possess a well-developed Golgi complex and numerous mitochondria and dense-cored

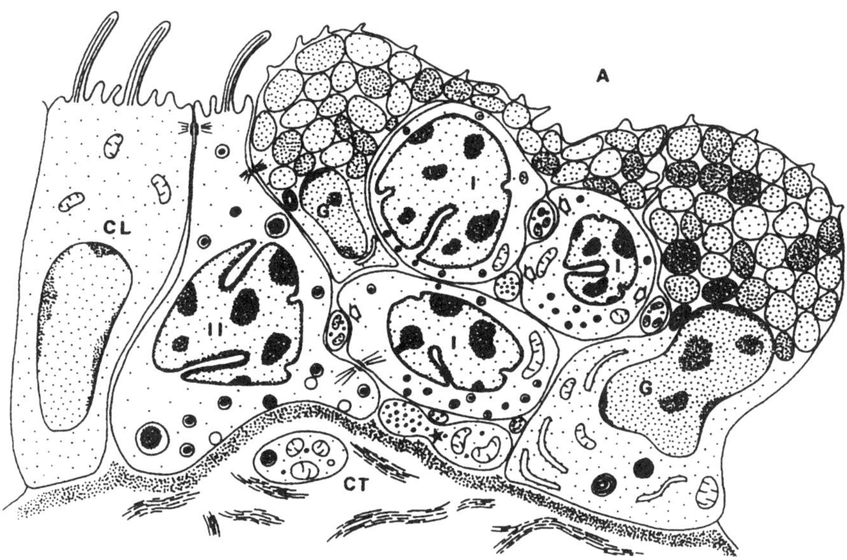

Figure 3 Schematic diagram of open-type neuroepithelial body in salamander lung. Type I cells (I) have small, dense core vesicles and are surrounded by goblet cells (G). Type II cells (II) have large, dense core vesicles, are in contact with the respiratory surface of the airway (A) by a single modified cilium, and are surrounded by ciliated cells (CL). Innervation is in the base (star). (From Ref. 30.)

vesicles (20,55,99). Falck-Hillarp fluorescence indicates that branchial neuroepithelial cells contain significant concentrations of monoamines (19,20,55) and immunocytochemical studies have identified the major monoamine contained within these cells as serotonin (99). Two populations of branchial neuroepithelial cells have been identified in bowfin based on immunoreactivity for either serotonin or met- and leu-enkephalin (97), but the physiological significance of this difference has yet to be determined. A neuronal function for gill neuroepithelial cells in bowfin was confirmed by the presence of neuron-specific enolase (31). However, the presence of some of the other major transmitters (i.e., dopamine, acetylcholine) found in mammalian glomus and pulmonary neuroepithelial cells and thought to be involved in O_2 chemoreception has not been examined.

Although the morphology and location of branchial neuroepithelial cells are suggestive of an O_2-sensitive chemoreceptor, histological and cytochemical data are supported by only a few physiological studies. An O_2-sensory function is suggested by the fact that hypoxia causes degranulation of the dense-cored vesicles and reduces monoamine content (55) similar to the degranulation seen in mammalian type 1 cells during hypoxia (24). Afferent O_2-sensitive activity has been recorded from the gills of fish (14,62) but it cannot be determined from these studies if the activity is from branchial neuroepithelial cells. Recent, preliminary studies on neuroepithelial cells in primary culture from channel catfish gills demonstrate O_2-sensitive currents similar to those of mammalian glomus and pulmonary neuroepithelial cells (61).

Goniakowska-Witalińska (30) has proposed a three-tiered scenario of the evolution of neuroepithelial cells in vertebrates based on studies in extant species. The first type are the most primitive and are solitary noninnervated neuroepithelial cells. These may be of the "closed type," deep in tissue, or the "open type," in contact with the ventilatory flow. This is the predominant pattern seen in larval *Salamandra* lungs and the respiratory swim bladders of the actinopterygian fishes *Polypterus* (bichir) and *Amia* (bowfin). The next step is innervation of the solitary neuroepithelial cells. Again, these may be the closed type as seen in the amphibian *Triturus* or open in the sarcopterygian lungfish *Protopterus*. The most advanced form of neuroepithelial cells are the innervated clusters or neuroepithelial bodies. Closed types are found in the lungs of most amphibia and open types are in some amphibia and the higher vertebrates such as mammals. Since the carotid and aortic arches are derived from gill arches, it has been suggested that the branchial neuroepithelial cells of fish represent the evolutionary precursors of mammalian carotid and aortic body glomus cells (13).

III. Physiology

A. Stimulus Specificity and Mechanism of Transduction

The linkage between changes in arterial Pa_{O_2} and chemosensory discharge in this tissue has remained elusive. Several hypotheses exist, which are not mutually exclusive, and which are thoroughly outlined in other chapters in this volume. What has received less attention is the correlation between hypoxic ventilatory responses, chemosensory discharge, and the O_2 content or saturation of the blood. While the O_2 saturation of the blood is normally tightly linked to the arterial O_2 partial pressure, this is not always the case and many vertebrates show a much stronger correlation between the threshold for the hypoxic ventilatory response and O_2 saturation than with Pa_{O_2}. Thus, in fish, ventilation rate varies inversely with blood O_2 content, independently of partial pressure, indicating that arterial receptors respond to the rate of delivery of O_2 to the receptor site (74,82), and throughout the air-breathing vertebrates there are strong indications that this is also the case. It has been argued that in animals where adaptive changes occur in the shape of the ability of hemoglobin to bind O_2, O_2 saturation would be a better indicator of the O_2 homeostasis of arterial blood than Pa_{O_2} (94). Thus, in turtles and hibernating ground squirrels there are reductions in the Pa_{O_2} threshold of the hypoxic ventilatory response accompanying the temperature-induced increase in the ability of hemoglobin to bind O_2 as body temperature decreases (29,63). This is also the case with animals that exhibit an increase in hemoglobin-O_2 binding as a consequence of chronic exposure to hypoxia as occurs in animal burrows (6) or at altitude (5,10). In all cases, if the hypoxic ventilatory response is plotted against the % Hb saturation of arterial blood, ventilation begins to increase when saturation falls below 80–85% (Fig. 4) (29,94). It has been further shown that many species (ectotherm and endotherm) behaviorally reduce their body temperatures under hypoxic conditions, increasing oxyhemoglobin affinity such that levels of O_2 saturation are maintained despite the reduction in Pa_{O_2} (38,95).

At present, only the aortic bodies of mammals have been shown to respond to changes in blood O_2 content per se, although this does not seem to elicit any reflex ventilatory response (36,51). The responses of O_2 chemoreceptors to changes in O_2 saturation in most species of vertebrates have not been studied. Several hypotheses have arisen from recent research to explain chemotransduction in the carotid body. These hypotheses are not mutually exclusive. One group of researchers (3,28,57) suggest that hypoxia leads to a reduction in the binding of O_2 to a heme protein receptor (Fig. 4c). This molecule is believed to be an H_2O_2-producing NAD(P)H oxidase that uses cytochrome b_{558} as a central electron carrier. The H_2O_2 is scavenged by glutathione peroxidase, thereby decreasing the ratio of reduced glutathione to oxidized glutathione. During hypoxia when O_2 levels are reduced, less H_2O_2 is produced leading to higher levels of reduced glutathione. This in turn leads to closing of potassium channels that are sensitive to the level of reduced glutathione that reduces K^+ conductance. As less K^+ leaves the cell, positive charge builds up inside causing depolarization of the cell. This leads to increased uptake of extracellular calcium by calcium channels whose opening or closing is dependent on the transmembrane difference in voltage. The increase in internal calcium concentration, in turn, promotes neurotransmitter release (2). This does not appear to be the complete story, however, because many of the components of this complex enzymatic chain of events are soluble cytoplasmic factors, yet hypoxia still increases the closing of K^+ channels in isolated patches of receptor cell membranes. Thus, whether this proposed mechanism is the basis of O_2 chemotransduction or simply

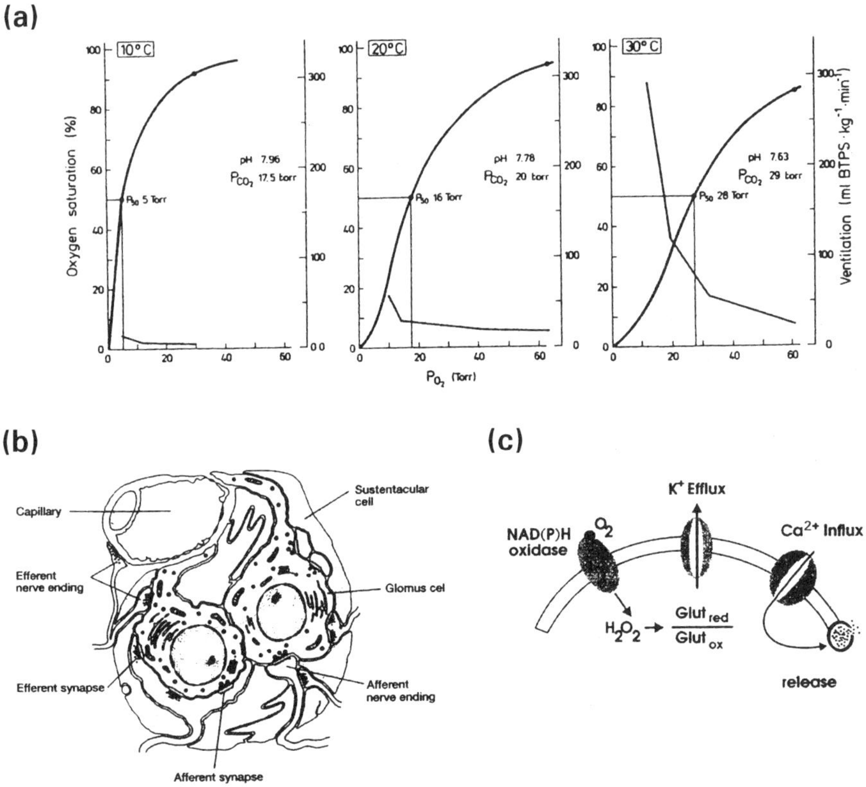

Figure 4 (a) Relationship between the oxygen saturation of whole blood (left axis) and the level of ventilation (right axis) and the partial pressure of O_2 for the turtle *Pseudemys scripta* at different temperatures. Note that the ventilatory response to decreasing levels of O_2 begins to increase significantly when the blood begins to desaturate below 50%, regardless of the temperature. (From Ref. 29.) (b) A schematic diagram of the glomus cell complex comprising the mammalian carotid body. This complex is associated with all respiratory-related O_2 chemoreceptors. (c) A schematic diagram of one theoretical model of the O_2-sensing mechanism in the carotid body. This model could account for the link between the hypoxic ventilatory response and the level of saturation of the arterial blood with oxygen (see text for details).

a modulating element of the transduction cascade remains to be resolved (11,28,32,66,96).

One of the appeals of the hypothesis that a heme protein serves as the O_2 receptor in chemoreceptor cells is the possibility that this heme protein would share similar properties and be modulated in a similar fashion as the hemoglobin in any given species. If this were the case, binding properties of receptors would shift in parallel to the hemoglobin-binding properties with changes in temperature, organophosphates, or other factors that alter the relationship between blood P_{O_2}

and O_2 content or saturation and shift the position of the O_2 equilibrium curve of the blood (9). This would maintain a tight coupling between receptor activation, blood O_2 saturation, and ventilation despite the shift in the relationship between arterial O_2 content and P_{O_2}. In other words, even though the receptor only responds to changes in local P_{O_2}, the receptor-binding properties might change in such a way under these conditions that receptor discharge only begins to increase significantly at the P_{O_2} at which hemoglobin begins to desaturate.

Another possible explanation of the observed correlations between cardio-respiratory changes and changes in arterial O_2 content that occur independent of changes in arterial P_{O_2} stems from the relationship between local tissue (and receptor cell) P_{O_2} and arterial P_{O_2}. P_{O_2} at the receptor site (which could be the cell surface or in the cytosol) is a function of mean capillary P_{O_2}, which, in turn, is a function of blood O_2 affinity and the change in O_2 content due to local O_2 consumption as blood travels along the capillary (21). The change in O_2 (which will in part determine the mean capillary P_{O_2}) will be a function of metabolic rate and the rate of O_2 delivery (blood flow) to the tissue. One of the appeals of this explanation is that it can account for the existence of distinct populations of receptor cells; aortic chemoreceptors that respond to changes in O_2 saturation (which well suits their primary role in regulating blood volume and O_2-carrying capacity) (51) and carotid chemoreceptors (which do not).

This same line of argument would apply to P_{O_2}-sensitive chemoreceptors in the venous circulation as has been suggested to be the case for amphibians and turtles (91). In these animals two receptor populations with different reflex roles have been proposed to exist to explain why there are no ventilatory responses to reductions in blood O_2 content, whereas heart rate and pulmonary blood flow increase following reductions in blood O_2-carrying capacity due to anemia, inhalation of CO, or nitrite infusions (89–91). O_2-sensitive chemoreceptors have been identified on the pulmocutaneous artery in amphibians as well as on the pulmonary artery in turtles (42,43,87), and blood in these vessels is predominantly venous systemic blood although the exact composition depends on the level of left-to-right intracardiac shunt. This led to the suggestion that perhaps O_2-sensitive chemoreceptors on the pulmonary artery are responsible for cardiovascular control while the arterial chemoreceptors are responsible for ventilatory control (91). Recent experiments designed to test this hypothesis have been equivocal, however, and while not ruling out this possibility, indicate that there are receptors in both the carotid and pulmonary circulations that have overriding effects on the cardiovascular system (92).

Whatever the mechanistic basis behind the apparent ability of some species to distinguish changes in arterial O_2 saturation independent of changes in O_2 partial pressure, the trend that appears is less a phylogenetic trend than an adaptive trend. In fish, ventilation rate varies inversely with blood O_2 content, independently of partial pressure (74,82). In amphibians and reptiles, animals with intracardiac shunts, changes in shunt fraction appear to be tightly related to changes in O_2 content while changes in ventilation are less so. All air-breathing vertebrates that undergo broad changes in body temperature show much tighter correlations between changes in O_2 delivery than in Pa_{O_2} per se. It is tempting to speculate that while the O_2-sensing mechanism may be the same for all receptor cells, their locations have evolved in such a way to make the responses of some more sensitive to changes in O_2 delivery than others. This ability appears to be present in all circumstances where it would appear to be of benefit.

B. Pharmacology

The link between O_2 sensing by glomus cells and activity in primary afferent nerves is believed to be one or more of the many neurochemicals identified in the glomus cells. The precise nature of this chemical link continues to elude physiologists and virtually all work has focused on mammals. A variety of different neurochemicals have been localized to mammalian carotid body glomus cells. Most of these alter afferent neural activity and/or cardioventilatory reflexes and many are released in response to hypoxia (see Ref. 72 for review). There is, however, no clear consensus on the role of any single chemical, much less so many neurochemicals, in O_2 chemoreception. Many of the same neurochemicals have been localized in O_2-sensitive cells in lower vertebrates. Most of these comparative studies have used neurochemical localization as a method for identifying putative O_2 receptor cells. Few have examined their effects on resting cardioventilatory variables, effects on hypoxic reflexes, or effects on O_2 chemoreceptor neural activity. As a result, there is little comparative data that provide information about how these chemicals are involved in O_2 chemoreception. Only one study has examined the effects of different neurochemicals on O_2 chemoreceptor control of cardioventilatory control in a nonmammalian vertebrate (15). Thus, the results of this study will be discussed in light of current theories on the pharmacology of O_2-sensitive chemoreceptors.

Of all the putative neurotransmitters tested on branchial O_2 receptors in rainbow trout (Table 1), acetylcholine was the most powerful stimulant of afferent activity and showed effects similar to NaCN and hypoxia. The effects of nicotine versus muscarine indicate that cholinergic stimulation of branchial O_2-sensitive chemoreceptors involves nicotinic receptors as in cats and rats (15). However, in early studies cholinergic blockers appear to have little effect on the response to hypoxia in mammals (24,72). There are significant species differences within mammals in the response to acetylcholine. The most notable difference is that between cats/rats and rabbits. In cats, rats, and rainbow trout, acetylcholine is stimulatory and this stimulation is mediated by nicotinic receptors. In rabbits, however, acetylcholine is inhibitory, an effect that appears to be due to the fact that cholinergic receptors in the rabbit carotid body are primarily muscarinic (69). Both cats and rabbits are able to respond effectively to hypoxia; thus, the significance of this difference and the mechanistic implications that it has on O_2 chemotransduction are unknown. The "acetylcholine hypothesis" has been revived by Fitzgerald and Shirahata (26) and strongly supported by Zhang et al. (100), who present convincing evidence that increased afferent neural activity during hypoxia is the result of acetylcholine release from type 1 cells. Thus, the cholinergic receptors on type 1 cells may be involved in negative feedback on neurotransmitter release. This is one

Table 1 Summary of the Effects of Different Pharmacological Agents on O_2-Sensitive Afferent Neural Activity Recorded from Cranial Nerve IX from the First Gill Arch of Rainbow Trout

Agent	Dose	Effect
NaCN	25 μg	Strong stimulation
Acetylcholine	100 nmol	Strong stimulation
Nicotine	100 nmol	Strong stimulation
Muscarine	100 nmol	Weak stimulation
Atropine	100 nmol	Inhibits effects of acetylcholine, nicotine, and muscarine but not hypoxia
Norepinephrine	5–1000 nmol	No effect
Epinephrine	500 nmol	No effect
Isoproterenol	100 nmol	No effect
Propranolol	500 nmol	Partially inhibits effects of NaCN and hypoxia
Dopamine	100 nmol	Weak stimulation followed by weak inhibition
Serotonin	100 nmol	Weak stimulation followed by weak inhibition

Source: Ref. 15.

area where further experimentation, especially in nonmammalian vertebrates, should provide important data on the pharmacological control of O_2-sensitive chemoreceptor activity.

The effects of other neurochemicals on branchial O_2 chemoreceptor activity in rainbow trout show similarities and differences to carotid body O_2 receptors. Although carotid body chemoreceptors are stimulated by epinephrine and norepinephrine, these neurochemicals appear to have no effect on branchial O_2 receptors. Dopamine, the predominant catecholamine in mammalian type 1 cells, also modulates O_2 receptor activity in trout gills (15). However, there has been no immunohistochemical demonstration of dopamine in the gills or neuroepithelial cells of fish. These results suggest that the neurotransmitter content of nonmammalian O_2 sensors should be examined to determine if there are any phylogenetic trends that may help resolve the functional significance of the various neurotransmitters in O_2 chemoreception.

IV. Environmental Adaptations

A. Magnitude and Variability of the Hypoxic Ventilatory Response

The relationship between ventilation and arterial P_{O_2} is relatively flat over the physiological range but substantial increases in ventilation do occur as hypoxia becomes more severe. The amount of variability in this response from species to species, however, is enormous. Much of this variability undoubtedly reflects adaptation to differing environmental, behavioral, or physiological demands. Thus the blunted hypoxic ventilatory responses seen in fossorial (burrowing) birds and mammals (7,8) are believed to be an adaptation that minimizes the use of increased air convection as a mechanism for dealing with the chronic hypoxia (and hypercarbia) of the burrow atmosphere. For much of the data, however, it is hard to draw such firm conclusions. Figure 5 illustrates the range of variability seen in the hypoxic ventilatory responses of various species of amphibia and reptiles. In general, large falls in inspired O_2 partial pressure are required to produce relatively small increases (2–3×) in levels of minute ventilation. The relative contributions of changes in the components of ventilation, respiratory frequency and tidal volume, to these changes are also highly variable. Thus, a fall in inspired O_2 tensions to 40 mmHg results in an elevation of ventilation to roughly 4 times resting levels, due exclusively to increases in breathing frequency, in the Nile crocodile (*Crocodilus niloticus*) but to only twice resting levels, due exclusively to increases in tidal

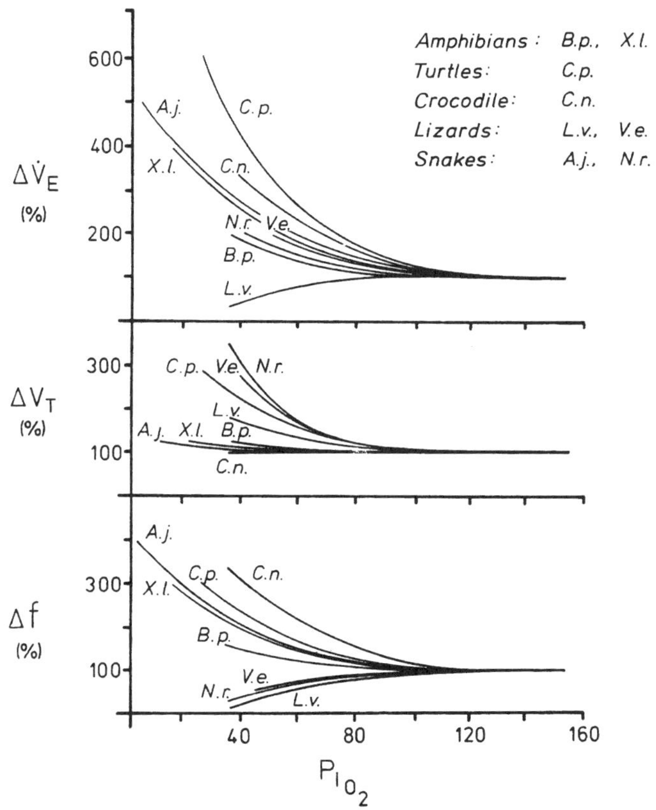

Figure 5 Relationship between the percent change in ventilation ($\Delta\dot{V}_E$) tidal volume ($\Delta\dot{V}_T$), and breathing frequency (Δf) and the partial pressure of O_2 in inspired air (PI_{O_2}) in various species of reptile. (B.p. = *Bufo paracnemis*, X.l. = *Xenopus laevis*, C.p. = *Chrysemys picta*, C.n. = *Crocodilus niloticus*, L.v. = *Lacerta viridis*, V.e. = *Varanus exanthematicus*, A.j. = *Acrochordus javanicus*, N.r. = *Natrix rhombifera*. (From Ref. 64.)

volume (breathing frequency actually decreases!), in the diamondback water snake (*Natrix rhombifera*). Much of this variability reflects the manner in which these measurements were made and reported in some species while much must represent adaptation to differing constraints placed upon the respiratory systems of these animals.

Perhaps most intriguing is the fact that ventilation does not increase and may even fall as O_2 levels decline in some species. In most air-breathing vertebrates, Pa_{O_2} is normally maintained within certain "homeostatic" limits by altering ventilation to match O_2 consumption. It is becoming increasingly clear, however, that just as Pa_{O_2} is a function of the level of ventilation of an animal, O_2 consumption is also a function of Pa_{O_2}. When O_2 becomes limiting, metabolism is reduced. Although this inter-relationship has been documented in all air-breathing vertebrates (see Ref. 39 for review), it is most evident from the data collected on reptiles. In this group, many studies have documented a "critical P_{O_2}" below which further reductions in ambient P_{O_2} lead to a depression of metabolism. There appears to be an inverse correlation between the critical P_{O_2} and the hypoxia tolerance of each species.

The data further suggest that this critical P_{O_2} is closely correlated with the partial pressure of O_2 at which arterial blood is 50% saturated. This correlation suggests that the critical P_{O_2} is a function of a decrease in the O_2-transport capacity of the blood. When blood (hemoglobin) leaving the gas exchange surface is no longer fully saturated with O_2, and thus the O_2 transport capacity of the blood falls, so too does metabolism. Finally, an inverse correlation also exists between the magnitude of the ventilatory response to hypoxia and the level of critical P_{O_2}. The greater the hypoxic ventilatory response, the more inspired O_2 levels must fall to produce any given change in Pa_{O_2} (the harder an animal breathes in response to a fall in O_2 in the environment, the less effect it will have on the arterial blood). Thus, the greater the hypoxic ventilatory response and the higher the oxyhemoglobin affinity, the lower the critical P_{O_2} (i.e., the greater the fall in inspired O_2 levels required to reduce Pa_{O_2} to a point where the O_2 transport capacity of the blood is reduced) and the greater the hypoxia tolerance of a species.

This is illustrated for three species with different degrees of hypoxia tolerance in Figure 6 (turtle > snake > lizard). The turtle, *Chrysemys picta*, possesses a brisk hypoxic ventilatory response and a high-affinity hemoglobin (29). The lizard, *Lacerta viridis*, on the other hand, possesses a very meager hypoxic ventilatory response and a lower-affinity hemoglobin (70). Values for the snake, *Natrix rhombifera*, lie in between (34,35). As levels of environmental O_2 fall, ventilation increases in an attempt to maintain levels of O_2 consumption in the turtle and snake. Thus the ratio of ventilation to O_2 consumption increases. The lizard, on the other hand, reduces O_2 consumption as levels of environmental O_2 fall and breathes only hard enough to maintain the new, reduced levels of O_2 consumption. Thus in this instance, the ratio of ventilation to O_2 consumption remains relatively constant. This reflects the fact that there are multiple strategies for dealing with decreases in environmental O_2 and will serve to remind us to take care in how we evaluate differences in the magnitude of the hypoxic ventilatory response. A blunted or

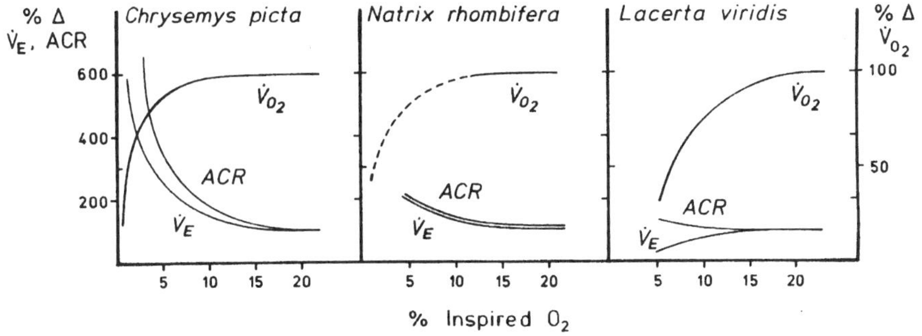

Figure 6 Relationship between the percent change in minute ventilation ($\Delta\dot{V}_E$), air convection requirement (ΔACR) (left axis), and oxygen consumption ($\Delta\dot{V}_{O_2}$) (right axis), and the percent oxygen in inspired gas. (Data from Refs. 35,70.)

reduced ventilatory response to reductions in inspired O_2 does not necessarily represent a reduced degree of adaptation to hypoxia.

B. Changes Associated with Air Breathing: Air-Breathing Fish

The atmosphere is a rich and stable source of O_2, compared to water, and the ability to breathe air has evolved multiple times in fishes. It is believed that the ability to breathe air marked the first step in the evolution of terrestriality (33). Studies on air-breathing fishes provide clues into the physiological changes that may have occurred in cardiovascular and ventilatory control systems in vertebrates during the transition from water to air breathing. Most studies have focused on two major groups of air-breathing fishes. The first group are the actinopterygian air breathers such as *Lepisosteus* (gar) and *Amia* (bowfin). These fish are facultative air breathers and have well-developed gills. Their gills are smaller when compared to similar-size water breathers, which is presumed to reflect a decreased dependence on aquatic respiration. The air-breathing organ in these fishes is a modified swim bladder, not a true lung (33). The second group of fish are the sarcopterygian air breathers, the lungfishes. In contrast to actinopterygians, the sarcopterygians have highly reduced gills (some arches lack filaments altogether), are obligate air breathers, and breathe air with true lungs (33).

Air-breathing fish have two functional sets of O_2-sensitive chemoreceptors in the gills innervated by cranial nerves IX and X as in water breathers (60,80). However, the reflex responses elicited in response to O_2-sensitive chemoreceptor stimulation differ because these animals have two very different sets of respiratory organs. The respiratory mode (water vs. air breathing) is variable and changes to meet different environmental and/or physiological demands. Gill ventilation (water breathing) is regular and rhythmic in the facultative actinopterygian air-breathing fishes but air breathing is sporadic and occurs as single breaths at irregular intervals (60,80).

In the spectrum of air-breathing fish, bowfin are more similar to unimodal water breathers than other air breathers. Their gills are not reduced to the extent they are in other air-breathing fishes, they do not have vascular shunts, and they rely on aquatic ventilation most of the time (18,73). Although the respiratory swim bladder is used for gas exchange, it is also still important for control of buoyancy. Some air breaths are not for gas exchange but to maintain neutral buoyancy by keeping the respiratory swim bladder inflated (37). Gill ventilation is stimulated by both internal and external NaCN, as in the majority of water breathers, and only external NaCN stimulates air breathing (60).

Gar have reduced gills and rely on air breathing more than bowfin (78). Their responses to O_2-chemoreceptor stimulation are also different. Stimulation of external O_2 receptors in gar with cyanide inhibits gill ventilation and stimulates air breathing (79). It has been hypothesized that this inhibition of gill ventilation limits the loss of O_2 from the gills during aquatic hypoxia seen in lungfish and bowfin (18,79,80). Stimulation of internal receptors with NaCN or by ventilating the respiratory swimbladder with N_2 increases both air and gill ventilation in gar. Smatresk et al. (79) suggest that this pattern of response is because internal O_2 receptors set the level of hypoxic drive and external O_2 receptors control the balance between air and water ventilation.

The gills of obligate air breathing sarcopterygians are reduced to the point that they are essentially useless for aquatic O_2 exchange (33,50). Air breathing is stimulated by both internal and external cyanide in the African lungfish (50) and Australian lungfish (46). In the South American lungfish (*Lepidosiren*), however, aquatic hypoxia had no effect on ventilation (76). Aerial hypoxia stimulates ventilation indicating that in this species ventilation is controlled primarily by internal receptors. Isovolemic anemia to reduce blood O_2 content by 50% did not affect ventilation indicating that the proximal stimulus for the ventilatory response to hypoxia is P_{O_2}, not O_2, content. The cardiovascular responses of South American lungfish to hypoxia appear to be much more similar to the hypoxic reflexes of tetrapods than of fish (76).

The phylogenetic trend in air-breathing fishes is a reduced sensitivity to aquatic hypoxia that parallels reduction and internalization of the gills along with their O_2 receptors. Although much is known about the reflex responses of air-breathing fish to O_2 chemoreceptor stimulation, no one has studied these receptors directly in these animals.

V. Conclusion

The ability to sense and respond to hypoxia is present in all taxa and represents a fundamental physiological requirement for O_2 at the subcellular level. An examination of the phylogeny of O_2 chemoreception reveals interesting trends as well as contradictions. Although there is extensive literature on the reflex effects of O_2-chemoreceptor stimulation, there are few neural or cellular studies in organisms other than mammals. Through comparative studies it may be possible to determine what physiological changes have occurred during the evolution of terrestriality and integrate this knowledge into a comprehensive understanding of O_2 chemoreception and cardioventilatory control.

References

1. Abdel-Magied EM, King AS. The topographical anatomy and blood supply of the carotid body region of the domestic fowl. J Anat 1978; 126:536–546.
2. Acker H, Xue D. Mechanisms of O_2 sensing in the carotid body in comparison to other oxygen sensing cells. News Physiol Sci 1995; 10:211–216.
3. Almaraz L, Gonzalez C, Obeso A. Effects of high potassium on the release of [3H]dopamine from the cat carotid body in vitro. J Physiol 1986; 379:293–307.
4. Berger PJ, Gibbins IL, Hards DK, Crosby LJ. The distribution and ultrastructure of sensory elements in the baroreceptor region of the truncus arteriosus of the lizard *Trachydosaurus rugosus*. Cell Tissue Res 1982; 226:389–406.
5. Black CP, Tenney SM. Oxygen transport during progressive hypoxia in high-altitude and sea-level waterfowl. Respir Physiol 1980; 39:217–239.
6. Boggs DF, Birchard GF. Relationship between hypoxic ventilatory response and hemoglobin O_2-affinity in birds. Physiologist 1982; 25:325–325.
7. Boggs DF, Kilgore DL. Ventilatory responses of the burrowing owl and bobwhite to hypercarbia and hypoxia. J Comp Physiol 1983; 149:527–533.
8. Boggs DF, Kilgore DL, Birchard GF. Minireview: Respiratory physiology of burrowing animals and birds. Comp Biochem Physiol 1984; A77:1–7.
9. Boggs DF. Hypoxic ventilatory control and hemoglobin oxygen affinity. In: Sutton JR, Houston CS, Coates G, eds. Hypoxia and the Brain. Burlington: Queen City, 1995.
10. Brooks JG, Tenney SM. Ventilatory response of llama to hypoxia at sea level and high altitude. Respir Physiol 1968; 5:269–278.
11. Bunn HF, Poyton RO. Oxygen sensing and molecular adaption to hypoxia. Physiol Rev 1996; 76:839–885.
12. Burleson ML, Smatresk NJ. Effects of sectioning cranial nerves IX and X on cardiovascular and ventilatory reflex responses to hypoxia and NaCN in channel catfish. J Exp Biol 1990; 154:407–420.
13. Burleson ML, Smatresk NJ, Milsom WK. Afferent inputs associated with cardioventilatory control in fish. In: Hoar WS, Randall DJ, Farrel AF, eds. Fish Physiology, Vol. XII, Part B: The Cardiovascular System. New York: Academic Press, 1992.
14. Burleson ML, Milsom WK. Sensory receptors in the first gill arch of rainbow trout. Respir Physiol 1993; 93:97–110.
15. Burleson ML, Milsom WK. Cardio-ventilatory control in rainbow trout. I. Pharmacology of branchial, oxygen-sensitive chemoreceptors. Respir Physiol 1995; 100:231–238.
16. Butler PJ, Taylor EW, Short S. The effect of sectioning cranial nerves V, VII, IX and X on the cardiac response of the dogfish, *Scyliorhinus canicula*, to environmental hypoxia. J Exp Biol 1977; 69:233–245.
17. Cutz E, Jackson A. Neuroepithelial bodies as airway oxygen sensors. Respir Physiol 1999; 115:201–214.
18. Daxboeck C, Barnard DK, Randall DJ. Functional morphology of the gills of the bowfin, *Amia calva* L., with special reference to their significance during air exposure. Respir Physiol 1981; 43:349–364.

19. Donald J. Comparative study of the adrenergic innervation of the teleost gill. J Morphol 1987; 193:63–73.
20. Dunel-Erb S, Bailly Y, Laurent P. Neuroepithelial cells in fish gill primary lamellae. J Appl Physiol 1982; 53:1342–1353.
21. Eyzaguirre C, Fitzgerald RS, Lahiri S, Zapata P. Arterial chemoreceptors. In: Shepherd JT, Abboud FM, eds. Handbook of Physiology. The Cardiovascular System III. Part 2. Bethesda, MD: American Physiological Society, 1983.
22. Eyzaguirre C. Some thoughts about mechanisms involved in carotid body chemoreception. Prologue. In: Gonzalez G, ed. The Carotid Body Chemoreceptors. Austin, TX: Landes Bioscience, 1997.
23. Eyzaguirre C. Electrical coupling between glomus cells and nerve terminals. In: Lahiri S, Prabhakar NR, Foster RE, eds. Oxygen Sensing: Molecule to Man. London: Plenum, 2000:349–357.
24. Fidone S, Gonzalez C. Initiation and control of chemoreceptor activity in the carotid body. In: Cherniack NS, Widdicombe JG, eds. The Respiratory System, Handbook of Physiology. Sect. 3, vol. II. Bethesda, MD: American Physiological Society, 1986:247–312.
25. Fitzgerald RS, Lahiri S. Reflex responses to chemoreceptor stimulation. In: Fishman AP, Cherniak NS, Widdicombe JG, Gdiger SR, eds. Handbook of Physiology, Section 3, The Respiratory System, Vol. II, Control of Breathing, Part 1. Bethesda, MD: American Physiological Society, 1986.
26. Fitzgerald RS, Shirahata M. Acetylcholine and carotid body excitation during hypoxia in the cat. J Appl Physiol 1994; 76:1566–1574.
27. Fritsche R, Nilsson S. Cardiovascular responses to hypoxia in the Atlantic cod, *Gadus morhua*. Exp Biol 1989; 48:153–160.
28. Ganfornina MD, López-Barneo J. Single K^+ channels in membrane patches of arterial chemoreceptor cells are modulated by O_2 tension. Proc Natl Acad Sci USA 1991; 88:2927–2930.
29. Glass ML, Boutilier RG, Heisler N. Ventilatory control of arterial P_{O_2} in the turtle *Chrysemys picta belli*: effects of temperature and hypoxia. J Comp Physiol B 1983; 151:145–153.
30. Goniakowska-Witalińska L. Neuroepithelial bodies and solitoary neuroendocrine cells in the lungs of amphibia. Microsc Res Tech 1997; 37:13–30.
31. Goniakowska-Witalińska L, Zaccone G, Fasulo S, Mauceri A, Licata A, Youson J. Neuroendocrine cells in the gills of the bowfin *Amia calva*: an ultrastructural and immunocytochemical study. Folia Histo Cyto 1995; 33:171–177.
32. Gonzalez C, López-López JR, Obeso A, Perez-Garcia MT, Rocher A. Cellular mechanisms of oxygen chemoreception in the carotid body. Respir Physiol 1995; 102:137–148.
33. Graham JB. Air-breathing Fishes. New York: Academic Press, 1997.
34. Gratz RK. Ventilation and gas exchange in the diamondback water snake, *Natrix rhombifera*. J Comp Physiol 1978; 127:299–305.
35. Gratz RK. Ventilatory response of the diamondback water snake *Natrix rhombifera* to hypoxia, hypercapnia and increased oxygen demand. J Comp Physiol 1979; 129:105–110.
36. Hatcher JD, Chiu LK, Jennings DB. Anemia is a stimulus to aortic and carotid chemoreceptors in the cat. J Appl Physiol 1978; 44:696–702.
37. Hedrick MS, Jones DR. Control of gill ventilation and air breathing in the bowfin *Amia calva*. J Exp Biol 1999; 202:87–94.

38. Hicks JW, Wood SC. Temperature regulation in lizards: effects of hypoxia. Am J Physiol 1985; 17:R595–R600.
39. Hochachka PW, Guppy M. Metabolic Arrest and the Control of Biological Time. Cambridge, MA: Harvard University Press, 1987.
40. Ishii K, Ishii K. The chemoreceptors of amphibians. In: Paintal AS, ed. Morphology and Mechanisms of Chemoreceptors. New Delhi: Vallabhbhai Chest Institute, 1976:265–274.
41. Ishii K, Ishii K. Electrophysiological aspects of reflexogenic area in the chelonian, *Geoclemmys reevesii*. Respir Physiol 1985; 59:45–54.
42. Ishii K, Ishii K, Kusakabe T. Chemo- and baroreceptor innervation of the aortic trunk of the toad *Bufo vulgaris*. Respir Physiol 1985; 60:365–375.
43. Ishii K, Ishii K, Dejours P. Activity of vagal afferent fibers innervating CO_2-sensitive receptors in the tortoise, *Testudo hermanni*. Jpn J Physiol 1986; 36:1015–1026.
44. Jia X, Burggren W. Developmental changes in chemoreceptive control of gill ventilation in larval bullfrogs (*Rana catesbeiana*). I. Reflex ventilatory responses to ambient hyperoxia, hypoxia and NaCN. J Exp Biol 1997; 200:2229–2236.
45. Jia X, Burggren W. Developmental changes in chemoreceptive control of gill ventilation in larval bullfrogs (*Rana catesbeiana*). II. Sites of O_2-sensitive chemoreceptors. J Exp Biol 1997; 200:2237–2248.
46. Johansen K, Lenfant C, Grigg GC. Respiratory control in the lungfish, *Neoceratodus forsteri* (Krefft). Comp Biochem Physiol 1967; 20:835–854.
47. Jones DR, Milsom WK. Peripheral receptors affecting breathing and cardiovascular function in non-mammalian vertebrates. J Exp Biol 1982; 100:59–91.
48. Kameda Y. Electron microscopic study on the development of the carotid body and glomus cell groups distributed in the wall of the common carotid artery and its branches in the chicken. J Comp Neurol 1994; 348:544–555.
49. Kusakabe T, Powell FL, Ellisman MH. Ultrastructure of the glomus cells in the carotid body of chronically hypoxic rats: with special reference to the similarity of amphibian glomus cells. Anat Rec 1994; 237:220–227.
50. Lahiri SE, Szidon JP, Fishman AP. Potential respiratory and circulatory adjustments to hypoxia in the African lungfish. Annu Rev Physiol 1970; 29:1141–1148.
51. Lahiri S, Milligan E, Nishino T, Mokashi A, Davies RO. Relative responses of aortic body and carotid body chemoreceptors to carboxhemoglobinemia. J Appl Physiol 1981; 50:580–586.
52. Laurent P. Action du pH et de la P_{O_2} sur le potentiel de membrane des cellules de l'épithélium récepteur dans la pseudobranchie d'un poisson téléostéen. Rev Can Biol 1969; 28:149–155.
53. Laurent P, Rouzeau JD. Action de la P_{O_2} sur l'activité nerveuse afférente provenant de la pseudobranchie d'un téléostéen. J Physiol Paris 1969; 61(Suppl 1):145.
54. Laurent P, Rouzeau JD. Afferent neural activity from the pseudobranch of teleosts. Effects of P_{O_2}, pH, osmotic pressure and Na^+ ions. Respir Physiol 1972; 14:307–331.
55. Laurent P. Gill internal morphology. In: Hoar WS, Randall DJ, eds. Fish Physiology, Vol. X. Gills, Part A, Anatomy, Gas Transfer, and Acid-Base Regulation. Orlando, FL: Academic Press, 1984:73–183.
56. Le Douarin N. The Neural Crest. London: Cambridge University Press, 1982.
57. López-Barneo J, López-López J, Ureña J, Gonzàlez C. Chemotransduction in the carotid body: K^+ current modulated by P_{O_2} in type I chemoreceptor cells. Science 1988; 241:580–582.

58. Malvin GM. Gill structure and function: amphibian larvae. In: Wood SC, ed. Comparative Pulmonary Physiology: Current Concepts, Vol. 39, Lung Biology in Health and Disease. New York: Marcel Dekker, 1989:121–151.
59. McDonald DM. Peripheral chemoreceptors: Structure-function relationships of the carotid body. In: Hornbein TF, ed. Regulation of Breathing, Part I. New York: Marcel Dekker, 1981:105–319.
60. McKenzie DJ, Burleson ML, Randall DJ. The effects of branchial denervation and pseudobranch ablation on cardio-ventilatory control in an air-breathing fish. J Exp Biol 1991; 161:347–365.
61. Mercer SE, Wilk-Blaszczak MA, Burleson ML. A comparative model for the investigation of the electrophysiological basis of oxygen sensing. FASEB J 2000; 14:A45(72.10).
62. Milsom WK, Brill RW. Oxygen sensitive afferent information arising from the first gill arch of yellowfin tuna. Respir Physiol 1986; 66:193–203.
63. Milsom WK, McArthur MD, Webb CL. Control of breathing in hibernating ground squirrels. In: Heller HC, Musacchis XJ, Wang LCH, eds. Living in the Cold: Physiological and Biochemical Adaptations. New York: Elsevier, 1986.
64. Milsom WK. Control and co-ordination of gas exchange in air breathers. In: Boutilier RG, ed. Advances in Comparative and Environmental Physiology. Vol. 6, chapt. 10. 1990:347–400.
65. Milsom WK. Cardiorespiratory stimuli: receptor cell versus whole animal. Am Zool 1997; 37:3–11.
66. Milsom WK. Phylogeny of respiratory chemoreceptor function in vertebrates. Zoology (Jena) 1998; 101:316–332.
67. Milsom WK, Reid SG, Rantin FT, Sundin L. Extrabranchial chemoreceptors involved in respiratory reflexes in the neotropical fish: *Colossoma macropomum* (the tambaqui). J Exp Biol 2002; 205:1765–1774.
68. Miyoshi M, Hashimoto Y, Kon Y, Sugimura M. Distribution of epithelioid cells in the wall of the chicken aorta and their functional role as chemoreceptors. Anat Rec 1995; 242:302–309.
69. Monti-Bloch L, Eyzaguirre C. A comparative physiological and pharmacological study of cat and rabbit carotid body chemoreceptors. Brain Res 1980; 193:449–470.
70. Nielsen B. On the regulation of respiration in reptiles. II. The effect of hypoxia with and without moderate hypercapnia on the respiration and metabolism of lizards. J Exp Biol 1962; 39:107–117.
71. Nye PCG, Powell FL. Steady-state discharge and bursting of arterial chemoreceptors in the duck. Respir Physiol 1984; 50:335–350.
72. Prabhakar NR. Neurotransmitters in the carotid body. In: O'Regan R, et al, eds. Arterial Chemoreceptors: Cell to System. New York: Plenum Press, 1994:57–69.
73. Randall DJ, Cameron JN, Daxboeck C, Smatresk N. Aspects of bimodal gas exchange in the bowfin *Amia calva* L. (Actinopterygii: Amiiformes). Respir Physiol 1981; 43:339–348.
74. Randall DJ. The control of respiration and circulation in fish during hypoxia and exercise. J Exp Biol 1982; 100:275–288.
75. Rogers DC. The structure of the carotid bifurcation in the lizards *Tiliqua occipitalis* and *Trachysaurus rugosus*. J Morphol 1967; 122:115–130.
76. Sanchez A, Soncini R, Wang T, Koldkjaer P, Taylor EW, Glass ML. The differential cardio-respiratory responses to ambient hypoxia and systemic hypoxaemia in the South

American lungfish, *Lepidosiren paradoxa*. Comp Biochem Physiol Part A 2001; 130:677–687.
77. Saunders RL, Sutterlin AM. Cardiac and respiratory responses to hypoxia in the sea raven, *Hemipterus americanus*, an investigation of possible control mechanisms. J Fish Res Bd Can 1971; 28:491–503.
78. Smatresk NJ, Cameron JN. Respiration and acid-base physiology of the spotted gar, a bimodal breather. III. Response to a transfer from fresh water to 50% sea water, and control of ventilation. J Exp Biol 1982; 96:295–306.
79. Smatresk NJ, Burleson ML, Azizi SQ. Chemoreflexive responses to hypoxia and NaCN in longnose gar: evidence for two chemoreceptor loci. Am J Physiol 1986; 251 (Regul Integr Comp Physiol 20):R116–R125.
80. Smatresk NJ. Respiratory control in the transition from water to air breathing in vertebrates. Am Zool 1994; 34:264–279.
81. Smith FM, Jones DR. Localization of receptors causing hypoxic bradycardia in trout (*Salmo gairdneri*). Can J Zool 1978; 56:1260–1265.
82. Smith FM, Jones DR. The effect of changes in blood oxygen-carrying capacity on ventilation volume in the rainbow trout (*Salmo gairdneri*). J Exp Biol 1982; 97:325–334.
83. Straus C, Wilson RJ, Remmers JE. Oxygen sensitive chemoreceptors in the first gill arch of the tadpole, *Rana catesbeiana*. Can J Physiol Pharmacol 2001; 79:959–962.
84. Sundin L, Reid SG, Kalinin AL, Rantin FT, Milsom WK. Cardiovascular and respiratory reflexes: the tropical fish, traira (*Hoplias malabaricus*) O_2 chemoresponses. Respir Physiol 1999; 116:181–199.
85. Sundin L, Reid SG, Rantin FT, Milsom WK. Branchial receptors and cardiorespiratory reflexes in a neotropical fish, the tambaqui (*Colossoma macropomum*). J Exp Biol 2000; 203:1225–1239.
86. Van Lommel A, Bollé T, Fannes W, Lauweryns JM. The pulmonary neuroendocrine system: the past decade. Arch Histol Cytol 1999; 62:1–16.
87. Van Vliet BN, West NH. Functional characteristics of arterial chemoreceptors in an amphibian (*Bufo marinus*). Respir Physiol 1986; 88:113–127.
88. Verna A. The mammalian carotid body: morphological data. In: Gonzalez C, ed. The Carotid Body Chemoreceptors. Austin, TX: Landes Bioscience, 1997:1–29.
89. Wang T, Branco LGS, Glass ML. Ventilatory respnses to hypoxia in the toad, *Bufo paracnemis*, before and after a decrease in haemoglobin oxygen-carrying capacity. J Exp Biol 1994; 196:1–8.
90. Wang T, Hicks JW, Milsom WK. Changes in arterial O_2 content affect cardiac shunt but not ventilation in turtles. Physiologist 1996; 39:7–27.
91. Wang T, Krosniunas EH, Hicks JW. The role of cardiac shunts in the regulation of blood gases. Am J Zool 1997; 37:12–22.
92. Wang T, Taylor EW, Reid SG, Milsom WK. Interactive effects of mechano- and chemreceptor inputs on cardio-respiratory outputs in the toad. Respir Physiol. Submitted.
93. West NH, Van Vliet BN. Sensory mechanisms regulating the cardiovascular and respiratory systems in amphibians. In: Feder M, Burggren W, eds. Environmental Physiology of the Amphibians. Chicago: University of Chicago Press, 1992.
94. Wood SC. Cardiovascular shunts and oxygen transport in lower vertebrates. Am J Physiol 1984; 247:R3–R14.
95. Wood SC. Interactions between hypoxia and hypothermia. Annu Rev Physiol 1991; 53:71–85.

96. Wyatt CN, Weir EK, Peers C. Diphenylene iodonium blocks K^+ and Ca^{2+} currents in type I cells isolated from the neonatal rat carotid body. Neurosci Lett 1994; 172:63–66.
97. Zaccone G, Lauweryns JM, Fasulo S, Tagliafierro G, Ainis L, Licata A. Immunocytochemical localization of serotonin and neuropeptides in the neuroendocrine paraneurons of teleost and lungfish gills. Acta Zool 1992; 73:177–183.
98. Zaccone G, Fasulo S, Ainis L. Distribution patterns of the paraneuronal endocrine cells in the skin, gills and the airways of fishes as determined by immunohistochemical and histological methods. Histochem J 1994; 26:609–629.
99. Zaccone G, Fasulo S, Ainis L. Neuroendocrine epithelial cell system in respiratory organs of air-breathing and teleost fishes. Int Rev Cytol 1995; 157:277–314.
100. Zhong H, Zhang M, Nurse CA. Synapse formation and hypoxic signalling in co-cultures of rat petrosal neurones and carotid body type 1 cells. J Physiol 1997; 503(3):599–612.

AUTHOR INDEX

Italic numbers give the page on which the complete reference is listed.

C

D

E

F

G

H

M

O

P

Q

R

W

SUBJECT INDEX

A

D

E

F

G

H

I

K

L

M

N

R